T0321247

DEVELOPMENTS
IN THE SCIENCE
AND TECHNOLOGY
OF COMPOSITE
MATERIALS

**EUROPEAN ASSOCIATION
FOR COMPOSITE MATERIALS**

**GERMAN AEROSPACE
RESEARCH ESTABLISHMENT**

DEVELOPMENTS IN THE SCIENCE AND TECHNOLOGY OF COMPOSITE MATERIALS

FOURTH EUROPEAN CONFERENCE ON COMPOSITE MATERIALS

SEPTEMBER 25-28, 1990
STUTTGART - F.R.G.

EDITORS: J. FÜLLER, G. GRÜNINGER, K. SCHULTE (DLR)
A.R. BUNSELL, A. MASSIAH (EACM)

WITH THE SPONSORSHIP
OF THE EUROPEAN ECONOMIC COMMUNITY

ELSEVIER APPLIED SCIENCE: LONDON AND NEW YORK

Copies of the publication may be obtained from :

EUROPEAN ASSOCIATION FOR COMPOSITE MATERIALS
2 Place de la Bourse - 33076 Bordeaux Cedex - France

ELSEVIER SCIENCE PUBLISHERS LTD
Crown House - Linton Road - Barking Essex IG11 8JU - England

Sole Distributor in the USA and CANADA
ELSEVIER SCIENCE PUBLISHING CO. INC.
655 Avenue of the Americas - New York NY 10010 - USA

C EACM 1990

ISBN 1-85166-562-5

Printed in France

SCIENTIFIC COMMITTEE ECCM-4

President : Mr. G. GRÜNINGER
German Aerospace Research Establishment (DLR)
Federal Republic of Germany

• SELECTION PAPERS COMMITTEE

WEST-GERMANY

F.J. ARENDTS	University of Stuttgart
N. CLAUSSEN	Technical University Hamburg-Harburg
R. CUNTZE	MAN-Technologie
G. EHRENSTEIN	Südd. Kunststoffzentrum
K. FRIEDRICH	Technical University Hamburg-Harburg (now University of Kaiserslautern)
B. GEIER	DLR
G. GRÜNINGER	DLR
K.P. HERRMANN	University of Paderborn
K.U. KAINER	T.U. Clausthal-Zellerfeld
H. KELLERER	MBB
R. KOCHENDÖRFER	DLR
H. KRAUSS	DLR
M. NEITZEL	BASF (now University of Kaiserslautern)
P. PETERS	DLR
R. PRINZ	DLR
K. SCHULTE	DLR
G. ZIEGLER	University of Braunschweig
G. ZIEGMANN	AKZO

FRANCE

C. BATHIAS	C.N.A.M.
A.R. BUNSELL	Ecole Nat. Supérieure des Mines de Paris
F. DE CHARENTENAY	P.S.A.
A. HORDONNEAU	Aérospatiale
J.F. JAMET	Aérospatiale
P. LAMICQ	SEP
C. LE FLOCH	Aérospatiale
R. NASLAIN	Lab. des Composites Thermostructuraux, Pessac
J.M. QUENISSET	Lab. de Chimie du Solide Talence
M. SUERY	Inst. Nat. Polytechnique Grenoble

GREAT BRITAIN

M.G. BADER	University of Surrey
P.W.R. BEAUMONT	University of Cambridge
T.W. CLYNE	University of Cambridge
J.H. GREENWOOD	ERA-Technology

B. HARRIS	University of Bath
D. HULL	University of Cambridge
F.R. JONES	University of Sheffield
F.L. MATTHEWS	Imperial College of Science Technology and Medicine

BELGIUM

A.H. CARDON	University of Brussels
N. SPRECHER	Owens Corning Fiberglass
I. VERPOEST	Katholic University Leuven

DENMARK

H. LILHOLT	Riso National Laboratory

ITALY

A. SAVADORI	Enichem

THE NETHERLANDS

J.A.N. SCOTT	Shell Research

SWEDEN

R. WARREN	Chalmers University of Technology

SWITZERLAND

P. DAVIES	Ecole Polytechnique de Lausanne

• ORGANIZING COMMITTEE

A.R. BUNSELL	EACM
A. MASSIAH	EACM
D. DOUMEINGTS	EACM
J. FÜLLER	DLR
G. GRÜNINGER	DLR
K. SCHULTE	DLR

• CONGRESS SECRETARIAT

H. BENEDIC	EACM
C. MADUR	EACM
N. TARDIEU	EACM
S. BOHLKE	DLR

PREFACE

The European Conference on Composite Materials (ECCM-4) will be held for the first time, in Germany after the successes of previous meetings in France and England. The meeting will take place in Stuttgart which is capital of Baden-Württemberg and a centre for new technologies in Germany. Amongst these new technologies, composite materials play a dominant role and it is the aim of the conference to promote scientific discussion of these materials.

Polymer matrix composites are well established and lie at the centre of interest so that a great number of contributions forms on plastic matrix and high temperature resin matrix composites. New developments in the area of reinforcement fibres will be discussed in a special section of the poster session.

Metal matrix and ceramic matrix composites as well as carbon fibre reinforced carbon are strong candidates for future structural materials. These classes of composites receive wide interest at the conference.

The conference organisers received more than 250 abstracts, from which about 160 contributed papers from 20 countries were accepted. In addition to the 80 oral presentations five invited papers on topics of special interest will be given. The recycling problem of fiber reinforced composites will be discussed in a plenary paper.

In the name of all those who were involved in preparation and organisation of this conference, we hope that fruitful discussions but also the social gathering will contribute to further steps in deepening the European cooperation in this fascinating composite research field.

Dr. A.R. BUNSELL
President of
the European Association
for Composite Materials

Prof. Dr. G. GRÜNINGER
President of the
Scientific Committee ECCM-4

VORWORT

Nach erfolgreichen Tagungen in England und Frankreich findet die Europäische Konferenz für Faserverbundwerkstoffe (ECCM-4) das erste Mal in Deutschland statt. Die Tagung wird in Stuttgart durchgeführt. Stuttgart ist die Metropole des Landes Baden-Württemberg und ein Zentrum für neue Technologien in Deutschland. Faserverbundwerkstoffe haben dabei immer eine hervorragende rolle gespielt. Die wissenschaftliche Beschäftigung mit diesen Materialien stellt die diesjährige Tagung in den Vordergrund.

Schwerpunkte liegen auf dem Gebiet der Polymermatrix-Verbundwerkstoffe. Den in Zukunft immer wichtiger werden den Thermoplastmatrix-Systemen und den Hochtemperatur-Kunstoffen wird besondere Beachtung geschenkt. Entwicklungen auf dem Gebiet der Verstärkungsfasern werden in der Poster Session dargestellt.

Faserverbundwerkstoffe mit Metall- und Keramikmatrix, sowie kohlenstoffaserverstärkter Kohlenstoff stehen an der Schwelle zur anwendung. Dem ausgeprägten wissenschaftlichen Interesse an dieser Werkstoffgruppe entspricht deren besondere Berücksichtigung.

Die Organisatoren der Konferenz erhielten mehr als 250 Abstracts, von denen 160 aus etwa 20 Ländern akzeptiert wurden. Zusätzlich zu den etwa 80 Vorträgen werden 5 eingeladene Vorträge zu Themen von besonderem Interesse gehalten. Die Probleme des Recyclings werden aufgegriffen und einen Schwerpunkt der Diskussion bilden.

Im Namen all jener, die an der Vorbereitung und Durchführung dieser Konferenz beteiligt waren, hoffen wir, daß die rege Diskussion und Auseinandersetzung ein weiterer Schritt zur vertieften Kooperation in Europa auf dem Gebiet der Faserverbundwerkstoffe sein wird.

Dr. A.R. BUNSELL
Präsident der
Europäischen Vereinigung
für Faserverbundwerkstoffe

Prof. Dr. G. GRÜNINGER
Präsident des
wissenschaftlichen
Komitees ECCM-4

TABLE OF CONTENTS

FATIGUE/*ERMÜDUNG* :

INTERFACES/*GRENZFLÄCHEN* :

METAL MATRIX :

CERAMIC MATRIX :

CARBON-CARBON COMPOSITES :

DELAMINATION :

FIBRES :

ANALYSIS/*ANALYTISCHE METHODEN* :

PROPERTIES/*EIGENSCHAFTEN* :

SESSION CHAIRMEN

- **FABRICATION**
  Dr. J. KRETSCHMER, Daimler-Benz AG
  Prof. F.L. MATTHEWS, Imperial College

- **FATIGUE/ERMÜDUNG**
  Prof. F.L. MATTHEWS, Imperial College
  Prof. B. HARRIS, University of Bath

- **INTERFACES/GRENZFLÄCHEN**
  Dr H. LILHOLT, Riso National Laboratory
  Prof. A. KELLY, University of Surrey

- **METAL MATRIX**
  Prof. M.G. BADER, University of Surrey
  Mr. M. ABIVEN, Aérospatiale

- **CERAMIC MATRIX**
  Prof. M. VAN DE VOORDE, Commission des Communautés Européennes
  Prof. G. ZIEGLER, University of Bordeaux I

- **CARBON-CARBON COMPOSITES**
  Dr. W. HÜTTNER, Schunk-Kohlenstofftechnik

- **DELAMINATION**
  Prof. J.L. LATAILLADE, University of Bordeaux I

- **CONSTITUENTS/KOMPONENTEN**
  Dr. A.R. BUNSELL, Ecole Nationale Supérieure des Mines de Paris

- **IMPACT AND ENERGY ABSORPTION/SCHLAGVERHALTEN UND ENER GIE ABSORPTION**
  Prof. I. VERPOEST, Catholic University of Leuven

- **NON DESTRUCTIVE TESTING/SERSTÖRUNGSFREIE PRÜFUNG**
  Mr. C. LE FLOCH, Aérospatiale

- **ANALYSIS**
  Prof. M. NEITZEL, BASF, now University of Kaiserslautern
  Prof. S.W. TSAI, University of Stanford

- **APPLICATION/ANWENDUNG**
  Prof. S.W. TSAI, University of Stanford
  Prof. I. CRIVELLI-VISCONTI, University of Naples

- **PROPERTIES/EIGENSCHAFTEN**
  Prof. F.J. ARENDTS, University of Stuttgart
  Dr. A. SAVADORI, Enichem

- **THERMOPLASTICS**
  Dr. J.A.N. SCOTT, Shell Research
  Prof. K. FRIEDRICH, T.U. Hamburg-Harburg, now University of Kaiserslautern

PLENARY PAPERS

- **"PRODUCTION, PROPERTIES AND APPLICATIONS OF METAL MATRIX COMPOSITES"**
  Prof. G.A. CHADWICK, Hi-Tec Metals, Southampton, Great Britain

- **"ASPECTS FOR RECYCLING - PLASTICS ENGINEERING PARTS"**
  Prof. Dr. A. WEBER, BASF, Ludwigshafen, West-Germany

- **"MICROMECHANICS AS A BASIC MILESTONE IN FIBRE REINFORCED POLYMERS AND METALS MACROMECHANICS UNDERSTANDING"**
  J.P. FAVRE, ONERA, Chatillon, France

- **"POTENTIAL AND DESIGN ASPECTS OF CERAMIC MATRIX COMPOSITES"**
  R. KOCHENDÖRFER, DLR, Stuttgart, West-Germany

- **"GLASS REINFORCED THERMOPLASTIC MATRIX COMPOSITES (GMT) : TECHNOLOGY AND APPLICATION"**
  Dr. F. ROSSI, FIAT AUTO, G. MOLINA, Centro Ricerche FIAT, Torino, Italy

PRODUCTION, PROPERTIES AND APPLICATIONS
OF METAL MATRIX COMPOSITES

G. CHADWICK

Hi-Tec Metals R&D Ltd
Unit B4 - Millbrook close - Chandlers Ford - Eastleigh
SO5 3BZ Southampton - Great Britain

ABSTRACT

This paper presents a brief review of the pressure casting of
MMCs. These melt production processes can accommodate reinforcements
of continuous fibre, short fibre, particulate and hybrid fibre-
particulate types and the reinforcement phase can be distributed
uniformly throughout the composite structure or, using preform
technology, in selected regions of the casting. Some mechanical
properties of squeeze cast MMCs are discussed, mainly from the
standpoint of the conflicting requirements of strength and toughness.

1 INTRODUCTION

Fundamental research on the structure/property relationships of
metal matrix composites is now flourishing in academic establishments
across Europe (1) whilst at the same time industrial organisations are
assessing and developing the viable manufacturing routes and exploring
possible applications for this new type of material. Figure 1 shows
the variety of manufacturing processes available for MMC production
from both the liquid state and the solid state. It is clear that the
liquid state processing routes are far more numerous than solid state
routes, reflecting a cost differential in favour of the liquid metal
processes. Of these latter processes, the products are either near-
net-shape or in more massive form for subsequent solid state processing.
This article deals with near-net-shape processing of MMCs using liquid
metal casting routes and especially those in which an applied pressure,
be it positive or negative, is deliberately manipulated to positive
effect during the infiltration of a reinforcement preform and during
subsequent solidification of the matrix phase.

3

Some of the earliest work on the production and testing of metal matrix composites used liquid metal infiltration techniques under vacuum. Copper-tungsten composites /2/ were made by sucking copper upwards through a stack of tungsten rods contained in a silica tube which had a constriction at its top to serve as a choke, by solidification, to the flow of copper when infiltration was complete. Similar techniques are still being used today /3,4/: evacuated moulds containing the reinforcement phase are filled with liquid metal to yield complex shaped composites. Sometimes an additional positive pressure is applied after vacuum infiltration in order to prevent, or more usually to minimise, the formation of contraction cavities. These "low pressure" infiltration processes usually operate at a few atmospheres pressure at most; that is, <1 MPa. In contrast, "high pressure" die casting /5/ and compocasting /6/ utilise pressures of 50-80 MPa. Squeeze casting /7/, at the extreme, uses applied pressures typically in the range 50-200 MPa in order to produce large, full integrity, pore-free metal matrix composites.

All these processes have one thing in common: the metal matrix and the reinforcement phase are generally not in thermodynamic equilibrium. Thus, the physical and chemical parameters affecting the manufacture of these type of metal matrix composites have universal influence.

2 PREFORM INFILTRATION

A basic requirement of composite theory and practice is the intimate contact and efficient bonding between the matrix phase and the reinforcement phase in order to promote effective load transfer from the matrix to the reinforcement. A low wetting angle between the phases will aid infiltration, but too low a wetting angle might create such effective load transfer so as to endow the composite with a low fracture toughness. On the other hand, too high a wetting angle will increase the difficulty of preform infiltration, especially for dense preforms, and residual porosity in the MMCs will be more probable. In order to . balance out the two conflicting requirements of good load transfer and high toughness, much effort is being expended on finding suitable surface coatings for fibres to optimise load transfer and toughness and to minimise chemical interactions.

Theoretical and experimental studies /8,9,10/ have been made on the thermodynamics and kinetics of fibre preform infiltration. In essence, the total pressure differential, ΔP, required to infiltrate a preform is dependent on four separate terms:

P_o, the pressure on the liquid metal at the surface of the preform;
P_g, the gas pressure within the preform;
ΔP_γ, the Gibbs-Thomson pressure at the liquid front,
and ΔP_η, the pressure drop due to flow through the porous preform.

$$\Delta P = P_o - P_g = \Delta P_\gamma + \Delta P_\eta$$

The pressure to infiltrate a preform, then, depends on the wetting angle; the preform density and geometry; the back pressure in the preform; and the applied pressure, which can vary over orders of magnitude according to the casting technique used.

4

The temperatures of the fibre preform, the mould, and the liquid metal are also important parameters in securing proper infiltration. Liquid metal entering the preform freezes onto the fibre surface and chokes the flow of liquid by diminishing the effective pore size and increasing the viscous drag. The lower the preform temperature the thicker is the solidified layer on the individual fibres of the preform and the shorter is the infiltration distance. Since latent heat is evolved during the metal deposition within the preform, metals with high latent heats will penetrate further into the preform than metals with low latent heats. Penetration of the preform ceases when the velocity of the liquid metal reaches a critical value, at which point the fibres ahead of the infiltration front extract sufficient latent heat to solidify the metal front entirely. Based upon this analysis, the minimum pressure required to infiltrate a preform ideally composed of uniaxial fibres is in the region of 5-10 MPa. This analysis has so far neglected the presence of air in the preform which can be compressed to very high pressures if it is not allowed to escape, thus necessitating very high values of P_0 to effect preform infiltration. Venting of internal preforms can be problematical; and the mass production of MMCs under vacuum is not foreseen to be industrially economic.

At the present time it appears that most low and medium pressure casting processes leave residual porosity in the reinforcement preform due to capillarity effects, gas entrapment, or solidification contraction effects, either acting singly or in concert. The only casting process that has been shown to date to be capable of producing, and guaranteeing, porosity-free MMCs is squeeze casting.

3 SQUEEZE CASTING OF MMCs

The principles of squeeze casting /11/ have been understood for many decades and the process has been in use for 50 years or more. It is only recently, however, that squeeze casting has been used for the manufacture of metal matrix composites, but already the process has been commercialised for the production of fibre reinforced piston /12/. The process is illustrated schematically in Figure 2. The magnitude of the pressure required to avoid porosity is governed by both the casting and the preform geometries. Some reported research has been conducted with applied pressures below the critical level to achieve full infiltration of the preform /13/, and possibly even below the level to avoid porosity in the matrix phase alone /14/.

After filling the mould, in air, with a metered amount of liquid alloy, pressure is applied across the entire surface of the casting by the punch. Liquid metal is pushed into the preform capillaries by the applied pressure but against the back pressure arising from the entrained air within the preform. Since the preform is often placed, as shown, at the base of a solid mould and since the application of pressure to the melt effectively seals the system and prevents the air from escaping, the pressure of the gas increases during the infiltration process. To reduce the volume of the trapped air to, say, one-thousandth of its original volume, an applied pressure of 1000 atmospheres, or 100 MPa, will need to be exerted on the melt by the punch of the squeeze casting machine. This is where the idealised analyses of threshold pressure requirements for preform infiltration are seen to be

5

inapplicable to the more general, practical case. Squeeze casting under vacuum does not reflect industrial reality.

A further problem arises because both oxygen and nitrogen react with aluminium to form very stable compounds; only the oxide reaction has been considered to date /15,16/. The oxide-coated aluminium alloy will deposit its surface oxide onto the preform fibres as it passes, with a wetting angle of ∿180°. The freshly exposed clean aluminium will spontaneously form more oxide to be layered onto the preform, and so on. Thus, for alumina preforms, the infiltration process is not so much one of forming aluminium/alumina interfaces but one of creating amorphous alumina/crystalline alumina interfaces. The argument is equally valid for all types of reinforcement phase.

Although large scale (interfibre) porosity is not observed in properly squeeze cast MMCs it is entirely possible for some gas to be trapped between the deposited oxide layer and the reinforcement phase, particularly if the latter has an irregular surface contour. Nonetheless, squeeze casting is considered to be the most effective and efficient route for the large-scale production of full-integrity MMCs.

4 MECHANICAL PROPERTIES

The mechanical properties of MMCs are inevitably a compromise between the properties of the matrix and reinforcement phases. For melt produced composites such compromises are compounded because the phase mixtures are thermodynamically unstable and interfacial reactions occur rapidly to degrade the interfaces and diminish their load bearing capabilities. Any binder phase in the reinforcement preform can complicate matters further. Much attention is now being focussed on interfacial monitoring and control in MMC systems. Interfacial stability can be improved either by creating a stable microenvironment around each fibre using barrier coatings (e.g. surface treated SiC fibres) or by altering the macroenvironment of the reinforcement phase by changing the chemistry of the matrix phase (e.g. adding at least 7% Si to the aluminium phase for SiC particulate composites). The problems of microporosity, gas entrapment, and of oxide or nitride formation at the fibre-matrix interfaces have not been addressed.

Research on the mechanical properties of composite materials has been restricted by the limited availability of a small variety of reinforcement materials, and by the limited amounts of MMC material that have been available to researchers for testing. This is not to say that little research has been performed, but that work has been selective and specific rather than comprehensive and progressive. Except for the early work on model systems, most recent mechanical property investigations have been carried out using MMC material produced using some form of squeeze casting process or an alternative, less efficient, lower pressure process.

It is clear that the composition and properties of the matrix phase affect the properties of the composite both directly, by normal strengthening mechanisms, and indirectly, by chemical interactions at the reinforcement/matrix interface. In this section we consider first the properties of some pure aluminium composites in which interfacial effects are minimised, before examining the properties of more complex composite couples.

6

4.1 Aluminium-Saffil Composites

Short-fibre alumina (Saffil[R]) preforms have been used to produce MMC test pieces having fibre loadings of 0.05-0.30V_f in 0.05V_f increments. The preforms, having a 2-D random array of fibres 3μm ∅ x 3mm long, were squeeze infiltrated at a pressure of 120 MPa with commercially pure (>99.9%) aluminium. Figure 3 shows the modulus of the composites measured in compression on samples of dimensions 20x10x10mm, the long axis taken in the layer plane of the preform. The calculated modulus of the composites is also shown in the figure. Excellent agreement is observed between the empirical and theoretical values.

Compression testing and toughness testing have also been conducted on the composites. Figure 4 shows that both the compressive yield stress and ultimate failure stress increase with increasing fibre loading. Examination of the failed specimens shows that interfacial debonding is prevalent in all of these composites, consistent with the low measured work of adhesion in the alumina-aluminium system. In contrast, the impact and fracture toughness values of these pure aluminium matrix MMCs both decrease with increasing reinforcement loadings, as shown in Figure 5 for fracture toughness. In these composite couples, increased modulus and strength are paid for by a decreased toughness.

In order to establish the effects of alloying elements on the fracture toughness, alloys of Al-2Mg and Al-4Zn-2Mg were used to squeeze infiltrate saffil preforms at fibre loadings of 0.10, 0.15 and 0.25 V_f. The results for alloys infiltrated at a common 120 MPa are plotted in Figure 5 and show that for the unheat-treated alloys, the Al-2Mg matrix improves the fracture behaviour whereas the Al-4Zn-2Mg matrix lowers the fracture toughness relative to the pure aluminium matrix composites at low V_f but increases it at high V_f. The effect of Mg in the binary alloy can possibly be rationalised in terms of an increased work of adhesion between fibre and matrix which decreases the amount of delamination around fibres normal to the principal stress axis. The zinc containing alloy behaves differently, presumably due to enhanced segregation effects around the fibre interfaces particularly in the low V_f composites.

Other experiments have been conducted with Al-5Mg and 7010 aluminium alloys as matrices for 0.15V_f saffil MMCs: tensile, fracture toughness and crack propagation rate tests have been conducted to determine how composite properties are altered with changing matrix phase. Table I presents the results of this study. It is obvious that the heat treated high strength 7010 alloy MMC shows little ductility compared with the unheat-treated Al-5Mg matrix composite, a feature attributed to the inability of the high strength alloy to plastically deform sufficiently to blunt the cracks formed by fibre breakage. This is supported by the observation that the 7010 matrix phase exhibits a much flatter and more brittle appearance than the Al-5Mg matrix alloy.

In three-point bend tests, both 7010 and Al-5Mg composite materials exhibit low fracture toughness values, being just over one-half of the respective monolithic matrix values. Critical crack lengths in the composite materials are very small compared with those of the matrix phases alone, being only 1.5mm for the MMC compared with >15mm for the

7010 alloy system. Crack propagation rate curves show that the rate of crack growth is lower in 7010 than in Al-5Mg despite the fact that the latter shows larger transgranular facets than the former. Fatigue cracks in the Al-5Mg composite operate on different macroscopic planes, these planes being linked together by high-energy-absorbing shear cliffs: in the 7010-saffil MMC these macroscopic shear cliffs are absent. This difference in behaviour manifests itself in a higher crack growth rate in the 7010-saffil MMC than in the Al-5Mg-saffil MMC. The K_q values calculated from the crack propagation tests agree reasonably well with those obtained from the three-point bend tests.

These experiments indicate that overall better engineering properties are obtained with the softer Al-5Mg matrix phase than with the hardened 7010 matrix alloy. In particular, fracture toughness is generally found to be very low in high strength matrix composites. Despite this fact, it is common for the heat treatment schedules of MMCs to be optimised for highest strength, notwithstanding the prime requirement of producing tough composites for high tensile load-bearing applications.

4.2 Cast SiC particulate MMCs

The ambivalence referred to above can be illustrated with the Duralcan-type SiC particulate materials now available in commercial quantities for near-net-shape casting of engineering components. Research work at Hi-Tec Metals R & D on the type of material has shown that a heat treatment different from that used to maximise tensile strength should be selected if high toughness is a desired property. The recommended heat treatment of A357-20% SiC_p is that which gives the maximum tensile properties of the matrix phase alone; but this heat treatment schedule gives a composite fracture toughness value of only 19.7 $MPam^{\frac{1}{2}}$. It can be seen from Table II that this is 25% less than the figure for the as-cast composite material and 25% less than the fracture toughness of the heat-treated matrix phase alone. Over-ageing the composite (250°C/2hr) produces a toughness of 25.8 $MPam^{\frac{1}{2}}$ whilst retaining the high modulus, but the strength drops to less than that of the A357-T6 alloy. By solutionising and naturally ageing, the strength of the composite can be increased to above 300 MPa whilst at the same time achieving a toughness of 36.4 $MPam^{\frac{1}{2}}$. Annealing the composite increases the toughness still further but the tensile properties are reduced to extremely low values. Clearly it is possible to optimise to a certain extent the toughness, strength and stiffness of this type of particulate MMC by selecting the appropriate heat treatment for the particular combination of properties required.

5 SUMMARY

Liquid-phase processing of MMCs is the industrially favoured route for the mass production market due to its cheapness and versatility. Squeeze casting at pressures in the region of 100 MPa can produce components having localised reinforcements emplaced at specific locations in the casting and being guaranteed porosity free. Casting at lower pressures, especially for complex shapes or for shapes having large ratios of depth to width, can leave remnant porosity in the composite region which will lead to an ensuing diminution in mechanical properties. In all casting processes care has to be most critically

8

exercised to prevent the entrapment of gas within the fibre reinforcement phase.

Metal matrix composites containing high-modulus small-diameter ceramic fibres appear to be of inherently low toughness. Increasing the toughness by applying coatings to the fibres is possible but not yet practicable for the cheaper alumina preforms. Changing the composition of the matrix phase can produce tougher composites, as shown above for the Al and Al-5Mg matrix saffil reinforced composites. Such improvements are derived empirically, however, and alloy optimisation tests are time-consuming and expensive.

It has not yet been possible to obtain from a fundamental point of view any significant understanding or predictive capability of the fracture behaviour of these type of MMCs. Basic values of interfacial shear strengths are difficult if not impossible to obtain with short-fibre 2-D random ceramic preforms. The application of the newly developed fibre push-out tests to metal-based systems could be useful in providing fundamental data for mechanical property evaluation and interpretation.

6 ACKNOWLEDGEMENTS

The author would like to thank all his colleagues at Hi-Tec Metals R & D for providing the information recorded in this paper.

7 REFERENCES

1. "Metal Matrix Composites" E-MRS Symposium, Strasbourg, May, 1990. In press. Elsevier 1990.

2. R.W. Tyson, Ph.D. Thesis, University of Cambridge, 1964.

3. F. Folgar, J.E. Widrig and J.W. Hunt, "Design, fabrication and performance of fibre FP/Metal matrix composite connecting rods". 32nd SAMPE Symposium, April 1987. pp 1559-1567.

4. R.L. Trumper, P.J. Sherwood and A.W. Clifford, in "Materials in Aerospace" Royal Aeronautical Society Conference, London 1986, Vol.II pp 249-289.

5. N.W. Rasmussen, P.N. Hansen and S.F. Hansen, "High pressure die casting of fibre-reinforced aluminium alloys" E-MRS Symposium B. Strasbourg 1990.

6. F.A. Girot, L. Albingre, J.M. Quenisset and R. Naslain, "Rheocasting Al matrix composites". Journal of Metals, Nov. 1987.

7. G.A. Chadwick, "The squeeze casting of light alloys and composites", E-MRS Symposium 1985, pp 213-219. Les Editions de Physique. Les Ulis. France.

8. H. Fukunaga, "Squeeze casting processes for fibre reinforced metals and their mechanical properties". Cast Reinforced Metal Composites. ASM 1988. pp 101-108

9. A. Mortensen, V.J. Michaud, J.A. Cornie, M.G. Flemings and L. Masur, "Kinetics of fibre preform infiltration". Ibid. pp 7-13.

10. J.M. Quenisset, R. Fedou, F. Girot and Y. lePetitcorps, "Effect of squeeze casting conditions on infiltration of ceramic preforms". Ibid. pp 133-138.

11. G.A. Chadwick and T.M. Yue, "Principles and applications of squeeze casting" Metals and Materials, 1989 pp 6-12.

12. R. Munro, CIMAC Conference, Oslo, 1985.

13. H. Fukunaga, K. Goda, Y. Kurita and Y. Fujita. "Strength and reliability of silicon carbide reinforced aluminium composites by squeeze casting including a vacuum system". ICCM VI, 1987, Volume 2, pp 362-371.

14. J. Dinwoody and I. Horsefall. "New developments with short staple alumina fibres in metal matrix composites" Ibid pp 390-401.

15. T.W. Cline and J.F. Mason. "The squeeze infiltration process for fabrication of metal matrix composites". Met. Trans., 1987, 18A 1519-1530.

16. S. Schamm, J.P. Rocher and R. Naslain. "Physicochemical aspects of the K_2ZrF_6 process for the infiltration of SiC (or C) preforms by liquid aluminium". EACM3, Bordeaux 1989. Elsevier. pp 157-163.

Table I Properties of a low strength and a high strength aluminium alloy and their saffil reinforced composites.

	0.2%YS MPa	UTS MPa	El at UTS %	E GPa	K'_q MPa m$^{\frac{1}{2}}$	K''_q MPa m$^{\frac{1}{2}}$
7010	490	550	10.5	70	33	>30
7010 MMC	-	220	<0.2	85	21	13
Al-5Mg	122	282	13.8	70	>50	>46
Al-5Mg MMC	155	276	2.0	85	29	14

K'_q obtained from shallow-notched three-point bend tests
K''_q obtained from asymptotic value of crack growth rate curves

Table II Properties of A357 and Duralcan alloys.

	0.2%YS (MPa)	UTS (MPa)	El%	E (GPa)	K (MPa m$^{\frac{1}{2}}$)
A357-F	106	214	10.3	$\sim$70	41.7
A357-T6$_1$	296	347	8.5	$\sim$70	26.7
A357-T6$_2$	302	359	9.6	$\sim$70	24.7
A357-15SiC$_p$-F	132	193	1.5		25.3
A357-20SiC$_p$-F	144	218	2.1		25.2
A357-15SiC$_p$-T6$_1$	351	380	0.6	93	17.9
A357-20SiC$_p$-T6$_1$	369	398	1.3	105	19.7
A357-20SiC$_p$-T6$_2$	377	408	1.4	105	21.8
A357-20SiC$_p$-Stab.	217	274	3.4	105	25.8
A357-20SiC$_p$-Sol.	227	319	4.3	105	36.4
A357-20SiC$_p$-Ann.	149	232	5.4	105	>50

F = as cast
T6$_1$= 525°C/24hr; w.q.; 170°C/8h.
T6$_2$= 540°C/24h; w.q.; RT/24h; 160°C/24h.
Stab. = 540°C/12h; w.q.; 250°C/2h. Sol. = 540°C/12h; w.q.
Ann. = 540°C/72h; air cool.

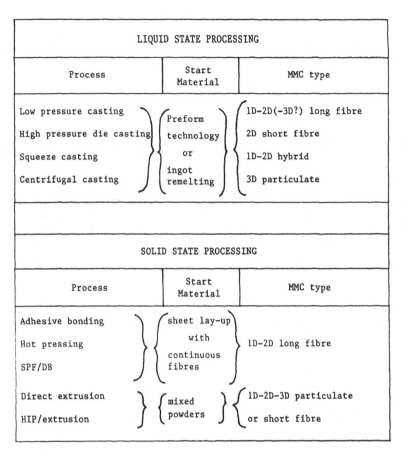

LIQUID STATE PROCESSING		
Process	Start Material	MMC type
Low pressure casting		1D-2D(-3D?) long fibre
High pressure die casting	Preform technology or ingot remelting	2D short fibre
Squeeze casting		1D-2D hybrid
Centrifugal casting		3D particulate

SOLID STATE PROCESSING		
Process	Start Material	MMC type
Adhesive bonding		
Hot pressing	sheet lay-up with continuous fibres	1D-2D long fibre
SPF/DB		
Direct extrusion	mixed powders	1D-2D-3D particulate
HIP/extrusion		or short fibre

Figure 1 Compilation of MMC processing routes.

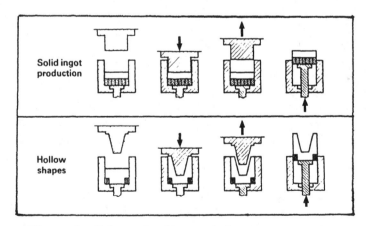

Figure 2 Schematic diagram of squeeze casting process.

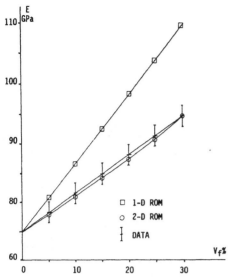

Figure 3 Experimental modulus values of
aluminium-saffil MMCs compared
with rule-of-mixture predictions.

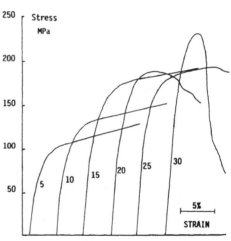

Figure 4 Compressive stress-strain curves of
aluminium-saffil MMCs.

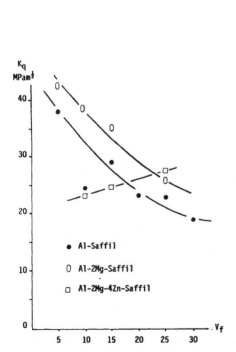

Figure 5 Fracture toughness values of
aluminium-base-saffil MMCs.

ASPECTS ON RECYCLING
PLASTICS ENGINEERING PARTS

Prof. Dr. A. WEBER

BASF AG
Abt. KTE - Postfach - 6700 Ludwigshafen - West-Germany

1. Introduction

In a previous study (1), an attempt was made to describe the role
of plastics as indispensable automotive engineering materials in
the light of the demands imposed on modern automobiles. The present
paper is intended to show the other side of the coin, i. e. the
disposal of plastics parts from abandoned vehicles. The subject has
hit the public eye, but often only individual aspects are placed in
the forefront. A matter-of-fact contribution to the discussion is
therefore essential.

2. Initial situation

According to the Plastics Institute in Aachen (2), the average life
of a motor vehicle in the period from 1975 to 1995 has been ten
years. Over the same period, the proportion of plastics in automobiles has risen from 5.2 % to 13.0 %. In a presentation by the Prognos-Institut in Basle (3), a figure of 10 % was quoted for the
weight fraction of plastics in automobiles for 1995, if reticence
were shown; and of 19 %, if the rate were forced. The latest information from Volkswagen (4) includes a figure of 14 % for the proportion of plastics in 1995. The average weight of a vehicle is
about 1 000 - 1 200 kg. The number of vehicles registered over the
last ten years in West Germany and due for scrapping in 1990 is 2.6
million. In 1984, it was 1.7 million cars and station-wagons. Consequently, the amount of plastics to be disposed of from scrapped
vehicles - 77 kg/automobile - in West Germany is about 200 000
tons. The export of used cars may cause these figures to be altered
slightly. It can be expected that about 252.000 tons will have to
be disposed of in 1995 (3). Since the situation in Europe is similar to that in Germany, the corresponding figure for 1990 would be

about 778 000 tons of plastics.

A breakdown of the materials used in automotive engineering into the various classes is shown in **Fig. 1.**

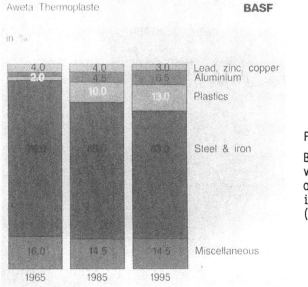

Fig. 1

Breakdown of the various classes of materials used in automobiles (1965 - 1995)

According to Volkswagen research (4), an average car weighs 1 150 kg. It contains 540 kg of rolled steel, 130 kg of cast iron, 75 kg of forgings, 80 kg of light metals, 120 kg of plastics, 90 kg of rubber, 50 kg of glass, 35 kg of paint, and about 65 kg of other materials such as wood, fibrous substances and leather. It is evident that the growing plastics fraction will increase the problems confronted by the shredder industry, which disposes of about 85 % of scrapped vehicles. The economic viability of shredders depends on the recovery of metals from the automobiles. It is current practice to dump or lay aside for disposal plastics and other organic and inorganic waste which together amount to about 25 % of the weight. Plastics are only a part, and their mass fraction is 120/1150 = 10.4 %. The other nonmetallic materials - rubber, paint, glass, wood, fibrous substances and leather - account for about 25 - 10.4 = 14.6 %. This figure is augmented by the oil and other utilities remaining in the vehicle and by the dirt deposited on the bodywork, which amounts to a substantial contribution, as is evident from the calorific value of the shredded waste.

Dumping sites are becoming few and far between and more expensive. Many have already shut down. In 1975, there were about 4 500 dumping sites in West Germany, but now there are only about 350. Dumping costs in 1985 were 95 DM/ton, and a figure of 110 DM/ton is expected for 1995.

The mass flow rates of various materials through shredders in West Germany are shown in **Fig. 2**, which was compiled by Menges (5) from statistical data for 1986 provided by the German Association of scrap merchants (Bundesverband der Deutschen Schrottwirtschaft e.V., Düsseldorf).

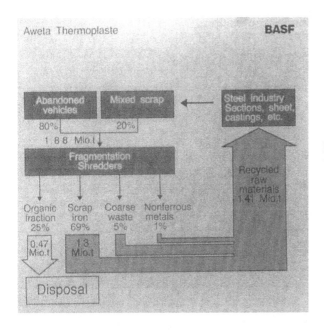

Fig. 2

Mass flow rates through the shredder (5) (1986)

What is referred to as the organic shredder fraction actually contains large amounts of glass and other inorganic matter. In 1986, the weight of plastics in abandoned vehicles was 115 000 tons, or only about 25 % of the figure for the organic fraction, i. e. 470.000 tons,which in turn was 25 % of the total weight of scrap.

2.1. Future applications for plastics in automobiles

As has been mentioned before (1), the automobile and plastics industries have entered into a close partnership aimed at further innovation. The life cycle of important classical applications for plastics is shown in **Fig. 3** together with some new development projects in automotive engineering. It can be seen from this diagram that many applications that offer bright prospects in terms of volume are still in the introductory and growth phases; examples are cylinder head covers and various parts in the bodywork and engine compartment. Thus, there is no doubt that the significance of plastics in automotive engineering will continue to grow in terms of both quality and quantity, with the result that the question of disposing of this plastics fraction will become more and more urgent.

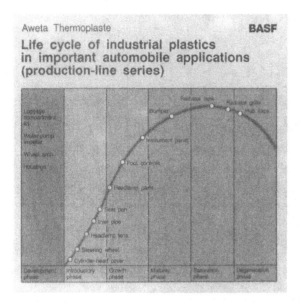

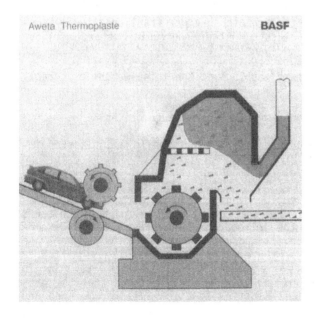

Fig. 4
Schematic diagram
of a shredder

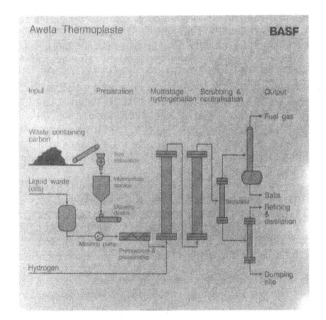

Fig. 5

Operation of a
shredder

2.2 Concepts for disposal

The shredder represents the end of the road for many consumer
durables. They fragment electrical equipment, industrial installa-
tions, and - most of all - abandoned vehicles. After shredding,
the materials are separated and recycled.

The shredder **(Fig. 4)** is a hammer mill in which the flattened
automobile bodies together with the engines, transmission systems,
axles and other scrap are shredded into pieces about the size of
man's hand in a matter of minutes.

Before shredding, the batteries are removed from the automobiles,
and the petrol tanks are emptied. The lubricants remain. In some
cases, assemblies that are still usable are salvaged before shredding, i.e. nonferrous metal parts; and in others, the engine and
transmission systems are missing.

Metal scrap is classified in magnet separators and settling ba-
sins, and the organic fraction is windsifted in cyclones **(Fig. 5)**.
The light fluff contains plastics, textiles, rubber, residual
paint and oil, dirt, and inorganic substances such as rust and residual glass.

Shredder scrap is classified as waste, the disposal of which demands special treatment, and it is likely to remain so. The reason
for this classification is not the plastics fraction but the oils.
Plastics scrap can never fall under this category.

In view of the increasing shortage of dumping sites and the rising costs of dumping, the shredder industry (there are about 45 shredders in West Germany) fears that it can no longer operate economically. Consequently, a joint project has been devised by the branches of industry represented in the raw materials committee of the German confederation of the automobile industry*, i.e. the automobile industry itself, the chemical industry, and steel manufacturers. The result of their effort has been a research programme on recycling plastics and disposing of shredder waste. All the industrial groups concerned have organized a workshop to coordinate the detail tasks within the scope of the Working Group 10 of the automotive engineering research association**.

The federation of plastics manufacturers*** has assumed the task of investigating all the questions that relate specifically to plastics. In October 1989, it presented a waste disposal concept at the K'89 Plastics Fair. The subjects brought up in this concept are being dealt with by three working groups:

1. material recycling of plastics waste

2. pyrolysis, hydrolysis, alcoholysis and glycolysis for recycling the high-molecular-weight building blocks in waste consisting of mixed plastics

3. energy recycling

*   Verband der Automobilindustrie e.V. (VDA)
**   Forschungsvereinigung Automobiltechnik (FAT)
***   Verband Kunststofferzeugende Industrie (VKE)

2.3   Initial results and current situation

As can be seen from **Fig. 6**, the research programme on recycling plastics and disposing of shredder scrap has focused attention on individual problems.

2.3.1   Designing motor vehicles to facilitate recycling of plastics

The automotive engineering research association has commissioned a company, Dr.-Ing. h. c. F. Porsche, to devise a scheme for designing automobiles that can be easily dismantled. The aim of the project was to submit suggestions for a production-line car in the medium price bracket that would allow plastics parts and assemblies to be readily dismantled and subsequently sorted into pure fractions that could be recycled.

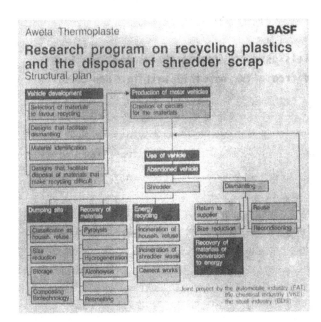

Fig. 6

A production-line vehicle currently on the market had to be analyzed to derive guidelines of a general nature that would lead to a design that could be readily dismantled. The vehicle studied contains 97 kg of all kinds of plastics, 63 % of which could be sorted; 46 % of the total, i.e. 73 % of the fraction that could be sorted, was accounted for by parts that could be separately dismantled and consisted solely of a given type of thermoplastic. As was to be expected, the bumpers, polyethylene fuel tanks, and polyolefin interior trim could be dismantled particularly easily and rapidly; 30 % of the total could be dismantled in about 20 minutes. The total time required for dismantling all the plastics parts and assemblies was 74 man minutes, and another 41 man minutes were required for dismantling these parts themselves and sorting them into the individual classes. Thus the grand total was 115 man minutes. Accordingly, the time required for removing each kilogram of plastics from the vehicle was 46 seconds, and a total of 71 seconds was required for dismantling and sorting.

A breakdown of the thermoplastic parts and assemblies that could be dismantled within 20 minutes is presented in **Table 1**, and the sequence in which they were dismantled is evident from **Table 2**. The results will be published by FAT (automotive engineering research association) in the near future (6).

Table 1

Fractions of materials used to produce the assemblies that could be dismantled from a VW Passat 83 within a period of twenty minutes

Polypropylene	15.4 kg
Polyethylene	8.7 kg
ABS/ASA	5.4 kg
Polyoxymethylene	0.4 kg
Polyamide	0.07 kg

Table 2

Aweta Thermoplaste   **BASF**

Suggested sequence for dismantling the various assemblies

Part or assembly	Cumulative time for dismantling and taking a part (sec)	Cumulative mass (kg)
1. Front bumper	90	5.598
2. Rocker panel	120	6.692
3. Rear bumper	270	12.370
4. Map pocket on front door	330	14.272
5. Sidewipe strip	410	16.156
6. Fuel tank	830	24.675
7. Rear shield	860	25.244
8. Door window seal	900	25.956
9. Tool box	910	26.128
10. Roof lining	920	26.298
11. Air filter	1010	27.725
12. Cover for windscreen wiper linkage	1020	27.860
13. Expansion tank for cooling water	1060	28.374
14. Front wheel arch shell	1180	29.844
15. Radiator cover	1195	30.017

In the light of this initial analysis, the automobile and plastics industries intend to devise designs that will facilitate dismantling from combinations of materials that permit recycling. An example is a bumper system that was presented by BASF at the K'89. It consists of a polypropylene sheath, an expanded polypropylene core and a glass-mat-reinforced polypropylene beam. Basic rules for designing engineering parts that permit recycling have already been published*. It has also been suggested that existing West German recommendations on identifying plastics parts (7) be revised and enlarged.

* VDI-Richtlinie 2243 E

In many cases, the concept of automobiles designed to facilitate dismantling conflicts with the principle of light and composite structures. The question also arises whether it would be in the interests of technological progress and the human race in general to design a highly sophisticated product, the automobile, not from the point of its optimum use but from the aspect of optimum recycling of its individual parts. It would appear worthwhile to find a sound compromise with the aim of restricting dismantling to assemblies that are readily accessible in any case. A realistic basis would be to recover and recycle about 30 % of the plastics used in automobiles. Weight-saving necessitates anisotropic structures of great strength, as can be realized by adopting the principles of lightweight construction, and effectively contributes towards the economics of automotive engineering and protection of the environment (cf. (1) 3.1.1).

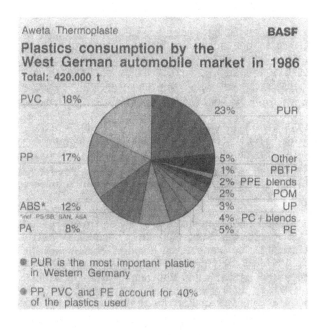

Fig. 7

However, light structures need not necessarily be easy to dismant-
le. In fact, sandwich structures and highly integrated assemblies
point to the opposite. A reliable ecological balance ought to be
struck between the energy that can be saved by weight-saving and
that which can be achieved by recycling the materials. As is evident from **Fig. 7,** the plastics used in automobiles can be broken
down into a few predominant types. Polyolefins hold out the best
prospects for recycling. This fact leads to the demand that the
plastics used in automobiles should be restricted to ten classical types (4). However, all the parts in the engine compartment,
assemblies, electronic units and other highly technical parts cannot be classified in this category.

2.3.2   Reuse of worn parts

The second circuit in **Fig. 6** concerns the reuse of worn parts,
which is an established tradition in automotive engineering and is
carried out on an industrial scale. Ever since motor vehicles have
existed, replacements have been offered for engines, transmission
systems, axles and electrical equipment. Engines reconditioned by
the manufacturer are encountered in everyday practice. Replacements for plastics parts are also quite common. Big scrap dealers,
e.g. Kisow in Norderstedt, offer facilities for dismantling plastics bumpers and other parts or assemblies in order to gain spare
parts (8).

2.4   Recycling plastics removed from abandoned vehicles

2.4.1   Regrinding pure sorts of plastics scrap

Plastics scrap that is intended to be recycled must be reduced in
size. If necessary, it has to be sorted beforehand, and impurities
have to be removed, before it can be sold. The secondary market
for regrind is well-established and is in the hands of private dealers. Clean regrind consisting of only the one type of plastic
can bring prices of up to 70 % of the virgin quality.

In 1988, 9.2 million tons of plastics were produced in West Germany, and the corresponding figure for plastics scrap was about 2.5
million tons, of which about 500 000 tons was recycled by about
200 companies (9). This figure includes the clean scrap recovered
from automobiles. No figures are available for the amounts of material recycled by processors.

Four major West German chemical manufacturers have investigated
the possibility of finding applications for this secondary material. Means have been sought for recycling high-molecular-weight
polyethylene fuel tanks (BASF), impact-modified polypropylene bumpers (Hoechst), polypropylene battery casings and tanks for winds-
creen wash assemblies (Hüls), and glass-reinforced nylon and ABS
wheel hub caps (Bayer). The studies are to be extended to include
other large plastics parts.

The first results in the studies on the individual recycling projects are now available. Differences were revealed in the ease with which the individual plastics could be recovered. Plastics fuel tanks, all of which were unfluorinated and ten years old, presented difficulties because of the petrol that they had absorbed. They thus had to be degased beforehand. The main difficulty presented by the wheel caps was the paint, and a large outlay on cleaning was required for the bumpers. Batteries presented the least problems.

It is thus already apparent that the economic aspects of recycling can by no means be left out of consideration and that recycling at any price is no solution to the problem. On the other hand, no difficulties exist in processing the secondary material. No deterioration in quality was observed in processing the regrind obtained from plastics fuel tanks; even the mechanical properties were not impaired. The strength of articles obtained from recycled wheel caps was 90 - 95 % of that achieved with the virgin material.

The experiments are still in progress.

Volkswagen are erecting a plant in Leer, in which experiments will be performed in recycling readily accessible parts of scrapped vehicles, e.g. bumpers, fuel tanks, and trim. The installation includes a shredder. BMW are pursuing activities of this nature in Wackersdorf, and other concerns are preparing similar projects.

The studies embrace dismantling techniques, the outlay on cleaning, and size reduction and sorting procedures. Information must be obtained on the attainable properties of the recovered material and, in particular, on the means of processing it.

The long-term targets are closed circuits; for instance, a fuel tank should become a fuel tank again by the techniques described. Obviously, this does not succeed in all cases.

Recycling quotas of 30 % are aimed at, and recycling of materials and energy must go hand in hand. Since the recovered material must satisfy quality specifications, in conformance with standard practice in the automobile industry, the product streams of recycled material must be kept under strict control by the automobile manufacturer.

Close cooperation in these experiments is maintained between automobile manufacturers and the plastics industry, i.e. between the processors and the raw materials manufacturers. However, the infrastructure is undoubtedly the crucial factor in recycling materials. It already commences in channelling the abandoned vehicles to the scrap heap and separating them according to their make and model.

Plans entertained by German chemical concerns on a joint venture for recycling plastics waste are in the same direction but, transgress the limits set by the recovery of plastics from abandoned vehicles. Several companies have established central services departments dealing with plastics and the environment. The task of these departments is to devise solutions for disposing of and recycling plastics waste and to realize pilot projects. The most difficult task in this connection is again presented by the logistics, i.e. the planned collection of plastics waste and sorting it into the various types. A new field of activity has thus been opened up.

2.4.2 Recycling mixed plastics scrap

Contaminated and mixed plastics scrap can also be recycled, but an attendant deterioration in quality can frequently be expected (10). Articles with thick walls can thus be produced: flower tubs, park benches, fences, acoustic walls, etc. However, the market for these cannot be expanded indefinitely, and this also applies in some cases to plastics scrap consisting of only the one sort. It is precisely the automobile industry that is seldom prepared to use secondary qualities, because each processing cycle entails a reduction in the material's quality.

2.4.3 Fundamental aspects on simple recycling

Organic substances are subject to degradation and thus cannot be recycled indefinitely. The outcome of each cycle is again scrap until, finally, the quality has detriorated to such an extent that no further use is possible. The final repose is the rubbish dump, to which two-thirds of household and industrial scrap, including that from shredders, still finds its way today. Only one-third of the refuse is incinerated, although this is precisely the most efficient means of waste disposal, because the ash occupies only 10 % of the original volume. From the ecological aspect, there is no point in expending energy in purifying contaminated plastics scrap, because it entails the expenditure of energy and other utilities that themselves require to be protected, e.g. air and water. Thus, the one solution would raise fresh problems, because energy must be expended once again in purifying the polluted air and water. These aspects demonstrate how imperative it is to recycle energy.

2.4.4 Recycling by chemical processes

1. Pyrolysis of mixtures of plastics scrap

An industrial-scale demonstration plant for the pyrolysis of plastics has been in existence in Ebenhausen, near Ingolstadt, since the early 1980s. It was promoted by the plastics industry and is now operated by Asea Brown Boveri (ABB). The pyrolysis reaction breaks down all kinds of plastics into their chemical molecular building blocks. The plastics do not burn, but undergo destructive distillation with the exclusion of oxygen **(Fig. 8)**. A gas is formed that allows the plant to be operated without external energy. Other products are pyrolysis oil and soot, which can also be marketed, and slag. The excess gas can be led into the municipal sy-

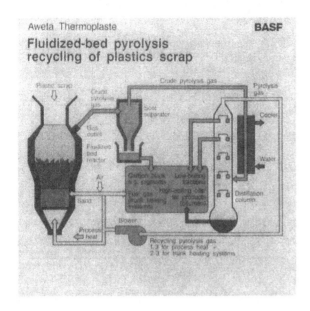

Awela Thermoplaste   **BASF**

Fluidized-bed pyrolysis
recycling of plastics scrap

Fig. 8

stem, and the fractionated pyrolysis oil can be used for the production of plastics. Pyrolysis has already been adopted today for
the main component of plastics waste derived from scrapped automobiles, i. e. polyolefins. However, since much more than 90 % of
the crude oil produced is still being consumed for heating or as
engine fuels, there is no point in squandering resources by converting plastics into oil and gas, most of which will be burnt in
any case. It is better from the ecological aspect to chose the direct route of converting "white coal" into energy than the indirect route of decomposing it into a combustible fuel by a technique that causes pollution.

2. Hydrogenation of plastics scrap

Union Rheinische Braunkohlen-Kraftstoff AG in Wesseling has demonstrated another process for recycling mixed plastics scrap by breaking it down into its molecular building blocks. A current of hydrogen is passed at elevated temperatures and moderate pressures
over the plastics scrap and converts it into oils, i. e. hydrocar-
bons **(Fig. 9)**. The yield is very high, but the process itself is
still in its infancy. If the price of crude oil rises substantially, say by a factor of 3, in the next ten years, the process ought
do have developed sufficiently to allow a reliable means of waste
disposal.

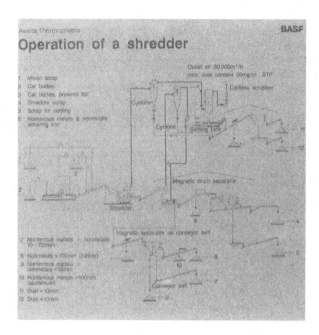

Fig. 9

3. Hydrolysis of plastics scrap

Polyurethanes, polyesters, and polyamides are accessible to hydro-lysis. In other words, they are dissociated at high temperatures in the presence of water, or the condensation reaction that led to their formation is reversed. Dihydric alcohols may be substituted for the water, in which case the term glycolysis applies. This process has already been adopted in a study project performed by the Aalen Engineering College. In this case, polyols were libera-ted from pure polyurethane scrap obtained in the production of au-tomobile seats. The polyols, in turn, were reconverted into polyu-rethane foam.

2.5 Energy recycling with particular reference to shredder scrap

1. Energy recycling

The estimate for the volume of plastics scrap contained in the au-tomobiles abandoned in West Germany in 1990 is 200 000 tons. In the ten years that follow, it will increase to 400 000 tons 500.000 tons but will represent only 10 % of all the plastics wa-ste that accumulates in Germany, the current figure for which is about 2.0 million tons, of which 0.7 million tons is incinerated and 1.4 million tons is dumped. Thus, the question of plastics scrap in abandoned automobiles - no matter how significant it is - is merely the tip of the iceberg. The philosophy adopted by the plastics industry for plastics waste as a whole, or even for all products, is simply to avoid it, to reduce it, or to make use of

28

it. All the aspects mentioned up· to now, each and severally, fall
under these three headings. Energy recycling is definitely part
and parcel of waste disposal. It relieves the load on dumping sites and, at the same time, preserves natural raw materials resources to the benefit of the environment as a whole.

Organic, i.e. combustible, waste ought not to be dumped at all. In
contrast to inorganic waste, it contains energy that, if dumped,
is withheld from us today but has actually been expended at the
cost of the current and coming generations. If this energy were
tapped, fossil fuels, the reserves of which are finite, would be
saved **(Fig. 10).**

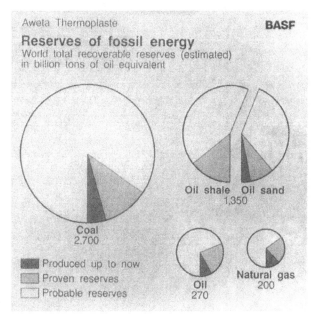

Fig. 10

As has already been explained, recycling materials merely implies
postponement of the inevitable hour and cannot proceed ad finitum,
as is the case in carbon cycle **(Fig. 11)** (11), (12), in which pho-
tosynthesis converts solar energy into biomass at a rate of 100
billion tons a year. Civilization, i.e. living beings and the eco-
nomy, make use of only 3 % or three billion tons of this biomass.
Fossil fuels, i.e. coal, oil and gas, can be regarded as biomass
that has been formed in the biosphere over millions of years. The
stocks that have thus accumulated - about 1 500 billion tons - are
used up at a rate of eight billion tons (0.5 %) a year. The rate
of accumulation does not keep pace with the rate of withdrawal,
and it follows that reserves of fossil fuels are finite and must
therefore be used sparingly. Hence the most urgent ecological engineering task is to reduce energy consumption to a minimum. This
aim can be served by energy recycling. A breakdown of the total
energy consumption reveals that the chemical industry is extremely
thrifty and accounts for only 7 % of the total oil consumption

(Fig. 12). It converts 4 % of the total oil consumed into plastics. And, in contrast to the 80 % that is consumed as fuel for heating and propulsion, all this 4 % can still be converted into energy after the plastics have served their purpose as engineering materials.

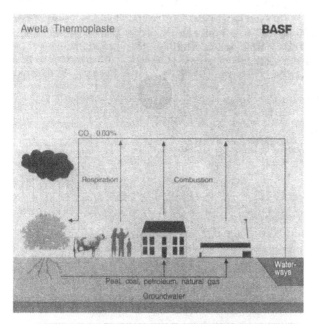

Fig. 11

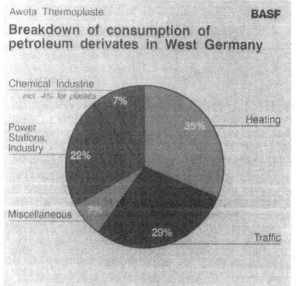

Fig. 12

The enthalpy of plastics corresponds to that of petroleum. Even glass-reinforced unsaturated polyester resins, i.e. SMC/BMC, contain energy, viz. about 10 000 MJ/kg. Household refuse contains about 7 % of plastics waste, which is regarded as a welcome additional source of fuel in incinerators. By virtue of their high calorific value, plastics supply 30 % of the heat required for incineration. Without this contribution, much household refuse could no longer burn independently and would require the support of valuable primary resources in order to burn. The combustion of plastics represents a credit note to reduce the costs of waste disposal.

All that need be said here on the problems caused by heavy metals, halogens, and dioxin is that the difficulties involved have now been mastered and that no environmental hazard need be feared from these substances. After all, the majority of them are not inherent components of plastics, which consist predominantly of nothing more than carbon and hydrogen. In fact, these substances are additives that have been deliberately introduced and could be substituted by others that have less impact on the environment. The quality of plastics as fuels is superior to that of natural resources, i.e. coal and oil. More knowledge on the subject and a basis for optimization can be expected from experimental and pilot plants that are in the planning stages.

Experiments were carried out in a Hamburg incinerator in order to study the combustion and emission characteristics, particularly of dioxin (PCDD/F), under defined operating conditions and controlled fuel input. The results (13) reveal that the emission of PCDD/F from the incinerator could not be changed at all by altering the fuel. This applies particularly to adding given amounts of copper and PVC to the fuel and to sorting plastics from household refuse. In other words, the PVC fraction in household refuse did not have any effect on the formation of polychlorinated dibenzodioxins (PCDD) and dibenzofurans (PCDF).

The costs involved in incinerating PVC, i. e. for the off-gas scrubbing and neutralization caused by the formation of hydrogen chloride, are compensated by its calorific value (14). Thus, plastics on the whole, even including PVC, do not make any recognizable contribution by themselves to the formation of dioxin.

Dioxin formation is suppressed below the natural risk level by optimized design of the combustion chamber and by controlling the combustion and other parameters. The studies mentioned indicate that precautionary measures ought to be adopted more for nonplastics (15). Up to 15 % of plastics may be contained in the waste that is burnt in modern incinerators. There is no point in removing plastics scrap from refuse in general; it would merely entail the expenditure of more energy or further air and water pollution. On the contrary, plastics represent a welcome additional source of energy that allows virginal fossil fuels to be saved by supporting combustion in the incinerator (16).

The heavy metals present in the plastics waste and in printed matter largely originate from paints and inks and are disposed of in the furnace slag. In certain processes, they are embedded in the vitrified slag or are bonded in a cement slurry and can thus be dumped. Heavy metals are not leached by moisture.

Another effective means of disposing of slags and off-gas scrubbing residues containing heavy metals is to embed or fuse them in a plastic.

2. Special incinerators for plastics waste obtained in industry
   Power stations fired on plastics

Table 3

Calorific values of plastics and fuels

	(MJ/kg)
Polyethylene (PE)	43.3
Polypropylene (PP)	44.0
Polystyrene (PS)	40.0
PVC	18.0 - 26.0
Thermosetting plastics	ca. 20.0
Petroleum/fuel oil	42.0
Coal	29.0
Wood	15.0 - 17.0
Paper	13.0 - 15.0

Table 3 shows that plastics are high-energy fuels. It would be an obvious step to exploit the energy by burning plastics together with low-energy waste. For instance, the incineration of sewage sludge together with plastics waste represents a method of disposal that is congenial to the environment. Other methods present growing difficulties. Formerly, the sludge from water treatment plants was spread on cultivated soil and was alowed to seep in; but this gave rise to ecological problems, because sludge contains heavy metals that originate from various sources. If the sludge is incinerated, the heavy metals are included in the ash, in which they may be vitrified or bonded in concrete and thus safely disposed of. An example of an application for this method of disposal would arise if the yield of heavy metals were too lean to allow recovery by metallurgical means.

The West German plastics industry offers municipalities and public utility companies cooperation in erecting a plant on a suitable site for demonstrating the utilization of plastics and sewage sludge for heating. A plan exists for a plant of this nature and merely has to be tried out. Its realization necessitates financial contribution from the Federal Government and support from the politicians responsible. No time should be lost in authorizing the plant and allowing capital investment plans to be drawn up. Plastics manufacturers are confident of support and have already taken concrete steps in finding a suitable site and realizing a demonstration plant (17).

Within the next few years, BASF will plan and project (18) a power station that is to be fired by plastics (18). The aim is to demonstrate that the pollutant load caused by incinerating plastics waste from shredders can be kept well below the permissible level. The installation is to be designed to cope with 30 000 tons a year, and it is planned to incinerate the shredder waste together with household refuse and sewage sludge. A preliminary study of the project has been made by Fichtner, Stuttgart, a well-known engineering consultant,on behalf of the association of plastics manufacturers (Verband Kunststofferzeugende Industrie/VKE).

The results of this study, which is entitled "Thermal utilization of plastics-rich fractions in refuse" (19), are to be taken into consideration in drawing up the plans for the BASF plastics-fired power station. The study is also part of a research programme that concerns recycling plastics and the disposal of shredder waste and has been jointly sponsored by the automobile industry (VDA/FAT), the plastics industry (VKE/GKV), the scrap recycling industry (BDS and DSV), and the steel industry.

3. The disposal of SMC/BMC waste

In West Germany, cured thermosetting plastics scrap is rated as household refuse or as industrial waste similar to household refuse and can thus be disposed of on normal dumping sites or in incinerators. Hence no special measures are required for the disposal of SMC waste. Experiments performed by BASF (20) have proved that the incineration of SMC and BMC does not raise any fresh problems on waste disposal.

Waste glass-reinforced unsaturated polyester composites also do not present any special problems. It can be seen from **Fig. 13** that all the methods of disposal that have been described here also apply to SMC. If SMC or GRP waste is incinerated, the glass fraction can be recovered. Comminuted GRP waste can be recycled in the form of a filler for GRP. If care is taken to ensure that the glass fibres retain their length, they will also exert a reinforcing effect. Work is in progress on processes of this nature.

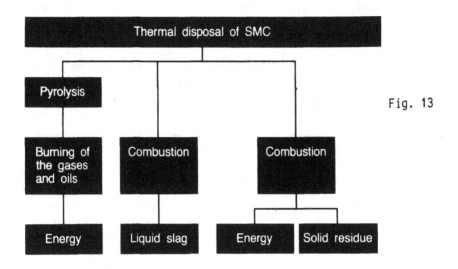

Fig. 13

4. Special recycling techniques for advanced composites

Recycling advanced composites is particularly worthwhile because of the price of the reinforcing materials. The matrix can be removed by mild oxidation or by solvents so that only the rovings and the nonwoven or woven reinforcement remain. In a process developed by MBB, the reinforcement recovered is chopped into fibres of about 5 mm length by rotary cutters fitted with blades that intersect at right angles. The fibres thus obtained can be used for reinforcing thermoplastics. The high price that can be obtained from these regenerated, extremely strong fibres justifies the complicated procedure.

The author wishes to extend his cordial thanks to Dr. Alfred Hauß and Dr. Dieter Müller, BASF AG, Ludwigshafen, for the valuable information and suggestions that contributed towards the compilation of this paper.

References

(1) Weber, A.: "Die technischen Gesichtspunkte zur Rolle der Kunststoffe im Automobilbau", Kunststoffe, March 1990

(2) Müller, H. & E. Haberstroh: "Verwendung von Kunststoff im Automobil und Wiederverwendungsmöglichkeiten", IKV Aachen; FAT Publication No. 52, Frankfurt/Main 1986

(3) Müller, K & S.S. Klockhow: "Entwicklungstendenzen und Recycling-Probleme bei der Substitutuion von Stahl im Automobilbau", Prognos-Institut, Basle 1987

(4) Steiert, H.: "Schon bei der Konstruktion des Automobils muß an die Wiederverwertung seiner Bestandteile gedacht werden", Blick durch die Wirtschaft, Frankfurt/Main, 13 November, 1989

(5) Menges, G. et al.: "Recycling des Kunststoffanteils von PKW", Kunststoffe 78 (1988), 7, 573 et seq.

(6) Mast, P. & K. Strobel: "Demontagefreundliche Gestaltung von Automobilen", FAT Publication, in print

(7) Verband der Automobilindustrie (VDA) 260: "Kennzeichnung von Kunststoffteilen", August 1984

(8) Wutz, M.: "Recycling - ein Kunstruktionsziel im zukünftigen Fahrzeugbau", 2nd Autec 1988 - Congress, 19 - 21 October 88, Sindelfingen

(9) Emminger, H.: "Verwertung von Kunststoffen aus Hausmüll - Möglichkeiten und Probleme ", Paper presented at the 4th International Recycling Congress, Berlin 1984, Published as a reprint by the Verband Kunststofferzeugende Industrie (VKE), Frankfurt/Main 1, Karlstr. 21, 11/84

(10) Schönborn, H.-H. et al.: "Wiederverwerten von Kunststoff-Abfällen unter besonderer Berücksichtigung des Hausmülls", Documentation on studies by Süddeutsches Kunststoff-Zentrum (SKZ), Würzburg; LG-Stiftung (Landes-Giro-Kasse) Natur und Umwelt, Stuttgart 1988

(11) Menges, G. et al.:

"Recycling von Kunststoffen in Altautos"; Paper presented at the 1st International Plastics Forum, Salzburg 1985. Also published under the title, "Fünf Thesen zum Recycling von Kunststoffen aus Altautos", Kunststoffe 79 (1989) 3, 195 - 198

(12) Hauß, A.F.:

"Kunststoffe, Umwelt u. Recycling", Techn. Rundschau Bern, 35 (1987), 26 - 30 and personal communication to the author

(13) Martin, Johannes J.E. and M. Zahlten:

"Betriebs- und Inputvariationsversuche an einer Müllverbrennungsanlage - Ergebnisse und Ausblick -"; EF-Verlag für Energie- und Umwelttechnik GmbH, Berlin. Reprint from Abfallwirtschaftsjournal 5/89

(14) VKE:

"HCL-Emissionen aus der Müllverbrennung und PVC", page 29; Published by the Verband Kunststofferzeugende Industrie, Frankfurt/-Main, Karlstr. 21, 1986

(15) FAT:

Publication No. 72; Untersuchung über das Emissionsverhalten der Leichtmüllfraktion aus Autoshredderanlagen beim Verbrennen

(16) Dirk, E.:

"Thermische Verwendung von Kunststoffen in Haushaltsabfällen"; Frankfurt 1988; The written version of this presentation can be obtained from the Verband Kunststofferzeugende Industrie (VKE), Frankfurt/Main, Karlstr.21

(17) VKE:

Annual Report 1988/89; Verband Kunststofferzeugende Industrie (VKE), Frankfurt/Main, Karlstr. 21, 1989

(18) Newspaper "Mannheimer Morgen", 14 June, 89
Newspaper "Die Rheinpfalz", 27 June, 89

(19) VKE Publication:

Thermische Verwertung kunststoffreicher Abfallfraktionen", Frankfurt/Main, July 1989

(20) König, U., Dorn, I. & G. Heinrich:

"Zur thermischen Entsorgung von SMC-Abfällen"; Paper presented at the 22nd AVK Meeting in Mainz, May 1989

MICROMECHANICS AS A BASIC MILESTONE IN THE FIBRE REINFORCED POLYMERS AND METALS MACROMECHANICS UNDERSTANDING

J.P. FAVRE

ONERA
Materials Department - BP 72 - 92322 Chatillon - France

ABSTRACT

Examples are developed that illustrate how micromechanics can give support to more comprehensive models of the composites mechanical behaviour including quantitative aspects. After the importance of intercorrelating the various methods of measuring interface parameters has been emphasized, the contribution of micromechanics to the prediction of the tensile strength of unidirectional silicon carbide/metal composites and the residual strength of notched carbon/resin laminates are discussed.

INTRODUCTION

It is obvious that all composites are concerned with the micromechanical approach outlined in the above title and the development of the fibre/reinforced ceramics for some advanced programs has recently enlightened the importance of the micromechanical aspects in the damage process.

But fibre/reinforced polymers and metals have in common that, in a number of mechanical situations, the fibres break first, which is the opposite situation of that encountered in reinforced ceramics. The load has then to be transferred to the remaining intact fibres by the organic or metallic matrix via the interface. That there is a profound identity in the behaviour of both kinds of composites is attested by the many early workers, interested in the fundamentals of the reinforcement concepts, indifferently using or thinking of metals or polymers.

While micromechanics has always been called upon to provide ma-cromechanics with simple theoretical models of fibre/matrix interaction, three reasons emerged that explain the renewed interest for an "experimental" micromechanics :

- a large variety of fibres and resins or metals is now available for composite making. However, selecting the proper combination by testing composites and laminates covering all the processing conditions and mechanical situations requires an huge effort. There is an evident need of selection criteria based on the most elementar composite entities i.e. the fibre, the matrix and the interface.

- a great concern has been always put on the latter.While fibres and matrix can be tested independently, interface testing requires that both constituents must be associated. But due to the complexity of the laminate response, informations about the interface characteristics are very difficult to extract from standard mechanical tests on composites. There is a constant need for mechanically "simple" interfaces with the materials in their appropriate form, that is thin brittle fibres, properly cured or moulded resins, properly processed metals.

- even though the main features of the composite effective behaviour are now well described in their broad lines (the mechanics of laminates, based on the "ply", pays no attention to the micromechanical processes, but number of "black boxes" parameters that are finally called upon practically incorporate the corresponding, yet undifferentiate information), there is a general recognizing that damage initiation and development have to be first studied at the micro scale before to be introduced into integrated formal descriptions. To speak shortly, more micromechanics is needed to implement a more reliable macromechanics.

I - Micromechanics does not give just anything

It is well known that composite fracture properties cannot be simply derived from the independantly measured constituents characteristics : the fibre "efficiency" (fibre-to-composite strength ratio) depends upon a lot of parameters /1/ and the matrix tensile or shear yield stress is far from being arrived at when neat resin or metal properties are roughly compared to the average composite properties /2,3/. In addition to the inherent fibre or matrix properties and the actual stress state (thermal and mechanical), the modes by which stress is transferred and the corresponding limits are a critical step. A lot of micromechanical analogs has been developed so far that try to put more "science" into what is commonly called "interface".

Interface is not considered here as the series of layers of chemical species surrounding the fibre that is actually observed. What is in view is a description of the fibre-to-matrix or the matrix-to-fibre interaction in terms of mechanical parameters as stress and strength or energy, i.e. quantities likely to be incorporated into mechanical models. We assume that there is no a priori identity between these interface parameters and the constituents properties that can be independantly determined by appropriate methods. On the same way, even if correlations can be found sometimes, interface parameters and the consequent composite properties cannot be simply derived from an physico-chemical analysis of the fibre surface or the matrix because mechanics is at stake.

It is suggested here that from the moment when the interface parameters are measured in a consistent way and can be readily related to each other, when these parameters, introduced into reliable models, ultimately give the effective properties of the macrocomposite with a good approximation, thus micromechanics does not give just anything.

II - Interface parameters in relation to each other

The parameters are acceded to by different ways that have been reviewed at IPCM'89 in Sheffield /4/ .The first task is that the results of the various methods are to be related to each other in order to make sure that relevant figures are obtained. The relation between pull-out tests and fragmentation tests for carbon/epoxy systems is an example of such a relation /5/.

As pointed out formerly by Piggott /6/, the pull-out test seems to be better described in terms of energy instead of strength. However, as usual for materials properties, both concepts may be turned to a convenient description of the interface properties and, most frequently, the pull-out test has been analyzed in terms of debonding strength. In the model of fragmentation process presented in /5/, two successive regimes of stress building at the fibre short pieces extremities, namely elastic stress transfer and friction, combine to give an overall interfacial shear stress that is actually measured in the fragmentation test. When compared with the Fraser-Di Benedetto original model /7/, or the more recent one by Netravali et al /8/, our model has something new in that it is the very strength calculated from pull-out tests which is taken as the debonding criterion to tilt from one regime to the other. If completed by the determination of the friction stress of the previously debonded fibre (as Piggott does, cf /6/) , the model thus show that pull-out tests and fragmentation tests join together into a quantitative description of the interfacial transfer.

More relations between the data of various methods (including micro-indentation) are expected to be available in the next few years, possibly as a result of a recently initiated round-robin on interfacial testing methods.

III - Micromechanical computation of tensile properties

Several models have been reported in the last ten years to describe the tensile strength of unidirectional composite according to a stochastic mode of fracture taking into account the distribution of the fibre characteristics and the fibre-to-fibre stress transfer /9-11/. Contributions have been also brought by directly observing the progressive fracture of randomly /12/ or regularly arranged /13/ bunches of fibres.

As for any other composite, the tensile strength of silicon carbide/ titanium or /aluminium alloys is controlled to a large extent by the characteristics of the fibre/matrix interface. For metal matrix composites, these characteristics depend very much on the chemical interaction that takes place during processing and on the stresses induced on cooling due to the large difference in coefficients of thermal expansion as well as the high temperature of the forming process. The fragmentation test was applied on single SiC fibres/metal alloy specimens as a function of the temperature to determine quantitatively the interface ability for stress transfering. A single parameter τ, similar to that of the preceding section, is calculated where the ultimate interface bonding strength, the after-debonding friction stress and the plastic deformation of the metal matrix are combined. From the analysis of the fragmentation data, it was concluded that the properties of the composite are likely to be limited by the interface properties for titanium, by the matrix plastic deformation for aluminium /14/. Similar results have been obtained independantly by the same fragmentation test for titanium /15/ and aluminium respectively /16/.

Finally, τ was directly introduced into a model of stochastic rupture of the SiC/metal unidirectional composite. The approach is justified as the fibre volume fraction in the material is small enough so that the conditions around each fibre are about the same as in the single fibre model. The evolution of the tensile strength vs temperature up to 450 or 750°C for SiC/Al or SiC/Ti respectively was then predicted and agrees very well with the experimental tensile data of the corresponding composites /14/.

Obviously, conditions are far from being so ideal with irregularly close packed thin glass or carbon fibres and fibre-to-fibre transfer rules have to be deeply modified. Such a study is now in progress with cured impregnated tows as models of multifilament unidirectional composites. This is an important topic since fibre characteristics are currently given by the manufacturers on the basis of tow testing with a resin whose properties are optimized to give the best performance.

IV- Laminates hole sensitivity

As far as strength is concerned, the multilayer laminates are very sensitive to the stress concentrations around a hole. The commonly applied elastic analysis fails to predict the residual strength of notched laminates because the influence of the micromechanical damage is ignored. Of particular signification is the example of the contribution of this damage to the residual tensile properties of carbon fibre reinforced thermosets or thermoplastics : in contradiction with the expected trend, the "tough" thermoplastics show some deficiencies in key properties compared to the relatively "brittle" thermosets /17/.

As depicted qualitatively in /18/, the elastic stress transfer is likely to be limited in thermoset systems by the ultimate interfacial shear strength. Fibre/matrix debonding then follows with a lot of friction, resulting in a spreading out of the overstressed region over a wider area and a final de-creasing of the local stress concentrations. Jacques had previously pointed out that micromechanical fragmentation data in terms of the interfacial transfer shear stress might be used to predict the damage tolerance of various laminates /19/.

In contrast, due to the high quality of bonding, the thermoplastic matrix (PEEK) yields rather than debonds. Around the hole in the laminate, the overstresses cannot be relieved so easily, resulting in a rapid failure of the material. Owing to the early "necking" of neat PEEK samples /20/, the fragmen-tation process cannot be brought to its normal end (saturation) and there was a lack of quantitative data to verify that transfer is limited by the matrix plastic deformation. However, there is enough experimental evidence that carbon/PEEK interface does not break until the matrix yields.

REFERENCES

1 - Bader M. G, Science & Eng. of Comp. Mat.,1,1(1988)1-11.

2 - Tirosh J., Katz E., Lifschuetz G & Tetelman A. S, Eng. Fract. Me-chanics,12(1979)267-277.

3 - Nicholls D.J. in "Composite Materials : Testing & Design (7th Con-ference)"ASTM STP 893, ed. by J. M. Whitney (ASTM, Philadelphia,1986) 109114.

4 - Favre J.-P. in "Interfacial Phenomena in Composite Materials'89", ed. by F. R. Jones (Butterworths, London, 1989) 7-12.

5 - Favre J.-P., Sigety P. & Jacques D., J. of Mat. Science (in press).

6 - Piggott M. R., Comp. Science & Tech. 30(1987)295-306.

7 - Fraser W. A., Ancker F. H, DiBenedetto A. T & Elbirli B., Polymer Comp. 4, 4, (1983)238-248.

8 - Netravali A. N, Henstenburg R. B., Phoenix S. L. & Schwartz P., Polymer Comp. 10, 4(1989)226-241.

9 - Barry P. W., J. of Mat. Science 13(1978)2177-2187.

10 - Ochiai S. & Osamura K. in "Composites'86 : Recent Advances in Japan and the United States", Proceed. Japan-US CCM-III, Tokyo, 1986, ed. by K. Kawata , S. Umekawa & A. Kobayashi, 751-759.

11 - Manders P. W., Bader M. G. & Chou T. W., Fibre Science & Tech. 17 (1982)183-204.

12 - Wadsworth N. J. & Spilling I., Brit. J. Appl. Phys. (J. Phys. D), ser.2, 1 (1968)1049-1058.

13 - Wagner H. D. & Steenbakkers L. W., J. of Mat. Science (in press).

14 - Molliex L, Thesis Ecole Centrale de Paris (1990).

15 - Le Petitcorps Y, Pailler R. & Naslain R., Comp. Science & Tech. 35(1989)207-214.

16 - Hamann R., Fougères R., Rouby D., Fleischmann P., Gobin P. F., Lonca-Hugot F. & Boivin M., Mémoires & Études Scient. Revue de Métall., décembre 1989, 789-797.

17 - Carlsson L. A., Aronsson C. G. & Bäcklund J., J. of Mat. Science 24(1989)1679-1682.

18 - Vautey P. & Favre J.-P., Comp. Science & Tech. (in press).

19 - Jacques D, Thesis Institut Polytechnique de Lorraine (1989).

20 - Vautey P., Mérienne M.-C, Cottenot C & Favre J.-P., cf ref.4, 53-56.

POTENTIAL AND DESIGN ASPECTS OF CERAMIC MATRIX COMPOSITES

R. KOCHENDÖRFER

DLR - Institute of Structures and Design
PO Box 80 03 20 - 7000 Stuttgart 80 - West-Germany

ABSTRACT

The characterizing properties of ceramic matrix composites differ from those of fiber reinforced polymers or monolithic ceramic materials. These properties are outlined in principal, as are the consequences for design aspects and applications, in order to show the potential of ceramic matrix composites.

INTRODUCTION

"Ceramic matrix composites will be able to solve all the problems of bulk ceramic materials. They will add to the inherent bulk-ceramic properties high tensile strength and high fracture toughness, thus increasing damage tolerance and reliability of ceramic structures". These were the expectations of numerous designers and even the hope of material scientists. However, with increasing experience it became obvious that ceramic matrix composites (CMC) have to be treated as a special class of materials with characteristic properties matching neither those of bulk ceramics, nor the widely used polymer matrix composites (PMC), nor metal matrix composites (MMC). For people dealing with carbon reinforced carbon (C/C), this was no surprise. The experience of those few experts could be totally applied to CMC's and actually today most people categorize C/C as

CMC's. Composite people will recognize a high degree of similarity in material behaviour, and in fact, CMC's offer the highest temperature stability of all matrix systems (polymer, metal, glass, ceramic). Bulk ceramic experts are confronted with the lowest communality in material behaviour, both as to manufacturing parameters and material characterisations.

In the following, only those ceramic matrix composites are discussed which have continuous fiber reinforcement with fibers suitable for textile fabrication processes which are applicable for manufacturing temperatures exceeding 1000°C, e.g., carbon, Nicalon, Tyranno or silicon nitride fibers embedded into carbon or silicon carbide matrix systems.

1. CHARACTERIZING PROPERTIES OF CERAMIC MATRIX COMPOSITES

The main reason to shift from bulk ceramic to fiber reinforced ceramic is to achieve higher fracture toughness. Due to lack of plasticity, dominating properties like long critical crack length, low crack growth velocity, crack stopping effects, crack splitting, crack bridging by fibers and fiber pull-out effects will increase fracture toughness.

Polymer matrix composites are characterized by these properties, offering (like ductile metals) a "forgiving material". This material character comes from strong and/or stiff fibers which are embedded in a weak matrix (low strength, low stiffness, low temperature stability). To achieve a perfect interface between fiber and matrix, the fiber surface is activated to ease chemical reactions with resin constituents. Despite a strong bond, cracks will stop and deviate at the fiber/matrix interface due to the drastic difference in stiffness.

For ceramic matrix composites the situation has to be totally different, because the stiffnesses of matrix and fibers have the same order of magnitude.

Fig. 1 lists the key words characterizing polymer matrix composites and bulk ceramic materials as well as ceramic matrix composites. Some of these aspects will be highlighted in the following comments.

1.1 Interface aspects

In the case of a perfect interface, the stress level within the ceramic matrix would be as high as within the ceramic fibers and a propagating crack would not even notice a change in material character when penetrating from the matrix into the fiber. Thus, CMC materials characterized by a perfect fiber/matrix interface would offer the same brittle nature as bulk ceramics in general, without any noticeable increase in fracture toughness. At the other extreme, where there is no fiber/matrix interface, the matrix would hardly be stressed at all. In the case of a totally fiber-dominated material, even if the matrix is completely cracked or porous, the fibers will carry the load and the matrix has just to keep them in place. This type of material would offer high values of fracture toughness, dominated by friction effects, but due to other requirements it would not be a suitable engineering material.

Thus, a certain amount of fiber/matrix interface strength between 0% and 100% is required to realize damage tolerant ceramic matrix composites, the lower the bonding strength the higher the fracture toughness.

From the quality assurance point of view this is a very critical issue. People dealing with metal matrix or glass matrix composites are well aware of the problems with guaranteeing and qualifying a certain interfacial bonding condition by manufacturing process parameters and by nondestructive testing methods.

1.2 Material purity aspects

Material scientists of bulk ceramics are optimizing their material to be of the utmost purity (clean room processing), small in grain size (submicron powders) and perfect in grain boundary (no or low sintering aids). No defects, no porosity, no cracks should be detectable.

Ceramic composites with high values for damage tolerance and fracture toughness are characterized by totally different properties. In general, a large variety of constituents is involved. The ceramic fibers themselves are characterized by a high amount of impurity, today. To achieve weak fiber/matrix bonding and especially oxidation resistance over a wide temperature range, a variety of coating materials as well as glassy phases have to be used which inevitably increase overall material impurity.

1.3 Material defect aspects

Due to different thermal elongations of fibers, surface coatings, and matrix, and due to volume change and shrinkage impediment during the manufacturing process, microcracks and porosity within ceramic matrix composites cannot be avoided; but even if they could, they should not be avoided. Cracks and voids "soften" the ceramic matrix, reducing its overall stiffness and the inherent stress level within the material. Thus, a microcrack pattern and a certain amount of porosity is favourable and can even be tailored to achieve high fracture toughness.

However, these matrix cracks should not initiate fiber cracks. Again, a weak fiber/matrix interface is a suitable means to overcome this problem.

2.   FIBERS AND MATRIX SYSTEMS FOR CMC's

There are not many fibers available on the market which are suitable as reinforcements for CMC's. Carbon fibers offer the highest thermal stability within an inert atmosphere. However, in an oxidative environment pronounced degradation starts as early as at 450°C, i.e., oxidation protection rather than thermal stability is the outstanding problem with using carbon fibers.

The opposite is true for ceramic fibers. Silicon carbide as well as silicon nitride fibers are not pure fibers, Fig. 2. At high temperatures - especially at temperatures exceeding the fiber manufacturing temperature - the large number of fiber constituents will lead to fiber strength degradation due to diffusion processes or change in microstructure, Fig. 3. The fibers are not homogeneous along the fiber length. Thus, in carbon rich fiber spots local oxidation may occur, reducing the fiber strength values even after only a short time of high temperature exposure in air, Fig. 4. It is indicated in Fig. 3 and Fig. 4 that today's ceramic fibers are not suitable for longtime temperature application above 1000°C.

The most common matrix systems are carbon for C/C composites and silicon carbide for carbon fibers and silicon carbide fibers, respectively (C/SiC, SiC/SiC). To impregnate the fibers with the C or SiC matrix mainly chemical vapour infiltration (CVI) and liquid infiltration processes are applied. These processes are quite well documented in literature [3,4,5,6,7,8,9]. In general, the parameters of the CVI process are not constant over the processing time and the infiltration process has to be repeated several times. Thus, the material quality is not constant over the wall thickness.

3.   DESIGN ASPECTS OF CERAMIC MATRIX COMPOSITES

The statements of the previous sections characterize the specific nature of CMC's, whose material character has to be taken into account in component design. Fig. 5 summarizes representative values of today's materials from the world's leading companies, Société Européenne de Propulsion (SEP), and Aérospatiale (AS). After 10 to 15 years of development the processes are close to maturity, so improvements of more than 20% are not likely in future.

3.1 Strength considerations

Obviously, the failure strength of CMC's is not higher - it is even lower - than for bulk ceramics, Fig. 5. However, this is only true if we compare values gained with small specimens. For large structures this is different. This can be easily explained considering the interface and material defect aspects outlined in sections 1.1 and 1.3.

Given the fact that a microcrack pattern, weak fiber matrix bonding conditions and a certain amount of porosity are the failure dominant factors even on a microstructure level, the failure strength of a small volume specimen has to be the same as that of a large structure, following the weakest link theory. This can be summarized in a global statement. Contrarily to bulk ceramics, CMC's show no or a negligable size effect, and the internal stress level is reduced to be almost zero due to a pronounced microcrack pattern.

3.2 Strain and modulus considerations

The failure strain of damage tolerant CMC's is almost an order of magnitude larger than that of the relevant monolithic matrix system, Fig. 6. This corresponds with a non-linear elastic stress/strain behaviour, e.g. during increasing stress level non-reversible mechanisms occur within the material, like debonding, crack propagation, increasing crack pattern, local fiber fracture and pull-out effects, etc. Thus, also the stiffness or the modulus of the material decreases with increasing stress level applied to the material. As a consequence of the CMC character it is not surprising that already the modulus of the as-fabricated CMC material is lower than that of the corresponding monolithic matrix system.

As for PMC's, the values of stiffness and failure strength of a CMC material depend on the number of fiber orientations realized within the material. A technically feasible material needs at least two fiber directions in plane (2D or 0°/90°). If the weaving process realizes a certain fiber link through the thickness of the fabric, this is called a 2,5D arrangement. 3D means two directions in plane plus one direction perpendicular to the planes. A quasi-isotropic, fiber dominated material can sum to 6D reinforcements. The higher the number the

lower the modulus and failure strength as well as the
degree of anisotropy, Fig. 5.

3.3 Shear strain considerations

The interlaminar shear values are the most critical
mechanical values for CMC's. The shear strength depends
on the void content, fiber/matrix bonding and microcrack
pattern. Although also valid for PMC's, this is much
more pronounced for CMC's. Crack propagation between the
layers needs a minimum of energy; thus, this crack path
is the favourite within ceramic matrix composites, and
consequently the most likely failure mode even due to
secondary stresses in compression and load bearing
tests. Under bending conditions the amount of shear
deformation and shear failure has to be carefully
analyzed. Again, the higher the number of fiber directions (1D, 2D, 3D), the more the fibers dominate and the
more the premature failure or delaminations due to shear
will be reduced.

3.4 Considerations concerning fracture toughness

Fracture toughness is a criterion for crack propagation. Well established methods are available for
homogeneous brittle materials like cast iron, glass or
monolithic ceramics. In general, for fiber reinforced
materials the applicability or modifications of those
methods are still in discussion. This is also true for
CMC's. No doubt, it is possible to measure for CMC's
high stress intensity factors by selecting a proper
fiber orientation and loading condition [10,11]. However, in reality the crack will propagate at a minimum
level of energy dissipation which may lead to a crack
path not coinciding with that of the test. There is
still a lot to do to understand crack propagation
mechanisms in a material with a complex microcrack
pattern.

3.5 Considerations concerning mechanical cycling

As outlined in section 3.2, with increasing stress
level an increasing number of nonreversible mechanisms
change the material's behaviour. By reducing this
maximum stress level and cycling the load within this
lower stress field, we may find a stress level where
obviously no change in material behaviour occurs even
for high cycle fatigue testing. This can be defined as a
maximum design stress level which can be as high as 75%
of the material failure stress as indicated in Fig. 7,
[10]. For tension/tension fatigue tests in one of the
fiber directions (R = 0.1) these excellent fatigue
values correspond with the results of PMC's; however,
more information is needed to explain the good tension/

compression test results (R = -1). According to the experience with PMC's, fiber buckling, edge effects and delamination should influence the test results more than is indicated in Fig. 7, due to weak fiber/matrix bond, porosity and low shear values. Most probably all three effects will dominate the fatigue behaviour in bending even more than in tension condition.

3.6 Considerations concerning high temperature stability

CMC's will never be high purity materials, as indicated in section 1.2. Thus the temperature stability is not expected to be as high as for monolithic high purity ceramics.

Short time tensile test values for SiC/SiC materials decrease beyond 1000°C to 1100°C, whereas C/SiC values drop below RT strength level after exceeding 1700°C, Fig. 8. For longtime applications in oxidative atmosphere two phenomena limit the useful lifetime, fiber degradation and oxidation. Concerning carbon fiber reinforced ceramic composites (C/C, C/SiC), the material lifetime depends mainly upon the quality of the oxidation protection schema, optimized and tailored for each specific application. Thus, a global figure cannot be stated. The lifetime of ceramic fiber reinforced ceramic composites will be limited mainly due to the pronounced fiber degradation which occurs slightly above 1000°C. If this degradation is limited to certain spots along the fiber axis (section 2) a SiC/SiC material, Fig. 10, can be assumed to be an aligned short fiber composite with thermally stable SiC fiber sections in a SiC matrix, showing no degradation of bending strength even after 500h of exposure at 1100°C in air. Contrarily to that, the strength of the reinforcing Nicalon fibers drops drastically in air as well as in argon due to this local fiber spot degradation, Fig. 10.

3.7 Considerations concerning nondestructive testing (NDT)

NDT or quality assurance aspects for CMC's are not resolved today. The damage tolerance behaviour of CMC's is mostly dominated by manufacturing parameters (section 1) which will certainly not be revealed by the manufacturing company. Because of the existent microcrack pattern a number of NDT methods are not successfully applicable, e.g., penetration test, ultrasonic test, etc. There is no proved concept available to quantify the amount of fiber/matrix interfacial bond or the effects due to oxidative degradation by NDT methods. There are still a lot of questions open concerning effects of closed and open porosity or critical crack size within a microcrack pattern, etc.

4. Potential of Ceramic Matrix Composites

Despite a number of open questions CMC components are already successfully integrated in technical systems, and the applications will grow in future due to the unique properties of this class of material.

- C/C materials took a certain market share from graphite as heating or tooling devices in the nonoxidative high temperature regime. Rocket nozzles proved the high temperature, short time applicability of C/C material, even in an oxidative environment [10]. C/C braking systems are a market which may be expanded by higher oxidation resistant C/SiC materials not only for racing cars and airplanes.

- C/C with oxidation protection was the only material available and suitable for the wing leading edge structures and the nose cone of the US Space Shuttle [12].

Replacing the carbon matrix of the C/C material with a silicon carbid matrix results in an increase of oxidation resistance and at the same time in an improvement of the mechanical values. Thin walled structures operating in oxidative environments became feasible, like a Ramjet nozzle, Fig. 9, or flaps for airbreathing engines [14]. However, these are still limited life structures due to the oxidative attack on the carbon fibers.

Replacing the carbon fibers of the C/SiC material by nonoxidating ceramic fibers on the basis of silicon carbide or silicon nitride a further increase in oxidation stability is possible at a somewhat lower temperature limit. The aim of ceramic fiber developing companies is obviously to achieve a higher long time temperature stability for the fibers in future.

Already today ceramic matrix composites are unique materials which alone can simultaneously meet such design requirements as

- large size
- light weight
- thin walled
- highly integrated
- damage tolerant

structures for long time temperature applications up to 1000°C as well as for limited life structures up to 1800°C.

REFERENCES

[1] Kochendörfer, R.: Monolithic and Fiberceramic
     Components for Turboengines and Rockets.
     AGARD Conference Proceedings No. 449,
     72nd Specialist Meeting, 3-5 Oct 1988,
     Bath, England

[2] Kochendörfer, R.: Heiße Tragende Strukturen
     aus Faserverbund-Leichtbauwerkstoffen,
     DGLR-Jahrbuch 1987

[3] Hüttner, W.: Parameterstudie zur Herstellung von
     C/C-Körpern nach dem Flüssigimprägnierverfahren.
     Dissertation Universität Karlsruhe, 1980

[4] Fitzer, E.: The Future of Carbon-Carbon Composites.
     3rd Distinguished Lecture on Materials Technology,
     Southern Illinois Univ., Illinois/USA (1986)

[5] Gadow, R.: Die Silizierung von Kohlenstoff.
     Dissertation, Universität Karlsruhe (1986)

[6] Caputo, A.J.; Lackey, W.J. and Stinton, D.P:
     Development of a new faster Process for the
     Fabrication of Ceramic Fiber-Reinforced Ceramic
     Composites by CVI. Ceramic Eng. Sci. Proc. 6 (1985)

[7] Boisvert, R.P.; Diefendorf, R.J.:
     Siliciumcarbid-Verbundwerkstoffe aus polymeren
     Precursoren. Proceedings, Int. Kongress
     Hochleistungsverbundwerkstoffe für neue Systeme,
     Verbundwerk '88, Wiesbaden

[8] Krenkel, W.; Hald, H.: Liquid Infiltrated C/SiC,
     An Alternative Material for Hot Space Structures.
     Proceedings Int. Conference on Spacecraft
     Structures and Mechanical Testing, Nordwijk,
     19-21 Oct. 1988

[9] Schäfer, W.: Continuous Fiber Reinforced Silicon
     Carbide, A High Temperature Material for Aerospace
     Applications. Proceedings Verbundwerk,
     Wiesbaden 1990

[10] Heraud, L.: Main Characteristics and Domains of
     Application of C/SiC and SiC/SiC Ceramic Matrix
     Composites. Proceedings, Verbundwerk,
     Wiesbaden, 1990

[11] Lamicq, P.J.; Bernhart, G.A.; Dauchier, M.M.
and Mace, J.G.:
SiC/SiC Ceramic Composites. American Ceramic Soc.
Bull. 65, pp. 336 (1986)

[12] Curry, D.M.; Scott, H.C.; Webster, C.N.:
Material Characteristics of Space Shuttle
Reinforced Carbon-Carbon.
SAMPE Vol. 24, p. 1524 ff, May 1979, San Francisco

[13] Hald, H.: Szasz, P.; Dittrich, K.:
Development and Real-Test of a Ramjet Nozzle Made
of Liquid-Silicon-Infiltrated C/SiC. Proceedings
ESTEC, Nordwijk/Netherlands, 21-24 March 1990

[14] Mestre, R.: Utilisation des Composites Hautes
Temperatures dans les Turboreacteures. In [1]

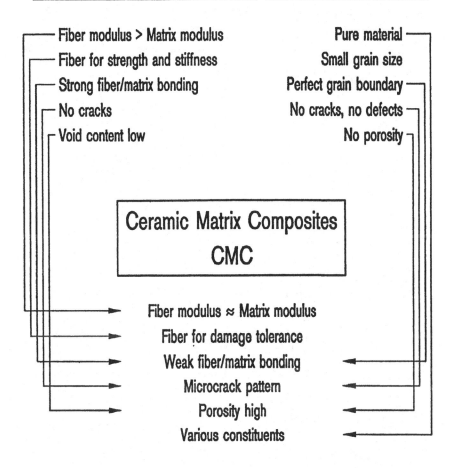

Fig. 1: Characterizing material properties for PMC's
monolithic ceramic materials and CMC's

Fiber	Manufacturer	Composition	Number of Filaments	Diameter μm	Price DM/kg [*]
Carbon HT	Torayca	C 95%	1000 – 12000	7	80/200
Carbon HM	Torayca	C 100%	1000 – 12000	7	250/800
Carbon UHM	Tonen Corp.	Precursor: Pitch	3000	10	5000/–
Nicalon NL-202	Nippon Carbon	SiC 65.3% SiO$_2$ 23% C 11.7%	500	15	1000/2500
Tyranno	UBE	Si 48% C 28% O 18% Ti 4%	400 – 1600	8 – 10	1200/2000
Si$_3$N$_4$	Tonen Corp.	Si 59.8% N 37.1% O 2.7%	-	10	–/–

[*] Price of Roving/Fabric

Fig. 2:   Fibers as reinforcements of ceramics

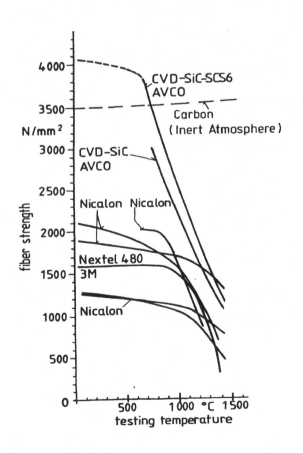

Fig. 3:

Fiber strength as a function of testing temperature (short time values), [1]

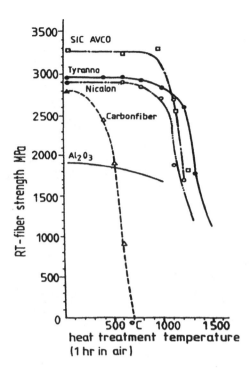

Fig. 4: RT tensile strength after 1h exposure
in air at heat treatment temperatures [2]

		2D C/C	3D C/C	2D C/SiC	3D C/SiC	2D SiC/SiC
Manufacturer		Schunk	Aerospatiale	SEP	SEP	SEP
Process		Precursor	CVI	CVI	CVI	CVI
Tensile Strength	MPa	150 – 200	110	350	80	200
Young's Modulus	GPa	70 – 90	31	90	75	230
Elongation at Break	%	0.2 – 0.4		0.9	0.5	0.3
ILSS	MPa	8 – 12	56	35	50	40
Compression Strength	MPa		76	420 – 580	650 – 740	420 – 580
Flexural Strength	MPa	150 – 200		500	300	300
Therm. Expansion	10^{-6} 1/K	0.8 – 6.9	0.2	3 – 5	1.7 – 2.3	1.5 – 3
Coefficient of Emmision			0.9	0.8	0.8	0.8
Thermal Conductivity	W/mK	15 – 80	4.7 – 22	6.5 – 14	13 – 17.5	10 – 19
Thermal Capacity	J/kgK		840	620	620	620
Porosity	%	5 – 8		10	12	10
Fiber Content	Vol.%	55 – 65		45	24	40
Density	g/cm³	1.55 – 1.65	1.6	2.1	2.3	2.5
Max. Temperature	°C	2000	2000	1600 – 1800	1600 – 1800	1000 – 1200

Fig. 5: Mechanical properties of ceramic matrix
composites at RT

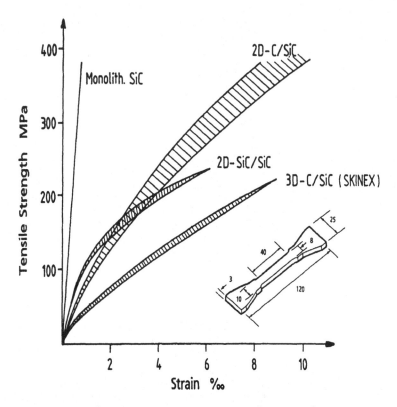

Fig. 6: RT tensile strength of ceramic matrix composites [10]

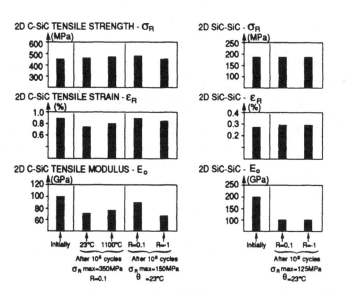

Fig. 7: 2D C-SiC and 2D SiC-SiC fatigue behavior [10]

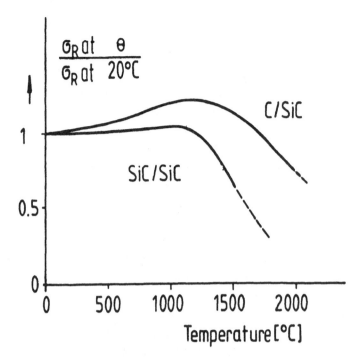

Fig. 8: Tensile strength of C/SiC and SiC/SiC material at test temperatures related to the RT values

Fig. 9: Hot test of a C/SiC Ramjet Nozzle [13] (length 220 mm, diameter 75mm, wall thickness 1.7 mm, internal pressure 7 bar)

57

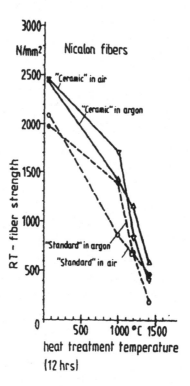

RT-tensile strength
of SiC-fibers after
12 hrs exposure in
air and argon

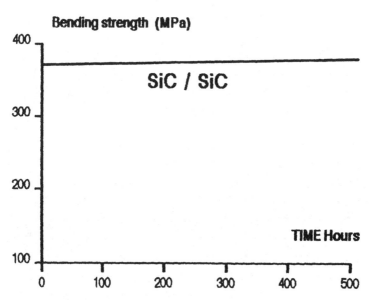

Fig. 10: Degradation of bare Nicalon fibers and of
SiC/SiC material, reinforced with Nicalon
fibers [1,11]

GLASS REINFORCED THERMOPLASTICS MATRIX COMPOSITES (GMT): TECHNOLOGY AND APPLICATIONS

F. ROSSI, G. MOLINA*

FIAT AUTO
Corso Agnelli 200 - Torino - Italy
*CENTRO RISERCHE FIAT
Strada Torino 50 - 10043 Orbassano - Torino - Italy

ABSTRACT

The best exploitation of peculiar weight and specific strength characteristics of GMTs requires increasingly detailed knowledge of their behaviour during molding. The comparison between two components of the same shape and size, the first one ribbed and the second one flat, is the starting point of a direct analysis of process' conditions.

A systematic approach based on design of experiments techniques is used for the optimization of molded part quality.

I - INTRODUCTION

At the moment widescale use of compression molded thermoplastic components is made in both the automotive industry in particular and the transport industry in general to obtain components that require high mechanical resistance and resilience.

However present information on the behaviour of these composites during molding and distribution of the values of performance characteristics in the molded objects is still quite fragmentary.

On this point a first tests' series was performed using a circular flat mold to make a preliminary classification on the molding behaviour of different types of glass reinforced thermoplastics, /3/, /4/.

Purpose of this paper is to continue this kind of analysis taking now into account a completely 3D shape with two different internal profiles (ribbed or flat).

The final target is,in any case,a complete definition of
a test methodology based on statistical design techniques,
which is the basis to build a systematic comparison among
different materials and different settings of the process
parameters .

II - MOLDING BEHAVIOUR OF GMT . METHOD OF INVESTIGATION

For the analysis of the technological molding parameters
reference will be made below to a commercial semi-finished
part, obtained with the traditional dry process, with
polypropylenic matrix and reinforced with 40% by weight of
continuous random glass fibers . To do this way we take the
position of the production engineer who considers the raw
or semi-finished material on the point of view of the final
performances only, without taking into account the previous
manufacturing process of the composite .
About molding the correct use of the material is
obviously based on as accurate as possible assessment of the
values to be assigned to the typical process parameters of
the technology examined . Use in this phase of the
statistical techniques of parameter design makes it possible
to achive this target with a reduced number of tests .
We can summarize the procedure followed in application
of parameter design methodology to the study of compression
molding as follows :

A. The experimental inquiry was made considering two
components of the same shape, size and thickness (external
dimensions 400x400x85 mm) , one with ribs and bosses on the
internal side and one without (see fig. 1) .

B. The objective functions (or quality characteristics)
for the system examined were defined as follows :
1. Maximum average molding pressure. The value for each
experiment was obtained as average during three tests.
Having assigned the piece to be molded minimization of this
characteristic must be aimed at (smaller the better) .
2. Aesthetic appearance of the surface . The evaluation
was made with a Diffracto D-Sight examination consisting in
computer processing of the image of the piece filmed with a
TV camera . The numerical value obtained (average on three
measurements) is of the "smaller the better" type .
3. Ultimate flexural strength (test ASTM D 790). Two
pieces were examined for each experiment with 10 recordings
on each (flat box) or 12 recordings (ribbed box, see fig.2).
The quality function applied is of the type $Q = \mu / \sigma$ (where
μ is the average value of the 2x10 or 2x12 data
distribution and σ is the standard deviation) . In order to
maximize μ and to reduce σ (limit the dispersion of the
results) the Q function must be maximized (larger the better
type) .
4. Flexural modulus (test ASTM D 7.90) . The quality
function used is of the type $Q = \mu / \sigma$ (see point above) .
5. Distribution by weight of the glass fibers . The

evaluation was made on 28 samples for each non-ribbed piece and on 34 samples for each ribbed part (measurements obtained by ashing the samples at 500 °C for 8 hours) . Two component were tested for each experiment (see fig.2). Quality characteristic $Q = \mu / \sigma$ (larger the better) .

C. On the basis of the process considered, the control factors were selected representative of the technology (see also tab. 1) :
1. Average mold temperature (80 °C and 95 °C)
2. Compression speed during the molding phase (2.4 mm/s and 4.8 mm/s)
3. Number of sheets used in each test (i.e. initial dimensions of the sheet : level 1 = 200x300x3.7 mm and level 2 = 200x200x3.7 mm) . In each experiment a preheating temperature of 210 °C was used .

D. Having defined the problem addressed , the experimental matrix was selected . Considering 3 control factors with 2 levels it is possible to execute a full factorial design with 8 experiments only . In this way the statistical design is able to measure the effects due to the main factors and all the effects due to all their interactions without the aliasing effects that are always present in fractioned plans .
The factorial design plan was executed with 3 replications for each experiment and, of course, the plan was executed separately on the flat box and on the ribbed box .

III - EXAMINATION OF THE RESULTS

The final results of the statistical analysis (ANOM and ANOVA) performed on the 5 quality characteristics are given in tab. 2 .
Starting from these data we can draw the following considerations .

3.1. Flat box

A. The optimal setting for the case of the unribbed box is :

T2 , S1 , N1

B. The most important factor is, no dubt, the number of sheets used . This parameter is in fact significant for all 5 quality functions considered, with a best setting corresponding to level 1 . The larger the initial dimensions in plan of the sheet (and therefore, the lower the flow) the better the final performance in terms of homogeneity of phisical and strength characteristics .

C. It should be noted that the temperature, considered singly, does not appear to have any effects on the distribution of the strength characteristics .

D. The interactions among the control factors are of little importance (both for first and second order) .

3.2. Ribbed box

A. The optimal setting , giving priority to good distribution of the mechanical strength characteristics, is:

T1 , S1 , N1

B. The number of sheets must be placed at level 1 also in this case . It is therefore confirmed that high flow values in the mold are damaging also in the presence of ribs .

3.3. Effects of the ribbing

The comparison between the values of the characteristics examined, obtained with and without ribs, leads to the conclusion that :
A. The presence of ribs and/or bosses causes a considerable increase in the average molding pressure .
B. The ribbed areas cause a considerable worsening of the aesthetic quality on the whole (read-out effects) .
C. The distribution of the ultimate flexural strength, flexural modulus and glass content are also worse for ribbed components .

IV - CONCLUSIONS

A. Experimental design techniques is able to give us a methodology for GMT composites' testing . This method is suitable to study the effects due to different kind of shapes as the case of the presence of ribs .
B. The results obtained could be used to make a comparison, in terms of performances, among different types of reinforced thermoplastic composites .
C. The evaluation of parameters effects and the definition of parameters best setting it is also possible .

REFERENCES

1 - Whitney J.M. , I.M. Daniel and R.B. Pipes , "Experimental Mechanics of Fiber Reinforced Composite Materials" , SESA Monograph No.4 (1982)
2 - Tomkinson - Walles G.D. , "Performance of Random Mat Reinforced Thermoplastics". Proceedings S.P.I. 44th , 11-F (February 1989)
3 - Folonari C.V. , Molina G. ,"Mouldability and performance of Glass Reinforced Polypropylene". Proceedings ATA-MAT 89 , C2.6 (June 1989)
4 - Lepore S., Molina G., Rossi F.S., Serra R.,"Compression Moulding behaviour and performance of Glass Reinforced Thermoplastics". Proceedings S.P.I. 45th , 18-E (February 1990)

CONTROL FACTORS		LEVELS	
		1	2
T –	Mold temperature (°C)	80	95
S –	Compression speed (mm/s)	2.4	4.8
N –	Sheets No. (without ribs)	4	6
N –	Sheets No. (with ribs)	5	7.5

Tab. 1 – NUMERICAL VALUES FOR THE CONTROL FACTORS.

	Factors Interactions / Quality Functions	MOLD TEMP. T	COMPR. SPEED S	SHEETS No. N	1st LEVEL INTERACTIONS T*S	T*N	S*N
1st STATISTICAL DESIGN — BOX WITHOUT RIBS	MOLDING PRESSURE	2	1	1	2*1		
	DIFFRACTO INDEX	2		1			
	FLEXURAL STRENGTH		1	1	1*1		
	FLEXURAL MODULUS			1			
	GLASS CONTENT			1			(2*1)
2nd STATISTICAL DESIGN — BOX WITH RIBS	MOLDING PRESSURE	2	1	1			
	DIFFRACTO INDEX					(2*1)	
	FLEXURAL STRENGTH			1	1*1		
	FLEXURAL MODULUS	1		1	1*1		
	GLASS CONTENT	2	2				

( ): 80% < confidence limit < 90%

Tab. 2 – "ANOM" AND "ANOVA" FINAL RESULTS.
PARAMETERS SETTING TO OPTIMIZE THE QUALITY FUNCTIONS
(90% CONFIDENCE LIMIT).

Fig. **1** - Molded boxes in GMT (40% continuous random glass fibers).
Component with ribs and flat box.
Dimensions: 400 x 400 x 85 mm
Thickness : 3.0 mm
Ribs thickness (average) : 1.5 mm

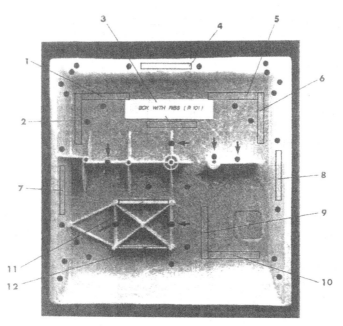

Fig. **2** - Positions of the specimens used for flexural
tests (rectangular shapes) and glass fiber
content measurements (black circles).

Flexural test :
   flat box   : positions 1 ÷ 10
   ribbed box : positions 1 ÷ 12

Glass content :
   flat box   : 28 points
   ribbed box : 34 points

The six arrows show the six last points used for
the ribbed part only.

64

FABRICATION
FERTIGUNG

Chairmen : **Dr. J. KRETSCHMER**
Daimler-Benz AG
Prof. F.L. MATTHEWS
Imperial College

A NOVEL METHOD AND MATERIAL
FOR THE PROCESSING OF
GLASS MAT REINFORCED THERMOPLASTICS (GMT)

M. ERICSON, L. BERGLUND

Linköping Institute of Technology
Composites Group - Dept of Mechanical Engineering
58183 Linköping - Sweden

ABSTRACT

A novel processing method for the fabrication of glass mat reinforced thermoplastics, GMT, has been developed. Commercially available GMT's are delivered in the form of semi-manufactured sheets or webs, whereas this novel method utilizes fibre roving and thermoplastic powder to build a preform which is heated and moulded directly into a component. The present work describes the basic concept of the process and a laboratory scale processing apparatus. Experimental results on creep modulus are discussed and compared with data for commercially available materials.

INTRODUCTION

Glass mat reinforced thermoplastics, GMT, have been commercially available since the early 1970's /1,2/. The materials have many applications in the automotive industry where the short moulding times makes the process attractive.

A number of different ways to produce GMT's are now available. These commercial techniques all have in common that the materials are delivered in the form of semi-manufactured laminates, plates or webs. These are heated and moulded by a hot stamping technique similar to compression moulding. The oldest method, and still the most common /2,3/, is based on the impregnation of a glass mat by a polymer melt.

A glass mat is fed to a laminator together with a resin film and also molten resin from a large sheet extruder. The so called laminator is basically a belt press with heating and cooling zones. The resin film melts in the heating zone and wets the fibre mat due to the applied pressure. The resulting composite is then cooled, trimmed and cut into rectangular blanks.

More recently, paper making technology has been used to mix reinforcing fibres and polymer powder in an aqueous slurry /2,3/. The mixture is then drained, dried and consolidated. Some manufacturers do not consolidate their laminates. Instead a special heating technique is used.

Both the existing techniques for the production of GMT components use semi-manufactured sheets as starting materials for the moulding operation. This increases the material cost. In order to decrease the component cost, a novel method to produce GMT components was developed. This method utilizes the constituent fibre and polymer matrix, in the form of roving and powder, respectively, to build a porous preform which is heated and moulded. By this concept, the raw materials are processed only once to form the finished component. In the present work, this new method and material for the production of components from glass mat reinforced thermoplastics is described. The tensile modulus of the material is compared with the modulus of commercially available materials.

I - DESCRIPTION OF THE PROCESS

The basic principle of the process can be described as follows. Fibre roving is cut and mixed with a thermoplastic powder and the fibre/powder mixture is combined into a preform. The preform is then heated in a through-air heater and moulded into a component by hot stamping. A laboratory scale processing apparatus consisting of three units was built; a preform fabrication apparatus, a preform heating apparatus and a modified moulding press.

1.1 Constituent materials

For the present study we used a glass fibre roving and a polypropylene powder. The roving was a Scandinavian Glasfiber roving, type RPA38 20/9 EC12-600, an E-glass silane sized 600 tex roving used for winding, pultrusion and weaving of thermoset matrix composites. The polymer was a polypropylene homopolymer powder, type GY 545M, from ICI Petrochemicals and Plastics Division (Great Britain).

1.2 Preform fabrication

The fibre and the thermoplastic powder are combined into a porous preform in the preform fabrication apparatus, see Figure 1. The roving from the glass fibre bobbin (a) is cut in a rotating cutter (b) into chopped fibre bundles. The bundles are thrown into a T-connection (c) where they are mixed with the thermoplastic powder supported by the powder feeder (d), a powder container with an ejector driven by the pressure P_2 and a fluidized bed produced by the pressure P_3. An air flow extender (e) accelerates the mixture and forces it through a PVC-tube (f). While transported in the tube, the fibres and the powder are electrostatically attracted to each other. The fibre/powder mixture is blown to a mould (g) made from a perforated metal sheet where it results in the formation of a preform. The fibre and powder is kept in place by a vacuum P_4 which has to produce approximately the same air flow through the preform as the flow coming out of the PVC-tube. This air flow was achieved by the use of an industrial vacuum cleaner. The preform net used was circular with a diameter of 145 mm with a 50 mm high border.

The cutter used was an Applicator (Sweden), type 8200, the powder feeder used was a 5 litre Michaelsson (Sweden) and the air flow extender used was a Vortec (USA) Transvector, type JET 905 B. The roving and the processing parameters in the preform fabrication apparatus will determine fibre content, fibre length and fibre orientation distribution as well as fibre bundle configuration. These preform processing parameters are the speeds of the cutter, the roving and the fed powder. In our case the 600 tex roving was cut into 25 mm long bundles which disintegrated when the air flow extender, (e) in Figure 1, was passed.

1.3 Preform heating

After the preform fabrication, the preform needs to be heated. In the preform heating apparatus, see Figure 2, the preform net with the preform is heated by hot gas. The gas, by a pressure P, is heated by a Karl Leister (Switzerland) 5 kW electric gas heater, see Figure 2 (a). The heater controls the temperature T_1 by an electronic temperature control integrated with the heater. A conical steel hood (b) connected to a steel pot with holes just above the bottom (d), leads the hot gas through the preform (c). Surprisingly little polymer powder was lost in this operation. The temperature T_2 is measured by a thermocouple. Without the steel pot in Figure 2, the surrounding air will cool the lower side of the preform and prolong the heating. In the preform heating apparatus , nitrogen (N_2) at a pressure of 0.1 MPa, and a temperature T_1 of 220 °C, see Figure 2, was used. Nitrogen had to be used as heating medium, since the polypropylene powder suffered from oxidation when air was used. The heating time was chosen so that as the temperature T_2 reached 200°C, the heating was stopped and the preform was taken out of the preform net and charged into the mould.

1.4 Moulding

When the preform has been heated, the next step is the moulding operation. The heated preform was moved to the press and put in the mould cavity. The mould at 25-35 °C was then rapidly closed and the material flowed to fill the mould. In order to have a laboratory press which could be closed fast enough, a Moore (Great Britain) "Jubilee" hand-operated 50 tons press was modified. The hydraulic cylinder was changed to an Enerpac (USA) 50 tons cylinder. The cylinder was supported by an Enerpac 700 bar hydraulic pump. The mould used was a matched die steel mould for cylindrical discs with a diameter of 180 mm. The time needed for the transport from the heater to the mould, closure of the mould and maximum pressure build up (20 MPa) was less than 15 seconds.

II - MATERIAL PROPERTIES

Tensile creep modulus tests were done on specimens from eight manufactured discs. One rectangular specimen (30 mm by 160 mm) was cut centrally from each disc. Average thickness of the samples was 2.3 mm. A constant load was applied to the specimen and the displacement, measured by an extensometer, was recorded after 100 seconds. Each specimen was loaded three times at different load levels and the creep modulus was calculated from the isochronous stress-strain curve. By burning off the matrix from all tested specimens in an oven (five hours at 550 °C) the fibre weight fraction was determined. The fibre volume fraction for each specimen was calculated using densities of 0.90 g/cm^3 and 2.56 g/cm^3 for the matrix and fibre, respectively.

From the appearance of the preform and mechanical tests, it was found that the fibre orientation distribution was random in the plane. The modulus of the material will therefore primarily depend on the fibre content. In Figure 3 the creep modulus versus fibre content is shown. The drawn lines represent the creep modulus of two commercial GMT's which were processed in the same mould and tested in the same manner /4/. The modulus of the new material is similar to the two commercial materials. The scatter in the data is believed to be due to variations in the local fibre content and/or fibre orientation. As is apparent in Figure 3, the fibre content varied from 16 to 21 percent by volume. When efforts were made to fabricate discs with higher fibre content, a tendency for uneven fibre distribution was observed with regions of very low matrix content.

CONCLUSIONS

A novel concept for the fabrication of glass mat reinforced thermoplastics has been described. This method utilizes the constituent fibre and polymer matrix, in the form of roving and powder, respectively, to build a porous preform which is heated and moulded. Circular discs were moulded using glass fibre roving and polypropylene powder. The material showed similar tensile creep modulus as commercially available GMT's, 4-7 GPa depending on the fibre content. The suggested concept lowers the cost for GMT's and the preform technique allows more complex geometries to be manufactured.

ACKNOWLEDGEMENTS

Financial support from the National Board for Technical Development and many helpful suggestions from Michael Jander, Owens Corning, and Åke Nylinder, Volvo Car Corporation, are gratefully acknowledged. Materials were kindly supplied by Owens Corning and ICI.

REFERENCES

1 - **Six, J.**: GMT- ein Werkstoff und eine Verarbetungstechnologie setzen sich durch. 21. Öffentliche Jahrestagung der Arbeitsgemein-schaft Verstärkte Kunststoffe e.V. Mainz November 1987. Proceedings. pp. 22.1-22.8.

2 - **Bigg, D.M. ; Preston, J.R.** : Stamping of Thermoplastic Matrix Composites. Polymer Composites. 10 (1989) 4. pp. 261-268.

3 - **van Damme, P.A.**: Economic composites for mass production in the automotive industry. 3rd International Conference on Fibre Reinforced Composites. University of Liverpool, March 23-25, 1988.Proceedings. London (Plastics and Rubber Institute) 1988. pp.1/1-1/14.

4 - **Ericson, M.; Berglund, L.**: Deformation and fracture of glass mat reinforced polypropylene. To be published.

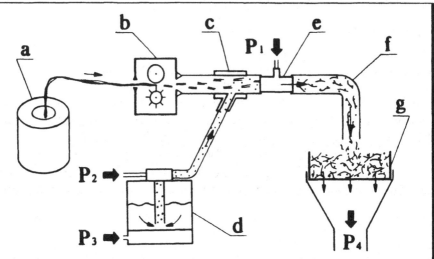

Figure 1. Preform fabrication apparatus. a) fibre roving bobbin, b) rotating cutter, c) T-connection, d) powder feeder, e) air flow extender, f) PVC-tube and g) preform net .

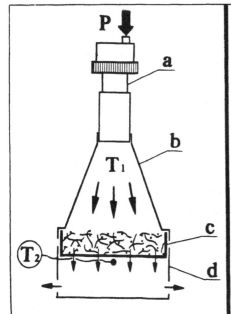

Figure 2. Preform heater. a) electric air heater , b) steel hood, c) preform net and d) steel pot.

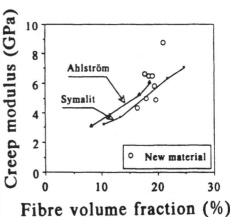

Figure 3. Creep modulus (100s) of the new material compared to two commercially available GMT's, Symalit GM PP made by Symalit AG in Switzerland and Ahlström RTC-C made by A. Ahlström Corporation in Finland.

ELECTRON BEAM CURING OF COMPOSITES

D. BEZIERS, B. CAPDEPUY, E. CHATAIGNIER

Aérospatiale
BP 11 - Issac - 33165 St Médard en Jalles - France

ABSTRACT

Since 1980, **AEROSPATIALE** has developped a new curing process of composite using electron beam and X-rays with the first objective to manufacture filament wound motor cases. To industrialize this process studies have been performed : formulations of matrix, synthesis of a specific carbon fiber sizing, caracterization of composite on samples and on full scale structure, settlement of procedure associating electron beam and X-rays. These results led to decide the setting up of industrial facilities, so in the mid 1991 **AEROSPATIALE** will able to cure large structures (maximum length : 10 m maximum, diameter : 4 m).

1 - INTRODUCTION

AEROSPATIALE supplies cases for solid propellant motors for the French National Deterrent Force.
These cases are made in filament wound composite, mainly carbon fiber since the 80's. These composites used epoxy resins as matrix and therefore need heating to cure them. In order to cancel this thermal treatment, research was initiated in 1979 to select a new process.
The first results demonstrated us that a combination of electron beam and X rays is the best adapted to our products and we decide to develop a research activity mainly oriented towards wound carbon fibers composites.
AEROSPATIALE has been comitted since 1979 to the development of this process referred to as radiation curing (or E.B.C. - electron beam curing - when only electrons are involved), with emphasis on two points :

- development of a process combining electrons and X-rays, and of the relevant equipment and implementation,
- research into basic products - new resin formulations, characterization of suitable fibre/resin systems, new adhesives and application technologies (WET filament winding, prepreg, layup).

2 - PRESENTATION OF RADIATION CURING

2.1 Advantages of radiation-curing

The use of the combined process, electron and X-rays provides the following advantages :
- easy to apply,
- very short curing time for substantial penetration capacity, due to use of $\bar{e}$,
- penetration greater than our needs achievable by increasing curing time, due to use of X-rays,
- curing with very low temperature increases, limiting stresses of thermal origin,
- resin pot life much longer than the manufacturing time for filament of thermal structures,

and with only 2 disadvantages :
- cost of equipment,
- special chemistry required.

2.2 Development of the process and definition of the relevant equipment

AEROSPATIALE patent VS 4689 4887, VS 4789 505.
This process requires the use of electron accelerators which, at an industrial level, exist in several forms depending on their characteristics. They are principally characterized by two parameters :
- the energy which determines the penetration of the radiation (electrons or X-rays). This is expressed in M.e.V. (10^6 electron volts),
- the power in Kw, directly linked to the exposure time.

The electrons and X-rays are obtained from the same accelerator. The production of X-rays also requires a conversion target placed beneath the electron output window. This target, which is endependent of the accelerator, is placed in position mechanically when the thickness of product to be crosslinked exceeds the possibilities of electron penetration. The parameter which determines the limit of electron curing is the surface mass ($\Sigma e \times p - e$ and are thicknesses and specific mass of different materials).

The use of X-rays, although it does make it possible to crosslink considerable thicknesses, nonetheless substantially increases the curing time : the X-rays dose rate is very low, and the distribution of the radiation is not highly directive.

These considerations determine the choice of characteristics for the accelerator.

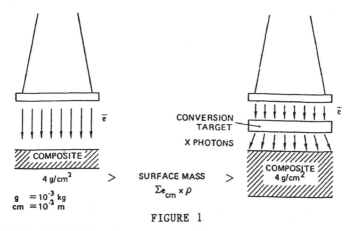

FIGURE 1

3 - RADIATION CHEMISTRY

This process requires special chemistry, and **AEROSPATIALE** has devoted its efforts since 1979 to developing this within the framework of its composite application for filament-wound motors.

Such research has involved the formulation of resins based either on commercially marketed substances or on original syntheses produced in collaboration with French universities. These resins were then characterized using Kevlar, and subsquenty carbon, fibers.

The principle of crossliking under radiation is the following :

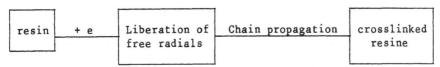

| resin | + e | Liberation of free radials | Chain propagation | crosslinked resine |

This mecanism involves to have polymers with double bonds at the chain ends. One of the consequence of this fact, is that the crosslinking of the resin does not implicate any hardening agent.

3.1 Resins

3.1.1 Acrylic resins

The most commonly used resins for radiation are mainly ethylene unsaturated polymers (with photoinitiators for UV radiation). They are :

1) Acrylate epoxides
   - either in a difunctional form : glass transition 150°C (302°F).
   - or in a tetra functional form : glass transition $\leqslant$ 200°C (392°F).

2) Acrylate polyesters

3) Acrylate urethanes

The properties of these resins are roughly equal to those of their thermal equivalents.

Formulations are based on mixtures of these various types of acrylic resins with matrices having the required properties, as a compromise for motors.

3.2 Resin characterizations

These were carried out essentially as part of the applications, i.e. in response to the specifications for filament-wound motors.

tg $\leq$ 150°C (302°F) resins

		ΔL/L (%)	σ (MPa)	E (MPa)	tg (°C)	(°F)
Heat reference resins	X	4–6	85	3000	145 –	293
	Y	10	60	2500	100 –	212
ē resins	A	5–6	65	3200	120 –	248
	B	3	75	3200	120 –	248
	C	13	50	2100	85 –	185

tg 200°C (392°F)

		ΔL/L (%)	σ (MPa)	E (MPa)	tg (C°)	(F°)
Heat reference resin	Z	4 – 5	90	3000	175 –	347
ē resins	D	2	65	–	175 –	347
	E	2 – 3	60	3000	180 –	356

3.3 Fiber/matrix interface

The first characterization we have done on composites acrylic matrix-carbon fiber (HERCULES IM6) give low results on shear stress with or without epoxy based commercial sizing. To solve this problem we have imagined to put a specific sizing on the fiber based on a monomer with a double fonctionality : $F_1 - R - F_2$. The first one F_1 must be able to react with the fiber and the second one F_2 with the acrylic matrix.
Analysis of the fiber surface by ESCA technic give a majority of hydroxyl functions so F_1 can be isocyanate function. To react with the matrix acrylic function can be suitable. So we have synthetized a monomer which general formula is $CH_2 = CH - R - NCO$. The improvment, given by this way, had been demonstrated by two tests. In a first elemental test we have measured pull out stress of elementary fiber embedded in pure resin. The observed value has increased from 40 MPa to 100 MPa. So we have applied this new sizing to composites, the interlaminar shear strength in greatly improved (near twice) 50 MPa versus 25 MPa. This value is equivalent to that obtained with conventionnal epoxy resin.

3.4 Composite characterizations

Evaluations are carried out using characterization test samples specific to filament winding. These are essentially :

- N.O.L. rings for interlaminar shear stress,
- 304 mm (12") diam. vessels representative of the stress on composites within a motor (biaxial stresses) : determination of hoop and helix stresses.

Class tg < 150°C (302°F)

Carbon composite		σ_F ** (MPa)	τ * (MPa)
Heat reference prepreg		54	4500
Wet		55	3500
e wet	A	50	4400

*σ_F : hoop filament stresses

**τ : shearing stress on N.O.L.

3.5 Accelerated ageing assessment

Assessment on 304 Wmm (12") diam. vessels in accordance with motor specifications (heat and humidity) demonstrates no effect on performance.

4 - REAL SIZE DEMONSTRATION

After the laboratory work on test pieces, demonstrations were carried out on a scale roughly corresponding to real structures.

This structure has a diameter of 1150 mm (45.3 in.). It is a filament-wound vessel roughly corresponding to the second stage of a strategic missile.

The main results obtained were as follows in comparison with the same structure with heat curing :
- hoop filament stress = 4100 MPa (583 ksi)/3700 MPa (526 ksi),
- performance factor K * = 46 km (1.81 x 10^6 inches)/41 km (1.62 x 10^6 inches).

5 - BRIEF DESCRIPTION OF THe UNIPOLIS PRODUCTION UNIT

This unit comprises 3 main modules :
- a high energy (10 M.e.V.) and high power (20 Kw) electron accelerator,
- infrastructures for the installation of the equipment and the biological protection of the environment from the radiation,
- an operating mechanism which positions the electron beam with regard to the structure.

The purpose of the equipment installed is to make it possible to cure cylindrico-spherical structures with the following characteristics :

- diameter between 100 mm (4") and 4000 mm (157"),
- length between 1500 mm (60") and 10500 mm (413").

In addition, the curing of a structure of the maximum dimensions must not take more than 8 hours.

6 - CONCLUSION

AEROSPATIALE has demonstrated the interest (good performance, lower costs) of a new electron beam curing process in the field of filament wound structure.

The industrial unit will be operationnal in the 1991. To en large the field of application of this process research are already on the way and we can mentionne :

- new high temperature resins (Tg > 300°C),
- rubber for thermal insulation,
- laminates,
- multidirectionnal composites.

EXPERIMENTAL INVESTIGATION OF THE SUPERPLASTIC FORMING TECHNIQUE USING CONTINUOUS CARBON FIBER REINFORCED PEEK

S.G. PANTELAKIS, D.T. TSAHALIS*, T.B. KERMANIDIS*, S.R. KALOGEROPOULOS

Hellenic Aerospace Industry Ltd
Tanagra - PO Box 23 - 32009 Schimatari - Greece
*University of Patras - Greece

ABSTRACT

Experiments were performed with continuous carbon fiber reinforced PEEK, using the diaphragm forming technique for the purpose of improving the control of the process of part formation. Both UD and woven C/PEEK prepregs were employed with a single diaphragm of UPILEX The effect of temperature, pressure and time on the quality of the formed parts was studied. The most important result of the study is the demonstration that very short times over which the forming hydrostatic pressure is applied, of the order of few seconds, lead to less uneven thickness distribution and less fiber wash.

1. INTRODUCTION

Thermoplastic composites have been emerging as very promising aerospace materials because of their desirable properties such as high service temperatures and high impact resistance, not to mention their highly desirable process advantages, e.g., repairability by remelting and short cycle times with hot forming techniques. Therefore it is not surprising that their behaviour and forming techniques have attracted attention in recent years /1-7/.

One of the most promising advanced composites is PEEK reinforced with carbon fibers, i.e., C/PEEK. Some

79

results on forming techniques using this composite have been reported recently /2,3/.

However, the application of hot forming techniques, such as the diaphragm forming technique, originally developed for metallic materials, to C/PEEK has been exploited in a rather limited fashion /1,4,5,6/. Consequently, the present state of affairs concerning the diaphragm forming technique is that, - despite some recent efforts /5/, only qualitative control of the process can be achieved.

The present work aims to contribute towards the improvement of the control of the process of part formation using continuous carbon fiber reinforced PEEK with the "superplastic forming technique". The experimental results are presented.

2. FORMING TECHNIQUE AND MATERIALS

The experimental setup employed is shown schematically in Figure 1. The process involves heating in a "female" mould the composite material, over which a backing material is placed. Then pressure is applied on the top surface of the backing material with Argon gas (introduced through the top half of the mould) and vacuum is drawn from the bottom "female" part of the mould.

Two different types of C/PEEK material were used, namely, C/PEEK prepreg and C/PEEK woven. The stacking sequence for the C/PEEK prepreg was (45°,90°, - 45°, 0°)s. For the trials two shapes were used, namely, an "ashtray-like" and an "ellipse-like" cover.

3. RESULTS AND DISCUSSION

Initially a parameter study was carried out in order to determine empirically the suitable process parameters for forming the selected parts, i.e., temperature, pressure, time. For these tests, the cooling rate, which influences strongly the material properties /7/ and to a lesser extent the forming process itself, was kept constant at 4° C/min.

After some experimentation with several Al foils, Kapton and the Thermoplastic foil UPILEX R, it was found that the use of the thermoplastic foil leads to significant shorter process cycles, reduced required forming pressures, better formed parts and better surface finish.
The process cycle empirically derived as being the most

suitable for forming the sellected parts was for both types of the material: Temperature=370°, Pressure=10 bar, process cycle duration= 90 min. (within this time are included: forming, consolidation and cooling time). In Photo 1, samples of the produced parts are shown.

Although the parts produced were satisfactory, it was found that the achieved thickness distribution was not uniform and substantial fiber wash was observed. Especially for the ashtray-like part, due to its double curvatures, the thickness variation reached values up to 30%.

4. PROCESS IMPROVEMENT EXPERIMENTS

In order to quantify the process, with the final aim of achieving fine control over the distribution of the thickness, various superplastic forming trials were performed with C/PEEK prepreg with the experimental set up and the forming procedure described earlier.

4.1 Experiments

The mould used was the Ashtray-like. The specimens were prepared by stacking in UD-Mode 12 layers of Uni-directional C/PEEK pre-preg. The single diaphragm used was UPILEX R 125.

The uneven distribution of the thickness and fibre wash are the result of the flow of the matrix. In order to determine the detail characteristic of the matrix flow during forming the following information are required:

a. If there is any relative displacement between corre-sponding points on top and bottom free surfaces of the composite.

b. Quantification of the displacement of the matrix in directions transverse and parallel to the fibres as well as for in between directions.

c. The relative displacement of the "middle" layer of the composite with respect to corresponding points on the free surfaces.

In order to achieve objectives a) and b) the following technique was devised: After some experi-mentation it was found that graphite adheres to the matrix and does not deteriorate during forming. Accordingly concentric circles centered at the center of the mould were drawn on both free surfaces of the

specimen (Fig. 2a). In order to achieve objective c)
metal specks were implanted at the middle plane of the
specimen, i.e. between layer 6 and 7 at the same radii
as the circles of graphite. X-Raying of the specimen
before and after formation in conjuction with
information from objectives a) and b) provides the
required data.

4.2 Experimental results and discussion

From the performed trials the following results were
obtained:
a. The displacement of the matrix in the fiber
   direction is almost zero. The maximum matrix displacement occurs in the direction transverse to the
   fibers and reaches values up to 80%. The graphite
   circles deform to ellipse-like shapes that conform
   to the shape of the mould. The major axis of the
   ellipse is transverse to the fiber direction and its
   minor axis is parallel to the bibers and almost
   equal to the radius of the initial circle (Fig. 2b)
b. In support of the experiments, an analytical model
   was developed of the initial stages of the forming
   process [8]. The predictions of the model point to
   the direction of application of the hydrostatic
   pressure over the shortest possible time interval in
   order to minimize the uneven thickness distribution
   and fiber wash. How short this time interval can be
   is determined by the strain rate sensitivity of the
   diaphragm material.

Accordingly, experiments were performed with the
time to reach the forming pressure of the order of few
second, with the smallest time achieved equal to 2
seconds. It was found that shorter times lead to less
uneven thickness distribution and less fiber wash. The
possible influence of the differential stretching
between UPILEX R and matrix remains a subject of investigation.

In Fig. 3, characteristic results from these experiments are shown. Specifically, in Fig. 3a the relative
ratio of the matrix displacement, 1, over the diameter
of the initial circle, 1o, is plotted. The thickness
distribution along a crss-section transverse to the
fibers, is shown in Fig. 3b.

The predictions of the above referenced model and
the findings of the present experiments are indirectly
supported by the forming trials of /6/. In those
trials, the total forming time was reduced to 60 sec
when Upilex diaphragms were used.

c. No apparent differential movement of the curvilinear displacements of the upper and lower surfaces due to differential matrix movement was observed. The relative displacement of the middle layer of the middle layer of the composite with respect to corresponding points on the free surfaces is limited.

References:

1. Cattanach, J.B. and A.J. Barnes, "Forming Fibre-Plastic Composites", U.S. Patent No. 4657717, April 14, 1987.

2. Kempte G. and Krauss H. (1988): Proceedings of 16th Congress of the International Council of the Aeronautical Sciences (ICAS), Jerusalem/Israel, August 28 - September 2, (1988, Vol. 2, pp. 1780-1800.

3. Nagumo T., Nakamura H., Yosshida Y. and Hirakoka K. (1987): Proceedings of the 32nd International SAMPE Symposium, April 6-9, 1987, p.p. 369-407.

4. Paipetis S.A., Pantelakis S.G., Papanikolaou G.D., Pissinou G. and Schulze V.: Proceedings of the 2nd COMP'88, Patras, Greece, August 22-27, 1988, in print.

5. Smiley A.J. and Pipes R.B. (1988): Journal of Thermoplastic Composite Materials, Vol. 1, pp. 298-321.

6. Ostrom R.B., Kock S.B., Wirz-Safranek D.L.: Sampe Quarterly 21 (1988), pp. 39-45.

7. Vautey P.: Sampe Quarterly 21 (1990), 2, pp. 23-28.

8. Tsahalis, D.T. Pantelakis, S.G. and Schulze V., accepted for presentation at the Symposium on the Processing of Polymers and Polymeric Composites of the ASME 1990 Annual Winter Meeting, Dallas, Texas, November 1990.

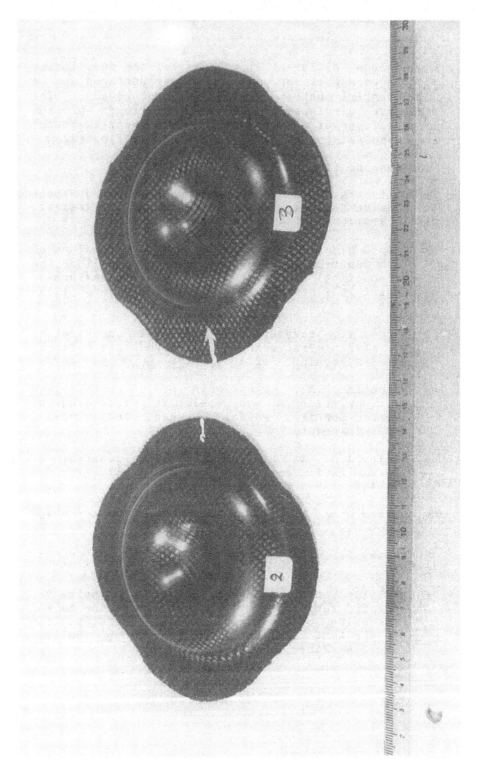

Photo 1. Sample products using the superplastic forming technique

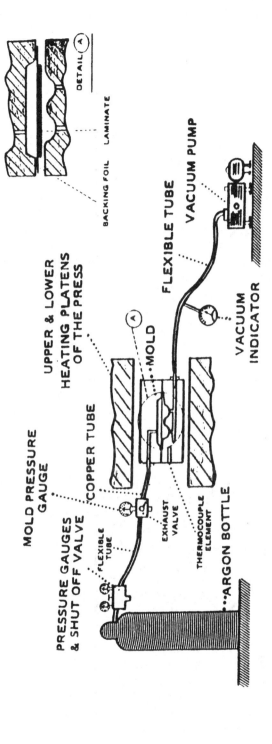

Figure 1. Schematic Layout of the Experimental Set - Up for the Superplastic Forming

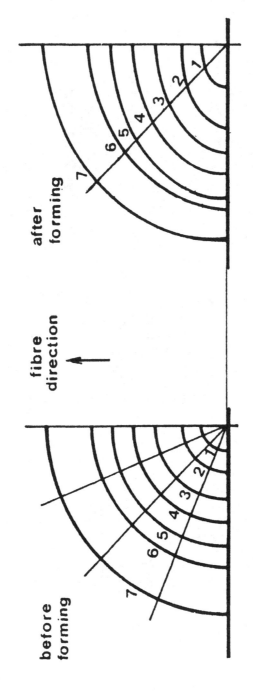

before forming

fibre direction →

after forming

*Numbers correspond to concentric circles

Figure 2. Experimentally derived matrix displacement (quadrant)

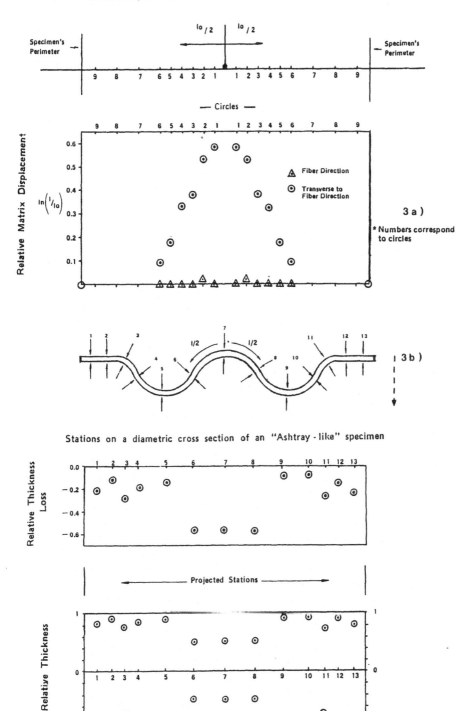

Figure 3.   Matrix Displacements

LASER PROCESSING OF THERMOPLASTIC MATRIX FILAMENT WOUND COMPOSITES

F. FERRARO, G. DI VITA, M. MARCHETTI*, A. CUTOLO**, L. ZENI**

CIRA Via Maiorise - 81043 Capua - Italy
* University of Rome - Via Eudossiana 16 - Rome - Italy
**University of Naples - Piazzale Tecchio - Naples - Italy

ABSTRACT

The aim of this paper is to present our investigation on chemical and physical interaction between laser beam and tows of thermoplastic material for filament winding.

We carried out an experimental evaluation on the effect of process parameters such as the laser wave length, the interaction time, the irradiated energy, the laser spot width.

1 INTRODUCTION

Filament winding is a very interesting technique in advanced composites industry. The automation of the process provides very high quality products allowing low manufacturing cost /1,2/.

Considerable enhancements could be achieved by use of thermoplastic matrix in filament winding process /3/.

Recent investigations showed performance improvements in moisture absorption, no evidence of thermally induced micro-cracking and minimum outgassing /4/.

The main relevant benefits of thermoplastic, compared to thermoset matrix, lie in the significant reduction of manufacturing costs and preliminary treatments as well. In fact refrigeration of prepreg, additional bagging, handling, and curing cycle steps are not required.

Taking advantage of thermoplastic's melt and remelt characteristics a large amount of heat can be supplied while winding: this permits laying the fiber down in non-geodesic patterns (Fig. 1) and in complex shapes with concave surfaces /5,6/.

If enough heat and pressure are added, the part will be in-situ consolidated during the winding process. This application involves the solution of technological problems such as the correct heating and melting of the matrix and the consolidation mode /7/.

Among the available heating techniques we pointed our attention on the laser beam method (Fig. 2). This form of heat supply is theoretically ideal becouse a large amount of energy can be concentrated in a small area.

2 EXPERIMENTAL

Tests were carried out, among the commonly used lasers in classical manufacturing technologies, to evaluate the best material-wave length combination. At Edinburgh Instruments laboratories, a CO_2 laser (10.6 μ) gave unsatisfactory results. Successively, repeating the trial at Laser Optronic, the same result was obtained, while a Nd-Yag (1.06 μ) gave better performances.

2.1 First experience

A set of measurements on a stationary sheet of thermoplastic material was led.

The laser was a 12 W Q-switched Nd-Yag delivering a pulse train with adjustable repetition frequency and with a pulse length of about half a microsecond, so that we was able to control the average power and the energy transfered to the sample. The laser was driven, through an appropriate trigger signal, to the acousto-optic cell of the Q-switching. The interface card was formed by a Camac Controller Interface Mod. C111 and an IBM PC-AT Single Crate Board. We used the computer to set the beam parameters and its interaction time with the sample.

We carried out tests varying interaction time and laser average power. This permitted to bound a set of process parameters giving the required objectives: no combustion of the resin, no damage on the back side of the composite sheet.

2.2 Second experience

An experimental equipment was performed (Fig. 3) to check out practicability of thermoplastic filament winding by a laser heat source.

We used a 400 W maximum average power Nd-Yag laser.

On the basis of the previous experience, a set of basic process variables (winding speed and laser power) was selected.

In this way we identified a wide region of appropriate operating conditions to avoid bad welding or resin burning (Fig. 4).

We remark that *PPS* thermoplastic resin seems to have a better behaviour to laser processability compared to *APC2*.

Furthermore, the importance of a uniform energy distribution was evidenced. In fact it was not possible to get a correct welding along the whole tow cross section using a 3 mm. spot, as the gaussian distribution of energy give rise to an analogous temperature profile and to a narrow welding area. Using the central region of a larger spot (5 mm), providing the same specific power, a more uniform and suitable temperature distribution was obtained.

3   CONCLUSIONS

The feasibility of thermoplastic filament winding by use of the Nd-Yag laser beam heating technique was proved.

Inferior and superior limits in both winding speed and average laser power have been located.

We verified by an analitical model that there is a constant temperature difference between the front and the back side of the processed tape. For this reason it will be possible, on the basis of the experimental data, to fix work parameters so as to obtain, on the exposed surface of the tape, the appropriate welding temperature. At the same time a lower temperature, on the not exposed surface, will be available. In this way the tow will not bond to the consolidation rollers.

The next step will be an improvement of the system by use of optic fibers. They allow to transfer the laser beam directly to the delivery arm, as well as to get a higher uniformity of energy distribution.

ACKNOWLEDGEMENTS

The authors wish to express their gratitude to Dr. L. Grisoni and Mr. F. Bianchi of Laser Optronic, for a most valuable collaboration.

REFERENCES

1. M. Marchetti, D. Cutolo, G. Di Vita, "Design of domes by use of the filament winding technique", in ECCM-III, pp.401-408, Proceedings of the 3rd European Conference on Composite Materials, Bordeaux, France, 1989.

2. M. Marchetti, D. Cutolo, G. Di Vita, "Filament winding of composite structures: validation of the manufacturing process", in ICCM-VII, Proceedings of the 7th International Conference on Composite Materials, Guangzhou, China, 1989.

3. M. W. Egerton, M. B. Gruber, "Thermoplastic filament winding demonstrating economics and properties via in-situ consolidation" , pp.35-46, Proceedings of the 33th International SAMPE Symposium and Exhibition, Anaheim, California, U.S., 1988.

4. E. M. Silvermann, R. A. Griese, W. F. Wright, "Graphite and kevlar thermoplastic composites for spacecraft applications" , pp.770-779, Proceedings of the 34th International SAMPE Symposium and Exhibition, Reno, Nevada, U.S., 1989.

5. G.M. Wells, K.F. Mc Anulty, "Computer aided filament winding using non-geodesic trajectories", in ICCM-VI, pp.1.161-1.173, Proceedings of the 6th International Conference on Composite Materials, London, UK, 1987.

6. G. Di Vita, M. Farioli, M. Marchetti, "Process simulation in filament winding of composite structures", in CADCOMP 90, pp.19-37, Proceedings of the 2nd International Conference on Computer Aided Design in Composite Material Technology, Brussels, 1990.

7. S. M. Grove, "Thermal modelling of tape laying with continuous carbon fibre-reinforced thermoplastic", Composites, Vol. 19, no. 5, pp. 367-375, Sept. 1988.

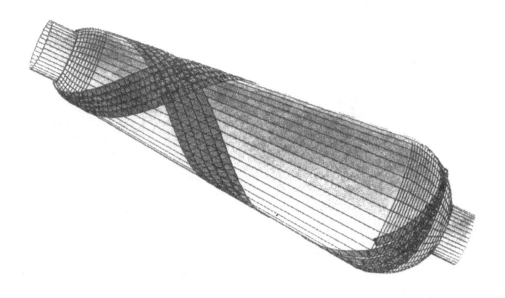

Fig. 1 — Simulation of a non-geodesic tape winding on a n axisymmetric mandrel.

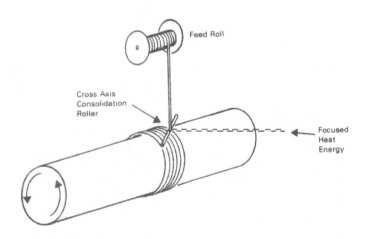

Fig. 2 — Thermoplastic filament winding with continuous consolidation.

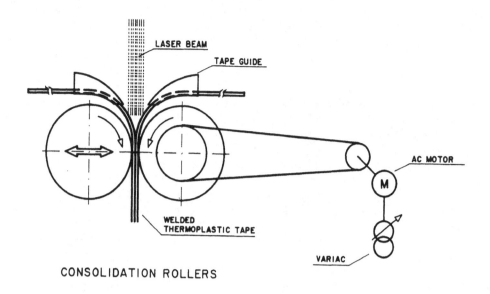

Fig. 3 — Sketch of the experimental equipment.

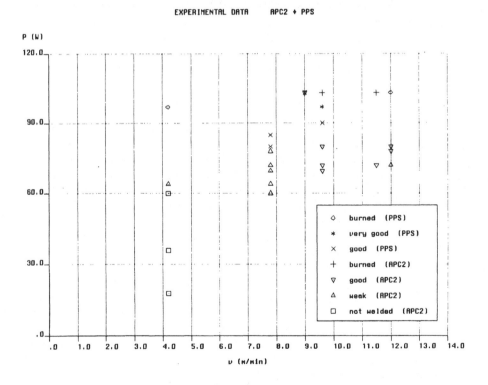

Fig. 4 — Feeding speed - Power beam working points.

FABRICATION OF CONSTRUCTIONAL COMPOSITE MATERIALS BASED ON POWDER POLYMER BINDERS

V.A. DOVGYALO, O.P. YURKEVICH

Metal-Polymer Research Institute
Kirov Str. 32a - 246652 Gomel - USSR

Composite materials, based on continuous fiber fillers and polymer binders stand among the most durable contemporary constructional materials (according to their specific characteristics). Application of a wide range of thermosetting and thermoplastic polymer binders and reinforcing fillers(carbon-, glass-, organic fibers and their combinations) promote achieving polymer articles of various exploitation properties. For all this, the range of their properties depends on the method of combining different components at the stage of prepreg formation.

Two methods are mainly used today to combine components : in solution and in the polymer binder melt. The method of combining in melt is quickly developiing nowadays because of its oecological and technico-economical advantages. Prepreg drying and recuperation of solution are eliminated in this kind of technique. Binders in their initial state are presented as fiber, film, or powder materials in these processes.

The choice of initial polymer and the corresponding technique is determined primarily by structural-mechanical characteristics of the filler and rheological properties of the binder melt which is thermally treated while producing prepregs and their articles.

The technique for producing composite materials, based on powder binders are today the cheapest and the most available among the solutinless technological processes for producing thermosetting and thermoplastic polymers, and a wide range of woven and unwoven fillers. The powder state of polymers has the following characteristics firstly, the particles sizes are comparable with those of different structural elements (both, monofibers and complex threads) of the fillers ; secondly, the particles have large specific surface values (as compared with other types of polymer binders). These peculiarities hinder the process of continuous polymer matrix formation in the filler bulk, when using conventional treatment means of polymer components : pressure, temperature and thermal treatment duration. Polymer binders belong to highly ohmic dielectrics, capable to acquire and retain electgric charges. This permits to use such effective means as interaction of electric fields and components charges to influence the process of combining components.

Suc interaction can be carried out by applying electrically charged binder particles on the filler in the external field of hiqh voltage. This can be realized in apparatus of the chamber type with electrode system for creation electric field. The continuous filler is being pulled through the interelectrode space in the chamber, or over the chamber filled with binder in the fluidized state. It determins the following stages of the operating process :

1. Charging of the binder particles in the fluidized layer (static electrization of the particles in their contact interaction, or ionic charging in corona field, depending on the system parameters).
2. Selection of the particles of the proper size, charge value and polariry from the fluidized layer, their orientation and directed removal with determined intensity in the electric field of the proper voltage.
3. Distribution and arrangement of the charged particles in the bulk and on the reinforced filler surface according to their charge, field valtage, structure and nature of the filler.
4. Retaining of the charged binder particles on the filler surface in the process of prepreg formation.
5. Prepreg layers fixing during production of articles.

The mentioned peculiarities of the operating process make it possible to vary the operating scheme, to chose the most optimum one, taking into consideration structural-morphological and electro-physical properties of the components, to solve interesting technological problems and ensure great advantages :

1. The possibilities to use as binders both thermosetting and thermoplastic polymers.
2. Oecological safety of the process.
3. Simplicity in adjustment of the equipment when substituting one type of binder by another.
4. One layer at a time application of binder compound on the reinforcing layer.
5. Binder spreading on the surface or in the bulk of the reinforcing filler.
6. Creation one of the binder components concentration gradient along the prepreg thickness or its length.
7. A wide range prepreg composition change without changes in productivity of the process.

Electric effect on the process of prepreg creation results in 5 to 80% of binder concentration, high degree of evenness in thickness, increased density of particles arrangement in the deposited layer with regulated process intensity.

Alongside with technological aspect of electric fields and components charges interaction, it is important to mention physico-chemical aspect. Particularly, its influence on the structure formation processes in the binder layers boundary with the filler which results in changes of composite materials quality. It is evident, that the binder electric charge value effects the kinetics of thermosetting polymer binder (polyaminoimide) freezing in the carbon-filled system and influences the carbon-reinforced plastics properties. Two effects are observed in thermoplastics during relaxation processes in the electrically deposited polymer layers : firstly, polymer vetrification temperature increase as a respond to the layer charge increase ; secondly, changes in concentration dependence of carbon-filled system vetrification temperature. These effects determine the influence of polymer charge value on the carbon-reinforced plastics properties.

When using thermo-resistant thermoplasts with rigid chains and high melt viscosity, it is not sufficient in many cased to apply only processes, maintaining the most favourable conditions as to wet, spread and saturate the filler bulk with binder melt. To achieve articles of high properties it is necessary to use some additional methods, particularly - chemical modification of thermoplastics by doping with other polymers or oligo-mers. Injection of powder thermosetting compounds of the similar chemical structure is one of the effective means for thermoresistant thermoplastics modification. These powder com, pounds can dissolve polymer in the melt with the subsequent formation of homogeneous system of low melt viscosity at the initial stage of materials formation. Then the compounds structuralize linear polymer thus forming a matrix of specific structure and properties.

In conclusion, we must note, that the developed technique field of application is not restricted only by production of constructional materials. It can be effectively used in creation of antifriction band materials, foil-coated dielectrics, insulating plastics and laminated systems based, on manycomponent binders.

PROCESSING CONCEPT FOR TUBES MADE OF LONG FIBRE REINFORCED THERMOPLASTICS

W. HOLSTEIN, J. NICK

HMS - Antriebssysteme Gmbh
Kurfürstenstr. 13-14 - 1000 Berlin - West-Germany

ABSTRACT

The processing of carbon-fiber reinforced thermoplastics to little, high-loaded tubes in the scope of medical technique is subject of this here-shown development. It is a modification of the well known 'blow-up-technique', which utilizes the plastic deformation of an inner mold by simple thermodynamic processes for heating and pressing of the thermoplastics.

I-EINLEITUNG

Die Firma HMS Antriebssysteme GmbH ist auf dem Gebiet der Hochleistungsfaserverbundwerkstoffe tätig. Die Konstruktion von 'Bauteilen aus diesen Werkstoffen und die Entwicklung von optimierten Fertigungsverfahren bilden dabei den Schwerpunkt, die Produktion von Kleinserien vervollständigt die Bandbreite des Unternehmens.

Im Rahmen einer Auftragsproduktentwicklung aus dem Bereich der Medizintechnik sollte ein Fertigungsverfahren für dünnwandige, hochfeste und hochsteife Röhrchen aus biokompatiblem Material entworfen und erprobt werden.

Als Werkstoff kommt der Thermoplast PEEK (Polyether-Ether-Keton) von ICI Fiberite mit Carbon-Langfaser-verstärkung zur Anwendung, der unter der Produktbezeichnung APC-2 (Aromatic Polymer Composite) im Handel ist. Dieses 'prepreg tape' hat einen Fasergewichtsanteil von 68 % und ist bei Temperaturen von 380° - 400°C zu verarbeiten.

Ziel der Entwicklung ist es, Bauteile herzustellen, deren mechanische Kennwerte denen des Grundwerkstoffes nicht oder nur unwesentlich nachstehen. Das Fertigungs-verfahren soll zudem einfach und kostengünstig sein und reproduzierbare Qualitäten ermöglichen.

II-PRINZIP DES FERTIGUNGSVERFAHRENS

Der unten beschriebenen Vorgehensweise liegt der Gedanke zugrunde, die bekannte Blow-up-Technik für den geforderten Temperaturbereich zu modifizieren. Das Wesentliche dabei ist, daß das formgebende (Innenrohr) und das pressende (Schlauch) Werkzeug in einem Teil zusammengefaßt werden, d. h. der Kern, der mit dem Werkstoff belegt wird, dient gleichzeitig dazu, das Bauteil in die Außenform zu pressen.

III-VERFAHRENSBESCHREIBUNG

In ein an den Enden verschlossenes dünnwandiges Metallrohr wird eine Flüssigkeit gefüllt, die im oben beschriebenen Temperaturbereich verdampft ist. Um dieses Röhrchen wird anliegend das faserverstärkte Thermoplast-Prepreg gelegt. Beides wird in ein Stahlrohr geschoben, das als Innenmaße die Außenkontur des zu fertigenden Bauteils hat.

Die Form wird vor dem Einschieben des Innenteils mit Inert-Gas gespült, um zu verhindern, daß während des Umformprozesses Werkstoffschädigungen durch Oxidation entstehen. Aus dem gleichen Grund geschieht das Erwärmen in einer evakuierten Heizvorrichtung.

Während der Erwärmung verdampft die Flüssigkeit im inneren Röhrchen, Druck baut sich auf und es kommt zu einer elasto-plastischen Verformung des Werkzeugs (Innenrohr) und damit zu einem Anpressen des Thermoplastes in das Außenrohr.

Nach Erreichen der gewünschten Temperatur wird entsprechend den Verarbeitungshinweisen des Werkstoff-herstellers abgekühlt. Durch die bleibende Verformung des

Innenrohrs kann der notwendige Anpreßdruck durchgehend gehalten werden. Das Entformen des Bauteils geschieht durch das einfache Herausziehen aus der Außenform und durch Verjüngen des Querschnitts des inneren Röhrchens durch leichtes Strecken. Dadurch ist es problemlos für die Herstellung mehrerer Bauteile verwendbar.

Die Vorteile dieses Fertigungsverfahrens lassen sich nur durch die genaue Abstimmung der einzelnen Parameter verwirklichen. Diese sind:

- Geometrie

- Werkstoffauswahl für die Formteile

- Temperaturverlauf beim Aufheizen und Abkühlen

- Art und Menge der Flüssigkeit im Innenrohr

Die äußeren Abmessungen müssen in der Weise aufeinander abgestimmt sein, daß sich einerseits die Packung Innenform-Prepreg noch in das Außenrohr schieben läßt, andererseits der Werkstoff des Röhrchens durch zu große plastische Verformung nicht reißt. Wichtig ist es auch die Prepregs, die in Plattenform geliefert werden und in Querrichtung leicht brüchig sind, für ein möglichst dichtes Anliegen am Innenrohr stufenweise vorzuformen. Dies geschieht in mehreren Arbeitsschritten durch Umformen der Einzelschichten auf immer kleinere Rohrdurchmesser.

Der Werkstoff des Außenrohres ist Stahl, der, entsprechend großzügig dimensioniert, beim auftretenden Innendruck nur kleine Dehnungen aufweist. Für das Innenrohr wird eine Aluminiumlegierung mit hoher Duktilität ausgewählt. Die durch den Innendruck bewirkten Zugspannungen müssen vom Material aufgenommen werden können. Außerdem muß darauf geachtet werden, daß das erhitzte Gas nicht durch den Rohrwerkstoff diffundieren kann.

Die Flüssigkeit muß während des Aufheizens und Abkühlens entsprechend den Verarbeitungsvorschriften für das Thermoplast den erforderlichen Druck aufbringen.

Entscheidend dabei ist auch die Wahl und die Menge des verdampfenden Mediums in Abhängigkeit vom Volumen des inneren Rohres. Aufschluß darüber erhält man in den entsprechenden Dampftabellen.

IV-ZUSAMMENFASSUNG

Die für den Anwender wichtigsten Vorteile dieses Herstellungsverfahrens liegen in der Einfachheit der Vorgehensweise, dem geringen Kostenaufwand für das Werkzeug und einer hohen Produktvariabilität. Bauteile mit zylindrischer oder sich konisch verjüngender Form lassen sich einfach herstellen.

Durch den problemlos einstellbaren Preßdruck lassen sich die geforderten hochwertigen Bauteile herstellen. Dem Konstrukteur ist durch die hohe Qualität des umgeformten Thermoplasten die Möglichkeit gegeben mit den im Katalog des Herstellers angegebenen Werkstoffwerten zu arbeiten und durch die Wahl der Faserbelegung das Produkt den Anforderungen anzupassen.

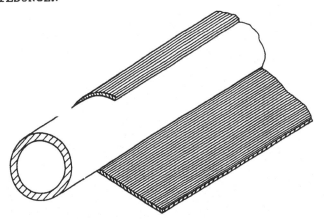

Abb. 1: Langfaser Composite (folienartig) um Vorformrohr

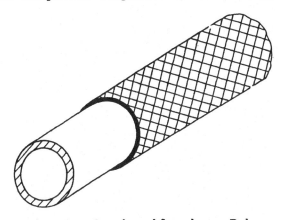

Abb. 2: +/- 45 -Gewebeschlauch um Rohr

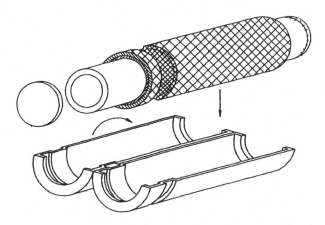

Abb. 3: Teilbare Form und Innenrohr (offen) mit Prepeg-
        Lagen

ADHESION AND WELDING OF CONTINUOUS CARBON-FIBER REINFORCED POLYETHER ETHERKETONE (CF-PEEK/APC2)

G. KEMPE, H. KRAUSS, G. KORGER-ROTH

DLR - Institute of Structures and Design
Pfaffenwaldring 38-40 - 7000 Stuttgart - West-Germany

ABSTRACT

Structures made of more than one component must be joined together by suitable bonding techniques, enabling high loads without damage to the fibers. For joining thermoplastics both adhesion and welding are suitable techniques. In this paper the advantages and disadvantages of these different bonding techniques are discussed.

1. INTRODUCTION

The investigations were done with continuous carbon fiber reinforced Polyetheretherketone (CF-PEEK). Possible joining methods for thermoplastic matrix materials are:

- Adhesion with and without surface treatment
- Welding with heat (mold, heating plate, heating element, hot gas, laser)
- Welding with friction (ultrasonics, vibration, rotation)
- Induction welding

Tensile shear tests were conducted so that identical test conditions could be used for comparing all joining processes. Fig. 1 shows the sample geometry. The shear stress values thus calculated are suitable for comparative evaluation even if they cannot necessarily serve as design values. For further comparisons the samples welded in the mold were taken as a reference standard, with a shear stress value of 100 %.

2. ADHESION

Bonding thermosets with adhesives is a familiar known method. For joining thermoplastics with adhesives some pretreatment is required. The investigation was divided into "Adhesion without Surface Activation" and "Adhesion with Surface Activation".

2.1 ADHESION WITHOUT SURFACE ACTIVATION

Based on the manufacturer's recommendation, tests were conducted with the following adhesives:

- FM 300 (Cyanamide)
- EA 929 (Hysol)
- Araldit AV 119 (Ciba Geigy)
- Agomet F 310 (Agomet Adhesives)
- Agomet F 330 (Agomet Adhesives)

Pretreatment of the samples was performed as follows:

- Degreasing
- Slight roughening of adhesive surface
- Cleaning
- Thin application of adhesive

The adhesive was cured according to appropriate manufacturer's instructions. The bonded CF-PEEK plates were cut in accordance to the sample shape shown in Fig. 1, and the notches were milled. Only samples from those bonded with FM 300 withstood the processing. Differences in shear stress values depend on cleaning agent (ca.20%) and the kind of roughening the adhesive surface received (ca. 100 %). These values aren't representative and too poor for joining primary structures. For such kind of structures a surface activation is required.

2.2. ADHESION WITH SURFACE ACTIVATION

The following surface activations are possible:

- flame scarfing
- etching with chromic acid
- plasma treatment

For further investigations specimens were plasma treated and bonded with the adhesives FM 300 (epoxy adhesive film) and EA 929 (thixotropic paste). In addition to the tensile shear tests the influence of the delay between plasma treating and adhesion and the aging of the adhesion was investigated. In contrast to the samples not plasma treated, those specimens treated withstood the process of cutting and milling the notches.

The investigations about the delay between plasma treating and adhesion showed a decreasing shear stress, Fig. 2. The samples bonded 72 hours after plasma treating attained only 70 % of the shear stress value of the reference sample. But the values of the samples bonded immediately after plasma treating, conditioned in an environmental chamber (rel. humidity 85 %, temperature 70° C, time 2160 h) and moistened in water only showed a decrease of 10 %. These investigations aren't completed yet.

3. WELDING

Thermoplastics may be replasticized which makes them suitable for welding techniques. Therefore in our institute welding techniques were compared to adhesive bonding techniques.

3.1. WELDING WITH HEAT

Welding with heat involves the generation of heat by heating elements (e.g. cartridge heaters, heating foils) or by heat transfer through radiation.

3.1.1. JOINING IN THE MOLD

Thermoplastics offer the advantage of resoftening and welding, and this property is employed when joining components at appropriate pressures. Relatively high levels of dimensional accuracy can be obtained if a mold is used. If the component has hollow spaces, supporting cores must be inserted since the entire component softens. In general, the process of joining in the mold renders good and reproducable values.

3.1.2. HOT PLATE WELDING

In order to avoid the heating of the entire component with fiber-reinforced thermoplastics, an electrically heated plate (heated "mirror") was positioned between the surfaces to be welded. This heating plate can be made according to the contour of the component concerned. The joining zone is heated with contact or radiation heat. After obtaining the necessary plastification the heating plate is removed and the parts are pressed together.

3.1.3. WELDING WITH A HEATING ELEMENT

The use of a heating element embedded in the joining surface provides the advantages of welding with a heating plate, without the danger of damaging the welding surface by fiber and matrix-sticking. At DLR work was done on a process not requiring the use of other materials. Particularly, in the case of CF-PEEK/APC 2 the good insulation properties of the matrix material and the good electrical conductivity of the carbon fibers make it possible to produce a heating element, Fig. 3, which corresponds to the properties of the component's composite material.

3.2. WELDING WITH FRICTION

The usual welding processes

- Ultrasonic welding
- Vibration welding
- Rotation welding

all employ the principle of heat produced by friction. Introduction of required energy is achieved in different ways.

3.2.1. ULTRASONIC WELDING

In ultrasonic welding facilities high frequency electric vibrational energy -supplied by a generator- is converted into mechanical energy by an ultrasonic converter. The energy introduced is converted via molecular and interfacial friction into the heat required for fusion. The results of the tensile shear tests, Fig. 4, show a decrease in tensile shear strength with increasing power, due to damages to the welding area.

3.2.2. VIBRATION WELDING

Vibration welding is a linear friction process. The required plasticization is achieved by pressure and forced friction in the joining zone. For the welding process one part is held in fixed position. The second part vibrates according to the set amplitude. Energy conversion takes place at the location where it is needed. One prerequisite is that the joining seam allows a straight frictional movement along one axis. PEEK-Foil as a welding aid is used to get the same levels of the thicknesses of the samples and to avoid fiber damages in the joining area.

3.2.3. ROTATION WELDING

Rotation welding is a type of friction welding. This process is suitable for bonding rotationally symmetrical formed parts. One part rotates and the second part is held in place. After rotation is stopped the weld "freezes". The rotational speed and friction time have an important effect. A conical surface weld was chosen to maintain the favourable properties of continuous reinforcement. The cone had an angle of 7°. The study showed that it was necessary to clamp both tubular pieces and that the welding point had to be supported not only on the inside by a centering mandrel but also on the outside.

3.3 INDUCTION WELDING

The EMAWELD process /1/ is a type of induction welding. In this process a welding aid made of EMAWELD material with an application of magnetically activatable powder is positioned between the two components to be joined. The powder particles are embedded in the same or in a similar thermoplastic from which the components are made. A high-frequency magnetic field causes the particles to be heated through hysteresis and eddycurrent losses, raising the welding aid to melting temperature and thus producing a weld. Welding tests using plates produced by DLR (for 2 welding tests) and made of CF-PEEK (APC 2) were conducted by Emabond Incorporated.

4. CONCLUSION

For comparing the different joining techniques they must be subdivided into processes, Fig. 5,

- which weld the entire surface
- which only perform spot welding
- which can damage the interface.

The results of different bonding techniques are shown in Fig. 6. The best results are obtained with those methods which employ heat (mold, heating plate, heating element and the EMAWELD process). With these welding methods fiber orientation and wall thickness of the components are retained.

The application of methods which weld by friction is possible, however, these produce lower shear stress values. The fiber orientation at the interface does not remain unchanged, the fibers shift partly together with the matrix material. The best results with this welding category can be obtained with vibration welding.

Adhesive bonding of CF-PEEK without surface treatment is problematic. Tests in a low pressure gas plasma show an improvement of the surface activation whereby good results are obtained in large joining areas. Compared with the welding techniques joining structures with adhesive avoids additional thermal stress.

LITERATURE:

Emabond   Eletromagnetische Schweißverfahren für
Incorporated   die Verbindung von thermoplastischen
                 Kunststoffteilen; 12.1985

Kempe, G.   Manufacturing Processes and Molding of
Krauss, H.   Fiber-Reinforced Polyetheretherketone
                 (PEEK); ICAS-86-4.6.1, London/UK,
                 Sept. 1986

Kempe, G.   Fügeverfahren   für   kontinuierlich
Krauss, H.   verstärkte Thermoplaste
                 Seminar: Faserverstärkte Hochleistungs-
                 thermoplaste, Umsetzung und Verarbeitung
                 zu Sturkturkomponenten (ETH), Zürich/CH,
                 Sept. 1989

Hudis, M.   Plasma Treatment of Solid Materials,
                 in: Hollahan, J.R; Bell, A.T.
                 "Techniques and Applications of Plasma
                 Chemistry", New York: John Wiley + Sons

Rasche, M.   Klebetechnik, Oberflächentechnik, Nieder-
                 druckplasmatechnologie
                 Mitteilung Ingenieurbüro M. Rasche,
                 Berlin/D

Liston, E.M.   Plasma Modification of Surfaces to
Rose, P.W.   Enchance Bonding, Painting, Printing
                 and Wetting, ISPC-7, Paper No. P-4-5
                 Eindhoven, July 1985

Grünwald, H.   Plasmapolymerisation von Acrylnitrid,
                 Propylenoxid und Tributylzium-Methacrylat
                 sertation Universität Tübingen, 1987

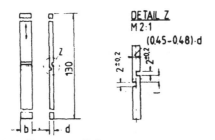

LAMINATE WIDTH   b : 12.0 mm
LAMINATE THICKNESS   d : 4.0 mm
NOTCH-DISTANCE   l : 3.0 mm

Fig. 1:   Specimen for Tensile Shear Test

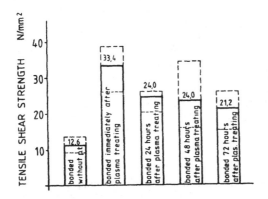

Fig. 2:   Influence of Surface Activation with
        Oxygen

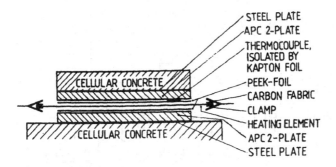

Fig. 3:   Heating Element, Schematic Test Set-Up

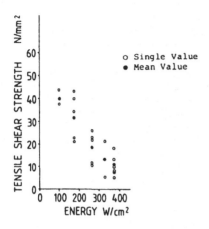

Fig. 4: Ultrasonic Welding, Test Results

| | BONDED AREA: | |
	NOT DAMAGED	DAMAGED
AREA- BONDING	MOLD HEATING MIRROR HEATING ELEMENT INDUCTION ADHESIVE	VIBRATION ROTATION — — —
SPOT- BONDING	—	ULTRASONIC

Fig. 5: Bonding Techniques, Influence on Bonding Area

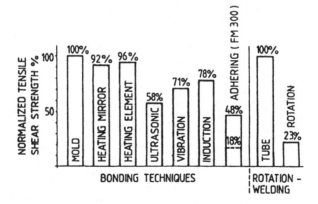

Fig. 6: Comparison of Different Bonding Techniques, Test Results

112

AN EXTENDED MODEL FOR THE FORMING SIMULATION OF FABRIC REINFORCED THERMOPLASTIC PREPREGS

J. MITSCHERLING, W. MICHAELI

IKV - Pontstr. 49 - 5100 Aachen - West-Germany

ABSTRACT

Deep drawing of continuous fibre reinforced thermoplastic prepregs is a promising new composite production process. Geometric models are used to simulate the draping configuration of fabrics to complex shaped tool surfaces without wrinkling or bridging. An extended model of the deep drawing process taking into consideration the forming stages, friction and fabric shearing forces, too, is presented in this paper.

INTRODUCTION

A production process for composite parts of increasing importance is the forming of continuous fibre reinforced thermoplastic prepregs. These are available in the form of semifinished lamina plates, single plies which must be stacked to lamina and several covoven and comingled materials.
Various forming processes are used to produce parts, partly wellknown from corresponding forming processes of sheet metal and thermoplastic plates:
deep drawing, hydro-, rubber- and diaphragm forming etc.
The prepregs being used for high performance composites in most cases up to now textiles with long straight fibres, such as UD-layers and fabrics are prefered in this context.
Parallely to the process technology simulation methods for the forming process have been developed, /1 bis 6/.

I - THE GEOMETRIC FABRIC MODEL

Unlike in e.g. random mats or knitted materials in fabrics in the direction of warp and weft neither flow nor strain can occur to a considerable amount. The drapability of the fabrics to double curved surfaces is merely based on the shearing of warp and weft fibres.

Hence in the forming simulation fabrics are modeled as lattices, the vertices of which are posed on the tool surface in a single step keeping constant distances a and b of the warp resp. weft yarn paths. Only rhombic deformations of the lattice squares are allowed.

By fixing the path of one warp and one weft fibre on the surface the position of the other yarn paths is uniquely determined. Each point is calculated from the before calculated neighbour lattice points. (fig.1)

$$(1) \quad \| P_{ij} - P_{ij\text{-}1} \| = a, \quad \| P_{ij} - P_{i\text{-}1j} \| = b$$

The nonlinear equation system is solved by a Newton method.

In the resultant fabric position the extent of the local shear deformation can be observed. Depending on the construction of weave and fibre density there is a limit of deformation which can not be exceeded without wrinkling. Also areas of fibre bridging can be detected. (e.g. /6/)

Different strategies are used to contrain the first warp and weft yarn paths: Often symmetry reasons call for laying them into the tool's symmetry planes, /1,2/. Geodethic paths can be presumed /4/, or the total extent of fabric deformation is minimized, /3/.

The programs based on the geometric model are an aid to define theoretically admissable fabric layups rather than a real process simulation. The various possible results are influenced by the user through more or less comfortable means of fixing the first warp and weft path.

How the process has to be performed to achieve the calculated fabric layup can not be derived direcly. Contrariwise it can not be predicted how a fabric is deformed under certain process conditions.

Furthermore a pseudo simulation of the process stages by stepwise increase of the tool height shows that even in that case where the finally resulting layup does not wrinkle during the forming regions of not admissable deformation can occur. (fig.2)

An extended forming model to calculate the forming stages from a starting position of the fabric to the finished part taking into consideration the working forces, too, is presented in the following chapter.

II - THE EXTENDED FORMING MODEL

2.1. The calculated quantities

The fabric is modeled as freeform surface parametrized in warp and weft direction. Membrane stresses, friction and pressure on the tool surface are related to the local coordinate system given by the accompanying tripod. (fig. 3)
Bending is presumed neglectable, which is a reasonable assumption for the molten flexible prepregs during forming. Whether theory has to be modified for thicker multiplied lamina future will show.
The main differential equations are derived from the momentum- and force equalities by the means of the Gauß Theorem:

$$(2) \quad \tau_{uv} - (\cos \alpha) \sigma_v = \tau_{vu} - (\cos \alpha) \sigma_u =: B$$

This equation describes the symmetry of the shear stresses expressed in the nonorthogonal local coordinate system (e_1, e_2).

$$A_u := [\sigma_u - (\cos\alpha)\tau_{vu}] \qquad A_v := [\sigma_v - (\cos\alpha)\tau_{uv}]$$

$$(3a) \quad
\begin{aligned}
(\sin^2\alpha)\tau_{wu} = {} & \left((\cot\alpha)\, \frac{d\alpha}{du} + \frac{\cos\alpha}{\sin^2\alpha}\, \frac{d^2g}{du^2}\, e_2\right) A_u + (\cot\alpha)\, \frac{d\alpha}{dv}\, B \\
& - \frac{1}{\sin^2\alpha}\, \frac{d^2g}{dv^2}\, e_1 A_v - \frac{d}{du}\, A_u - \frac{d}{dv}\, B
\end{aligned}$$

$$(3b) \quad
\begin{aligned}
(\sin^2\alpha)\tau_{wv} = {} & (\cot\alpha)\, \frac{d\alpha}{du}\, A_u + \left((\cot\alpha)\, \frac{d\alpha}{dv} + \frac{\cos\alpha}{\sin^2\alpha}\, \frac{d^2g}{dv^2}\, e_1\right) A_v \\
& - \frac{1}{\sin^2\alpha}\, \frac{d^2g}{du^2}\, e_2 A_u - \frac{d}{du}\, B - \frac{d}{dv}\, A_u
\end{aligned}$$

$$(3c) \quad -(\sin^2\alpha)\sigma_w = \frac{d^2g}{du^2}\, e_3 A_u + 2\, \frac{d^2g}{dudv}\, e_3 B + \frac{d^2g}{dv^2}\, e_3 A_v$$

The prepreg temperature influences the material properties concerning friction and fabric deformation and, therefore, has to be calculated. This will not be explicitly mentioned in the further text.

2.2 The fabric deformation theory

The assumptions of pinned joints in the warp and weft intersections anu or inextensible fibres is kept. The corresponding equations (4) after discretization by linear finite elements (dirac functions for higher level derivations) lead to the geometric algorithm (1) explained above. In this sense the geometric algorithm is the partial solution of the extended model.

$$(4) \quad \left\| \frac{dg}{du} \right\|^2 = \left\| \frac{dg}{dv} \right\|^2 = 1$$

The description of the fabric shear deformation behaviour is based on papers deriving the shear stress-deformation relation of (not matrix impregnated) fabrics by micromechanical models. /7/

These contributions show that the forces required for shearing a fabric is indirectly depending on the biaxial stresses through the compression force between weft and warp fibre in the yarn intersections. A linear statement proved to be sufficiently exact for describing the shear curves.

The deformation angle is nonlinearly influencing the shear stress, especially limiting the deformation angle which can be achieved without wrinkling.

Because of the viscous thermoplastic matrix a dependancy on the shear rate is expected, too.

The exact deformation function is presently examined in shearing tests:

$$(5) \qquad \tau_{uv} = f(\alpha, \frac{d}{dt} \alpha, F_c (\sigma_u, \sigma_v))$$

2.3 Friction between prepreg and tool

The tool movement is input as a geometric contraint to the prepreg configuration in the different forming stages.

Provided that the prepreg does not affect the tool shape during forming, non-rigid tools such as hydroforming tools or diaphragms can be calculated, too. The shape and the movement of the tool surfaces is calculated seperately in that case. (s. /5/ for diaphragm forming)

$$(6) \quad G_1 (g_x (u,v,t), g_y (u,v,t), t) \le g_z (u,v,t) \le G_2 (g_x (u,v,t), g_x (u,v,t), t)$$

During forming the contact areas between prepreg and tool resp. blank holders are determined. (fig.4)

Friction is constitutive for the forming process, because, unlike for plastic materials, the material required to fill the curved regions of the part can not be delivered by thinning, but must be totally pulled out of the peripheral blank holders.

In the contact areas a friction law is valid, the exact form of which is presently determined in friction tests. The molten matrix embedding the fabric causes a mixture of Coloub and laminar friction. Simple sliding of the prepreg on the surfaces has to be distincted from pulling the prepreg out of the clamping blank holders.

(7) K_1, K_2 : $|\, \tau_w \,| \le f_{st}\,(\sigma_w)$ $V_{rel} = 0$, (sticking)

$\tau_w \;\; = -\, f_{sl}\,(\sigma_w, V_{rel})V_{rel}$ $V_{rel} > 0$, (sliding)

$K_{1,2}$: modified f_{st}, f_{sl}

$\overline{K}$: $\sigma_w = \tau_w = 0$

V_{rel} is the relative velocity of the prepreg on the tool surface

2.4. Numerical solution

The principal proceeding to the numerical solution of the equations is similiar to the simulation of deep drawing sheet metal, especially the consideration of contact and friction.
The differential equations are discretized by finite elements.
Starting from an initial prepreg configuration in each time step the contact areas are determined. The inequality constraints (6) are forced by a penalty funtion. Prepreg geometry and stresses are iteratively calculated by turns.
The implementation of the presented model is in hand, and the practical verification in experiments is planned for the future.

ACKNOWLEDGEMENT

The authors would like to thank the DFG for their supporting this work, and Bayer AG for their material donations.

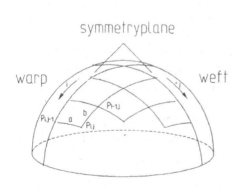

Fig. 1 - Geometric fabric model

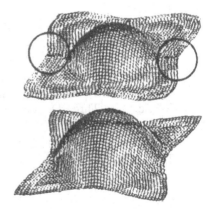

Fig. 2 - Wrinkling in earlier stages of deep drawing

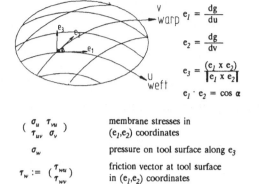

$$e_1 = \frac{dg}{du}$$

$$e_2 = \frac{dg}{dv}$$

$$e_3 = \frac{(e_1 \times e_2)}{|e_1 \times e_2|}$$

$$e_1 \cdot e_2 = \cos \alpha$$

$\left(\begin{smallmatrix} \sigma_u & \tau_{vu} \\ \tau_{uv} & \sigma_v \end{smallmatrix} \right)$ membrane stresses in (e_1, e_2) coordinates

σ_w pressure on tool surface along e_3

$\tau_w := \left(\begin{smallmatrix} \tau_{wu} \\ \tau_{wv} \end{smallmatrix} \right)$ friction vector at tool surface in (e_1, e_2) coordinates

double contact   no contact   single contact
$K_{1,2}$ $\bar{K}$ K_1 bzw. K_2

Fig. 3 - Extended forming model

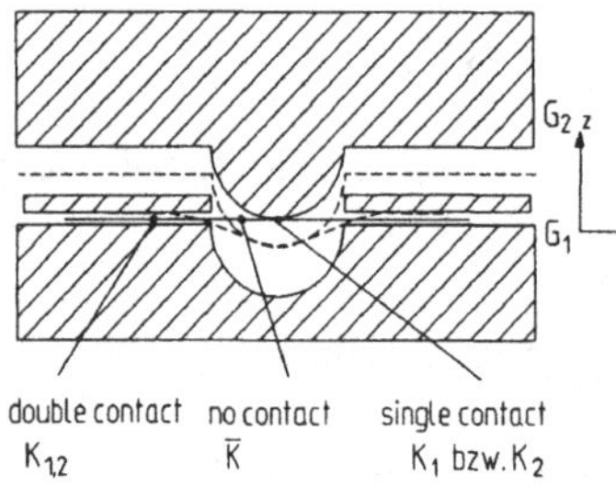

Fig. 4 - Contact areas between prepreg and tool

REFERENCES

1. - Robertson R.E., Hsiue E.S., Yeh G.S.Y., Continuous fiber rearrangements during the molding of fiber composites, Polymer Composites, Vol. 5, No. 3, July 1984
2. - Smiley A.J., Pipes R.B., Fiber Placement During the Forming of Continuous Fiber Reinforced Thermoplastics, Society of Manufacturing Engineers, Technical papers, EM 87-129
3. - Bergsma O.H., Huisman J., Deep Drawing of Fabric Reinforced Thermoplastics, CADCOMP-Conference, April 1988, Southampton, Springer
4. - Heisey F.L., Haller K.D., Fitting woven fabric to surfaces in Three dimensions, Journal of Textile Inst., 1988, No. 2.
5. - Smiley J.A., Diaphragm forming of Carbon Fiber Reinforced Thermoplastic Composite Materials, CCM Report 88-11, May 1988
6. - Van West B.P., Keefe M., Pipes R.B., The Draping of Bidirectional Fabric Over Three-Dimensional Surfaces, Univ. of Delaware, 1989
7. - Kawabata S., Niva M., Kawai H., The FINITE-Deformation Theory of Plainweave Fabrics, Part III: The Shear Deformation Theory, J. Text. Inst., Vol. 64, Manchaster, 1973

A NOVEL TOOL TO DETERMINE THE OPTIMUM FABRICATION PARAMETERS OF THERMOPLASTIC CFRP LAMINATES

P.W.M. PETERS, A. LYSTRUP*, S.-I. ANDERSEN*

DLR - Institute of Materials Research
PO 906058 - 5000 KÖLN 90 - West-Germany
*Riso - Postbok 49 - Roskilde - Denmark

ABSTRACT

A device is described which enables the application of non-uniform pressure on stacked plies of carbon fibre thermoplastic prepreg in an autoclave. The device consists of a pressure plate on top of the stacked plies; the plate is larger than the plies and thus bends over them leading to a non-uniform pressure distribution with a high pressure at the edge and low pressure in the middle of the laminate.

The pressure plate and laminate were chosen to be circular. Three different ways of production were used:
(a) the vacuumbag is vented to atmosphere.
(b) the stacked plies are evacuated at the circumferential edge only.
(c) the stacked plies are evacuated through a centre hole in the plies and the pressure plate as well as through the edge.

Ultrasonic C-scan investigations of the laminates produced show hardly no difference between the first two methods. The last method, however, leads to a clearly improved compaction. This indicates that the free volume between the different plies should be kept open to vacuum as long as possible. Furthermore, from the shape of the not fully compacted area in the centre of the produced laminates it could be concluded, that the flow in a direction transverse to the fibre at the surface of the different plies controls the consolidation process.

INTRODUCTION

Fibre reinforced thermoset matrix materials have found numerous applications in aerospace and other structural components. One of the advantages in their use is their rather easy processability, which makes it possible to fabricate complex-shaped structures. The use of thermoplastic matrix systems has recently been encouraged in order to overcome the brittle behaviour of thermosetting CFRPs especially under impact. The fracture strain of a thermoplastic matrix generally is an order of magnitude higher than that of thermosetting systems, so that impact behaviour and other matrix dominated properties can be improved considerably. Processability of thermoplastic matrix systems is however more difficult, because the viscosity generally is one or two orders of magnitude larger than that of thermosetting systems. Further, the processing temperature generally is 100°C to 200°C higher. The processing window of fibre-reinforced thermoplastics is generally determined experimentally by fabricating laminates under a range of temperatures and pressures: a time-consuming process. The present work is based on a special tool, designed to produce a varying pressure inside the laminate from one location to another.

The non-homogeneous pressure distribution within the vacuum bag on the stacked plies is realized by choosing the plate on top of these plies to be somewhat larger than the plies themselves (figure 1). In this case the top plate bends over the plies leading to a non-uniform pressure distribution with a larger pressure at the edge. The influence of this pressure distribution on compaction is investigated.

MATERIAL AND EXPERIMENTS

Laminates were produced from APC-2 (AS-4 carbon fibre-reinforced polyetheretherketon supplied by ICI). The quality of consolidation largely depends on that of the prepreg. Good quality prepreg has a constant (mean) thickness over the width and length of the prepreg and the roughness should be small. For a characterization of the prepreg quality see /2/.

The pressure distribution on the stacked plies as a result of the oversize of the pressure plate can be calculated with the aid of the theory of plates on an elastic foundation /1/. This is done elsewhere /2/. For ease of calculation the shape of the produced laminates were choosen to be circular. According to the calculation the pressure distribution was adequate by chosing the following dimensions, diameter of laminates 200 mm, diameter of pressure plate 240 mm with a thickness of 8 mm (material: construction steel St 50) under an autoclave pressure of 2 bar.

Three different setups were used. At first three circular laminates $(0)_7$, $(0/90/0/\overline{90})_s$ and $(0/+45/90/-45)_s$

were produced with the vacuum bag vented to atmosphere. Another laminate $(0/90/0/\overline{90})_s$ was produced under an autoclave pressure of 1 bar and with vacuum pressure in the vacuum bag. In this case vacuum has access to the stacked plies only through their circumferential edge.

Finally a plate $(0/90/0/\overline{90})_s$ was produced under the same conditions as the last plate, but now with a 2-mm centre hole both in all individual prepreg-plies as well as in the pressure plate. This allowed the plies to be evacuated also through the centre hole.

The plates were consolidated at a temperature of 380°C. Vacuum (or ambient pressure) was maintained during the whole cycle. The pressure (1 or 2 bar) was applied, when the consolidation temperature is reached. The temperature is reduced (at of rate of about 20°C/min) after 5 min of constant pressure in the autoclave and the pressure is relieved when the temperature is below 260°C.

The produced plates were investigated in a waterbasin by C-scan with two sensors, one on either side of the plate, making use of an Echograph 1054 ultrasonic inspection equipment (Company Deutsch). For an extensive description of the evaluation of the ultrasonic signal see /2/.

RESULTS

The C-scan pictures of the first series of produced laminates are presented in figure 2. Under the applied pressure distribution the unidirectional laminate (0_7) shows a rather good compaction, whereas the $(0/90/0/\overline{90})_s$ and $(0/+45/90/-45)_s$ laminates show poor compaction in the centre of the laminate. Material flows because of the pressure gradient to the centre of the plate. At the edge, however, there is a disturbing radial flow to the outside. This is extreme in the case of the unidirectional plate. After consolidation the originally circular plies (diameter 200 mm) become deformed to an ellipse with the larger axis $l = 212$ mm lying transverse to the fibre direction.

The laminate $(0/90/0/\overline{90})_s$ shows an area of poor compaction in the middle of an approximate square with the diagonals in the 0°- and 90°-directions. The laminate $(0/+45/90/-45)_s$ shows a rather circular shape of the poor compacted area.

The laminate $(0/90/0/\overline{90})_s$ produced with vacuum access only through the prepregs edges only shows a minor reduction of the poor compacted area. If, however, vacuum is also applied through the centre hole of the pressure plate and the plies, the compaction of the laminate is improved. Figure 3 shows a C-scan of this laminate. The area of poor compaction is small and its extension in the 0°-direction is greater than that in 90°-direction. (The deflection of the plate is not influenced by the 2 mm hole /2/).

DISCUSSION

Before these particular results of consolidation are discussed some general remarks concerning the production of laminated plates from thermoplastic prepreg have to be made. In the ideal case of prepregs with a uniform thickness and no roughness (surface is mirror like) the prepregs have to be put in contact at an adequate temperature to produce a bond between the plies by linking the amorphous chains. In reality, however, the following non-homogenities have to be considered:
- the prepregs are not of a constant thickness
- the prepregs are rough and
- the press plates (of a press) are not parallel or flat.

It is assumed that the prepreg layers consist only of a single piece of prepreg. That is, no overlapping or gap between the prepregs are considered.

In the case of an autoclave the last mentioned influence of the press plates is irrelevant. If we assume a completely flat tool and a flat rigid bottom plate only a non-homogeneous thickness and the roughness of the prepreg can influence the quality of compaction.

At the beginning of compaction the rough surfaces make contact at the thickest locations with a free volume between the plies, the size of which depends on the value and irregularity of the roughness. During compaction there is a flow of material from the areas where the prepregs touch each other to areas where they do not. As long as the roughness is small (and there is no variation in mean thickness) the amount of flow necessary for compaction is small. Flow can take place unhindered under vacuum conditions (in an autoclave, generally) or hindered by ambient pressure (in a press, generally).

If a laminate is produced under ambient pressure it is important to keep the free volume between the separate plies open to the atmosphere. This is the case for a good quality prepreg with a regular distributed but small roughness. As soon as the roughness is irregular it is likely that during compaction the free volume between the different plies can be cut off from the atmosphere. Thus on further compaction these areas have to be closed against a gas pressure that builds up during further compaction and thereby increasing the pressure needed for producing a good quality laminate. Exemples of this phenomenon are the $(0/90/0/\overline{90})_s$ - and $(0/+45/90/-45)_s$ - laminates of which the C-scan pictures are shown in figure 2.

Air and gasses that are expected to be generated by the prepreg during the heat-up process are vented to atmosphere. At an early stage, however, (probably only shortly after the consolidation temperature is reached), the edge of the circular plate is compacted, leading to a sealing off of the inner free volume. Under the autoclave pressure of 2 bar, the consolidated area grows to the centre of the plate centre also forcing eventually formed gasses and entrapped air towards the centre. Thus in the laminate centre compaction gets more difficult because of the lower stress and the opposing pressure of the entrapped and compressed gasses. By drilling a centre hole through all of the different plies and through the press plate, formed and entrapped gasses are able to escape through the centre. This leads to clearly improved consolidation in the centre of the plate (see figure 3). One reason for its lack of full compaction can be that the hole in the stacked plies (or press plate) has been blocked with or sealed off by matrix material before all the gasses and air escaped through the hole.

The further question arises: why is the consolidation under an angle of $\pm$ 45° to the fibre directions better than in one of the fibre directions? The answer comes from the consolidation experiment of the unidirectional laminate. This laminate showed extreme outward flow perpendicular to the fibre direction as a result of the free edge. Thus the flow is strong in a direction perpendicular to the fibre caused by a movement of the fibres in that direction whereas in the fibre direction flow is restricted by the presence of the fibres themselves. In this case only the matrix will flow (tunnel flow) /3/.

In the $(0/90/0/\overline{90})_s$ -laminate at an angle of 0°, 90°, 180° and 270° to the $\overline{x}$ direction material flow (figure 2 b) has to occur along one of the fibre directions of the plies, which is restricted as mentioned above. At an angle of 45°, 135°, 225° and 315° the radial flow in the different plies is equal and controlled by the radial component of transverse flow (in respect of the fibre direction of the separate plies). If we compare the shape of the poor compacted area (ultrasonic signal < 600 mV, white area) of both $(0/90/0/\overline{90})_s$ -laminates (figures 2b, 3), it is not directly clear why the shape of the area of the poorest compaction changes from a square to a rectangle with the longer axis in the 0°-direction. If we consider the shape of the poor compacted area with only one threshold level smaller than good quality material (1200 mV < ultrasonic signal < 1000 mV, dark grey area), then again the shape is a square, with one of the diagonals in the 0°-direction. Thus, it is clear, that the shape of the not fully compacted material in the centre is similar; only the poorest compacted area is of rectangular shape. One reason for this shape can be the presence of more layers (4) in the 0°-direction in comparison with the number of layers (3) oriented in the transverse direction. As the free volume between the separate prepreg plies is filled mainly by local transverse flow at the surface of the respective plies it can be expected that at 90° and 270° there is greater flow (in the transverse direction of the four 0°-plies) than at 0° and 180° (in the transverse direction of the three 90°-plies). Thus in the 90°- and 270°-directions the material should theoretically be more effectively compacted. The difference in compaction is larger can be reasonably expected, however. This could be caused, as mentioned before, by closure of the centre hole rim before all gasses could be removed.

Further the poor compacted area of the $(0/+45/90/-45)_s$-laminate is smaller in comparison with the $(0/90/0/\overline{90})_s$-laminate (figure 2), because the difference in fibre orientation of the different plies is always 45°. Thus transverse flow (which is mainly responsible for filling up the free volume between the plies) is less restricted than in the case where the fibres of different plies cross under right angles.

CONCLUSIONS

A novel tool which makes it possible to apply a non-uniform pressure on stacked plies of fibre-reinforced thermoplastic prepregs during fabrication of laminates has been developed, and a series of laminates produced by this tool has been studied to gain a better understanding of the consolidation process during the fabrication of laminates. During laminate fabrication from thermoplastic prepregs the flow of material and thereby the degree of consolidation is controlled by local pressure differences between areas of contact of the different plies and areas of non-contact. Entrapped air and gasses hinder this flow by the formation of an opposite pressure difference.

Consolidation takes place by local flow of material mainly at the surface of the different plies. If the prepreg has a uniform thickness and a uniform roughness the amount of lateral flow integrated over the complete prepreg surfaces is zero. In the present investigation the pressure gradient towards the laminate centre induces a flow in this direction. The fibre orientation of the different plies with respect to this radial flow clearly influences the degree of compaction. If the radial flow is in one of the fibre directions then the flow is restricted, whereas transverse to the fibres, flow (and thus compaction) is easy. This leads to a typical shape of the poor compacted area in the centre of the fabricated laminates. From the results it can be concluded that under normal process conditions (uniform pressure on the laminate) consolidation preferably takes place by local flow at the surface of the different plies transverse to the direction of fibres. This flow, and thus compaction, is reduced by neighbouring plies if their fibres cross at a right angle. Differences in the fibre orientation of neighbouring plies of 90° should therefore be prevented.

If a laminate is consolidated by using a uniform pressure distribution on top of the laminates the quality of the laminate is strongly dependent on the smoothness and thickness variation of the prepreg plies. Smooth and uniform thick prepregs produce good quality laminates, prepregs that are not uniform thick and rougher have a tendency to entrap air and gasses; thereby, pressure is locally built up inside the laminate, preventing it from being fully consolidated.

If a laminate is consolidated under a non-uniform pressure distribution at its surface entrapped air and gasses inside the laminate are forced towards areas of lower pressure, good quality laminates can then be produced only if it is possible to vent or evacuate these gasses and air pockets from the low pressure areas.

REFERENCES
1. Hetényi, Beams on elastic foundation.Ann Arbor: The university of michigan press, 1946.
2. Peters, P.W.M., Lystrup, Aa., Andersen, S.I., to be published in Composites Manufacturing.
3. Lee, W.I. and Springer, G.S., J. Comp. Mat., 21 (1987), 1017-1055.

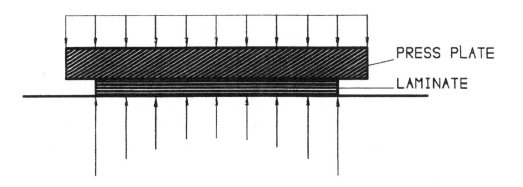

Fig. 1:   Pressure distribution in laminate as a result of the overbending press plate.

Fibre direction

Fig. 2a: laminate 0_7

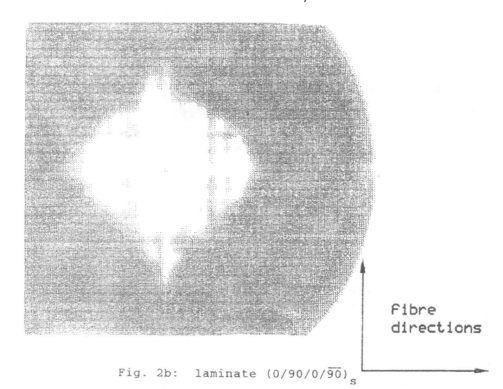

Fibre
directions

Fig. 2b: laminate $(0/90/0/\overline{90})_s$

126

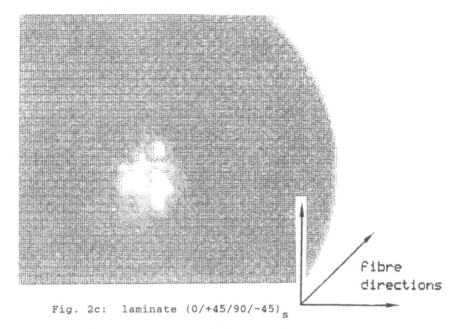

Fig. 2c: laminate $(0/+45/90/-45)_s$

Fig. 2:   C-Scan of laminates produced at an autoclave
           pressure of 2 bar with the vacuumbag vented   to
           atmosphere.

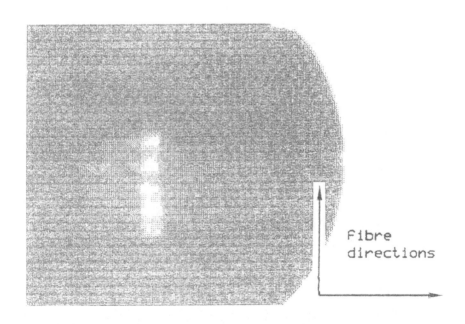

Fig. 3:   C-Scan of the laminate $(0/90/0/\overline{90})_s$ produced at
           an autoclave pressure of 1 bar and with   vacuum
           access   at   the   centre   of   the   press   plate
           (through a 2 mm hole) and at the laminate edge.

STAMP FORMING OF CONTINUOUS FIBER/POLYPROPYLENE COMPOSITES

M. HOU, K. FRIEDRICH

Institute for Composite Materials
Postfach 30496 - University of Kaiserslautern
6750 Kaiserslautern - West-Germany

ABSTRACT

A right angle experimental mold is employed for anisothermal stamp forming of continuous CF/PP composites. Results about the effect of stamping time on the part shape, fiber movement at the bend range and mechanical behaviour of the part are presented. Experimental results showed that (a) a useful stamping time was 15 seconds, (b) fiber movement at the exterior bend range was much more intensive than that at the interior bend range, and (c) two modes of failure- transverse matrix cracking and interlaminar delamination-were observed.

INTRODUCTION

In a previous investigation /1/, a right angle tool was employed for experimentally studying of the anisothermal stamp forming of CF/PP composites. Optimum processing conditions, such as stamping temperature, stamping velocity and pressure in relation to the properties of stamped parts were determined. It is very important to further develop the advantage of the short cycle processing time of this technique and to determine the fiber waviness induced during the stamp forming from a flat layup. These and other issues, e.g. concerning the effect of stamping time on the springback phenomenon of the stamped bend part are content of the present study. Fiber movement at bend range were investigated with Cu tracer wires embeded in preconsolidated laminate samples. Finally, the mechanical behavior of stamped parts was studied by bending tests.

EXPERIMENTAL

The material used in this study is continuous carbon fiber (AS4) reinforced polypropylene prepreg with approximately 20% fiber volume fraction, known as PLYTRON, manufactured by ICI,U.K. The experimental set-up is shown in Fig. 1. The manurfacturing process of preconsolidated flat laminates and the stamp processing conditions are explained in detail in reference /1/.

Cu tracer wires of diameter 0.063 mm were placed in the preconsolidated flat laminate with a lateral distance of 2.5 mm between each other prior to stamp forming. The stamped bend with Cu tracer wires was studied by X-ray analysis. Mechanical behavior of the bend samples is performed on a bending test set-up, as shown in Fig. 2. The test velocity is 5 mm/min.

RESULTS AND DISCUSSIONS

Effect of stamping time on part shape

Final part angles as a function of stamping time under two levels of stamping pressure are shown in Fig. 3. All angles are smaller than the actual angle given by the forming tool. In fact, it is well known that laminates exhibit a phenomenon referred to as "reverse springback" after forming. The extent of springback is mainly dependent on the applied time of loading. Further, the final part angle will be influenced by two factors, i.e. the wrinkling of fiber in the bend range /2/ and the thermal properties of the composite /3/. It should be noted, in addition, that shorter stamping times result in more irregular cross sections of the bend sides(Fig. 4). This effect leads to more extensive waviness of fibers and should therefore result in a further reduction of the part angle. As stamping time is above 15 seconds, the final part angle approaches a constant value. Under this condition, a higher stamping pressure causes more transverse flow of matrix between the fibers which seems to result in a greater movement of fibers near the bend region, thus giving birth to smaller final part angle.

Fiber alignment at bend range

By X-ray analysis of bend laminates with tracer wires, the Cu absorb more of the radiat waves and thus appears as dark lines on the film. Comparing with the original flat laminate, the variation of distance between Cu tracer wires at bend range give a qualitative idea of fiber misalignment during processing. A typical histogram for distance distribution of Cu tracer wires at the exterior bend range is shown in Fig. 5. The abscissa indicates the distance between the fibers and the ordinate the percentage of fibers at this distance respectively. A dramatic disorder in fiber distance with increasing stamping pressure is visible. Further it was detected that the fiber misalignment at exterior bend range is larger than that at interior range for a given stamping pressure.

Failure modes of stamped bend parts

Bending tests were performed with three stacking sequences in the laminates, i.e. $[0]_6$, $[0,90,0]_s$ and $[0,90_2]_s$ respectively. A schematic load-time diagram of bend test is shown in Fig. 2. Fig. 6 illustrates that the failure shear stress (defined as the load at flexural failure devided by the cross section of the bend side) is highest for the unidirectional $[0]_6$-laminates. Two modes of failure were observed: (a) transverse matrix cracking of $[0,90_2]_s$ laminated bends due to the weak tensile strength of $90°$-plies, and (b) delamination of $[0]_6$ bends due to the interlaminar shear stresses. Depending on the thickness of $90°$ layup at bend region, which strongly depends on the processing parameters, some $[0,90,0]_s$ laminated bends failed in delamination without experiencing transverse matrix cracking.

CONCLUSIONS

In conclusion, the following findings of this study should be emphasized:
1. A useful stamping time was found to be above 15 seconds. From this time and above, a uniformity in the desired cross section of the bend side and a constant final bend angle could be achieved.
2. Fiber waviness at the exterior bend range was greater than that at the interior range, which would be increased with increasing stamping pressure.
3. Two modes of failure in the bend region were observed (a) transverse matrix cracking and (b) interlaminar delamination, both depending on laminate stacking sequence and stamping pressure. Highest failure strengths of the bends were measured for unidirectionally laminated bends.

ACKNOWLEDGEMENTS

Thanks are due to ICI, U.K. for providing the testing material, and to the German Science Foundation (DFG FR 675-7-1) for financially supporting these studies.

REFERENCES

1- Hou M., Friedrich K., Anisothermal Stamp Forming of

Thermoplasitc Matrix Composites, paper submitted at the Third International Symposium: Advanced Composites in Emerging Technologies, COMP 90, Univ. of Patras, Greece (1990)

2- Zahlan N. and O`Neili J.M., Design and Fabrication of Components: the Springforward Phenomenon, Composites, 20 (1989) 77-81

3- Kim C.G. et al, Spring-in Deformation of Composite Laminate Bends, private communication with Korea Institute of Machinery and Metal, Changwon, Korea

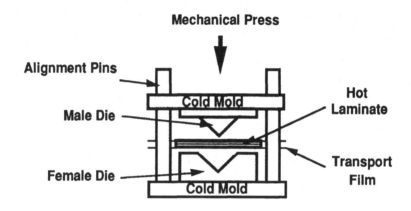

Fig. 1 - Experimental set-up

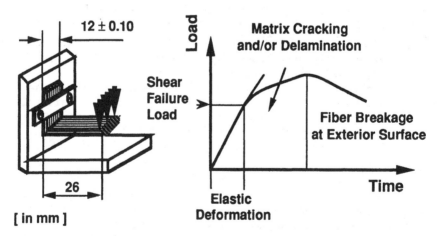

Fig. 2 - Bending test set-up and schematic load-time diagram of bending test

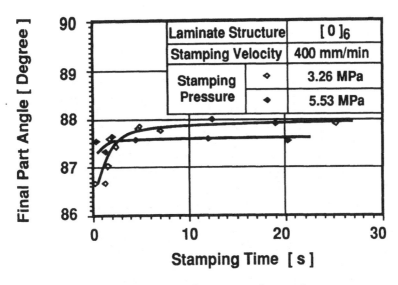

**Fig. 3 - Effect of stamping time on
final part angle**

Stamping Time [s]	Stamping Pressure	[MPa]
	3.26	5.53
0.4		
20.0		

**Fig. 4 - Effect of stamping time on
cross section of bend side**

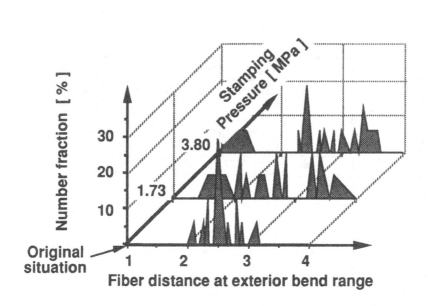

Fig. 5 - Histograms for distance distribution
of Cu tracer wires

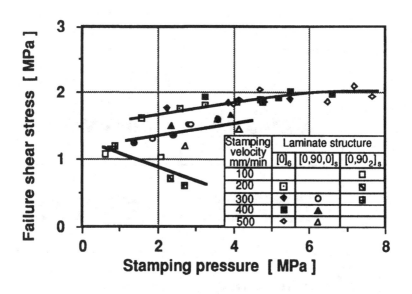

Fig. 6 - Effect of stamping pressure and
velocity on failure shear stress

AUTOMATIC LAYING OF UNIDIRECTIONAL PREPREG TAPES ON COUMPOUND SURFACES

A. FALCHI, A. GREFFIOZ, E. GINSBACH

Brisard Group - Composites Department
BP 25 - 12700 Capdenac - France

ABSTRACT

In order to meet the repetability and reliability required to tape-lay aircraft structures, the BMO Group has developed an automated tape layup system composed of two machines working in succession.
The first machine cuts the tape courses for complete part and store them in tape reel cassette.
The second machine lays the cuts from the cassette onto the mold at high speed.
Examples of productivity and accuracy achieved by this technology are given in this paper.

INTRODUCTION

Such strides are being made in developping and applying advanced composites that projects labelled futuristic are now becoming industrial facts.

At the present·stage, the increasing needs in high performance composites sructures made of unidirectional pre-impregnated fibers urge for a radical industrialization.

In response to the need for an alternative to hand layup, with respect to repeatability and reliability required to manufacture an aircraft wing, the BMO Group has developed jointly with AEROSPATIALE and DASSAULT/BREGUET, the automated tape laying system called ACCESS-ATLAS.

I - TWO STAGE SYSTEM - AN ALTERNATIVE SOLUTION

Any tape placement machine to be truly economic must not only be accurate and fast, but also 100% dependable in all its functions. Because the solutions to tape shear and tape tail control dependability were extremely difficult if impossible in practice, a new philosophy emerged splitting these functions into two machines.
One to perform all the tape preparation tasks such as cutting angle and measuring course length. The other to perform only one function, that of laying the tape on the tool or mold.

II - STAGE I - ACCESS - TAPE PREPARATION MACHINE

Access is an acronym for "advanced Composite Cassette Edit Shear System"

Access processes pre-impregnated tape material from original supply spools by :
- Continuously removing the tape from the original backing paper
- Cutting each tape course to length and angle
- Repositioning the course on a new backing paper
- Re-winding the course into a dispensing cassette for the stage II.

The knife device, under computer control is able to cut virtually any angle, either positive or negative, coumpound angle and curves as well. Typically the entire laminate stacking sequence for a part is programmed into ACCESS, starting with the last course first and working in reverse such that when the tape cassette thus prepared is placed on the stage II machine, ATLAS,the proper course sequence is achieved.

III - STAGE II - ATLAS - TAPE PLACEMENT MACHINE

ATLAS is an acronym for "Advanced Tape Laying System"
The prepared tapes are dispensed onto the surfaces of tools with the advantage that the machine does not pause for shearing and perform no function other than high speed actual tape laying. The main feature of ATLAS is the simplicity and light weight of its components to further improve speed by reducing inertia.
It allows to install dual tape laying heads for product versatility and six axes of computer controlled motion for growth potential and contour tape laying.

IV - AUTOMATION OF THE PROCESS - PALS SOFTWARE

From the part definition to total completion of the part, all the functions of the machine are under computer control (fig. 2).
To propose a free man's hand system able to master key issues of tape laying process, BMO has developped the sophisticated software PALS. PALS is an acronym for Program of Automatic laying System. It is specially dedicated to :

- Translate a mould surface definition input file into strip cutting and laying part-programmes.
- Laying tape on a dual curvature surface , with no gaps or overlaps.
- Automating and surpressing the process, printing out a log book.

V - EXAMPLES OF USE

5.1. Flat layup

The tespiece shown in fig. 3 illustrates the capability of the machine to overcome important variations in thickness. Such a component is theoretically impossible to layup as this involves deviating the fibers in the thick places. As a matter of fact, the deviation is ignored as ATLAS can force the fibers along a path, programmed along the geodesic line (i.e. programmed as a flat path).
In fig. 3, the change in thickness from area n° 1 (40 layers) to aera n° 2 (140 layers) is 15 mm over a 125 mm distance that is a 12% slope.
This component has been made within the prescribed tolerances, the gap between two adjacent cuts never exceeding 0.25 mm.

5.2. Developable surface layup -

This is one step further in layup difficulty as compared to the previous example.
For a developable surface like that of fig. 4 defining the geodesic paths amounts to defining straight paths on the developed surface.
A developed surface is not the projection of a curved surface onto a plane. It is the initial surface, flattened as it were, and PALS software provides for the calculation automatically (see fig. 2).

This technology can be extended to some cases of non-developable surfaces. It has been successfully applied in industrial occurences like upper and lower wing panels of French-Italian civil aircraft ATR 72.

5.3. Complex surface layup

A simple example of non-developable surface is a sphere. To layup tape on a spherical surface, two technologies are applicable (fig. 5):

- Geodesic path
Each cut is laid-up along a geodesic path, that is, in the case of a sphere, a great circle.
Geodesic path layup involves, in order to avoid overlapping of prepreg, having split spindle-shaped cuts.

- Natural path
A first cut (neither split nor sindle-shaped) is laid-up along a geodesic path.
The other cuts are laid-up in succession so that the edge of each layer is parallel to that of the previous one until the corrugation limit of the prepreg is reached (acceptable equivalent radius is 20m with ATLAS*.
Beyond this threshold, the next cut is split and laid-up along the geodesic path. Then, the following cuts are laid-up parallel as before.

With the "geodesic path" method, the fibers are placed on the mould in the optimum direction but numerous splits, detrimental for the performance of the material as well as for the scrap rate, are required.
With the "natural path" method, the scrap rate is significantly lower as most of the cuts are full width but the direction of the fibers is not that of the geodesic path of the sphere.

VI - MACHINE PRODUCTIVITY

With the 2-phase concept, tape cutting and layup are separated. As tape cutting is performed on ACCESS while tape layup is in process, productivity depends on tape laying speed of ATLAS only.
Fig. 6 shows the productivity of the ACCESS-ATLAS system in a real industrial application to a series of aircraft wing panels. The curves show how machine productivity changes in relation with the length of the cuts to be laid-up.
Curve (A) represents productivity in the case of one-direction layup that is with tape laying head at a constant angular position.
Curve (B) for reciprocating layup, that is when the head has to rotate each time a cut has been laid-up.
It can be noticed that one meter is a critical length of cut beyond which reciprocating layup is more productive than one-direction.
Point M of curve B shows maximum machine productivity, i.e.
135 m^2 per hour as obtained for a structure with 8 m. lay cuts, laid-up in the reciprocating way.

VII - CONCLUSION

Productivity and accuracy achieved by this tape laying technology open the door to industrial scale production of high-performance composites structures.

REFERENCES

1 - Greffioz A., Système de nappage pour surface à double courbures, 4eme Journées d'Automne, I.N.E.R.N., FRANCE (octobre 1989).
2 - Goldsworthy W., Two stage placement system, "progress in advanced materials and processes" (G. Bartelds, ed.) (1985).

TAPE LAYUP SYSTEM
ACCESS

Width of prepreg	25 - 150 mm
Prepreg reel	500 mm dia. 300 m of tape approx.
Cutting system	dual, mechanical cutter type
Elimination of scraps	automatic
Detection of flaws in material	by flags on prepreg support tape
Prepreg feedrate	30 m/mn
Cutting velocity	15 m/mn

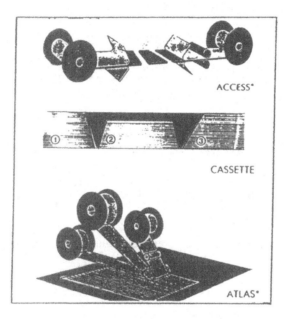

TAPE LAYUP SYSTEM
ATLAS

Working capacity	4 m x 7 m (4 x 12 m optional)
Vertical stroke (Z)	500 mm
Rotational stroke of heads (C)	$\pm$ 200°
Number of tape layup heads	2
Swivel stroke of heads (A and B)	$\pm$ 30°
X and Y axis motion speed	30 m/mn
Z axis motion speed	10 m/mn
A and B axis motion speed	1800°/mn
C axis motion speed	6000°/mn
Width of prepreg	25 - 150 mm

Fig. 1. Technical description of the two stage System

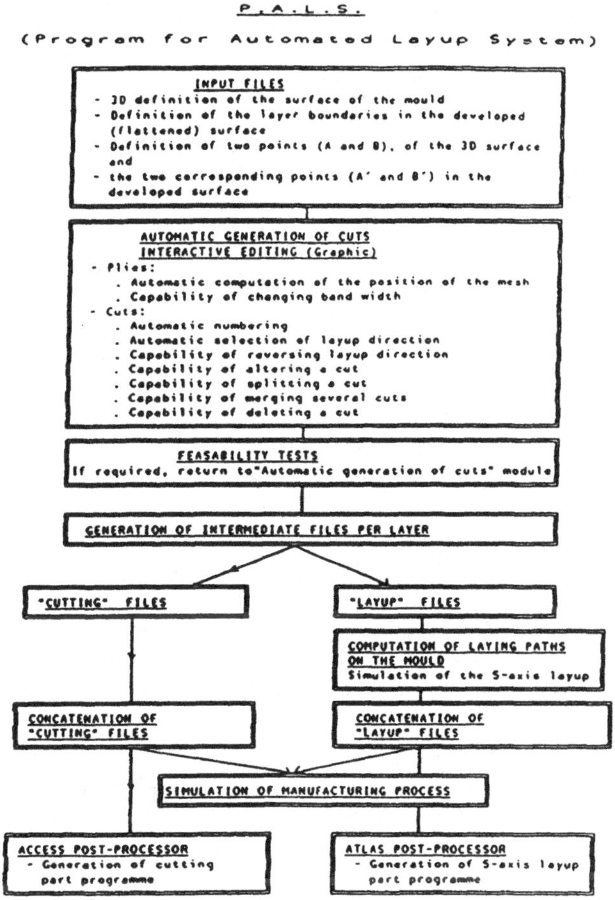

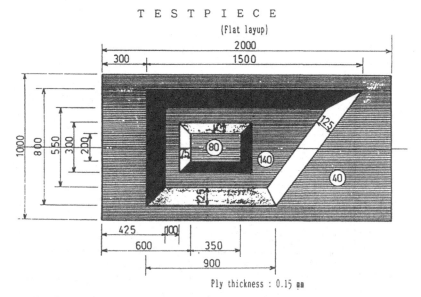

P.A.L.S.
(Program for Automated Layup System)
INPUT FILES
- 3D definition of the surface of the mould
- Definition of the layer boundaries in the developed (flattened) surface
- Definition of two points (A and B), of the 3D surface and
- the two corresponding points (A' and B') in the developed surface
AUTOMATIC GENERATION OF CUTS
INTERACTIVE EDITING (Graphic)
- Plies:
. Automatic computation of the position of the mesh
. Capability of changing band width
- Cuts:
. Automatic numbering
. Automatic selection of layup direction
. Capability of reversing layup direction
. Capability of altering a cut
. Capability of splitting a cut
. Capability of merging several cuts
. Capability of deleting a cut
FEASABILITY TESTS
If required, return to "Automatic generation of cuts" module
GENERATION OF INTERMEDIATE FILES PER LAYER
"CUTTING" FILES
"LAYUP" FILES
COMPUTATION OF LAYING PATHS ON THE MOULD
Simulation of the 5-axis layup
CONCATENATION OF "CUTTING" FILES
CONCATENATION OF "LAYUP" FILES
SIMULATION OF MANUFACTURING PROCESS
ACCESS POST-PROCESSOR
- Generation of cutting part programme
ATLAS POST-PROCESSOR
- Generation of 5-axis layup part programme

Fig. 2. Description of P.A.L.S. Software

TESTPIECE
(Flat layup)
2000
300
1500
1000
800
550
300
200
125
125
75
80
140
40
125
425
100
600
350
900
Ply thickness : 0.15 mm

Fig. 3. Example of flat layup

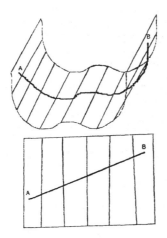

Fig. 4. Tape layup on developable surface

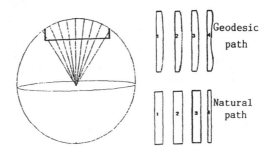

Fig. 5. Tape layup on a sphere

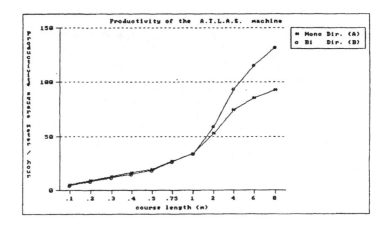

Fig. 6. Productivity of the Atlas machine versus course length

PERFORMANCE ANALYSIS OF A SMC PANEL MOULDING

A.F. JOHNSON, G.D. SIMS, A. HARRISON*

*National Physical Laboratory
TW11 0LW Teddington - Great Britain
*Ford Motor Company Ltd
Research and Engineering Centre - Laindon SS15 6EE - Great Britain*

ABSTRACT

Design analysis calculations were carried out on a vehicle bonnet panel moulded in a sheet moulding compound (SMC) and the results compared with the measured panel performance under flexural and torsional loads. The design analysis was based on both finite element calculations and simplified methods for ribbed panel design developed by the authors and described in /1/. Both design methods showed reasonably good agreement with measured panel properties, which gives confidence in these procedures for large ribbed SMC composite panels.

I - INTRODUCTION

This paper describes a case study on a vehicle bonnet panel moulded in a sheet moulding compound (SMC), in which the material properties were determined and then design analysis predictions were compared with the measured performance of the panel under flexural and torsional loads. Details of the panel geometry are given in Section 2, along with a description of the materials test programme in which the variability in properties due to moulding conditions was assessed. In Section 3 design analysis calculations are given for the panel, based on finite element analysis (FEA) and on simplified design procedures for panels developed by the authors and described in /1/. Section 4 describes the complete panel loading tests and discusses the validity of the simpler design methods and FEA for the design of large ribbed SMC composite panels.

II - PANEL GEOMETRY AND MATERIALS PROPERTIES

The component considered is a prototype car bonnet which is compression moulded in SMC. It is essentially a rectangular panel with slight curvature whose upper (exterior) surface is smooth with ribs moulded on the lower (interior) surface. The rib pattern consists of a pair of cross ribs on the main panel with additional longitudinal and transverse ribs reinforcing the front edge panel. Dimensions are marked on a schematic diagram of the bonnet in Fig 1. Measurements were made of panel thickness and rib geometry. The thickness varied from 2.8–4.2 mm over the panel with a mean value of 3.6 mm, which is the value used in the design calculations. The main cross ribs had a rib height of 15 mm and width of 4 mm. There are also additional ribs stiffening the front edge. The bonnet has fixing points for the hinges on the rear corners, and for the lock assembly on the stiffened front edge, with bump stops in the front corners. For the loading tests it is usually supported on the hinges and bump stops so that it is the distances between these fixing points which define the plate span for the design calculations.

The panel is made from sheet moulding compound (SMC) and fabricated by hot press moulding between matched metal dies. For such compression moulded components, material properties may vary with flow conditions in the mould, and hence with position in the moulding. In order to obtain appropriate materials property data for design, small plate and beam test specimens were cut from a bonnet moulding and tested in flexure. The measured values of flexural modulus E and strength σ are summarised in Table 1. In addition Poisson's ratio was measured in tension and a mean value $\nu = 0.3$ was obtained. In the table the orientation referred to is relative to the longitudinal panel axis. Mean values are given based on testing 213 flexure beam specimens cut from different parts of the moulding. There were variations in materials properties with position in the moulding and the measured coefficients of variation are shown in Table 1. The variability in strength properties is much higher than that in modulus. The effect of flow during moulding has led to slight anisotropy in the panel, but this is not considered significant for design purposes. Thus the SMC material is assumed to be approximately isotropic in the plane of the panel and on taking the mean of the flexural data in Table 1, the basic materials data for design are:

$$E = 10.2 \text{ GPa}, \quad \nu = 0.3, \quad \sigma = 130 \text{ MPa} \tag{1}$$

Note that these measured values are close to the typical values quoted in the literature for SMC, see /2/, showing that in this case the literature data would be a suitable basis for design analysis.

III – PANEL DESIGN ANALYSIS

The bonnet panel must satisfy several design requirements, including vibration and flutter, frontal impact, flammability and structural integrity requirements. In this paper only panel stiffness and strength under flexural and torsional loads are considered. The flexural load case is a transverse load of 88N applied to the centre of a fully supported panel, and for the torsional case a panel with three supported corners is loaded by a transverse load of 102N at the free corner.

3.1 Simplified design procedures

In order to use the design procedures of /1/ the SMC panel is first idealised into simpler beam and plate elements. In this case the subpanel of interest is the rectangular panel between the hinge points and bump stops, ABCD in Fig 2, which is stiffened by the diagonal cross ribs. Using the design procedures of /1/, this panel is idealised further as a flat orthotropic panel with equivalent flexural rigidities D_1, D_2, D_3 and torsional rigidity T. The stiffening effect of the ribs depends on the rib angle θ with respect to the plate longitudinal 1–axis (Fig 1), the flexural rigidity of the unribbed plate of thickness h, $D = Eh^3/12(1 - \nu^2)$ and a rib stiffness parameter S, and is given by

$$D_1 = SDc^4 + 2Ds^2c^2 + Ds^4, \qquad D_2 = SDs^4 + 2Ds^2c^2 + Dc^4,$$

$$D_3 = D + 3D(S - 1)s^2c^2, \qquad T = D(1 - \nu) + 2D(S - 1)s^2c^2 \tag{2}$$

where $c = \cos\theta$, $s = \sin\theta$. A design formula for the rib stiffness factor S is given in /1/, /3/ which shows the dependence on rib height and rib thickness. Using geometry and materials data relevant to the SMC subpanel ABCD, it is found that $S = 9.6$, $\theta = 39^\circ$, hence the effective orthotropic panel stiffnesses are:

$$D_1 = 181 \text{ Nm.}, \quad D_2 = 103 \text{ Nm}, \quad D_3 = 313 \text{ Nm}, \quad T = 210 \text{ Nm} \tag{3}$$

which should be compared with the flexural rigidity of the unribbed SMC panel $D = 44$ Nm. Note that the cross–ribs have significantly enhanced the torsional rigidity, as expected, but also increased the panel flexural rigidities D_1, D_2.

The effect of load and fixing conditions on the SMC panel are now assessed by analysing the equivalent orthotropic plate with stiffness properties (3). The design procedures in /1/ for transverse loading of a rectangular orthotropic plate of dimensions a, b are based on a design formula for the maximum plate deflection w

$$w = \alpha P a^2/12D_2, \tag{4}$$

where P is the applied load and α a dimensionless stiffness parameter which depends on geometry a/b, plate edge conditions, load position and the plate material properties through the ratios D_1/D_2, D_3/D_2. Tables of values of α for different conditions are given in /1/, /4/. Here the panel ABCD is modelled with simply supported edges along AB, CD and with AD, BC assumed to be clamped where the heavily ribbed front edge subpanel and the hinge brackets provide considerably more restraint. Unfortunately the relevant value of α for these mixed edge conditions is not included in /1/. However a value of $\alpha = 0.112$ is given in /1/ for a fully supported orthotropic panel, and a value $\alpha = 0.028$ may be obtained from the design formula in /4/ for a fully clamped panel. These values may be used to calculate upper and lower bounds for the panel deflection. These calculated bounds based on (4) with a load $P = 88N$ are shown in Table 2. In the supported case the calculated deflection of 7.3 mm is nearly twice the panel thickness and is corrected in the table for large deflection effects using the methods of /1/. The large difference between the predicted deflections shows the significance of edge conditions on panel stiffness. It is considered here that the partially restrained panel should be closer to the clamped case rather than the supported case.

Torsional loading of a rectangular orthotropic plate twisted at its corners, with free edges, is also discussed in /1/ where it is shown that the corner deflection w is given by (4) with

$$\alpha = 3(b/a)(D_2/T), \tag{5}$$

from which the corner deflection under a 102N load predicted in Table 2 is obtained. The simplified design methods also enable maximum stresses in the panel to be estimated and values of flexural stress $\sigma = 21$ MPa and shear stress $\sigma = 24$ MPa were calculated for the flexural and torsional loading cases. These values are well below the strength data for SMC given in Table 1, showing that the panel design is stiffness limited.

3.2   Finite element analysis

The FEA calculations were carried out on MSC NASTRAN FE software using the bonnet model shown in Fig 2. This consists of 80 shell elements for modelling the bonnet surface. Coordinates were taken from actual component drawings, and hence the FE model includes the shallow curvature effects as indicated in the figure. The bonnet ribs were modelled using off-set beam elements to connect the appropriate nodes in the FE mesh, and are marked by the heavier lines on Fig 2. Compared with the simplified geometrical modelling used for the simplified design procedures it is clear that FEA allows a more precise modelling of bonnet curvature effects and edge conditions. However, it should also be pointed out that the setting up of the FE model is time consuming and the computations expensive.

The analysis used the measured materials property data given in (1). The FEA calculations were carried out under centre point flexural load and corner torsion load. As boundary conditions, it was assumed that the bonnet had free edges and was supported or fixed at the hinge points B, C and the bump stops A, D marked on Fig 2. For the flexural load case the bonnet was fixed at A, B, C, D and centrally loaded at E by a load of 88N. For the torsional analysis, points A, B, C were fixed with the load of 102N applied at D. The results of the analysis in the form of predicted loading point deflections are summarised in Table 2.

IV – DISCUSSION OF MEASURED AND PREDICTED PANEL PERFORMANCE

Loading tests were carried out on the bonnet panel under the following conditions. For the flexural test a panel was placed on four supports at the points A, B, C, D of Fig 2 and loaded at the point E over a 75 mm diameter disc by adding loads in increments of 2.5 kg. Displacements were measured with a displacement transducer mounted underneath the panel, beneath the loading disc. The torsion test was accomplished by clamping the hinge positions B, C and simply supporting the bonnet at A. Weights were suspended at D and the vertical displacement measured. From each test a linear load–displacement curve was obtained, from which the deflections corresponding to the design loads of 88N in flexure and 102N in torsion were taken. These measured deflections are given in Table 2.

A deflection of 2.15 mm was measured under the 88N centre point load. This may be compared with predicted values of 2.65 mm by FEA and the upper and lower bound values of 4.7 mm and 1.8 mm from the simplified design procedure. Apart from the upper bound value based on the unrestrained simply supported edges, there is reasonably good agreement between the predicted values and the measured performance in the flexural load case. Note that the design specification for the bonnet is a limiting deflection of 6.3 mm under the 88N centre load, showing that the SMC panel is well within this specification.

In the case of the torsional load, the agreement between measured and predicted values is not as good as in flexure, but still satisfactory. Thus the measured corner deflection under the 102N load was 127 mm, compared with a predicted value of 104 mm from the analysis above and the much lower value of 43.6 mm from FEA. Factors not taken into account in the calculations, such as the very large deflections and the curvature should both lead to an increased torsional stiffness and reduce the predicted deflection of 104 mm. As expected, the FEA prediction which includes these features is lower, although it is significantly below the measured value. We note that the bonnet design requirement is for a 127 mm deflection under a 267N corner load. It is apparent from the test results that the bonnet is much less stiff in torsion than the specification requires, and it would be necessary to more than double the torsional rigidity to meet this requirement. This could be achieved by increasing the panel thickness or by increasing the number or size of the diagonal ribs.

V – CONCLUSIONS

The main points to emerge from this SMC panel case study are:

1. The panel moulding had materials properties which were essentially isotropic in the panel plane and measured mean values of modulus and strength were close to typical data given in the literature for SMC /1/, /2/.

2. The methods for design analysis developed in /1/ give reasonably good agreement with measured behaviour of a loaded SMC panel, and therefore provide a firm basis for preliminary design analysis of such components under short term loads.

3. A much more detailed FEA study of the panel did not give any better prediction of bonnet performance under test than the simpler design methods.

4. With the simpler design methods it is possible to quickly assess several factors in the design, such as the effect of materials properties, shape, edge and fixing conditions, rib design etc. It is found that in general

material properties have less influence on panel performance than geometrical factors such as panel thickness and rib height. It is also found that the assumptions about edge fixing conditions have considerable effect on panel stiffness and that large deflection effects may need to be taken into account.

REFERENCES

1. A.F. Johnson and G.D. Sims, Design Procedures for Plastics Panels, (1987) National Physical Laboratory, Teddington.

2. A. Kelly (ed.) Concise Encyclopedia of Composite Materials, (1989) Pergamon.

3. A.F. Johnson and G.D. Sims, Composite Polymers, 2 (1989) 89–112.

4. A.F. Johnson and G.D. Sims, in "Proc 2nd Int. Conf. Composite Structures", (I. Marshall, ed.), (1983) 302–325, Applied Science.

Table 1 Measured flexural properties of the SMC panel material

Loading Direction	Modulus GPa	Coefficient of Variation %	Strength MPa	Coefficient of Variation %
0°	10.5	8.6	134	18
90°	9.8	8.6	126	21

Table 2 Calculated and measured values of panel deflection

Loading	Design Method	Predicted deflection (mm)	Deflection corrected for large deflection effects (mm)	Measured deflections (mm)
Panel flexure 88N centre load	Eqn (4) with α from /1/ Table 5.3 for SUPPORTED edges	7.3	4.7	2.15
	Eqn (4) with α from /4/ for CLAMPED edges	1.8	1.8	2.15
	FEA		2.65	2.15
Panel torsion 102N Corner load	Eqns (4),(5)	104	–	127
	FEA	–	43.6	127

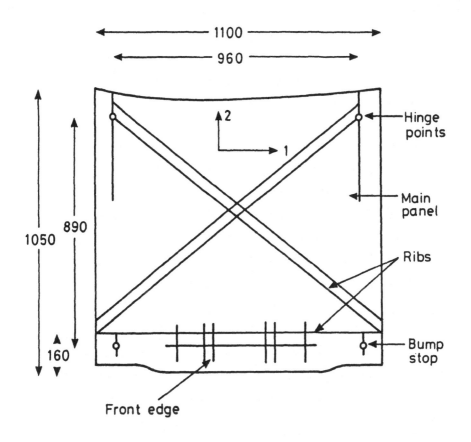

Fig 1   Schematic diagram of bonnet panel and rib system.

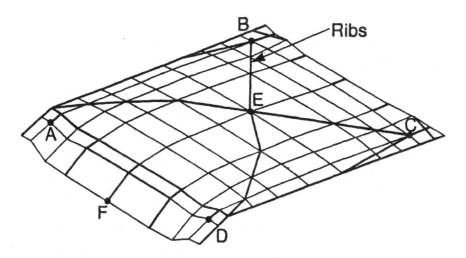

Fig 2   Panel elements as used in FEA.

INJECTION MOLDING OF LONG-FIBER-REINFORCED THERMOPLASTICS

B. SCHMID, H.-G. FRITZ

IKT
Böblingerstr. 70a - 7000 Stuttgart 1 - West Germany

ABSTRACT

Long-fiber-reinforcements in thermoplastics and their lowcost processing on injection molding machines is a combination that advances into areas of use, which were previously reserved for metals. This paper discribes the possibilities for optimum utilization of fiber reinforcement. Considerations involving processing and machinery to achieve a minimization of fiber length reduction represents a key aspect for long fiber reinforced thermoplastics.

Langfaserverstärkte Thermoplaste wurden Anfang der achziger Jahre als Ergänzung bzw. Alternative zu kurzfaserverstärkten und glasmatten-verstärkten Thermoplasten entwickelt. Der entscheidende Unterschied zu den gebräuchlichen thermoplastischen Verbunden ist in der Faserlänge des Langfasergranulates zu sehen. Die konstante Faserausgangslänge beträgt zwischen 5≤l≤12 mm. Dieser Länge stehen ca. 350 µm mittlere Faserlänge bei Kurzfaserverstärkungen und mehreren Zentimetern bei Glasmattenverstärkungen gegenüber.

Die Herstellung dieses Verbundwerkstoffes konnte durch die Optimierung des Pultrusionsverfahrens realisiert werden (Bild 1). Diese relativ kostenintensive Technologie stellt hohe maschinen- und verfahrens-technische Ansprüche. Als Hauptvorteile dieser Materialien gegenüber anderen thermoplastischen Verbundsystemen sind zu nennen: hoher E-Modul, verbessertes Impact-Verhalten und eine im Vergleich zu Wickel- und Prepregverfahren kostengünstige Verarbeitung im Spritz-gießprozeß (Bild 2).

Um Langfaserstrukturen in Spritzgießteile einbringen zu können, d.h. verbesserte Materialeigenschaften zu erhalten, werden in einer Arbeitsgruppe am Institut für Kunststofftechnologie Alternativen zum kostenintensiven Pultrusionsprozeß getestet. Die Zugabe von Schnittfasern zum Matrixgranulat mit anschließender Direktverarbeitung auf Spritzgießmaschinen ist kostengünstig und somit für teure Verstärkungsfasern wie C-Fasern geeignet /1/. Als Variante für Spritzgießmaschinen mit Entgasungsaggregat bietet sich die Zugabe der Schnittfasern in die bereits erschmolzene thermoplastische Matrix an. Voraussetzung für diese Technologie ist eine gute Rieselfähigkeit und eine exakte Dosierung der Schnittfaser, um einen konstanten Fasergewichtsanteil zu gewährleisten. Dieses Verfahren erscheint geeignet, erhöhten Faserbruch in der Plastifizierphase zu umgehen.

Die Einarbeitung von Endlosfasern in Thermoplastschmelzen mittels Entgasungsaggregate wurde bereits mit Erfolg praktiziert /2/. Mit diesem Verfahren können Faserlängen realisiert werden, wie sie bei GMT-Bauteilen vorliegen. Die derzeit auf diesem Sektor laufenden Untersuchungen behandeln den Einfluß der effektiv notwendigen Faserlänge auf die Bauteileigenschaften; ferner werden Probleme der Fasercluster-Auflösung und der Bauteileigenschaftsanisotropie in Abhängigkeit von Faserlängenspektren analysiert.

Beim Spritzgießen langfaserverstärkter Thermoplaste tritt in Abhängigkeit von Art und Beschaffenheit der Formmasse, den Verarbeitungsbedingungen sowie der Geometrie des Plastifizieraggregates und der Kavität eine Faserlängenreduktion ein, die die mechanischen Werkstoffkennwerte i.a. negativ beeinflußt. Um diese Faserlängenverkürzungen imzuge der Verarbeitung zu minimieren, wurde ein Versuchsprogramm zum Studium der Haupteinflußgrößen für den Faserabbau durchgeführt. Im Rahmen dieser Grundlagenuntersuchungen wurde die lokale Faserschädigung des thermoplastischen Faserverbundwerkstoffs auf dem Weg der Trichteröffnung bis zum Werkzeughohlraum quantitativ ermittelt. Das dabei verwendete, pultrudierte langfaserverstärkte Polyamid 6.6 enthielt 30 Gew. % Glasfasern (16 Vol. %), welche eine einheitliche Ausgangslänge von 10 mm aufwiesen. Weiter wurde die Faserlängenreduktion auch im Anguß und über dem Fließweg in der Kavität erfaßt. Für diese Untersuchungen wurde im Hinblick auf die realisierten Strömungsformen ein sog. Topfwerkzeug eingesetzt: In dem zentral angespritzten Topfboden ist eine zweidimensionale, überlagerte Scher-/Dehnströmung, in der Topfwandung hingegen mit guter Nährung eine eindimensionale Scherströmung verwirklicht.

Zur Erfassung der Faserlängenreduktion in der Plastifiziereinheit einer Schneckenkolben-Spritzgießmaschine wurde der Spritzgießzyklus nach Beendigung der Nachdruckphase unterbrochen und die Schnecke bei Betriebstemperatur aus dem Zylinder gezogen. Aufgrund dieser Vorgehensweise konnten Proben längs der Schnecke entnommen und hinsichtlich ihrer Faserlängenverteilung analysiert werden (Bild 3).

Als erster signifikanter Einflußparameter auf die Faserlängenverkürzung wurde die Schneckenumfangsgeschwindigkeit näher analysiert. Untersuchungen mit kurzglasfaserverstärktem PA 6.6 von Filbert /3/ ergaben eine zunehmende Faserschädigung mit wachsender Schneckenumfangsgeschwindigkeit. Dieser Zusammenhang konnte bei der Spritzgießverarbeitung von langfaserverstärktem PA 6.6 bestätigt werden (Bild 4). Bis zu einer Schneckenlänge von 14 D konnte kein signifikanter Faserabbau festgestellt werden. Mit Beginn der Kompressionszone ab 15 D tritt eine progressiv ansteigende Faserlängenreduktion ein, die sich auch im Bereich der Ring-Rückströmsperre fortsetzt. Bemerkenswert ist, daß an der Schneckenspitze drehzahlabhängig noch 30 bis 50% Langfaseranteile ($1,0 \leqslant l \leqslant 10$mm) im Plastifikat vorliegen.

In einer weiteren Versuchsreihe wurde der Einfluß der Schneckengeometrie bei n=konst. auf die Faserlängenverkürzung untersucht. Im Vergleich zur konventionellen Dreizonenschnecke (l_E=10D, l_K= 6D, l_M=4D) bewirkt eine relativ flach geschnittene Meteringzone aufweisende PVC-Schnecke (l_E=10D, l_K=8D, l_M=2D) eine deutlich höhere Faserschädigung.

Über die Bauart und geometrische Gestaltung der Rückströmsperre wird ebenfalls auf die Faserlängenverkürzung Einfluß genommen. Eine neuentwickelte Zentralkugel-Rückströmsperre (KRSP), die sich durch strömungsgünstige, kleine Widerstandsbeiwerte aufweisende Massekanalgestaltung auszeichnet, läßt hohe Massedurchsätze ohne jede weitere Faserlängenreduktion in dieser Funktionszone zu. Herkömmliche Ringrückströmsperren bewirken aufgrund scharfkantiger Masseumlenkungen und enger Strömungsquerschnitte eine nennenswerte Faserschädigung.

Der Staudruck stellt einen weiteren Einflußparameter für den Faserabbau dar. Ausgehend von der staudruckfreien Betriebsweise bewirkt eine Anhebung des im Hydrauliksystem gemessenen Staudrucks auf p_{St}= 10 bar unter sonst identischen Versuchsbedingungen (Standardschnecke, KRSP, n= 50 min^{-1}) bereits eine Abnahme des Langfaseranteils von 50% auf 20%. Zurückzuführen ist diese verstärkte Faserlängenabnahme auf die staudruckbedingte Reduktion des spezifischen Massedurchsatzes und damit verbundene Steigerung der dem Plastifikat aufgeprägten Gesamtscherdeformation.

Eine systematische Variation des Plastifizierzylinder-Heizprogrammes erbrachte weitere Aufschlüsse bezüglich des Problems "Minimierung des Faserbruchs während der Plastifizierphase". Einem von der Trichteröffnung zur Schneckenspitze hin ansteigenden Temperaturprofil wurde ein degressiv verlaufendes entgegengestellt. Ergänzend wurden alle fünf Heizzonen des Zylinders auf einer einheitlich vorgegebenen Solltemperatur gehalten. Wie die Erfahrung zeigt, sind unzulässig hohe Scherkräfte eine Ursache des Faserbruches während der Plastifizierphase. Da diese der Viskosität direkt proportional sind, empfiehlt sich deren Absenkung über eine Temperaturanhebung. Die Versuche zeigten, daß das Zylinderheizprogramm so zu wählen ist, daß der Plastifiziervorgang auf einem kurzen Schneckenabschnitt der Plastifizierzone konzentriert und anschließend eine weitere steile Temperaturanhebung vorgenommen wird.

Um die rheologischen Vorgänge beim Werkzeugfüllvorgang mit langfaserverstärkten Polymerschmelzen besser verstehen und beschreiben zu können, wurden kapillarrheometrische Untersuchungen durchgeführt. Ziel dabei war, geometrieunabhängige rheologische Stoffwertfunktionen mittels eines Hochdruckkapillarrheometers unter Verwendung eines Satzes Kreiskapillaren mit konischem Einlaufwinkel (α=30^0) zu ermitteln. Die zunächst aufgenommenen scheinbaren Fließkurven und Viskositätsfunktionen zeigten noch eine starke Abhängigkeit von der Variablen "Düsenradius". Mit Hilfe einer modifizierten Auswertetechnik gelang es schießlich diese in geometrieinvariante Stoffwertfunktionen zu überführen. Dabei zeigte sich, daß aufgrund von Faserorientierungsvorgängen im Vorlageraum und Düsenbereich das strukturviskose Verhalten stärker als beim ungefüllten Matrixwerkstoff ausgeprägt ist. Durch die Bestimmung von Faserorientierung und Faserlängenverteilung im Düsenbereich sowie mit Hilfe rasterelektronenmikroskopischer Aufnahmen, konnten die ermittelten Ergebnisse auch phänomenologisch erklärt werden.

Die bislang gefundenen Ergebnisse zur optimalen Verarbeitung langfaserverstärkter Thermoplaste belegen, daß mit kleinen Schneckenumfangsgeschwindigkeiten (- vergrößertes Plastifizieraggregat), einer Zentralkugel-Rückströmsperre in Verbindung mit einer Standard-Dreizonenschnecke (Gangtiefenverhältnis h_1/h_3 = 2), einer staudruckunabhängigen Betriebsweise und einem ansteigenden, aber insgesamt hoch liegenden Zylindertemperaturprogramm eine schonende Materialplastifizierung möglich ist. Dabei erzielbare Langfaseranteile von 45 bis 55% des Gesamtfaserzusatzes bewirken eine signifikante Verbesserung der Durchstoßfestigkeit und ein deutlich isotroperes Verhalten der daraus spritzgegossenen Formteile (Bild 5).

1- Kompalik, D.; Schmid, B.: Kunststoffe 78 (1988), 308-311
2- BAYER AG: Europ. Patentschrift 0 124 033 B1
3- Filbert, W.: Plastverarbeiter 22 (1971), 413-417

An dieser Stelle sie dem Wirtschaftsministerium Baden-Württemberg, der Dt. Forschungsgemeinschaft, den Firmen Krauss-Maffei Kunststofftechnik GmbH, BAYER AG, ICI PLC U.K. und der Sigri GmbH für die geleistete Unterstützung gedankt.

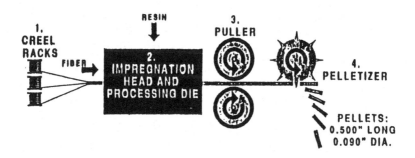

Bild 1: Schematische Darstellung des Pultrusionsprozesses

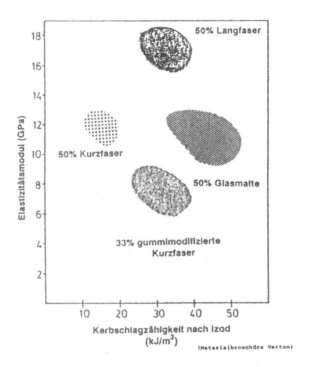

Bild 2: E-Modul und Impact-Verhalten div. thermoplastischer Verbunde

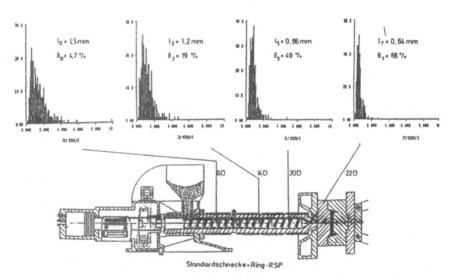

Bild 3: Faserlängenverkürzung im Plastifizieraggregat (n=100 min^{-1})

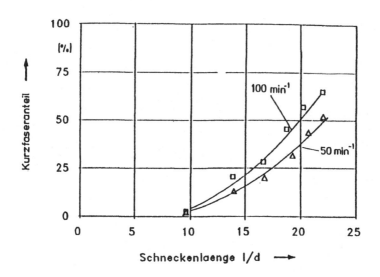

Bild 4: Einfluß der Schneckenumfangsgeschwindigkeit auf die
Faserreduktion von langfaserverstärkten Thermoplasten

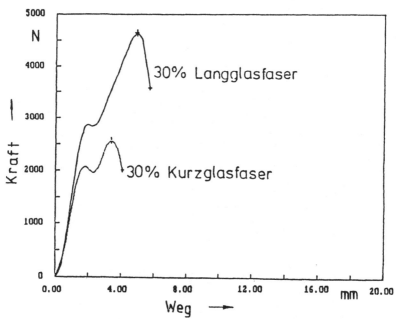

Bild 5: Kurz- und langglasfaserverstärkte Thermoplaste im Vergleich
während des Durchstoßversuches

154

FATIGUE
ERMÜDUNG

Chairmen : **Prof. F.L. MATTHEWS**
Imperial College
Prof. B. HARRIS
University of Bath

RESIDUAL STRENGTH AND TOUGHNESS OF DAMAGED COMPOSITES

B. HARRIS, A. CHEN, S.L. COLEMAN, R.J. MOORE

School of Materials Science
University of Bath - BA2 7AY Bath - Great Britain

ABSTRACT

A study has been made of the effect of prior damage on the tensile strength and toughness of a number of carbon and glass fibre reinforced plastics materials. This prior damage was introduced by tension fatigue cycling, by transverse compression, by repeated impact, and by stress corrosion. The effects of the damage were assessed by measuring the un-notched and notched tensile strengths. Acoustic emission monitoring and microstructural studies were carried out in support of the mechanical tests. Microstructural damage can result in independent changes in the notched and un-notched strengths of a composite, although the pattern of changes is not simple.

I   INTRODUCTION

In a recent survey of the tensile strength and toughness of a wide range of fibre composites containing several different types of constituents and of many kinds of lay-up Harris *et al*/1/ have shown that there is a linear relationship between the strength and the candidate fracture toughness, K_Q, of the form:

$$K_Q \approx 64\sigma_f \; . \; . \ldots \ldots \ldots \ldots \ldots \ldots 1)$$

(with stress in GPa and K_Q in MPa$\sqrt{m}$). Other recent papers/2,3/ show similar results. This is contrary to the common belief that the strength of composites is independent of fracture toughness/4,5,6/. It was concluded that the use of the critical stress intensity concept has no meaning for fibre composites on account of their complexity. But this raises interesting questions about the degree of interdependence of the notched and un-notched strengths of fibre composites and the extent to which there is scope for modifying the toughness and strength of a composite independently, a concept commonly accepted as a vital part of the design of a component to meet specific requirements. We now present a study of the effects of prior structural damage on the strength/toughness relationship.

II   EXPERIMENTAL DETAILS

The materials were autoclaved, non-woven carbon and glass fibre reinforced epoxy laminates of [($\pm45,0_20$)$_2$]$_S$ construction manufactured from prepregs, and a hot-pressed woven-roving glass/epoxy laminate. They were subjected to several methods of inducing prior damage, and the effects of this damage were assessed by means of measurements of the un-notched and notched tensile strengths. Notched strengths were measured on single edge-notched test pieces with a notch-depth ratio, 2a/W, of 1/3. Some of the experiments were monitored by acoustic emission (AE) methods with the aid of an AETC amplitude analysis system and data acquisition equipment which has been fully described elsewhere/7,8/.

2.1   Methods of Introducing Damage.

Fatigue: Samples of CFRP were subjected to repeated tension cycling in servo-hydraulic machines under load control at 5-10Hz (R = +0.1).

Transverse compression: The autoclaved CFRP and GRP composites were subjected to damage by transverse loading in compression to various fractions of the fracture load for this mode of deformation in a plane strain compression jig. The damage zones were created in the centres of specimen test lengths. For the notched strength measurements, the edge notches were cut along the centre-lines of the damaged zones.

Repeated impact: These experiments were carried out on the thicker WR GRP laminates. Prepared samples were held against a rigid anvil bolted into a pendulum impact machine and were damaged by repeated impacts from the tup which had a 20mm diameter striker of semi-cylindrical shape in place of the usual impacter. The line of contact of blows delivered by the striker was perpendicular to the sample length. Typically, the threshold for barely visible impact damage (BVID) was 25J. The residual toughness of this thicker laminate was assessed by means of notched Charpy impact tests rather than notched tensile strength tests.

Stress corrosion: Strips of GRP laminate were bent elastically in simple jigs to a radius corresponding to a maximum tensile strain of 0.2%, approximately a tenth of the flexural failure strain. They were then immersed in 5N HCl solution for various times during which acoustic emission monitoring indicated the occurrence of substantial amounts of fibre damage, possibly with some resin cracking.

III   EXPERIMENTAL RESULTS

Fatigue damage: For cyclic (peak) stress levels between about 65% and 95% of the mean monotonic tensile strength, reductions in Young's modulus and strength, following initial small increases of one or two percent, are of the order of only 20% even after 10^6 cycles. No significant reductions occur until well beyond 10^4 cycles, even though in the early stages of life cycling is accompanied by a continuing increase in AE activity. Residual stiffness and strength vary together in such a way that the tensile failure strain of the material remains unchanged by the fatigue damage. The fractional reductions in notched strength were markedly greater than those in un-notched strength for similar lives at a given stress.

Figure 1 shows the effect of cycling at different peak tensile stress levels on the ratio of the notched to un-notched tensile strengths, σ_N/σ_f, which is also effectively the ratio K_Q/σ_f. It can be seen that there is no change in the value of this ratio over the first 10^4 cycles, whatever the cyclic stress. Beyond this, however, the more rapid deterioration in the residual notched strength results in a downward shift of the σ_N/σ_f ratio, this reduction occurring at about the same number of reversals for cyclic stresses of 67% and 77% of the failure stress, but earlier for 93%. The simplest explanation is that while the random distributions of broken fibres in the 0° plies (evidence of which is also provided by edge replication) must affect both the notched and un-notched strengths equally, the resin or interface cracks in the 45° plies appear to facilitate the propagation of a pre-cut notch, either by providing additional easy crack paths or by effectively "sharpening" the tip of the precut notch.

Transverse Compression Damage: The damage induced by transverse compression resulted in some minor permanent changes in shape of the sample cross section in the damage zone. In the CFRP, microstructural examination revealed matrix plastic deformation, with local bending of fibres, matrix cracking and substantial fibre breakage in both 0° and 45° plies, some of the former being directly attributable to the effects of matrix cracks in neighbouring plies. By contrast, the GRP laminates showed no visible signs of either fibre breakage or matrix cracking.

Figure 2a shows the effect of transverse compression on the un-notched and notched strengths of the GRP laminate. The mean undamaged strength and K_Q values for the GRP were 529MPa and 36MPa$\sqrt{m}$, respectively. For transverse compression loads close to failure (90% of maximum) the strength is only marginally reduced, while beyond a damage level of about 50%, the notched strength appears to rise by some 30%. Thus although the initial value of $67\mu m^{-\frac{1}{2}}$ for the ratio K_Q/σ_f is close to the value of 64 given by equation 1, it is about 50% greater than this (about 100) after the most severe compression damage. Thus, the strength and fracture toughness appear to change relative to each other in the opposite manner to that implied by equation 1.

The CFRP laminate was much less strongly affected by transverse compression. Figure 2b compares the fractional changes in notched strength for both GRP and CFRP. The apparent toughness of the CFRP remains unaffected until transverse damage levels as high as 95% of the failure level. The value of K_Q for the undamaged CFRP was 67MPa$\sqrt{m}$, and the small variations of K_Q with damage below 90% all fall within the undamaged scatterband. The measured tensile strength for 91% damage was 0.72GPa and the ratio K_Q/σ_f was therefore increased, by the highest levels of compression damage, to $92\mu m^{-\frac{1}{2}}$ from a starting point of 62 for undamaged material.

Repeated impact damage: Figure 3 shows fractional changes in fracture energy and un-notched strength of the WR GRP as a function of cumulative sustained impact energy. This material was little affected by repeated impacts, despite a marked stress whitening and loss of translucency in the damage zone and visible damage to 0° fibres in the surface layers of the laminate.

Stress Corrosion Effects: Although AE monitoring gave clear evidence of damage to the GRP laminate exposed to HCl, the changes in strength and apparent toughness brought about by corrosion damage were less marked than those caused by transverse compression, as figure 4 shows. The changes in strength, again only marginally outside the undamaged scatter band, indicate a small overall decrease, while the corresponding notched strength values show no significant net change. The K_Q/σ_f ratio therefore shows only a slightly rising trend with increasing damage level, as indicated by the least squares line in figure 4.

IV. DISCUSSION

A "map" of the effects observed in these experiments is given in figure 5, which is a plot of all measured K_Q and σ_f data pairs together with the line representing equation 1. The effects of damage are not as simple as might have been expected. First, the net reduction in K_Q/σ_f for the CFRP as a result of cycling, over the range of stresses and cycles used, is small (about 11%), although both strength and toughness are affected by the fatigue damage. Second, transverse compression of CFRP results in a reduction in σ_f without any significant change in K_Q, with the result that the K_Q/σ_f ratio rises. Third, transverse compression of GRP samples results in little change in σ_f, but a marked upward shift in K_Q, with an accompanying net upward shift in K_Q/σ_f. Finally, stress corrosion of the GRP results in no significant change in either K_Q or σ_f. These several trends are illustrated by the arrows in figure 5. Repeated impact to high cumulative levels of absorbed impact energy also result in no net change in the toughness/strength ratio of the woven roving glass/epoxy laminate (these results are not included in figure 5).

V. CONCLUSION

Under various circumstances, the effects of microstructural damage may result in independent changes in the notched and un-notched strengths of a composite. However, there appears not to be a simple pattern of changes. The toughness to strength ratio, K_Q/σ_f may rise or fall, depending on the material and the damaging conditions, as shown in figure 5. However, the results for all of the tests presented here still fall within the 90% confidence limits for the relationship of equation 1 identified in the survey of Harris et al[1]. These new results therefore answer no questions about the significance of the 1mm "critical defect size". The only significant observed change in the ratio K_Q/σ_f, by a factor less than two for a very heavy compression load exerted over an area of some 400mm^2, suggests that the size of the damage zone itself bears no relation to the fracture mechanism.

VI. ACKNOWLEDGMENT

AS Chen wishes to acknowledge the financial support of British Council for the period of his research program at the University of Bath.

VII. REFERENCES

1. B Harris, SE Dorey and RG Cooke (1988), Comp Sci and Technol, 31, 263-277.
2. RC Wetherold and MA Mahmoud, (1986), Mater Sci and Eng, 79, 55-65.

3. FJ Belzunce, A Gutiérrez and J Viña, (1989), Proc 6th Meeting of El Grupo Español de Fractura, Sevilla, Spain, March 1989, 211-216.

4. C Zweben, (1971), J Mech Phys Solids, 19, 103-1165.

5. *idem*, (1969), J Comp Mater, 3, 7136.

6. SS Wang, JF Mandell and FJ McGarry (1975), in Fracture Mechanics of Composites, ed GP Sendeckij, ASTM STP 593, 36-85.

7. FJ Guild, FJ Ackerman, MG Phillips and B Harris, (1983), Proc First International Symposium on Acoustic Emission from Reinforced Composites, 34San Francisco, (SPI, New York), paper A-4, 1-14.

8. FJ Guild, MG Phillips and B Harris, (1985), J Mater Sci Letters, 4, 1375-1378

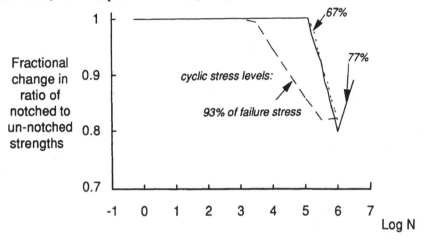

Figure 1. Change in ratio of normalised notched and un-notched strengths following repeated tension cycling of [(±45,0₂0)₂]s CFRP laminates (R = +0.1)

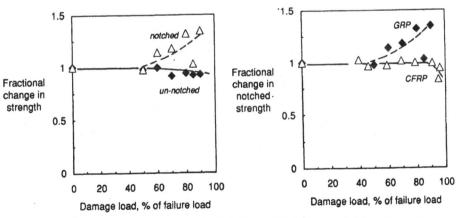

Figure 2a. Variation of the notched and un-notched tensile strengths of [(±45,0₂0)₂]s GRP laminate as a function of transverse damage load (strengths are normalized with respect to undamaged values and the transverse damage loads with respect to the transverse failure load).

Figure 2b. Comparison of the effects of transverse damage on the notched strengths of [(±45,0₂0)₂]s GRP and CFRP laminates.

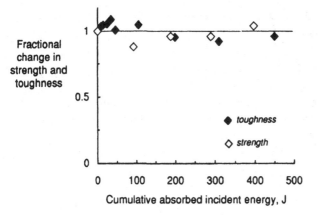

Figure 3. Effect of single-sided repeated impact on the work of fracture and un-notched tensile strength of WR glass/epoxy laminate.

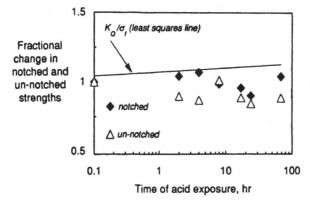

Figure 4. Effect of stress corrosion on the notched and un-notched strengths of GRP laminate (normalized strength data).

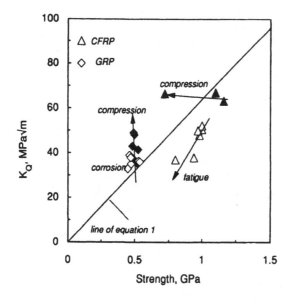

Figure 5. Changes in the K_Q/σ_f ratio as a consequence of prior damage induced in various ways.

CHARACTERIZING PURE MODE I DELAMINATION USING MULTIDIRECTIONAL DCB SPECIMENS FOR GLASS/EPOXY

X.J. GONG, M. BENZEGGAGH, A. LAKSIMI, J.M. ROELANDT

Université de Technologie de Compiègne
BP 649 - 60206 Compiègne - France

ABSTRACT

In order to etablish DCB as a general quantitative method, an advanced understanding on this configuration is necessary. Using different angle ply laminates of glass/epoxy, we have tried to explain the effect of lay-up of laminates upon the pure mode I critical energy release rate. The results obtained revealed that in the multidirectional DCB specimen, three dimentional effects are too strong to be negnected. Application of the compliance method requires a more exact theoretical analysis. GIc measured in this study seems indenpendent of specimen rigidity, but varies with curvature couplings ky/kx and kxy/kx.

INTRODUCTION

Fiber-reinforced epoxy structures have been widely used in industries by their excellent mechanical properties in the fiber direction. However, the interlaminar resistance under mode I loading (peeling mode) is so poor that the delamination becomes a major cause of their premature structural failure. An understanding of this fracture behavior is therefore of great importance in the design and testing of composite materials.

Until recently, the double cantilever beam (DCB) configuration was used extensively in determining interlaminar fracture toughness. The critical energy release rate G_{IC} was determined as the criterion for the onset of delamination. To this end, many analytical models and experimental methods were presented in the literature /1,2,3,4,5/. But most of them were limited to an unidirectional composite. In fact, the composite material structure commonly used in industry consists of stratifications of the type [±θ]. Contrary to a unidirectional composite a delamination defect can exist and propagate between any group of angle plies. In this case, the strain and stress fields as well as the fracture mechanisms become very complex /6,7,8/. In the literature, several investigations exist using angle ply DCB specimens . However the results found are often incomplete and even contrary one another.

A.S.D. Wang /9/ revealed that the G_{IC} depends greatly on the angle between adjacent fiber orientation and the direction of delaminations. According to the DCB tests performed by Culen and William, 0°/0° interface delamination value GI_c was very different from that of the 90°/90 interface one for the same material (135 J/m^2 and 228 J/m^2 respectively). Results presented in the references /10,11,12,13,14,15/ appear to support this point of view. On the contrary, works of H. Chai /16/, R.L. Rankumar and J.D. Whilcomb /17/ showed that G_{IC} is independent of both the orientation of adjacent plies to the delamination and stratification of the laminate. Moreover, experimental and analytical methods proposed in these works were not always satisfactory, nor were interpretations always convincing.

In this paper, in order to establish DCB test as a general quantitative method, different symmetrically stratified laminates were chosen so that the crack growth direction is no longer parallel to the fiber orientation of adjacent plies, and all of specimens may be strictly loaded in the condition of pure peeling mode. After having measured GIC for each layup laminate, we have tried to answer the questions as to what are the main processes of energy dissipation during the delamination and how the material parameters can influence the crack resistance.

I - EXPERIMENTAL METHOD

1.1. Material and specimen fabrication

The material used in this study is an epoxy M10 (VETROTEX) reinforced with 52% by volume of E glass fibers. The DCB specimens, their dimensions shown in fig.1, were prepared from 16 quasi-unidirectional preimpregnated plies (lightly woven). The initial cracks were formed by inserting a teflon film in mid-plane during mondling. Seven different lay-up were fabricated, as follows :

$[0_3/90]_{2S}$, $[\pm\theta_2]_{2S}$ (θ = 15, 30, 45, 60, 75) $[90/0_3]_{2S}$

to fulfill the following conditions:
-two identical arms of specimens to ensure pure peeling mode during loading;
-symetric arms to eliminate the coupling of bending and extension;
-varied fiber orientations to examine the effects of lay-up on G_{IC}.

The mechanical properties of these laminates are listed in table 1.

1.2. Data reduction and testing procedure

Experimentally the critical energy release rate G_{IC} was determined by the compliance method as follows:

$$GIC = \frac{Pc^2}{2B}\frac{dc}{da} \qquad (1).$$

where compliance c is measured during tests. Then a law of c versus the initial crack length a is established for each lay-up using the formula :

$$c = \alpha\, a^n \qquad (2).$$

where α and n are two constants of the laminate considered, which are found empirically. Replacing c in the (1) by (2),we obtain:

$$GIC = \frac{Pc^2}{2B} \; \alpha \; na^{n-1} = \frac{Pc^2 \; c}{2B \; a} \tag{3}.$$

It is important to recall that this experimental method should only be used when self-similar crack growth occurs. In other words the crack front dose not change shape. In reality because of the presence of coupling of D12 abd D16n D26 for angle laminates, crack front contour may change. This report assumes that in the initial defect plane the unit new surface created by crack growth would cause the same compliance change despite of this situation. Thus the compliance method can be used. Moreover, the appearance of fiber bridging during crack growth would make measuremented crack length increments difficult /18/. Hence the calibrations of compliance has to be performed by using a series of specimens with different initial crack length.

As a result, six or seven specimens were tested for each lay-up laminate. The crack was separated at a constant rate of 2 mm/min in a INSTRON machine while the load P, the deflection δ, the reponse of a gauge, and acoustic emission signals were recorded continuously. With the help of this information, a linear part of load-deflections curve allows us to determine compliance. The onset point was characterised by appearance of the first acoustical emissions, as well as a sudden change in the slopes of curves P-δ and P-deformation curves /19/. This provides an objective method to determine the critical values.

II - THEORETICAL ANALYSIS

2.1. Simple linear beam theory

Consider an symetrical anitropic cantilever beam having a rectangular cross section. A uniform force is applied to the end of the specimen to subject it to bending.

The basic differential equation for the deflection of a composite beam undergoing pure bending is :

$$\frac{d^2W}{dx^2} = \frac{M}{E_{fx} \; I} \tag{4}.$$

where   w : deflection of the beam
        M : moment $M = Px$
        E_{fx}: effective bending modulus along eam axi
        I : moment of inertia $I = \frac{Bh^3}{12}$, B, h are the width and thickness of an arm of DCB

Integrating equation (4) with the associated boundray conditions, and considering the deflection of a DCB at the end as equal to twice that of one beam, we can obtain the compliance of a specimen :

$$c = \frac{2w}{p} = \frac{8 \; a^3}{E_{fx} \; B \; h^3} \tag{5}.$$

Substituting the relationship (5) into (1), G_{IC} can be calculated by :

$$G_{IC} = \frac{12 \ Pc^2 \ a2}{E_{fx} \ b^2 \ h^3} \tag{6}.$$

2.2. Finite element analysis

For a multidirectional DCB specimen, the strain and stress fields are too complex to find an exact solution. For this reason, finite element analysis is very efficient.

In this work, two dimensional quadrilateral elements was used in creating the mesh the DCB ply by ply. Near the crack tip relatively smaller grids were used to describe more accurately the larger stress gradient. Suppose that a very small crack extension in mid-plane occurs under a critical load Pc. G_{IC} can be then calculated by the following equation :

$$G_{IC} = \frac{1}{2} \{U\}^T \frac{\Delta[K]}{\Delta A} \{U\} \tag{7}.$$

where : U : displacement field of the specimen under the critical force Pc

ΔA : vitural crack extension

ΔK : variation in the rigidity matrix cause by ΔA

In this analysis, the assumptions were made :
-each ply is homogeneous and anisotropic
-the composite exhibits linear elastic behavior
-crack propagation is self-similar
-along the thickness of the specimen, the distribution of stress, strain and displacement are the same in plane stress.

III - RESULTS AND DISCUSSION

3.1. Analytical and experimental compliances

To determine G_{IC} by the application of the compliance method depends greatly on how to calibrate the compliance. It is evident that if the law of compliance includes certain error, its derivative could significantly amplify this error.

Fig. 2 present the variation of compliances for each lay-up laminate versus initial crack length, determined by three method : experimental measurement, simple beam theory and finite element analysis.

Herein, we note three important points.
First, experimental values are always higher than analytical ones and this difference increases with inceasing initial crack length; secondly, as adjacent ply angle varies from 0° to 45° or from 90° to 45°, effects of coupling D_{12} and D_{16} (see table 1) become stronger, experimental and analytical points separate more. That is logical,because that is more complex than the one plane problem as assumed in the analytical calculations. the third point is surprising. The simple beam theory coincides better with the experimental results than the finite element method. As seen, in the case of multidirectional DCB configuration, a better three-dimensional finit element model has to be developed to improve the calculated results.

3.2. Correlation of G_{IC} determined by three methods

Herein GIc^{exp-c} measured by applying the compliance method to the experimental value (3) and GIc^{bt-c} using beam theory (6) were compared with GIc^{fe} found by the virtural extension method of finite element analysis (7) (see Fig.3). It can be seen that the G_{IC} determined by three method agree for small $D_{12}^2/D_{11}D_{22}$ and $D_{16}^2/D_{11}D_{22}$ laminates. However stronger these couplings become, greater the scatter. this confirms that three dimensional effects are not negligible.

3.3. Influence of lay-up on G_{IC}

In general a crack dose not maintain its initial plane in a multidirectional DCB test, but shifts interfaces as it grows, especially for the laminates whose fiber reinforced angle is important. Once the crack tip shifts from the mid-plane of beam, instead of a pure mode I, the specimen is undergoing mixed mode fracture. This case will be discusse later. For the moment, we focus on the onset of delamination. Indeed the critical point detected, as described above, signifies damage initiation at the crack tip without macroscopic crack opening. Hence G_{IC} actually represents the critical mode I value.

In order to examine the effect of lay-up upon G_{IC}, the experimental results were plotted versus different fiber orientations as shown in Fig.4. Althrough G_{IC} for this material varies less than shown in references (9) and (10), it shows a marked tendency, similer to certain material properties Ky/Kx and Kxy/Kx (see Fig.4).

From lamination theory, the relationship between the loads applied and deformations is expressed by :

$$\left\{ \begin{array}{c} N \\ - \\ M \end{array} \right\} = \left[\begin{array}{c|c} A & B \\ \hline B & D \end{array} \right] \left\{ \begin{array}{c} \varepsilon^\circ \\ - \\ K \end{array} \right\} \tag{8}.$$

where : N, M : vectors of forces and moments

ε°, K : vectors of neutral plane strain and curvature

A,B,D : extensional, coupling and bending stiffness matrix respectively

In this test symmetrical DCB was assumed to be subjected to a pure bending following the beam axis, that is to say B = 0, My = 0 and Mxy = 0. But D16 ≠ 0, D26 ≠ 0 in general. Then, equaton (8) can be reduced to :

$$\left\{ \begin{array}{c} Mx \\ 0 \\ 0 \end{array} \right\} = \left[\begin{array}{ccc} D_{11} & D_{12} & D_{26} \\ D_{12} & D_{22} & D_{26} \\ D_{16} & D_{26} & D_{66} \end{array} \right] \left\{ \begin{array}{c} kx \\ ky \\ kxy \end{array} \right\} \tag{9}.$$

From (9), We can obtain :

$$\frac{ky}{kx} = \frac{D_{16}\,D_{26} - D_{12}\,D_{66}}{D_{22}\,D_{66} - D_{26}^2}$$

$$\frac{kxy}{kx} = \frac{D_{12}\,D_{26} - D_{16}\,D_{22}}{D_{22}\,D_{66} - D_{26}^2} \tag{10}.$$

and

$$E_{fx} = \frac{12}{h^3} D_{11} \left[1 + \frac{D_{12}}{D_{11}} \frac{ky}{kx} + \frac{D_{16}}{D_{11}} \frac{kxy}{kx} \right] \qquad (11).$$

For two orthotropic laminates $[O_3/90]_s$ and $[90/O_3]_s$, $D_{16} = D_{26} = 0$,

$$E_{fx} = \frac{12}{h^3} D_{11} \left[1 - \frac{D_{12}^2}{D_{11} D_{22}} \right] \qquad (12).$$

Substituting this last formula into (5), an exact plane stress solution of theory is obtained/20/. These material properties were also listed in table 1. GIc shows little relationship to E_{fx}.

In reality, under test condition, instead of a free pure bending, curvatures ky and kxy at the ends of the beam were constrained. In other words, in addition to the principle bending, at the crack tip the beam also tended to bend in the direction virtical to the beam axis, and to twist around the beam axis as well. Pure mode I would be a combinason of these three different peeling modes. It is possible that these two last modes produce supplemental stress fields and damage zones during the onset of delamination. Since linear elastic material behaviour was assumed, energy absorbed by the specimen could be a superposition of them. By conjunction with the variations of ky/kx and kxy/kx, it is plausible to suppose a relationship between G_{IC} and them :

$$G_{IC} = G_{IC}^x + F1 \left(\frac{ky}{kx} \right)^n G_{IC}^y + F2 \left(\frac{kxy}{kx} \right)^m G_{IC}^{xy} \qquad (13).$$

where G_{IC}^x- effective critical energy release rate caused in x direction bending without Ky and Kxy. In pratical, it never happens, because ky$\neq$0

G_{IC}^y- effective critical energy release rate caused only in y direction bending,it never happens, because kx$\neq$0

G_{IC}^{xy}- effective critical energy release rate caused only by twisting

Now, if we assume that $G_{IC}^x = G_{IC}^y = G_{IC}^{xy}$ is true for the composite tested, then we

obtain :

$$G_{IC} = \left[1 + F1 \left(\frac{ky}{kx} \right)^n + F2 \left(\frac{kxy}{kx} \right)^m \right] G_{IC}^* \qquad (14).$$

In the case of isotropic material and unidirectional composites, G_{IC} remains a constant, because of their invariant ky/kx value and zero kxy/kx.

Since the sign of kxy/kx may not change material behavior, n=1 and m=2 are assumed, and we apply model (14) by rewriting the expression as:

$$G_{IC} = \left[1 + \alpha \left(\frac{ky}{kx} \right) + \beta \left(\frac{kxy}{kx} \right)^2 \right] G_{IC}^* \qquad (15).$$

With $\alpha=-1$, $\beta=1$, $G_{IC}^*=140(J/M^2)$ determined empirically.

G_{IC} were recalculated by this model and plotted also in the fig.3. They agree very well with the experimental results, the only exception being the $[\pm 75_2]_{2s}$. It must be noted that the experimental value of this lay-up is doubtable, because damage due to bending on the face of the specimen under compression was observed during tests.

The present work is a very begining. For attempting to explain the main processes of energy dissipation during the onset and propagation of delamination and to confirm the effect of layup on GIc, on the one hand, it does require richer experimental data and a lot of further researches, such as morphology examination on the delamination surface in microscopic level, and on the other hand we need find and confirm a suitable theoritical analysis for multidirectionnal DCB test as to associate with the experimental results.

CONCLUSIONS

For a multidirectional DCB configuration test, three dimensional effets are too important to be negnected. Therefore two dimensional theoretical analysis would be not enough to correlate the experimental results.

Althrough the critical energy release rate G_{IC} for pure opening mode is independent of the effective bending modulus, it appears to vary with the stratification, and to a greater extent with the level of coupling curvatures ky/kx and kxy/kx.

REFERENCES

1 -De Charentenay F.X., Harry J.M., Prel Y. and Benzeggagh M.L., Effect of defects in composite material, ASTM.STP 836 (1980) pp.84-103
2 Benzeggagh M.L., Prel Y. and De Charentenay F.X., Fifth International Conference on Composite Materials, ICCM-V (1985.) pp.127-139
3 -Hwang W. and Han K.S.,Journal of Composite Materials,23(April 1989) pp.386-431
4 -Hu M.Z., Kolsky H. and Pipkin A.C., J. Comp. Mat., 19 (May 1985) pp.235-249
5 -Aliyuand A.A., Daniel I.M., ASTM.STP 876 (1985) pp.236-248
6 -Hunston D.L. and Bascom W.D., Comp. Tech. Reviex (1983) pp.118-119
7 -Kageyama K., Kobayashi T. and Chou T.W., Composites, 18, n° 5 (Novembre 1987) pp.393-399
8 -MohamadS., El-Zein and Kenneth , Reifsnider L., Journal of Composites Technology & Research, 10, n° 4 (Winter 1988) pp.151-155
9 -Wang A.S.D., Composites Technology Review, 6, n° 2 (Summer 1984) pp.45-62
10-El Senussi A.K. and Webber J.P.H., Composites, 20, n° 3 (May 1989) pp.249-259
11-Marom G., Romain I, Harrel H. and Rosensatt M., ICCM-VI, 3, pp.65-273
12-Miller A.G., Hertzberg P.E., and Rantala V.W., 12th Nationale SAMPE Technical Conference (Octobre 7-9 1980) pp. 279-293
13-Wilkins D.J., Eisenmann J.R., Camin R.A., Margolis W.S. and Benson R.A., ASTM. STP 775 (1982) 168-183
14-Jordan W.M. and Bradley W.L., ASTM.STP 937 (1985)
15-Nicholls D.J. Gallagher J.P., J. of Reinforced Plastics and Composites, 2 (January 1983) pp.2-17
16-Chai H., Composites,15, n° 4 (Octobre 1984) pp.277-290
17-Ramkumar R.L. and Whitcomb J.D., ASTM.STP 876 (1985) pp.315-335
18-Johnson W.S. and Mangalgiri P.D., Nasa Technical Memorandum 87716 (Avril 1986)
19-Benzeggagh M.L., Gong X.J., Roelandt J.M., ICCM-VII (1989) pp.210-216
20-Davidson B.D.and Schapery R.A.,J.of Comp. Mate., 22(July1988), pp.640-656

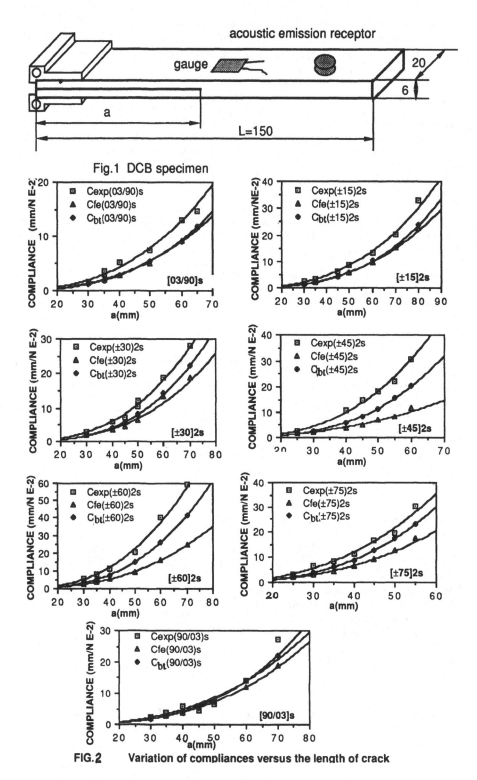

Fig.1 DCB specimen

FIG.2   Variation of compliances versus the length of crack

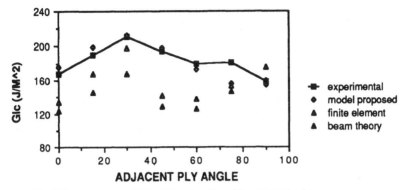

Fig.3 Comparaison of Glc determined by different methods

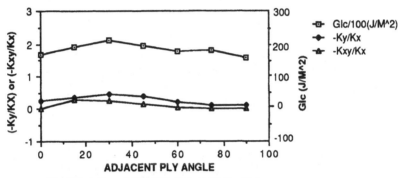

Fig.4 Variation of Glc ,(-Ky/Kx) and (-Kxy/Kx)

TABLE1. LAMINATE CONSTANTS								
	U.D.	[03/90]s	[±15]2s	[±30]2s	[±45]2s	[±60]2s	[±75]2s	[90/03]s
Exf(GPa)	36,2	35,82	31,99	23,03	15,57	12,02	10,844	21,53
Eyf(GPa)	10,6	11,01	10,84	12,02	15,57	23,03	31,99	25,55
Gxyf(GPa)	5,6	5,6	6,709	8,95	10,08	8,95	6,709	5,6
NUxyf	0,0761	0,2505	0,3466	0,4529	0,3904	0,2364	0,1175	0,1085
NUyxf	0,26	0,077	0,1175	0,2364	0,3904	0,4529	0,3466	0,1288
D16/D11	0	0	0,0596	0,1094	0,1293	0,0967	0,0374	0
D16/D22	0	0	0,1787	0,2141	0,1293	0,0494	0,0125	0
D16^2	0	0	0,0107	0,0234	0,0167	0,0048	0,0005	0
(D11*D22)								
D12/D11	0,076	0,077	0,1191	0,2438	0,4087	0,4772	0,3568	0,1288
D12/D22	0,26	0,2506	0,3568	0,4772	0,4087	0,2438	0,1191	0,1085
D12^2	0,0198	0,0193	0,0425	0,1164	0,1671	0,1164	0,0425	0,014
(D11*D22)								
Ky/Kx	-0,26	-0,251	-0,347	-0,453	-0,39	-0,236	-0,118	-0,109
Kxy/Kx	0	0	-0,275	-0,25	-0,142	-0,068	-0,027	0

WIND TURBINE ROTOR BLADES UNDER FATIGUE LOADS

C. KENSCHE, H. SEIFERT

DLR - Institute for Structures and Design
Pfaffenwaldring 38-40 - 7000 Stuttgart - West-Germany

ABSTRACT

To investigate fatigue life of wind turbine rotor blades DLR operates a 100 kW wind energy converter with 25 m diameter rotor at its test site in Southern Germany. Additionally, fatigue tests on Gl-Ep material and full scale components are carried out at the laboratory. Results of these investigations are presented.

1. Measurements of Fatigue Loads on Rotor Blades

Wind turbines and their components have to be designed for a life time of more than 20 years under severe meteorological conditions. For fatigue investigations and to verify the load assumptions, the DLR-developed three bladed wind turbine DEBRA-25 (Figure 1) has been successfully operated and continuously observed since 1984 at the wind test site *Ulrich Hütter* in Southern Germany [1]. During more than 17,000 grid connected operation hours about 510 MWh energy were delivered to the local utility. This downwind machine has a rotor diameter of 25 m. Rated power is achieved at 10.5 m/s averaged wind speed. At higher wind speeds pitch control is active and by turning the blades to feathering position electrical power output can remain constant.

Prior to manufacture of the glass-epoxy (Gl-Ep) prototype blades several static and dynamic tests on a test rotor blade and on components, including the prestressed bolt load introduction system, were carried out at the DLR-laboratory [2],[3]. When assembling the turbine at the test site, a data acquisition system for long time observation and for short term measurements, especially for the rotor blade loads, was applied.

Since commissioning the DEBRA-25 raw data of nearly all typical operation modes and wind speeds could be recorded. In this time the temperature varied between -30 to +40°C and the maximum measured wind speed was 43.5 m/s. Loss of power output due to heavy rain has been investigated as well as the additional loads on rotor blades caused by heavy icing during operation. Another example for an extreme meteorological situation is the operation at very high wind velocities. Figure 2 shows the reaction of the wind turbines supervisory system to a gust starting at 28 m/s and ending at 40 m/s within 1 second. This controlled shut down to idling led to the highest loads on the blade root that were ever measured during the observation time.

As a result of statistical measurements and evaluations during one year, Figure 3 shows typical operation details. The percentages of stationary operation modes e.g. running in partial power mode (LII) with fixed pitch angle and the transient modes e.g. emergency stops (Em stop) related to the time measured (right side of Figure 3) and extrapolated to 100 per cent technical availability (Rayleigh 4.3 m/s) and to higher yearly averaged wind speeds are shown (Rayleigh 6.5 m/s). Calculating the rotor revolutions, using the 6.5 m/s-Rayleigh distribution of wind speed, and a simplified method of designing load spectra, leads to $3.6 \cdot 10^8$ load cycles that fits very well to the predicted $4 \cdot 10^8$ load cycles of the DEBRA-25 load assumptions [1], [3].

The investigations proved two advantages of the DEBRA design. These can be seen in Figure 4 where the averaged blade root bending moments and their fluctuations during the Operation Mode II are related to the wind speed. Due to the cone angle centrifugal forces induce negative bending moments whereas aerodynamic loads induce positive bending moments (see principle sketch at the right side of Figure 4), thus the averaged loads are minimized in that range of operation where the turbine runs most time. The same is valid for Operation Mode I [4]. The other advantage is that, due to the pitch control mechanism, the measured bending moments at the fatigue critical section of load introduction to the hub do not grow significantly with increasing wind speeds within the designed operation limit of 25 m/s. The exceeding of this limit, shown in Figure 4, was due to test purposes. Even the operation at more than 30 m/s results in relatively low averaged root bending moments and also the maximum amplitudes only reach 55 per cent of the design load [1], [2].
The light weight design of the rotor blades diminishes the deterministic in-plane loads due to gravity. The stochastic loads due to the wind variations mainly characterize the out-of-plane bending moments of the blades.

Using Rainflow-counting algorithm and combining the measured raw data with the operation statistics a range spectra for blade root bending moments was evaluated (Figure 5) [5]. Results of these investigations were submitted to an expert group of the International Energy Agency (IEA) and influenced the standardized wind turbine load sequence called WISPER [6].

2. Fatigue Tests in Laboratory

Rotor blades as components of wind turbines usually have a certified life time of 20 years. The design strain level is quite conservative, i.e. 0.3 per cent in the tension, 0.2 per cent in the compression zone. These low values lead to an unnecessarily high mass of the blades and in consequence of nacelle, drive train, tower and foundation. Thus, higher design strain levels should be considered. In [7], [8] a way of evaluating the life time of sailplanes depending on the working stress level in the spar caps of the wings was tried. A programme was proposed to improve the life prediction method on the basis of the Relative Miner Rule usable for gliders and for wind turbine rotor blades. For this purpose fatigue tests on coupon specimens and large scale components are carried out at the DLR within a CEC-project [9].

For the $\varepsilon - n$ curves GI-Ep specimens of typical rotor blade materials were used. One aim was to investigate the fatigue limits up to 10^8 load cycles. Figure 6 shows the $\varepsilon - n$ curves for two different lay-ups and two stress ratios. The values for the laminate with $\pm45°/0°$-lay-up and the low temperature-low pressure prepreg system M9/M10 (Ciba Geigy) tested with a stress ratio of R = -1 seems to be lower than that for the pure UD-material with the cold curing resin system L20/SL (Bakelite). The M9/M10-specimens had steel tabs and a dogbone shape (see Figure 7), whereas the other ones had GRP-tabs.

Different sizing (I 550 and I 700 from Interglas) was used on the UD-material. No significant difference could be observed in the fatigue results. The specimens with load cycles less than 10^8 at a strain level of 0.4 per cent did not fail in the measured cross section but in the grips or due to malfunction in the control system of the servo hydraulic cylinder, respectively. Besides the constant amplitude tests, which are still ongoing, the WISPER [6] and its modified derivation WISPX (ommission of the lower WISPER-ranges) is applied to enable a life time prediction of the specimens depending on the load level. One cycle of this standard simulates two months of a fictive wind turbine operation. For specimens with a maximum strain level of 1.0 per cent until now more than 1000 WISPX cycles have been performed. These tests are still being carried out on the tension part of the standard load sequence and represent the lower spar cap of the rotor blade. Tests on the compression side will be the next step of our investigations.

However, the fatigue behaviour of full scale components will differ from that of the specimens. Thus, for a better service life evaluation large structures including load introduction were also investigated. Single step tests and the WISPER/WISPX-standard were used. The 3 m long spar beams are derived from the rotor blades of the AEROMAN 12.5/40 wind turbine manufactured by MAN. Normally they are designed conservatively according to the certification rules. However, for testing purposes a thin-walled test cross section at r = 1 m was built in to achieve a maximum strain of about 0.6 per cent at the design load [10]. All structures are sufficiently equipped with strain gauges in the cross section. They are cantilever mounted in a fatigue testing rig for static and fatigue tests and loaded at the 2.5 m position in bending by a hydraulic cylinder.

The simultaneous fatigue loading of three components is carried out to save time (see Figure 8). Before starting this the global stiffness and the damping behaviour are investigated, generally. The structures are loaded first with the design load causing a bending moment of 18.75 kNm at the blade root. While the deflection of all tested spar beams remained nearly the same a large scatter could be observed at the maximum strains (Figure 9). Also the frequency response is measured by means of forced vibrations by electro-dynamic excitations with constant energy input. The logarithmic damping decrement is evaluated according to the half-bandwidth method [11]. Figure 10 shows the damping of six components in the virginal state [12]. For better comparison of the damping behaviour the logarithmic decrement is normalized at constant amplitude.

Two components were tested statically, one of them failed due to buckling [10]. To prevent this a foam core was applied in the thin-walled area at the other beams. Unfortunately a debonding and thus instabilities occured at some of them after the first loadings, nevertheless. The fatigue investigations were run at constant amplitude tests with strain levels up to 0.5 per cent and using WISPER with maximum strain level of 0.6 per cent. After surviving more than 240 sequence cycles without damage the strain was increased up to 0.75 per cent at those components which were well manufactured, i.e. which showed linear behaviour of the strains. Another increase of the strain level after survived 360 cycles of the WISPER standard is planned. However, the other spar beams failed either due to severe manufacturing defects or due to local buckling in the measurement cross section. E.g. spar beam P05 failed in the upper spar cap after a time history of WISPER cycles represented more than 60 years of operation. The failure seemed to result from intralaminar shear stress in the top of the local buckle observed during the maximum loadings. The stiffness check did not show any signs of fatigue. The deflection and also the natural frequency remained whithin the scatter of measurement accuracy. Figure 11 shows constant deflection and frequency while the logarithmic decrement increases during the service life but decreases later. Even after the occurance of the first failure the damping did not exceed the maximum level achieved before.

3. Summary

For the design of economic medium and large sized wind turbines the composite structures of rotor blades have to be optimized and the life time has to be predicted more exactly. Thus, evaluations of site measurements under all operating conditions with different types of wind turbines have to be carried out. Parallel laboratory tests on materials and structures are necessary. Both site and laboratory tests showed that the load assumptions can be improved and that the design strain level can be increased. Especially in the laboratory it could be proven that contrary to small scale specimen, where stiffness degradation is often observed shortly before failure, in large components a fatigue failure may not be detectable except by inspection by eye. However, the quality control in the blade factory has to be of higher standard because for well manufactured components no real fatigue limit has been reached up to now even at maximum strain levels of more than 0.6 per cent.

Bibliography

[1] H. Böhnisch, H. Hald, Ch. Kensche, A. Kußmann, J.P. Molly, H. Seifert *Entwicklung, Bau und Betrieb einer 30/100 kW-Windkraftanlage.* DFVLR-Mitteilung 88-06 (1988)

[2] H. Hald, Ch. Kensche *Development and Tests of a Light Weight GRP-Rotor Blade for the DFVLR-100 kW WEC.* Windpower '85, Conference. American Wind Energy Ass., San Francisco, California/USA, Aug. 27-30, 1985.

[3] Ch. Kensche, H. Seifert *Fatigue Tests on the DEBRA 25 Rotorblade and Fatigue Monitoring in Operation.* IEA-Workshop on "Fatigue in Wind Turbines", Harwell/U.K., March 21-22, 1988.

[4] H. Böhnisch, H. Seifert *Analysis for Fatigue Relevant Operation Data from the DFVLR 100 kW Turbine.* European Community Wind Energy Conf. and Exhibition, June 6-10, 1988, Herning/Denmark.

[5] P. Schanz *Belastungsanalyse an den Rotorblättern des Windenergiekonverters DEBRA-25* DLR IB 435-88/07 (1988)

[6] A.A. ten Have *Wisper: A Standardized Fatigue Load Sequence for HAWT-Blades.* European Community Wind Energy Conference (ECWEC '88) Proceedings 1988, Herning/Denmark

[7] Ch. Kensche *Fatigue of Composite Materials in Sailplanes and Rotorblades.* 19. OSTIV-Congress, Rieti, August 1985

[8] Ch. Kensche *Testing and Certification of Sailplane Structures.* Institution of Mechanical Engineers 126/86, 1986

[9] Ch. Kensche *Research Programme for Fatigue-Critical Structures of Small and Medium-sized Wind Turbines.* European Community Wind Energy Conference (ECWEC '88) Proceedings 1988, Herning/Denmark

[10] C. Schloz *Experimentelle Untersuchung und Berechnung eines Windkraft-Rotorblattholms aus Glasfaserverstärktem Kunststoff (GFK) unter Biegebelastung* DLR IB 435-89/08 (1989)/ Studienarbeit Universität Stuttgart

[11] H. Georgi *Damping Effects in Aerospace Structures* AGARD Conference Proceedings No.277, 1979

[12] T. Kalkuhl *Dämpfungscharakteristiken von GFK-Rotorblättern unter verschiedenen Randbedingungen* Studienarbeit, Universität Stuttgart 1990

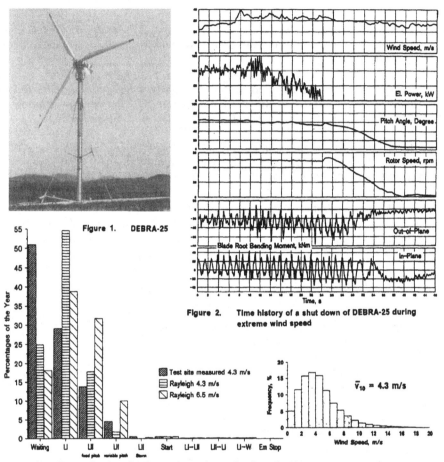

Figure 1.   DEBRA-25

Wind Speed, m/s

El. Power, kW

Pitch Angle, Degree

Rotor Speed, rpm

Out-of-Plane

Blade Root Bending Moment, kNm

In-Plane

Time, s

Figure 2.   Time history of a shut down of DEBRA-25 during extreme wind speed

Test site measured 4.3 m/s
Rayleigh 4.3 m/s
Rayleigh 6.5 m/s

$\bar{v}_{10}$ = 4.3 m/s

Figure 3.   Measured and predicted percentages of operation time of stationary and transient operation modes. Frequency distribution of measured wind speed at DLR-test site.

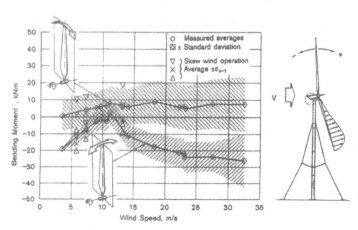

◇ Measured averages
⊠ ± Standard deviation
▽) Skew wind operation
✕ } Average ±δₙ₋₁
△

Figure 4.

Averaged blade root bending moments and its standard deviation in-plane and out-of-plane at Operation Mode II with 50 rpm rotor speed. Pitch control starts at 10.5 m/s wind speed.

178

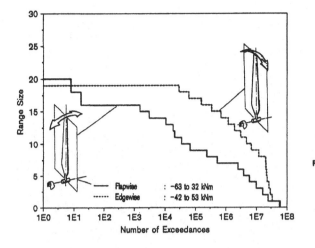

—— Flapwise	: –63 to 32 kNm
------ Edgewise	: –42 to 53 kNm

Number of Exceedances

Figure 5. Rainflow counted range spectra of root bending moments extrapolated to one year. Short time measurements combined with operation statistics.

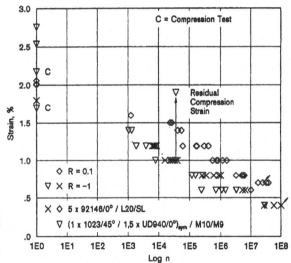

C = Compression Test

Residual Compression Strain

◇ R = 0.1

▽ X R = –1

X ◇ 5 x 92146/0° / L20/SL

▽ (1 x 1023/45° / 1,5 x UD940/0°)$_{sym}$ / M10/M9

Log n

Figure 6. $\varepsilon - n$ curves for GI-Ep used in wind turbine rotor blades

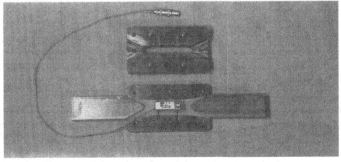

Figure 7. Specimen with antibuckling guide and extensometer

179

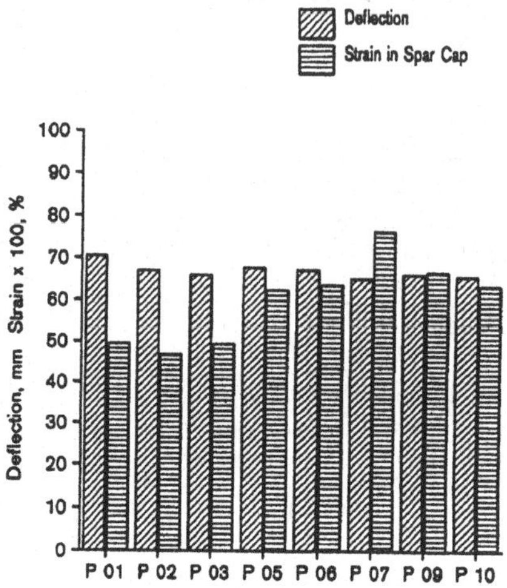

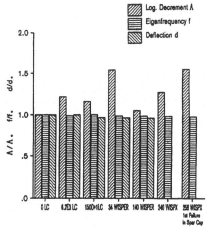

Figure 9. Deflection and maximum strain of spar beams under 18.75 kNm bending moment

Figure 8. Rotor blade components in fatigue testing rig

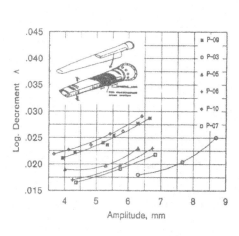

Figure 10. Damping of components before design loading

Figure 11. Damping and global stiffness of P05 during life time and after first failure in upper spar cap

180

IS CRACK CLOSURE DUE TO FATIGUE LOADING CAUSING MORE DAMAGE IN CARBON FIBRE REINFORCED EPOXY COMPOSITES ?

M. WEVERS, I. VERPOEST, P. DE MEESTER

Katholieke Universiteit Leuven
De Croylaan 2 - 3030 Leuven - Belgium

0. ABSTRACT

This tension-tension fatigue study on crossplied laminates, revealed that the fatigue stress ratio R has a considerable influence on the fatigue damage development. It has been found that for stress ratios smaller than 0.5 additional damage is introduced due to crack closure. A micromechanical explanation will be presented to model this crack closure phenomenon in $(0°_2, 90°_2)_s$ laminates.

1. INTRODUCTION

The fatigue damage behaviour of carbon fibre reinforced epoxy laminates is characterized by the initiation and growth of specific damage modes: damage in the epoxy matrix, damage of the carbon fibres, damage in between the individual laminae and damage of the interface between the fibre and the matrix.

A fatigue study carried out in tension revealed that beside the fatigue stress amplitude σ_a, the fatigue stress ratio R has a considerable influence on the damage development in crossply carbon-epoxy laminates.

It was found that a crack closure phenomenon, which is well known for metals but has not yet been recognized for composites, will depend upon the value of the R-ratio. The exact value of the stress at which crack closure takes place can be deduced from the hysteresis-curves and the graphs of the acoustic emission activity in function of the stress.

This crack closure phenomenon introduces additional damage at low stress levels i.e. shear fractures of the 90° plies, micro-delaminations between the 0° and 90° plies and initiation of fibre breakage in the 0° plies.

This surplus of damage in crossply carbon-epoxy laminates fatigued with low R-ratios, is well documented using the edge replica technique, the X-ray radiography and the deply technique.

2. MATERIAL AND TEST CONDITIONS

Carbon fibre reinforced epoxy (cfre) prepreg layers, Fibredux 920C-TS-5-42 supplied by Ciba Geigy Co., were laminated into a $(0°_2, 90°_2)_S$ stacking sequence and cured in a pressclave resulting in a 1.16 mm thick laminate. Specimens, 19.8 x 128 mm^2, were cut from this laminate with a diamond saw. The edges of the laminate were polished up to 1 μm enabling damage studies at the edges with the optical microscope and the replica technique.

The fatigue tests were carried out on a servo hydraulic fatigue testing machine, Schenck Hydropuls 25 kN, in an air conditioned room at 20±2°C and 45±5% relative humidity. The frequency of the sinusoïdal fatigue cycle was very low. For a maximum fatigue stress of 70% or 75% of the UTS, test series with different $R=\sigma_{min}/\sigma_{max}$ ratios have been carried out.

3. DAMAGE STUDY

3.1. Damage assessment techniques.

Three different nondestructive test techniques were used to monitor the damage initiation and development. These techniques are described in previous papers[1,2,3] and will only briefly be mentioned here.

The *acoustic emission technique* was used to follow the initiation and growth of the distinct damage modes during fatigue. For this paper we will focus on the stress level at which the AE-events of the damage modes are generated.

At regular intervals, *replicas* were taken of the edges of the specimens. They were investigated under the light microscope, so that maps of matrix cracks and delaminations could be drawn. The replica, taken at the end of the fatigue test, was also studied with the scanning electron microscope.

X-ray radiographs were taken at the end of each test in order to visualize the internal damage state of the specimen. A solution of zinciodide was used as an enhancing agent penetrating under a low strain.

On top of this, *longitudinal stiffness degradation measurements* were carried out using a dynamic extensometer, Instron 2620μ-601, with a gauge length of 100 mm. The hysteresis curves taken at several time intervals were carefully studied and a correlation with the fatigue damage was investigated.

3.2. Description of the fatigue damage.

On the X-ray radiograph taken at the end of the fatigue tests the total length of matrix cracks as well as the total number of matrix cracks was counted. For the tests run at R<0.5 much more and much longer matrix cracks were found then for the tests run at R≥0.5 for which the number of fatigue cycles was even higher. It is thus clear that the fatigue damage state of the specimens tested at R≥0.5 was less severe than the specimen tested at R<0.5.

The main damage mode seen on the radiographs was matrix cracking, but some matrix cracks were ravelled. This ravelled outlook indicates the presence of incipient delaminations (type 1) which follow the matrix cracks.

In Fig. 1a and 1b the fatigue damage developed at the edges, as revealed by the replicas taken at the end of the test, is shown. For a specimen tested at R≥0.5 (Fig. 1a) we notice the presence of full thickness matrix cracks in the 90° layers, small straight 90° cracks not fully grown over the thickness of the specimen and very small edge delaminations (type 2) between the 90° and 0° layers. The damage state in a specimen tested at R<0.5 is more severe (Fig. 1b). More full thickness and small matrix cracks, sometimes at an angle to the other matrix cracks, can be found, whereas the edge delaminations (type 2) have grown over a longer length along the interfaces between the 0° and 90° plies.

The results of the AE-measurements will be discussed on the basis of the graph of the AE-activity versus the stress level. On Fig. 2a and 2b it is obvious that in the upper stress region ($\sigma > 0.7\ \sigma_{max}$) the AE-activity is very similar for both test series. However note the difference of the scale on the Y-axis. In the lower stress range of the R≥0.5 tests (Fig. 2a), the activity is peaking at the stress reversal. However, this activity remains much lower than the AE-activity in the lower stress range of the R<0.5 tests (Fig. 2b). In other words, the majority of the AE-signals in the R<0.5 tests are generated in the lower third of the stress cycle.

Finally the hysteresis curves for each test series have been compared. These curves reveal also information on the fatigue damage development. On the curves for the R<0.5 tests (Fig. 3) one can see that the upper part of the σ-Є curves shifts parallel to itself but the lower part becomes convex towards the end of the fatigue test. For the curves of the R≥0.5 tests (Fig. 3) only a parallel shift of the whole curve is measured. From studies on metals it is well known that the shape of the σ-Є curve can give an indication of crack closure.

4. MICROMECHANICS OF THE FATIGUE DAMAGE MODES

The micromechanics of the formation of small cracks at an angle to the full thickness matrix cracks and the formation of type 1 delaminations and longer type 2 delaminations in the $(0°_2, 90°_2)_s$ laminates, will now be discussed. In our opinion the concept of **crack closure** provides a sound explanation for this additional damage found in the specimens tested at R=0<5.

During the growth of 90° cracks, material debris is formed between the crack faces. This excess material causes compressive forces in the 90° plies for perpendicular cracks (Fig. 4a) or sliding forces for inclined matrix cracks (Fig. 4b) when the crack closes down. As the shear strength is low, matrix cracks parallel with the plane of maximum shear stress can be formed. It can be demonstrated that for a similar reason, larger delaminations at the edge (type 2) and delaminations (type 1) inside the laminate will be formed along these 90° cracks. This is due to the τ_{xy} and σ_z stresses in the interface between the 0° and 90° plies caused by crack closure of the 90° cracks. For the tests at R≥0.5 the crack will stay open and no delaminations or shear cracks can be formed in the low stress level.

All these additional damage modes, formed during the lower part of the sinusoïdal load function, will cause acoustic emission at low stress levels; Fig. 2b confirmed this. The crack closure phenomenon also explains the convex part of the hysteresis curves. At crack closure there are compressive forces in the 90° plies around the matrix crack, but the 0° plies experience hereby tensile stresses. So deforming again from σ_{min}, smaller additional loads are needed to obtain any deformation than when the closed crack was not present or already open.

5. CONCLUSIONS

The tension-tension fatigue behaviour of $(0°_2,90°_2)_s$ laminates for different stress ratios R is documented in this paper. The fatigue damage studies performed, indicated a difference in damage state which is quite important in view of residual mechanical properties of these laminates. Crack closure, which is a well known phenomenon in fatigue of metals, plays an important role in the fatigue damage development of this type of laminate, fatigued at a stress ratio R<0.5. A revolutionary mechanistic model was worked out to explain the crack closure phenomenon. This research work highlights the importance of carefully designing composite structures to be used in fatigue environments even if the fatigue loads are only tensile because one has to take into account the possibility of crack closure. It is evident from our results that crack closure in crossply carbon-epoxy laminates degrades the mechanical properties of the composite even more by causing additional damage in the composite.

6. REFERENCES

[1] Wevers, M., Verpoest, I., Aernoudt, E. and De Meester, P., In: Verbundwerkstoffe Technologie und Prüfung, Deutsche Gesellschaft für Metallkunde E.V., 1985, p. 143-158.

[2] Wevers, M., Verpoest, I., Aernoudt, E. and De Meester, P., Journal of Acoustic Emission, 4, 2 and 3 , 186-190 (1985).

[3] Wevers, M., Identification of fatigue failure modes in carbon fibre reinforced composites: part I and II, PhD-thesis, Dept. MTM, Katholieke Universiteit Leuven, 7 May 1987.

FIGURES:

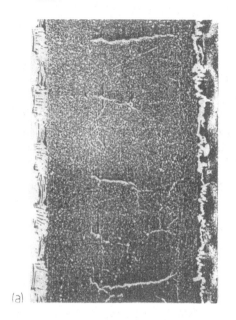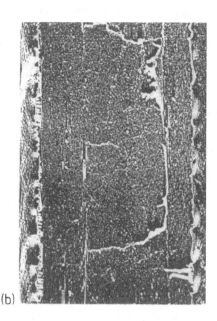

(a)   (b)

Fig. 1: Fatigue damage on the replicas of the edges of the $(0°_2, 90°_2)_s$ specimens, taken at the end of the fatigue tests respectively with R=0.5 (a) and R=0.03 (b), σ_{max}=70% of the ultimate tensile strength and a frequency of 3 Hz.

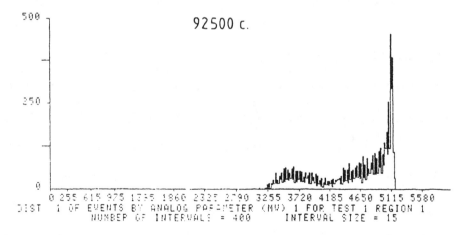

Fig. 2a: The number of AE-signals in function of the stress level for a fatigue test with R=0.6, σ_{max}=75% of the ultimate tensile strength and a frequency of 1Hz.

186

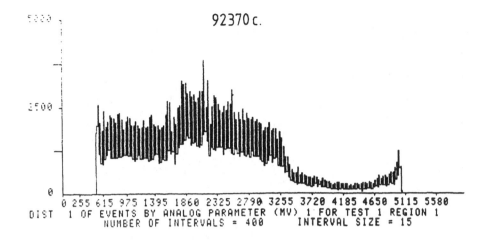

Fig. 2b: The number of AE-signals in function of the stress level for a fatigue test with R=0.1, σ_{max}=75% of the ultimate tensile strength and a frequency of 1 Hz.

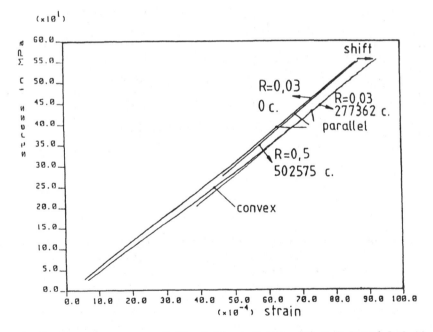

Fig. 3: Hysteresis curves of the fatigue tests with R=0.5 and R=0.03, σ_{max} = 70% of the ultimate tensile strength and a frequency of 3 Hz.

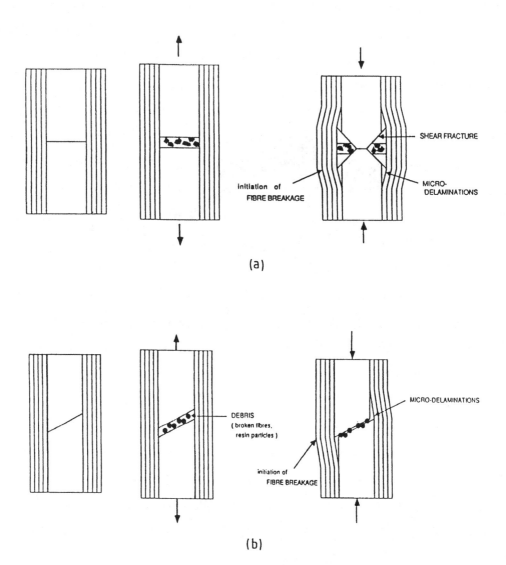

Fig. 4: Model for the damage modes caused at crack closure of the perpendicular matrix cracks (a) and the inclined matrix cracks (b) in the 90° plies.

DAMAGE RATES FOR INTERLAMINAR
FAILURE OF FATIGUED CFRP LAMINATES

R. PRINZ

DLR - Institute for Structural Mechanics
PO Box 3267 - 3300 Braunschweig - West-Germany

Abstract

For the determination of *strain energy release rates* and *fracture toughness* of graphite/epoxy continuous fiber laminates of unidirectional stacking order were used whose central plies were separated prior to curing. In the course of cycling loading a lateral crack developed in the area of the separations from which two delaminations initiated at each of the interfaces of the separated and the continuous plies. For the *stiffness/delamination-length relation* and the *energy release rate G_{II}* a simple equation was derived. The *cyclic energy release rate* ΔG_{II} as a function of the delamination (crack front) propagation rate da/dN in correlation with the *critical buckling state* (CBS) of separated plies can be used for fatigue life estimations or of residual compressive strength calculations.

Introduction

Laminates, in general, are assemblies of unidirectional plies oriented under 0°, ± 45° and 90° with respect to the main load path. The nature of final failure of the specimen depends on the kind of cyclic loading. At the stress ratio of $R = \sigma_l/\sigma_u = 0.1$ (σ_l and σ_u lower and upper stress respectively) fatigue failure is usually due to an increasing number of lateral cracks in the 90°-plies (first ply failure, FPF, and characteristic damage state, CDS, [1]) followed by interfacial delamination, 0°-fiber breaks in different locations and massiv degradation by matrix cracks due to shear loading in the 0°-plies in longitudinal direction.

In contrast to that tension-compression fatigue tests with different stress ratios $R = \sigma_l/\sigma_u$ indicate that in graphite-fiber-reinforced resins fatigue damage manifests itself most severely at $R = -1$, [2], [3]. The fatigue strength losses is due to delaminations which develop during the compression as well as the tension load phase at comparatively small stress amplitudes, [2]. After a period of sta-

189

ble delamination growth the separated plies tend to buckle and the subsequent instable progression of the delaminations will eventually lead to failure, the residual life depending on the strength of the remaining intact cross-sections. Based on these recognitions an model of delamination growth was developed with the aim of predicting the number of cycles at which buckling of delaminated plies is imminent, [2], [3]. The analysis method which has been applied in a number of cases is sound and is considered a valuable improvement of the qualitative and quantitative comprehension of the damage mechanisms in CFRP.

1. Energy Release Rate

The formation and propagation of cracks in laminates is accompanied by energy releases which also cause a reduction of the laminate stiffness. By establishing a relationship between the extent of damage and the change of stiffness, the energy release rate G and the stress intensity factors K can be determined. For the experimental determination of energy release rates different methods have been employed for mode I and mode II and for mode I/II loading [3]. It is apparent that the majority of the methods utilize very simple test configurations like *double cantilever beam tests* (DCB) or *end notched flexure tests* (ENF) which do not accurately reflect the damage mechanisms occurring in realistic environments. Especially the cyclically changing conditions at the crack front of delaminations subjected to tension-compression loading and the unstable growth of delaminations due to buckling of delaminated plies are rarely taken into account. For that reason, new test configurations were developed which are denoted as *transverse crack tension test* (TCT) and *transverse crack tension compression test* (TCTC).

From the experimental evaluation of the stiffness $C \sim E$ and the delaminated area A and on the assumption that the delaminations grow at constant force F, the energy release rate G can be formulated as, [4],

$$G = \frac{1}{2} \frac{d(1/C)}{dA} \cdot F^2 \;=\; G_I + G_{II} + G_{III}. \tag{1}$$

The total energy release rate G for one delamination front, in general, is composed of three contributions: G_I release rate due to tension, G_{II} release rate due to shear in the delamination plane and G_{III} release rate due to shear transverse the delamination plane. For problems at hand (TCT-tests), the latter is of no consequence. The individual contributions of G_I and G_{II} are not readily separable. A numerical approximation is possible by a finite element approach which allows, by appropriate coupling of elements, the introduction of either a mode I or a mode II displacement and the simultaneous suppression of the other. Compared to the the energy release rate G_{II} the rate of G_I is about 1% and therefor negligible. The corresponding stress intensity factors K can be obtained from the well-known relationships for orthotropic materials [6], [3].

2. Energy Release Rates for TCT-Tests

For the determination of critical energy release rates (fracture toughness) unidirectional test specimens were used whose central plies, in accordance with figure 1, were separated prior to curing. Such laminates are referred to as $[0_n, \emptyset_m]_s$, the slash indicating the cut. In the course of cycling loading a lateral crack developed in the area of the separations from which two partial delaminations initiated at each of the interfaces of the separated and the continuous plies. The two delamination which were observed at specimen edges by a microscope, grew at approximate uniform growth rate, but the positions of the four delamination fronts are shifted by some mm. After predetermined cycle numbers extent of delaminations was recorded by C-scans, examples of which are given in Figure 1, by X-ray examinations or by grid reflection method.

During cycling stiffness E was continually registered, so that a clear connection between the size of delamination, a and the corresponding stiffnesses E could be established in the form of

$$\frac{E_o}{E} = 1 + \frac{t_c}{t_l} \cdot \frac{E_o}{E_l} \cdot \frac{a}{2L} \qquad (2)$$

where E_o denotes the initial stiffness of the test specimen with separated plies, $E_l = E_x$ the stiffness of the fault-free laminate, L the specimen reference length, t_c the thickness of separated plies, $t_l = t - t_c$ the thickness of continuous plies and t the total thickness of the laminate. Figure 2 shows strong relation of the specimen stiffness rate E_o/E as a function of delamination length a. The initial stiffness E_0 has to be measured during the first loading prior to formations of delaminations. Introducing equation (2) into equation (1) under consideration of stiffness $C = E \cdot b \cdot t/L$, volume $V = b \cdot t \cdot L$ and $dA = b \cdot da$, the strain energy release rate assumes the simple form

$$G_{II} = \frac{1}{2} \cdot V \cdot \sigma^2 \cdot \frac{d(1/E)}{dA} = \frac{1}{4} \cdot \frac{\sigma^2}{E_l} \cdot \frac{t_c}{t_l} \cdot t . \qquad (3)$$

The results obtained from this approach compare very favorably with the finite element calculations so that the energy release rate G_{II} needed for this type of delamination can be calculated rather accurately from equation (3).

3. Tests and Results.

In order to determine the delamination growth rate da/dN constant amplitude tests and cyclic shedding load tests with increasing and decreasing amplitudes were conducted. This tests run with a load amplitude frequency of 5 Hz and a stress ratio of $R = 0.1$ or $R = -1$. The test specimens made of T300/914C and M40/Code49 fiber-matrix material were preconditioned at environmental conditions of the laboratory with a temperature of 20°C and 40% relative humidity. Tests results of both materials are depicted in figure 3a for a stress ratio of $R = 0.1$. For the cyclic energy release rate ΔG_{II}, a logarithmic linear relationship between the delamination (crack front) propagation rate da/dN and the ratio ΔG_{II} in the range of stable delamination grow was observed.

The comparison of results from constant load amplitude tests with $[0_2, \emptyset_2]_s$-specimens and from load shedding fatigue tests with $[0_4, \emptyset_4]_s$-specimens indicates that the delamination growth rate da/dN is not influenced by the specimen thickness and by increasing or decreasing load amplitudes. TCT-test results compare very well with conventional mode II ENF-tests as can be seen in figure 3b. There exists a threshold G_{IIth} below which the delamination does not develop and a critical value of G_{IIc}, above which the growth rate is unstable.

TCT-tests are relatively simple and should be considered for the characterization of toughness sensitivity of fiber-reinforced resin systems in general.

In corresponding TCTC-tests with a stress ratio of $R = -1$ ultimate failure of the test specimens was caused by buckling of the delaminated sections during compression load cycle. In a number of tests load cycling was terminated at various cycle numbers before failure and residual static compression strength was recorded. According to figure 4 linear relationship was found between the length of delamination a and the compressive strength (static or cyclic loading) of the test specimens. The solid lines in figure 4 represents the theoretically computed buckling strength of knife-edge supported delaminated orthotropic plates [7]. The test results fit the experimental data quite well.

4. Fatigue Life and Residual Strength.

The description of delamination growth in cyclically loaded *unidirectional and multidirectional* laminates may be based on formulations used in classical fracture mechanics [5]. By distinguishing between the mean values of the growth components a_i and b_i in the x- and y- directions of the ith-ply, the approximate size of the delaminated area $A_i = a_i \cdot b_i$ is governed by the equation, [3],

$$\frac{dA_i}{dN} = \hat{c} \cdot f(\Delta G_i)^n = \bar{c} \cdot \varepsilon^n \cdot A_i^m , \qquad (4)$$

where ΔG_i is the amplitude of the energy release rate and ε the stress σ induced strain; $c(\hat{c}, \bar{c})$, n and m are experimentally determined values. For delamination growth with $b_i = const$, the area A_i in the preceding equation can be replaced by $a \cdot b$. Upon integration and substitution of the limits for $N = 0$ with $a = a_0$, a cycle number N_S for damage S and a cycle number N_B for *critical buckling state* (CBS) can be obtained, which corresponds to a delamination length a_S and a_B respectively, [3]:

$$N_S = \frac{a_S^{(1-m)} - a_0^{(1-m)}}{(1-m) \cdot c \cdot \sigma^n} \quad , \quad N_B = \frac{a_B^{(1-m)} - a_0^{(1-m)}}{(1-m) \cdot c \cdot \sigma^n} . \qquad (5)$$

Buckling of a delaminated section, or of the complete laminate, occurs when the delaminated area of the plies reaches a critical buckling dimension. Because of the complexity of the issue a precise assessment of the buckling load is not feasible, but the possibility of an approximation will be demonstrated by an example. Assuming an internal delamination of length a and width b the critical buckling load F_B, and therefor the critical delamiation length a_B, is

$$F_B = K_B \frac{\pi^2 D_{11}}{b^2} = \sigma_B \cdot t_d , \quad a_B = \sqrt{2.69 \frac{\pi^2 D_{11}}{F_B}} . \qquad (6)$$

$F_B = \sigma_B \cdot t_d$ is the load intensity per unit width, D_{11} the bending stiffness, and t_d the thickness of the delaminated section. K_B is the buckling factor, figure 5, which depends on the aspect ratio a/b, the boundary conditions and the stiffness properties of the delaminated sections. a_B follows on the minimum value of $K_B = 0.83$ for the aspect ratio of $a_B/b = 1.8$. This model was be applied successfully for the estimation of residual strength and load cycles to failure of delaminated TCTC-specimens. To facilitate the comparison with residual strength tests results of *multidirectional* laminates the following dimensionless formulation is introduced, [3].

$$S = f\left(\frac{N_S}{N_B}\right) = \left[\frac{a_S}{a_B}\right]^n = \left[\frac{(\sigma_{c0}/\sigma_{cres})^{\frac{1-m}{2}} - 1}{(\sigma_{c0}/\sigma_l)^{\frac{1-m}{2}} - 1}\right]^n, \qquad (7)$$

where σ_{c0} is the initial and σ_{cres} the residual compressive strength, respectively. σ_l is the lower compressive stress of cyclic loading. As the number of test results is as yet scant, $f \cdot (N_S/N_B) = (N_S/N_B)^n$ was introduced as a potential function. The associated scatter range and curves of equal survival probabilities $P_s = 90$ %, 50 % and 10 % for the damage state S are shown in figure 6.

5. Summary.

Analytical and experimental investigations on unidirectional CFRP specimens with transverse separations of central plies (TCT- and TCTC-tests) were reviewed for characterization of mode II delamination fracture in CFRP. A model for estimation of fatigue life based on mode II cyclic energy release rate was developed, which inlude buckling failure in cyclic tension-compression loading and which can be used for approximate calculation of multidirectional laminates damage growth also.

6. Bibliography

[1] R.D.Jamison, K.Schulte, K.L.Reifsnider and W.W.Stinchcomb : *Characterization and Analysis of Damage Mechanisms in Tension-Tension Fatigue of Graphite/Epoxy Laminates.* Published in Wilkins,D.J.: Effects of Defects in Composite Materials. ASTM STP 836 (1982) pp. 21 - 55.

[2] Prinz,R. : *CRP Damage Mechanics under Fatigue Loading.* European Space Agency ESA-TT-1093 (March 1987) 287 p., Translation of : Schadensmechanik kohlenstoffaserverstärkter Kunststoffe bei Schwingbelastung. DFVLR-Mitteilung 87-08, 1987, 246 p.

[3] Bergmann, H.W. and R.Prinz : *Fatigue Life Estimation of Graphite/Epoxy Laminates under Consideration of Delamination Growth.* Journal for Numerical Methods in Engineering 27 (1989) pp. 323 - 341.

[4] Irwin,G.,R. : *Fracture.* In S.Flügge : Handbuch der Physik, Band VI, Elastizität und Plastizität. Springer Verlag, Berlin - Göttingen - Heidelberg, 1958, pp. 551 - 590, esp. p. 563.

[5] Paris, P.C. and F.Erdogan : *A Critical Analysis of Crack Propagation Laws.* Trans. ASME, Ser. D, Journal of Basic Eng. 85 (1963) pp. 528 - 534.

[6] Sih,G.C.; P.C.Paris and G.R.Irwin : *On Cracks in Rectilinearly Anisotropic Bodies.* International Journal of Fracture 1(1965) pp. 189 - 203.

[7] Shivakumar,K. and J.D.Withcomb : *Buckling of Sublaminate in a Quasi-Isotropic Composite Laminate.* Journal of Composite Materials 19(1985) pp. 2 -18.

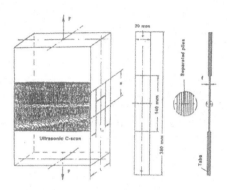

Figure 1. TCT-Test specimen with central separated plies and delaminations, emanating from the transverse crack tips, shown as ultrasonic-C-scan of a $[0_2, \emptyset_2]_s$-laminate

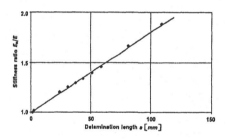

Figure 2. Comparison of the observed delaminations length with the measured and calculated stiffness ratio E_0/E (equation 2).

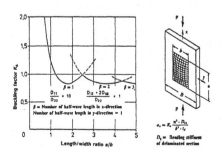

Figure 5. Critical Buckling load of separated plies.

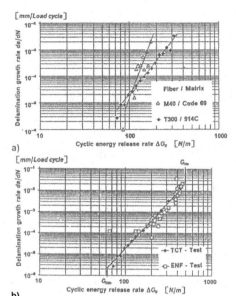

Figure 3. Delamination growth rate da/dN versus cyclic energy release rate ΔG_{II}.

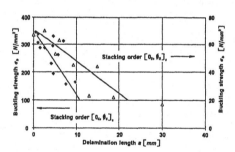

Figure 4. Residual compressive strength versus delamination length. Stacking order $[0_2, \emptyset_2]_s$ and $[0_4, \emptyset_4]_s$. Lines show the theoretical buckling strength of the simply supported plate.

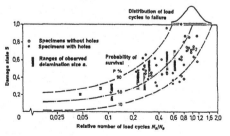

Figure 6. Damage state S of delaminated multidirectional laminates versus load cycle ratio N_S/N_B

194

THE MECHANICS OF FATIGUE DAMAGE IN STRUCTURAL COMPOSITE MATERIALS

P.W.R. BEAUMONT

Cambridge University - Engineering Department
Trumpington Street - CB2 1PZ Cambridge - Great Britain

Abstract

Static and cyclic loading, impact and environmental attack all contribute to the accumulation of *damage* in composite laminates. The damage can take many forms: delamination and splitting during load cycling, matrix cracking during thermal fatigue, and so on. With this diversity of damage mechanisms, it is no wonder that the variability in their static strength is significantly enhanced by service in the field. We recognise, therefore, that damage is progressive and is accompanied by gradual changes in strength and stiffness of the laminate. In other words, residual strength and stiffness and life-time are part of the same design phenomenon.

1. Introduction

We can assess the effect of damage on the strength and stiffness of a composite experimentally. One way is to measure residual strength or stiffness after the composite has sustained damage, by cyclic loading, for instance. An S-N curve can be drawn through the locus of data points where the residual strength curve equals the applied stress. Catastrophic failure occurs when the level of damage exceeds some critical amount; alternatively, the component breaks when the applied stress reaches the degraded strength of the composite. If the application is critical, a distribution of reduced strengths can be plotted against number of load cycles and probability of survival curves drawn through the data.

Although there exists vast collections of mechanical property data of this kind which is useful, empirical design does have its limitations. While a simple rule may *describe* a composite's response to a tensile stress at room temperature, it provides no guidance in *predicting*, for example, the composite's behaviour under a cyclic-stress at elevated temperature or in wet environments because the empirical information contains no knowledge about mechanisms.

195

An understanding of the modes of failure and their dominance under service conditions is important in engineering design because, for a particular application, they determine the "design allowables", the initial cracking stress, the composite's resistance to impact damage, and the damage accumulation rate which affects damage tolerance, residual strength and stiffness, and hence, the life-time of the component.

Microscopic modelling can provide insight into the mechanisms of failure and *in parallel* with empirical methods they can be used to develop equations having the predictive power of microscopic modelling with the precision of ordinary curve-fitting. This is called "model-informed empiricism" which has led to the development of a new branch of mechanics called "damage mechanics".

2. Damage Mechanics of Fatigue Failure

The concept of damage tolerant design requires that a structure retains adequate strength and stiffness after sustaining damage by impact or fatigue.

One way forward is to consider the microstructural changes during damage growth and the effect they have on the residual properties. This approach is preferred since a prediction of damage growth and life-time includes the actual failure processes. The current level of damage can be used to determine the residual strength and stiffness and remaining life-time of a damaged structure.

2.1 A Damage-based Fatigue Model

We know that cyclic loading of composite laminates causes damage, e.g., splitting and delamination, matrix cracking, and eventually fibre breakage, and that such damage accumulates with time. It is reasonable to assume, then, that failure will occur when some critical level of damage in the composite is exceeded. The physical nature of this damage is likely to depend on the material and load history (1-3). If the damage growth-rate depends on the cyclic stress range, $\Delta\sigma$, and the load ratio R, and the current "value" of damage, D, then, provided all other variables (temperature, frequency, etc) do not change, then (1):

$$\frac{dD}{dN} = f(\Delta\sigma, R, D) \qquad (1)$$

It follows that the fatigue life-time is simply the number of load cycles it takes to raise the initial damage state, D_i, usually equal to zero unless some prior damage, like impact, was caused during manufacture or installation, to the final or critical level of damage, D_f:

$$N_f = \int_{D_i}^{D_f} \frac{dD}{f(\Delta\sigma, R, D)} \qquad (2)$$

The problem, of course, is that we do not know the function f. However, if damage affects the stiffness, E, of the laminate, and it usually does, we can write:

$$E = E_0 \, g(D) \qquad (3)$$

where E_0 is the undamaged modulus. By differentiating eqn. (3) and combining with eqn. (1) we have

$$\frac{1}{E_0}\frac{dE}{dN} = g'\left[g^{-1}\left(\frac{E}{E_0}\right)\right] f\left[\Delta\sigma, R, g^{-1}\left(\frac{E}{E_0}\right)\right]$$

where g^{-1} is the inverse of g:

$$D = g^{-1} \left(\frac{E}{E_o} \right)$$

To obtain the function f we must first establish the function g(D). This can be done experimentally; it will depend on the laminate properties. Having gathered data, then, for E/E_o with load cycles N, and knowing g(D), we can evaluate f:

$$f(\Delta\sigma, R, D) = \frac{1}{g\left[g^{-1}\left(\frac{E}{E_o} \right) \right]} \left(\frac{1}{E_o} \right) \frac{dE}{dN} \qquad (4)$$

To do so, the right-hand side of the equation is evaluated for a range of $\Delta\sigma$ values at constant E/E_o and R, for a range of R at constant $\Delta\sigma$ and E/E_o and for a range of E/E_o at constant $\Delta\sigma$ and R.

2.2 Fatigue Damage Growth in [0/90/±45]$_s$ CFRP

A quasi-isotropic carbon fibre-epoxy laminate in tensile cyclic loading fails by matrix cracking of the 90° ply followed by the 45° ply; delamination between the 45° and 90° and $90^o/-45^o$ ply interfaces; and finally by fibre fracture. In high-cycle fatigue (long fatigue lives), the dominant mode of failure is by delamination (1).

Ignoring other forms of damage, we can define a damage parameter, D, as the normalised delamination area, A/A_o, where A is the actual delamination area and A_o is the total area available for delamination. A simple model for the loss of modulus with delaminations is:

$$E = E_o + (E^* - E_o)\, A/A_o \qquad (5)$$

where E^* is the modulus which corresponds to total delamination of the laminate, i.e. where $A/A_o = 1$ and $E/E_o = 0.65$. It follows that:

$$D = 2.857\, (1 - E/E_o) \qquad (6)$$

Thus:

$$\frac{dD}{dN} = f(\Delta\sigma, R, D) = -2.857 \left(\frac{1}{E_o} \frac{dE}{dN} \right) \qquad (7)$$

The right-hand side of eqn. (7) can easily be evaluated experimentally.

Fig. 1 shows the damage rate, dD/dN, determined experimentally as a function of $\Delta\sigma$ (R = 0.1). Clearly, there are 3 regimes of fatigue behaviour:

(1) there is a power law relation between dD/dN and $\Delta\sigma$ over most of the stress range;

(2) the damage-rate is much higher than expected from a power law at stresses approaching the ultimate strength; essentially we have a static failure; and

(3) there is an apparent threshold stress of 250 MPa below which no damage occurs.

197

It is interesting to note that this is the stress which corresponds to the first cracking stress (the "design ultimate") observed in a monotonic tensile test.

2.2.1 The terminal damage

Finally, to predict fatigue life, we need to know the critical or terminal value of D. We can determine D_f as follows: in a monotonic tensile test, assuming no stiffness reduction:

$$\varepsilon_c = \frac{\sigma_{TS}}{E_0} \tag{8}$$

After cycling at some maximum stress, σ_{max}, the increased strain is:

$$\varepsilon = \frac{\sigma_{max}}{E} \tag{9}$$

Since failure occurs when the maximum strain during the fatigue cycle equals ε_c, by equating eqns. (8) and (9) and substituting for D_f from eqn. (6), we have, at failure:

$$D_f = 2.857 \left(1 - \frac{E_f}{E_0} \right) = 2.857 \left(1 - \frac{\sigma_{max}}{\sigma_{TS}} \right) \tag{10}$$

For power law damage growth (from fig. 1):

$$\left(\frac{dD}{dN} \right) = 9.2 \times 10^{-5} \left(\frac{\Delta\sigma}{\sigma_{TS}} \right)^{6.4} \quad \text{(for R = 0.1)} \tag{11}$$

2.2.2 Prediction of S-N curves

We can now substitute for $f(\Delta\sigma, R, D)$ from eqn. (11) into eqns. (1) and (2) and integrate to obtain N_f as a function of $\Delta\sigma$.

Figure 2 shows the theoretical S-N curve of the composite laminate for R = 0.1. The life-time prediction is shown by the solid lines, for $\sigma_{TS} = 550$, 600 and 650 MPa. Good agreement between theory and experiment indicates consistency in our treatment of fatigue damage. Clearly, the low cycle fatigue behaviour is highly sensitive to changes in tensile strength.

The extension of the model to include random load cycling follows in a logical way (4).

3. Fatigue Damage Mechanics and Notch Strength

Fatigue damage in a notched cross-ply carbon fibre-epoxy laminate consists of splits in the 0° ply growing from the notch tips, delaminations between the 90° and 0° plies, and transverse ply (matrix) cracking in the 90° plies over the regions bounded by the splits and the specimen edges (3) (Fig. 3). During load cycling, damage grows in a self-similar manner; hence damage can be characterized by a single dimension, e.g., split length l.(Fig. 4). In other words, fatigue damage consists of several interacting, planar cracks whose formation and growth depend on the properties and behaviour of the matrix.

3.1 Fatigue damage growth in $(90/0)_s$ CFRP

For multiple cracking that constitutes damage propagating from a notch in a composite laminate, the strain energy release rate ΔG offers a simple method of analysing the driving forces for crack propagation (3). The crack growth-rate equation is of the form:

$$\frac{da}{dN} = A(\Delta G)^{m/2} \qquad (12)$$

However, this is insufficient for describing the present example of combined split and delamination growth. As l increases, the associated delamination area grows according to l^2, which implies an increasing resistance to further crack advance. It would be more appropriate, therefore, to use an equation of this form:

$$\frac{dl}{dN} = A\left[\frac{\Delta G}{G_c}\right]^{m/2} \qquad (13)$$

where G_c is the current critical strain energy release rate or apparent toughness for damage growing under monotonic loading.

Damage growth in static loading is governed by (2):

$$P^2 = \frac{2}{(\delta C/\delta l)}\left[G_s t + G_d l \tan \alpha\right] \qquad (14)$$

where the applied stress, σ_∞, on the specimen is directly proportional to the applied load, P. $\delta C/\delta l$ is specimen compliance change with split length, and G_s, G_d are the energies per unit area for splitting and delamination.

The parameters l, t, and α are defined in Fig. 4.

Now the relationship between ΔG and the change in compliance is well known:

$$\Delta G = \frac{(\Delta P)^2}{2t} \frac{\delta C}{\delta l} \qquad (15)$$

$\Delta P = (2t)(W/2)(\Delta \sigma_\infty)$ for one half thickness (two ply) quadrant of the specimen of width W and $\Delta \sigma_\infty$ is the applied cyclic stress range. Since the apparent toughness of the composite is:

$$G_c = G_s + G_d \left(\frac{l \tan \alpha}{t}\right) \qquad (16)$$

eqn. (13) becomes:

$$\frac{dl}{dN} = A\left[\frac{1/2(\Delta P)^2(\delta C/\delta l)}{G_s t + G_d l \tan \alpha}\right]^{m/2} \qquad (17)$$

Values of G_s and G_d have already been determined by Kortschot and Beaumont (2) for this material to be 158 and 400 Jm^{-2} respectively. The delamination angle, α, has been measured at about 3.5°, the initial split length,l_o, can be determined using the analysis of Kortschot and Beaumont (2) for quasi-static damage growth. The constants A and m can be obtained by experimental calibration. For carbon fibre-epoxy, m = 14. Finally, it is necessary to evaluate $\delta C/\delta l$, for which a finite element model is required (2).

Figure 5 shows split growth data for several (90°/0°)$_s$ specimens subjected to three different peak cyclic stresses (R = 0.1). The split lengths are normalised with respect to the notch size, a. Also shown is the integrated form of the crack growth equation for these stress levels. A good fit is obtained using A = 5.0 x 10^{-5} and m = 14. Likewise, a similar model can be applied to a quasi-isotropic laminate (90/+45/−45/0)$_s$. Furthermore, choosing the appropriate values for G_s, G_a and α for the carbon fibre-PEEK system (5), eqn. (17) accurately predicts fatigue damage growth in this example also (Fig. 6).

3.2 Residual Strength

The model described by Kortschot and Beaumont can be used to predict the residual strength of fatigue damaged material . It consists of a two-part Weibull model to account for the statistical variability of laminate strength. The Weibull parameters are found by performing strength tests on unnotched unidirectional 0° specimens. Essentially, the Weibull model reduces to (2):

$$\sigma_{0f} = \sigma_o\left[\frac{V_o}{0.0018\,(l/a)\,8(a^2 t_0)}\right]^{1/\beta} \qquad (18)$$

σ_{0f} = failure strength of a volume V of the 0° ply in the cross-ply,
σ_o, V_o, reference stress and volume from experiments on unnotched 0° specimens.
σ_o = 1.88 GPa, V_o = 7.4 mm^3. β = Weibull modulus = 20 from unnotched 0° strength tests, t_0 is 0° ply thickness.

Residual strength, $\sigma_{\infty f}$, can therefore be calculated as a function of split length. The model assumes that laminate fracture occurs when the longitudinal tensile stress in the 0° ply of the cross-ply laminate exceeds its strength:

$$\sigma_{\infty f} = \sigma_{0f}/K_t$$

where the notch tip stress concentration factor K_t is $f(l/a)$ (Fig. 7) (2).

Figure 7 shows the residual strength data plotted as a function of split length. The predicted residual strength dependence on damage is superimposed. Clearly, Kortschot and Beaumont's model accurately predicts the residual strength when applied in conjunction with the growth-law described above. The notch tip blunting effect is the dominant factor governing residual strength, with residual strength increasing throughout the duration of the fatigue test (Fig. 8).

4. Final Comments

The models developed in this work are fairly simple in form and compare well with the data available. Most of the parameters which appear in the models have a clear physical meaning and can be determined experimentally.

Further experimental work is needed to investigate the range of conditions (temperature, humidity, strain-rate, state of stress, etc.) over which the models apply. Given the general nature of the approach it may also be possible to extend the work to various lay-ups and geometrical configurations.

Acknowledgements

Much of the work on which this paper is based was performed by members of my research group of the Cambridge University Engineering Department. It is a special pleasure to acknowledge the enjoyable collaboration with Dr Anoush P. Poursartip, Dr Mark T. Kortschot and Dr Mark Spearing. The work has benefitted greatly from the many invaluable contributions of Professor Michael F. Ashby.

The financial support of the British Science and Engineering Research Council, the British Council and the European Space Agency is gratefully acknowledged.

References

1. A.P. Poursartip, M.F. Ashby and P.W.R. Beaumont (1986): Fatigue Damage-Mechanics of CFC Laminates; Part 1, Composite Science and Technology, **25**, 193.

2. M.T. Kortschot and P.W.R. Beaumont: Damage Mechanics of Carbon Fibre Composite Materials, ESA Workshop, Advanced Structural Materials: Design for Space Applications, 23-25 March 1988, ESTEC, Noordwijk, The Netherlands. (To be published in Comp. Sci. and Tech. (1990)).

3. S.M. Spearing, P.W.R. Beaumont, and M.F. Ashby (1989): Fatigue Damage Mechanics of Notched Graphite-Epoxy Laminates. Presented at the ASTM Meeting "Fracture and Fatigue of Composites", Orlando, Florida (USA). (November 6-9).

4. A.P. Poursartip and P.W.R. Beaumont (1986): Fatigue Damage-Mechanics of CFC Laminates; Part 2, Composite Science and Technology, **25**, 283.

5. S.M. Spearing and P.W.R. Beaumont (1990): in preparation.

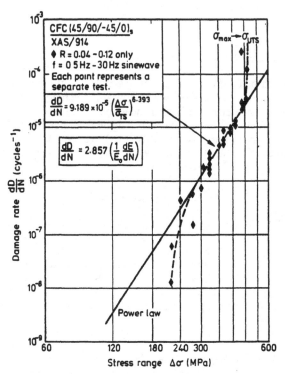

Fig. 1   Damage growth-rate as a function of cyclic stress range for a carbon fibre-epoxy laminate (1). The solid line is represented by eqn. (11).

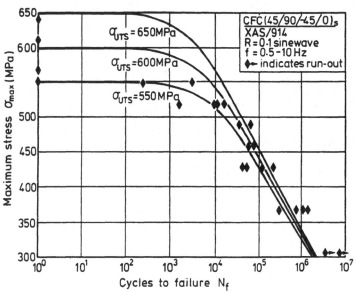

Fig. 2   Predicted S-N curve for a carbon fibre-epoxy laminate together with superimposed experimental data of measured life-time (1). The solid curves are determined by combining eqns (1), (2), (10), (11) together with appropriate experimental data.

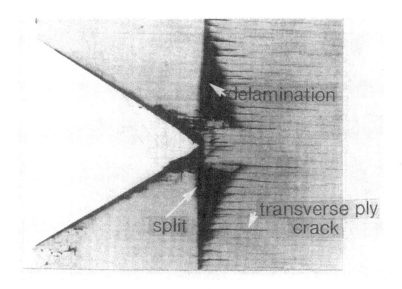

Fig. 3   A photomicrograph showing splitting, delamination and transverse ply cracking in a $(90/0)_s$ carbon fibre-epoxy laminate (2).

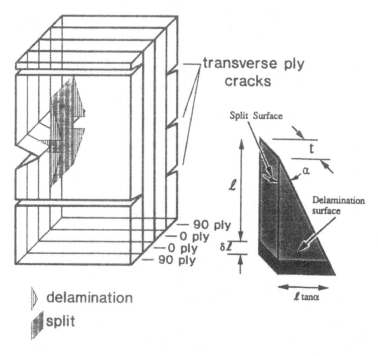

Fig. 4   A model of the various forms of notch tip damage with the critical failure parameters defined (2).

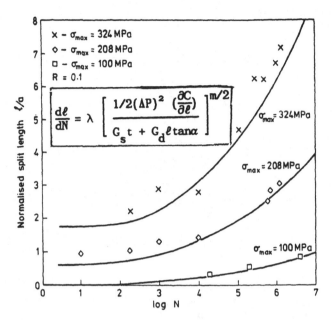

Fig. 5   Predicted damage growth (eqn. 17) with load cycles compared with experimental data for a $(90°/0°)_s$ carbon fibre-epoxy laminate (W = 24 mm; a = 4 mm). Each data point refers to a different specimen (3).

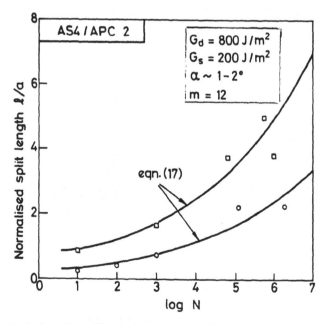

Fig. 6   A similar plot to Fig. 5 of experimental data and theoretical prediction for the carbon fibre-PEEK laminates (AS4/APC2) (5).

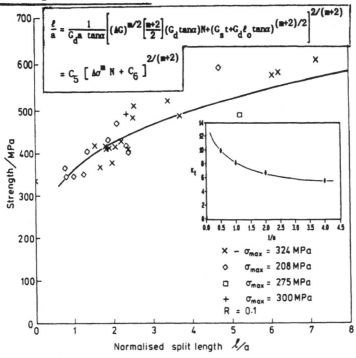

Fig. 7   Predicted residual strength as a function of dynamic split length together with experimental observation (W = 24 mm; a = 4 mm; R = 0.1) (3).

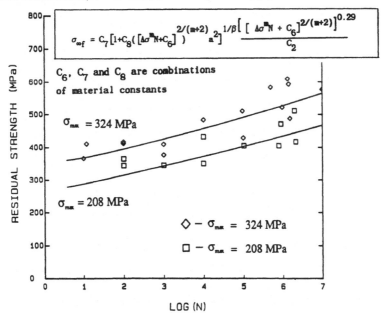

Fig. 8   Predicted residual strength as a function of load cycles (N) together with experimental observations (W = 24mm; a = 4mm; R = 0.1) (3).

FATIGUE DAMAGE ANALYSIS OF NOTCHED CARBON/EPOXY COMPOSITE BY SPATE AND FINITE ELEMENT METHOD

K. KAGEYAMA, M. KIKUCHI*, K. NONAKA**

University of Tokyo, 7-3-1 Hongo - Bunkyo-ku - 113 Tokyo - Japan
*Science University of Tokyo - Noda - Chiba 278 - Japan
**Mechanical Engineering Lab. - Tsukuba - Ibaraki 305 - Japan

ABSTRACT

Damage-threshold analysis of notched $[0/90]_{2s}$ and $[90/0]_{2s}$ carbon/epoxy laminates is carried out by using thermoelastic technique (SPATE) and a three dimensional finite element method. Fatigue strength and damage extension is examined experimentally and the process of transverse lamina cracking is analyzed numerically in the framework of Linear Elastic Fracture Mechanics.

INTRODUCTION

Fracture behavior of carbon composites is very complicated because of their inhomogeneity and extreme orthotropy. O'Brien [1] has pointed out that edge delamination plays an important role in fatigue strength of quasi-isotropic laminates. In the case of notched laminates, transverse lamina crack and local delamination may initiate from the highly stressed regions and effect upon the fatigue strength.

A method of stress pattern analysis by thermal emission (SPATE) has been developed [2] based on the thermoelastic effect. The method has been successfully applied to carbon composites by authors [3]. When fatigue damages accumulate in the composites, it may changes the stress distribution. In the present paper, the three dimensional finite element method and the thermoelastic technique are combined and applied to fatigue damage analysis of notched carbon/epoxy laminates based on Linear Elastic Fracture Mechanics.

I - THERMOELASTIC STRESS ANALYSIS

1.1. Principle

When elastic body is compressed and expanded adiabatically, the temperature increases and decreases, respectively, because of the thermoelastic interaction. Cyclic load is applied to the specimen under the adiabatic condition, and the temperature change is detected with exactly the same frequency of the applied load. Stress patterns are obtained as the temperature distribution by scanning on the surface of the specimen.

1.2. Thermoelastic Effects of Orthotropic Plate

In the case of orthotropic materials, a linear differential relation has been obtained between the sum of the principal stresses and the temperature. The equation cannot be applied to the unidirectionally reinforced composites because of their extremely orthotropic properties. The thermoelastic effect of orthotropic plates has been investigated by authors [3], and a linear relation is obtained between the temperature change, ΔT, and the change of the longitudinal and transverse stresses, $\Delta\sigma_L$ and $\Delta\sigma_T$.

$$\Delta T = -\frac{T}{\rho C_\sigma} (\alpha_L \Delta\sigma_L + \alpha_T \Delta\sigma_T) \tag{1}$$

Where, T is the absolute temperature, ρ is the density, and C_σ is the specific heat at constant stress. α_L and α_T are the thermal expansion coefficients in the longitudinal and transverse directions, respectively. In the case of an unidirectionally reinforced carbon/epoxy lamina, α_T is much larger than α_L in the absolute value. Consequently, the first term in the parentheses can be neglected.

$$\Delta T = -K_m \cdot T \cdot \Delta\sigma_T , \qquad K_m = \frac{\alpha_T}{\rho C_\sigma} \tag{2}$$

K_m is the thermoelastic constant. Eq. (2) indicates that transverse stress component can be detected by the thermoelastic technique.

1.3. SPATE System

The thermal emission from the specimen is detected and analyzed by SPATE. Data acquisition is under computer control and scans are produced in a raster-like manner as shown in Fig. 1. Stress detail to 0.5 mm dimension is resolvable. A correlation technique has achieved the extremely high detection sensitivity of temperature [2]. The resolution of temperature is 0.001 K.

The thermoelastic constant, K_m, of an unidirectionally reinforced carbon/epoxy lamina is 2.5×10^{-11} Pa^{-1}, and the value is about seven times larger than steel. The sensitivity of 0.001 K is equivalent to stress of 0.14MPa.

II - EXPERIMENTS

2.1. Materials and Specimen

Material tested is carbon/epoxy $[0/90]_{2s}$ and $[90/0]_{2s}$ laminates which are made from Toray P305 prepregs (T300/#2500). The nominal thickness of the lamina and laminate are 0.25 mm and 2.0 mm, respectively. The nominal fiber volume fraction is 60 %. Tensile coupons with circular hole are machined as shown in Fig. 2.

2.2. Test Method

The specimen was mounted in a test frame by using wedge action tension grips, and tension-tension cyclic load is applied to it by an electrohydraulic testing machine. The load frequency was 10 Hz, which satisfy the adiabatic condition. Ratio of the minimum and maximum loads, R, was chosen to be 0.1 and kept constant during the test. S-N behavior up to 10^7 cycles was measured and the extension of fatigue damages was investigated by SPATE during the fatigue test.

III - THREE DIMENSIONAL FINITE ELEMENT ANALYSIS

A three dimensional finite element code for a super computer (CRAY X-MP/216) has been developed and applied to fatigue damage analysis of composite laminates because the fracture processes of delaminations and transverse lamina cracking are essentially three dimensional. An isoparametric element with 20 nodes and a singular element of crack tip are used for the analysis.

Transverse cracks, which propagate from a circular notch in a $(90/0)_{2s}$ carbon/epoxy laminate, are analyzed in the present paper. Lamina cracking is modeled as shown in Fig. 3, and three dimensional stress intensity factor of the transverse lamina crack is calculated by using connected extrapolation method proposed by Kitagawa[4].

Numerical results of K_I of a propagating transverse crack in the surface lamina is shown in Fig. 4. K_I decreases with the crack propagation under fixed load condition, and thus the transverse lamina crack growth is stable.

IV - RESULTS

4.1 S-N curves

S-N curve is shown in Fig. 5. The maximum stress, σ_{max} , is measured as the average stress on the unnotched ligament area, that is, $\sigma_{max} = P_{max}/[B (w\text{-}d)]$. The fatigue strength, σ_f, of $[0/90]_{2s}$ and $[90/0]_{2s}$ laminates are 580 MPa and 450 MPa, respectively, at $N = 10^7$ cycles. $[0/90]_{2s}$ laminate has longer fatigue life than $[90/0]_{2s}$ laminate, though no remarkable difference is observed in the static strength.

4.2 Fatigue Damage Extension

Transverse lamina cracks of $[90/0]_{2s}$ laminate are detected by SPATE at σ_{max} of 70 MPa which value is about 15% of σ_f. Fatigue damage initiates at very low stress level. Moreover, transverse lamina cracking is stable and the cracks propagate with increasing the applied load. The load history of $[90/0]_{2s}$ laminate is shown in Fig. 6. Points indicated by arrows represent the near threshold levels, at which the transverse lamina crack propagation rate is too small to be detected by SPATE. Examples of thermoelastic stress patterns at the near threshold levels are shown Fig. 7 (a)-(b). The transverse cracks on the surface lamina can be clearly detected as the unstressed regions.

Thermoelastic patterns of $[0/90]_{2s}$ laminate are shown in Fig. 8 (a)-(b). Transverse lamina cracks is not well detected by SPATE, because the transverse lamina cracks in the surface ply propagate parallel to the loading direction, though the pattern indicate that the stress concentration decreases with extension of transverse lamina crack.

The crack length of $[90/0]_{2s}$ laminate is measured from the thermoelastic stress patterns, and K_{max} is calculated from Fig. 4. A relation between K_{max} and the crack length is shown in Fig. 9. K_{max} is nearly constant with crack growth and the average value is obtained to be 0.68 MPa$\sqrt{m}$.

CONCLUSIONS

A hybrid experimental/computational method is applied to fatigue damage analysis of notched carbon/epoxy laminates. The results are summarized as follows.
(1)Fatigue strength and S-N behavior are different between $[0/90]_{2s}$ and $[90/0]_{2s}$ notched laminates, though no remarkable difference is measured in the static strength.

(2) The extension of fatigue damages has been investigated experimentally by using SPATE. Transverse lamina crack from notch is clearly detected in the [90/0]$_{2s}$ laminate at the first stage of fatigue failure.

(3) Experimental and numerical results indicate that the transverse lamina cracking is stable. K_{max} is nearly constant independent of crack length at near threshold level.

ACKNOWLEDGMENT

This work is a part of MITI program of Basic Technology for Future Industry and financially supported by the Agency of Industrial Science and Technology, Ministry of International Trade and Industry, Japan.

REFERENCES

1 - O'Brien, T. K., NASA Tech. Memorandum, 100548 (1988).
2 - Webber, J. M. B., Proc. 1st Int. Conf. Stress Analysis by Thermoelastic Technique, (1984), 1-8.
3 - Kageyama, K., Ueki, K. and Kikuchi, M., Proc. VI Int. Congr. Exp. Mech., (1988), 931.
4 - Kitagawa H. et al., Trans. JSME, Ser. A, 50-450 (1984), 129.

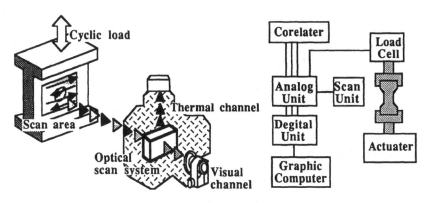

Fig. 1 - SPATE System

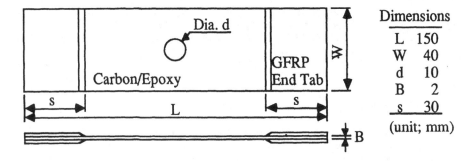

Fig. 2 - Notched tension specimen

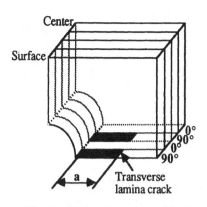

Fig. 3 - Finite element model of transverse lamina cracks

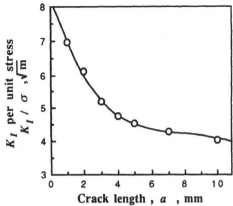

Fig. 4 - Numerical relation between K_I and crack length

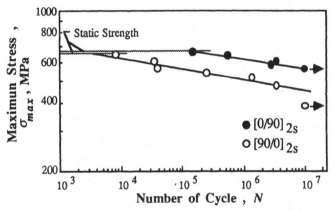

Fig. 5 - S-N curves of [0/90]2s and [90/0]2s laminates

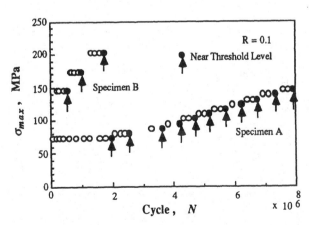

Fig. 6 - Loading history

(a) Before fatigue test   (b) σ_{max} = 145 MPa

Fig. 7 - SPATE patterns of damaged [90/0]$_{2s}$ laminaie

(a) Before fatigue test   (b) σ_{max} = 200 MPa

Fig. 8 - SPATE patterns of damaged [0/90]$_{2s}$ laminaie

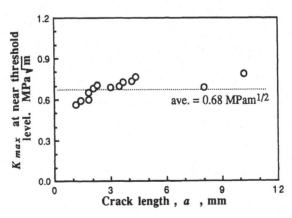

Fig. 9 - K_{max} versus crack length curve at near threshold level

THE EFFECT OF RESIN FLEXIBILITY ON THE CREEP BEHAVIOUR OF FILAMENT-WOUND GLASS-REINFORCED RESIN PIPES

I. GHORBEL, D. VALENTIN*, M.-C. YRIEIX**, P. SPITERI**

Ecole des Mines - Centre des Matériaux - BP 87 - 91003 Evry Cedex - France
CNRS URA 866 - 91003 Evry Cedex - France
**EDF - BP 1 - 77250 Moret sur Loing - France*

ABSTRACT
This study deals with the long term behaviour of filament wound glass fiber/resin pipes internally pressurized. The samples under study were exposed to water at 60°C for 3000 hours before being tested in creep. It has been seen that the life time of the pipes is governed by the critical creep strain to failure of the liner. Moreover the use of a more flexible resin affects the life time , the damage kinetic's, and the failure mechanisms.

I - INTRODUCTION

This paper describes the long term behaviour of internally pressurized filament-wound glass reinforced polyester and vinylester pipes. Several investigations of the effect of the resin flexibility on the short term behaviour of tubes have been reported /1, 2, 3/. The general conclusions were that resin flexibility affects both the load at which transverse cracking starts and the type of cracking behaviour. The purpose of this study is to examine the effect of resin flexibility on the time-dependence of the deformation behaviour and to evaluate the damage growth of filament wound pipe samples internally pressurized with water at 60°C.

II - MATERIALS AND METHODS APPARATUS

The winding angle of the pipes is 55°, the internal diameter is about 100 mm, the thickness is approximately 5 mm and the length is 700 mm. Internal pressurisation tests have been conducted on tubes with the closed end testing procedure after exposure to water at 60°C for 3000 hours. These pipes had an integral liner of mat material with resin. They are fabricated with two different resin matrices. The polyester resin is reinforced by E-glass fiber when the vinylester resin is reinforced by ECR glass fiber.

The unreinforced vinylester resin has a lower Young's modulus (E = 3125 MPa) than the polyester resin (E = 3576 MPa) and higher elongation to break (ϵ_r = 4,6% instead of 0,71% for polyester resin). The glass content and other physical caracteristics of both composites are given in table 1. The axial (ϵ_x) and hoop (ϵ_ϕ) deformations are recorded with the help of a specific extensometer /4/. The damage growth is measured by the loss of modulus during the creep tests through unloading and loading path /5/.

III - CREEP BEHAVIOUR

Polyester specimens were tested at up to four different pressure levels when vinylester pipes were only tested at up to three pressure levels. The range of the applied pressure "P" is about 0,5 to 2 of the short term damage threshold "P_D". The different pressure levels "P", the ratio of "P" to "P_D" called "R", and the initial damage present in the pipes before creep tests are given in Table 2. According to experimentally obtained creep data, the strain value in the hoop direction "$\epsilon\phi$" of both materials is not significant. Therefore, only axial deformation " ϵ_x " will be reported in this paper. The creep strain data, for the different pressure levels and for each material are represented in figure 1 and 2. We have observed linear correlation of strain versus time when plotted on Ln/Ln scales. Hence the material creep response time-dependence is in the form of a power law function. The long term behaviour can then be described by the following equation :

$$\epsilon_x(t) = \epsilon_x(o) + \alpha \, t^\beta \quad ; \; t \text{ in secondes}$$

The instantaneous strain "$\epsilon_x(o)$", the amplitude of the creep strain "α" and the time exponent "β" are the creep parameters. They are obtained by using a least squares best fit regression. Their values are given in table 3. According to these results, the time exponent is independant of the pressure level but a function of the material. Hence more flexible the resin is, higher the time exponent is (β = 0,2 for polyester pipes instead of 0,28 for vinylester pipes).

The creep strain amplitude response pressure-dependence is not linear.Despite of the low experimental points number the relationship between "α" and "R" can be expressed using a power function of the form :

$$\alpha = \Upsilon . R^m$$

Υ and m are materials constants. They were found by using a least squares best fit regression and represented in table 4. According to these results, we can see that the α-R dependency is almost linear for polyester pipes whereas it is nonlinear for vinylester ones. It is important to note, that "R" can be either above as below 1 for polyester pipes whereas it is greater than this value for vinylester pipes. We think that α depends probably on both pressure and damage levels. The creep equation can therefore be expressed as :

$$\epsilon_x^c = \epsilon_x(t) - \epsilon_x(o) = \Upsilon R^m \, t^n \; ; \; t \text{ in secondes}$$

IV - FAILURE MECHANISMS

Some creep curves were followed up to complete failure, while others were terminated after various durations for microscopic examination of the samples. The time of rupture " T_R " of the pressurized pipes is given in table 5. Experiments leading to complete failure rupture of the internally pressurized pipes shows that long term ultimate failure depends on the flexibility of resin and the stress level.

According to experimental results dealing with the failure mechanisms we cannot extend the short-time behaviour to long-term behaviour.

Short term failure occurs by the weeping of the pressurising fluid in glass reinforced polyester and by bursting in ECR reinforced vinylester /6/. Under creep the failure process is associated with weepage for polyester pipes while both weeping and bursting are reported for vinylester tubes. We note that weeping occurs for values of "R" lower than that required for bursting. Obviously, after a 3000 hours exposure to water at 60°C, the fiber/resin interface for both materials was affected. Also it is possible that the strength of ECR glass fibers decreases due to the bad quality of the interface and because of ageing. So if the interface stresses due to the applied pressure are less than the ECR glass fiber/vinylester bonding strength creep failure occurs by weeping else bursting is reported.

V - DAMAGE GROWTH

According to experimental results, we note, during creep tests on glass reinforced polyester and vinylester, at different pressure levels (table 2), the damage accumulation due to intralaminar cracking can reach a saturation limit " P_c " (figures 4 and 5). This saturation limit of damage can be reached before leading to weepage of both materials. The time-dependence damage can be described using a function of the form :

$$D = D_c - B* \ EXP \ \left(- \frac{t}{c} \right)$$

D_c is a material constant, B a function of the applied pressure and C a function of the initial damage D_0 , found using a least squares best fit regression and represented in table 6. Hence, the damage accumulation rate " $\dot{D}$ " can be written as follows :

$$\dot{D} = \frac{D_c - D}{C}$$

We note that the creep damage kinetic's above is similar to the damage model developped in (7). It is of interest to examine the pressure and the initial damage dependence of the damage law functions found for the studied pipes. Hence it has been seen that for different pressure levels the " D_c " is equal to 30% for

215

E/polyester pipes whereas it is about 12% FOR ECR/vinylester pipes. It was found that the greater the resin flexibility the lower is the saturation limit. Moreover the relationship between "B" and the "R" parameter can be described for E/polyester pipes by the following equation :

$$B = 62 \exp (- 1,2R) \quad \text{with } R = P/P_D$$

while for ECR/vinylester pipes we cannot expressed "B" as a function of "R" because we have only two experimental points. However we observed that for the same "R" the greater the flexibility of the matrix, the lower its B values. Finally we note that the "C" parameter depends on the initial damage present in the structure. When the range of applied creep pressure is not associated with initiation of cracks, then the "C" is great and the damage growth is slow. If creep tests were conducted on damaged pipes the rate of damage evolution is rapid. The use of a more flexible matrix involves according experiments less values of "C".

VI - CONCLUSION

According to this study dealing with long term behaviour of polyester and vinylester pipes, the final failure (weepage) of the tubes is governed by the critical strain to failure of the liner. So, it seems difficult to predict the lifetime of the pipes using short term tests.

On the other hand, the use of a more flexible resin affects both the critical time required to failure occuring by bursting or weeping and the damage kinetic's. It is observed from experiments conducted on both materials at the same ratio "R" that "C" decreases with an higher resin flexibility.

The presence of the liner with the same resin as matrix in filament wound layers governs the complete failure by weeping of both pipes under short and long term behaviour.

The bursting of tubes with the more flexible matrix can be described in terms of both the flexibility of the resin and the bonding and fiber strengths.

REFERENCES

1 - Legg H.J., and Hull D.,"Effect of resin flexibility on the properties of filament wound tubes". Composites October 1982, pp. 369-376.
2 - Carswell W.S., "The behaviour of glass filament wound pipes under internal pressures". Proc. 2nd. Int. Conf. on Composite Materials, pp. 472-483.
3 - Garrett K.W., and Bailey J.E., "The effect of resin failure strain on the tensile properties of glass-reinforced polyester cross-ply laminates". J. of Mat. Sci., 12, (1977), pp. 2189-2194.

4 - Ghorbel I, Valentin D., Yrieix M.C. et Grattier J, "Evaluation of the influence of rheological properties of the matric on the acoustic emission and damage accumulation in GFRP tubes" accpted to bepublished in Composite Science and Technology.

5 - Lemaitre J., and Chaboche J.L., "Aspect phénoménologique de la rupture par endommagement". Journal de Mécanique Appliquée, Vol. 2, n° 3, 1978.

6 - Ghorbel I., Valentin D., Yrieix M.C., Grattier J., "Influence du vieillissement sur le comportement au perlage des tubes verre-résine". Développement in the Science and Technology of composite materials, ECCM3, 20-23 mars 1989, Bordeaux-France, pp. 635-642.

7 - Oytana C., "Mechanics and mechanisms of damage in composites and multimaterials". MECAMAT INTERNATIONAL SEMINAR 1989, Damage Indicators - Part I.

Résin	ϵ % matrix	glass fiber used	Number of layers	V_f %	V_v %	Tg (°C)
polyester	0.7	E	8	49	2	132
vinylester	5	ECR	10	54	1	110

Table 1 : Caracteristics of dry pipes.

Material	P bars	R	D^o_x %
E glass/ polyester	30	0,79	0
	45	1,18	5
	60	1,58	16
	75	1,97	26,5
ECR glass/ vinylester	130	1,08	7
	145	1,21	11
	160	1,33	15

Table 2 : Mechanical conditions before creep tests

Material	P bars	α % s.$^{-\beta}$	β	ϵ_x (o) %
E glass/ polyester	30	7,9 E-3	0,18	0,03
	45	11,5 E-3	0,22	0,08
	60	19,4 E-3	0,20	0,243
	75	23,3 E-3	0,24	0,22
ECR glass/ vinylester	130	11,5 E-3	0,27	0,184
	145	16,2 E-3	0,27	0,194
	160	23,3 E-3	0,28	0,205

Table 3 : Creep parameters.

Material	γ	m
E glass/polyester	0,01	1,24
ECR glass/vinylester	0,09	3,48

Table 4: fitted parameters of the creep strain amplitude versus "R".

Material	P (bars)	failure mechanisms	t_R (hours)
E glass/ polyester	30	Weeping	> 5600
	45	"	> 3500
	60	"	> 1000
	75	"	35
ECR glass/ vinylester	130	Weeping	163
	145	"	15
	160	bursting	3

Table 5 : Creep mechanisms and time of failure of pressurized pipes.

Material	D_c (%)	B (%)	C (hours)	P (bars)
E glass/ polyester resin	33,63	23,78	2490	30
	27,56	14,83	80	45
	29,77	7,96	77	60
	32,6	6,70	21	75
ECR glass/ vinylester resin	11,38	6,33	65,70	130
	12,87	4,92	1,46	145

Table 6 : The parameters of the damage law.

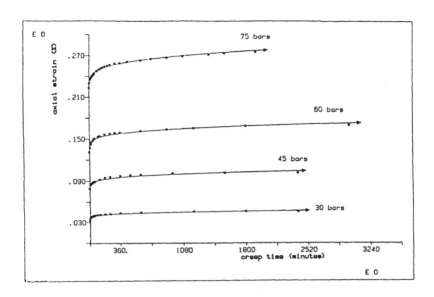

Fig. 1 : Glass E/polyester creep curve.

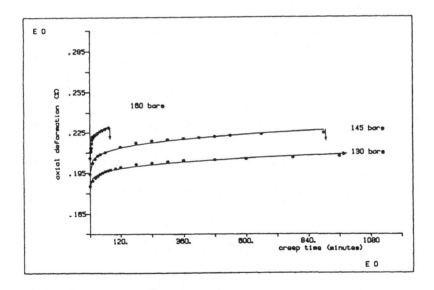

Fig. 2 : Glass ECR/vinylester creep curve.

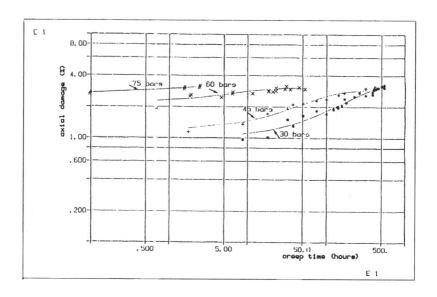

Fig. 3 : Glass E/polyester damage growth.

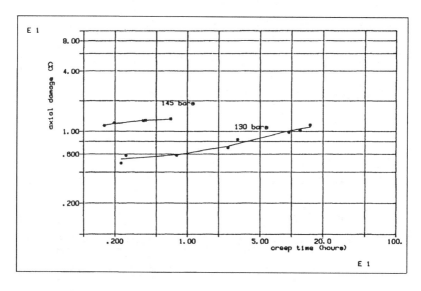

Fig. 4 : Glass ECR/vinylester damage growth.

COMPRESSIVE FATIGUE DAMAGE IN T800/924C CARBON FIBRE-EPOXY LAMINATES WITH AN OPEN HOLE

C. SOUTIS, N.A. FLECK

Cambridge University Engineering Department
Trumpington St. - CB2 1PZ Cambridge - Great Britain

SUMMARY

The compression fatigue behaviour of T800/924C $[\pm 45/0_2)_3]_s$ carbon fibre-epoxy laminates containing a single hole has been studied. The fatigue damage consists of fibre microbuckling, ply cracking and delamination. Fatigue lives exceed 10^6 cycles when the maximum compressive load in the fatigue cycle is less than 85% of the static strength. After fatigue cycling at lower peak loads the residual strength of the specimen increases: splitting and associated delamination reduce the stress concentration at the hole edge. At peak fatigue loads equal to 90% of the static strength fatigue failure occurs as a result of microbuckle initiation and growth.

I   INTRODUCTION

When carbon fibre reinforced epoxy is subjected to compressive loading, the fibres are the principal load bearing elements and must be supported from undergoing a microbuckling type of failure. This is the task of the matrix and the fibre matrix/interface, the integrity of both being of far greater importance in compression than in tension. Because of the greater demand on the matrix and the interface in compressive loading, compressive fatigue loading generally has a greater effect on the strength of composite panels than tensile loading, Curtis [1], Rosenfeld et al. [2].

In the present paper we describe the fatigue compressive behaviour of the $[(\pm 45/0_2)_3]_s$ T800/924C CFRP laminate with a single hole. The aims of the study were to examine the various damage mechanisms which occur in fatigue, and to assess their effects on residual strength.

II   EXPERIMENTAL PROCEDURE
2.1   Material and Specimen Configuration

The material used was Toray T800 carbon fibres in a Ciba-Geigy 924C epoxy resin. The preimpregnated tapes were laid-up by hand into 1m by 0.3m panels using a $[(\pm 45/0_2)_3]_s$ stacking sequence and cured at the Royal Aerospace Establishment, Farnborough. Following cure, the laminates were inspected ultrasonically to establish specimen quality and void content.

Specimens of dimensions 245mm by 50mm were cut from the 3mm thick panels, and reinforcement tabs of aluminium were bonded giving a gauge length of 115mm. Circular holes of 5mm diameter were drilled at the centre of the specimen using a tungsten carbide bit to minimise fibre damage at the hole boundary.

In order to obtain the static strength of the multidirectional material, five unnotched specimens of dimensions 245mm by 30mm were cut and end tabs affixed.

2.2   Mechanical Tests

Static compressive tests were performed on a screw driven machine at a crosshead displacement rate of $0.017mm\ s^{-1}$. The specimens were loaded by shear action using compressive wedge grips. The average measured value of static compressive strength was used to select the maximum stresses imposed during cyclic loading. Compression-compression fatigue tests were run on a servo-hydraulic testing machine under constant amplitude, load controlled, sinusoidal axial loading at a frequency of 10Hz. At this frequency hysteretic heating effects are negligible. Maximum load levels were equal to 75, 85 and 90% of the compressive strength of the notched specimens, and a load ratio R(= minimum load/maximum load) of 10 was used.

For all static and fatigue tests an anti-buckling device was employed in order to prevent column buckling. Its window size was 90mm long by 38mm wide, allowing local microbuckling associated with delamination around the hole to occur, but restraining the specimen from general buckling. Teflon tape was used on the inside surfaces of the fixture to reduce friction.

2.3   Damage Monitoring Techniques

The damage that occurred during fatigue loading was monitored by using penetrant-enhanced X-ray radiography. Care must be taken using this technique because the presence of penetrant may affect the subsequent damage growth [3].

Some of the damaged specimens were sectioned with a diamond circular saw and sections were polished for examination by light microscopy. These section studies are useful both for identifying the particular ply interface at which delamination occurs and for locating matrix cracking.

The effect of fatigue damage on the stiffness of single hole specimens was measured throughout the fatigue test by periodically recording the elongation of the specimen using a displacement clip gauge. The elongation was measured over 70mm gauge length along the centre line of the specimen and symmetrically disposed with respect to the hole.

III   TEST RESULTS AND DISCUSSION
3.1   Static Compression Results

The mean failure stress of the unnotched $[(\pm 45/0_2)_3]_s$ laminate is 810 MPa, with a standard deviation of 80MPa, and the elastic modulus in the loading direction is 85GPa. The large scatter in strength is probably due to imperfect manufacturing conditions of these laminates resulting in fabrication defects and non-uniform laminate thickness. The average failure strain is 1.1%; this is almost equal to the failure strain of the 0° laminate indicating that the 45° plies have little influence on the response of the 0° plies. The failure mode observed is fibre microbuckling [4,5]. The 0° fibres break at two points creating a kink band inclined at an angle $\beta = 5° - 15°$ to the horizontal axis. A schematic representation of the buckling mode is shown in Fig. 1, together with the geometric parameters that define it.

The effect of a single 5mm hole is to reduce the compressive strength of the multidirectional composite laminate by more than 45%. The average failure stress is 430 MPa and the mean remote failure strain is 0.55%; stress values are based on the gross sectional area.

The notched specimens exhibit less than 5% scatter in strength. They fail from the hole and the fracture surface is approximately at right angles to the loading axis. The governing failure mechanism is fibre microbuckling in the 0° layers. Fibre buckling initiates at the edges of the hole at high compressive loads, at 85%-90% of the fracture strength. It extends stably under increasing load before becoming unstable at a critical microbuckle length of 2-3mm. The 0° plies carry most of the load and hence it is the failure of these laminae which results in laminate fracture. Matrix cracking and delamination also occur but are thought to be secondary modes of damage. A typical x-ray radiograph of static loading induced damage is shown in Fig. 2. For further details on static testing see Soutis-Fleck [5].

3.2   Damage Development in Fatigue
i) σ_{max} = 320MPa

Figure 3 shows a typical sequence of X-ray photographs of the

223

damage development in compression-compression fatigue at
σ_{max} = 320 MPa, i.e. 75% of the static notched strength. In the
initial radiograph the dark circular band around the hole shows
drilling damage. Within the first 100 cycles 0° ply cracks appear
at the top and bottom of the hole, figure 3a. Splitting

tangential to the hole in the 0° layers occurs after about 10^3
cycles. Splitting between the 0° fibres extends in length with
the number of cycles and damage is restricted to between the two
longitudinal splits tangent to the hole, figure 3b. Many small
matrix cracks in the ± 45° layers exist along these splits.

At about 10^5 cycles, delamination develops at the 45/0° ply
interfaces. Delamination initiates in areas which have seen
extensive cracking and therefore have high interlaminar stresses

associated with the cracking. It becomes more evident at 10^6

cycles, figure 3c. At 10^6 cycles, in the region between the 0°
splits above and below the hole, delamination grows
longitudinally. This delamination is arrested in lateral growth

by the presence of the 0° tangent cracks. After 6×10^6 cycles
the specimen remained intact and the test was stopped. The growth
rate of the splits tangential to the hole was estimated from the
series of radiographs and found to decrease with increasing split
length, Figure 4.

Transverse sectioning and microscopic examination enable
individual matrix cracks in the 0° layers to be observed. A
matrix crack in a single ply constrained on both sides by ± 45°
plies does not grow directly into the neighbouring layers since
this would require fracture of the ± 45° fibres. The interfaces
between plies of dissimilar orientations arrest the crack, usually
by delamination. Figure 5 shows a photograph of a typical cross

section of a specimen cycled at σ_{max} = 320MPa for 10^6 cycles. The
matrix cracks within the 0° layers and associated delaminations
are evident.

The fatigue damage described above tends to reduce the
notched specimen effectively into two unnotched laminates on each
side of the hole, separated by damaged material. The residual
compressive strength of the specimen is 506MPa. This corresponds
to a 15% increase in strength. Other workers have made similar
observations. For example Whitcome [6] inferred from his
experimental work (tensile and compressive fatigue behaviour of
carbon-epoxy laminates) that matrix splitting can reduce the notch
stress concentration significantly, leading to an increase in the
residual strength of the damaged laminate. He measured residual
strengths which were greater than or equal to the orginal strength
of the composite laminate. Soutis, Fleck and Smith [3] have
performed a 2-D finite element calculation of the damaged specimen
in an attempt to account for the increase in residual strength due
to splitting and to delamination. The predicted increase in
strength by 16% are in good agreement with the measured increase
in strength of 15%.

Compression fatigue loading for 10^6 cycles at the stress σ_{max} = 320 MPa (75% of static notched strength) causes a decrease of approximately 5% of stiffness over the clip gauge-length. This is consistent with the observation that the damage is restricted to the vicinity of the hole. Similar measurements were made by Black and Stinchcomb [7]. They investigated the damage development mechanisms in $[\pm 45/0_2/\pm 45/0_2/\pm 45/0/90]_{2s}$ notched graphite/epoxy laminates during constant amplitude compressive fatigue loading and found that the longitudinal stiffness does not undergo large changes during cyclic loading.

ii) σ_{max} = 365MPa

Damage development at the stress level of σ_{max} = 365 MPa (85% of static failure strength) is qualitatively similar to that observed at the lower load level but is more severe. Figure 6 shows damage after 10^6 cycles. The residual compressive strength of this specimen is 520MPa, which corresponds to a 20% increase in strength over the undamaged specimens with 5mm diameter holes. This increase is attributed to the redistribution of stress around the hole associated with splitting of the 0° plies and with delamination.

iii) σ_{max} = 387MPa

At a maximum cyclic compressive stress σ_{max} = 387MPa (i.e. 90% of the nominal notched ultimate strength), a specimen failed in fatigue after 6×10^4 cycles. Figure 7 illustrates damage after 3×10^4 cycles. Fibre microbuckling develops on the first loading cycle: it occurs in the 0° layers at the edges of the hole, which is the location of the highest in-plane compressive stresses. The buckled zone propagates by fatigue across the specimen width until catastrophic failure occurs. It appears that the stress relieving effects associated with splitting and delamination are not sufficient to arrest the microbuckle growth. The post-failure appearance of the specimen is similar to those tested under static loading.

IV CONCLUDING REMARKS

Compression-compression fatigue damage in notched $[\pm 45/0_2)_3]_s$ composite laminates consists of matrix cracking, delaminations and, at high compressive loads, fibre microbuckling. In single-hole specimens at a stress level σ_{max} less than 85% of the static notched compressive strength, damage in the form of intra- and interlaminar cracking extends along the longitudinal axis of the laminate parallel to the applied load and is well contained between axial splits at the edges of the hole. This type of damage lowers the concentration of stress around the hole. The residual strength is greater than the notched static strength.

225

The longitudinal stiffness undergoes a small reduction in value
during cycling loading; this is associated with the fact that the
damage is restricted to the vicinity of the hole. When the
maximum compressive load in the fatigue cycle equals 90% of the
notched compressive strength, fibre microbuckling extends in the
transverse direction and causes fracture of the laminate in less

than 10^5 cycles.

V   ACKNOWLEDGEMENTS

The authors are grateful for funding from the Procurement
Executive of the Ministry of Defence, under a joint SERC/MOD
contract. We would like to thank Dr P T Curtis of the Royal
Aerospace Establishment, Farnborough for many helpful discussions.

VI   REFERENCES

1. CURTIS, P.T. RAE TR-86021, U.K., 1986.
2. ROSENFELD, M.S.,GAUSE, L.M., ASTM STP 723, 1981, pp. 174-196.
3. SOUTIS, C., FLECK, N.A. and SMITH, P.A. Submitted to
    Int.J.of Fatigue, Feb. 1990.
4. SOUTIS, C., Ph.D. Thesis, Cambridge University Engineering
    Department, England, 1989.
5. SOUTIS, C. and FLECK, N.A. J. Composite Materials, May 1990.
6. WHITCOMB, J.D. ASTM STP 723, 1981, pp. 48-53.
7. BLACK, N.F., STINCHCOMBE, W.W. ASTM STP 813, 1983, pp.
    95-115.

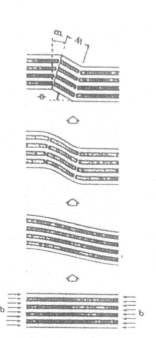

Fig.1 Geometry of fibre microbuckling mode
β= boundary orientation angle
φ= inclination angle
w= microbuckle width.

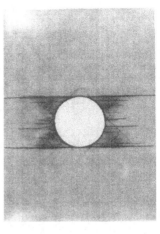

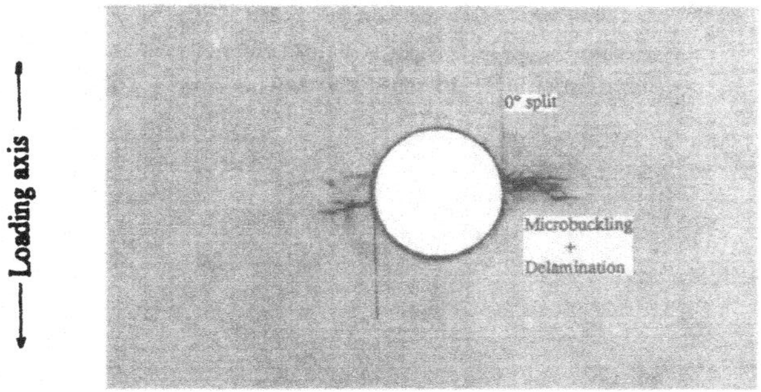

← Loading axis →

Fig.2 Fibre microbuckling emanating from a circular hole at 95% of failure load (d=5mm).

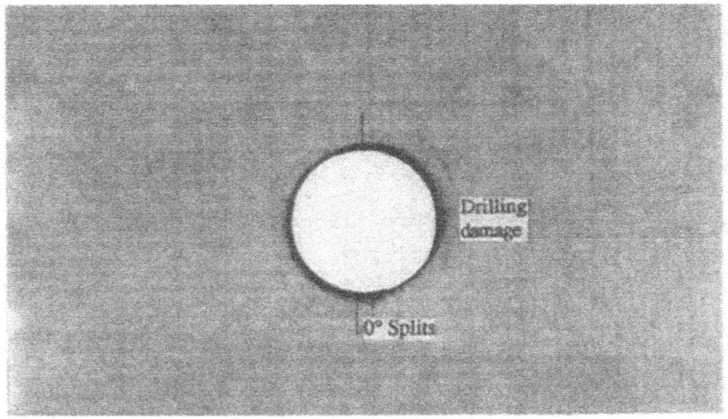

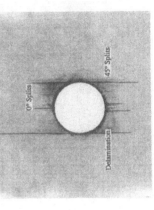

Fig.3a) X-ray radiograph showing compression-compression fatigue damage in a specimen with a 5mm diameter hole, cycled at σ_{max}=320MPa (75% of failure stress) after N=100 cycles.

Fig.3b) Fatigue damage after N=10⁵ cycles.

Fig.3c) Compression-compression fatigue damage after 10⁶ cycles (stress level=75% of failure stress).

227

Fig.5 Transverse section of compression-compression fatigue damaged laminate. σ_{max}=320MPa, N=106, ply thickness=0.125mm. The arrows marked 'D' denote delamination, while the arrows marked 'S' denote splitting. Ply thickness=0.125mm.

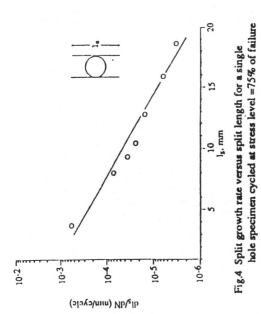

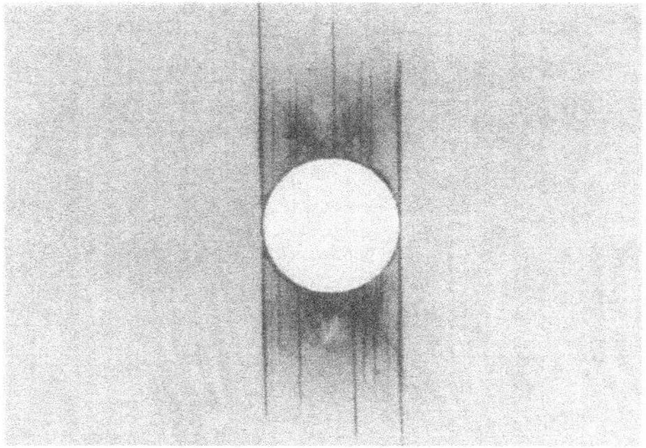

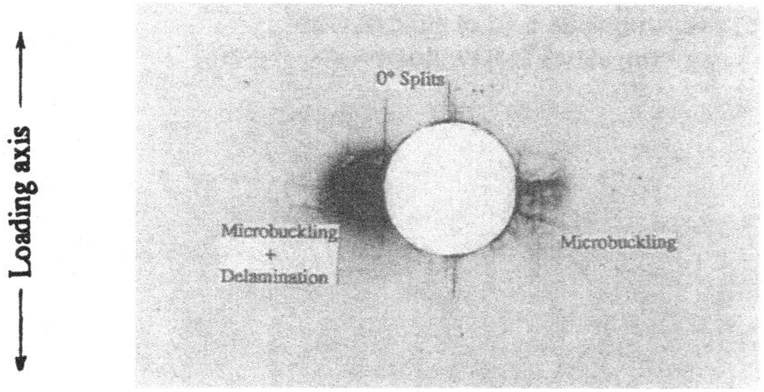

Fig.7 Compression-compression fatigue damage in a specimen cycled up to 10^4 cycles at σ_{max}=387 MPa (90% of failure load).

← Loading axis →

Fig.4 Split growth rate versus split length for a single hole specimen cycled at stress level =75% of failure stress.

Fig.6 Compression-compression damage in a specimen cycled up to 10^6 cycles at σ_{max}=365MPa (85% of ultimate stress).

228

MONITORING ANOMALIES IN COMPOSITE COMPONENTS BY MEANS OF VIBRATION ANALYSIS

K. EBERLE

Institute for Computer-Applications
7000 Stuttgart 80 - West-Germany

ABSTRACT

The production on a large scale of fiber-reinforced composite components requires procedures for quality control which garantee that properties vary within tolerable limits. Defects which arise during manufacturing process or service life have an effect on vibration behaviour of the component. Therefore vibration analysis methods can be used for monitoring different levels of defects between components. This paper reports experiences in applying a simple method to characterize production variations of SMC components in terms of frequency and damping variations.

MEASURING METHOD

The method applied for monitoring anomalies is derived from the decay experiment /4/ which often is used to determine resonant frequencies and damping coefficients. In general the object is excited to vibrate in one mode and after the exciter has been turned off, the decay rate is calculated from vibration measurements. Using impulse technique instead of harmonic excitation, it causes an object to vibrate in a multimode state, but still one vibration transducer is sufficient to obtain decay characteristics. The transducer's signal (acceleration, velocity, sound) summing up all vibration modes after an exciting pulse, is digitized and transformed to the frequency domain by means of Fast Fourier Transformation (FFT).

The results of each transformation are depicted in an amplitude spectrum with sharp narrow peaks indicating resonant frequencies. Shifting a window along the time axis of the transducer's signal and repeating FFT successively provides a set of amplitude spectra from which finally damping coefficients can be evaluated. If damping in terms of logarithmic decrement is plotted versus frequencies, characteristic values of all excited modes are combined in one diagram (Fig.2), which in the following is called damping spectrum. To apply the method successfully, two basic requirements have to be met:

- Approximate linear damping behaviour of the investigated component
- Elastic suspension elements operating with negligible damping or friction.

In general, reliability of results is determined by the quality of boundary conditions which can be verified by reproducible damping values due to repeated excitations. Resolution and accuracy depend on other parameters, for instance on location of excitation, sampling frequency, window size or time shift increment. But as long as different components are compared under the same test conditions, these parameters affect the results in the same way and can be ignored. Thus the limitations of the method are to work only comparatively and to require a master spectrum, which refers to a "normal part". Normality or standard properties of a component can be achieved by testing with other NDT-methods for example.

APPLICATION TO SMC COMPONENTS

A GFRP-component just produced on a large scale has been chosen to test whether the method described is suitable as a quality control method in the production line. The component is part of a truck carriage body and fabricated in SMC-technique. Fig.1 outlines the object as a slightly curved plate with stiffening ribs on the inner side.

Firstly, from testing a sample of normal components, reliable and reproducible test conditions has been established and a master damping spectrum was obtained. Next samples of components were measured, which had been produced anomalously. Due to arbitrary variations of manufacturing process parameters like prepreg quantity, curing cycle or fiber volume fraction, properties within a sample were constant, but varied globally or locally between samples.

By comparing damping spectra of composite components with different properties significant differences could be detected. Fig.2 gives a good example, how frequencies and logarithmic decrements are affected by prepreg systems, having different resin types and fiber volume fractions. The shaded rectangles indicate higher fiber volume fraction, causing lower damping values (-8%) and increased frequencies (+4,5%) within the analysed frequency range (0 to 1000 Hz).

While this variation has a global effect on the damping spectrum, an other comparison indicates the damping spectrum to be selectively affected (Fig.3): In components produced from the same material, but cure time being 30% shorter with respect to normal duration of 120 s, vibration modes in the range 400 Hz to 550 Hz are excited, which are not included in the master spectrum.

Similar effects to damping spectra arise from local defects. For example a damaged rib in one of three components in sample B (Fig.5) causes damping values of particular modes (294 Hz, 568 Hz) to increase. The assumption that friction in the crack surface contribute to material damping can be verified according to mode shape diagrams (Fig.1) of related frequencies. In the damaged area all affected modes have a strong gradient of displacement that activates additional damping.

With respect to experiences of this investigation, it can be stated that the method is suitable to detect anomalies or defects in composite components by means of characteristic variations in damping spectra. Using a microphone as vibration transducer instead of a accelerometer and its time consuming application has further advanced the method towards a procedure for on-line quality control.

VERIFICATION AND IMPROVEMENT BY COMBINED FE ANALYSIS

A combined finite element analysis of the SMC plate completed the experimental investigations. In Fig.1, depicting the mode shape of 568 Hz, the idealisation is additionally outlined. The model's characteristic data are 1495 elements (TRUMP, BECOSX, BETACX) and 4700 degrees of freedom. Results of eigenfrequency analysis based on material data of specimen, which had been extracted from a plate, agreed well to the results of damping spectra and served for calibrating the FE model. Advantages to be achieved by combined combutations are:

- Mode shapes can be calculated and plotted on the computer rather than evaluating them from extensive measurements.

- Boundary conditions are easy to optimize using calculated mode shapes.

- Reference damping spectra can be obtained calculating time history response.

Mode shapes are helpful tools to estimate accuracy of measurements or to verify special effects of local defects and distributed anomalies on damping spectra. In addition they simplify the procedure necessary to find optimal places on an object for suspension (negligible motion) and excitation (large displacements). This is important in attempting to excite a great number of modes with few pulses.

By the means of FE computations effects of anomalies or defects on dynamic behaviour of components can be simulated supporting assumptions concerning damping mechanisms in vibration modes. Finally it can be stated that a combination of computation and measurements improves the choice of experimental parameters as well as the applicability of the method.

ACKNOWLEDGEMENTS

The author wish to acknoledge the support of Mercedes Benz AG (ZWT/WN Stuttgart) who has initiated and sponsored this investigation.

REFERENCES

1 - Schütze, R.,Zerstörungfreie Prüfung von Verbundstrukturen, Strukturmechanik Kolloquium DFVLR, Braunschweig 1979

2 - Teagle, P.R., The quality control and non-destructive evaluation of composite aerospace components, Composites, Vol.14, No.2, April 1983

3 - Argyris,J.H., Eberle,K., Faust,G., Ickert,K., Kirschstein,M., Schadensmechanik von kohlefaserverstärkten Kunststoffen, ISD-Report Nr.322, Stuttgart 1984

4 - Eberle,K., Dämpfungsverhalten geschädigter Faserverbundbauteile, DGLR-Symposium "Entwicklung und Anwendung von CFK-Strukturen", Berlin 1984

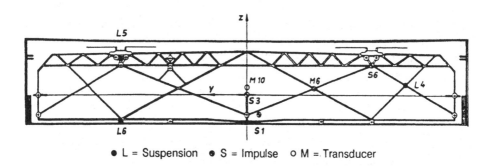

• L = Suspension   • S = Impulse   o M = Transducer

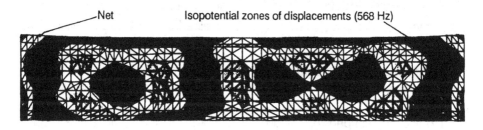

Figure 1 - SMC component, FE net and mode shape

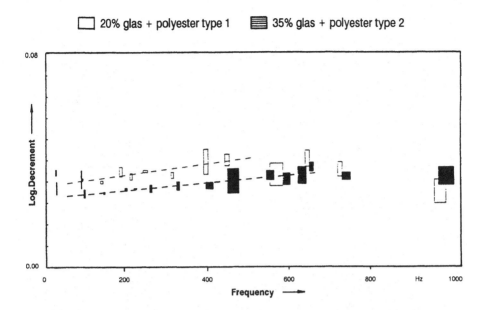

Figure 2 - Effect of different materials on damping

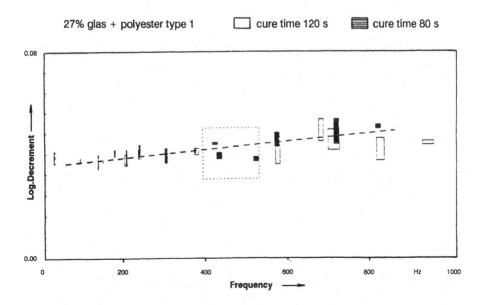

Figure 3 - Effect of variations in cure time on damping

Figure 4 - Component B1 with damaged rib

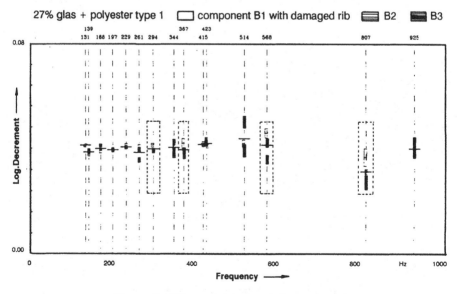

Figure 5 - Effect of damaged rib on damping

INTERFACES
GRENZFLÄCHEN

Chairmen : **Dr. H. LILHOLT**
Riso National Laboratory
Prof. A. KELLY
University of Surrey

INTERFACES IN AL$_2$O$_3$ FIBRE REINFORCED ALUMINIUM ALLOYS

G. NEITE, H.J. DUDEK*, A. KLEINE*, R. BORATH*

Metallgesellschaft AG
Central Laboratory - Reuterweg 14
6000 Frankfurt/Main 1 - West-Germany
German Aerospace Research Establishment (DLR)
Institute for Materials Research - Postfach 90 60 58
5000 Köln 90 - West-Germany

ABSTRACT

The fibre-matrix interface in aluminium alloys reinforced by alumina and alumosilicate fibres, respectively, is analysed using analytical transmission electron microscopy, X-ray photoelectron spectroscopy and Auger electron microanalysis. The interfaces in alumina fibre reinforced alloys show high stability with only negligable indications of a reaction after processing. After long term heat treatments at high temperatures spinel growing is observed. In the alloys reinforced with alumosilicate fibres, which are amorphous thermal treatments at high temperatures can lead to a crystallisation and transformation of the fibres.

INTRODUCTION

In recent years several new metal matrix composite materials were developed /1,2/. A large group of these composites are short fibre reinforced aluminium alloys produced by squeeze casting techniques /3,4/. For the reinforcement alumina, aluminium-silicon mixed oxides and carbon fibres are used. One major problem in all metal matrix composites is the interaction between fibres and matrix during processing and application of the composites. In this paper results of microanalytical, microscopical and surface analytical investigations on the fibre-matrix interface in the alumina and alumosilicate, respectively, fibre reinforced piston alloy AlSi12CuMgNi will be discussed.

237

EXPERIMENTAL

The composites were produced by squeeze casting techniques. The starting materials are short Al_2O_3-fibres from ICI (Saffil[R]) bonded with silica to form a preform with a fibre content of about 20 vol.-% and a common AlSi12CuMgNi piston alloy /2,5/. Additionally some castings are produced with about 10 vol.-% alumosilicate fibres (48% Si_2O, 52% Al_2O_3). The preforms were infiltrated with the liquid alloy applying a pressure of up to 110 MPa. The short fibres have a random planar orientation in the composite.

The interface analyses were performed mainly using the analytical transmission electron microscope (ATEM) Philips 430 attached with a Tracor 5000 EDX- and a Gatan 601 EELS-system. The transmission electron microscopy samples were prepared by dimple grinding and ion milling. For surface analysis a Perkin Elmer PHI 590A Auger microprobe and a PHI 550 ESCA equipment were used. This system is connected with an "in situ" equipment allowing tensile tests under UHV conditions /6/.

RESULTS

Analytical transmission electron microscopy

In the as cast alumina fibre reinforced AlSi12CuMgNi two kind of interface phases were observed. In most cases smooth interfaces decorated with some single crystallites were found (Fig. 1a). The crystallites are only a few 10 nanometers in size. A few percent of the fibres show interface layers with varying thicknesses between 10 and 100 nm (Fig. 2a). The crystallite size of this interface phase is of only a few nanometers. The identification of the crystal structure of this phase by electron diffraction methods was not yet possible /7/. EDX analyses gave a segregation of magnesium, silicon and pollution elements like molybdenum, iron, and lead in the interface layer. EELS analyses result in a high oxygen content /8/. A typical linescan from an interface with magnesium segregation is shown in Fig. 2b.

A thermal treatment of the composite at 500°C for 60 days results in forming of crystallites in the surface of the alumina fibres (Fig. 1b, /5/). By EDX analyses in the transmission electron microscope and convergent beam electron diffraction these crystallites were identified as $MgAl_2O_4$-spinels /9,10/. These results imply that the small crystallites at the interfaces of as cast composites shown in Fig. 1a are spinels.

The situation changes completely when alumosilicate fibres are used as reinforcement. In the as cast condition of the composite these fibres consist of two different regions as is shown in Fig. 3a. The fibres are covered with fine grained material with grain sizes of a few nanometers, while the fibres are amorphous. There is a sharp transition of this shell to the fibre, while the transition to the matrix is less sharp. EDX analyses of the fibres gave nealy equal amounts of aluminium and silicon, as is expected from the fibre composition. Analyses of the shell gave approximately 20 mass-% Al, 65 mass-% Si and 15 mass-% Mg. With the used EDX detector system oxygen cannot be measured.

A thermal treatment of the alumosilicate composite for 2 houres at 500°C leads to strong reactions between fibres and matrix. The fibre structure is changed leaving only the inside regions of the fibres unreacted (Fig. 3b). The reaction zone is crystalline. EDX analyses gave a significant increase of the aluminium and magnesium content (Fig. 3c).

X-Ray Photoelectron Spectroscopy

The chemical composition of the surface of the fibres can be analysed by X-ray photoelectron spectroscopy. Table 1 summarizes uncorrected results for differently treated Saffil[R] alumina fibres. Besides high concentrations of contamination elements like carbon and oxygen an identification of the silica binder and its distribution on the fibres surfaces is possible. On the surfaces of the fibres with a high binder content no aluminium is identified, indicating a complete surface coverage with silica. The fibres with the low binder content are only partly covered with silica.

Scanning Auger Microanalysis

Scanning Auger microanalysis is an excellent tool to investigate fracture surfaces. Contamination problems can be overcome by "in situ" experiments. After an "in situ" tensile test the fracture surface of a composite was analysed by AES (Fig. 4). The crack has propagated through many fibre-matrix interfaces. These interfaces were analysed by AES /6,8/. The magnesium content in the fracture surface is much higher than in a standard alloy with a three times higher magnesium content (Fig. 4b). This can be explained by magnesium enrichment in the fibre-matrix interfaces bare layed by the crack. Contrary to the results of analytical TEM the "in situ" AES analysis shows a magnesium segregation on the surface of all fibres. This can be explained either by the higher sensitivity of AES to magnesium monolayers on the fibres surfaces in comparison to analytical TEM or by preferential crack propagation in interfaces with magnesium segregations.

CONCLUSIONS

The present investigations show a segregation of magnesium and pollution elements in the fibre-matrix interface. The growth of only small crystallites on the fibres in the alumina fibre reinforced AlSi12uMgNi has no significant effect on the mechanical properties /11/. After a long term exposure at 500°C spinels grow. They can partly destroy fibres. But even a 60 days heat treatment shows only a small drop in mechanical properties.

In the alumosilicate fibre reinforced AlSi12CuMgNi strong reactions in the fibre-matrix interface can take place. Obviously a reaction during processing between the silica binder on the fibres and the matrix leads to diffusion of aluminium and magnesium into the binder forming a reaction shell around the fibres. After heat treatments at 500°C magnesium diffuses into the amorphous fibres leading to a crystallisation of the outer parts of the fibres. An aluminium-silicon-magnesium mixed oxide is formed. The strength of this newly formed mixed oxide fibre is yet unknown. Influences on the composite strength were not yet investigated.

Tabelle 1: XPS-element surface concentrations from differently treated Saffil[R] alumina fibres in at.-%

	Al	Si	O_2	C
1.5 mass-% binder	2	5	80	13
6 mass-% binder	-	4	56	40
Removed from composite	6	-	48	46

100 nm   TEM bright field image   TEM dark field image   200 nm

a) Smooth interface with small spinel crystallites; observed at most fibres. as cast condition

b) Spinel crystallite growing into an alumina fibre; heat treated condition

Fig.1: Interface decorated with spinel crystallites in alumina fibre reinforced AlSi12CuMgNi

DLR

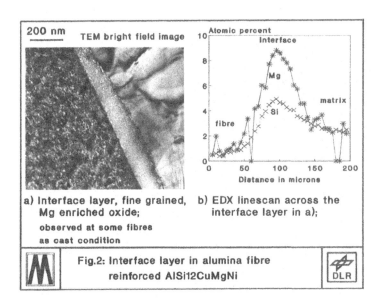

200 nm TEM bright field image

a) Interface layer, fine grained,
   Mg enriched oxide;
   observed at some fibres
   as cast condition

b) EDX linescan across the
   interface layer in a);

**Fig.2: Interface layer in alumina fibre
reinforced AlSi12CuMgNi**

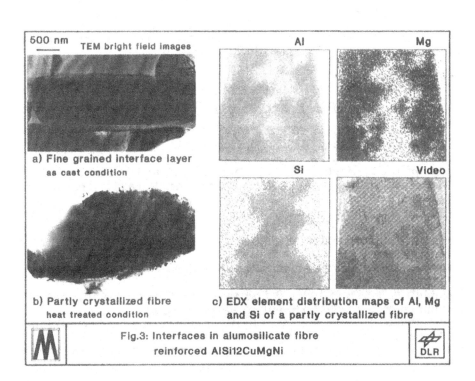

500 nm TEM bright field images

a) Fine grained interface layer
   as cast condition

b) Partly crystallized fibre
   heat treated condition

c) EDX element distribution maps of Al, Mg
   and Si of a partly crystallized fibre

**Fig.3: Interfaces in alumosilicate fibre
reinforced AlSi12CuMgNi**

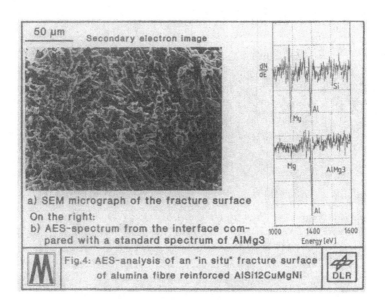

a) SEM micrograph of the fracture surface

On the right:
b) AES-spectrum from the interface compared with a standard spectrum of AlMg3

Fig.4: AES-analysis of an "in situ" fracture surface of alumina fibre reinforced AlSi12CuMgNi

Acknowledgement

This work was financially supported by the German Government (BMFT), Reference No. 03 M 1016 5.

REFERENCES

1 -Bunk W. und Schulte K., Mat.-wiss. u. Werkstofftech. **19** (1988), 391
2 -Lux J., Aluminium **63** (1987), 932
3 -Clyne T.W., Bader M.G. Cappleman G.R. and Hubert P.A., J. of Mat. Sci. **20** (1985), 85
4 -Masur L.J., Mortensen A., Cornie J.A. and Flemmings M.C., Metall. Trans. **20A** (1989), 2549 and Metall. Trans. **20A** (1989), 2535
5 -Neite G., Matucha K.-H., Dudek H.J. and Bunk W., Proc. Euromat 1 (1989), Aachen 22.-24.11.1989
6 -Mucha H. and Dudek H.J., ECCM 4, Stuttgart (1990), this proceedings
7 -Dudek H.J., Inst. Phys. Conf. Ser. No. 93, Vol. **2** (1988), 437-438
8 -Dudek H.J., Mat.-wiss. und Werkstofft. **21** (1990), 48
9 -Kühnen M., Diplomarbeit FHS-Aachen (1989)
10-Suganuma K., Okamoto T., Hyami T., Oku Y., and Suzuki N., J. **Mat.** Sci. **23** (1988), 1317
11-Internal report, Central Laboratory, Metallgesellschaft AG

TEM CHARACTERISATION OF FIBRE-MATRIX INTERACTIONS IN LIGHT ALLOY MMC CONTAINING ALUMINA FIBRES

S. FOX, H. FLOWER, D. WEST

Imperial College - Dept of Materials
Prince Consort Road - SW7 2BP London - Great Britain

ABSTRACT

This paper reviews the literature, published over the last two decades, on the TEM characterisation of interfacial reaction zones resulting from the use of chemically reactive matrices/alloying additions in aluminium and magnesium based composites containing alumina. The results of a systematic study, carried out by the authors, on aluminium alloy and magnesium based metal matrix composites, reinforced with alumina fibres, are then presented and the importance of fibre microstructure is highlighted.

I-LITERATURE REVIEW

Thermodynamic considerations suggest that the use of lithium or magnesium containing matrices may result in the direct reduction of alumina /1/. Figure 1 shows that both metals are able to reduce alumina and silica over the temperature range encountered during composite fabrication by casting (typically 700-850 °C). Silica is included in the diagram as it may be present as a fibre addition or preform binder material. Other oxides (ZnO, FeO, NiO and CuO) are not shown as they all have less negative free energies of formation than silica.

The solid lines in Figure 1 show the relative stabilities of alumina and magnesia per mole of oxygen, assuming both products and reactants at unit activity. The broken lines show the effect of decreased magnesium activity on the free energy of formation of magnesia, calculated from the relevant thermodynamic data /2/ using standard thermodynamic relationships. The driving force for the reduction of alumina is clearly reduced and the crossover between the stabilities of the two oxides is correspondingly brought to a lower temperature. A magnesium concentration of 1 at.% (0.9 wt.%) resulting in such a low activity that the relative stabilities of magnesia and alumina are reversed (compared to unit activity) over the entire temperature range considered.

The rate at which alumina reacts with the melt will be strongly influenced by the rate of diffusion of reactive species into the fibre. Figure 2 shows a compilation of experimentally determined values for the lattice diffusivity of aluminium and oxygen in alumina as a function of temperature /3/. Extrapolation of the aluminium diffusivity to an expected upper limit casting temperature of 800°C gives an approximate value for the diffusivity of 10^{-24}cm^2/s. Simple $x=\sqrt{Dt}$ calculations, assuming a solidification time of 60 seconds suggests an interaction depth of 1nm.

The fibres referred to in this paper are listed in the table below which also shows some typical fibre parameters.

Manufacturer	Trade Name	Alumina Phase	Grain Size (nm)	Fibre Diameter (μm)
E.I. du Pont	FP	alpha	500	20
I.C.I.	RF	delta/theta	50	3
I.C.I.	RD	eta	5	3
I.C.I.	SD	delta/theta	50	3
I.C.I.	LD	eta	5	3
Nippon L.M.	--	-	50	3

The literature on FP/Al-Mg based composites contains widely differing reports of large interaction zones in the matrix /4/, and fibre reaction zones ranging from 50nm /5/ to 200nm /6/ in thickness, depending on the exact processing conditions and magnesium concentrations used. In the RF/Al-Mg composites, reaction zones ranging from less than 1nm /7,8/ to up to 25nm /9/ have been reported, with a further increase to 1um in the RD fibres /9/. A reaction layer of 100nm has also been reported in the Japanese fibre in an AA6061 (Al- 10wt.% Mg, 0.6wt.% Si, 0.25wt.% Cr, 0.2wt.% Cu) matrix composite /10/.

The situation is similar with magnesium base composites, with reaction zones of 100 to 500nm being observed /11-14/, depending on processing conditions and alloy additions.

In Al-Li based FP fibre composites there is general agreement that the overall reaction zone thickness is dependant on processing conditions with reaction zones of between 100 and 500nm being reported /15-20/. However, the reaction zone was of non-uniform thickness in all cases. Additional thermal exposure, after fabrication, also resulted in reaction zones of up to 6um being observed /20/.

II-EXPERIMENTAL PROCEDURE

This study is concentrated on the binary Al-Mg and Al-Li, and Mg-based systems as matrices. The experimental composites containing RF, SD and LD fibres, with either resident silica or fugitive methyl methacrylate (mma) binders, were produced by I.C.I. Runcorn, U.K.. The composites were fabricated by pressurised liquid metal infiltration at I.C.I. using a range of superheats starting from the minimum necessary for satisfactory infiltration (700°C) to excessive melt superheat to promote interaction between the fibre and the melt (850°C). A pressure of 75MPa was aplied and infiltration completed in 2-5 seconds, solidification then occurred in 1-2 minutes.

Specimens were examined by light and scanning electron microscopy. Thin foils for TEM were produced by mechanical dimpling and ion beam thinning. Elemental concentration profiles, for Al, Mg and Si, were determined by stepping a 100nm diameter beam at intervals of 100 or 200 nm across the different fibre-matrix interfaces using X-ray EDS.

III-RESULTS

3.1 Aluminium Alloy Matrices

TEM examination of Al- 10 wt.% Mg matrix composites showed both the RF and SD fibres to consist mainly of delta alumina crystallites, of the order of 50nm in diameter, with approximately 5% residual porosity /21/. Casting at different melt superheats did not result in any marked changes in the internal microstructure of these fibres. However, the LD fibres, which contained finer eta-alumina crystallites of the order of 5-10nm in diameter, and approximately 30% intentional porosity, showed very different microstructures. The initial fine, lamellar structure of these fibres was gradually replaced with a coarser, more equiaxed structure with increasing melt superheat.

X-ray EDS analysis profiles from the RF fibres showed no obvious temperature dependence with regard to magnesium penetration. Interaction depths ranged from 100 to 500nm, the maximum penetration being observed with a melt temperature of 800°C. All of the SD composites showed a low level (0.5 to 1 wt.%) of penetration at the fibre centre and a progressive increase in the level of magnesium at the fibre surface, with increasing melt temperature. Differences in the behaviour of the RF and SD fibres, which have similar microstructures, are thought to be due to differences in fibre volume fraction (20 and 55% respectively) and binder materials (silica and mma respectively). The more porous LD fibres showed higher levels of magnesium penetration at the fibre centre than the SD fibres, and maximum overall melt penetration with a melt temperature of 750°C, as shown in Figure 3.

Observations of lithium depletion from the matrix of a composite made with Al- 3 wt.% Li (approx. 10 at. %) and modifications to the fibre microstructure indicate that lithium behaves in a similar manner to magnesium.

3.2 C.P. Magnesium Matrix

The concentration profiles in Figures 4 and 5 demonstrate the increased reactivity of elemental magnesium. The profiles were determined from two composites, both cast at 700°C, containing LD fibres with slightly different microstructure/porosity distributions. Figures 6 and 7 show the corresponding as-cast microstructures, the overall fibre volume fractions being approximately 50 and 25% respectively. Figure 4 shows a large magnesium peak at the fibre surface (corresponding to the reaction zone in Figure 6) with significant magnesium penetration through the fibre. Figure 5 shows complete fibre reaction ie. the fibre is now MgO. These fully reacted fibres, which were also observed in the 50% fibre volume fraction composite, probably encountered pure magnesium early on in the infiltration process whereas the partially reacted fibres were probably in contact with liquid Mg-Al, due to the reduction of alumina elsewhere and hence show similar behaviour to the Al-Mg alloy in Figure 3.

IV-DISCUSSION

In all cases the extent of the reaction zone and also of magnesium penetration is much larger than predicted by simple diffusion calculations based on lattice diffusivity. It is therefore likely that other diffusion processes are taking place. Indeed, the grain boundary ($2 \times 10^{-17} cm^2/s$, assuming a 50nm crystallite size) and surface diffusivities ($10^{-15} cm^2/s$) of aluminium in alumina are more consistent with the observed penetration depths, being several orders of magnitude higher than the calculated lattice diffusivity. The presence of an interconnected pore structure in the LD fibres will therefore increase the propensity for melt penetration into the fibre.

The complete penetration of the porous eta-alumina fibres by magnesium contrasts with the reported formation of surface reaction layers, extending approximately 100nm into the fully dense alpha-alumina fibres. This may be explained by a combination of two factors. The alpha-alumina fibres do not contain the short circuit diffusion paths described in the eta-alumina fibres. Also the slightly greater stability of the alpha alumina may decrease the driving force for the reduction of alumina by magnesium to form magnesia. For example the free energies of formation of alpha and gamma alumina at 827°C are -885 and -876 kJ/mole oxygen respectively /22/. This small reduction in the thermodynamic driving force may be sufficient to inhibit reaction when the activity of magnesium is reduced by the presence of aluminium, as has been shown in Figure 1.

CONCLUSIONS

1. The results have demonstrated some of the effects of melt superheat, alloy chemistry and fibre microstructure. They show that some of the previous reports indicating negligible interaction (in Al-Mg based composites) are clearly incorrect.

2. Melt penetration as well as reaction zone formation must be considered in relation to fibre-matrix interaction. It is only in fully dense fibres, with relatively coarse crystallites that penetration and reaction zone formation are of equal importance.

ACKNOWLEDGEMENTS

The financial support of SERC and MOD and the provision of squeeze cast composites by I.C.I. are gratefully acknowledged.

REFERENCES

1. ET Turkdogan, "Physical Chemistry of High Temperature Technology" Academic Press: London (1980)
2. R Hultgren, "Selected Values of the Thermodynamic Properties of the Elements" ASM (1973) p180
3. E Dorre and H Hubner, "Alumina: Processing, Properties and Applications" Springer-Verlag: Berlin (1984)
4. CG Levi, GJ Abbaschian and R Mehrabian, Met. Trans. 9A (1978) p697

5 A Munitz, M Metzger and R Mehrabian, Met. Trans. 10A (1979) p1491

6. WD Johnston and IG Greenfield, in Proc. Int. Adv. Cast Reinf. Met. Comp. (eds. SG Fishman and AK Dhingra) ASM (1988) p355
7. GR Cappleman, JF Watts and TW Clyne, J. Mat. Sci. 20 (1985) p2159

8. TW Clyne, MG Bader, GR Cappleman and PA Hubert, J. Mat. Sci. 20 (1985) p85
9. J Dinwoodie and I Horsfall, in Proc. 6th Int. Conf. on Comp. Mats. (eds. FL Mathews et al) EAS: London 1987 p2.390
10. K Suganuma, T Okamoto, T Hayami, Y Oku and N Suzuki, J. Mat. Sci. 23 (1988) p1317
11. RA Page, JE Hack, R. Sherman and GR Leverant, Met. Trans. 15A (1984) p1397
12. JE Hack, RA Page and R Sherman, Met. Trans. 16A (1985) p2069

13. A McMinn, RA Page and W. Wei, Met. Trans. 18A (1987) p273

14. A Magata and IW Hall, J. Mat. Sci. 24 (1989) p1959

15. AK Dhingra, Phil. Trans. Royal Soc. Lon. A 294 (1980) p411

16. ibid p559

17. IW Hall and V Barailler, in Proc. ECCM1 (eds. AR Bunsell et al) EACM (1985) p589
18. IW Hall and V Barailler, Met. Trans. 17A (1986) p1075

19. J England and IW Hall, Scripta Met. 20 (1986) p697

20. WH Hunt, in "Interfaces in Metal Matrix Composites" (eds. AK Dhingra and SG Fishman) AIME (1986) p3
21. S Fox and HM Flower, in Proc. EUREM88 (eds. PJ Goodhew and HG Dickinson) Inst. of Physics (1988) Vol. 2 p435
22. DR Stull, H Prophet et al, "JANAF Thermochemical Tables" 2nd edition NSRDS NBS 37 (1971)

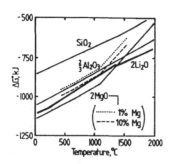 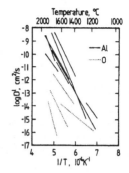

Figure 1 Free energy of formation Figure 2 Lattice diffusivity (Al₂O₃)

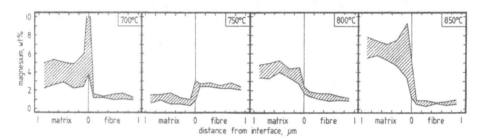

Figure 3 Magnesium concentration profiles in LD/Al-10 wt.% Mg composite

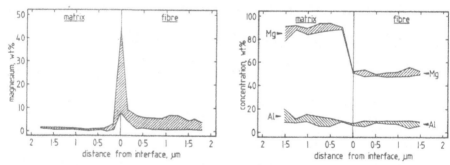

Figure 4 LD/C.P. Mg conc. profile Figure 5 As Fig.4, modified LD fibre

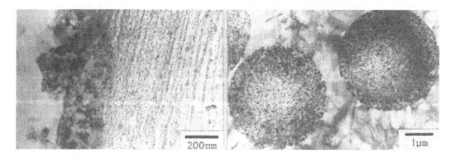

Figure 6 TEM of interface, Fig. 4 Figure 6 TEM of fibres as in Fig. 5

CHARACTERIZATION OF THE SQUEEZE CAST
M6-ALLOY AS41 WITH AL$_2$0$_3$ FIBERS

H. TELLESBÖ, S. GULDBERG*, H. WESTENGEN*, T.E. JOHNSEN*

Senter For Industriforskning
Forskningsv. 1 - Oslo - Norway
Norsk Hydro - 3901 Porsgrunn - Norway

Abstract

The magnesium alloy AS41 reinforced with Al_2O_3 have been through extensive microstructure investigations, mechanical and physical testing. A reinforcement of 20 % fibers raises the ultimate tensile strength (UTS) by 40 % at roomtemperature (RT) and 100 % at 250 °C. The measured elastic modulus is dependent on the test method and the test parameters selected. Reinforced AS41 has a steady state creep rate in an order of magnitude lower than unreinforced AS41 at 250 °C and 25 MPa load. Thermal expansion is measured to be 18×10^{-6} °C^{-1} for reinforced material. The thermal conductivity of the composite is in the range 50 -55 W/m°C at RT increasing to above 60 W/m°C at 500 °C as determined by a heat pulse method.

1. Introduction

The object of this work has been to investigate the potential of a Mg-based alloy for use in composites at elevated temperatures for applications in automotive engine parts. The alloy AS41 (~4 Al, ~1% Si) is selected due to its good high temperature properties especially for creep /1/, and the relative low cost. The composite material of this alloy reinforced with Al_2O_3-fibers has been produced by squeeze casting. In this paper the results of the evaluating experiments are reported, i.e. characterization of the microstructure, mechanical and physical testing.

2. Material

Preforms of δ-Al_2O_3 Saffil-fibers with an aspect ratio of 3:69 were used as reinforcement. The fiber density was 20 %, and the fibers were oriented randomly in planes parallel to the floor of the die (2D) (see Figure1). 1.2 wt% SiO_2 was used as a binder. Two different geometries of the preform, circular and square were tried in the prelimenary experiments. The circular preforms had a diameter close to 100 mm and a height of 15 mm. The square ones had dimensions of 69x69x17 mm^3. Large series of composites were produced at a larger press using square preforms with dimensions of 100x100x20 mm^3 in order to have enough test materials.

3. Casting parameters

The composites were in a prelimenary study produced by squeeze casting using a cylindrical die of diameter 10 cm. A preform temperature of 800 °C and a tool temperature of 300-350 °C were used. The investigations showed that a melt temperature of about 770 °C combined with a solidification pressure of 75 MPa eliminated visible pores in the material. If the pressure was higher than 75 MPa, the melt temperature could be less than 770 °C and vice versa to give a dense material. The height of unreinforced metal above the preform had to be more than 1.5 cm to avoid deformation of the preforms near the sides of the die. The process parametres in the later castings were the same with exceptions of a higher pressure and a slighter lower melt temperature.

4. Microstructure

A careful characterization of the microstructure in the composite specimens has been carried out by optical microscopy and Scanning electron microscopy (SEM). Mn-rich particles are observed in a layer just above the reinforced part of the casting. Primary Mg_2Si-particles are found in the reinforced part with the highest concentration near the bottom of the casting. Eutectic particles of Mg_2Si are common both in the reinforced and unreinforced part. Another phase, $Mg_{17}Al_{12}$, are present, mainly near the bottom of the casting. Oxide films are found at the corners and the sides of the preforms. In the case of circular preforms some cracks are observed in the fiber preform.

The location of Mn-particles, oxide films and Mg_2Si-particles give an indication of the melt flow through the preform during infiltration. Standard AS41 alloy contains some manganese to combine iron into primary Mn-Fe-particles. These particles have a diameter of about 10 µm, and are filtered from the melt as it is squeezed into the preform. The Mn-rich particles are therefore located in the regions where the melt has entered the preform. The oxide films are on the other hand located in the regions where the melt has flowed out of the preform. The SiO_2-binder is dissolved by the melt, and a melt with high Si-consentration is transported towards the bottom. Mg_2Si-particles are formed by reaction between the melt and the binder. The cracks found in the circular preforms are probably due to the infiltration path of the melt. Such preforms are probably only filled from the top. The square preforms, on the other hand, are infiltrated both from the sides and the top of the preform so that stresses are balanced. The test specimens for mechanical and physical testing are machined from the castings with the square preforms in order to avoid cracks.

5. Testing

5.1 Hardness

For unreinforced AS41 the Vickers hardness value in HV_1 [kg/mm^2] is about 50. HV_1 for the composite was 120. There is a slight increase in hardness towards the bottom of the composite as a result of the precipitated Mg_2Si and $Mg_{17}Al_{12}$.

5.2 Ultimate tensile strength

The UTS is measured in directions parallel with the fiber plane See Figure1). The tensile tests have been carried out at RT, 150 °C, 200 °C and 250 °C for both unreinforced and reinforced material. The tensile test results are shown in Figure 2. The points in Figure 2 represent the mean value of the two or three best test parallels. The curves of reinforced and unreinforced AS41 show similar trends. The curve for reinforced material is 80 MPa higher than the unreinforced curve at room temperture. At 250 °C the difference between the two curves has increased to 100 MPa, i.e. 100 % increase in tensile strength by reinforcing AS41 with fibers. The fracture elongation for reinforced material is about 1% at all temperatures.

5.3 Elastic modulus

Young's modulus has been investigated using both tensile testing, rheological three point bending /2/ and vibrating reed. The material has been tested both perpendicular to and in fiberplane using the latter two methods. The former test method has only been carried out in fiberplane. All measurements of the reference material and the composite have been carried out at RT. The results of the three different tests are summarized in Table 1. The tensile test results show that the elastic modulus of the composite material decreases considerably with increasing applied tensile strength compared with the reference material. This can be explained by internal stresses in the material caused by differences in the thermal expansion of the fibers and the matrix during cooling of the composite. The results from tensile testing are measured by cycling between 0 and 40 MPa. The measurements are taken during reloading. This gives an increase in Young's modulus from 49.8 GPa for unreinforced AS41 to 77.7 GPa for AS41 with fibers. The dynamic three point bending tests are calibrated by setting the elastic modulus of pure magnesium to 45 GPa. The tests are run with a small preload and a small sinoidal oscillating load. The measurements have been carried out at different frequencies of the oscilltaing load, from 6×10^{-2} Hz - 32 Hz. The elastic modulus in Table 1 is given for a frequency of 2 rad/s (6×10^{-1} Hz) and a strain rate of 10^{-3} s^{-1}. The elastic modulus is calculated from the force needed to oscillate the material with the given frequency and strain rate. The results show that the elastic modulus increases sligthly with increasing frequency. The elastic modulus measured in this way gives 59 GPa for unreinforced material and 81-83 GPa for reinforced material dependent of the orientation of the fibers. These values are slightly higher than those measured by tensile testing. This can to some degree be explained by the lower applied forces and less relaxation when using the three point bending method. The same phenomenon is mentioned by Busk /3/. The vibrating reed experiments have been carried out by sending sound waves towards the test pieces and measuring the resonance frequency. The resonance frequency are around 1000 Hz for both reinforced and unreinforced material. This testing method gave lower values of the elastic modulus than the other two methods The measured value of unreinforced AS41 was 38.8 GPa. The values of reinforced AS41 were 51.7 - 58.0 GPa. The conclusion is that the elastic modulus is dependent on the testing method.

5.4 Creep properties

Creep properties have been tested at 150, 250 and 300 °C. At 150 °C and tensile stress of 50 MPa both unreinforced and reinforced AS41 have no visible creep strain during 100 hours. This material behaviour is quite good compared with the value of 0.2 % creep strain of an AS41 material reported at 150 °C, 40 MPa and 100 hours in Busk /3/. The difference is probably due to larger grain size and thereby longer diffusion paths in our material than in the material reported by Busk. The results of the creep measurements carried out at 250 °C and 25 MPa load are shown in Figure 3. The creep properties of reinforced and unreinforced AS41 are compared to the most common magnesium alloy, AZ91. The steady state creep rate of reinforced AS41, unreinforced AS41 and AZ91 are as follows: 1.4×10^{-9} s^{-1}, 12.5×10^{-9} s^{-1} and 20.2×10^{-9} s^{-1}. At 300 °C the composite material withstood a load of 25MPa in 1000 hours without failing. The conclusion is that the reinforced AS41 has excellent creep properties compared with the other magnesium material tested.

5.5 Stress relaxation

Testing of stress relaxation has been carried out at 150 °C on sylinderical test pieces of diameter 12 mm and height 5 mm and with an applied load of 80 MPa. After 100 hours the remaining load is respectively 80 % and 90 % for unreinforced and reinforced material. For unreinforced high pressure die cast AS41 the corresponding stress relaxation is reported to be 50 % /3/. This lower value can be explained, as for creep strain, by a smaller grain size.

5.6 Thermal expansion

The thermal expansion is measured by means of a Perkin-Elmer TMA7. The specimens are heated from 10 - 300 °C at a rate of 2 °C/sec. The expansion coefficient is measured in two directions; one parallel with the fiberplane and another perpendicular to fiberplane. For unreinforced AS41 the thermal expansion coefficient is 24×10^{-6} °C^{-1}. Corresponding values for the reinforced material are 21×10^{-6} °C^{-1} and 18×10^{-6} °C^{-1} perpendicular to fiberplane and with fiberplane respectively.

5.7 Thermal conductivity

The thermal conductivity is measured indirectly using a heat-pulse method. The conductivity was calculated from the heat diffusivity values of the material. The conductivity for the composite material is in the range 50-55 W/m°C at RT increasing to above 60 W/m°C at 500 °C.

6.Conclusion

A proper designed tool is required in order to avoid cracking during infiltration of melt into the preform. An substatial increase in UTS , 40 - 100 %, is obtained by adding fibers.The measured elastic modulus is sensitive to the testing method. Thus a description of testing method is required. The elastic modulus of the reinforced AS41 alloy is 10 - 15 % higher than the modulus of commercial aluminum alloys when measuring by the three point bending method.The creep properties and stress relaxation properties are increased by fiber reinforcement. Reinforced AS41 has a steady state creep rate in an order of magnitude lower than unreinforced AS41 at 250 °C and 25 MPa load. The thermal expansion coefficient is reduced by the additon of fibers to a level below that for unreinforced aluminum alloys. The thermal conductivety is only slightly reduced by fiber addition.

The test results of the mechanical and physical properties of the composite investigated, indicate that the AS41-alloy has a high potential as base alloy in composites.

7. References

/1/ "Magnesium casting alloys"
Adr.: Norsk Hydro,Magnesuim Division, Bygdøy alle 2, P.Ø.Box 2594 Solli, N-0203 OSLO 2, NORWAY.

/2/ Tor Einar Johnsen and Tone D. Selnæs: "Elastic modulus measurements in metal matrix composites". To be published in "Fundamental Relationships Between Microstructure and Mechanical Properties of Metal Matrix Composites", Proceedings of a Symposium held at the TMS Fall Meeting, Indianapolis, Indiana, 2.-5. October 1989.

/3/ Robert S. Busk: "Magnesium products design", page 240-241 and 274.
   Marcel Dekker, Inc., 270 Madison Avenue, New York,New York10016,
   1987.

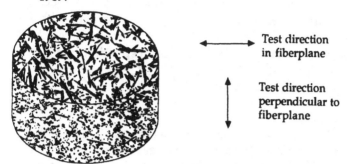

Figure 1.
The orientation of fibers in the composite. The two test directions are indicated.

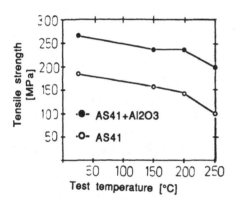

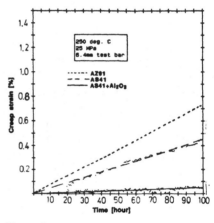

Figure 2.
Tensile strength [MPa] as a function of
temperature [°C] for AS41 and reinforced
AS41.

Figure 3.
Creep strain [%] as a function of time
[hours] for reinforced AS41 compared
with unreinforced AS41 and AZ91.

Table 1
The elastic moldulus for unreinforced and reinforced AS41 measured with three different
methods.

| Material | Test method | Youngs modulus [GPa] | |
		Test direction perpendicular fiberplane	Test direction parallel with fiberplane
AS41	Tensile testing		50
AS41	Three point bending		59
AS41	Vibrating reed		39
AS41+Al2O3	Tensile testing		77
AS41+Al2O3	Three point bending	83	81
AS41+Al2O3	Vibrating reed	52	58

THE INTERFACIAL ROLE OF SILICA ON CERAMIC FIBRE - REINFORCED ALUMINIUM ALLOY COMPOSITES

T. STEPHENSON, Y. LEPETITCORPS, J.M. QUENISSET

Laboratoire de Chimie du Solide
Université de Bordeaux I - 351 cours de la Libération
33405 Talence - France

Silica present on the surface of ceramic fibres plays a critical role in the mechanical behaviour of composites due to fibre-matrix (F/M) interaction. This study has examined reaction kinetics and diffusion phenomena in the $SiO_2/Al(Mg,Cu)$ system between 400-800 °C. Magnesium strongly affects the rate of reaction. Apparent activation energies in the solid-liquid state were 8-74 kcal/mol depending on alloy composition. Analysis by TEM showed limited reaction between silica fibres and metal matrices for short liquid infiltration times. The F/M interface was found to be mechanically discontinuous.

INTRODUCTION

The implementation of metal matrix composites has lagged behind initial expectations because of the formation of brittle phases at the F/M interface which degrades ceramic fibre strength and impairs composite ductility. The interfacial bonding between the reaction zone and the fibre strongly affects overall composite strength. For instance, a weak bond can act to deflect or arrest crack propagation, whereas a strong bond acts as a stress concentrator.

The nature and thickness of the phases produced from F/M interaction are therefore extremely important in maintaining composite ductility, while permitting sufficient load transfer from the matrix to the fibres. The formation of an interphase in aluminium alloys reinforced by ceramic fibres is strongly dependent on the processing conditions. These phases can be predicted thermodynamically, however kinetic phenomena associated with phase nucleation and solid state diffusion are important criteria for the non-equilibrium conditions encountered in composite processing.

Ceramic fibre reinforcements typically have a surface layer rich in silica. This layer may either be inherent, as in the case of silicon carbide, or introduced during fibre manufacturing to control grain size, phase stability or enhance wetting properties in the case of alumina or carbon fibres. The reaction of silica with aluminium and its alloying elements is thus the controlling factor in the chemical and resulting micromechanical behaviour at the F/M interface of these composites.

KINETICS AND PHASE IDENTIFICATION

Reaction kinetics and diffusion phenomena have been studied in the $SiO_2/Al(Mg,Cu)$ system between 400-800 °C. The nominal alloy compositions are noted in Table 1.

A model system was chosen for solid-liquid kinetic experiments where a fused silica rod (Φ = 5 mm) was immersed into liquid alloys (670-800 °C) protected by a purified argon atmosphere to prevent excessive loss of Mg /1/. Reaction times were limited to between 5-30 mn. The samples were solidified, sectioned radially and analysed by x-ray μprobe and x-ray diffraction. The interphase in $Al-Mg/SiO_2$ composites consisted exclusively of $MgAl_2O_4$ spinel except in high Mg-content alloys (10 at%), when a thin (1-2 μm) MgO phase appeared at the unreacted SiO_2/oxide product interface. This agrees favourably with some authors' work on $Al_2O_3-SiO_2$ fibres /2,3/. $Al-Cu/SiO_2$ composites were markedly less reactive . The reaction zone consisted of Al_2O_3 with small amounts of $CuAlO_2$ or $CuAl_2O_4$.

Arrhenius plots in Fig. 1 show the change in reaction rate as a function of the alloying element and its concentration in the matrix. Apparent activation energies were determined for each system and are listed in Table 2. The value for Al/SiO_2 is 53.5 kcal/mol which is higher than previously reported results (38-44 kcal/mol) /4/. This difference is possibly due to the choice of atmosphere (air, vacuum), which has a strong effect on the oxide film thickness on the liquid metal surface.

In the solid state, samples were prepared by hot pressing aligned 50 μm SiO_2 filaments (Heraeus) between thin discs of the above alloys at 500 °C for 2 h. These "composites" were then heat treated (400-600 °C) under flowing argon. The conditions necessary to weld the alloy discs together resulted in a substantial reaction with Al-Mg alloys. Elemental profiles performed by x-ray microanalysis showed that the filaments were largely transformed into $MgAl_2O_4$ spinel. Heat treatment above 550 °C further changed the filament composition to MgO. Al and Al-Cu alloys exhibited a sharp interface with SiO_2, only forming an alumina reaction product above 550 °C (12 μm after 100 h).

Samples were prepared for transmission electron microscopy analysis by infiltrating silica wool ($\Phi=9\mu m$) with the same alloys at 700 °C (contact time =1 mn). Discs ($\Phi=3mm$) were prepared from sections cut perpendicular to the infiltration direction, mechanically polished and dimpled, then ion (Ar) beam milled (cold stage, θ = 10-15 °). Semi-crystalline regions were often found indicating products of the reduction of SiO_2 fibres. Analysis by STEM and EDS revealed the simultaneous presence of Si, Al and Mg or Cu. In an Al-10%Cu/SiO_2 composite where a distinct interphase could be identified, the separation of the Al_2O_3-SiO_2 interphase from the fibre was evident. The role of the interface between the interphase and the fibre constitutes the second part of this study.

MECHANICAL BEHAVIOUR OF THE INTERPHASE/FIBRE INTERFACE

Adhesion tests in metal matrix composites which determine F/M interfacial shear stress are either difficult to realise or interpret due to phenomena such as residual thermal stresses at the interface, plastification and work hardening of the matrix. Fragmentation tests performed on SiO_2-covered SiC monofilaments in aluminium matrices /5/ indicated that load transfer at the F/M interface is such that the related shear stress is higher than the yield strength in shear (σ_E /2) of the matrix. As a consequence, the fragmentation model proposed by Fraser et al. /6/ is no longer valid in this case.

A simple test is needed to provide a measure of F/M bonding for metal matrices having a low σ_E. The toughness of the F/M interface can be considered as a quantitative measure of interfacial bonding in terms of G_{Ic}, the energy required to separate the two surfaces. Stress concentration is necessary at the interface to facilitate F/M debonding and may be accomplished using bi-material notched specimens.

In order to establish a coherent bond between aluminium and oxide materials at relatively low temperatures, the alumina film on the metal surface must be ruptured so that clean metal is in contact with the glass or ceramic surface. Attempts to create diffusion couples between planar fused silica and aluminium alloys via solid-liquid processing have been unsuccessful. An "all or nothing" type of interaction occurs depending on the processing conditions. When sufficient time is allowed for a reaction to take place, the interface ruptures either during growth of the reaction zone or upon solidification.

Rupture may occur during interphase growth because of the difference in molar volume between silica and the oxide products of reaction. Examination of SiO_2/Al alloy composites by optical microscopy (Fig. 2) reveals extensive radial cracking of the oxide interphase. These oxides are typically very fragile such that microcracks are able to initiate from defects at the fibre surface. M.

Sakai et al. /7/ found the tenacity K_{Ic} (CHV) of polycrystalline $MgAl_2O_4$ to vary between 1.29-1.95 MPa m$^{1/2}$. It is reasonable to assume that oxygen is rigid compared to the mobility of the metal cations, and that all oxygen in the interphase originates from silica. A simple calculation based on an equivalent number of oxygen atoms in a given volume of silica and an oxide product phase, shows that, for example, the spinel $MgAl_2O_4$ has a 27% smaller volume (Table 3).

This difference gives rise to circumferential and longitudinal tensile stresses at the silica/interphase boundary. As the interphase continues to grow, the replacement of thin silica layers of the fibre progressively relieves the tensile stress in the interphase, although previously initiated microcracks continue to grow. A critical thickness of the interphase is reached at which a radial tensile stress develops at the interphase-fibre interface allowing microcrack propagation and debonding occurs.

Aluminium has been bonded to fused silica and alumina in the solid state between 520-640 °C /8/. As the metal must be deformed plastically to form a bond, the couples were prepared by sandwiching aluminium discs between fused silica bars. Similar samples have been prepared in this laboratory at 630 °C under purified argon flow for 1-3 h. Fracture has been observed to take place at the reaction zone-silica interface. This implies weak bonding between the interphase and silica. Thus, in the absence of residual thermal stresses, debonding should occur at the interphase/fibre interface during deformation of the composite and prevent crack tip growth.

CONCLUSIONS

Silica is very reactive in contact with aluminium alloys in the liquid state, particularly those containing magnesium. Apparent activation energies determined between 670-800°C were 8-74 kcal/mol depending on alloy composition. Kinetic analysis and confirmation by TEM indicate little interaction for the short contact times encountered in the infiltration of ceramic fibre preforms. Subsequent annealing above 450 °C results in significant fibre degradation in the case of Al-Mg alloys. It is unlikely that aluminium alloys reinforced by ceramic fibres containing silica have a mechanically continuous F/M interface. Load is transferred from the matrix to the fibres by frictional forces resulting from residual stresses induced by solidification of the metal and molar volume changes with the growth of the oxide reaction zone.

REFERENCES

1. T. Stephenson, Y. LePetitcorps, J.M. Quenisset, Proceedings E-MRS Symposium B: Metal Matrix Composites, May 29-June1 1990, Strasbourg, France ,to be published

2. G. R. Cappleman, J.F. Watts, T.W. Clyne, J. Mat. Sci, 20 (1985) 2159-2168

3. A. Munitz, M.Metzger, R. Mehrabian, Met. Trans. A, 10A (1979) 1491-1497

4. K. Prabriputaloong, M.R. Piggott, J. Amer. Cer. Soc., 56 (1973) 184-185

5. T. Stephenson, Y. LePetitcorps, R. Naslain, Revue de Metallurgie, No.9 (1988) 492

6. W.A. Fraser, F.H. Anker, A.T. Dibenendetto, 30th Tech. Conf. 1975, Reinf. Plast./Comp. Instit., The SPI Inc., Section 22-A

7. M. Sakai, R.C. Bradt, A.S. Kobayashi, Nippon Ser. Kyo 96 [5] (1988) 525-31

8. J.T. Klomp, A.J.C. Van de Ven, J. Mat. Sci. 15 (1980) 2483-2488

FIGURES AND TABLES

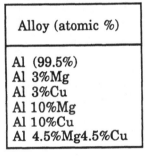

Alloy (atomic %)
Al (99.5%)
Al 3%Mg
Al 3%Cu
Al 10%Mg
Al 10%Cu
Al 4.5%Mg4.5%Cu

Table 1. Alloy compositions. Major impurities: Fe <0.2 mass%; Si <0.1 mass%.

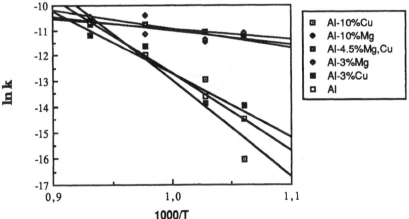

Figure 1. Arrhenius plots for silica/aluminum alloys (670-800°C).

Table 2. Calculated apparent activation energies.

Alloy/Silica	Ea/2 (kcal/mol)
pure Al	53.5
Al 3%Mg	11.0
Al 3%Cu	49.4
Al-10%Mg	15.1
Al 10%Cu	73.9
Al 4.5%Mg4.5%Cu	8.0

Table 3. Relative molar volume changes for oxide product phases.

Phase/Structure	Oxygen Density ($Å3$/atom O)	Volume Change (V-Vo)/Vo (%)
Silica amorphous	22.67	Vo
alumina (trigonal)	16.34	-27.9
alumina (fcc)	20.54	-9.4
alumina (triclinic)	15.74	-30.6
$MgAl2O4$ (fcc)	16.5	-27,0
MgO (fcc)	18.7	-17.4
$CuAl2O4$ (fcc)	16.49	-27.3

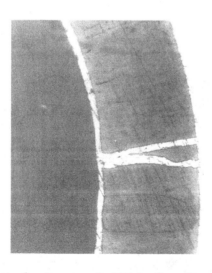

Figure 2. Optical micrograph (75x) of a silica (left)/Al-10%Cu sample (800 °C 15mn) showing crack initiation and phase separation at the spinel/silica interface.

MICROSTRUCTURAL STUDIES OF ALUMINA FIBRE REINFORCED ALUMINIUM ALLOY

M. YANG, V.D. SCOTT

University of Bath - School of Materials Science
BA2 7AY Bath - Great Britain

The microstructure of alumina fibre reinforced Al-7wt%Si alloy has been studied. The eutectic structure which characterised this alloy was changed by the presence of fibres, with silicon and intermetallic phase tending to segregate on the fibres. The coarse silicon precipitates exhibited twinning but no epitaxial relationship with the aluminium, whilst fine silicon precipitates had a cube-cube orientation with the aluminium lattice. Lath-like intermetallics, $FeSiAl_5$ and $FeSi_2Al_4$ with monoclinic and tetragonal structures, existed in equilibrium with a defined epitaxial relationship. Chemical reaction between fibres and matrix was not observed, although magnesium segregation at the fibre surface occurred. Dislocation substructures in the aluminium matrix were also studied.

1. Introduction

This study is a part of ongoing research programme aimed at optimising the properties of ceramic fibre reinforced aluminium alloys manufactured by the liquid metal infiltration method. It involves a detailed study of the microstructure of the composite, particularly the matrix and interface and a parallel assessment of its mechanical behaviour. Here, we report an investigation of an Al-7wt.% Si alloy reinforced with alumina fibres.

2. Experimental Procedure

The composite used in this study consisted of 41vol.% continuous Safimax δ-alumina fibres unidirectionally aligned in an Al-7Si-0.2Mg alloy. It had been produced by means of liquid metal infiltration of a fibre preform. Polished surfaces of the composite were analysed using optical and scanning electron microscopy (SEM) combined with energy dispersive spectrometry (EDS). Specimens for transmission electron microscopy (TEM) were prepared by slicing 3mm diameter discs from the bulk sample. The disc was dimpled, and ion milled at a voltage of 5kV until perforation. TEM examination was carried out at 200keV using instrument fitted with a high angle, thin window EDS detector.

3. Results and Discussion

3.1 Matrix microstructure

Optical microscopy of the composite revealed that the fibres were not perfectly aligned, fairly large regions of unreinforced matrix separating the layers of fibre which composed the preform. Infiltration was good with a very low level of porosity. Coarse silicon lamellae were observed in fibre-free regions, which appeared to grow radially from fibres. Where fibres were closely packed, silicon particles were smaller, their size roughly equating with the inter-fibre spacing. Observations of HF acid etched surfaces showed the majority of fibres and intermetallics residing at the boundary of the primary aluminium grains. The grain size was reduced by more than 20 times when compared with that of the unreinforced alloy cast under similar conditions. Hence not only do fibres have a pinning effect on the growth of aluminium dendrites, but also they act as favourable nucleation sites for the growth of silicon particles.

TEM observations show that silicon lamellae are characterised by twinning, Fig. 1, diffraction analysis indicating {111} habit planes with twins related by a rotation of 180° about the [110] axis. The lamellae have their longitudinal axis parallel to [112] and within the {111} plane. Fine silicon precipitates of polyhedral morphology are present in a very low density. An example in Fig. 2 indicates that the silicon particle shares a common [111] zone axis and has a cube-cube orientation relationship with the aluminium matrix. The 4% lattice misfit between aluminium and silicon gives rise to some dislocations in the surrounding matrix.

Fig.3 shows crystals bridging two fibres. The interface between the crystals and fibres is well defined without any evidence of chemical reaction. The crystals labelled A and B with the common planar interface were found by EDS to contain iron, silicon and aluminium, crystal A having the higher iron : silicon ratio. Lattice fringe images, taken from the marked area in Fig. 3, are presented in Fig. 4. The lattice spacings of A and B were measured to be 1.04nm and 0.47nm, respectively. Diffraction results combined with EDS data indicated that phase A was $FeSiAl_5$ with a monoclinic structure (a = 0.612nm, c = 4.15nm and β =91^0), and phase B was $FeSi_2Al_4$ with a tetragonal structure (a = 0.609nm and c = 0.944nm). They are aligned epitaxially as follows;

$$(001)_{tet} \, // \, (001)_{mono} \text{ and } [100]_{tet} \, // \, [100]_{mono}$$

corresponding to a misfit of 0.5% at the interphase plane. Ledges of several atomic layers were found at the interface between the two crystals. It may be argued that the tetragonal phase formed initially, and then, upon cooling the composite from the infiltration temperature, underwent a diffusional transformation to the monoclinic form by a ledge mechanism. Given adequate time at temperature, it is likely that all the tetragonal phase would have been converted into the monoclinic form.

3.2 Interface structure

Evaluation of the fibre/matrix interface of the composite is complicated by the multiphase nature of the matrix. The majority of silicon and iron-rich intermetallics precipitated directly on fibres and the resultant interfaces were well defined with no evidence of chemical reaction. Some interfacial porosity was present, resulting from poor metal infiltration of closely packed fibres. There was also some evidence of fibre/matrix decohesion induced by excess shrinkage of the matrix upon cooling the composite from the fabrication temperature.

Fig. 5a shows an interface in which dislocations in the aluminium matrix are generated from the fibre surface. The presence of the dislocations has made it difficult to discern any phase segregated at the interfacial region, but there appears to be some fine-scale structural transition adjacent to the fibre. Whilst electron diffraction analysis did not reveal any new phase, EDS results showed a detectable increase in the magnesium content, Fig. 5b. This result is consistent with previous observations/1/ whereby magnesium had segregated on the fibre surface as complex oxide compounds. This segregation is considered to be beneficial from the processing point of view, since it reduces the surface tension of the molten aluminium alloy and promotes the wettability.

3.5 Dislocations

Dislocations are present, particularly in the regions of the aluminium matrix close to fibres and precipitates. The dislocation dipoles and loops, Fig. 6a, are characteristic of strain hardening, whilst network and low angle boundaries, Fig. 6b, are indicative of thermal recovery processes. Hence, many dislocations produced by thermal contraction stresses in the early stage of the cooling process are able to climb and interact to reduce the overall dislocation density and associated energy. As the composite temperature falls further, thermally activated processes will be restricted and dislocation density will increase more rapidly. An estimate of the dislocation density in areas such as that shown in Fig. 6a was ~ 4 x 10^7mm^{-2}, typical of small strains ~0.5%. Taking thermal expansion coefficients of 23 x 10^{-6}per ºC and 8 x 10^{-6} per ºC for aluminium and alumina fibre, respectively, a misfit parameter of 0.3% between the fibre and aluminium matrix is estimated by assuming below a temperature of ~200C no stress relaxation can occur. Hence, the experimentally measured value of the plastic strain does not disagree with that calculated.

4. Conclusions

Fine diameter alumina fibre preforms can be infiltrated by molten Al-Si alloy with the aid of external pressure to give a composite with only very limited porosity. The eutectic structure which characterised this alloy was modified by the presence of fibres, with silicon and intermetallic phases tending to segregate on the fibres. The resultant interface formed between the fibre and second phase was well defined with no evidence of chemical reaction. The small amount of magnesium added to the alloy was found to segregate at the fibre surface, possibly dissolved in the surface regions. This segregation is considered to have a beneficial effect on the infiltration process by reducing the surface tension of the molten alloy and enhancing the wettability. Limited porosity at the interface was observed and evidence of fibre matrix decohesion was found. Finally, the morphology and crystallography of second phases were established.

References

1. T.W.Clyne, M.G.Bader, G.R.Cappleman and P.A.Hubert. J. Mat. Sci., 20 (1985) 85-96.

Acknowledgment to SERC and MOD for support.

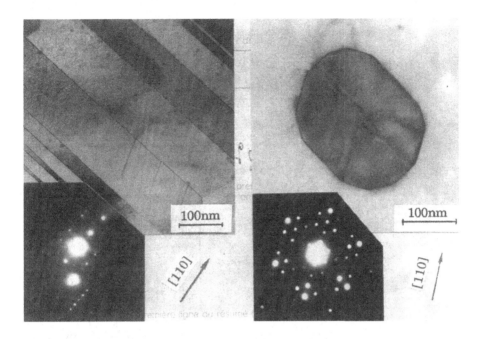

Fig. 1 Silicon twins with {111} habit planes, TEM.

Fig. 2 Silicon and aluminium sharing the common [111] zone zxis, TEM.

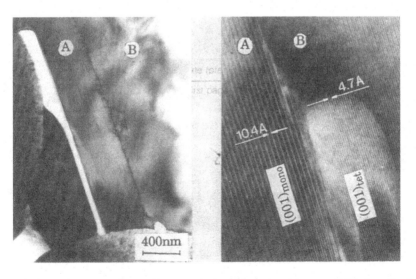

Fig. 3 Iron-rich intermetallics adjacent to fibres, TEM.

Fig. 4 Lattice image of the marked area in Fig. 3, TEM

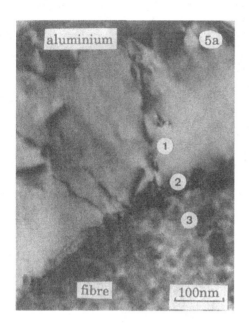
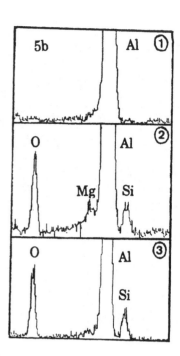

Fig. 5a Interface between fibre and aluminium matrix.
5b EDS data showing increase in Mg level at the interface region.

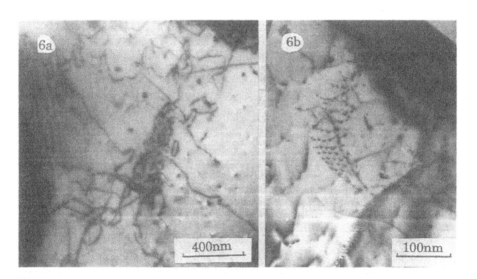

Fig. 6a Dislocation loops and dipoles.
6b Recovery structure with dislocation networks and low-angle grain
boundaries.

ON INTERFACES IN C-FIBRE REINFORCED MAGNESIUM ALLOYS

A. KLEINE, H.-J. DUDEK, G. ZIEGLER*

DLR - Institute for Materials Research
PO Box 90 60 58 - 5000 Köln 90 - West-Germany
*University of Bayreuth
P.B. 101251 - 8580 Bayreuth - West-Germany

Abstract

Transmission electron microscopy (TEM) and energy dispersive X - ray micro analysis (EDS) have been applied to characterize the fibre - matrix interfaces in P100 and M40 graphite (Gr)/AZ61 Mg alloys respectively. In both composites areas of exfoliation between fibre and matrix were observed. Coarse precipitates averaging a few μm in size and enriched with Al were formed at the interfaces could mostly be identified as $Mg_{17}Al_{12}$. Reaction products or reaction zones were not observed in both composites.

1. Experimental

The AZ61 (Mg 93 At%, Al 5 - 6 At%, Zn 0.5 - 1 A%) alloy reinforced with P100 (Union Carbide) and M40 (Toray) graphite fibres was processed by ONERA / Chatillon by pressing metallic alloy foils and fibre layers at 630 °C and 13.5 MPa processed by liquid hot pressing method. Fabrication details will be discussed elsewhere / 1 /. The composites had a fibre volume fraction of 50% with fibres oriented mainly in the pressing plane. The interface analysis was performed mainly using an analytical TEM operating at 300 KV (Philips 430) attached with a Tracor 5000 EDS - System. Dimple grinding prior to ion milling was applied to prepare thin TEM foil /1, 3, 4/. The fibre axes of the TEM specimens were orientated parallel to the foil surface.

2. Results

2.1 P100 Gr / AZ61 Mg Composite

Fig. 1 a shows a typical micrograph of the composite. The exfoliation between matrix and fibre was observed in different areas of the sample. It is assumed
that this phenomenon is related to poor wetting and
bonding between the fibre and the matrix / 2 / or to
preparation effects / 3, 4 /. In case of good wetting
and bonding between fibre and matrix a nearly smooth
interface was observed with some dislocation adjacent to
the interface (Fig. 1 b). Precipitates formed at the
interface shows a strong Al signal in the EDS - spectra
and were identified as $Mg_{17}Al_{12}$ by Selected Area- (SAD)
and Convergent Beam Electron (CBED) -Diffraction (Fig.
2). EDS - spectra (Fig. 3 a) very often prove a small
increase of Al content in the interfacial region in comparison to the baseline matrix (Fig. 3 b). There are no
extensive reaction zones or reaction products between
fibre and matrix observed.

2.2 M40 Gr / AZ61 Mg Composite

Exfoliation between fibre and matrix could also be
observed in some areas of these samples. In case of
good wetting and bonding between fibre and matrix a
smooth interface was observed and no reaction zone could
be detected. The matrix adjacent to the fibre often exhibits dislocations as well as planar or lath shaped
features which vary in diffraction contrast during specimen tilt. The micrograph in Fig. 4 is typical for
situations where the distance fibre / fibre is less than
the average diameter of one fibre. It is assumed that
these planar or lath shaped features are secondary phase
precipitates because the EDS - analysis shows much
higher Al content for some of these objects compared to
the AZ61 matrix. Precipitates of the $Mg_{17}Al_{12}$ - phase
(Fig. 5) and some coarse particles containing Al and Mn
were observed at the fibre matrix interface. Qualitative X - ray mappings (Fig. 6) performed in the TEM in
some instances display a small enrichment of Al at the
interface.

3. Conclusion

Microstructural TEM analysis shows that areas of no
wetting and bonding between fibre and matrix exist in
both composites. At present it is not clear whether this
exfoliation is related to a) the limited wetting and
bonding of molten Mg alloys with graphite fibres / 2 -
6 /, b) to the large differences of thermal expansion
between matrix, precipitates and fibres or c) due to

preparation effects /3/. Segregation of Al and impurity
elements in the interfacial region probably occur during
processing which favors the formation of coarse precipitates and an Al - enrichment at the interfaces in both
composites. Similar results were obtained by /3,4 /. It
is assumed that the fibres act as nucleation centers
where equilibrium metallic phases are formed. The number
of coarse precipitates (e.g. $Mg_{17}Al_{12}$) in the interfacial region is higher in the P100 /AZ61 - than in the
M40 / AZ61 - composite. On the other hand the planar
features and the dislocation structures typical for the
M40 /AZ61 composite could not be found in the P100 /
AZ61 composite. The formation of lamellar or spherical
shaped precipitates of the equilibrium $Mg_{17}Al_{12}$ phase in
the matrix region normally observed in Mg - Al - Zn
alloys / 7 / were not detected in both composites.
Currently it can not be excluded that these different
microstructures develop because of different fabrication
conditions and / or because of special treatments or
properties of the fibre materials. In both composites no
reaction products (e.g. Mg_2C_3, Al_4C_3) or reaction zones
between the fibre graphite and the matrix constituents
elements could be detected.

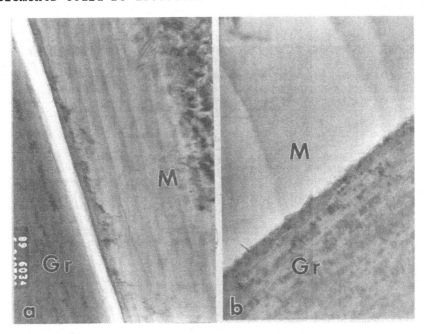

300 nm   400 nm

Fig. 1 a) and b) TEM bright field micrographs of P100
Gr / AZ61 composite (Gr = Graphite; M = Magnesium alloy)

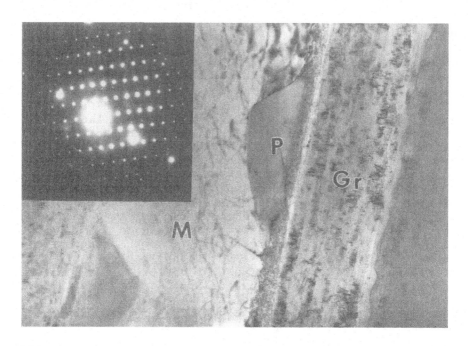

300 nm

Fig. 2) Interfacial region in P100 Gr / AZ61 Mg alloy.
The precipitate (P) is identified as $Mg_{17}Al_{12}$ (see SAD
pattern; zone axis is parallel [0 1 1]).

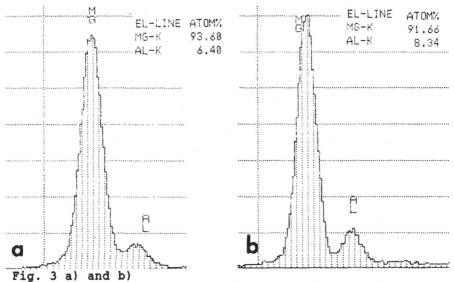

EL-LINE	ATOM%
MG-K	93.60
AL-K	6.40

EL-LINE	ATOM%
MG-K	91.66
AL-K	8.34

Fig. 3 a) and b)
EDS - spectra obtained from the matrix (a) and the
interface (b).

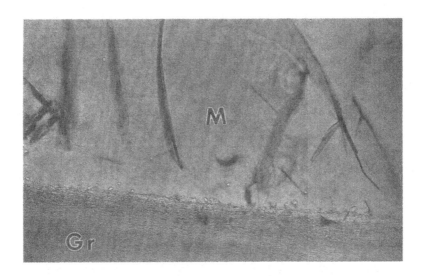

600 nm

Fig. 4) TEM bright field micrograph of M40 Gr / AZ61 composite (Gr = Graphite; M = Magnesium alloy).

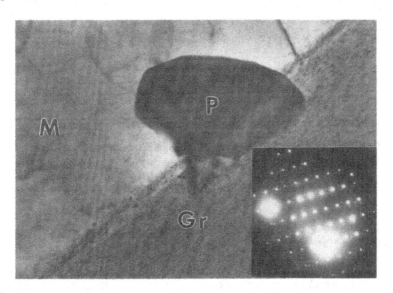

150 nm

Fig. 5) Interfacial region in P100 Gr / AZ61 Mg alloy. The precipitate (P) is identified as $Mg_{17}Al_{12}$ (see SAD pattern; zone axis is parallel [0 1 1]).

Fig. 6
EDS qualitative X - ray mapping scans covering inter-
facial regions in M40 Gr/ AZ61 composite. The mappings
shows an enrichment of Al in the interface region.
(Mg : Magnesium, Al : Aluminium, P : Particle in Fig. 5
Gr : Graphite)

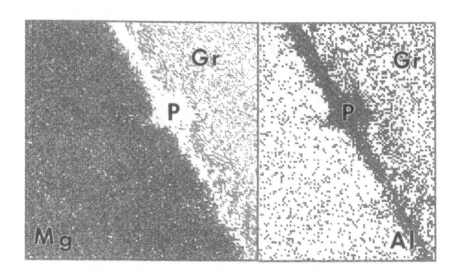

4. References

/ 1 / R. Mevrel et. al. in preparation
/ 2 / M.H. Richman et. al., Army Materials and
Mechanics Research Center, Watertown; Massachusetts;
1973, Distributed by: National Technical Information
Service, U.S. Department of Commerce
/ 3 / S.P. Rawal, L.F. Allard, 6th Intern. Conf.
on Composite Materials and 2nd European Conf. on
Composite Materials (ICCM & ECCM), London 1987,
169 - 182
/ 4 / S.P. Rawal, L.F. Allard, M.S. Misra, Interfaces in metal - matrix composites; New Orleans Loui-
siana, USA; 2 - 6 March 1986, 211- 225 Warrendale 1986
/ 5 / H.A. Katzman, Jour. Mat. Sci. 22 (1987)
144 - 148
/ 6 / A.P. Levitt, E. di Cesare, S.M. Wolf, Metallurgical Transactions (3), 1972, 2455
/ 7 / G.W. Lorimer, Magnesium Technology, The
Institute of Metals, 1987, 47 - 53

METAL MATRIX

Chairmen : **Prof. M.G. BADER**
University of Surrey
Mr. H. ABIVEN
Aérospatiale

MEASUREMENT AND ANALYSIS OF THE STIFFNESS AND BAUSCHNIGER EFFECT DURING LOADING AND UNLOADING OF A CARBON FIBER AND ALUMINUM MATRIX COMPOSITE

P. FLEISCHMANN, Y. KAGAWA*, R. FOUGERES

INSA de Lyon - GEMPPM UA CNRS 341 - Bât. 303
69621 Villeurbanne - France
*University of Tokyo - Inst. of Industrial sc.
22-1 Roppongi - Tokyo 106 - Japan

ABSTRACT

Plastic behavior of a randomly distributed short carbon fibers pure aluminum matrix composite has been investigated by means of mechanical measurements during cyclic tension tests and Acoustic Emission (AE) measurements. It is shown that the plasticity is exhibited in two stages: the first is related to interactions between dislocations and fibers. Bauschinger effect measurements indicates that the mean internal stress increases with the plastic strain. However, a full understanding of plastic strengthening requires to take into account the local internal stress near the fibers. The second stage is clearly related to fiber breaks.

INTRODUCTION

We have studied a metal matrix composite (MMC) with a commercial pure aluminum matrix (A1100) containing randomly distributed short carbon fibers. In opposition to MMC with a dispersion strengthened matrix, where precipitates have also to be considered, this material could be treated as a pure two constituent composite. As in thge two phases aluminum alloys, a Bauschinger effect can exist in such material and the objective of this study is to perform tests in order to obtain experimental evidences of this effect. For this purpose, repeated tensile tests were performed. We measured with a good accuracy the stress, the strain and the time in order to calculate precisely mechanical parameters as the stiffness. These measurements are analyzed with hardening and stiffness models. A complementary measurement, Acoustic Emission (AE), leads to informations about the micro-deformation events in the sample.

I - MATERIAL

The Al-C composite with 30% vol. C fibers (T300) in a commercial pure Aluminum matrix (A1100) was prepared by squeeze casting technic /1/. The diameter of the fibers were 7 μm and their average length was 100 μm. The aspect ratio l/d, l being the length and d the diameter, was over 10 with a large distribution due to the length-scatter. The fibers can be considered as randomly distributed in a plane. In a previous paper /1/, it has be shown that the properties of the fibers are not significantly decreased after this treatment.

II - TESTING CONDITIONS

The size of the useful part of the sample is 200x4.9x2mm. The crosshead speed of the tensile machine was in all test 0.01 mm/mn. The strain was measured with a single strain gauge (dynamic 0-1.5%) glued on the surface of the sample. A HP 3457A multimeter (5 true digits) connected to a HP Vectra ES/12 computer digitalized strain, stress, AE event count with a sampling period of 0.77s. All results presented here were obtained by processing the measurements stored on the hard disc.

III - REPEATED TENSILE TEST

Repeated tensile tests were performed on the sample in order to measure the Bauschinger effect. The tensile test was interrupted when the total strain reaches a value near n*0.15% (n is the cycle number). After interruption,the sample is unloaded to zero stress and then reloaded. The crosshead speed remains constant during this test. Results are only given for the first 10 cycles. The fracture of the sample occurs at a stress of 120 MPa and at an estimated strain of 4%. Fig.1 presents the measured parameters versus strain. The stress-strain curve shows a large Bauschinger effect: by decreasing the load, reverse plasticity occurs before load zero. The acoustic emission event count exhibits clearly two stages: before a total strain of 0.6% (cycles 1 to 4) the AE event activity is very low but over this strain, the AE event count increases rapidly.

IV - DISCUSSION
4.1. Before the beginning of AE   (ε_p <=0.6% strain)

The measured behavior in repeated tensile tests, especially reverse plasticity, is very close to the behavior of strain hardening alloys with undeformable particles as described for instance by Brown and Stobbs /2/. If we consider that the observed effects are owing to similar reasons, the tensile stress σ, can be given by the following relation:

$$\sigma = \sigma_j + \sigma_{pd} + \sigma_f + \sigma_c + \sigma_i \qquad (1)$$

where: σ_i and σ_{pd} are the thermal stress component, corresponding to the effect of jog dragging and impurity-dislocation interaction, respectively. The values of these two components are generally small: about 1 MPa for the jog dragging and 2 MPa for the impurity-dislocation interaction as it has been estimated in the A1100 aluminum material. σ_f is due to the strengthening of the dislocation forest. An upper bound of σ_f can be deduced from the following considerations: in the course of the tensile test, the plastic deformation ε_p is given by

$$\varepsilon_p = s \Lambda b \, l \qquad (2)$$

with s the Schmidt factor(0.3),$\Lambda$the dislocation density and l the average distance of dislocation displacement. So, from the relation (2) σ_f is given by:

$$\sigma_f = (\mu b^{0.5} \varepsilon_p^{0.5})/(s^{0.5} l^{0.5})$$

with μ the shear modulus of the matrix. An upper bound σ_{fm} is obtained by taking into account that l is at least equal to the mean distance between carbon fibers in the slip planes of the matrix. With l=20 μm and μ=27 GPa σ_{fm} can be calculated. Its value is reported on Fig.4 far from ε_p=0 where the relation (2) is not suitable. σ_c is the Orowan stress due to the fiber bypassing by dislocations. This component is given by $\sigma_c = \mu b/sl$ and is 4MPa for l=20 μm. σ_i is the internal stress due to the fact that incompatible deformations between fibers and matrix have to be assumed. These deformations are induced by difference in thermal expansion coefficient or difference in plastic behavior between fibers and matrix. On Fig.3 the internal stress is represented in a simplified manner. At large distance from the fibers, σ_i is equal to the mean internal stress $<\sigma>$. According to Eshelby's approach, the local internal stress near the fibers is generally much higher than the mean stress. At low volume fraction of particles or fibers, it is considered that σ_i in (1) is given by $<\sigma>$. The calculation of $<\sigma>$ /3,4,5/ leads to a linear relation between $<\sigma>$ and the plastic strain ε_p, but important is that this mean stress is assumed to be not modified by an elastic unloading.

In our case, a low level of reverse plasticity occurs during unloading and gives us the possibility to measure the beginning of plasticity in reverse flow, easy to observe on the unloading stiffness curve (point B on the curve Fig.2). The loading stiffness curve gives us also the beginning of plasticity in forward direction of the load (point F) with a good accuracy. The mean stress is then obtained by the simple calculation :

$$<\sigma> = (\sigma_F + \sigma_B)/2 \qquad (3)$$

On Fig.4, we have plotted the variation of the calculated composite strengthening stress σ, as the sum of the different components occurring in equation (1) with $\sigma_i = <\sigma>$ ($<\sigma>$ being calculated according to relation (3) and experimental results) and $\sigma_f = \sigma_{fm}$. We can see that whatever the plastic strain, the calculated value σ is lower that the experimental value σ_t. We think that the difference $\sigma_t - \sigma$, is due to the fact that with a high volume fraction of fiber (30% in our case), the dislocation movement is also affected by the local internal stress in the vicinity of the carbon fiber (see Fig.3). Therefore the difference $\sigma_t - \sigma$ can be consider as a good order of magnitude for the local internal stress.

4.2 Behavior during unloading the sample

The stress field, acting on the dislocations, is related to the stress-strain curve during loading or unloading. The assumption is that the total strain is due to three parts/6/:
   - the pure elastic strain $\varepsilon_e = \sigma/E$ (E the Young modulus)
   - the anelastic strain ε_a due to the change in curvature of the dislocations segments.
   - the plastic strain ε_p^u during the unloading controlled by the dislocation motion in the stress field discussed previously added to a short range internal stress field due to jog dragging. This plastic strain is represented in the upper corner of Fig.1 and can be numerically calculated. As the interaction dislocation-jog motion is thermally activated, Hamel /6/ has shown that the plastic strain rate $\dot{\varepsilon}_p^u$ can be written as:

$$\dot{\varepsilon}_p^u = 2sb^2\, hv\,(\Lambda_f/\lambda_f)\exp(-\Delta F_1/kT)\sinh(s\,(\sigma-\sigma_i)\,bd\lambda_f/kT) \quad (4)$$

where b is the Burger's vector(2.86 A), v is the atomic vibration frequency(10^{13}Hz),ΔF_1 is the activation free energy, d is the activation distance (near b), h is the distance of advance per jump, Λ_f is the mobile dislocation density and λ_f is the distance between jogs.
As in the case of pure aluminum /6/, at the beginning of the unloading(upper value of the stress), the plastic strain rate is positive, it means that the dislocation motion remains during a short time in the forward direction as predict by relation (4). With an activation energy of 1.4eV corresponding to self diffusion of jog in pure aluminum /7/ and a free loop length of$\lambda_f = 1/\sqrt{\Lambda_f}$, the measured plastic strain

rate at the beginning of the unloading leads to a value of $(\sigma-\sigma_i)$ of 1 MPa. This is proof that the contribution of jog dragging to the total stress cannot explain the lack of stress previously observed and that the internal stress is much higher than the mean stress. We have again to consider that the internal stress near the fibers is much higher than the mean stress, that dislocation motion is affected by the local stress field near the fibers and so, can create a dislocation pile up against the fibers.

4.3. After the beginning of burst type Acoustic Emission.
    This stage is clearly related to fiber breaks which can then be optically observed and which are at the origin of the AE burst. It can be generally observed that one or two large slip lines are associated with each fiber break. So the fiber break releases the pile up of dislocations. We can observe numerous broken fibers which are not oriented in the tensile direction, and broken parts with very small length. All these results are inconsistent with the critical length predict by models from the shear lag theory. If we consider the stress field near the head of a dislocation pile up, as described in Stroh's theory /8/, a few number of dislocations can induced such fiber breaks.

CONCLUSION

    The studied material exhibits clearly two plastic deformation stages: the first is ruled by dislocation interactions with the fibers. The mechanical behavior is close to that observed in strain hardening alloys with undeformable particles. For instance, a true Bauschinger effect can be observed in repeated tension test. A second stage occurs after 0.6% total strain and is clearly connected with fiber breaks.

ACKNOWLEDGEMENT

    Experimental part of this study was done in the laboratory of Prof T. Kishi, RCAST, University of Tokyo with a support of Nippon Steel. Their help is gratefully acknowledged.

REFERENCES

1 - Y. Kagawa, S. Utsunomiya, Y. Kogo and M. Imaizumi
      6th ICCM and 2nd ECCM, Vol 2, p2.199
2 - L.M. Brown and W.M. Stobbs in Phil. Mag., 23, 1971, 1185
3 - L.M. Brown and D.R. Clarke in Acta Met., 25, 1977, 563
4 - O.B. Pedersen in Acta Met., 31, 1983, 1795
5 - H. Lilholt in Proc. of 9th Riso International Symposium
      on Metallurgy and Material Science, Riso, 1988
6 - A. Hamel, A. Vincent ,R. Fougeres Acta Met.,38,1990,301
7 - J. Friedel in Dislocations, Pergamon Press, Oxford 1964
8 - A.N. Stroh in Adv. Phys., 6, 1957, p418

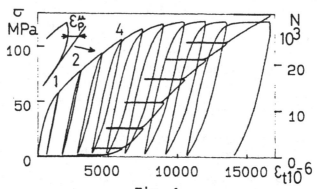

Fig. 1
Stress and AE event count versus strain (10 first cycles)

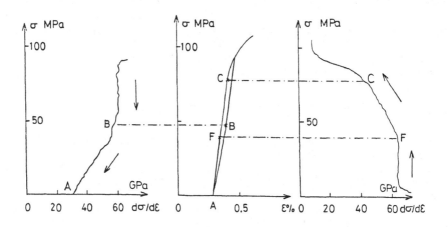

Fig. 2
Strain (center) and stiffness versus stress
left: unloading cycle 3, right: loading cycle 4

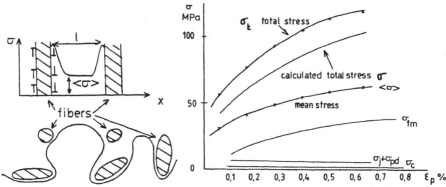

Fig. 3
Schematic representation of
a dislocation and of the stress
field between two fibers

Fig. 4
Calculated and measured
stress versus plastic
strain

MICROMECHANICAL MODELIZATION OF THE INFLUENCE OF MICROSTRUCTURE ON THE BEHAVIOR AND DAMAGE OF A G_r/AL COMPOSITE

P. BREBAN, D. BAPTISTE, D. FRANCOIS

Ecole Centrale Paris - Grande Voie des Vignes
92295 Chatenay Cedex - France

ABSTRACT

A new solid phase coextrusion process to manufacture metal matrix composites has been developed. As part of this study, the influence of following parameters on the behavior and damage mechanisms of the composite was investigated : aspect ratio of the fibers, orientation of reinforcement and local concentration distribution. This has been achieved through an equivalent inclusion analytical approach based on a micro-macro modelization. The prediction of the elastic constants, the coefficients of thermal expansion and the yield criterion can be determined from the microstructure. The influence of the distribution of fibers on the development of local plasticity has been demonstrated. It appears that the configuration of bundles of fibers has a dramatic effect on the propagation of damage which occurs at the fiber tip. An analytical criterion for damage initiation has also been suggested. These results are compared with in situ tensile tests performed inside a scanning electron microscope,where every stages of the failure mechanism can be observed.

INTRODUCTION

Most of discontinuous metal matrix composites (DMMC) studied at the moment are manufactured by three main technologies [1]: squeeze- casting, compo-casting, powder metallurgy. But the relatively high cost of these processes restricts the field of their application. To prevent this limitation, we have improved a solid phase coextrusion process in order to obtain in a single operation a discontinuous fibers composite. This technology has been developed with an aluminum matrix reinforced by T300 graphite fibers (diameter, $7\mu m$). The final product presents bundles of fibers well infiltrated by the matrix, but not evenly distributed in the aluminum. The mechanical properties are subsequently improved compared to the matrix. The longitudinal Young's modulus is 25% higher, and the strength to failure is two times more important.

In order to improve the mechanical properties by controlling the manufacturing parameters, the influence of the microstructure has been evaluated through an analytical modelization. The poor effect of bundles of fibers has been displayed. A damage initiation criterion based on physical observations has been suggested through the comparision of residual stresses and the local stress field around the fiber when loading the composite.

1 - MICROSTRUCTURE OF THE COMPOSITE

The material selected for the experiments is a 1050Al (A5) - 20vol.% T300 carbon fibers reinforced metal-matrix composite. It was extruded at low speed and high temperature according to the developed process. The material presents a good impregnation of fibers. The aspect ratio of fibers (length divided by diameter) is relatively small, about 5 in average. An important heterogeneity of local volume fraction can also be noticed with some areas almost free of reinforcement and others with more than 50vol.%. The last important point to be remarked is the large distribution of orientation of the fibers.

2 - ANALYTICAL MODELIZATION OF THE COMPOSITE BEHAVIOR

The important anisotropy of carbon fibers requires a three dimensional approach taking into account all the coefficients of the stiffness matrix. An equivalent inclusion method based on Eshelby's theory [2] has been chosen. Among different models inspired by the self consistent scheme, TANAKA and MORI's model [3] appears to give an explicit solution to the modelization of discontinuous metal matrix composites. An attempt is made to clarify and generalize this approach in term of a micro-macro relationship.

The macro scale corresponds to the elementary volume representative of the structure. This volume includes all microstructural parameters such as distribution of fiber length, disorientation of the reinforcement and local heterogeneity of volume fraction. The cell is submitted to an uniform macro-stress Σ. The presence of numerous fibers disturbs locally the uniform stress field. Thus the total stress field at any point of the space can be decomposed into the uniform and disturbed stress fields (σ).

The evaluation of σ requires to write the behavior of a single fiber surrounded with the others. Let us introduce at the micro scale a short fiber characterized by its stiffness, aspect ratio and orientation in a small volume of matrix. The constitutive equation completed with the equivalent inclusion theory provides a relation of localization where the eigenstrain is calculated from teh macro strain fields through the localization tensor which only depends on the characteristic parameters of this fiber and of the matrix.

The homogeneization consists in writing that the disturbed stress field vanishes in average over the entire volume of the composite. The problem can then be solved. The total strain field E associated to the uniform stress field Σ is determined. The linear relation between Σ and E gives the stiffness matrix of the composite.

3 - RESULTS

3.1. Effect of fiber aspect ratio

In the particular case where all fibers are well oriented and identical, the previous equations become very simple. This formulation is first used to predict the effect of variable aspect ratio on longitudinal and transverse Young's modulus for different composites. Figure 1 presents the curves obtained for a 2124Al-20vol.% SiCw composite (1-a) and the studied material (1-b). The difference of reinforcement between an isotropic fiber (SiC) and a greatly anisotropic one (Carbon - [4]) is shown on the two diagrams. An important point to be noticed is the saturation of the curve for a lower threshold in the case of carbon fibers. Beyond an aspect ratio of 15 no significant increase of the elastic properties has to be expected. The very low transverse modulus of T300 carbon fibers has an important impact on the longitudinal modulus of the composite. The influence of the properties of fibers appears to be predominant on the elastic characteristics of the composite.

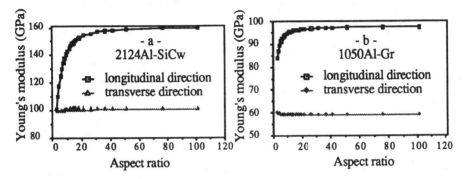

Fig. 1 - The variation of longitudinal and transverse Young's moduli with the aspect ratio of fibers for different composites with 20vol.% of fibers

3.2. Effect of fibers disorientation

Other developments of the same formulation can be made, such as the effect of scattered fiber aspect ratio [5], but the most important microstructural parameter is the disorientation of fibers. TAYA and al.[6] have studied the effect of a distribution of orientation on the longitudinal Young's modulus. The developed formalism allows the same study on all the coefficients of the stiffness matrix. The results show that this parameter cannot be neglected, excepted for narrow angle distributions (less than 20° of disorientation).

3.3. Effect of local volume fraction

The last microstructural parameter to be modelize is the effect of a local volume fraction distribution. This has been achieved through the introduction at the micro-scale of a term of short range interaction between the fibers (δ^*). δ^* corresponds to the difference between the eigenstrains calculated with the local volume fraction and the average one.

Figure 2 shows the evolution of longitudinal and transverse moduli with the main volume fraction, for a given distribution. All the points correspond to the same average fiber rate. The longitudinal modulus exhibits a small decrease when increasing the fiber volume fraction in the bundles. The effect on the transverse one appears to be neglectible. Nevertheless this parameter is less important than the disorientation of fibers.

3.4. Extensions of the modelization

An extension of the modelization was carried out by TAYA and al. [7] to calculate the coefficients of thermal expansion (CTE). The previous generalized formulation is also extended to thermal loadings. Thermal residual stresses have also been calculated. This model gives no influence of this stress field on the elastic constants but displays a translation of the threshold surfaces.

The last extension to be presented here is the calculation of the yield criterion. The experiments on DMMC have shown that strain localization involving local plasticity appears at very small stresses near fiber tips without generating a macro plasticity. So a yield criterion has to take into account the average stress over the matrix volume. An attempt was made by W. J. CLEGG [8] to modelize the tensile behavior of particle reinforced MMC with a similar approach, using a Tresca criterion for the matrix and involving elastic residual stresses. The aluminum is supposed here to plastify according to a Mises criterion. When the stress $\Sigma+<\sigma>_{D-\Omega}$ reaches the flow limit, the Hill criterion is assumed to be verified by the composite. Yield stresses in longitudinal and transverse directions for the same three different composites as previously described, have been calculated. The shape of the curves is identical to the evolution of the Young's moduli, with the same saturation for a threshold value of the aspect ratio. Results obtained for the 2124Al-20vol.% SiC are in good agreement with measurements made by CHAMBOLLE [9] with an average aspect ratio of the whiskers of 5. For longer fibers, this model seems to overestimate the axial flow stress, mainly because it neglects the initiation of local plasticity around the fibers due to residual stresses. The very small incidence of the aspect ratio on the transverse properties can be noticed. The influence of the matrix alloy which is much more important in this case. Figure 3 presents the influence of thermal residual stresses on the yield criterion. We can verify that the matrix is initially in tension at least in the fiber axis. The important coupling between longitudinal and transverse directions explains the behavior perpendicularly to the fibers.

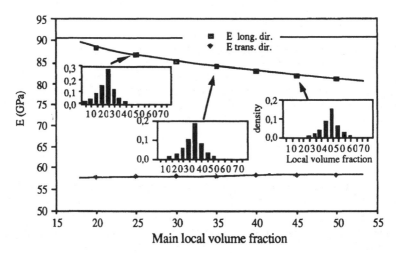

Fig. 2 - The variation of the longitudinal and transverse Young's moduli with a distribution of local volume fraction. Each case corresponds to the same average volume fraction (20%).

4 - APPROACH TO DAMAGE MODELIZATION

4.1. Experimental observations

In situ tensile tests on micro-samples have been performed in a SEM, in order to identify the physical parameters of damage. The sample was chosen for its large distribution of local volume fraction, at least at the surface. The different steps of damage can be listed : strain localization in the bundles of fibers, cavity nucleation at the fiber tips, cavity growth and coalescence, crack propagation with a plastic area in front of the crack tip.

We measured the load during the test which allows us to correlate the observations to the macro-tensile behavior (figure 4). Cavity nucleation just has already begon when the stress reached the macro-yield stress of the composite.

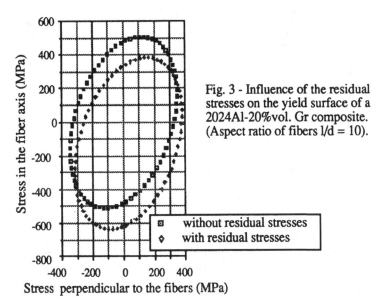

Fig. 3 - Influence of the residual stresses on the yield surface of a 2024Al-20%vol. Gr composite. (Aspect ratio of fibers l/d = 10).

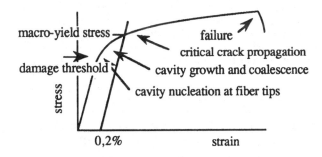

Fig. 4 - The different steps of damage propagation correlated to the macro behavior of the studied composite.

4.2. Analytical approach

We are able to calculate the residual stresses at the fiber interface. Fracture surfaces show that no reaction occured during the manufacturing process between carbon and matrix; so only residual stresses assure the binding of fibers. When an external load is applied to the composite, Eshelby's theory gives an evaluation of the interface stress field.

The damage threshold is supposed to be reached when the normal stress to the considered point of the interface is equal to the residual stress. A single criterion allows the modelization of an anisotropic phenomenon as presented figure 5. The comparision between the yield and damage surfaces presents a good agreement with figure 4. This approach also displays that areas with an important fiber volume fraction are more sensitive to damage.

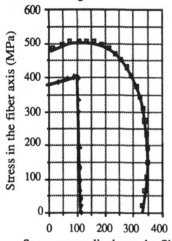

Fig. 5 - Comparision between the yield criterion and the damage threshold calculated by the analytical approach for a 2024Al-20%vol. Gr composite

◆ Damage threshold

▫ Yield surface

CONCLUSION

In this study, the influence of different microstructural parameters on the mechanical properties of a Gr-Al composite has been investigated. We have outlined the ill effect of carbon fibers anisotropy on the elastic characteristics of the composite. The choice of a strong matrix is more important for the increase of the flow stress. A fiber disorientation of less than 20° induces only a 2% drop of the stiffness, but transversely oriented fibers have a more considerable effect. A distribution of local volume fraction induces a small decrease of the elastic characteristics.

Taking into account the residual stresses generated by the cooling from the manufacturing temperature, a configuration of bundled fibers with little matrix inbetween has been shown to induce an earlier initiation of damage by cavity nucleation at the fiber tips. Damage tends to propagate in the direction of the crack that is to say perpendicularly to the fiber axis. The better configuration corresponds to homogeneously distributed fibers.

The developed models are included in a more general scheme to predict a damage probability of a structure.

Acknowledgments - This work has been supported by the Ets R. CREUZET for the manufacturing of the composites and the AEROSPATIALE, through a CIFRE contract.

REFERENCES

1 - M. G. McKIMPSON and T. E. SCOTT (1989), Mat. Sci. Eng. A107, 93-106
2 - ESHELBY J. D. (1957), Proc. Roy. Soc. London Series A 241, 376-396
3 - TAYA M. and MURA T. (1981), J. Appl. Mech. 48, 361
4 - KOWALSKI I. M. (1986), SAMPE J., 38
5 - TAYA M. and TAKAO Y. (1987), J. Comp. Mat. 21 , 140
6 - TAKAO Y., CHOU T. W. and TAYA M. (1982), J. Appl. Mech. 49, 536-540
7 - TAYA M. and TAKAO Y. (1985), J. Appl. Mech. 52, 806
8 - CLEGG W. J. (1988), Acta Met. 36 2141-2149
9 - CHAMBOLLE D., BAPTISTE D. and BOMPARD Ph. (1989), Proc. Mecamat 89 (to be published).

CORROSION STUDIES ON METAL MATRIX COMPOSITES

S. COLEMAN, B. McENANEY, V. SCOTT

University of Bath - School of Materials Science
BA2 7AY Bath - Great Britain

ABSTRACT

A double cycle polarization technique is described which provides important corrosion parameters such as pitting potential and protection potential. The parameters are assessed by comparison with open circuit potentials and gravimetric studies from longer term immersion tests. Corrosion in de-aerated NaCl solution of aluminium alloys reinforced with alumina, silicon carbide and carbon is studied. Second phases, often at the fibre-matrix interface, generally result in preferential attack. This dominates the corrosion behaviour of the composite with reinforcements playing a secondary role, for example, by acting as relatively inert physical barriers to corrosion pit propagation or by modifying matrix microstructure.

1. INTRODUCTION

Whilst corrosion properties of aluminium and its alloys are well understood, assessment of corrosion of composites based upon these metals has posed a number of problems. These can be related to complex microstructural features which are produced when incorporating ceramic fibres or particulates into the metal matrix by various manufacturing techniques. Such features include not only the matrix and fibre types, but also binder phases, matrix alloy impurities and fibre/matrix reaction products. This paper examines methods of measuring corrosion behaviour and, in particular, the possibility of employing an accelerated test - the double cycle polarization (DCP) technique. The results are compared with those given by natural immersion tests involving open circuit potential and gravimetric measurements.

2. MATERIALS

Two matrix metals have been studied, Al-7Si and Al-4Cu, with four types of reinforcement, carbon fibre, Nicalon fibre, Saffil fibre and silicon carbide particulate. Fibre-reinforced composites were manufactured by liquid metal infiltration and studied in the as-fabricated condition. The particulate reinforced material was produced by a powder route and supplied in the wrought condition. Details of materials are given in Table 1.

3. EXPERIMENTAL

Corrosion tests were conducted in de-aerated 3.5wt.% NaCl solution at 25°C.

3.1 Gravimetry

Specimens 20x15x5mm^3 in size were prepared to a 600 grit finish on all surfaces and immersed in saline solution for periods of up to 3 weeks. At the end of a test the corrosion products were removed by scrubbing with a bristle brush and immersing in 10% nitric acid for 10 minutes. The process was repeated until a reproducible weight loss figure was obtained.

3.2 Open circuit potential tests

Specimens were mounted in epoxy resin and metallographically polished to a fine finish. They were inserted into a PTFE holder and electrical contact made with an insulated copper wire. The open circuit potential (OCP) was recorded regularly over a period of 3 weeks.

3.3 Double cycle polarization tests

Specimens were prepared and fitted into a PTFE holder as for the OCP test. Measurements were carried out in an all-glass cell, electrode potentials being recorded using a saturated calomel electrode. A potential sweep rate of 20mV/min was used.

4. RESULTS AND DISCUSSION

4.1 Gravimetry

Weight changes recorded after 3 weeks immersion are given in Table 2. For the majority of systems studied a weight loss with a relative error of 10% was found. In the case of Al-7Si alloy reinforced with carbon or Saffil fibres substantial and irreproducible weight gains were measured. This was caused by the difficulty in removing entrapped corrosion product, as evidenced in Figure 1, which shows a cross-section from corroded carbon-fibre reinforced composite. The micrograph shows that corrosion has occurred deep in the composite due to penetration of saline solution, mainly down the fibre/matrix interface.

With regard to the unreinforced Al-7Si alloy the weight loss (6.2±0.5gm^{-2}) is approximately twice that of Nicalon reinforced Al-7Si alloy (2.5±10gm^{-2}). Since approximately one half of the composite consisted of reinforcement, the results imply, firstly, that the Nicalon fibre has not increased corrosive attack and, secondly, that the reaction is largely dominated by the metal matrix. Certainly, observation of isolated fibres immersed in saline solution showed no evidence of chemical attack.

At first sight the dominant role of the matrix is not so obvious in the experiments using Al–4Cu systems. For example, the addition of only ~20vol.%

reinforcement in the form of SiC particles has caused a significant reduction in the degree of corrosion, from $5.4\pm0.5 \text{gm}^{-2}$ to $1.2\pm0.1 \text{gm}^{-2}$. It is possible, however, that the different metallurgical conditions of the matrix (the unreinforced metal was cast whereas the composite was forged) may be responsible to a large extent for the results, the SiC particles themselves playing only a minor role in the corrosion.

4.2   Open circuit potential test

A typical plot obtained in an open circuit potential (OCP) test, in this case for Saffil reinforced Al-7Si alloy, is illustrated in Figure 2. For the first hour of immersion the potential remained steady at -765mV. It then dropped to a value of -969mV before rising to ~-840mV over the following seven days. The initial potential corresponds to the pitting potential, E_{pit}, and the quasi-static potential to the protection potential, E_{prot} (1). Values of E_{pit} and E_{prot} are given in Table 2. For for three of the five materials examined E_{prot} could not be defined because a quasi-static potential was not obtained over a three-week period. Values for the difference potential (DP = E_{pit}- E_{prot}) which defines the range of potential over which pit propagation can occur, are also quoted for the other two materials.

The data show that the pitting potential is most noble (more electropositive) for the SiC-reinforced Al-4Cu, at -748mV. With regard to composites based upon Al-7Si alloy, E_{pit} becomes more electronegative as the electrical conductivity of the fibre increases, values ranging from -760mV for Al-7Si through to -791mV for the carbon reinforced material. This suggests that pitting is initiated more readily in the presence of a more conducting fibre, probably by microgalvanic action between fibre and matrix. This is confirmed in Figure 3 which shows enhanced attack at the fibre/matrix interface in the Nicalon-reinforced composite. It was also noted that the degree of interface attack was less with the low conductivity Saffil fibre composite and greatest for the high conductivity carbon fibre material.

The presence of second phases is also likely to contribute to pitting corrosion. For example, it is well known that aluminium carbide, which hydrolyses readily, can form at the surface of carbon fibres (2) and in a material such as the one studied here, the carbide can extend up to 1μm into the surrounding aluminium matrix (3). Other second phases which are associated with corrosion attack are intermetallics, such as $FeSiAl_5$, which are picked up as contaminants during the manufacture of the composite by liquid metal infiltration. Hence the absence of these impurity phases in the wrought, SiC-reinforced, Al-4Cu composite might well account for its higher pitting potential.

Examination of DP values showed a clear distinction between unreinforced and reinforced materials, 328mV for unreinforced Al-7Si alloy and ~200mV or less for the composites. Little correlation was found between difference potential and electrical conductivity of fibre, but there appears to be a connection with geometry of the reinforcing material. For example, composites with continuous fibres had the lowest DP values, 113mV and 159mV for Nicalon and carbon fibres, composites containing discontinuous reinforcements (short fibre Saffil and SiC particulates) had values of 204mV and 200mV. It is argued, therefore, that the main effect of reinforcement on pit growth is to act as a physical barrier to their development.

4.3   Double cycle polarization test

An example of a double cyclic polarization (DCP) curve is illustrated in Figure 4, in this instance for Al-4Cu alloy reinforced with SiC particles. The arrows indicate the direction of applied potential scan and the solid line corresponds to the first potential scan which starts at point A. The pitting potential and the protection potential are also marked on the diagram. Potentials recorded during the tests are listed in Table 2.

All composite materials gave pitting potentials which were more electronegative than the respective unreinforced alloys. This indicates that pitting susceptibility is greater for reinforced materials, in accord with the findings from OCP tests, Figure 5. A difference of ~15mV was found in E_{pit} values obtained from OCP and DCP tests, OCP values being more electro-negative in all but one case. The more positive values recorded during DCP tests probably reflect the positive displacement of E_{pit} due to the potential scan rate used, 20 mV/min. For the two composites where it is possible to make a comparison, E_{prot} values obtained from OCP and DCP tests are in reasonable agreement. Also, there is a positive correlation between DP data obtained from the DCP tests and gravimetric results on those specimens where sensible weight losses could be measured, Figure 6.

5.   CONCLUDING REMARKS

A double-cycle polarization (DCP) technique has been successfully used to study the corrosion behaviour, in de-aerated 3.5wt.% NaCl solution, of aluminium alloys reinforced with Saffil, Nicalon and carbon fibres as well as SiC particulates. The DCP method yields important corrosion parameters, such as pitting potential and protection potential, which are in reasonable agreement with the results from OCP and gravimetric experiments.

The work shows that salt water attack is essentially galvanic and localized at microstructural constituents of the MMC, such as the fibre-matrix interface and second phases in the matrix. All MMC systems investigated showed these characteristics to some extent, but there were some important differences. As regards the role of the fibre, results indicate that the most electrically-conducting fibre, carbon, produced the most marked galvanic effect, with corrosion occurring preferentially at the fibre-matrix interface. A contributory factor here was aluminium carbide formed between metal matrix and fibre during manufacture of the composite. Saffil, an insulator, was essentially passive as regards corrosion, but all materials exhibit significant microgalvanic activity within the matrix, especially at iron-containing intermetallics of the type Al_5SiFe.

6.   ACKNOWLEDGEMENTS

To the SERC and MOD for support.

7.   REFERENCES

1. Smialowska, Z.S. and Czachor, M.J., British Corr. J., **4**, (1969) 138-145.
2. Clyne T.W., Bader M.G., Cappleman G.R. and Hubert P.A., J. Mat. Sci, **20** (1985) 85-96.
3. Ming Yang and Scott V.D., J. Mat. Sci., (1990)*in press.*

Matrix		Reinforcement			
Alloy	Composition wt.%	Type	Geometry	Diameter	Vol. fraction
Al-7Si	7.22Si 0.38Mg 0.05Ti	Carbon Nicalon Saffil	continuous continuous short, 100μm	8μm 15μm 3μm	50% 50% 18%
Al-4Cu	4.0Cu 1.5Mg 0.6Mn	SiC	particulate	3μm	20%

Table 1. Composition of materials.

Material	Gravimetry	OCP			DCP		
	Wt change $(g.m^{-2})$	E_{pit} (mV)	E_{prot} (mV)	DP (mV)	E_{pit} (mV)	E_{prot} (mV)	DP (mV)
Al-7Si	-6.2±0.5	-760	*	-	-743	-812	69
Al-4Cu	-5.4±0.5	-	-	-	-667	-746	79
Al-7Si/C	+4.0	-791	*	-	-748	-850	102
Al-7Si/Nic	-2.5±0.2	-778	*	-	-759	-800	41
Al-7Si/Saf	+97	-765	-840	75	-781	-819	38
Al-4Cu/SiC	-1.2±0.1	-748	-760	12	-726	-738	12

* no quasi-static potential found.

Table 2. Corrosion data obtained from gravimetry, OCP and DCP tests.

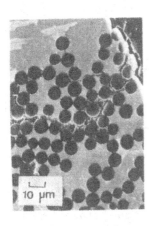

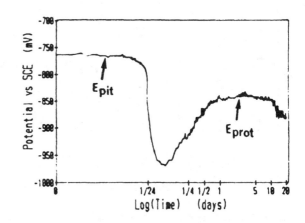

Fig.2. OCP curve; Saffil reinforced Al-7Si.

Fig.1. Corroded surface showing interface attack; carbon fibre reinforced Al-7Si.

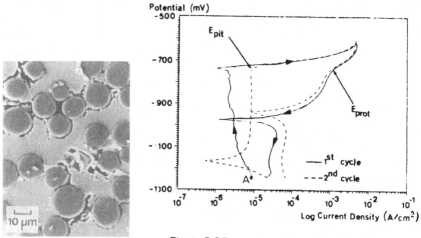

Fig.4. DCP curve for Al-4Cu alloy reinforced with SiC particles.

Fig.3. Enhanced attack at interface in Nicalon reinforced Al-7Si.

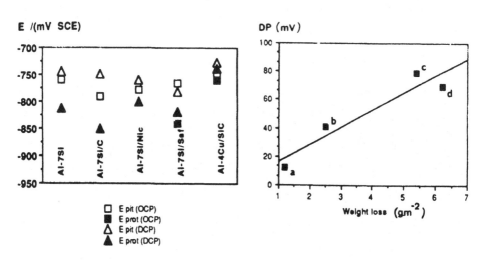

Fig.5. Pitting potentials and protection potentials for a range of materials.

Fig.6. Correlation between difference potentials and gravimetric data for a variety of materials (a) Al-4Cu/SiC (b) Al-7Si/Nic (c) Al-4Cu (d) Al-7Si.

CHARACTERIZATION OF ALUMINIUM OXIDE (FP) FIBRE REINFORCED AL-2,5L$_I$-COMPOSITES

K. SCHULTE, A. BOCKHEISER, F. GIROT*, G.K. KIM**

DLR - Institute for Materials Research
PO Box 90 60 58 - 5000 köln 90 - West-Germany
*Université de Bordeaux I
351 cours de la Libération - 33405 Talence - France
**CIMC - South Korea

ABSTRACT
Metal matrix composites based on Al_2O_3-fibre reinforcements are gaining increasing attention for structural application. However, the data base concerning its mechanical properties and microstructural characterization remains limited. For the present study an Al_2O_3-fibre (FP-fibre from Dupont) reinforced Al-2,5 Li composite has been investigated. Continuous fibre reinforcements, unidirectional in the 0° and 90°-direction were used for the investigation. Processing of the material was performed by vacuum melt infiltration at LCS, Bordeaux.
Transmission electron microscopy was performed in order to study the fibre/matrix interface. Precipitation free areas around the fibres containing any δ' particles, which in turn are present in the matrix, were found.
For mechanical characterization tensile and compressive tests were performed. In order to get sufficiently good results a special MMC-related test equipment was designed. The best results were achieved, when cylindrical hour-glass shaped specimens were used. Fatigue testing of the metal matrix composites showed, that with fibre reinforcement a pronounced improvement of the fatigue behaviour can be achieved.

INTRODUCTION
Conventional metallic materials have been tailored in the past close to their ultimate properties. New technological requirements ask for further improved materials. Metal-matrix composites (MMC) promise to reach this goal.

MMC can be described as materials whose microstructure comprise a continuous metallic phase (the matrix) into which a second phase has been artificially introduced during processing, as reinforcement. Presently the interest in MMC is primarily focused on light alloys reinforced with fibrous or particulate phases to achieve major jumps in selected mechanical properties or thermal stability. This new interest is mainly related to the fact that ceramic based reinforcement constituents became available, which are comparatively inexpensive. Al_2O_3- or SiC-based fibres, whiskers or particles, but also carbon fibres are used to reinforce aluminium, magnesium or titanium matrix alloys [1]. The present study will focus on α-Al_2O_3-fibre reinforced Al-2,5Li aluminium alloys.

EXPERIMENTAL

The material used in this study was supplied as vacuum infiltrated α-Al_2O_3 fibre (FP-fibre from Dupont, USA)/ Al-2,5Li matrix composite containing 30 vol.pct. of uni-directional fibres. The diameter of the polycristal FP-fibre is about 20 μm. The minimal fibre stress and the Young's Modulus is given with 1500 MPa and 380 GPa, respectively, while the fibre strain to failure does not exceed 0,4%. The density of the fibre is 3,9 g/cm^2. The 2,5 wt.pct. Lithium was added to the matrix aluminium in order to achieve improved wetting of the fibres and infiltration of the molten metal [2]. A positive secon-dary effect is, that the addition of Li increases the strength and decreases the density of the matrix.

The composite, produced at the University of Bordeaux (F) was available in the form of plates with the dimension 55 x 50 x 10 mm. For mechanical characterization tensile and compressive tests were performed on cylindrical hour glass shaped specimens, which were machined from the plates. The specimens had a thread on both ends and were gripped in the female screw such, that the load was intro-duced not only via the threads, but also by gripping stresses, as the female screw was cut into four sections with a slot in between, so that a mechanical gripping load could be born on them.

Tensile and compressive tests were performed on the unidirectional composite with fibres in the 0° and 90° direction. In addition, fatigue tests were made at a stress ratio of R = σ_u/σ_o = 0,1 respectively R = -1 (σ_u = minimum and σ_o = maximum stress in each load cycle).

Fracture surfaces were studied by extensive scanning electron microscopy (SEM) while the microstructure was examined by transmission electron microscopy (TEM). Samples for the TEM were prepared by mechanical thinning in a dimple grinder down to a thickness of 70 μm followed by ion milling until perforation.

RESULTS AND DISCUSSION

a) The Microstructure

Fig. 1 is a TEM-micrograph showing the microstructure of the Al-2,5 Li composite indicating the presence of a fibre/matrix reaction zone. Analysis of the reaction zone identified it as being primarily cubic $LiAl_5O_8$ spinel and $LiAlO_2$ [3-5] Within the matrix precipitation of δ'-phases (Al_3Li) the Al-Li system was observed giving rise to precipitation hardening. An important result of this investigation is, that between the reaction zone, which growth into the fibre by consuming parts of its volume, and the bulk matrix a precipitation free zone (PFZ) surrounds the fibre. Within the PFZ no δ'-particles can be identified, what means, that the formation of the reaction zone with its $LiAlO_2$ and $LiAl_5O_8$ reaction products absorbs Li from the surrounding matrix.

The formation of the lithium aluminate can be explained by a two step reaction, proposed by Dhingra [6]

$$6 \ Li + Al_2O_3 = 3 \ Li_2O + 2 \ Al \qquad and$$
$$Li_2O + Al_2O_3 = 2 \ LiAlO_2.$$

The formation of the spinel might follow the reaction

$$LiAlO_2 + 2 \ Al_2O_3 = LiAl_5O_8.$$

The reaction shows, that both the lithium of the matrix and the Al_2O_3 from the fibre are needed to form the reaction products.

The results show:

1. that the alumina grains close to the fiber surface have been completely surrounded by a thin layer of this reaction products. A strong bonding can be expected from this interface since the matrix not only adheres the fibers but also interconnects the first layers of alumina grains.

2. that the PFZ is probably due to
    a) a lack of lithium around the fiber, resulting from the interface creation, which impedes the precipitation of the δ'-phases.
    b) the fiber which acts as a dislocation trap, and is rapidly surrounded by a defect free zone in which the precipitates can not nucleate.

b) Mechanical Properties

In Table 1 are summarized the results for the mechanical properties in tension and compression for both 0° and 90° fibre orientation. The stress/strain behaviour is shown in Fig. 2a,b. The tensile properties achieved are similar to those shown in the literature elsewhere [4, 7], being slightly lower as predicted by the rule of mixtures. However, the strength (σ_B) in compression is considerably higher than in tension, which can also be observed for the deviation from the linear stress strain relation, σ_K. This, in general good compressive behaviour, can mainly be related to the fact, that the metal matrix itself has a relatively high Young's Modulus E, which allows to

better support the fibres and avoids their short wave
kinking [8].
In the transverse tests (90°-direction) due to the
special micromechanics situation, which is discussed in
more detail in [9], higher values are achieved in compression than in tension.
In Fig. 3 are shown the fracture surfaces from tensile
tests performed on a) [0°], longitudinal and b) [90°],
transverse test pieces. In the longitudinal test pieces
(Fig. 3a) fracture occured by only few pull out of
fibres, but the crack jumps from one fibre to the neighbouring one with extensive matrix deformation. The matrix
forms dimples around each broken fibre. The transverse
tensile test shows a good fibre matrix bonding which
results in fibre splitting (Fig. 3b), the bond strength
is higher as the transverse tensile strength of the
fibre.
Fatigue tests were performed on unidirectional specimens
with all fibres in the 0°-direction. In Fig. 4a and b is
shown the stress/strain behaviour after different numbers
of load cycles. In the case with σ_o = 410 MPa (R = 0,1),
which is about 75% of the fracture stress, the specimen
exceeds more than 4,7 million load cycles. At R = -1,
with a total stress variation of 600 MPA, the specimen
easily survives more than 3,7 million load cycles. The
reduction in the width of the hysteresis loops indicates
cyclic hardening. These results demonstrate the superior
fatigue behaviour of the observed MMC-composite. At these
high load levels conventional aluminum alloys would never
exceed these high numbers of load cycles.
Monolithic, non fibre reinforced aluminum alloys show
after fatigue loading a typical fracture morphology,
where fatigue crack propagation during each consecutive
load cycle, is indicated by striations. However, for
composite materials a comparable behaviour can not be
expected. In Fig. 5a is shown the fracture surface of a
fatigue test at R = 0,1. An overview of the fracture
surface with lower magnification does not show any
essential difference to a simple tensile test. Only at
high magnification, in a matrix rich area, parallel lines
can be observed, which have some similarities to fatigue
striations. Distinct differences, however, can be
observed on the fracture surface of specimens loaded at R
= -1. Because of tension/compression loading no dimples
are formed, but in matrix rich zones smooth scaled areas
are present showing parallel lines at high magnification,
indicating a prefered direction of crack propagation
(Fig. 5b).

SUMMARY
Static and fatigue properties of unidirectional fibre
reinforced aluminum alloys (FP-fibre/Al-2,5Li matrix)
were investigated. A fibre reinforcement led to an
increase in Young's Modulus. Extremely good compressive
properties could be achieved. Fibre reinforcement also
improves the fatigue behaviour, when compared to unreinforced aluminum alloys. During fatigue loading cyclic
hardening could be registered. On the fracture surfaces
of fatigue specimens, significant differences to those
from static tests could be found.

LITERATURE
[1] K. Schulte, W. Bunk: AGARD-Structures and Materials
     Panel Meeting No. 67, 2.-7. Oct. 1988, Mierlo, The
     Netherlands, Conf. Proceeding No. 444, New Light
     Alloys, pp. 26.1-26.16
[2] A.P. Maijdi, J.-M. Yang, T.-W. Chou: ASTM-STP 964
     (1988) 31-47.
[3] R.A. Page, G.R. Leverant: Proceedings ICCM-V, San
     Diego, CAL (1985) 867-886.
[4] W.R. Hunt: Proceedings of the AIME Symposium on
     Interfaces in Metal-Matrix Composites, New Orleans,
     Louisiana, March 2-6 (1986) 3-25.
[5] I.Hall, V. Barrailler: Proceeding ECCM-1, Bordeaux,
     F, (1985) 589-594.
[6] A.K. Dhingra: Phil.Trans.Roy.Soc. London, A 294
     (1980) 559-564.
[7] H.R. Shetty, T.-W. Chou: Metallurgical Transactions
     16 A (1985) 835-864
[8] A. Puck: Kunststoffe 53 (1963) 727
[9] K. Schulte, J.J. Masson: Proceeding ECCM-3,
     Bordeaux, F, (1989) 501-507.

Table 1: Mechanical properties after tensile and
          compressive testing (σ_K = knee in the
          stress-strain curve)

	[0°] Al_2O_3/Al-2,5Li		[90°] Al_2O_3/Al-2,5Li	
	Tension	Compression	Tension	Compression
σ_B [N/mm^2]	538	1460	177	225
ε_B [%]	0,36	1,09	1,2	7,4
E [GPa]	181	166	126	117
σ_K [N/mm^2]	155	291	91	103

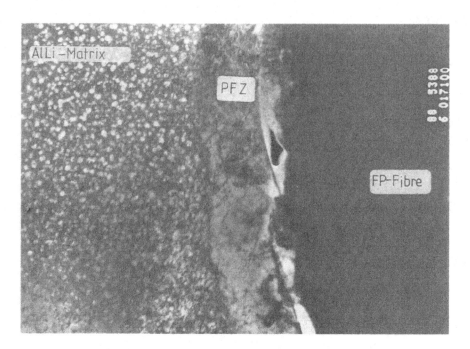

Fig. 1:　TEM micrograph of α-Al$_2$O$_3$ fibre/Al-2,5 Li matrix reaction zone.

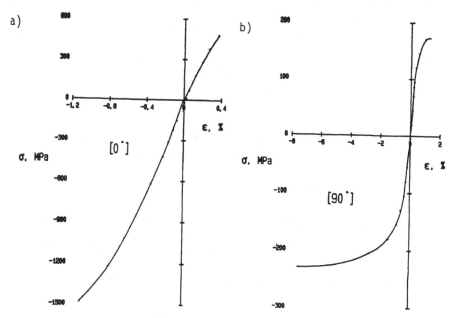

Fig. 2:　Stress/strain behaviour of Al$_2$O$_3$/Al-2,5 Li under tensile and compressive load, a) longitudinal [0°], b) transverse [90°].

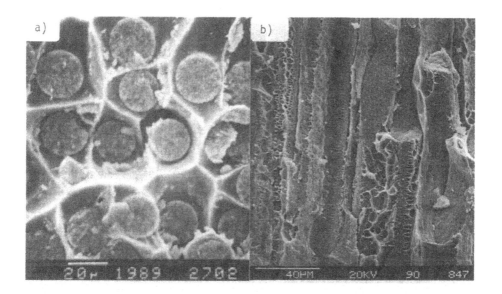

Fig. 3:   SEM-micrograph of the fracture surface after
          tensile testing
          a) longitudinal [0°], b) transverse [90°].

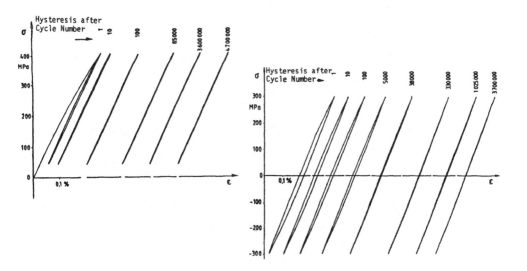

Fig. 4:   Stress/strain behaviour with hysteresis loops
          at various cycle numbers
          a) $R = 0,1$, $\sigma_o = 410$ MPa, $N > 4,3 \cdot 10^6$ cycles
          b) $R = -1$, $\sigma_o = 300$ MPa, $N > 3,7 \cdot 10^6$ cycles.

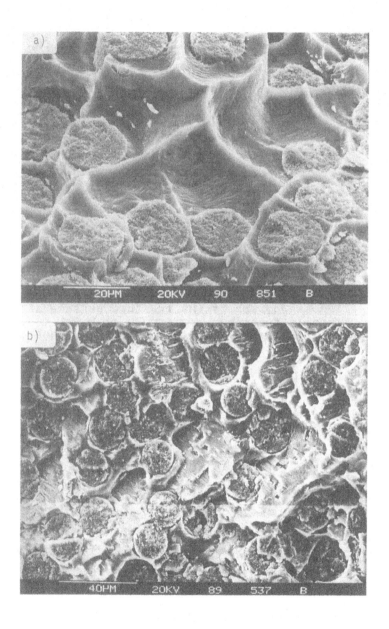

Fig. 5:   SEM-micrograph of the fracture surface after a
          fatigue test
          a) $R = 0,1$, $\sigma_o = 410$ MPa
          b) $R = -1$, $\sigma_o = 300$ MPa.

TENSILE AND FRACTURE TOUGHNESS BEHAVIOUR OF REINFORCED MAGNESIUM COMPOSITES

K. PURAZRANG, J. SCHROEDER*, K.U. KAINER*

Sharif University of Technology
PO Box 11365-8636 - Tehran - Iran
*Inst. für Werkstoffkunde - TU Clausthal
Agricolastr. 6 - 3392 Clausthal 2 - West-Germany

Abstract

Magnesium alloys were reinforced by alumina short fibres and silicon carbide particles using powder metallurgical or squeeze infiltration techniques. Tensile and fracture toughness tests were carried out at room and elevated temperatures. The effect of reinforcement and fabrication route on the mechanical properties are discussed. The reinforcement increases the tensile related properties and decreases the ductility and fracture toughness. A greater dependence of the mechanical properties on the fabrication route is observed. The P/M matrix alloy and composites are much stronger but more brittle than the squeeze cast alloy and composite.

I-INTRODUCTION

High strength magnesium composites are interesting materials for wide-spread use in engineering applications. Magnesium may be reinforced by continuous or discontinuous fibres, whiskers or particles. Studies of the mechanical properties have already revealed that composites with continuous fibres exhibits the best mechanical properties, at least in the direction of the fibres (S. P. Rawal et al. /1/, K. U. Kainer et al. /2/, E. G. Wolff et al. /3/). However, the anisotropic reinforced composites limits their application. In most practical applications a biaxial or triaxial stress states obtains in the components. MMC's with discontinuous fibre/particle reinforcement exhibits greater isotropy. This is the reason for increasing interest on discontinuous MMC's. The main objectives of present study was to compare the mechanical properties of a commercially available magnesium alloy and composites. The composite are: P/M AZ91+SiC$_p$, P/M AZ91+Al$_2$O$_3$-fibres and squeeze cast AZ91+Al$_2$O$_3$-fibres. Tensile and fracture toughness tests

301

were carried out at room and elevated temperature. The influence of the fabrication route on the mechanical properties is discussed.

II-EXPERIMENTAL PROCEDURE

The magnesium composites employed in this investigation were manufactured by powder metallurgical and squeeze infiltration techniques. Magnesium alloy AZ91 was used as matrix. SiC particles, 6 μm mean diameter and alumina short fibre, 3 μm mean diameter and length of about 150 μm were used as reinforcement. Magnesium alloy powder (particle size - 63 μm) was produced by machining and mixed with SiC particles or alumina fibres and consolidated by extrusion. Squeeze casting was used to produce composite plates 30-40 mm thick with discontinuous alumina fibres. Details of P/M and sq./c production have been published (J. Schröder et al. /4/, K. Purazrang et al /5/. The volume fraction of the reinforcing phase was 15 vol.%. Unreinforced material for comparison purposes was produced by the same techniques. Tensile and fracture properties of the different materials were determined. The fracture toughness at elevated temperatures was evaluated from the pick load of the load-time curves, which were found to be the same as the pick load of the load-displacement curves (K. Purazrang et al./6/). Scanning electron microscopy was used to analyse the fracture surfaces of the fracture toughness specimens and to examine the micromechanisms of crack initiation and crack propagation.

III-RESULTS AND DISCUSSION

Fig. 1 shows the micrographs of the matrix alloy produced by P/M and sq./c methods respectively. Whereas the P/M alloys showed a texture, the sq./c specimen was isotropic. The microstructure of the matrix composites is finer than that of the unreinforced alloy (Figs. 1 a' and 1c). The fibre promotes the nucleation of the $Mg_{17}Al_{12}$. Similar behaviour has been observed for materials produced by P/M methods (Figs. 1 b and 1 d). Fig. 2a shows that alumina fibre and SiC particles significantly increase the Young's modulus of the matrix alloy. The alumina fibres are more effective. The fabrication routes have little effect on Young's modulus. The tensile strength, the proof stress and the elongation to fracture were strongly effected by fabrication route. Fig. 2b and 2c shows that the tensile strength and 0.2 proof stress for P/M produced unreinforced Mg-alloy at room temperature were 50 % (UTS) and 200 % (0.2 proof stress) higher than sq./c. material. The improvement in tensile properties of the unreinforced P/M-material is induced by the extrusion process. On increasing the temperature, both tensile- and proof stress of sq./c alloy decrease gradually to lower values. The loss in strength was more pronounced for P/M produced matrix alloy. At 200°C both showed almost the same tensile- and proof stress. The effect of fabrication route and testing temperature on tensile properties of composites are similar to that on the unreinforced AZ 91. In squeeze cast produced composites there was less fibre damage which was reflected in a lower decrease in strength with increasing temperature. The temperature dependence of the ductility, measured as elongation to fracture, of P/M and sq./c matrix alloy and composites are shown in Fig. 2d. Reinforcement decreases substantially the ductility of the matrix

alloy of the both production routes. Whereas increasing testing
temperature improved appreciably the ductility of the matrix alloy,
the ductility increase of composite due to rising testing temperature
were only modest.

Fracture toughness tests were carried out at room and elevated temperatures using short rod specimens. The values of fracture toughness
of the matrix alloy and composites produced by P/M and sq./c method
obtained at room temperature are presented in Fig. 3a. The results
show that reinforcement decreases the values for the matrix alloy for
both production routes. The fracture toughness of the sq./c composite
was approximately 70% of that of the sq./c matrix alloy. Almost the
same relation was found between P/M composites and P/M matrix alloy.
The fracture toughness of the P/M matrix alloy was only 20 % of that
of the sq./c matrix alloy. The influence of production routes on the
fracture toughness of composites was slightly less than that for
matrix alloy. The values for the P/M composites were almost the same,
which indicates that the form of reinforcement has little effect. The
results of the tests at temperature up to 200°C are shown in Fig. 3b.
Al_2O_3 fibre sq./c composite revealed higher fracture toughness values
at elevated temperature than P/M SiC-particle composites. On increasing the temperatures the fracture toughness of sq./c. AZ 91 +
Al_2O_3 decreased slightly due to the change in interface properties.
The fracture toughness of P/M composites may be influenced by two
contrary parameters. Increasing testing temperatures have a stress
releasing effect which improves the toughness. On the other hand
microcrack formation at the interface between the particles and
matrix lowers the fracture toughness. These effects balance out.

Depending on fabrication route the fracture surface of the alloy
exhibited entirely different morphologies. Whilst the fracture
surface of the P/M produced alloy showed a preferentially orientated
fracture plane in the extrusion direction, random fracture planes
have been observed for sq./c matrix alloy. The fracture surface of
sq./c matrix alloy showed a mixed mode of cleavage and dimple ductile
fracture. The fracture surface of P/M AZ91 shows an intergranular
fracture. Some elongated dimples in the extrusion direction and
perpendicular to fracture plane have also been observed.

The P/M composites like the P/M matrix alloy also showed the orientation dependence on fracture surface (Figs. 4 b and 4c) . Whereas
reinforcement in P/M composites were orientated in the extrusion
direction the fibres in sq./c composites were randomly distributed
(Fig. 4a). In addition to the orientation differences the composites
revealed mainly a brittle fracture appearance with exception of
elongated dimples in extrusion direction on the plane perpendicular
to the fracture surface of P/M composites (Figs. 4b and 4c). It can
also be seen that the fibres of sq./c composite broke flush with the
fracture surface (Fig. 4a). Unlike sq./c composites there is little
evidence of a reaction zone at the interface between matrix and reinforcement in P/M composites. The fracture surfaces of composites
tested at elevated temperatures exhibited the same features as those
at room temperature. Increasing the testing temperatures results in
fiber/matrix debonding in sq./c composite and crack formation in the

303

extrusion direction in P/M composites. It appears that microcrack of
the damaged fibre in sq./c composite and crack formation mainly along
the elongated powder grain interfaces have been responsible for low
ductility and fracture toughness of composites at elevated temperatures. Microscopic examination of the crack path in controlled
fracture test specimens of fibre reinforced composites showed that
the tensile residual stresses at the fibre/matrix interface due to
differences in the coefficient of thermal expansion between matrix
and fibres facilitate the crack initiation and crack propagation.
However the nature of the crack path through the matrix, for instance
intergranular crack growth, indicates a matrix controlled fracture
mechanism (Figs. 5a and 5b). The crack pattern of the SiC particle
reinforced P/M composite showed also a matrix-controlled fracture
micromechanism. There is little evidence of particle cracking
(Fig. 5c).

CONCLUSIONS

Fabrication routes and types of reinforcements influence the tensile
properties both of the magnesium alloy and composites. The P/M matrix
alloy and composites posses much higher tensile strengths and proof
stress than sq./c matrix alloy and composite which is induced by
extrusion process. With increasing temperature a substitantially
decrease of the tensile strength and proof stress of P/M matrix alloy
and composites is observed, which leads to compareable strength at
200°C with sq./c.-materials. The fracture thoughness of the matrix
alloy and composites were also markedly affected by the fabrication
routes. Both matrix alloy and composites produced by P/M methods exhibited much lower fracture toughness at room temperature compared
with the sq./c matrix alloy and composite. The micromechanism of the
crack initiation and crack propagation in composites are mainly
controlled by the matrix of the composite.

REFERENCES

1 - S. P. Rawal. M. S. Misra and L. F. Allard, ICCM & ECCM Sec.
Europ. Conf. on Comp. Mat., 2.169
2 - K. U. Kainer, J. Schröder, B. L. Mordike, ECCM3 Third Furpo.
Conf. on Conf. Mat., 171-176
3 - E. G. Wolff, B. K. Min and M. H. Kural, Journal of Materials
Science 20 (1985), 1141-1149
4 - J. Schröder, K. U. Kainer and B. L. Mordike, ECCM3 Third
Europ. Conf. on Comp. Mat., 221-226
5 - K. Purazrang, K. U. Kainer and B. L. Mordike, Fracture toughness behaviour of magnesium alloy metal matrix composite produced by
infiltration technique, to be published
6 - K. Purazrang, K. U. Kainer and B. L. Mordike, The effect of
discontinuous alumina reinforcement content on stress-strain and
fracture toughness behaviour of magnesium matrix composites, to be
published

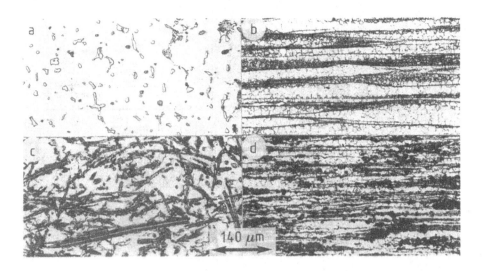

Fig. 1: Microstructure of unreinforced and reinforced AZ91.
  a) squeeze cast alloy   b) P/M produced alloy
  c) squeeze cast composite d) P/M produced composite

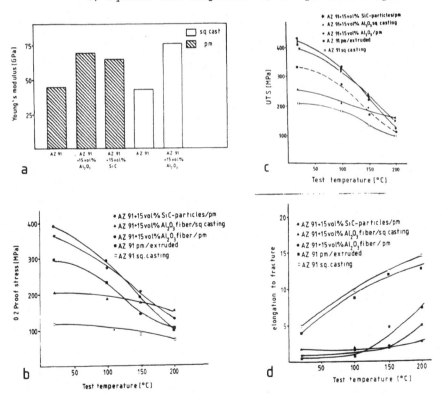

Fig. 2: Effect of alumina short fibre and SiC$_p$ reinforcement on
  Young's modulus (a), 0.2 proof stress^p (b), UTS (c) and
  elongation to fracture (d) of cast and P/M AZ 91 material.

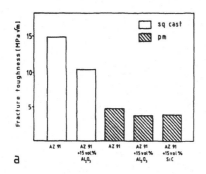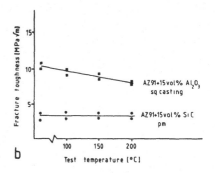

Fig. 3: Effect of fabrication routes and testing temperature
on fracture toughness of the AZ91 materials:
a) room temperature, b) temperature dependence.

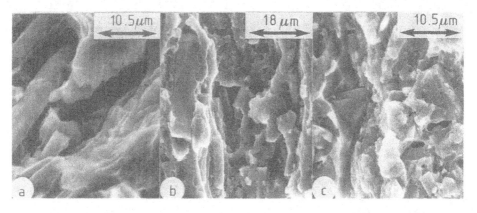

Fig. 4: SEM fractographs of different magnesium composites
a) sq./c. AZ91 + 15 vol.% Al_2O_3
b) P/M AZ91 + 15 vol.% Al_2O_3
c) P/M AZ91 + 15 vol.% SiC

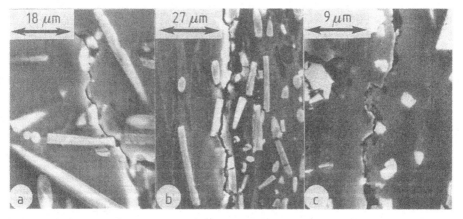

Fig. 5: Crack propagation in controlled fracture tests
a) sq/c AZ91 + 15 vol% Al_2O_3 fibres composite
b) P/M AZ91 + 15 vol% Al_2O_3 fibres composite
c) P/M AZ91 + 15 vol% SiC particle composite

306

THE CORROSION CHARACTERISTICS OF A MODEL C/AL METAL MATRIX COMPOSITE

C. FRIEND, C. NAISH*, T.M. O'BRIEN*, G. SAMPLE

Cranfield Institute of Technology
RMCS Shrivenham - SN6 8LA Swindon - Great Britain
*AEA-Technology Harwell Labs
Harwell OX11 0RA - Great Britain

ABSTRACT

This paper presents the results from a study into the corrosion characteristics of model carbon-fibre/pure aluminium MMC. These model MMC have been used to identify the fundamental characteristics which control the corrosion response of this type of material. The corrosion responses of these composites have been characterised by a number of electrochemical techniques including long-term immersion, potentiodynamic sweeps, and AC impedance measurements. Typical data is presented for all these techniques, and the characterisations derived from the different measurements are compared in the context of the overall corrosion resistance of these materials.

INTRODUCTION

Currently there is considerable technological interest in Metal Matrix Composites (MMC) since materials reinforced with high strength phases represent a unique method for tailoring a material's physical properties to suit particular applications. Currently most work on such systems has concentrated on optimisation of their primary mechanical properties, however, in many real applications additional secondary properties must also be considered. In most applications MMC will be subjected to various environments and their corrosion resistance is therefore an important material property. Little systematic data exists on the corrosion characteristics of MMC, however, the data from a number of unrelated systems indicates that generally one of two corrosion responses is observed.

1. MMC exhibit similar corrosion characteristics to those of the unreinforced matrix/1,2/.
2. MMC exhibit enhanced susceptibility to corrosion/3,4/.

Although these different responses can be observed, in practice it is often difficult to identify the fundamental processes which contribute to the corrosion response. This is because much of the work carried out previously has been conducted on commercial MMC systems. Where some mechanisms have been identified it has also been difficult to distinguish (i) the relative importance of these contributions and (ii) the appropriate corrosion measurement techniques for each of these processes. The solution to such problems is to carry out corrosion measurements on model MMCs. This paper presents the results from the initial stage of such a study. In this work MMCs have been fabricated with simple fibre geometries and microstructurally simple matrices. These model systems have been used to compare the corrosion responses derived from a number of standard electrochemical measurement techniques with the aim of identifying (i) differences between the characterisations and (ii) the fundamental processes which contribute to the corrosion response of these materials.

I - MMC FABRICATION

Three materials were investigated in this study. These were high modulus (HM) and high strength (HS) carbon-fibre reinforced 99.99% Al, and unreinforced 99.99% Al. The composite materials were produced by preform infiltration using unidirectionally oriented preforms containing a fibre fraction of 0.75. This high volume fraction was employed to maximise any possible effects resulting from the introduction of C-fibres into an aluminium matrix. Unreinforced aluminium was also cast under identical conditions for comparison with the composite materials. Encapsulated specimens were produced from these casings with exposed surface areas of 1cm². These were cut both parallel and normal to the fibre direction, and were abraided to 320 grit to obtain a reproducible surface finish.

II - ELECTROCHEMICAL MEASUREMENTS

The corrosion behaviour of all three materials were investigated in 3% NaCl solution. These were either saturated with air or deoxygenated using an argon gas purge. Two short-term techniques were employed (AC impedance and potentiodynamic sweeps) in combination with longer term electrochemical measurements. AC impedance measurements were carried out in deaerated solutions under potentiostatic control at the free corrosion potentials. These measurements were made using a Solartron 1186 Electrochemical Interface in combination with a Solartron 1250 Frequency Response Analyser (FRA) controlled by a Hewlett Packard 216 microcomputer. This computer-based system was used for both data acquisition and processing using Harwell AC Impedance software. The measurements were carried out over a frequency range of 1mHz to 100kHz. Electrochemical sweep measurements were carried out under deaerated conditions using Thompson Electrochem potentiostats and a Hewlett Packard 3497A data acquisition unit controlled by a 216 microcomputer. During such measurements the potentials of the samples were swept at a rate of 5mVmin⁻¹ from values close to their rest potentials to - 700mV (with respect to the Saturated Calomel Reference Electrode (SCE)). Once this final level was reached the samples were then 'inverse-swept' to their final rest potentials. Free corrosion potential-time measurements were conducted using a SCE, a high impedance digital voltmeter, and a high impedance buffer amplifier. These measurements were carried out over a period of 38 days under both aerated and deaerated conditions.

III - RESULTS AND DISCUSSION

Fig 1 shows Nyquist plots from AC impedance measurements on the three materials. Plots for the unreinforced aluminium showed a characteristic Randles' Circuit response with a low frequency tail. The HM composite tested over the same frequency range exhibited a similar response but with a slower reaction time constant, and the HS material showed an almost purely capacitive response. The free corrosion responses of these simple MMC were therefore significantly different from that of the unreinforced aluminium, and from one another. Further work is required to understand these differences however, the measurements do allow a semi-quantitative ranking of the three materials in order of their free corrosion rates. This can be carried out by comparing their low frequency real impedances. Such results are shown in table 1.

Fig 2 shows the results from potentiodynamic sweeps on the three materials. A crude ranking of the free corrosion rates can be obtained from such measurements by considering the corrosion currents associated with the passive regimes (table 2). These are interesting since they confirm the AC impedance measurements. This shows that both the short-term electrochemical techniques produce similar characterisations for the relative corrosion rates of the three materials. The potentiodynamic sweeps also produce additional information on the nature of the corrosion processes under imposed conditions. The most noticeable effect resulting from the presence of the fibres was the introduction of pitting at noble (more positive) potentials. This indicates that the presence of carbon fibres in pure aluminium produces an increased tendency for pitting corrosion, with the breakdown to pitting occurring at lower potentials in the MMC. Similar break-down potentials were observed in both MMC despite the fact that their general free corrosion rates were significantly different. The extent of pitting, as indicated by the difference in the hysteresis at positive pitting potentials, also varied with the type of reinforcement. Previous work on C-fibre reinforced MMCs has observed the susceptibility of these materials to pitting corrosion and discussed a number of possible mechanisms. These include galvanic coupling between the matrix and reinforcement[5], hydrolysis of Al_4C_3 interfacial reaction product[6] or a combination of both processes. In the present work there was no evidence of extensive Al_4C_3 formation which suggests that the enhanced pitting was associated with galvanic corrosion caused by the electrochemical difference between the C-fibre and aluminium matrix. However, further investigations are required to explain (i) the similarity in the pitting potential with both reinforcement types, and (ii) the variation in the extent of pitting in the two MMC.

The effect of immersion time on the free corrosion potentials under deaerated conditions is shown in fig 3. The HM composite consistently

309

exhibited a lower corrosion potential compared to the HS material. This agrees with the shorter-term electrochemical results which indicated a higher free corrosion rate for the HM composite, but an increased susceptibility to pitting in the HS material. Over the same time-scale the pure aluminium exhibited a highly variable free-corrosion potential, although its response was nearer to that of the HM composite. This again agrees with the short-term results. The response of the pure matrix was in general agreement with that observed in aluminium. This is usually explained in terms of the dissolution of the air-formed oxide film following short immersion times, with an associated increase in corrosion rate. However, at longer times growth of a protective solution-formed film occurs which returns the material to a more noble potential.

CONCLUSIONS

1. Short and longer-term electrochemical measurements have successfully identified differences in the corrosion behaviour of two model MMC compared to the unreinforced matrix.

2. The characterisations obtained from short-term perturbation techniques agree well with each other, and with longer-term immersion tests. This indicates that in these materials the short-term tests give a good indication of the longer-term behaviour.

3. Under imposed conditions potentiodynamic measurements produce additional data on the nature of the different corrosion processes.

4. Further work is required to fully understand the differences in behaviour between the model MMC materials.

REFERENCES

1 - Nath D., and Namboodhiri TKG., Composites, 19 (1988).
2 - Otari T., McEnaney B., and Scott VD., 'Cast Reinforced Metal Composites', Proc. Int. Symp. on Advanced in Cast Reinforced Metal Composites, Chicago, Sept 1988 (ed. S. G. Fishman and A. K. Dhingra), ASM Int. 1988 pp 383-389.
3 - Trzaskoma PP., McCafferty E., and Crowe CR., J. Electrochemical Soc. 130 (1983) 1804.
4 - Mansfield F., and Jeanjaquet S. L., Corrosion Science 26 (1986) 727.
5 - Aylor DM., and Moran PJ., Elecrochem. Science and Tech 132 (1985) 1277.
6 - Wu R., and Cai W., Proc. ICCM6 (Elsevier App Sci, 1986) Vol 2 pp 128-137.

TABLE 1. REAL COMPONENTS OF A.C.IMPEDANCE AT 1mHz

	Impedance (kΩ)	Corrosion Rate
75v/o HM-C MMC	10	Highest
99.99% aluminium	32	
75 v/o HS-C MMC	40	Lowest

TABLE 2. RELATIVE CORROSION RATES FROM POTENTIAL SCAN "PASSIVE CURRENTS"

	"Passive Current" (μAcm^{-2})	Corrosion Rate
75 v/o HM-C MMC	10	Highest
99.99% aluminium	10	
75 v/o HS-C MMC	1	Lowest

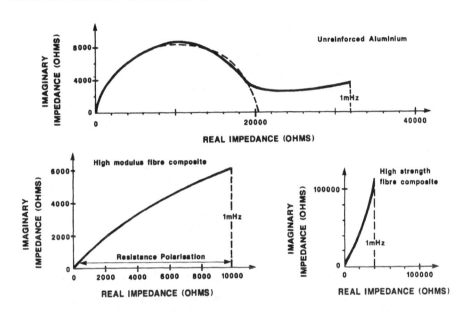

Fig. 1 NYQUIST PLOTS FROM A.C. IMPEDANCE MEASUREMENTS

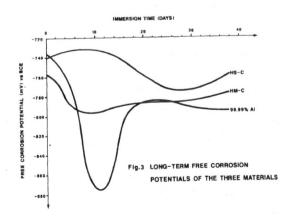

Fig.3 LONG-TERM FREE CORROSION POTENTIALS OF THE THREE MATERIALS

311

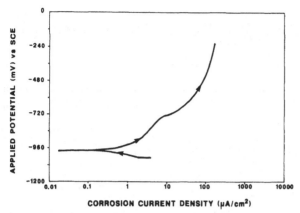

POTENTIODYNAMIC SWEEP FOR 99.99% ALUMINIUM.

ARROWS INDICATE THE SWEEP DIRECTION, SWEEP RATE 5mV min⁻¹

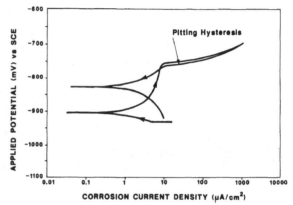

POTENTIODYNAMIC·SWEEP FOR HM–C.

ARROWS INDICATE THE SWEEP DIRECTION, SWEEP RATE 5mV min⁻¹

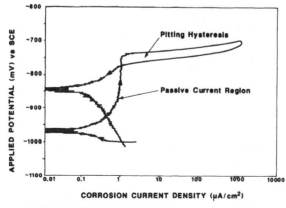

Fig. 2

POTENTIODYNAMIC SWEEP FOR HS–C.

ARROWS INDICATE THE SWEEP DIRECTION, SWEEP RATE 5mV min⁻¹

FIBRES-REINFORCED ZINC-ALUMINIUM MATRIX COMPOSITES FOR IMPROVED CREEP RESISTANCE AND FRACTURE TOUGHNESS

M.A. DELLIS, J.-P. KEUSTERMANS, F. DELANNAY, J. WEGRIA*

*Université Catholique de Louvain - Dépt. des Sciences
des Matériaux et Procédés - PCPM Reaumur - Place Sainte-Barbe 2
1348 Louvain-La-Neuve - Belgique
*Vieille Montage - a division of ACEC - Union Minière SA
Centre d'Etude des Applications - BP 1 - 59950 AUBY - France*

ABSTRACT

This work aim at an assessment of the effects of fibers reinforcement and the influence of manufacturing parameters on microstructure and properties of zinc-based alloys commonly used in the manufacturing of parts by pressure die casting. Microstructural observations were made on the squeeze infiltrated samples. The following properties have been investigated: young modulus, hardness, impact strength (Charpy), creep resistance and thermal expansion coefficient. The failure behaviour is espacially discussed.

INTRODUCTION

Among the zinc-based foundry alloys, the ZA family of alloys has been used increasingly during the past years. These alloys present some advantages in comparison to the ZAMAK family or the Al-based alloys, specially a high strength with a low casting temperature. Pressure casting of these alloys in hot or cold type of casting machines makes possible high production rates at a moderate cast.

The mechanical properties of the ZA alloys are generally considered satisfactory froom room temperature up to about 100°C. Above this temperature, the applications are restricted owing to the decrease of the creep resistance. It is known that metal matrix composites (MMC) usually exhibit improved creep properties with respect to the unreinforced matrix /1/. Examples of the application of the composite concept to zinc-based alloys are fairly scarce in the literature /2,3/. These studies have been often concentraded on wetting and interfacial rections with little emphasis on the achievable properties. The aim of the present work is to explore the possibilities of producing affordable zinc-matrix composites which could enhance the maximum temperature of application of zinc-based foundry alloys. In order to assess a range of properties as wide as possible, we chose to compare three types of reinforcing fibres: alumina (Saffil), Carbon and a chromium-rich steel (Fecralloy). It was also expected that fibre reinforcement could reduce the thermal expansion coefficient which may be beneficial for precision casting.

I EXPERIMENTAL

The composites were prepared by the squeeze infiltration of a preheated fibre preform under a pressure of 20 MPa. Two types of matrix alloys (ZA8 and ZA 27) were combined with three types of reinforcements: preforms of δ-Al_2O_3 short fibres (Saffil), felts of carbon fibres (Le Carbone) and felts of Fecralloy fibres (Bekaert S.A.). These reinforcements were selected both for the range of properties that they span and for their potential availability at a moderate price. Tables 1 and 2 give the compositions and characteristics of the alloys and fibres together with the temperatures of casting of the alloys and of preheating of fibres. The duration of the squeeze casting process from the casting of the liquid alloy to the quenching of the sample in cold water is about 3 minutes. The volume fraction of fibres was maintained at about 20% in all composites samples.

The microstructure of the composites was studied by optical microscopy and SEM. Brinell hardness was measured using a 2,5 mm dia. steel sphere. Young moduli were computed from the fundamental frequency of vibration of bar specimens (Grindo-Sonic). Charpy tests were conducted on unnotched 10x10x55 MM bars. Thermal expansion coefficients were measured on 3x3x30 mm bars in an Adamel-type dilatometer. Creep tests were conducted at 80°C under 70 MPa.

II-RESULTS AND DISCUSSION

2.1.Microstructure

Microscopic observations on polished sections of castings attest a very homogeneous dispersion of the fibres in the matrices with no significant trace of porosity. Down to the resolution of SEM imaging, no evidence was found of the presence of a reaction layer at the fibre-matrix interface in any of the composites.
Figure 1 illustrates the typical microstructure of the matrix around the fibres in a $ZA27/Al_2O_3$ composite. The fibres are surrounded by the most Zn-rich (brighest) phase wich solidifies the last whereas the Al-rich dendrites (dark phase) which solidify first have grown at some distance from the interface. This indicates that fibres do not act as nucleation sites for solidification Instead, as observed also in Al-base composites / /, fibres tend to segregate in the liquid. This phenomenon was very apparent in both alloy matrices in the presence of AL_2O_3 and Fecralloy fibres. It was less obvious in the presence of carbon fibres. The difference might be due to the lower specific heat of carbon or to a higher tendency to nucleate reaction products at the interface.

2.2.Mechanical properties

Table 3 presents the values of Young modulus, Brinell hardness, coefficient of thermal expansion and Charpy energy measured on the two matrix alloys and on three composites: ZA8/Fecralloy, ZA8/Carbon and ZA27/Saffil.

314

The Young modulus is increased in the presence of Fecralloy and
Saffil fibres whereas it is reduced in the presence of Carbon fibres.
In the cases of Fecralloy and Saffil fibres, the experimental values
have been compared to the theoretical values of the Young moduli
computed using the Halpin-Tsaï equations /4/. In order to account for
the random orientation, we used the relation

$$E = 3/8 \ E_L + 5/8 \ E_T$$

where E_L and E_T are the longitudinal and transverse moduli. The lower
stiffening effect observed experimentally suggests a weak interface
bonding with presumably some decohesion at the interfaces, espacially
for the fibres perpendicular to the loading direction. No model are
however presently avaible allowing to evaluate the extent of such
decohesions from Young modulus measurements. In the case of the
carbon fibres, one should also consider the very low transverse
modulus of the fibres.

The higher hardness of ZA27 in comparison to ZA8 is due to the
higher proportion of the eutectic phase. All three composites present
higher hardnesses than the matrix alloy. It is conspicuous that this
effect increases strongly when the size of the reinforcing fibres
decreases. This agrees with the theory which ascribe the main
contribution of the strengthening in metal matrix composites to the
dispersion hardening of the matrix by the reinforcing fibres or
particles /5/.

The most striking effect in Table 3 is the reduction of the
thermal coefficient of the composites with respect to that of the
matrix. This indicates that, in all three composites, the interface
bondings, though relatively weak, are still large enough to induce a
large constraint in the matrix so as to reduce significantly the
macroscopic strain. The highest effect (30%) is observed with the
Fecralloy fibres, although these fibres have the highest coefficient
of thermal expansion (see table 2). Quantitative interpretation of
these behaviours should take into account the size, aspect ratio and
orientation of the fibres and the relative values of thermal expansion
coefficient, young modulus and yield strength of both fibres and
matrices /6/.

The values of the Charpy energy at 20°C presented in table 3
indicate that the fibres cause a large decrease of the impact
toughness. The highest relative decrease is observed in the case of
the Saffil fibres. It is well known that MMC generally exhibit a
reduced toughness in comparison to the matrix alloy. This is due to
the occurence of damage in the form of fibre cracking and interfacial
debonding, the final rupture then proceeding by ductile tearing of the
metal matrix /7/. Figure 2 compares the fracture surfaces of the
ZA8/Fecralloy and ZA8/C. It appears that the matrices have failed by
a mixed brittle-ductile mechanism.(Room temperature is close to the
transition temperature of the ZA alloys). In the ZA27/Saffil
composite, the fibres appear to heave broken very near to the plane of
rupture of the matrix, with no apparent decohesion and pull-out. In
In the ZA8/Fecralloy and ZA8/Carbon composites, one observes instead
evidence of decohesion and pull-out. These phenomena are presumably
responsible of an increase of the work of rupture /8/,which might
explain why the decrease of the Charpy energy is limited to a factor
of 2 instead of a factor of 8 in the case of the Saffil fibres.

The Fecralloy fibres even show evidence of plastic stretching which may be a patent strengthening mechanism /8,9/. However, the contribution of such a mechanism may be limited in a Charpy test as the high strain rate and the constraint of the matrix may favor a brittle rupture of the Fecralloy fibres.

Figure 3 compares the creep curves at 80°C under 70 MPa for the ZA27 alloy and a ZA27/Al$_2$O$_3$ composite. It shows that the presence of Al$_2$O$_3$ fibres induces a two orders of magnitude decrease of the creep rate. It is usually interpreted as being due to the stabilization of a higher dislocation density, which brings about a drastic reduction of the power law (climb governed) creep. Fibre reinforcement thus appear a promising way for improving the high temperature properties of zinc-based alloys.

2.3.Comment about the role of interface bonding strength

The most promising properties which have been demonstrated in our zinc-based composites are the reductions of creep rate and thermal expansion coefficient. (high specific stiffness or strength is not a major issue for the applications of zinc-based alloys). the more crucial problem to be dealt with is the retention of adequate fracture toughness and ductility. Low modulus carbon fibres or ductile steel fibres ought to be prefered to brittle alumina fibres from this point of view. Optimum toughness also requires a proper tayloring of the interfacial bonding, which should be strong enough to avoid early damaging during straining, but weak enough to allow branching of cracks along interfaces and toughening by pull-out and, possibly, plastic flow of the fibres during crack propagation. The dependence of toughness and ductility on temperature should also be carefully assessed.
At the present stage of our investigation, a major unknown remains the actual value of the interfacial bonding strength. The absence of detectable reaction layer and the observation of fibre pull-out on fracture surfaces suggest that, at least in the case of carbon and Fecralloy fibres, the bonding may be more mechanical than chemical. In such a case, it should depend largely on the roughness of the fibre surface and on the residual thermal stresses in the composites. Our future work will focus on modifying bonding strength through thermal processing.

CONCLUSIONS

1.This work demonstrates the preparation of sound fibres reinforced zinc-alloy matrix composites with no reaction products at the fibre-matrix interface.
2.These composites exhibit significantly lower creep rates and thermal expansion coefficients than the matrix alloys.
3. The main problem remains the retention of adequate fracture resistance and ductility. Carbon or steel fibres appear more advantageous than alumina fibres from this point of view, but interface bonding should be taylored for optimum properties.

ACKNOWLEDGEMENTS

The authors are indebted to Mr Demaegd of Bekaert S.A.,Zwevegem, Belgium, for the supply of the Fecralloy fibres.Financial support for this work was partly provided by the Service de Programmation de la Politique Scientifique Belgium (Pôle d'attraction interuniversitaire). J.P. Keustermans acknowledges a grant of IRSIA, Belgium. M.A. Dellis acknowledges a grant of Vieille Montagne, a division of ACEC Union Minière S.A..

REFERENCES

1.H. Lilholt and M.Taya, "ICCM 1987",(W.C. Harrigan Jr, J. Striffe, A.K. Dhingra ed.), The Metall. Soc., (1987), vol 2, 2.234
2.A Madronero, Rev. Metal. Madrid, 21 (6), (1985), 346
3.J.A. Cornie, R. Guerriero, L. Meregalli, I. Tangerini, "Cast Reinforced Metal Matrix Composites", conference proceedings,(S.G. Fishman , A.K. Dhingra ed.),(sept 1988), 155
4.R.L. Mehan, Journal of Mater. Sci. 13 (1978), 358
5.R.J. Arsenault and N. Shi, Mater. Sci. Eng., 81, (1986), 175
6.D.Masutti, J.P. Lentz and F. Delannay, J. Mat. Sci.letters, 9, (1990),340
7.A.K. Vasudevan, O. Richmond, F. Zak and J.P. Embury, Mater. Sci. Eng. A, 107,(1989), 63
8.A.G. Evans, D.B. Marshall, Acta Metall. 37 (1989) 2567
9.M.F. Ashby, F.J. Blunt and M. Bannister, Acta Metall. 37 (1989) 184

Table 1 : Typical composition range of ZA8 and ZA27 (% in weight)

Alloy	Al	Cu	Mg	Zn	casting temp
ZA8	8-8,8	0,8-1,3	0,015-0,03	remainder	520 °C
ZA27	25-28	2-2,5	0,01-0,02	remainder	660 °C

Table 2 : Characteristics of fibers

	Saffil	Carbone	Fecralloy
Composition	96-97%Al2O3 (δ) 2,3% SiO2	C	20%Cr,5%Al, Fe
Diameter	3 µm	10 µm	22 µm
Length	500 µm	1 cm	1 cm
Preheating temp	1000 °C	500 °C	300 °C
Vol. fraction	~ 20 %	~ 20 %	~ 20 %
Thermal expans. coeff.(10-6/°C)	7	~ 0	11

Table 3: mechanical properties

	ZA8	ZA27	ZA8/Fercralloy	ZA8/C	ZA27/Saffil
Young Modulus(GPa)	89	89,2	92	78	94,6
Theoretical Modulus			105		114
Brinell hardness	85-90	110-120	96	132	177
Thermal expansion coefficient(10-6/°C)	29	28	20	25	22,5
Impact strength(J/cm2)	9	16	4	4,4	2

Fig.2a

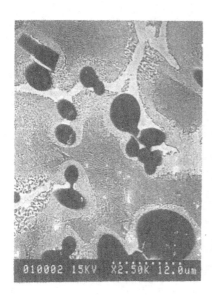

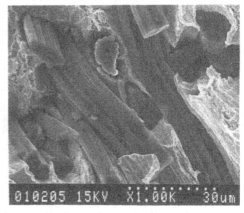

Fig.1

Fig.2b

Fig.1: SE micrograph of a polished ZA27/Saffil.

Fig.2: SE fractograph of a ZA8/Fecralloy (2a) and ZA8/C impact
specimen. (2b)

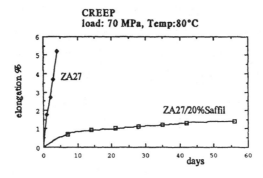

Fig.3: Comparison of the creep curve of a ZA27/Saffil composite with
that of the matrix alloy.

HIGH TEMPERATURE LOW CYCLE FATIGUE
OF A SIC/TI MATRIX COMPOSITE

W. WEI

MTU Motoren Und Turbinen Union GMBH
Dachauerstr. 665 - 8000 München 50 - West-Germany

ABSTRACT
Low cycle fatigue studies have been conducted on SiC continuous fiber
reinforced Ti-6Al-4V in air at temperatures up to 500 °C. Through the
use of beachmarks and fatigue striations, the progress of the crack
front could be retraced using the scanning electron microscope. The
results of these morphological studies indicate that during crack propagation, the crack bows out between the fibers analogous to dislocation bowing. At higher temperature there is an increased tendency for
interlaminar cracking and the appearance of stepped crack surfaces.
These observations have important consequences in the development of
fracture mechanics models for MMC life prediction.

INTRODUCTION

The proper design of high temperature components using continuous-fiber reinforced metal matrix composites (MMCs) presents new challenges
to currently available life prediction methods. The relatively wide
use of linear elastic fracture mechanics (LEFM) for the description of
the crack growth behavior of MMCs notwithstanding, these materials are
not and cannot be considered to be homogeneous either in the macroscopic or microscopic sense. In addition, crack growth and failure mechanisms vary strongly with fiber/matrix combination and interfacial properties [1-4]. Crack growth in as-received Ti based MMCs approximates
self-similar growth behavior, whereas Al based MMCs with weaker interfaces tend to fail through delamination or so-called H-type behavior.

The purpose of this paper is to present results of a morphological
study of the low cycle fatigue (LCF) behavior of a SiC continuous-fiber reinforced Ti alloy. It will be shown that, although crack growth
in this material approximates self-similar behavior, care must be

taken in applying traditional methods such as LEFM for fatigue life prediction of these inhomogeneous materials.

EXPERIMENTAL PROCEDURE

LCF testing was conducted on SiC (SCS-6) continuous fiber reinforced Ti-6Al-4V (Textron). The fatigue loading consisted of a load controlled trapezoidal waveform (1 sec rise / 1 sec hold at maximum stress / 1 sec fall / 1 sec hold at minimum stress) with a stress ratio of $r=\sigma_{min}/\sigma_{max}=0.1$. Maximum stresses used were 1000 and 1100 MPa. Beachmarks were created on the fracture surfaces to facilitate the observation of crack growth [6].These were produced using a higher frequency (2 Hz) triangular waveform with r=0.7 at a mean stress of 0.85 of maximum. The two waveforms were applied alternately in blocks of 1000 cycles trapezoidal waveform and 3000 cycles triangular waveform to create an array of beachmarks.

Rectangular specimens, 10 cm long by 1 cm wide were cut from 2 mm thick plate using a diamond cutoff wheel. The specimen edges were intentionally left in the as-cut, unpolished state in order to enhance crack initiation. Fatigue crack growth could then be obtained within reasonable test times. The specimens were tested at temperatures up to 500 °C in the as-received or a heat treated condition.The heat treatment involved annealing in air at temperatures up to 500 °C for ca. 500 hours in order to simulate operating conditions [5]. Heat treated specimens were tested at room temperature or at the corresponding annealing temperature.

Fracture surface analysis was conducted using scanning electron microscopy (SEM). The progress of the crack could be followed using the beachmarks in combination with the usual fatigue striations found on the fracture surface of the matrix material. Crack initiation sites could be located through the curvature of the beachmarks.

RESULTS

A typical fracture surface from the LCF tests is shown in Fig. 1. A relatively flat crack surface with a small amount of fiber pull-out can be seen. A number of crack initiation sites were found (arrows), mostly at the intersection of fibers with the specimen edge. Except for the fiber pull-out, this crack could be considered to have grown in an approximately self-similar manner. However, there is evidence of a step in the crack (delamination) and fiber splitting.

The progress of the corner crack can be followed using beachmarks and river lines, Fig. 1 . The beachmarks show that the crack front was not perfectly circular as would be assumed upon first observation. Evidence of crack front bowing through the fibers is highlighted by the beachmarks. This bowing is similar to that of dislocation bowing through microstructural obstacles. In addition, it appears that the crack "flowed" around the fibers in a manner similar to the flow of a fluid around cylinders in a flow channel as seen by the river lines. This type of flow behavior summarized schematically in Fig. 2.

It was further observed that the extent to which the crack front in-
teracted with the fiber/matrix interface depended upon heat treatment.
For the as-received condition, the fibers were well bonded to the
matrix, and high magnification views of fatigue striations near the
interface showed a complex interaction of the crack with the fibers.
For specimens annealed for long times at relatively high temperatures,
the fibers had reacted strongly with the matrix, resulting in interfacial separation [5]. The crack front then progressed past the fibers
with very little influence of the fibers on crack front curvature. A
tendency towards increased delamination was also observed for the heat
treated specimens. Also to be noted in the schematic diagram of Fig. 2
is that new cracks initiated on the "lee" side of the fibers, which
eventually joined up with the main crack as it passed.

DISCUSSION AND CONCLUSION

SEM studies of the LCF crack surfaces show that the crack growth of a
SiC/Ti MMC under low cycle fatigue loading exhibits a fairly complex
behavior. The crack front progresses through a field of fibers in a
manner similar to dislocation bowing, but also similar to a fluid
through cylinders. The interaction of the fibers with the crack front
depends on interfacial bonding, with heat treatment resulting in weaker bonding and less influence of the fibers at least on Mode I crack
growth (perpendicular to the loading direction).

It would appear that crack growth in MMCs such as seen in Fig. 1 can
be treated using conventional fracture mechanics methods. Many investigators and manufacturers of MMCs indeed use LEFM and the concept of
the stress intensity factor, K, and fracture toughness, K_{IC}, to describe the fracture and fatigue behavior of MMCs. While this may seem
to be a convenient way to present fracture and crack growth data, a
number of important assumptions upon which LEFM is based are not
necessarily fulfilled. These include the following:
  1. The material is homogeneous.
  2. The plastic zone size ahead of the crack tip is much smaller
     than the crack length.
  3. For fatigue loading, the fatigue life of a component is con-
     trolled by the initiation and growth of a single crack.
  4. The results of life prediction should be transferable to all
     materials and component geometries.

For MMCs the first assumption is clearly not true. For the second assumption, the plastic zone size cannot be defined, since, for example,
the plastic zone in the matrix material will be constrained by the interaction with a hard,elastically behaving fiber. Stereomicroscopy has
been used to show the complexity of this interaction [7]. An alternative to the plastic zone would be to define a damage zone size ahead
of the crack tip. By etching the matrix of a SiC(SCS-6)/Ti material
ahead of the crack tip, it was shown that the fiber coating was segmented in a large zone ahead of the crack tip [8]. The size of the
zone indicates that the second assumption may also be invalid.

The third assumption is not a fundamental assumption in the development of LEFM itself, but is the basis for the component life predic-

tion of monolithic metals. It is assumed that although a number of cracks may be initiated in a component, only one eventually grows sufficiently to control the failure process. An example of problems associated with applying this assumption to MMCs is shown in Fig. 3. The main fatigue crack which led to failure of a specimen tested at room temperature in the as-received condition is shown. As indicated in the accompanying tracing, the crack actually consists of two independent cracks, one which initiated from the rough specimen edge, A, and the other which initiated at an internal seam, B. Both cracks have approximately the same surface area, and appear to have grown simultaneously, meeting at point C. All of the specimens investigated contained several cracks.Cracks of approximately equal size were found at both edges,and internal cracks were also found originating from broken fibers or incompletely bonded seams between matrix foils. Thus, in all cases, several cracks controlled the failure life of the specimens,and life predictions based on the growth of a single crack are not valid.

A number of alternative solutions to life prediction of MMCs have been proposed in the literature. These include mixed mode LEFM [9], energy release rate models (strain energy release rate, G, or J-integral methods) [10-14], and damage accumulation models [4,15-16]. The energy based models seem to be the most reasonable for initial material comparisons, since the energy released in fracture is an intrinsic property of the material and is a direct measure of the failure behavior of the material. The contributions of the individual energy release mechanisms such as elastic and plastic strain energy, fiber pull-out, delamination, etc., cannot, however be easily separated.

All of the aforementioned methods have the disadvantage of depending on specimen and component geometry. None can be applied to the variety of failure mechanisms which have been observed for MMCs. As further progress in the development of MMCs is made, parallel developmental work will be required to reliably predict the behavior and life components constructed out of these materials.

ACKNOWLEDGEMENTS
The assistance of M. Schormair and J. Ritter, MTU München, in the conducting of the LCF tests is gratefully acknowledged. This work was funded internally by the MTU Motoren- und- Turbinen-Union GmbH.

REFERENCES
1. R.T. Bhatt and H.H. Grime, "Fatigue Behavior of SiC Reinforced Titanium Composites", NASA Technical Memorandum 79223 (1979).
2. D. Davidson et al, in "Mech. Behavior of Metal-Matrix Composites" (J. Hack,M. Amateau,eds.)(1983) 117-142, TMS-AIME, Warrendale,PA.
3. M. Gouda, K.M. Prewo and A.J. McEvily, in "Fatigue of Fibrous Composite Materials", ASTM STP 723 (1981) 101-115, ASTM,Philadelphia.
4. W.S. Johnson, in "Long Term Behavior of Composites", ASTM STP 813 (T.K. O'Brien) (1983) 160-176, ASTM, Philadelphia.
5. W. Wei, in "Fundamental Relationships Between Microstructure and Mechanical Properties of Metal-Matrix Composites" (P.K. Liaw and M.N. Gungor, eds.) (1990) 353-370, TMS of AIME, Warrendale, PA.
6. G.W. König and E.E. Affeldt, in "Proc. 2nd. Int. Conf. on Low-Cycle Fatigue and Elasto-Plastic Behavior of Materials" (K.T. Rie,

ed.) (1987) 673-679, DVM, Berlin.

7. D.L. Davidson, in "Proc. Fifth Int. Conf. on Composite Materials ICCM-V" (W. Harrigan,Jr. et al, eds.) (1985) 175-189, TMS of AIME, Warrendale, PA.

8. D.L. Davidson, in "Fundamental Relationships Between Microstructure and Mechanical Properties of Metal-Matrix Composites" (P.K. Liaw and M.N. Gungor, eds.) (1990) TMS of AIME, Warrendale, PA.

9. K.S. Chan and D.L. Davidson, Eng. Frac. Mech. 33 (1989) 451-466.

10. A. Daimaru, T. Hata and M. Taya, in "Recent Advances in Composites in the United States and Japan", ASTM STP 864 (J.R. Vinson and M. Taya, eds.) (1985) 505-521, ASTM, Philadelphia.

11. J.A. Nairn, in "Proc. American Society for Composites 2nd Tech. Conf." (1987) 58-66, Technomic Pub. Co., Lancaster, PA.

12. J.G. Morley, in "High-Performance Fibre Composites" (1987) Ch. 5, 151-171, Academic Press, London.

13. C. Chiang,W. Chen and K. Chen, Eng. Frac. Mech. 28 (1987) 301-307.

14. G. Newaz and J.-Y. Yung, Eng. Frac. Mech. 29 (1988) 483-495.

15. G. Dvorak and W. Johnson, Int J of Fracture 16 (1980) 16 585-607.

16. K.L. Reifsnider and W.W. Stinchcomb, in "Composite Materials: Fatigue and Fracture", ASTM STP 907 (H.T. Hahn, ed.)(1986) 298-313, ASTM, Philadelphia.

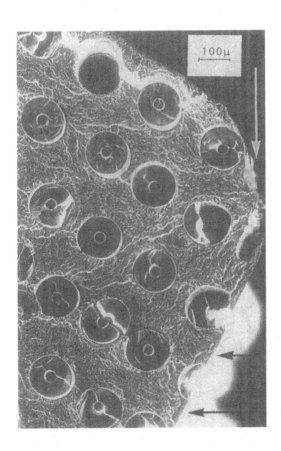

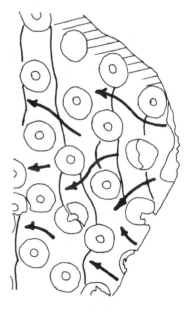

Fig. 1 - LCF crack in specimen tested at 400° C, annealed 506 hrs at 400° C, maximum stress = 1000 MPa.

⌢ beachmark
↢ crack "flow"
///// step (delamination)

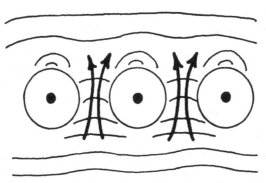

Fig. 2 - Schematic diagram of LCF crack propagation in
SiC continuous-fiber reinforced Ti-6Al-4V.

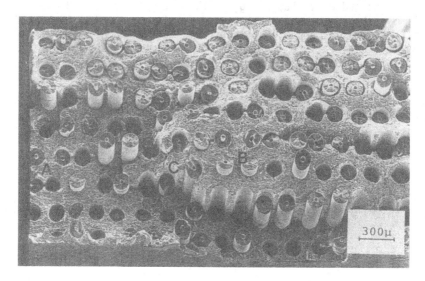

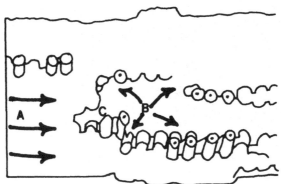

Fig. 3 - LCF cracks in specimen tested at RT, as-received
condition, maximum stress = 1000 MPa.

CONTINUOUS SIC FIBER REINFORCED METALS

M. MITTNICK

Textron Specialty Materials
2 Industrial Avenue - 01851 Lowell MA. - USA

ABSTRACT

Continuous silicon carbide (SiC) fiber reinforced metals (FRM) have been successfully applied on numerous aerospace development programs fulfilling primary design objectives of high specific strength over baseline monolithic materials. This presentation will review the current state-of-the-art in silicon carbide fiber reinforced metals through discussion of their application. Discussion will include a review of mission requirements, program objectives and accomplishments to date employing these FRMs.

INTRODUCTION

The term composite usually signifies a combination of two or more constituent elements (in this case a high-performance, low-weight filler embedded in a metal) to form a bonded quasihomogeneous structure that produces synergistic mechanical and physical property advantages over that of the base elements. Theoretically, there are three types of composites: (1) a particulate based material formed by the addition of small granular fillers into a binder that generally derives an increase in stiffness but not strength; (2) a whisker/flake filler that realizes a greater proportion of the filler strength due to its higher aspect ratio, and hence a greater ability to transfer load; and (3) a continuous fiber system (i.e., Fiber Reinforced Metal, FRM) that, due to fiber continuity, derives the full properties of the high-performance fiber (strength and stiffness).

There are various metal matrix systems commerciallyavailable today that conform to the three types of composites discussed above, the mechanical properties of these various systems vary significantly. The major import is the cleardistinction between continuous fiber system and the discontinuous (particulate and whisker/flake) systems. In the latter, only greater stiffness is generally realized, whereas in the former, both the modulus and the strength of the continuous reinforcing fiber are fully translated into the composite. In practice, if all of the filler strength is to be translated into the composite, then (a) the filler must have sufficient aspect ratio to transfer the load from "filler" to "filler" through the matrix, (b) the filler particles must be in close proximity to each other to avoid large matrix distortions (high volume loading), and (c) a large number of "filler termination sites" placed in close proximity must be avoided to preclude stress concentrations that overload the filler and matrix. In practice, it is very difficult to satisfy all of the requirements, and hence achieving the theoretical potential of the filler is close to impossible.

Although the mechanical properties of the continuous fiber composite systems (FRM) are superior to those of the discontinuous systems, the ease of providing material isotrophy is a significant advantage, for there are many stiffness-controlled applications that benefit from discontinuous reinforced materials. These applications generally involve complex geometries where it is difficult to position continuous fibers during the fabrication process. However, where there is sufficient volume of material to allow for orientation of fibers, the FRM systems will usually provide the most efficiency (weight savings).

SILICON CARBIDE FIBER

Since the advent of high-strength, high-modulus, low-density boron fiber, the role of fibers produced by chemical vapor depositing (CVD) in the field of high- performance composites has been well established. Although best known for its use as a reinforcement is resin-matrix composites,[1,2] boron fiber has also received considerable attention in the field of metal matrix composites.[3,4,5] Boron/aluminum was employed for tube-shaped truss members to reinforce the Space Shuttle orbiter structure, and has been investigated as a fan blade material for turbo-fan jet engines. There are draw-backs, however, in the use of boron in a metal matrix. The rapid reaction of boron fiber with molten aluminum[6] and long-term degradation of the mechanical properties of diffusion-bonded boron/ aluminum at temperatures greater than 480°C (900°F) preclude its use both for high-temperature applications and for potentially more economically feasible fabrication methods, such as casting or

326

low-pressure, high-temperature pressing. These draw-backs have led to the development of the silicon carbide (SiC) fiber.

Silicon Carbide Fiber Production Process - Continuous SiC filament is produced in a tubular glass reactor by CVD. The process occurs in two steps on a carbon monofilament substrate which is resistively heated. During the first step, pyrolytic graphite (PG), approximately 1 μm thick, is deposited to smooth the substrate and enhance electrical conductivity. In the second step, the PG coated surface is exposed to silane and hydrogen gases. The former decomposes to form beta silicon carbide (βSiC) continuously on the substrate. The mechanical and physical properties of the SiC filament are:

Tensile Strength = 3400 MPa (500 ksi)
Tensile Modulus = 400 GPa (60 msi)
Density = $3.045/cm^3$
Coefficient of
Thermal Expansion CTE = $1.5 \times 10^{-6}/^{\circ}C$
                 $(2.7 \times 10^{-6}/^{\circ}F)$
Diameter =   140 mm (0.0056 in.)

Various grades of fiber are produced, all of which are based on the standard βSiC deposition process described above where a crystalline structure is grown onto the carbon substrate. The βSiC is present as such across all of the fiber cross-section except for the last few microns at the surface. Here, by altering the gas flows in the bottom of the tubular reactor, the surface composition and structure of the fiber are modified by first, an addition of amorphous carbon that heals the crystalline surface for improved surface strength, followed by a modification of the silicon-to-carbon ratio to provide improved bonding with the metal.

Processing Considerations - As in any vapor deposition or vapor transport process, temperature control is of utmost importance in producing CVD SiC fiber. The Textron process calls for a peak deposition temperature of about $1300^{\circ}C$ $(2370^{\circ}C)$.

Temperatures significantly above this temperature cause rapid deposition and subsequent grain growth, resulting in a weakening of tensile strength. Temperatures significantly below the optimum cause high internal stresses in the fiber, resulting in a degradation of metal matrix composite properties upon machining transverse to the fiber.[7]

Substrate quality is also an important consideration in SiC fiber quality. The carbon monofilament substrate, which is melt-spun coal tar pitch, has a very smooth surface with occasional surface anomalties. If severe enough, the surface anomaly can result in a localized area of irregular deposition of PG and SiC which is a stress-raising region and a strength-limiting flaw in the fiber. The carbon monofilament spinning process is controlled to minimize these local anomalies sufficiently to guarantee routine production of high-strength > 3450 MPa (> 500 ksi) SiC fiber.

Another strength-limiting flaw which can result from an insufficiently controlled CVD process is the PG flaw.[8] This flaw results from irregularities in the PG disposition. Two causes of PG flaws are: (1) disruption of the PG layer due to an anomaly in the carbon substrate surface and (2) mechanical damage to the PG layer prior to the SiC disposition. PG flaws often cause a localized irregularity in the SiC deposition, resulting in a bump on the surface. Poor alignment of the reactor glass can result in mechanical damage in the PG layer by abrasion. A series of PG flaws result in what is called a "string of beads" phenomenon at the surface of the fiber. The mechanical properties of such fiber are severely degraded. These flaws are minimized by careful control of the PG deposition parameters, proper reactor alignment, and the minimization of substrate surface anomalies.

The surface region of Textron's SiC fibers is typically carbon rich. This region is important in protecting the fiber from surface damage and subsequent strength degradation. An improper surface treatment or mishandling of the fiber (e.g., abrasion) can result in strength-limiting flaws at the surface. Surface flaws can be identified by an optical examination of the fiber fracture face. These flaws are minimized by proper process control and handling of the fiber (minimizing surface abrasion).

Typical mechanical properties for the Textron CVD SiC fiber consist of average tensile strength of 3790 to 4140 MPa (550 to 600 ksi) and elastic moduli of 400 to 415 GPa (58 to 60 msi). A typical tensile strength histogram shows an average tensile strength of 4000 MPa (580 ksi) with a coefficient of variation of 15%.

Fiber Variations - The surface region of the SiC fiber must be tailored to the matrix. SCS-2 has a 1 μm carbon-rich coating that increases in silicon content as the outer surface is approached. This fiber has been used to a large extent to reinforce aluminum. SCS-6 has a thicker (3 μm) carbon-rich coating in which the silicon content exhibits maxima at the outer surface and 1.5 μm from the outer surface. SCS-6 is primarily used to reinforce titanium.

SCS-8 has been developed as an improvement over SCS-2 to give better mechanical properties in aluminum composites transverse to the fiber direction. The SCS-8 fiber consists of 6 μm of very fine-grained SiC, a carbon-rich region of about 0.5 μm, and a less carbon-rich region of 0.5 μm.

Cost Factors - From an economic standpoint, SiC is potentially less costly than boron for three reasons: (1) the carbon substrate used for SiC is lower cost than the tungsten used for the boron; (2) raw materials for SiC (chlorosilanes) are less expensive than boron trichloride, the raw material for boron; and (3) deposition rates for SiC are higher than those for boron, hence more product can be made per unit.

COMPOSITE PROCESSING

The ability to readily produce acceptable SiC fiber reinforced metals is attributed directly to the ability of the SiC fiber to (a) readily bond to the respective metals and (b) resist degradation of strength while being subjected to high-temperature processing. In the past, boron and Borsic^{TM} fibers have been evaluated in various aluminum alloys and, unless complex solid-state (low-temperature, high-pressure) diffusion binding procedures were adopted, severe degradation of fiber strength has been observed. Likewise in titanium, unless fabrication times are severely curtailed, fiber/matrix interactions produce brittle inter-metallic compounds that again drastically reduce composite strength.

In contrast, the SCS grade of fibers have surfaces that readily bond to the respective metals without the destructive reactions occurring. The results in the ability to consolidate the aluminum composites using less-complicated high-temperature casting and low-pressure (hot) molding. Also for titanium composites, the SCS-6 filament has the ability to withstand long exposure at diffusion bonding temperatures without fiber degradation. As a result, complex shapes with selective composite reinforcement can be fabricated by the innovative superplastic forming/diffusion bonding (SPF/DB and hot isostatic pressing (HIP)) process.

In the following discussion, further details of fabrication techniques will be discussed; however, first the production of intermediary products such as preforms and fabrics used in the component fabrication are described. These are required to simplify the loading of fibers into a mold and to provide correct alignment and spacing of the fibers.

Composite Preforms and Fabrics - "Green tape" is an old system consisting of a single layer of fibers that are collimated/spaced side by side across a layer, held together by a resin binder, and supported by a metal foil. This layer constitutes a prepreg (in organic composite terms) that can be sequentially "laid up" into the mold or tool in required orientations to fabricate laminates. The laminate processing cycle is then controlled so as to remove the resin (by vacuum) as volatilization occurs. The method normally used to wind the fibers onto a foil-covered rotating drum, overspraying the fibers with the resign, followed by cutting the layer from the drum to provide a flat sheet of "prepreg". "Plasma-sprayed aluminum tape" is a more advanced "prepreg" similar to "green tape" but replaces the resin binder with a plasma-sprayed matrix of aluminum. The advantages of this material are (a) the lack of possible contamination for resin residue and (b) faster material processing times because of the hold time required to ensure volatilization and removal of the resin binder is not required. As with the green tape system, the plasma-sprayed preforms are laid sequentially into the mold as required and pressed to the final shape.

"Woven fabric" is perhaps the most interesting of the preforms being produced, since it is a universal preform concept that is suitable for a number of fabrication processes. The fabric is a uniweave system in which the relatively large-diameter SiC monofilaments are held straight and parallel, collimated at 100 to 140 filaments per inch and held together by a cross-weave of a low-density yarn or metallic ribbon. There are now two types of looms that can be specifically modified to produce the uniweave fabric. The first is a single-arm Rapier-type loom capable of producing continuous 60 in. wide fabric with the SiC filament oriented in the "fill" (60 in width) direction. The other is a shuttle-type loom in which the SiC monofilaments are oriented in the continuous direction with the light-weight yarn a metal ribbon in the "fill" axis. The shuttle loom can weave fabric up to 6 in. wide. Various types of cross-weave materials have been used, such as titanium, aluminum, and ceramic yarns.

330

Processing methods - "Investment casting" is a fabrication technique that has been used for many years, but is still universally accepted as a very cost effective method for producing complex shapes.

The aerospace business has for some time rejected aluminum castings due to the low strengths that are typically achieved; however, with a material that is now fiber dependent and not predominantly matrix controlled, significant structural improvements have been derived so as to revive the interest in this low-cost procedure. The investment casting technique, sometimes called the "Lost Wax" process, utilizes a wax replicate of the intended shape to form a porous ceramic shell mold where, upon removal of the wax (by steam heat) from the interior, a cavity for the aluminum is provided. The mold includes a funnel for gravity pouring, with risers and gates to control the flow of the aluminum into the gage section. A seal is positioned around the neck of the funnel, allowing the body of the mold to be suspended into a vacuum (imposed through the porous walls of the shell mold), the total cavity is filled with aluminum.

The SiC fibers are installed in mold using the fabric described above by either first placing the fabric into the wax replica or simply splitting open the mold and inserting the fabric into the cavity after the was has been removed. At present, the latter approach is usually used, due to contamination and oxidation of the fibers during wax burnout. At some future date, the necessary techniques for including the fiber in the wax (thereby reducing the processing costs) will probably be developed.

"Hot molding" is a term coined by Textron to describe a low-pressure hot pressing process that is designed to fabricate shaped SiC-aluminum parts at significantly lower cost than the typically diffusion bonding, solid-state process. As stated previously, the SCS-2 fibers can withstand molten aluminum for long periods; therefore, the molding temperature can now be raised into the liquid-plus-solid region of the alloy to ensure aluminum flow and consolidation at low pressure, thereby negating the requirement for high-pressure die molding equipment.

The best way of describing the hot molding process is to draw an analogy to the autoclave molding of graphite epoxy where components are molded in an open-faced tool. The mold in this case is a self-heated, slip-cast ceramic tool embodying the

331

profile of the finished part. A plasma sprayed aluminum preform is laid into the mold, heated to a near molten aluminum temperature, and pressure consolidated in an autoclave by a "metallic" vacuum bag. The mold can be profiled as required to produce near net shape parts including tapered thicknesses and section geometry variations.

"Diffusion bonding of SiC/titanium: is accomplished by hot pressing (diffusion bonding) technology, using fiber preforms (fabric) that are stacked together between titanium foils for consolidation. Two methods are being developed by aircraft and engine manufacturers to manufacture complex shapes. One method is based on the HIP technology, and uses a steel pressure membrane to consolidate components directly from the fiber/metal preform layer. The other method requires the use of previously hot pressed SCS/titanium laminates that are then diffusion bonded to super-plastic forming operations.

This is typical of the first fabrication procedure noted above. The fiber preform is placed onto a titanium foil. This is then spirally wrapped, inserted and diffusion bonded onto the inner surface of a steel tube using a steel pressure membrane. The steel is subsequently thinned down and machined to form the "spline attachment" at each end. Shafts are also being fabricated for other engine fabricators without the steel sheath. The concept developed for superplastic forming of hollow engine compressor blades. Here, the SCS/titanium laminates are first diffusion bonded in a press. These are then diffusion bonded to form monolithic titanium sheets, with "stop-off" compounds selectively positioned to preclude bonding in desired areas. Subsequently, the "stack-up" is sealed into a female die. By pressurizing the interior of the "stack-up", the material is "blown" into the female die to form the desired shape, stretching the monolithic titanium to form the internal corrugations.

These processes typically require long times at high temperature. In the past, all of the materials used have developed serious matrix-to-fiber interactions that seriously degrade composite strength. SCS-6, however, due to its unique surface characteristics, delays intermetallic diffusion and retains its titanium at $925^{\circ}C$ ($1700^{\circ}F$).

COMPOSITE PROPERTIES

Since continuous SiC reinforced metals have been in existence for a relatively short period of time, the property data base has been developed sporadically over this period, depending on funded applications.

SiC/Aluminum - The most mature of the SiC reinforced aluminum (SiC/Al) consolidation approaches is hot molding, and therefore the greatest mechanical property data base has been developed using this material. The design data base for hot molded SCS-2/6061 aluminum includes static tension and compression properties, in-plane and interlaminar shear strengths, tension-tension fatigue strengths (SN curves), flexure strength, notched tension data, and fracture toughness data. Most of the data have been developed over a temperature range of -55°C to 75°C (-65°F to 165°F) with static tension test results up to 480°C (900°F). As can be seen from these data, the inclusion of a high performance, continuous SiC fiber in 6061 aluminum yields a very high strength 1378 MPa (+200 ksi) high-modulus 207 GPa (30 msi) anisotropic composite material having a density just slightly greater, 2.85 g/cm^3 (0.103 lb/in.^{3}) than baseline aluminum. As in organic matrix composites, cross or angle plying produces a range of properties useful to the designer.

The property data developed to date for investment cast SCS/aluminum have been limited to static tension and compression. Fiber volume fractions are lower (40% maximum) than the hot molded laminates (47% typical) due to volumetric constraints in dry loading the shell molds; however, good rule-of-mixture (R.O.M.) tensile strengths and excellent compression strengths (twice the tensile strength) are being achieved.

The use of 6061 aluminum as the matrix material and the capability of the SiC fiber to withstand molten aluminum has made conventional fusion melting a viable joining technique. Although welded joints would not have continuous fiber across the joint to maintain the very high strengths of the composite, baseline aluminum weld strengths can be obtained. In addition to fusion welding, traditional molten salt bath dip brazing has been demonstrated as an alternative joining method.

An important consideration for emerging materials is corrosion resistance. Testing has been performed on SCS-2/6061 hot molded material at the David W. Taylor Naval Ship R&D Center[9] under marine atmosphere, ocean splash/spray, alternate tidal immersion, and filtered seawater immersion conditions for periods of 60 to 365 days. The SCS/aluminum material performed well in all tests, exhibiting no more than

333

pitting damage comparable to the baseline 6061 aluminum alloy.

SiC/Titanium - SCS-6/Ti 6-4 composites were originally developed at high temperature. There has been a successful program to reinforce the beta titanium alloy 15-3-3-3 with SCS fiber and superior composite properties have been achieved at 1585 to 1930 MPa (230 to 280 ksi) tensile strengths[10]. Fabrication of titanium parts has been accomplished by diffusion bonding and HIP. The HIP technique has been particularly successful in the forming of shaped reinforced parts (e.g., tubes) by the use of woven SiC fabric as a preform. The high-strength, high-modulus properties of SCS-6/Ti represent a major improvement over B_4C-B/Ti composites in which the modulus of composite is increased relative to the matrix, but the tensile strength is not as high as would be predicted by the rule of mixture.

SiC/Magnesium and SiC/Copper - SCS-2 has been successfully cast in magnesium.[11] Under a recent Naval Surface Weapons Center (NSWC) Program,[12] development of SiC-reinforced copper has been initiated. At present, about 85% of R.O.M. strengths have been achieved at a volume fraction of 20 to 33%.

APPLICATIONS

The very high specific mechanical properties of SiC reinforced metal matrix composites have generated significant interest within the aerospace industry, and as a result many research and development programs are not in progress. The principal area of interest is for high-performance structures such as aircraft, missiles and engines. However, as more and more systems are developing sensitivities to "performance" and "transportation weight" other and less sophisticated applications for these newer materials are being considered. The following paragraphs describe a few of these applications.

"SiC/aluminum wind structural elements" are currently being developed. Ten foot long "Zee" shaped stiffeners are to be hot molded and then subsequently riveted to wing planks for full-scale static and fatigue testing. Experimental results obtained to date have verified material performance and the design procedures utilized.

"SiC/aluminum bridging elements" are currently being developed for the Army to be used for the lower chord and the king post of a 52m assault bridge. Future plans call for development of the top compression tubes for the new Tri-Arch Bridge being developed by Fort Belvoir.

"SiC/aluminum internally stiffened cylinders" are being developed using the previously discussed investment casting process. A wax replica is first fabricated that incorporates the total shape of the shell, including internal ring stiffeners and the end fittings. The fabric containing the SiC fibers is then wound onto the inner shell mold, the two halves of the shell are remated and sealed, and infiltration of the aluminum is then accomplished.

"SiC/aluminum fins" for high-velocity projectiles are in the process of evaluation.

"SiC/aluminum missile body castings" have been fabricated utilizing a unique variation of filament winding. An aluminum motor case is first produced in the conventional manner; this time, however, with significantly less wall thickness than normally required. This casing is then overwrapped with layers of SiC fibers, where each layer is sprayed with a plasma of aluminum to build up the matrix thickness. No final consolidation of the 90% dense system is required, for the hydrostatic internal pressure on the circular body imposes no (or very minimal) shear stresses on the matrix. It is hoped that further development of this technique will permit full consolidation of the matrix by vacuum bagging the total section and hot isostatic pressing.

"SiC-titanium drive shafts" are being developed and fabricated by the hot isostatic pressing process described previously. These are generally for the core of an engine, requiring increased specified stiffness to reduce unsupported length between bearings and also to increase critical vibratory speed ranges. SiC-Ti tubes up to 5 feet in length have been fabricated and have incorporated into their ends a monolithic load transfer section for ease of welding to the splined or flanged connections.

"SiC discs for turbine engines" are currently under development. Initially discs were made by winding SiC-Ti monolayer over a mandrel followed by hydrostatic consolidation (hot isostatic pressing). The concept now being developed utilizes a "doily" approach where single fibers are hoop wound between titanium metal ribbons to be subsequently pressed together in the axial direction, reducing the breakage of fibers and simplifying the production of tapered cross-sections.

"Selectively reinforced SiC-titanium hollow fan blades" are being developed.

"SiC/copper materials" have been fabricated and tested for high-temperature missile applications. Also, SiC/bronze propellers have been case for potential Navy applications where more efficient/quiet propellers are required.

FUTURE TRENDS

The SiC fiber is qualified for use in aluminum, magnesium and titanium. Copper matrix systems are under development and reasonably good results have been obtained using the higher temperature titanium aluminides as matrix materials. The SCS-6 fiber demonstrates high mechanical properties to above 1400°C (2550°F). It is natural, then, to project systems such as SiC-nickel aluminides/iron aluminide/ superalloys, etc., all of which on an R.O.M. basis at least, project very useful properties for "engine" and "hypersonic vehicle" applications. Work required in this area includes diffusion barrier coatings and matrix alloy modifications to facilitate high-temperature fabrication processes. Also required is the detailed investigation of any detrimental thermal/mechanical cycling effects that may occur as a result of the mismatch in thermal expansion coefficients between matrix and fiber.

REFERENCES

1.   DeBolt, H., "Boron and Other Reinforcing Agents," in
      Lubin, G., ed., Handbook of Composites, Van Nostrand
      Reinhold Co., New York, 1982, Chapter 10.

2.   Krukonis, V.J., "Boron Filaments", in Milweski, J.V., and
      Katz, H.S., ed., Handbook of Fillers and Reinforcements
      for Plastics, Van Nostrand Reinhold Co., New York, 1977,
      Chapter 28.

3.   McDaniels, D.L., and Ravenhall, R., "Analysis of
      High-Velocity Ballistic Impact Response of Boron/Aluminum
      Fan Blades", NASA TM-83498, 1983.

4.   Salemme C.T., and Yokel, S.A. "Design of Impact-Resistant
      Boron/Aluminum Large Fan Blades", NASA CR-135417, 1978.

5.   Brantley, J.W., and Stabrylla, R.G., "Fabrication of J79
      Boron/Aluminum Compressor Blades", NASA CR-159566, 1979.

6.   Wolff, E., "Boron Filament, Metal Matrix Composite
      Materials", AF33 (615)3164.

7.   Suplinskas, R.J., "High Strength Boron", NAS3-22187, 1984.

8.   Aylor, D.M., "Assessing the Corrosion Resistance of Metal
      Matrix Composite Materials in Marine Environments",
      DTNSRDC/SMME-83/45, 1983.

9.   Kumnick, A.J., Suplinskas, R.J., Grant, W.F., and Cornie,
      J.A., "Filament Modification to Provide Extended High
      Temperature Consolidation and Fabrication Capability and
      to Explore Alternative Consolidation Techniques",
      N00019-82-C-0282, 1983.

      Cornie, J.A., and Murty, Y., "Evaluation of Silicon
      Carbide/Magnesium Reinforced Castings", DAAG46-80-C-0076,
      1983.

11.   Marzik, J.V., and Kumnick, A.J., "The Development of
      SCS/Copper Composite Material", N60921-83-C-0183, 1984.

SIC-FIBRE REINFORCED TITANIUM ALLOYS : PROCESSING, INTERFACES AND MECHANICAL PROPERTIES

H.-J. DUDEK, R. LEUCHT, G. ZIEGLER*

DLR - Institute of Materials Research
PO Box 90 60 58 - 5000 Köln 90 - West-Germany
*University of Bayreuth - Box 10 12 51
8580 Bayreuth - West-Germany

ABSTRACT
For the development of SiC-fibre reinforced titanium
alloys MMC-suited processing techniques have to be
designed, and the problem of fibre-matrix interface has
to be solved. Hot pressing and hot isostatic pressing are
thought to be promising processing techniques on a laboratory scale, using diffusion bonding as basic mechanism.
The problem of interface reactions can be solved by applying protecting coatings based on carbon.

INTRODUCTION
Metal matrix composites with titanium based alloys and
SiC fibres as reinforcing components are proposed to
introduce new high-temperature materials with a high
specific strength [1]. Two main problems have to be
solved before an industrial application appears to be
possible: the development of MMC-adequate processing
techniques and the control of the interface interaction
during processing and application [2].

PROCESSING
Processing of metal matrix composites for
high-temperature applications such as titanium alloys
requires processing temperatures far below the melting
point of the matrix because of the technical problems
connected with temperatures higher than 1600°C, the
limited thermal stability of fibres and the interaction
between the ceramic fibres and the metal melt. Therefore,
mainly the hot pressing and hot isostatic pressing
techniques are used for the processing of SiC-fibre rein-

forced titanium alloys [3], both based on the diffusion bonding mechanism.
The hot pressing conditions for the MMC depend on the deformation behaviour of the matrix material as a function of temperature, and can be determined experimentally using a hot press with exactly defined parameters for temperature, pressure as well as deformation degree and rate, fig. 1, [3]. Composites with a high fibre volume fraction up to 30% and well consolidated (complete envelopment, aligned fibres, no fibre fracture) presently can only be obtained, if a sufficient workability of the matrix is reached below approximately 1000°C, fig. 2a.
Successful incorporation of the SiC-fibres into near-net-shaped parts, such as rings, tubes and shafts [4], is expected from processing composites using the hot isostatic pressing technique. High-quality tensile test samples of the composites were obtained by applying the following procedure: fibres are coated with the matrix material by PVD or sputtering and introduced into tubes made of the matrix material. Subsequently, the part is encapsulated in stainless steel tubes, outgassed at 450°C and HIPed. A fibre volume fraction of up to 45% was achieved in this way, fig. 2b.

INTERFACES
Uncoated fibres introduced in titanium-based alloys result in composites with reduced strength, compared with the matrix material. This strength reduction is caused by fibre-matrix interaction resulting in a fibre strength degradation [5]. The interaction in the interface can be successfully prevented by fibre coating, fig. 3, (Textron/AVCO SCS-6 SiC fibre). The coating consists in this case of a few layers of carbon with varying content of silicon carbide, which is distributed as small, approximately 10 nm large particles. During introduction of the fibres into the matrix the reaction between titanium alloy and the coating results in the growth of a reaction zone consisting mainly of titanium carbide and titanium silicides [6]. A one hundred nanometer layer consisting of approximately ten nanometers large titanium carbide particles adjacent to the unreacted coating acts together with the remaining carbon coating – as a diffusion barrier for titanium, preventing an interaction with the SiC-fibre surface [6,7].
The applicability of SiC-fibre reinforced titanium-based alloys is therefore restricted to the life time of the protecting coating on the fibres. Thermal treatment experiments were performed to evaluate the stability of the coating. It was demonstrated that up to 600°C the carbon coating on the SiC fibre remains stable [7], which is in agreement with the observation that no reduction of mechanical properties is observed up to this temperature [1]. The thermal treatment at higher temperatures results

in a gradual consumption of the coating. The upper limit for the thermal treatment observed using analytical transmission electron microscopy was 800°C/500 h [7].

MECHANICAL PROPERTIES
Using MMC-processing techniques which give well consolidated composites, and protecting the SiC fibres with thermally stable coatings, composite strength and Young's modulus values are obtained near the rule of mixture [3]. These results are reached preferentially by using the Ti6Al4V-alloy as matrix material. Based on the rule of mixture a prediction of the properties can be made for composites with different matrices, under the assumption that problems connected with MMC processing and with the interface interaction are solved. Following these considerations fig. 4 presents the temperature dependence for composites with the Ti6Al4V-alloy and Ti_3Al as a matrix.

CONCLUSIONS
Mechanical properties in agreement with the rule of mixture of SiC-fibre reinforced titanium alloy-based composites will be achieved if suitable MMC-processing techniques are developed and fibre-matrix interface problems are solved. On a laboratory scale, an adequate processing method was found for cylindrical tensile test samples using HIPing; however, for transferring this technique to large, complex-shaped parts some problems associated with fibre fraction have to be solved [4]. The coatings of the fibres (Textron SCS-6 SiC-fibres) presently available allow applications up to about 600°C. Nevertheless an improvement of the coating for applications at higher temperatures should be possible.

LITERATURE
[1] T.M.F. Ronald: Advanced Materials to fly high in NASP", Adv. Mat. & Proc. 5 (1989) 29-35.
[2] H.J. Dudek, A. Kleine, R. Borath, R. Leucht, H. Mucha: "Mikrobereichs- und Oberflächenanalytik im Grenzschichtenbereich von Metallmatrix-Keramik-faser-Verbundwerkstoffen", DGM-Symposium "Haftung bei Verbundwerkstoffen und Werkstoffverbunden", 21.-22.6.1990, Konstanz, in press.
[3] R. Leucht, H.J. Dudek, G. Ziegler: "Laboratory Scale Processing of SiC-Fibre Reinforced Ti6Al4V-Alloys", Proc. Conf. Fibre Reinforced Composite, Liverpool, 1990.
[4] R. Leucht, H.J. Dudek, G. Ziegler: "Processing of Parts Using Hot Isostatic Pressing", these proceedings.
[5] R. Leucht, H.J. Dudek, G. Ziegler: "SiC-faserver-stärkte Titanlegierung; Einfluß von TiC- und TiN-Zwischenschichten auf die Reaktionszone", Z. Werk-stofftechnik 18 (1987) 27-32.

[6] M. Lancin, J. Thibault-Desseaux, J.S. Bour:
     "Structure of the Interface in the SCS-6
     Fibres/Ti6Al-4V Composite", Microsc. Spectrosc.
     Electron. 13 (1988) 503-509.
[7] H.J. Dudek, R. Leucht, R. Borath, G. Ziegler:
     "Analytical Investigations of Thermal Stability of
     the Interface in SiC-Fibre Reinforced Ti6Al4V-
     Alloy", Progress in Materials Analysis, Vol. 13,
     1990, Microchim. Acta Suppl., in press.

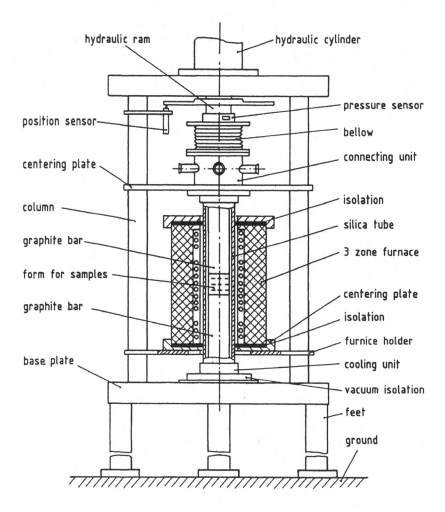

Fig. 1: Hot press developed for optimizing the
         processing conditions of SiC-fibre reinforced
         titanium based alloys and particularly by
         applying extremely slow deformation rates
         (superplastic deformation).

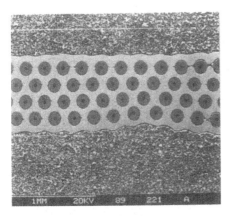 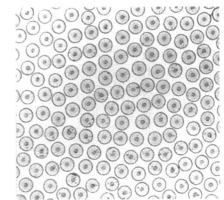

a) processed by hot pressing   b) processed by hot isostatic pressing

Fig. 2: Cross section of a SiC-fibre reinforced
Ti6Al4V-alloy.

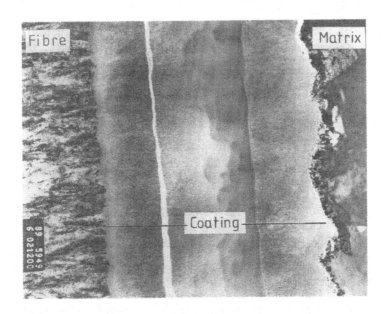

Fig. 3: Transmission electron microscopical image of
the interface region of a SiC-fibre (Textron/
AVCO SCS-6) reinforced Ti6Al4V-alloy, processed
by HIPing.

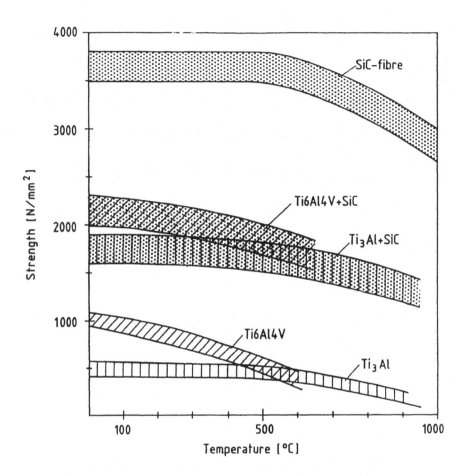

Fig. 4:   Tensile strength of SiC-fibre reinforced
           Ti6Al4V-alloy and the Ti₃Al compared with the
           strength of the fibers and the matrix material
           as a function of temperature. These data are
           obtained by evaluating the tensile strength
           under the assumption that MMC-adequate proces-
           sing techniques can be developed and the
           problem of interfacial interaction can be
           solved by developing thermally stable fibre
           coatings.

MANUFACTURE, MICROSTRUCTURE AND PROPERTY RELATIONSHIP FOR A FIBRE REINFORCED METAL

A. CHAPMAN, V. SCOTT, R. TRUMPER

University of Bath - School of Materials Science
BA2 7AY Bath - Great Britain

ABSTRACT

The microstructure which results from the combination of production route and component choice is shown to influence the mechanical performance of a Nicalon reinforced aluminium-7%silicon-0.4%magnesium alloy. The microstructure is very complicated with silicon particles and iron-containing intermetallic phases in the matrix forming bridges between fibres. Formation of a spinel phase is observed at the Nicalon/matrix interface and also some suggestion of a magnesium-enriched zone. Mechanical performance of the composite is much poorer than expected and the reason for this is explained in terms of the microstructure.

INTRODUCTION.

The current interest in fibre-reinforced metals (FRM) stems from the ready availability of commercial ceramic fibres which can be incorporated into light alloys to provide components with high specific stiffness and strength. Whilst the mechanical performance of the FRM will depend upon the components used and their interactions with each other, the effect of other microstructural features which result from the production route must also be considered.

This paper describes studies on a FRM consisting of an aluminium-7%silicon-0.4%magnesium alloy reinforced with Nicalon (Nippon Carbon Co Ltd) continuous fibres. The composite was produced by   liquid metal infiltration (LMI) of a preform in which glass weft binding held the Nicalon fibres in place.

EXPERIMENTAL

Optical studies were carried out on metallographically prepared specimens. These were cut from the bulk composite using a diamond saw, mounted in filled resin and ground on successively finer diamond slurries followed by a final polish using colloidal silica. Such surfaces were examined also by scanning electron microcopy (SEM); in addition, fracture and "deep etched" specimens were studied with this technique (deep etching is the selective removal of certain phases). A JEOL35C instrument was used, fitted with an energy dispersive spectrometer (EDS) for determining the chemical composition of the microconstituents. Supplementary data was obtained using a JEOL 8600 instrument equipped with a wavelength dispersive spectrometer (WDS).

Specimens were prepared for transmission electron microscopy (TEM) by slicing a thin (500μm) section with a diamond saw. A disc of 3mm diameter was either punched or cut out using a hollow drill from the slice. The disc was ground on silicon carbide paper until ~150μm in thickness, and dimples ground on both sides using a VCR model 500 machine to give a central thickness of typically 30 μm. Final thinning was carried out on a Gatan ion beam mill at an angle of 12º until perforation and further thinning at an angle of 8º to provide an enlarged thin region. Foils were examined on a JEOL 2000FX instrument at 200kV which was equipped with a thin window, high take-off angle EDS detector.

Mechanical testing in tension of aligned fibre composites was carried out on a Instron 1195 machine at a cross-head speed of 0.05mm min[-1].

A measure of the fibre/matrix interfacial strength was obtained using the micro-indentation technique described by Marshall /1/. The indenter was positioned on an individual fibre and a load applied to depress the fibre and allow indentation of the surrounding matrix. The stress was calculated using the equation given. Comparative indentation tests were carried out on commercially pure aluminium reinforced with Nicalon fibres.

RESULTS AND DISCUSSION.

Figure 1 shows the stress vs strain behavior of the composite when tested in tension; for comparison purposes, the response of the Nicalon fibre and experimentally obtained data for an unreinforced alloy are included in the diagram. The failure stress of the composite was 140MPa with an associated strain of 0.12%. By applying the rule of mixtures (ROM) to the alloy and fibre data at given strains, it was possible to predict a stress vs strain curve, see also in Fig.1.. The calculated failure stress using the ROM is 1.3GPa. The difference between measured and predicted data may be explained from an examination of the microstructure of the composite.

Even at low magnifications it can be seen that the microstructure is very complicated Fig.2.. In addition to poorly distributed Nicalon fibres in closely packed tows, there are a number of other features. Not surprisingly, the melt infiltration was poor where fibres were in contact and porosity is a common feature in these regions. Many particles of pure silicon are present which show a tendency to segregate to the fibres. By selectively removing the aluminium, the silicon was seen to form three-dimensional networks, some branches of which bridge fibres and tows Fig.3.. Intermetallic phases were present and found to contain iron, aluminium and silicon using electron-probe microanalysis (EPMA).

TEM of thin sections showed that the intermetallic phases also extended between fibres, and the corresponding electron diffraction data identified them as $FeSiAl_5$. The phase originated from iron impurities introduced during the LMI process.

Fig.2. shows the presence of a glass weft and this appears to present a problem of poor infiltration between its 5 μm diameter fibres.

Examination of the Nicalon/matrix interface revealed the presence of discrete particles. Using TEM combined with EDS, the discrete particles, Fig 4a., are seen to contain magnesium, aluminium and oxygen, Fig.5.. Electron diffraction data showed the particle was a spinel. The formation of this phase is thought to be due the reaction of the magnesium in the melt alloy with the silica known to be present /2/ in Nicalon fibre. In addition to the magnesium-rich particles, some evidence of a diffusion zone of magnesium was found at the interface. The significance of the zones is not fully understood and it is being further investigated.

Indentation tests indicated that the bond between Nicalon fibre and matrix was good. For the aluminium-silicon-magnesium composite the value for the interfacial shear stress was 150MPa, whereas for the commercially pure aluminium composite the figure was 30MPa. This would indicate that failure of the composite does not result from a weak interface. Supporting information is available in Fig.6., the fracture surface which exhibits little or no fibre pull-out. It also appears that the failure occurred along and within the glass weft. Using EDS, broken particles of silicon and iron intermetallic phases were identified on the fracture surface. Some plastic deformation of the aluminium matrix was evident, but in regions containing intermetallics the fracture path lay in the plane of the plate-like particle, Fig.7.. This is clearly a plane of weakness, especially as it extends across several fibres.

CONCLUSION.

The composite contains many microstructural constituents which result from the combination of fibre, matrix alloy and production route. These create a complex array of interfaces which can affect the mechanical properties of the material. However, the controlling factors which reduce the strength, are the presence of silicon, the iron containing intermetallics and the glass weft. The glass weft may be seen to improve the transverse properties but is thought to reduce the strength in the longitudinal direction.The result of the segregation and reaction of magnesium is not yet clear and will be investigated further.

Acknowledgment to the SERC and MOD for their support.

REFERENCES.

1. Marshall D.B. and Evan A.G.. J. Am.Ceram.Soc. 68 No 5 (1985) 225-231.

2. Clark T.J., Prack E.R.,Haider M.I. and Sawyer L.C.   Ceram.Eng.Sci.Proc. 8 (1987) pp.717-731.

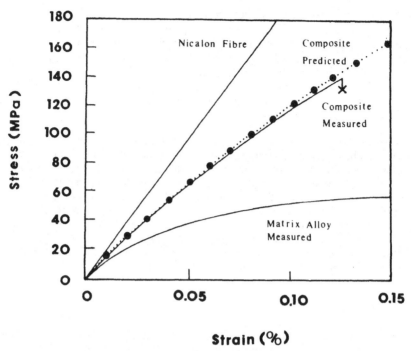

Strain (%)

FIGURE 1. TENSILE STRESS VS STRAIN FOR THE COMPOSITE AND IT'S CONSTITUENTS.

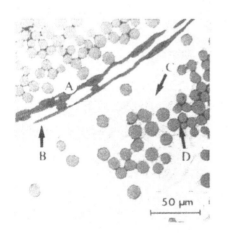

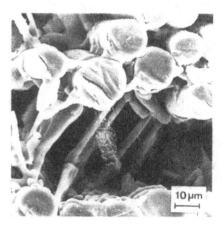

FIGURE 2. PREPARED SECTION OF FRM SHOWING A. GLASS WEFT, B. MODIFIED SILICON STRUCTURE, C. IRON RICH INTERMETALLICS AND D. POOR INFILTRATION OF FIBRES.

FIGURE 3. SILICON "BRIDGES" BETWEEN NICALON FIBRES, REVEALED BY SELECTIVE REMOVAL OF ALUMINIUM.

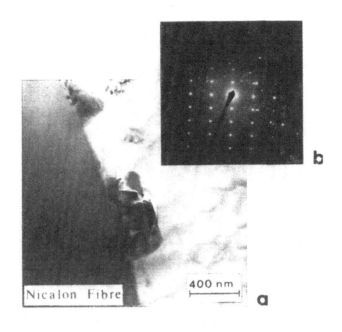

FRACTURE SURFACE OF THE FRM SHOWING A. PLASTICALLY

FIGURE 4.a. TRANSMISSION ELECTRON MICROGRAPH OF PARTICLE AT THE FIBRE
SURFACE. b. DIFFRACTION PATTERN TAKEN FROM THE PARTICLE.

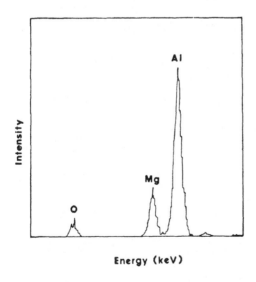

FIGURE 5. EDS TRACE TAKEN FROM THE PARTICLE IN FIGURE 4.a..

FIGURE 6. FRACTURE SURFACE OF THE FRM SHOWING A. PLASTICALLY DEFORMED ALUMINIUM, B. SHATTERED SILICON, C. SHATTERED INTERMETALLIC AND D. CRACK FOLLOWING THE GLASS WEFT.

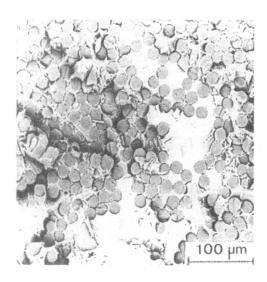

FIGURE 7. BACK SCATTERED ELECTRON IMAGE OF FRACTURE SURFACE SHOWING A LARGE REGION OF INTERMETALLIC EXTENDING ACROSS SEVERAL FIBRES.

INFLUENCE OF SAFIMAX FIBRES ON MICROSTRUCTURE AND MICROSEGREGATION IN AN ALUMINIUM ALLOY

Q. LI, D.G. McCARTNEY, A.M. WALKER*

University of Liverpool - Dept of Materials Science
PO Box 147 - L69 3BX Liverpool - Great Britain
*ICI Advanced Materials - PO Box 11
The Heath Runcorn Chesshire - Great Britain

ABSTRACT

The solidification behaviour of a fibre reinforced Al-6 wt.% Cu alloy containing 30 vol. % of 3μm diameter, semi-continuous, aligned alumina fibres has been studied. Results are presented to show the influence of fibres on microstructure and matrix microsegregation. The effect of total solidification time, θ_t, on solidification behaviour was examined for $1 < \theta_t < 520$s. When $\theta_t > 100$s the final matrix microstructure is non-dendritic and CuAl$_2$ is located principally at the fibre matrix interface. When $\theta \simeq$ 1s it was observed that the final microstructure is dendritic, with a periodic segregation pattern and the CuAl$_2$ is more dispersed. The matrix composition becomes more uniform and the minimum composition rises as θ_t increases. The microsegregation is analysed theoretically using a simple analytical model.

I - INTRODUCTION

The current extensive research on metal matrix composites (MMCs) has focussed, to a large degree, on their manufacture by solidification processing routes such as squeeze casting, pressure infiltration and slurry casting. In these processes the molten metallic alloy solidifies in the presence of inorganic ceramic fibres or particulate leading to mutual interactions. A principal reason for manufacturing MMCs is to obtain materials with mechanical properties superior to those of monolithic alloys, and clearly properties such as tensile strength and fracture toughness depend upon the overall microstructure of the MMC. Specifically, attention must be paid to both the microsegregation which occurs within the matrix alloy, and the nature and distribution of intermetallic phases which form. Indeed intermetallic phases around fibres have been shown /1/ to act as flaws which may substantially reduce tensile properties of fibre-reinforced metals. It is thus of great importance to understand in detail how microstructure develops as a results of processing, and it is with aspects of the solidification behaviour of fibre reinforced alloys that this paper is concerned.

Mortensen, Cornie and co-workers /2-6/ have reported, in some detail, microstructural development in fibre reinforced alloys. The present study is significantly different in that very much smaller diameter fibres were employed; 3μm in this case as opposed to 20 or 140μm /2-6/. Consequently average interfibre spacings in this work were $\simeq$ 6μm, and much smaller than in the previous studies /2-6/.

II - EXPERIMENTAL PROCEDURE

The fibre reinforced MMCs were initially prepared by squeeze infiltration of the matrix alloy into a binder-free ceramic preform. The matrix alloy was of composition Al- 6.0 wt.% Cu and was prepared from 99.995 wt.% Al and 99.9 wt.% Cu. The fibres were standard density Safimax fibres manufactured by ICI Advanced Materials, Runcorn, U.K. (Safimax is a semi-continuous, aligned, alumina fibre 3μm in diameter.) After removal from the squeeze casting die, composite bars were cut into shorter lengths to provide samples for remelting which were 15mm in diameter and 12mm in height. A small diameter thermocouple was inserted axially to a depth of approximately 6mm in each sample to determine temperature (T)/time (t) histories during remelting and resolidification.

In order to produce a series of specimens cooled under different conditions i.e. with different total solidification times θ_t, samples were reheated to 720°C and subjected to slow furnace cooling, air cooling - with different amounts of ceramic fibre insulation - and direct water quenching. The different values of θ_t obtained are listed in Table I. All values of θ_t were obtained from T $\underline{V}$ t thermocouple traces except for $\theta_t \simeq$ 1s. In this sample θ_t, for a water-quenched, unreinforced alloy, was estimated using secondary dendrite arm spacing measurements /3/, and θ_t for the composite assumed to be the same.

Samples were prepared for optical metallography using a standard colour etching technique /5,7/, and the micrographs in Fig. 1 were taken from such colour etched samples. Different shades of grey in the matrix indicate different copper concentrations. Electron probe microanalysis (EPMA) was used to measure matrix compositions. Suitable areas for microanalysis were located using the colour etch which was then polished off before the EPMA measurements were made. Operating conditions are described in detail elsewhere /7/.

III - RESULTS AND DISCUSSION

Five different values of θ_t were used as listed in Table I, and micrographs of transverse sections are shown in Fig.1. It is clear from Fig.1(a) that for θ_t = 520s the final microstructure as non-dendritic, and there is virtually no evidence of variations in copper concentration within the matrix. A small amount of the intermetallic phase $CuAl_2$ formed, and was located predominantly at the fibre/matrix interface. There is also apparently clustering and chaining together of fibres, with regions of $CuAl_2$ linking together neighbouring fibres. The solid α-Al appears to have avoided the alumina fibres as it has grown, and there is no evidence for the secondary dendrite arms normally seen in cast structures. Figs. 1(a)-(e) show that as the solidification time, θ_t, decreased there is increasing evidence - from the colour etching effects - for compositional variations within the matrix. Indeed in Fig.1(e), for which θ_t was $\simeq$ 1s, the matrix segregation pattern is similar to that commonly seen in cast-dendritic microstructures. In this sample it was found that the $CuAl_2$ phase was entrained within the α-Al matrix as well as being located at the fibre matrix interface. When $\theta_t \simeq$ 1s, the secondary dendrite arm spacing, λ_s, in a fibre free alloy was found to be $\simeq$ 6μm, whereas for θ_t = 520s, it was $\simeq$ 65 μm. Thus as θ_t decreased from 520s to 1s the ratio λ_f/λ_s (where λ_f is the average interfibre spacing assuming a random distribution) changed by an order of magnitude from $\simeq 0.1$ to 1.

Details of EPMA analyses are given in Table I. The sample with $\theta_t \simeq 1s$ was not analysed because of the relatively small spatial scale of compositional variations within its matrix. As can be seen from Fig.1(e) the periodicity of segregation is around 5μm which is only marginally greater than the resolution of EPMA. Hence meaningful composition profiles would not have been obtained. All the measured compositional variations provide quantitative support for the qualitative interpretation of the colour etched microstructures. Clearly there was very little compositional variation within the matrix for $\theta_t = 520s$, and the minimum copper content was 4.6 wt.%. The matrix region examined was relatively large and it is to be expected that smaller regions would be of even more uniform composition. As θ_t decreased the minimum copper concentration decreased, but even for $\theta_t = 110s$ this minimum value was well above that expected in an unreinforced Al-6 wt.% Cu alloy. This value would be expected /8/ to be approximately kC_0 where k is the partition coefficient and C_0 the alloy composition. Taking $k = 0.18$ /9/ gives $kC_0 \simeq 1.1$ wt.% Cu.

The present observations on microsegregation can be compared with previous EPMA measurements by Mortensen and co-workers /2,5/ who found, in fibre-reinforced alloys, reduced microsegregation similar to that reported here; albeit with much larger diameter fibres than those used presently. Hunt and McCartney /10/ and Mortensen et al. /5/ have both modelled solidification in narrow channels and examined the effect of channel diameter on microsegregation. They have shown that reduced compositional variations and increased minimum solute levels are to be expected when the solid phase grows in narrow channels because of the important contribution from solid-state diffusion during freezing. This is often termed back-diffusion and qualitatively explains the measurements in Table I.

The present geometrical arrangement of fibres is much more complex than the channels assumed in the mathematical models. Nevertheless, a semi-quantitative analysis of the degree of microsegregation in the present work can be obtained by using the Clyne-Kurz modification /11/ of the Brody-Flemings /12/ microsegregation model which incorporates back-diffusion of solute in the solid. Composition profiles cannot be predicted using such a model, only the fraction of eutectic in the final solidified sample. However, this parameter is obviously a measure of the degree of homogenization. Using the Clyne-Kurz model the eutectic fraction, f_e, is given by

$$f_e = 1 - \left(\frac{1}{1-2\Omega k}\right) \left[1 - \left(\frac{C_e}{C_0}\right)^{\frac{1-2\Omega k}{k-1}}\right] \tag{1}$$

where

$$\Omega = \left[\alpha(1-\exp(-\frac{1}{\alpha}) - 0.5\exp(-\frac{1}{2\alpha})\right] \tag{2}$$

and

$$\alpha = \frac{4D_s\theta_t}{L^2} \tag{3}$$

C_e is the eutectic composition, C_0 the alloy composition, k the partition coefficient, D_s the solid state diffusion coefficient and L the average spacing of the last liquid regions to solidify. In Fig.2 f_e is plotted against α for $C_0 = 2$ and $C_0 = 6$ wt.% Cu.

Taking the length, L, to be equal to λ_f then α can be estimated for $\theta_t = 1s$ and θ_t

= 520s. Assuming D_s to be 4.6 x 10^{-7} mm^2 s^{-1} /9/ then $\alpha = 0.05$ for θ_t = 1s, and for θ_t = 520s, α = 26.6. Thus, from Fig. 2, it is clear that f_e should be about 0.12 and 0.02 respectively for the two values. In other words when θ_t = 520s freezing conditions are predicted to give almost the equilibrium eutectic fraction whereas when θ_t = 1s the value of f_e predicted is that for non-equilibrium freezing, as given by the Scheil equation /8/.

In order for f_e to be reduced below the Scheil equation value, the matrix must contain a greater fraction of solute (from mass balance considerations). Thus this simple analysis predicts that when θ_t = 520s there should be a virtually homogeneous matrix of composition close to the solubility limit for Cu in Al as indeed occurs. Moreover, the observed changes in segregation profile, with decreasing θ_t, accord with the decreasing homogeneity of the matrix predicted by the model.

IV - SUMMARY

1. In an Al - 6 wt.% Cu alloy containing 30 vol.% of Safimax fibres microstructure and microsegregation vary significantly with solidification time θ_t.

2. When the final matrix microstructure is non-dendritic the last liquid to solidify is located predominantly around the fibres. Such a microstructure develops when the ratio λ_f/λ_s falls below 1, to a first approximation

3. The minimum matrix composition increases with increasing θ_t, and this is accompanied by a decrease in compositional variations within it.

4. The increasing homogeneity of the matrix, as θ_t increases, is in agreement with a simple analytical model in which the average spacing of the last solidifying liquid pools is set equal to λ_f.

REFERENCES

1. Trumper R. and Scott V., ECCM 3, eds. A.R. Bunsell, P. Lamicq, A. Massiah, Elsevier App. Sci., 1989, pp.139 - 144.
2. Cornie J.A., Mortensen A., Gungor M.N. and Flemings M.C., ICCM V, eds. W.C. Harrigan, J. Strife and A.K. Dhingra, TMS-AIME, 1986, pp.809-823,
3. Mortensen A., Gungor M.N., Cornie J.A. and Flemings M.C., J. of Metals, March 1986, vol. 38, pp. 30 - 35.
4. Cornie J.A., Mortensen A. and Flemings M.C., ICCM VI, eds. F.L. Matthews, N.C.R. Buskel, J.M. Hodginson and J. Morton, Elsevier App. Sci., 1987, pp. 2.297 - 2.319.
5. Mortensen A., Cornie J.A. and Flemings M.C., Metall. Trans. A, vol. 19A, 1988, pp.709-721.
6. Idem, J. of Metals, vol. 40, Feb. 1988, pp. 12-19.
7. Li Q.F., McCartney D.G., and Walker A.M., to be published.
8. Kurz W. and Fisher D.J., 'Fundamentals of Solidification', Trans Tech Publishers, 1986.
9. Kirkwood D.H., Mat. Sci. and Eng., vol. 65, 1984, pp. 101-110.
10. Hunt J.D. and McCartney D.G., Acta Metall, vol. 35, 1987, pp. 89-99.
11. Clyne T.W. and Kurz W., Metall. Trans. A, vol. 12A, 1981, pp. 965-971.
12. Brody H.D. and Flemings M.C., Trans. AIME, vol. 236, 1966, pp. 615 - 623.

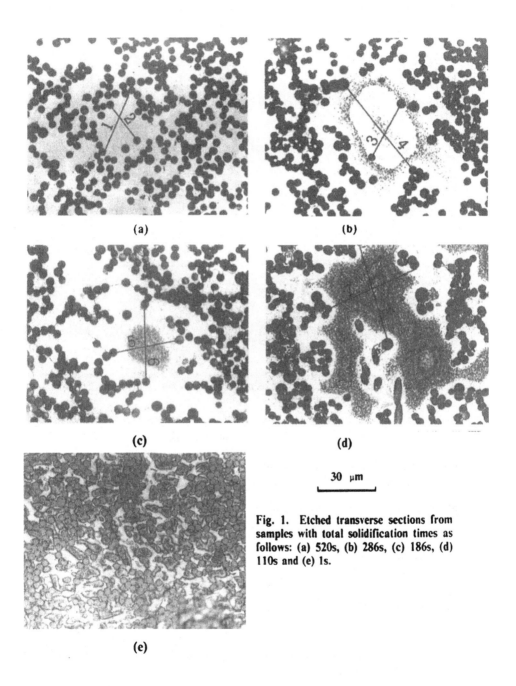

(a)

(b)

(c)

(d)

30 μm

Fig. 1. Etched transverse sections from samples with total solidification times as follows: (a) 520s, (b) 286s, (c) 186s, (d) 110s and (e) 1s.

(e)

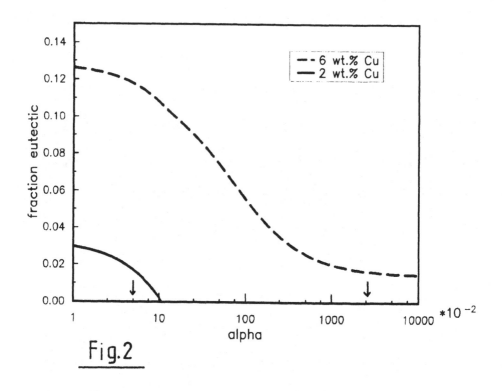

Fig.2

Table I. Matrix Copper Concentrations Measured by EPMA Along the Lines Shown in Fig. 1 Using Steps of 5μm. Compositions in wt.% Cu

θt (s)	Line No.	Distance along line (μm)						
		0	5	10	15	20	25	30
520	1	5.4	4.9	4.6	4.6	4.8	5.1	5.3
	2	5.4	4.9	4.6	4.8	5.0	5.2	-
286	3	3.6	3.2	2.7	2.6	2.7	3.0	-
	4	3.1	2.9	2.8	2.6	2.6	2.8	3.6
186	5	2.8	2.4	2.2	2.3	2.9	-	-
	6	2.7	1.9	2.0	2.5	3.1	3.6	
110	7	2.9	2.8	2.2	2.1	2.1	2.6	3.2
	8	3.7	2.5	1.9	1.8	2.0	2.4	3.2
1	No analysis values							

ORIENTATION OF SHORT REINFORCING FIBERS IN METAL MATRIX COMPOSITE MATERIALS

B. SPIGLER, M. SELA

Israel Institute of Metals - Technion Haifa
Technion City - 32000 Haifa - Israel

ABSTRACT

The strength and service behaviour of fiber reinforced Metal Matrix Composite Materials are known to depend strongly on the length and orientation of the individual fibers, as well as on the strength of the bonding between them and the metal matrix. A method is proposed for measuring the reinforcing effect of fibers, based on the measuring of the length and orientation of individual fibers by means of a quantitative Image Analysis system and on the calculation of the integral of their "reinforcing vector specific value". The advantages and limitations of the method are discussed.

INTRODUCTION

The improved strength of a Metal Matrix Composite Materials is obtained by combining the high strength of a hard ceramic reinforcing material with a relatively soft and plastic metal matrix. M.M. Composites reinforced with continuous unidirectional fibers show the highest strength /1/, but they are the most expensive and difficult to manufacture and very sensitive to fiber damage during secondary processing.

M.M.C. Composites reinforced with short-discontinuous fibers are less expensive and much easier to manufacture and much more apt to secondary processing, but they usually have an isotropic structure with randomly oriented fibers, which greatly reduces the strength of the composite. For this reason, efforts are made to obtain an unidirectional orientation of the fibers, either by using secondary processing methods such as extrusion, rolling etc., or by adequately varying the manufacturing process.

Kelly and Davis /2/ have shown that the strength of a composite reinforced with fully aligned short fibers is given by the equation:

$$\sigma_c = \sigma_f v_f (1 - \frac{l_c}{2l}) + \sigma_m (1-v_f)$$

where
σ_f = Fiber strength
v_f = Fiber volumetric ratio
l = Actual fiber length
l_c = The critical fiber length-defined as:

$$l_c = d_f \frac{\sigma_f}{\sigma_y}$$

where
d_f = Fiber's diameter
σ_y = Flow stress in matrix

For composites with isotropically (randomly) oriented fibers and random fiber length, the equation must use average values and becomes:

$$\sigma_c = \sigma_f v_f (1 - \frac{l_c}{2l_a}) \text{Cos } \theta^\circ_a + \sigma_m (12-v_f)$$

where
l_a = average fiber length
θ°_a = average fiber orientation angle

Since in most cases the fiber length and orientation are not dependent on each other, the real contribution off all fibers to the composite strength is given by a different equation, based on the measurement of each individual fiber:

$$F = f \sum_{\substack{l_{min} \\ \theta = 0^\circ}}^{\substack{\theta = 90^\circ \\ l_{max}}} n_i l_i \text{ Cos } \theta_i$$

where
F = Total contribution of the measured fibers to strength
f = Specific fiber / matrix reinforcement
l_i = Fiber length
θ_i = Fiber orientation

To perform such individual measurements on fibers through metallographic work, special techniques as well as adequate equipment have to be developed and experimentally evaluated.

I - EXPERIMENTAL APPROACH

The quantitative evaluation of fibers length and orientation is normally performed using two different metallographic methods.

a. Examination of individual fibers after deep etching of the metal matrix and exposing the fibers to a given depth underneath the (normally polished) surface
b. Measuring individual fibers laying (exposed) on the polished surface of an (usually opaque) metallographic specimen.

The first method has a number of significant advantages:

1. The length of the fibers (at least that part of the fiber laying below the surface up to the etching depth) can be directly measured or evaluated, as well as its orientation.
2. The fibers can be extracted from the composite etched surface, thus making accurate length measurement possible (but losing their initial orientation).

The main disadvantages of this method are its time consuming and destructive nature and its inadequacy to automatic measurements using quantitative image analysis systems.

The second method is limited to that part of the fibers laying on the polished surface and to a two-dimensional measuring of their length and orientation, requiring additional calculations for the evaluation of their three dimensional orientation, and making a complete length measurement impossible. Its main advantage is its adequacy to the use of computerized quantitative image analysis systems and the statistical reliability of the composite material's planar morphology evaluation.

Our present work has used the polished surface metallographic measurements method and focused on the measurement of fibers orientation in M.M.C.s made either using Powder Metallurgy or Squeeze Casting technology.

II - THE EXPERIMENTAL WORK

The experimental method developed for our investigation aimed towards two main goals:

a. The measuring of two and three dimensional orientation of individual fibers and its distribution over the specimen's polished surface.
b. The determination of the relationship between the orientation of fibers on different planes (x,y,z) of the composite material. The specimens were either Al.Alloy/SiC$_f$ (15-25%v$_f$) composite made by Powder Metallurgy or Al.Alloy/Al$_2$O$_3$ (20-25%v$_f$) composite made by squeeze casting.

The cylindrical specimens were cut along their longitudinal axis or perpendicular to it and automatically ground and polished, using a special procedure to avoid fibers damage or extraction. Three main axes were determined, x,y laying on the polished surface and z perpendicular to it. (See Figs. 1a,b).

After completion of the specimen's preparation, it was mounted on a Polyvar Optical Microscope connected to a Q 920 Cambridge Image Analysis System, for measurements. The measurement cycle included calibration, optical optimalization, image electronic enhancement, threshold determination for adequate fibers detection, semi-automatic manual fibers separation and editing and finally – measurement of each detected fiber and data storage. After the completion of a measuring cycle, a new field was automatically selected and put in focus and a new measurement cycle performed, and then another, up to a predetermined number of fields. The measuring process included the following operations:

1) Determination of length (l) and thickness/width (t) of each detected fiber to be used for further calculations.
2) Determination of fiber two dimensional (planar x,y) orientation of each detected fiber - measured from a given ($\theta^{\circ}_{xy} = 0^{\circ}$) axis. (See Fig. 2.)
3) Calculation of fibers inclination (θ°_{z}) using the measured dimensions (long and short diameters) of their elliptical cross-section laying on the specimen's polished surface. (See Fig. 3.)
4) Graphical display of the relationship between the orientation and "inclination" of each fiber (See Fig. 4.)

All data obtained for every individual fiber was stored in a "list" form to be used for damage assessment, strength calculations and preform evaluation.

III – RESULTS AND CONCLUSIONS

Results on fibers orientation and its dependency on preforms manufacturing and composites fabrication methods were obtained and their significance preliminary analysed. These results combined with damage evaluation and data on fibers "local" length and related to individual fibers may offer an efficient tool for the investigation of the relationship between microstructure, morphology and mechanical properties of metal matrix composite materials.

REFERENCES

1. Mel M. Schwartz, Composite Materials Handbook", McGraw–Hill Book Company, New York, N.Y., 1983.

2. A. Kelly and G.J. Davies, "The Principles of the Fiber Reinforcement of Metals", Met. Reviews 10(1965), p. 1.

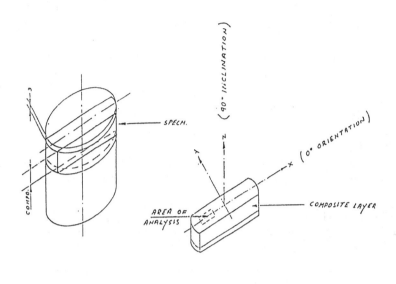

a.

b.

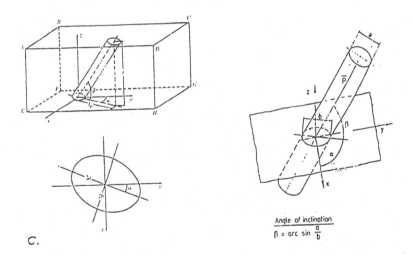

c.

$$\text{Angle of inclination}$$
$$\beta = \text{arc sin } \frac{a}{b}$$

FIG.1　a. LOCATION OF SPECIMEN IN SAMPLE.
　　　　b. ORIENTATION AND INCLINATION ANGLES, AND AREA
　　　　　　OF ANALYSIS.
　　　　c. DETERMINATION OF FIBER INCLINATION DEFINED BY
　　　　　　a AND b.

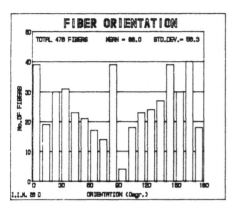

Fig. 2
Distribution of Fibers
Orientation in SPCM. No. 26D.

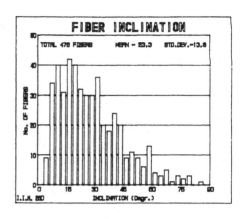

Fig. 3
Distribution of Fibers
Inclination in SPCM, 26D.

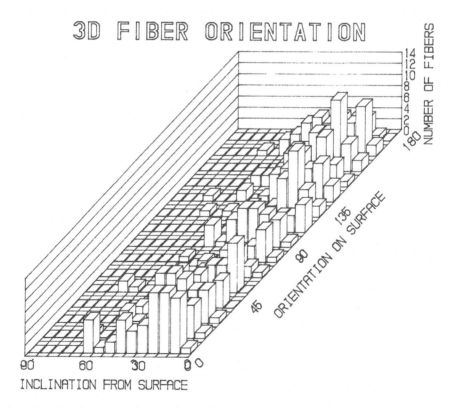

Fig. 4   3-Dimensional Display of Fibers Orientation
and Inclination in SPCM. No. 26d.

FATIGUE OF ALUMINA SHORT
FIBER / ALUMINIUM ALLOY COMPOSITES

Y. NISHIDA, I. SHIRAYANAGI, M. NAKANISHI, H. MATSUBARA

Government Industrial Research Institute - Nagoya
Hirate-Cho - Kitaku - Nagoya 462 - Nagoya - Japan

ABSTRACT

Cyclic bending fatigue tests were performed on alumina short fiber reinforced JIS AC8A alloys, to examine the effects of volume fraction of fibers and nonfibrous inclusions (shots) on the S-N curve. Fatigue limit stress rose with increasing volume fraction of fibers and the effect of shots was very small when shots content was less than 0.1%. Cyclic fatigue crack propagation was also measured on the composites. The sensitivity of the composites to pre-crack increased greatly with increasing volume fraction of fibers.

INTRODUCTION

Alumina short fiber reinforced aluminum alloy composites are attractive because of low cost. of fiber and have been actively investigated. However, high tensile strength has not been obtained for the composites, compared with whisker reinforced metals, although it has a good wear property. Then, the composites are regarded as a wear-resistance material and have been commercialized/1/. The properties of the composites have not been yet investigated sufficiently. There is a possibility to find some other characteristics. In addition, before the composites are put to practical use as a machine part, many kinds of properties must be made clear. In this study, fatigue tests were carried out to examine the influence of alumina fiber content and shots on S-N curve and fatigue limit stress. Tests for fatigue crack growth measurement were also made using CT specimens.

I - EXPERIMENTAL PROCEDURE

Alumina short fiber, which consists of 85% Al_2O_3 and 15% SiO_2 (produced by Denki Kagaku Kogyo Co. Ltd.,Japan) , was used for the fabrication of composites, volume fractions of which were 10%, 15% and 20%. The average diameter of the fiber was 3.6 μm. The composites were fabricated by squeeze casting and the matrix metal was JIS AC8A (Al-12%Si-1%Cu-1%Mg). Pouring temperature, mold temperature and applied pressure were 720°C, 300°C and 100 MPa, respectively. Nonfibrous inclusions(shots) are usually included with alumina short fibers, which were formed during production of fibers. Then, three shot content level preforms of alumina fibers were made, to examine the effect of shots on the properties of the composites. The shape of preforms were plate-like as shown in Fig.1(a). Since fibers tend to align in the plane parallel to horizontal plane during formation of preforms, fatigue specimens were cut out like Fig.1(a). The dimensions of the specimen are shown in the figure. The main part of bending fatigue testing machine is shown in Fig.1(c). The cycles of bending were 30Hz. The maximum stress (tensile stress and compressive stress) in the specimen occurs in both surfaces of the specimen. The maximum stress was taken as the stress of S-N curve, and the effect of the shots on S-N curve was examined for 0.1%, 1% and 5% shot content. The shot percentage is the volume fraction of shots in fibers, which are bigger than 93 μm in diameter. The content of shots was determined by BSI(British Standard).

Cyclic fatigue crack propagation was measured using CT specimen, which is shown in Fig.1(b), following ASTM Standard E647-83. Specimens were cyclically loaded at a load ratio (ratio of minimum to maximum loads) of 0.1 and a frequency of 29 Hz (sine wave). Electrical potential measurements across crack gage stuck on specimens were used to monitor crack lengths. Crack growth rates, da/dN, were determined over the range 10^{-9} to 10^{-6}m/cycle under computer controlled K-increasing conditions.

The optical microstructure of the composites was observed and fracture surfaces were also observed by SEM.

II - EXPERIMENTAL RESULTS AND DISCUSSION

2.1. Bending fatigue test

The optical microstructure of the composites is shown in Fig.2. The diameter of fibers is not uniform and some big fibers are involved. As fibers used in this experiment are transparent, matrix structure can be seen through fibers.

The S-N curve obtained for the case of 0.1 % shot content is shown in Fig.3(a). The fatigue limit stress increased greatly with increasing volume fraction, Vf, of alumina short fibers. The stress was 168 MPa for the case of 20% Vf, and it is 60% increase, compared with that of the matrix.

On the other hand, the results obtained for the effect of

shots on S-N curve is shown in Fig.3(b) and (c). When many shots are contained, the number of cycles to failure scattered and the S-N curve lowered. In the case of 0.1% shots, the scattering was almost the same, as can be seen in Fig.3(a), as that of matrix metal (small black dots in the figure). These results means that shots are in connection with the initiation of failure. To analyze quantitatively the influence of the amount of shot content on S-N curve, standard deviation of scatter of S-N plots was calculated and shown in Table 1. The standard deviation was obtained by the least square method on the assumption that S-N curve could be given in a linear form. When shot content of composites is 0.1%, the standard deviations are small and almost the same value as that of the matrix alloy. For composites with shot contents of 1.0% and 5%, the standard deviation is larger than that of the matrix alloy. Therefore, these results indicate that if shot content is lower than 0.1%, the influence of shot on fatigue is fairly small and can be neglected at least for this matrix alloy.

Fracture surfaces of fatigue specimens were also observed by SEM, and the possibility to find shots on the fracture surface was very high. It was, however, difficult to see whether shots were the origin of crack.

2.2. Fatigue crack propagation

Cyclic fatigue crack propagation data are plotted in Fig.4 as a function of the stress-intensity range ΔK for 10%, 15% and 20% alumina short fiber reinforced AC8A alloys composites, which have 0.1% shots, compared with matrix alloy. Taking account of no heat treatment for this specimen, the crack propagation rates of matrix alloy roughly agree with values in literatures /2/. As for composites, the rate became high. The rate for 20% fiber content composite is about 100 times higher than that of matrix alloy at the same ΔK. These results show that resistance to cyclic fatigue crack growth in composites is weakened with increasing fiber volume fraction.

III - CONCLUSION

Cyclic bending fatigue tests and cyclic fatigue crack propagation measurements were performed on alumina short fiber reinforced JIS AC8A alloy composites fabricated by squeeze casting. Shots involved with fibers have rigorous influence on S-N curve. However, it was made clear that if shot content is reduced less than 0.1%, the influence could be neglected for AC8A matrix composites.

REFERENCES

1 - T.Donomoto, N.Miura, K.Funatani, N.Miyake: SAE Technical Paper No.830252, 1983 International Congress.
2 - H.Egashira, I.Hirota, T.Kobayashi, S.Sakai: J.Japan Inst. Light Metals,Vol.39(1989),886.

(a)

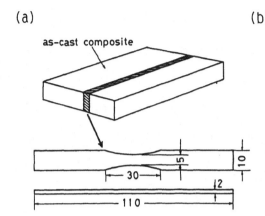

(b)

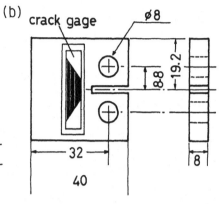

(c)

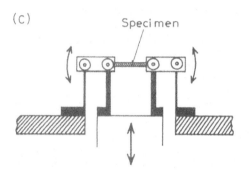

Fig.1 Dimensions of bending
fatigue specimen (a), CT
specimen (b) and main part of
bending fatigue test machine
(c).

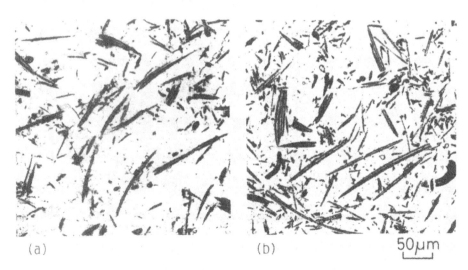

Fig.2 Optical microstructure of composites, (a) 10%fiber/AC8A alloy,
(b) 20% fiber/AC8A alloy.

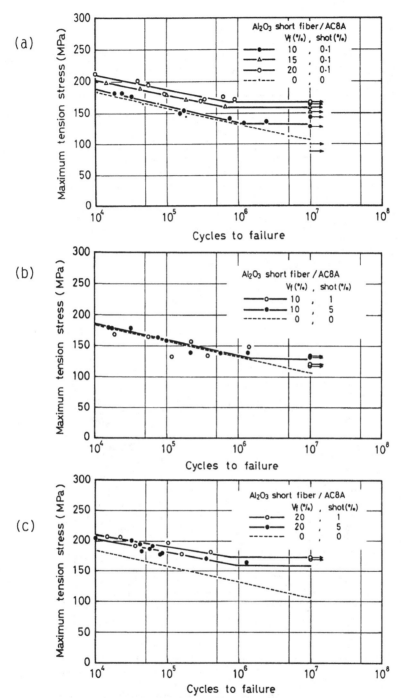

Fig.3 Relationship between maximum stress and number of cycles to
failure, (a) 0.1% shot content, (b) 10% Vf with different shot
content, (c) 20% Vf with different shot content.

Table 1 Degree of scatter of S-N plots due to shot content.

composite	shot content (%)	standard deviation (MPa)
10% Al$_2$O$_3$ fiber/AC8A	0.1	4.6
	1.0	13.4
	5.0	6.9
15% Al$_2$O$_3$ fiber/AC8A	0.1	1.4
20% AL$_2$O$_3$ fiber/AC8A	0.1	6.0
	1.0	7.3
	5.0	6.0
AC8A alloy	0	4.9

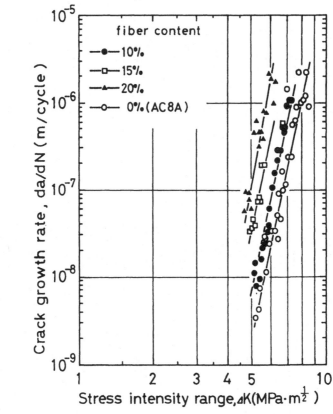

Fig.4 Cyclic fatigue crack growth rates as a function of the applied stress-intensity range for alumina short fiber reinforced AC8A alloy composites.

DEFORMATION MODE OF FIBRE METAL COMPOSITES UNDER IMPACT COMPRESSION

E. EL-MAGD, G. DACKWEILER

Lehr und Forschungsgebiet Werkstoffkunde
Augustinebach 4-22 - 5100 Aachen - West-Germany

ABSTRACT

The deformation mode and the behaviour of metal fibre composites copper/steel and silver/nickel with various fibre volume fractions and orientations of the fibres with respect to the impact compression direction were researched. Furtheron, a comparison was made between the deformation mode arising by impact and by quasistatic loading. Finite-element simulations are carried out in order to study the influence of effects of mass momentum of interia to the mechanical behaviour and the deformation mode.

INTRODUCTION

During compression of metal fibre composites damage occurs by buckling of the fibres, fracture of the matrix or by splitting off the interface matrix-fibre. This limits the formability of the material. This paper deals with the deformation mode and behaviour of metal fibre composites with various fibre volume fractions and orientations of the fibres with respect to the deformation description by impact compression.

I - EXPERIMENTAL RESEARCH

1.1 Methods of testing

Both testing fibre composites copper/steel and silver/nickel were made by a cover process, which can be used only by ductil materials. The fibre composite copper/steel (Cu/St) was consisting of copper matrix (SE-Copper) and continuous steel fibres (X2 CrNi 18 9), which has a medium diameter of 300 μm. Silver matrix (pure silver) and continuous nickel fibres (Ni 99,6) with a medium diameter of 30 μm were the components of the silver/nickel (Ag/Ni) composite. Researched were both materials with different fibre fractions of 0-100 % and different orientations with respect to the direction by impact compression under lubricating oil.

For the tests modified apparatus were used, using the Split-Hopkinson-principle for tension (Fig. 1a) and for compression (Fig. 1b) with strain rates ($\dot{\epsilon}$) up to 10^4 1/s. By those tests parameters of constitutive equations /1/

$$\sigma = \sigma_{f0} + H \epsilon + \eta \dot{\epsilon} \qquad (1)$$

were determined. These equations were implemented in a finite element program to simulate deformation processes. There were also made quasistatic compression tests on Cu/St to make a comparison to correspond impact tests.

After deformation, the specimen were measured and metallographically analysed. By the metallographical examination, vertical and horizontal cuts were made, with respect to the load direction. All micrographs have a scale from 5:1. The diameter of the Cu/St specimen before loading were 7 mm and the height 9 mm, from the Ag/Ni specimen 6 mm and the height 7.5 mm.

1.2 Test results

Fig. 2a shows the fibre deformation of a vertical micrograph in the middle of an unidirectional impact Ag/Ni specimen with a fibre fraction of 47 % by a drop height of 1 m with a 10 kg hammer. The picture demonstrates the symmetrical deformation of the fibres to the axle center.

The behaviour by impact compression (height 1 m, weight 10 kg) of an Ag/Ni specimen with 47 % fibre fraction and an orientation of the fibres from 30° to impact direction is displayed in Fig. 2b. The vertical cut point out the shear action of the matrix and the asymmetrical surface.

The symmetrical deformation of a Cu/St specimen with 45 % fibre part under impact compression (height 4.2 m, weight 10 kg) shows the micrograph of Fig. 3.

The top view of a Cu/St specimen with 45 % fibre part (height 2.3 m, weight 10 kg) and a 90° orientation of the fibres to the impact direction is presented in Fig. 4. The elliptical specimen form is traced back to the anisotropic condition, that is based on the fibre orientation.

The vertical micrograph in Fig. 5 shows a Cu/St 45 % specimen (height 2.3 m, weight 10 kg) by an orientation of 60° to the impact direction. The fibres left their original hexagonal order and some of them were deformed or in contact.

A quasistatic loading /2/ of a Cu/St 45 % specimen leads to deformation (Fig. 6), which is much different to the deformation found by impact compression (Fig. 3).

This different fibre behaviour leads to the assumption, that the high deformation speed by impact loading leads to effects of mass momentum.

II - FINITE ELEMENT COMPUTATION

2.1 Program description

To simulate the deformation process and to demonstrate the effects of high speed deformation, a two dimensional finite element program for transient dynamic problems was used /3/. It was applied to bodies loaded in axial direction using the principle of virtual work with the following equation:

$$\int_\Omega [d\,u]^T \,[b - \rho\,\ddot{u} - c\dot{u}] \,d\,\Omega + \int_\Omega [d\,\epsilon]^T \,\sigma \,d\,\Omega = 0 \qquad (2)$$

The FE-program used an explicit time step procedure and isoparametric 9-node cubical elements. The single scalar equations were decoupled by the diagonalizing of mass matrix M, so that no big global system of equations neededto be solved. The damping matrix C, which influence is very small in the case of metals, is neglected. The critical time step for this conditionally stable explicit procedure has to be worked out for each material and each constitutive equation seperately. The deformation process is computed using a plane strain problem with an idealization shown in Fig. 7.

The parameters of the constitutive equation for the matrix and the parameters for the fibres, which describe the material behaviour, used in the FE-program, were determined by the impact tension and compression test described above /4/.

2.2 Simulations

To demonstrate the influence of the deformation speed by impact compressed fibre composites, the FE-mesh (Fig. 7) is loaded with a linear time displacement function.

Fig. 8 shows the deformed mesh by a reduction of 40 % in 300 μs (a) and in 30 μs (b) without friction on the contact surface. It's seen, that the faster deformed mesh has a more symmetrical configuration.

To simulate a more experimental test, at the contact surface a linear friction-time function was implemented in the program. At the end of the deformation process the relation tangential stress to flow stress of the material was 0.1. This friction influence is demonstrated in Fig. 9a for a reduction of 40 % in 300 μs and Fig. 9b for a reduction of 40 % in 30 μs.

REFERENCES

1 - E. El-Magd, W.A. Ghany, M. Homayun: Structural Models for the behaviour of metals under dynamic loading

2 - E. El-Magd, D. Stöckel, E. Mokhtar: Knickverhalten beim Stauchen metallischer Faserverbundwerkstoffe, Z. Metallkunde 76(1985)4, S. 288-292

3 - D.R.J. Owen, E. Hinton: Finite Elements in Plasticity, Pineridge Press Limited, 1980

4 - E. El-Magd, W.A. Ghany, M. Homayun: Application of FEM to elasto-plastic wave propagation in metals, Eng. Comput., 1985, Vol. 2, June

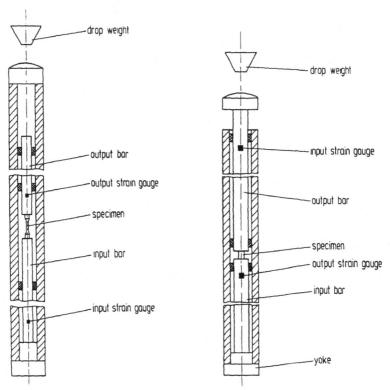

Fig. 1 - a) Impact tension apparatus   b) Impact compression apparatus

Fig. 2 - Vertical micrograph of an impact compressed Ag/Ni specimen with a fibre volume fraction of 47 %

Fig. 3 - Micrograph of an impact compressed Cu/St 45 % specimen

Fig. 4 - Top view of a Cu/St 45 % specimen with 90° orientation to the impact
direction

Fig. 5 - Vertical micrograph of a Cu/St 45 % specimen with orientation of 60° to
impact direction

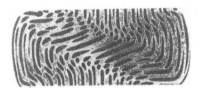

Fig. 6 - Quasistatic deformed Cu/St 45 % specimen

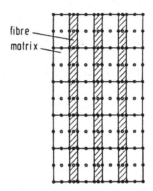

Fig. 7 - FE-idealization of a composit material by plain strain state

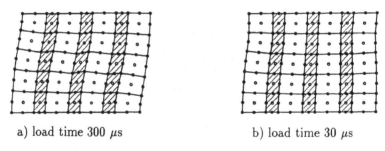

a) load time 300 μs   b) load time 30 μs

Fig. 8 - Deformed FE-meshs of composit materials to 40 %

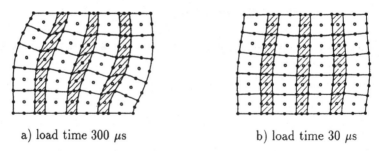

a) load time 300 μs   b) load time 30 μs

Fig. 9 - Deformed FE-mesh of composit materials to 40 % by friction

A CHARACTERIZATION OF THE FIBRE - MATRIX BOND IN SINGLE SIC FILAMENT - ALUMINIUM MATRIX COMPOSITES BY THEORITICAL AND EXPERIMENTAL APPROACH OF THE MULTIFRAGMENTATION

R. HAMANN, J. CHICOIS, P. FLEISCHMANN, R. FOUGERES, P.-F. GOBIN

GEMPPM - URA - CNRS 341
INSA Lyon - Bât. 303 - 20 Av. Albert Einstein
69621 Villeurbanne Cedex - France

ABSTRACT

The multiple fibre cracks phenomenon of a single filament Metal Matrix Composite was investigated all along a tensile test by using acoustic emission measurements. It appears that the same mean fibre critical length, from Kelly's model /4/, can be reached for very different composite strains, so this parameter cannot characterize entirely the fibre-matrix bond behaviour. After a theoritical analysis of the multiple fibre crack phenomenon, a new method is proposed for characterizing the fibre-matrix bond from a load transfer point of view.

INTRODUCTION

In order to process high-performance Metal Matrix Composites, a simple way to characterize the fibre-matrix bond from a load transfer point of view is needed. With such a test, process parameters can be easily supervised and improved. This paper presents a new method for characterizing the fibre-matrix bond quality. This method is first based on an experimental study of the multiple fibre crack phenomenon of a single SiC filament embedded in a 6061 aluminium alloy, by using Acoustic Emission (EA) measurements during a tensile test. Secondly, theoritical developments are made in order to predict the multiple fibre crack phenomenon by considering a perfect bonding. By comparing experimental and theoritical results, it appears that the bond quality can be determined all along the tensile test and not only in the ultimate state of the multifragmentation as in classical methods /1,4/.

I - THEORITICAL APPROACH

1.1 Calculation of internal stresses

First at all, the calculation of internal stresses due to the cooling of the specimen after the T6 ageing treatment, and of the evolution of the internal stress field with subsequent deformation during a tensile test, have been carried out /2/. Thermal induced internal stresses are calculated by taking in account the possible plasticity of the matrix at each step of the cooling. After cooling, the fibre is in a compression state, and a composite strain ε_{c0} of approximatively 0.2 % is needed to establish a tension stress field in the fibre. At a composite strain ε_c upper than ε_{c0}, the fibre stress field can be described according to a modified Voigt model, such as: $\sigma_f = E_f * (\varepsilon_c - \varepsilon_{c0})$ (1) where σ_f is the axial stress in the fibre and E_f the fibre Young modulus.

1.2 Prediction of multiple fibre cracks phenomenon

1.2.1 Load transfer model

The calculation of the fibre tensile stress has been carried out in the case of an infinite length filament. During the tensile test, the prediction of successive fibre cracks requires the use of a load transfer model between the matrix and the segments of the broken fibre. Different load transfer models have been proposed. The Cox' model /3/, is based on an elastic approach. As it was suggest by Kelly /4/, calculated shear stresses in the matrix near the fibre are too high for being consistent, so local plasticity of the matrix near the fibre ends must be taken in account. On another hand, experimental results show that much of the fibre cracks occur during the composite plastic flow. In consequence, a load transfer model considering a plastic behaviour of the matrix is needed. Such a model was reported by Ochiaï and Osamura /5/, but in their case the matrix yielding is not controlled by a general criterion, such as Tresca or Von Mises criterion. Thus, this model appears as not very suitable in our case. Mc Cartney /6/ has recently proposed a general approach of load transfer, with a possibility of frictionnal slip between the fibre and the matrix, but the matrix is still assumed to be elastic. In consequence, as a first approach, the Kelly's model, assuming that the upper limit of the matrix shear stress is the matrix shear flow stress, has been chosen in this study. Moreover it has been assumed that the axial stress in the segments of the broken fibre cannot be greater than the stress σ_f given by equation (1).

1.2.2 Fibre strength distribution

The distribution in failure stresses all along the single filament of the composite is also needed. Crack tests have been conducted on several set of uncoated fibres having differents lengths, in order to determine the three Weibull distribution parameters. Then, the distribution in failure stresses is computed by the following way: the fibre length is divided in a large number of small segments. For each of them, the failure stress is calculated from the uncoated fibre Weibull stress distribution by using the Monte-Carlo's method /1/.

During the computer simulation, composite strain steps are successively applied. Equation (1) and Kelly's model give the applied stress field all along the fibre. At each strain step, and for each small segments previously defined, a comparison between the applied and the calculated failure stress is made in order to determine the number of fibre cracks occuring during this strain step. Therefore, the "cumulative number of fibre cracks", Nc, is calculated by summing the crack numbers determined during the former strain steps. Then, a new strain step is applied and so on. Thus, from these calculations, Nc is determined as a function of the composite strain ε_c and this curve (Nc vs ε_c) will be used as a reference curve in the following part of this paper. This computed curve can be seen on figure 1, curve b. The dashed zones at the ends of curve b correspond to scattered zones due to fibre strength dispersion.

II - EXPERIMENTAL PROCEDURE

By using Acoustic Emission (EA) measurements, experimental curves of "cumulative number of fibre cracks" Nc versus "composite strain" ε_c have been obtained. First at all, it has been noticed a good agreement between EA event number and the actual number of fibre cracks observed after machining the composite.

Figure 2 shows an example of experimental (Nc vs ε_c) curve, with the corresponding stress-strain curve. Note that the first fibre cracks appear at the beginning of the composite plastic flow.

On figure 3, computed curve (A) and some experimental results (a-e) corresponding to different yield stresses $\sigma_{0.1}$ of the matrix can be seen. Let us note the following remarks:
- Thermal internal stresses have not a strong effect on the behaviour of the single filament composite.
- Difference between curves cannot be explained by difference in the intial yield stress of the composite matrix.
- The saturated number of fibre cracks Nt appears mainly independant on the plastic strain of the composite, that is to say that the mean critical length lc deduced from Kelly's model can be reached by different ways.

III - CHARACTERIZATION OF FIBRE - MATRIX BOND

As shown before, the mean critical length lc from the Kelly's model is not suitable to characterize alone the composite behaviour all along the tensile loading. A new method, based on comparison between computed and experimental curves (Nc vs ε_c) is proposed. In order to characterize the fibre-matrix bond three parameters are defined (figure 1):
- The gap dc between the first experimental crack and the beginning of the calculated curve. This parameter can be connected with a damage induced by the processing of the composite.
- The ratio of the experimental slope $P_{exp}=dNc/d\varepsilon_c$ to the calculated one P_{th}, (P_{exp}/P_{th}):
* A ratio having a value very lower than the unity can be considered as relevant from a very bad bonding with a very small load transfer.
* A ratio value near the unity indicates a good load transfer.
* A ratio value very higher than the unity, connected with a large gap dc, suggests a signifiant induced damage.
-The mean critical length lc from Kelly's model, for the bond under ultimate conditions.

CONCLUSION

Different experimentals results of multifragmentation are presented as well as their analysis by using a theoritical method proposed in this paper. It appears that the quality of a given fibre-matrix bond is very well understood by this approach all along the tensile loading.

REFERENCES

1 - Jacques D., Favre J-P., proc. JNC6, Paris, october 1988 - (Favre J-P. and Valentin D. ed.), (1988), 170-182, Amac, Paris.

2 - Hugot F., Boivin M., European journal of mechanics A: solids, 9, (1990), 175-188.

3 - Cox H-L., British J. of applied physics, 3, (1952), 72-79.

4 - Kelly A., Tyson W.R., in "High strength materials", (Zackay V.F. ed), (1965), 578-602, John Wiley & sons, New-York.

5 - Ochiaï S., Osamura K., Z. fur Metallkunde, 77, (1986), 249-254.

6 - Mc Cartney L.N., Proc. Roy. Soc. London A, A425, (1989), 215-244.

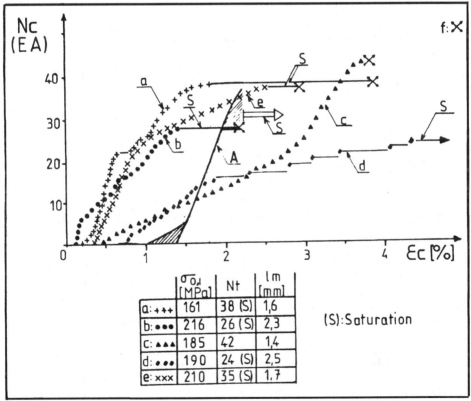

Fig. 3 - Comparison between computed (A) and experimental (a-e) (Nc vs ε_c) curves (see next page for legend).

$\sigma_{0,1}$: 0,1 % offset strain yield stress of the composite - lm: mean length of fibre segments at the end of the test - Nt: total number of fibre breaks - f: sample rupture.

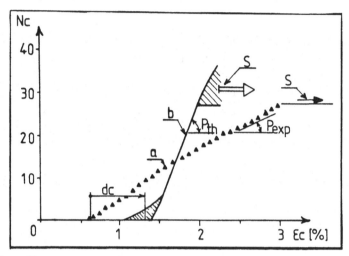

Fig. 1 - Parameters definition.
a: type of experimental curve - b: computed reference curve.

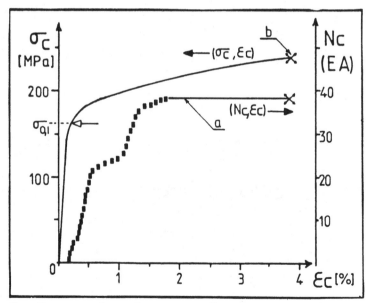

Fig. 2 - Stress-strain and corresponding (Nc vs ε_c) curves.
a: saturation: interruption of the multifragmentation - b: sample rupture.

SEVERAL RESULTS ON THE MECHANICAL BEHAVIOUR OF METAL MATRIX COMPOSITES WITH UNIDIRECTIONAL FIBERS OR NEEDLE SHAPED INCLUSIONS

M. TOURATIER, A. BEAKOU, J.-Y. CHATELLIER, L. CASTEX*, A. REMY**

Laboratoire Génie de Production - ENIT
Avenue d'Azereix - BP 1629 - 65016 Tarbes - France
*ENSAM - 13617 Aix-en-Provence - France
**ENSAM - 151 Bvd de l'Hôpital - 75640 Paris - France

ABSTRACT

This study was initiated to examine the dispersion of longitudinal and transverse waves in metal matrix composites , in order to obtain the dynamic elastic modulus and to evaluate various models for predicting the composite's macroscopic elastic constants from the properties of its constituents . The materials chosen for this investigation was alumina continuous fibers on the one hand and alumina short fibers on the other hand embedded in an aluminum alloy matrix. In more, some indications have been obtained for the residual stresses in the matrix , the acoustoelastic effect in the composite , the fatigue life of a specimen subjected to sinusoïdal tension-tension metal fatigue at the temperature of 120 °C .

INTRODUCTION

Fiber-reinforced composite materials have been extensively studied ,both experimentally and theoretically,/1/. Two types of fiber-reinforced composites occur in practice : continuous fiber (or wire) and short (or chopped) fiber. The former has been studied more thoroughly than the latter . Both of them are concerned in this study . Alumina and silica-carbide fibers embedded in an aluminum alloy matrix was particularly examined.

The aim of this work is to analyse the mechanical behaviour of metal matrix composites /2/ with long fibers or short fibers , i-e : macroscopic elastic constants deduced from ultrasonic-velocity measurements ; residual stresses in the matrix of the composite by X-rays ;

acoustoelastic effect in the composite ; fatigue life by sinusoïdal tension-tension loading at 120 °C.

The macroscopic elastic constants of composites with long fibers are predicted by an asymptotic expansion model , while those of composites with chopped fibers are predicted by a self-consistent model . Results from others models are shown for comparison.

I - MATERIALS AND SPECIMENS

Alumina (α type) long fibers embedded in an aluminum matrix composite specimens were fabricated from a liquid-infiltrated as cast cylinder of 25 mm diameter by Dupont de Nemours . An Al-2.5 % Li alloy was used as a matrix material by the manufacturer to promote bonding between the alumina fibers and the aluminum . The composite was manufactured in a constituent ratio of 50 % by volume of fibers which was unidirectional in orientation . Alumina (δ type) chopped fibers embedded in an aluminum matrix composite specimens were fabricated from squeeze-casting operation , the matrix material (aluminum alloy) had the following composition : Cu : 4.84 % , Mg : 1.57 % , Mn : 0.68 % , Ni:0.008% , Pb:0.024 % ,Fe : 0.25 % ,Sn : 0.017 % ; Ti : 0.0012 %; Zn : 0.013 %. The composite was manufactured in a constituent ratio of 22 % by volume of fibers which was randomly oriented . Silica-carbide (SiC) short fibers (or whiskers) embedded in an aluminum matrix composite specimens were obtained from squeeze-casting , the matrix material is the one described above , but the composite contain only 15 % volume Whiskers . All specimens used for the ultrasonic tests were parallelepipedic :4x17x17 mm^3 ; 8x17x17 mm^3 ; 10x17x17 mm^3 ; 17x17x17 mm^3 . The composite fatigue specimen was a standard traction specimen. The constituents properties for these composites are (E   Young's modulus , ν Poisson's ratio ; index f for fiber and m for matrix) :

Alumina (α type) long fibers reinforced composite :

$$E_m = 74 \, GPa \qquad v_m = 0.33 \qquad E_f = 380 \, GPa \qquad v_f = 0.25$$
$$\text{fiber diameter} : 15 \text{ to } 20 \times 10^{-6} \, m$$

Alumina(δ type) chopped fibers reinforced composite :

$$E_m = 75 \, GPa \qquad v_m = 0.30 \qquad E_f = 300 \, GPa \qquad v_f = 0.25$$
fiber diameter : 5 to 8 x 10^{-6} m ; fiber length : 500 x 10^{-6} m

SiC-Whiskers reinforced composite :

$$E_m = 75 \, GPa \qquad v_m = 0.30 \qquad E_f = 480 \, GPa \qquad v_f = 0.15$$
fiber diameter : 0.1 to 0.5 x 10^{-6} m ;fiber length : 50 to 200 x 10^{-6} m.

Figures 1 to 3 show the internal structure of the composites concerned in this study (increasely by 200)

II - DYNAMIC ELASTIC MODULUS : EXPERIMENT AND THEORY

Elastic modulus have been obtained using the classic stationary waves method , for instance ,see /3/,/4/, between 0.5 and 20 MHz . The method permits to deduce both geometrical dispersion of waves in the specimen and the elastic modulus from the measure of the wave velocity . The geometrical dispersion occurs if the wave velocity is depending of the wave frequency . Figure 4 shows the longitudinal and transversal waves dispersion in an isotropic specimen , and figure 5 , the wave dispersion, in long fibers reinforced composites . The elastic modulus are obtained when the wave velocity becomes constant , the wavelength being very large comparated to the diameter of the fibers.

The predicted elastic constants are deduced from homogenization techniques : asymptotic expansion for long fibers composites /2/,and self-consistent for chopped fiber composites or Whiskers composites /5/,/2/ . Experimental and numerical results are shown in tables 1 to 3 .

III - SOME REMARKS ON THE FATIGUE LIFE , RESIDUAL STRESSES AND ACOUSTOELASTIC EFFECT

3.1 Fatigue life

The specimen was a standard traction-sample , with a conventional reduction of the central section of length 32 mm which was of rectangular shape $8x5.7$ mm^2 . The specimen was subjected to sinusoidal tension-tension fatigue . Fatigue test was conducted by a loading of which the amplitude involves a maximum tension-stress of 205 MPa , and at the temperature of 120 ° C during the test . The rupture of the specimen near to the half length has been occured at 157000 cycles . The material tested was the Si-C Whiskers embedded in an aluminum alloy matrix(see "Materials and Specimens " title I) . Serrated end grips were used satisfactorily . Elastic modulus of this specimen measured after rupture and in the central region of the specimen were :
E = 96.7 GPa ± 2 GPa ; G = 37.2 GPa ±1 GPa

3.2 Residual stresses

Residual stresses have been evaluated by X-ray at the surface of parallelepipedic specimens with chopped fibers and whiskers embedded in an aluminum alloy matrix. Results concerning residual stresses in the matrix of composites , are showed in Table 4 : σ_n is the normal stress to the measure surface and σ_t the maximum shear stress in the measure surface .

3.3 Acoustoelastic effect

Parallelepipedic specimen has been simultaneously loaded by an uniform compression stress between 0 and 117 MPa , and by a transverse stress-wave at a frequency between 7.729 and 7.776 MHz for the alumina long fiber-reinforced composite . The results are shown in figure 6 .

CONCLUSION

Comparisons between experiment and theory are satisfactory for elastic modulus of long fiber reinforced-composite despite uncertainty of fiber volume fraction ,and are very satisfactory for elastic modulus of chopped fibers and whiskers reinforced aluminum alloy

It is important to note that the acoustoelastic effect is visible with these type of composite materials , i-e : alumina long fibers embedded in an aluminum alloy matrix. Results concerning fatigue life and residual stresses are not sufficient to obtain a significant interpretation.

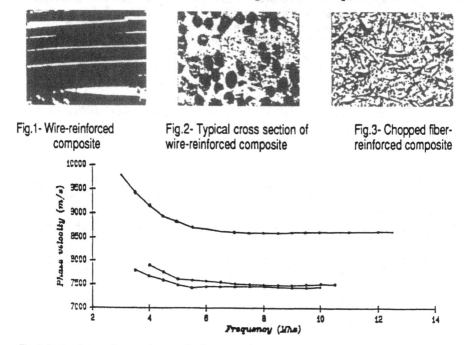

Fig.1- Wire-reinforced composite   Fig.2- Typical cross section of wire-reinforced composite   Fig.3- Chopped fiber-reinforced composite

Fig.4- Isotropic specimen : phase velocity versus frequency

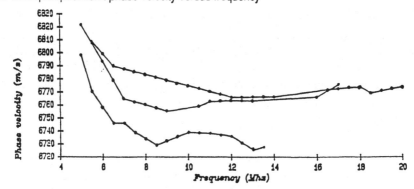

Fig.5- Wire-reinforced composite : phase velocity versus frequency

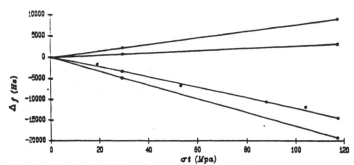

Fig.6 -Acoustoelastic effect in wire-reinforced composite

	E_L (GPa)	E_T (GPa)	v_{LT}	G_{TT}(GPa)	G_{LT}(GPa)
ASYMPTOTIC EXPANSION /2/	226.40	156.10	0.286	59.60	56.30
STATIC EXPERIMENT	207.00	144.00	0.244	48.00	48.00
WAVE EXPERIMENT	217.10	151.60	0.222	57.00	61.00
HASHIN /1/	226.40	139.50 149.10	0.286	52.90 58.10	56.30
DATTA-LEDBETTER /6/	226.40	144.10	0.286	52.90	56.30

Table 1- Macroscopic elastic modulus of wire-reinforced composite : comparison between experiment results and several model predictions (L longitudinal , T transversal direction) .

	E (GPa)	G (GPa)	v
WU FORMULA /5/	94.90	36.00	0.320
STATIC EXPERIMENT	110.00		
WAVE EXPERIMENT	94.40	36.00	0.314
MODEL 1 /7/	104.6	40.2	0.302
MODEL 2 /7/	97.5	38.7	0.260
HALPIN AND PAGANO /8/	84.5	35.8	0.179

Table 2- Macroscopic elastic modulus of chopped fiber-reinforced composite : comparison between experimental results and several model predictions.

	E (GPa)	G (GPa)	v
WU FORMULA /5/	90.40	34.20	0.320
STATIC EXPERIMENT	110.00	-	-
WAVE EXPERIMENT	91.40	34.50	0.325
MODEL 1 /7/	102.00	39.20	0.301
MODEL 2 /7/	98.10	40.00	0.225
HALPIN AND PAGANO /8/	84.9	35.50	0.198

Table 3- Macroscopic elastic modulus of Whiskers-reinforced composite: comparison between experiment results and several model predictions.

	AU4G1-Al2O3 (δ) SPECIMEN	AU4G1-Si C SPECIMEN	COMPOSITE SAMPLE
σn (MPa)	-196 ± 63	-90 ± 40	-379 ± 71
σt (MPa)	8.39 ± 5	-7.87 ±4.6	-12 ± 12
Intensity range	6.65	3.41	1.50
Peak breadth (10 $^{-9}$ m)	0.04887 ± 0.01954	0.04686 ±0.01698	0.03416 ±0.00127

Table 4- Residual stresses in the matrix of chopped fiber-reinforced composite

ACKNOWLEDGEMENT

This work was supported by the contract "DRET n°87/1163"
The authors are indebted to PSA for manufacturing the specimens used and M. Fournier (TURBOMECA-Bordes) for metal fatigue experiment .

REFERENCES

1 - Z.Hashin , J. Applied Mechanics , 50 (1983) 481-505
2 - M.Touratier ,DRET Technical Report n°87/1163 (1989)
3 - T.R.Tauchert , and A.N.Guzelsu , Transactions of the ASME , J. Applied Mechanics , (1972) 98-102
4 - J.Y.Chatellier and M.Touratier ,The Journal of the Acoustical Society of America , 83 , (1988) 109-117
5 - T.T.Wu ,Int. J. Solids Structures, 2 (1966) 1-8
6 - S.Datta ,H.Ledbetter ,Int. J. Solids Structures,19 (1983) 885-894
7 -R.M.Christensen ,Mechanical of Composite Material, (Wiley ed.N.Y.) (1979)
8 - J.C.Halpin and N.J.Pagano ,J. Composite Materials, 3 (1969) 720-724

SIC FIBERS AND PARTICLES REINFORCED
ALUMINIUM ALLOYS PRODUCED BY STIR CASTING

J. LECOMTE-BECKERS, D. COUTSOURADIS*, M. LAMBERIGTS*

University of Liege - Dept of Materials Science
Rue A. Stevart - Bât C1 - 4000 Liege - Belgium
*Centre de Recherches Métallurgiques
Rue E. Solvay 11 - 4000 Liege - Belgium

ABSTRACT

This study is concerned with the solidification processing of aluminium alloys reinforced by short fibers and particles of SiC. The reinforcements have been introduced and dispersed into aluminium alloys melts by the vortex method. The commercial alloys (2024, 319) containing elements such as Cu or Mg chosen to provide adequate wetting have been used and compared. The samples have been examined both in the as-cast or as-rolled and heat treated conditions (T6). Examinations of the mechanical properties by statistical analyses show that both SiC chopped fibers and particles increase significantly the mechanical properties of the alloys especially at high temperature.

I - INTRODUCTION

Fiber reinforced metal matrix composite (MMC's) are promising materials for future industrial applications where light weight structures, and engineerable mechanical and physical properties might be required /1/. Successful development of these applications could depend on successful progress in composite processing techniques. Although various processing techniques have been available for fibrous metal-matrix composite systems, casting and solidification processes have been drawing much attention recently since they offer near-net shape processing capability, selective reinforcement, controllable microstructures and perhaps most importantly lower cost compared to that of composite materials produced by other processing techniques. A potential technique to manufacture metal matrix composite is to stir the reinforcement into the metal melt. Several factors influence the final product, one very important being the metal melt'ability to wet the ceramic reinforcement. The term wetting is here used in a wide sens, including chemical reactions as well. For a given metal the wettability depends on many parameters :

1) type of reinforcements, their shape, size, surface roughness and surface chemistry of the outer atomic layers
2) alloys elements in the melt
3) gas environment of the reinforcements when injected into the melt
4) stirring, temperature and holding time in the melt.

We have studied MMC fabricated by stirring ceramic reinforcements, in particular SiC short fibers and particles into aluminium alloy melts. This study is concerned with the process used to provide adequate bonding between reinforcements and liquid matrix.

II - EXPERIMENTAL PROCEDURE

2.1. Materials

Two different aluminium alloys were used : an Al-Cu-Mg alloy (2024) which contains Mg, element susceptible to increase the wettability and an Al-Cu-Si alloy (A319) which contains Si, element which could delay the reinforcements reaction and dissolution with the aluminium melt.
The composition of these alloys is shown in Table I.

Typical properties of the Nicalon$^{(R)}$ SiC short fibers supplied by Nippon Carbon Co.Ltd. Japan are shown in Table II. The characteristics of the SiC particles are given in Table II.

2.2. Pretreatment of SiC fibers

The fiber, as-received, are in clumps and could not be dispersed uniformly into the melt in this state. Furthermore, the sizing on the as-received fiber led to non wetting and fiber rejection from the melt. To obviate these problems the as-received fiber had to be heated for 1 hour at a température of 600°C.

Moreover the SiC fibers when introduced in the melt had a tendency to floculate. This formation of flocules or aggregates of SiC in cast Al alloys is frequently observed /2/. To overcome these problems we have developed a specially designed mechanical device. The fibers pass through this device and are then introduced in the melt one by one.

2.3. Fabrication of the composites

The Al alloys were melted and degassed with dry argon at a temperature of 80°C above the liquidus. A vortex was created in the melt by using a quadruple bladed stirrer (to provide maximum down-wash) and the amount of fibers was added through the separating device. When particles were added instead of fibers the use of this separating devide was not necessary and the particles were put directly in the melt. This melt was contained in a silimanite crucible (charge : 1.5kg) and a shroud of dry argon gas was used over it during the addition of reinforcement. Melts containing 0, 2, 5, 10 V%SiC fibers and 0, 10, 20 V%SiC particles were prepared in this way and cast into a preheated copper plate die.

The samples were used in the as-cast state or subjected to subsequent heat rolling. These samples have been examined both in the as-cast or as-rolled and heat treated conditions (T6). The heat treating conditions are shown in Table IV.

III - RESULTS AND DISCUSION

The data for the mechanical properties of the two alloys with and without fibers and particles have been statistically treated by variance analysis. The effects of the different parameters have been studied.

The laminated samples possess better properties than the as-cast samples. The thermal treatment T6 increases the mechanical properties, which is the requested result, both with and without reinforcement. The mechanical properties are modified by the addition of reinforcements but the effects are different for the two alloys. The global results (in all treatments) are illustrated in Figs.1 to 8 for the yield strength (YS) and the ultimate tensile strength (R max), at 20 and 250°C.

The better results are obtained with 10% fibers or 10% particles. The use of too low fraction fiber (5%) introduces detrimental results as predicted by the mixing law. On the other side the use of a high fraction of reinforcement (20% particles) leads to other problems related to the soundness of the samples. The strengthening effect of the reinforcement (fibers or particles) is more marked at higher temperatures.

The distribution of the particles is not homogeneous due to the fact that we have not used our separating device for their introduction. Porosities often accompany the particles. This effect is rather marked for high volume percent which explains the bad results obtained.

The distribution of the fibers is more homogeneous and we find no porosity at the matrix-fiber interface. This proves the efficacity of our separating device. But although, in the as-cast state the reinforced matrix contains SiC fibers dispersed randomly in all directions, in the laminated state these fibers are broken in little pieces. This clearly show that with our rolling schedule the fibers must align in the rolling direction and are subjected to high shear strains which lead to the fiber breaking.

CONCLUSION

This research has lead to define the technological parameters to incorporate fibers or particles in aluminium alloys. The device and the procedure we have developed allow us to realise a composite by the difficult technique of incorporation in the liquid state. We have obtained strengthening effects both with fibers and particles with a more marked effect at high temperatures. One problem which still remains is the occurrence of porosity in the material.

The solution of this problem could be solved by the use of vacuum technique or squeeze casting.

ACKNOWLEDGMENT

The authors express their deep gratitude to Region Wallonne for financial support of this work.

REFERENCES

1 - Kelly A., Cast reinforced metal composites. Conference, ed. ASM International (1988), 1-5.
2 - Rohatgi P.K., Cast reinforced metal composites. Conference, ed. ASM International (1988), 85-91.

TABLE I - Alloys chemical composition

2024 : 4.1Cu - 0.3Mn - 1.6Mg
 319 : 3.6Cu - 6.6Si

TABLE II - Characteristics of SiC fibers

Mean fiber diameter :	15µm
Mean fiber length :	3mm
Cross section :	round
Filament per yarn :	500
Density :	2.55 g/cm³
Tensile strength :	2.4 - 2.9 GPa
Tensile modulus :	177-203 GPa

TABLE III - Characteristics of SiC particles

Mean size :	29µm
Minimum size :	16.5µm
Maximum size :	49µm

Table IV - Heat heating conditions (T6)

	2024	319
Solution heat treatment	2h1/2 at 493°C	12h at 500°C
Quenching	water (< 100°C)	water (< 100°C)
Aging	12h at 191°C	4h at 150°C

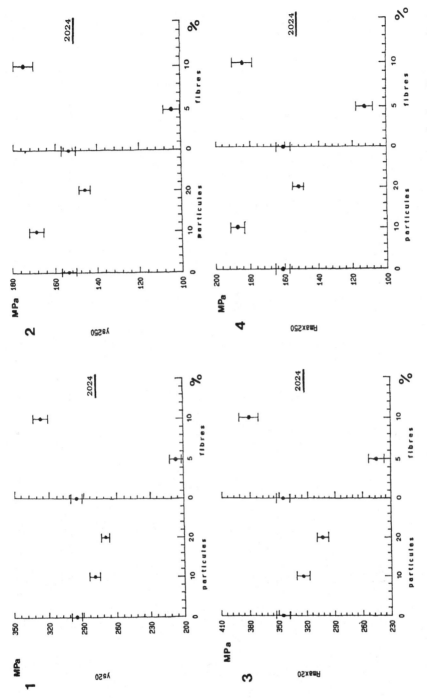

Influence of reinforcements on properties at 20 and 250°C : Standard Error Interval for Factor Means

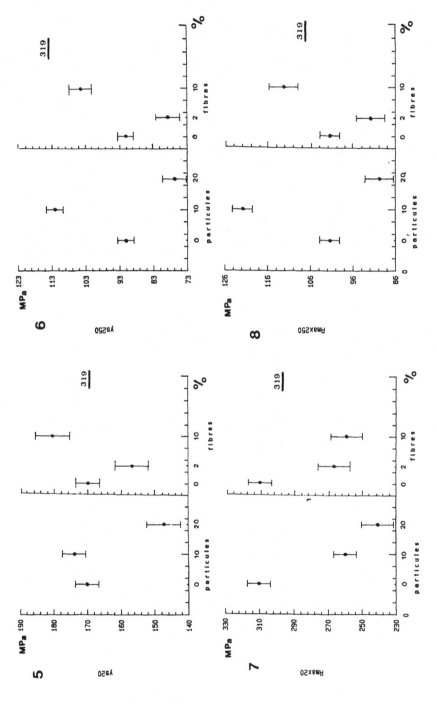

Influence of reinforcements on properties at 20 and 250°C : Standard Error Interval for Factor Means

PROCESSING OF PARTS MADE OF SIC-FIBRE REINFORCED TITANIUM USING HOT ISOSTATIC PRESSING (HIP)

R. LEUCHT, H.-J. DUDEK, G. ZIEGLER*

DLR-Institute of Materials Research
PO Box 90 60 58 - 5000 Köln 90 - West-Germany
*University of Bayreuth
Box 10 12 51 - 8580 Bayreuth - West-Germany

ABSTRACT
A hot isostatic pressing technique for processing
simple-shaped parts of metal matrix composites on a
laboratory scale is discussed. For reinforcing the
Ti6Al4V alloy, SiC-fibres (Textron/AVCO SCS-6) were used.
The processing of tubes, rings and shafts is presented.

INTRODUCTION
The hot isostatic pressing technique for processing metal
matrix composites [2] opens up the possibility of fabricating near-net-shaped parts by this method. Goals for
application are e.g. axially symmetrical parts in turbines.

PROCESSING TECHNIQUES
For processing simple-shaped parts by hot isostatic
pressing two different procedures were applied: tubes,
fig. 1, or rings, fig. 2, made of the matrix material
were grooved mainly using mechanical shaping or
electrical discharge machining, fig. 3. Grooving was
performed in such a way that the distance between the
fibres was twice the fibre diameter. Moreover, the depth
of the grooves was half the fibre diameter. On the
grooved surface, SiC-fibres were wound and the fibre ends
were fixed at the outside of the parts either
mechanically or by using a ceramic binder. After that,
the parts were coated with the matrix material using
electron beam evaporation or sputter deposition. As a
result, the fibres were completely enveloped by a matrix
layer of approximately 30 μm, fig. 4a. In a second step,

another layer of fibres was prepared and again coated with a 30 µm matrix layer. Using electrical discharge for grooving, the parts wound with the fibres were covered with a second tube or ring 1.5 mm thick, fig. 4b. The inner diameter of the second part should be as close as possible to the wound tube. The whole configuration was encapsulated in stainless steel tubes, outgassed at 450°C for two hours and HIPed at 900°C for 30 min at 1900 bar. Near-net-shaped parts were obtained after removing the stainless steel capsule by machining.

RESULTS AND DISCUSSION
Until now only rings and tubes with two layers have been processed, figs. 1 and 2. The fibres are well enveloped by the matrix, fig. 5a; however, during processing some of the fibres undergo fracture, fig. 5b.

Thus, cracking of fibres during HIPing is one of the main problems which have to be overcome for an industrial application of this technique. No fibre cracking has been observed in simple cylindrical samples [1] dimensioned 3.5 mm in diameter. If larger cylindrical parts, e.g. with a diameter of up to 30 mm, are processed without grooving the inner tube, promising results are expected, fig. 6a. However, in this case it has to be considered that it is difficult to wind fibres with equal distances on the matrix tube without grooving. When mechanical grooving is used, fig. 6b, cracking of fibres may take place during application of the gas pressure, particularly caused by the asymmetric arrangement of the inner grooved ring and the outer part and their differences in deformation resistance. The most promising technique, in our opinion, is the configuration in fig. 6c. However, the grooving of tubes and rings by electrical discharge machining is rather expensive.

LITERATURE
[1]   H.J. Dudek, R. Leucht, G. Ziegler: "SiC-Fibre Reinforced Titanium Alloys: Processing, Interfaces and Mechanical Properties", these proceedings.
[2]   R. Leucht, H.J. Dudek, G. Ziegler: "Laboratory Scale Processing of SiC-Fibre Reinforced Ti6Al4V-Alloys", Proc. Conf. Fibre Reinforced Composites, Liverpool, 1990.

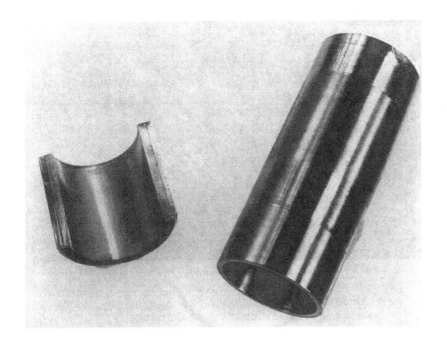

Fig. 1:   Processing of tubes by HIPing.

Fig. 2:   Processing of rings by HIPing.

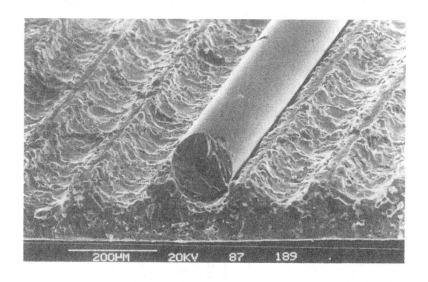

Fig. 3:   Grooved foil for MMC processing by using
           electrical discharge machining.

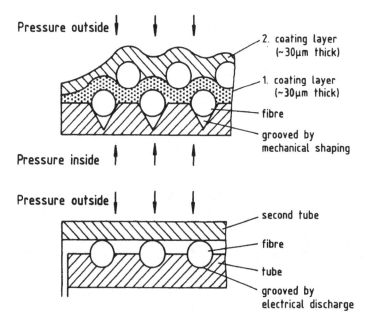

Fig. 4:
Different techniques for grooving and covering with
matrix material:
a) Grooving by mechanical shaping and matrix material
   deposition using physical vapour deposition.
b) Grooving by electrical discharge and covering with
   a tube consisting of the matrix material.

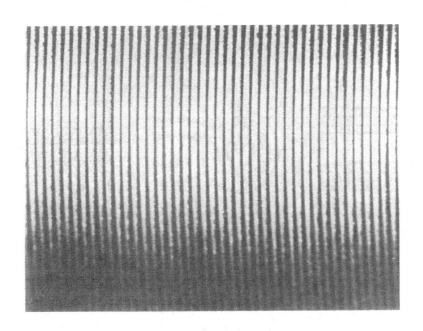

Fig. 5:   Consolidation of MMC parts during HIPing:
          a) Proper fibre envelopment,

b) Fibre fracture.

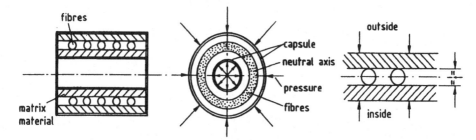

a) Without grooving the matrix cylinder,

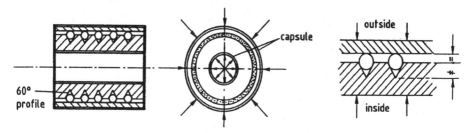

b) Using grooving by mechanical machining: V-shaped
   grooves, pressure from inside and outside

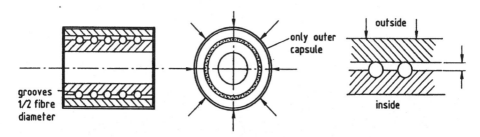

c) Using grooving by electrical discharge machining:
   round grooves, pressure applied only from outside
   due to encapsulation of the whole configuration.

Fig. 6:   Processing of MMC tubes by HIPing

WETTABILITY AND MECHANICAL PROPERTIES OF CVD SIC FILAMENT AND REINFORCED ALUMINIUM IN A VACCUUM SUCTION CASTING PROCESS

K. YU, V. DOLLHOPF*, R. KOCHENDÖRFER*

Institute of Aeronautical Materials CAE - 100095 Beijing - China
*Institute for Structures and Design - DLR
Pfaffenwaldring 38-40 - 7000 Stuttgart 80 - West-Germany

ABSTRACT

A vacuum suction casting process has been worked out for manufacturing CVD SiC filament reinforced aluminium. Three kinds of filaments were tested by using aluminium alloy Al-10%Si as a matrix. The wettability and the degradation of the filaments during the liquid infiltration and the mechanical properties of the composite rods were studied. Low infiltration temperatures were achieved due to good wettability under the specific conditions of the casting process. Uncoated SiC filament showed severe degradation after liquid infiltration, whereas SCS filaments exhibited an excellent protective ability against molten aluminium. The composite rods consisting of 50 vol.% SCS-6 filament possessed high strengths in tension as well as in compression.

INTRODUCTION

In recent years much interest has centred around the use of SiC reinforcement for metal matrix composites. The development of a manufacturing process for composites with SiC fibers is still receiving a large amount of attention. The purpose of this work is to develop a cost-efficient production method for SiC fiber reinforced aluminium. A specific vacuum suction casting process has been worked out to infiltrate CVD (chemical vapour deposition) SiC filaments with molten aluminium for manufacturing net shape composite samples with superior mechanical properties.

EXPERIMENTAL METHODS

The vacuum suction casting process was performed as follows. SiC filaments were inserted into a steel tube with one end sealed by an aluminium stopper and the other end connected to a vacuum system. The tube with the filaments was preheated to a high temperature and evacuated (3-5 x 10^{-2} mbar), after which the tube end with the stopper was put into molten aluminium. The stopper melted at once and the molten aluminium was sucked into the tube and infiltrated the filaments. After cooling down, the steel tube was etched away by nitric acid, leaving the SiC/Al rod intact.

Three kinds of CVD SiC filaments were used (SCS-2, SCS-6 coated with carbon core, both 140 μm in diameter from TEXTRON, USA; and uncoated with tungsten core, 100 μm in diameter from SIGMA, F.R. Germany). Cast aluminium alloy Al-10%Si was chosen as a matrix due to its high fluidity, low melting point and low probability of formation of Al_4C_3 in liquid state with SiC [1]. The results of infiltration were examined on micrographs of the cross sections of composite rods of 4mm diameter and 50 vol% filament showing different magnifications to evaluate wettability and the bonding between the filaments and the matrix. The composite rods (4mm in diameter) and the filaments extracted from them by etching the matrix with NaOH solution were tensile-tested using a specimen of 25 mm gauge length. Other properties were also examined.

RESULTS

The results of liquid infiltration with different combinations of two main technological parameters, the filament preheating temperature and the temperature of molten aluminium, which determine the conditions and results of wettability, infiltration, solidification and bonding, are summarized in Fig. 1. Low temperature is always recommended because there is less degradation of the fiber, if wettability and full infiltration can be achieved. Full infiltration was obtained at a temperature of molten aluminium as low as 700°C with filament preheating of 750°C, which is much lower than the wetting temperatures published in literature [1,2]. This is due to good wettability under the specific conditions of the casting process, namely, (1) no air between the filaments and adsorbed on the filament surface, which activated the filament surface, speeded up the aluminium flow and promoted the process of infiltration, (2) freedom from the interference of an aluminium oxide film due to the melting of the stopper in the depths of the molten aluminium and (3) the presence of atmospheric pressure.

A cross section of a 6 mm composite rod with 50 vol.% SCS-6 filament is shown in Fig. 2. The filament distribution on the whole can be considered satisfactory, but it is dense in some areas and sparse in others and usually gets a roll-structure. This is due to processing the filaments by hand and can be eliminated by using woven materials. The micrograph in Fig. 3 shows that the filaments are well infiltrated.

Fig. 4 shows the test results of the extracted filaments after casting at different temperatures of molten aluminium with the same filament preheating temperature of 750°C. The SCS filaments exhibited an excellent protective ability against the molten aluminium and could keep almost their initial strengths after casting, but the uncoated SIGMA filament showed severe degradation, lost about a half of its initial strength after casting and possessed only approximately 1500 MPa. It can be concluded that the uncoated SIGMA SiC filament is unsuitable for embedding into aluminium matrix by liquid infiltration.

Fig. 5 shows the tensile strength and modulus of ϕ 4mm composite rods cast at different temperatures of molten aluminium and the same filament preheating temperature of 750°C. In the tested temperature range of molten aluminium from 700 to 800°C the composite strengths dropped, as expected, with increasing temperature, but only very slightly. The highest tensile strength of 1700 MPa was obtained with the composite reinforced with 50 vol.% SCS-6 filament. SCS-2 reinforced aluminium exhibited a little lower strength, although the strengths of the extracted and as-received SCS-2 and SCS-6 filaments were quite similar. The 50 vol.% uncoated SIGMA filament reinforced aluminium had very low strength, even much lower than the 50% strength of the degraded filament, which indicated, presumably, a detrimental structure of the interface. As an intrinsic material characteristic, the tensile moduli were similar for all of them and ranged from 200-237 GPA, no matter whether the filaments degraded or not.

Using this process a unidirectionally reinforced tube with outside diameter of 20 mm and wall thickness of 2 mm was cast.

CONCLUSION

(1) CVD SiC filament can be easily wetted and infiltrated in the vacuum suction casting process due to specific favourable technological conditions. Wettability can be achieved at a temperature as low as 700°C

with Al-10% Si matrix. A temperature of 750°C for both filament preheating and molten aluminium is recommended and enough to guarantee full infiltration and a sound composite with good mechanical properties.

(2) The uncoated SIGMA filament degrades severely after liquid infiltration by aluminium and loses about half its initial strength, whereas the SCS filaments exhibit excellent protective ability against molten aluminium and remain almost the same as received. After embedding in the aluminium matrix, the SCS-6 filament seems to be the best one, and it gives the composite higher strength than does the SCS-2.

(3) The typical properties of 50 vol.% SCS-6 filament reinforced aluminium alloy Al-10%Si manufactured by the vacuum suction casting are as follows:

Tensile strength	1600-1700 MPa
Tensile modulus	208-220 GPa
Compression strength	1740-1810 MPa
Compression modulus	212-215 GPa
Shear strength (transverse)	240-270 MPa
Coefficient of thermal expansion (-100 to 300°C)	$2.6-3.6 \times 10^{-6}$ 1/°C

(4) The vacuum suction casting process studied here can create favourable conditions for liquid infiltration and provide good properties of manufactured composites and is suitable for producing net shape long parts. Using woven material of filament will further improve the properties of such composites and simplify the process. Its simple equipment, easy procedures and high productivity should make it a very cost-effective process for the manufacture of composite parts.

REFERENCES

1. R. Warren and C-H. Andersson
    Composites, 15 (1984), 101-111

2. S. Schamm, J. P. Rocher, R. Naslain
    ECCM 3 (1989) 157-163

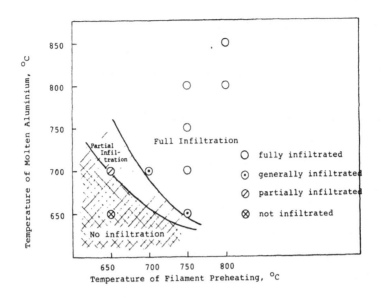

Fig.1 Effects of liquid infiltration at different temperatures

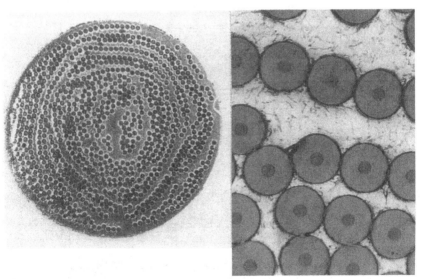

Fig.2 A cross section of fully infiltrated SiC/Al rod 6mm in diameter
with 50 vol% SCS-6 filament

Fig.3 Micrograph of the SiC/Al
rod (filament diameter 140 μm)

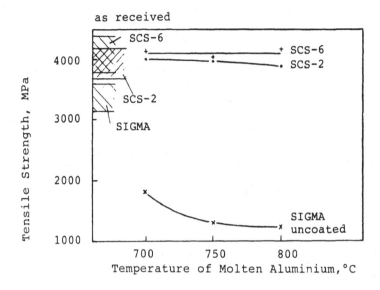

Fig. 4 Strength degradation of SiC filaments extracted
from matrix after casting versus temperatures of
molten aluminium (filament preheating 750°C)

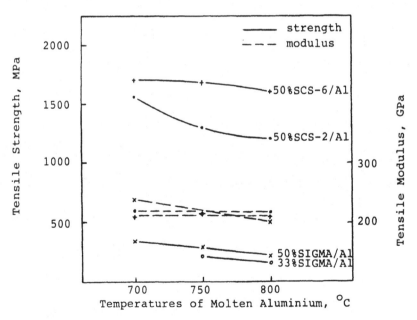

Fig. 5 Mechanical properties of CVD SiC/Al cast at
different temperatures of molten aluminium
(filament preheating 750°C)

CARBON-REINFORCED MAGNESIUM AND ALUMINIUM COMPOSITES FABRICATED BY LIQUID HOT PRESSING

M. RABINOVITCH, J.-C. DAUX, J.-L. RAVIART, R. MEVREL

ONERA
29 Avenue Division Leclerc - 92320 Chatillon - France

Abstract : An original process for fabricating magnesium- or aluminium composites reinforced by carbon fibres has been developed at ONERA. This process is based on the infiltration, under moderate pressure, of carbon tows by a liquid or semi-liquid magnesium or aluminium alloy. The composite materials obtained by liquid hot pressing have a large fibre volume fraction with a homogeneous fibre distribution, whereas operating in the semi liquid state provides a laminated material constituted of alternate unreinforced matrix and carbon-alloy composite layers. As an example of a metal matrix composite obtained by this process we present in detail the properties of a M40 carbon reinforced AZ61 magnesium alloy.

INTRODUCTION :

A study has been initiated at ONERA to develop magnesium- and aluminium-based metal matrix composites reinforced by carbon fibres. An equipment has been designed to fabricate these composites by liquid, or semi-liquid, hot pressing. Furthermore a low pressure CVD equipment has been constructed in order to deposit appropriate coatings on carbon fibres to optimize the properties of the fibre/matrix interface. An extensive study involving several fibre/matrix systems is under way to determine the influence of processing parameters and the nature of the fibre coatings on the properties of the composite material.
In this paper we describe the hot pressing technique, give some examples of materials obtained and focus in particular on the properties of a magnesium base alloy (AZ61) reinforced by high modulus M40 carbon fibres.

I - FABRICATION PROCESS :

In order to obtain a satisfactory infiltration of relatively compact fibre layers by liquid aluminium or magnesium, a hot pressing route has been selected involving moderate pressures, of the order of 10 to 20 MPa. Application of higher pressures, such as in squeeze casting processes, give generally very good infiltrations but may deteriorate or displace the reinforcing fibres or the preform. On the other hand, lower pressures are likely to cause infiltration defects such as pores, with deleterious effects on the mechanical properties of the composite.

In the process developed at ONERA a preform constituted of a sandwich comprising metallic alloy foils (aluminium or magnesium alloy) and unidirectional fibre layers is inserted into a tooling system having appropriate hermetic seals to avoid liquid metal leakages; this set is introduced between two heating plates (250 mm diameter) which are then placed in a press ; with this system, a temperature of 700 $^\circ$C can be reached within about 45 minutes and typical cooling rates are of the order of 10 $^\circ$C/mn. This equipment, which is practical and quite easy to operate, is controlled by a microcomputer, thus ensuring a good reproducibility of the processing conditions. It is to be reckoned though that for some applications, potential disadvantages may arise from the thermal inertia of the heating plates, too long a hold time at relatively high temperature (typically 720 s. above 550 $^\circ$C), and a solidification duration of 300 to 400 s. With such conditions, the matrix alloy is homogenized and as a consequence, its dendritic structure cannot be revealed. This fact was also noted by [1] in the case of a Al-4.5Cu (wt.%) alloy.

II - MATERIALS

Figure 1 illustrates a typical metal matrix composite tube obtained via the liquid route. After machining away its rounded edges, two unidirectional plates with dimensions of 80 x 38 x 1 mm^3 can be obtained. Figure 2 shows a transverse cross section of a plate constituted of AZ61 matrix (Mg-6Al-0.6Zn-0.2Mn, wt.%) reinforced with M40 fibres (Toray). This composite was obtained with an infiltration temperature of 635 $^\circ$C and a hydrostatic pressure of 14 MPa.

With an infiltration temperature lying between liquidus and solidus, a laminated material can be fabricated. An example of such a composite obtained by this semi-liquid route is shown in figure 3. The preform, constituted in this case of alternate layers of 5086 aluminium alloy foils and woven carbon fabrics (Brochier G884) has been pressed at a temperature of 620 $^\circ$C. At this temperature, the volume fraction of liquid in the alloy amounts to approximately 50%. Under the applied pressure, the liquid formed at the grain boundary in the metal is sqeezed out of the metallic foils and infiltrates the fabrics, and the remaining solid densifies by sintering. The material obtained presents a multilayered structure, constituted of alternate unreinforced alloy and composite layers.

III - PROPERTIES OF A M40/AZ61 COMPOSITE

It is known that pure magnesium does not react with high modulus fibres at temperatures typical of those adopted in our process (see for example [2]). However the presence of aluminium as alloying element in AZ61 which contains 6 wt.% Al, may induce a reaction at the fibre/metal interface and an investigation was initiated to study the properties of carbon reinforced AZ61 composites fabricated by the process described above. Another goal of this study was to evaluate the effects of a processing route carried out in a confined air atmosphere on such a material.

3.1. Microstructure of the composite
The transverse cross section represented on figure 2 shows that the fibre distribution is not ideal; each individual tow, which contains 3000 fibres, is still visible. Observation of longitudinal cross sections reveals slight local misalignment of the fibres due to the winding operation.

At higher magnification scarce isolated infiltration defects can be noted where fibres have remained in contact. Near the fibre/matrix interfaces, microscopic corrosion pits can be detected after the polished sample surface has remained in contact with ambient air for about ten hours. Previous experience with carbon/aluminium M40/Al composites lead us to attribute these pits to the hydrolysis of aluminium carbide particles. However, experiments carried out at D.L.R. [3] by transmission electron microscopy have not, up to now, revealed the presence of Al_4C_3. Work is still under way to clarify this point.

3.2. Tensile behaviour at 20°C of M40/AZ61
Tensile tests have been performed at room temperature on composite specimens having fibre volume fractions ranging between 40 and 45%, and a density of 1.8. The specimens are 80 x 8 x 1 mm³ testpieces which are held in the tensile machine by means of 3 mm duralumin end tabs. The tensile strength values corresponding to 35 specimens are reported in figure 4. They range between 600 and 1100 MPa with fifteen results exceeding 900 MPa and a *mean value of 853 MPa, representing 75% of the value calculated with the rule of mixture*. In this calculation the fibre strength is the value reported by the supplier and determined from tests carried out on resin-impregnated tows.

The dispersion in rupture strength is still too important but could certainly be narrowed by optimizing the successive steps of the process. It is to be emphasized that these results compare favorably with published results obtained on carbon/magnesium composites.

3.3. Fracture surface examinations.
All the fracture surfaces exhibit a macroscopic roughness (figure 5), with height variations of about 1 mm. At higher magnification two features can be noticed : fibre pull-outs on about 100 μm in zones where the local volume fraction is rather low, and numerous plateaux perpendicular to the applied stress surrounded by sheared regions in high volume fraction zones.

The presence of extracted fibres (pull out) on the fracture surface indicates that the first random fibre ruptures inside the material are accompanied by local interfacial decohesion. Stress concentration effects on the near neighbour fibres are therefore diminished and the propagation phenomena will start only at high stress levels. and simultaneously on several favourable sites. The weakened zones thus created at different locations will be joined together by shear ruptures at the end of the test. With such a scenario, one expects that a more uniform fibre distribution will significantly improve the tensile strength and narrow the dispersion in tensile strength.

CONCLUSION

An equipment has been designed for fabricating aluminium or magnesium metal matrix composites reinforced by long carbon fibres, by liquid or semi liquid hot pressing. Semi-liquid phase infiltration leads to a multilayered material having alternate unreinforced matrix-rich zones and composite layers. High tensile strengths (mean value : 850 MPa) have been obtained on M40/AZ61 magnesium-based composites fabricated via the liquid route.

ACKNOWLEDGEMENTS. Part of this work has been carried out within the framework of a EURAM project in cooperation with DLR (Köln) and Aérospatiale (Les Mureaux)

REFERENCES
[1] - **N.N. Gungor, J.A. Cornie, M.C. Flemings** : Solidification processing of an alumininum/alumina composite, in Cast Reinforced Metal Composites (Fishman S.G., Dhingra A.K. edit.), ASM International (1988), 39-46.
[2] - **P. Fortier** : Interaction chimique dans les composites à matrice métallique : systèmes Mg-C, Al-C-Si, Al-C-O-Si. Thèse (Université de Lyon I, 1988)
[3] - **J. Hemptenmacher, H. Kleine** : Private communication (1990).

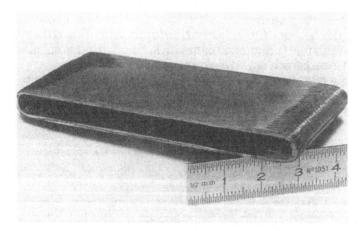

Fig. 1 - Composite tube obtained by liquid hot pressing.

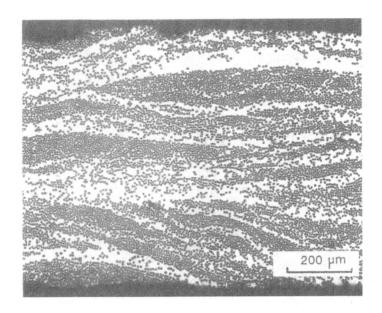

Fig. 2 - Transverse cross section of a M40/AZ61 composite.

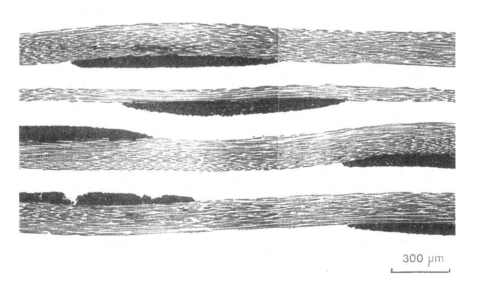

Fig. 3 - Aluminium alloy (5086) based composite fabricated according to the semi-liquid hot pressing route. Metallographic cross section showing the multilayered structure.

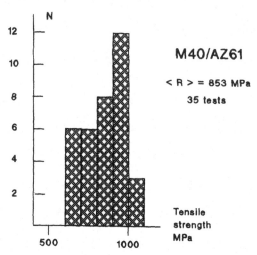

Fig. 4 - Histogram of tensile test results on AZ61 magnesium-based composite reinforced with M40 carbon fibres.

Fig. 5 - Typical fracture surface of a M40/AZ61 specimen ruptured at 1038 MPa.

CERAMIC MATRIX

Chairmen : **Prof. M. VAN DE VOORDE**
Commission of the
European Communities
Prof. G. ZIEGLER
University of Bayreuth

FAILURE OF CERAMIC MATRIX COMPOSITES
UNDER TENSILE STRESSES

J.J.R. DAVIES, R. DAVIDGE

AEA Technology
B 552 Harwell Laboratory - Didcot - Oxon OX11ORA - Great-Britain

ABSTRACT

A ceramic composite has been tested in tension and in 3 and 4-point flexure. The values of ultimate tensile strength from the different test techniques have been compared and it has been found that the strength varies with test type. Data have also been compared between test types. The Weibull modulus derived from specimen size variations is much lower than the modulus derived from within a single data set. These effects are explained by assuming that a fibre bundle-like failure mechanism is operating, but with complicating contributions from gripping effects in the tensile tests, and shear effects in the flexural tests.

1. INTRODUCTION

Ceramics matrix composites reinforced with strong continuous ceramic fibres have been developed to improve dramatically the fracture toughness of the matrix materials. This can induce a non-catastrophic fracture, should the ultimate fracture strength be exceeded, and thus give some measure of controlled failure during service. The microstructural effects that occur under stress are reasonably well understood [1] from a qualitative point of view. With uni-directional reinforcement the matrix first breaks up at intermediate stresses into a series of parallel blocks normal to the fibres. Then at higher stresses the reinforcing fibres begin to break followed by a controlled failure after the maximum stress has been reached where all the fibres are broken.

From an applications aspect it is essential to define quantitatively the various fracture processes because these control the safe regions of stress operation for a component. Failure may occur by a number of mechanisms including tensile (fracture of the fibres), shear (splitting between fibre layers) and compression (buckling). Here we are concerned primarily with tensile failure.

The tensile fracture strength can be measured by various mechanical tests, [2] particularly in bending and in uni-axial tension. It has been pointed out [3] that the numerical results obtained depend both on the nature of the test and the size of the specimen. It has been predicted particularly that strength should vary quite strongly with the specimen dimension parallel to the fibres and the current experiments have been designed to quantify these effects.

2.   EXPERIMENTAL

A fully crystalline hot-pressed glass ceramic matrix – Tyranno SiC fibre composite was produced as 100 mm² plates with unidirectional reinforcement and fibre volume fraction V_f 0.38 – 0.48. Specimens were machined to thickness with a 50 μm diamond finish equivalent surface grinder.

Flexural testing was performed on a jig with 10 mm diameter supporting rollers and a 25 mm diameter loading roller when used in 3-point flexure. For 4-point flexural tests, 6 mm diameter central loading points were used.

For tensile tests, the specimen configuration adhered as closely as possible to the "CRAG" [4] recommendations, as shown in Table 1. The gauge length and end-tab lengths were not strictly as recommended because gauge length size effects were of interest, and sufficiently large plates of material were not available. The end-tabs used to transmit gripping forces to the composite were 1.5 mm thick GRP, with stress-reducing contouring at the tab-ends.

Specimens were accurately aligned during stressing, and strain rates were kept approximately constant between specimens of different sizes to give a time to failure of around 2 minutes.

3.   RESULTS

Fig. 1 gives data for the strength of single fibres extracted from the matrix by dissolution. There is, as usually observed, a relatively large scatter in strengths (Weibull modulus m = 6.4). The mean strength is slightly below that in the original specification of the fibres indicating that the fabrication process has caused only modest fibre damage.

Fig. 2 presents strength data measured in 3-point bending for material tested with a range of support spans. There is a trend of steadily increasing strength with decrease in specimen span although data for the shorter span (25 mm) deviate from this trend. The inter-lamellar shear strength has been measured on separate tests using short 10 mm beams tested in 3-point flexure: the shear strength is 40 MPa. The anomalous behaviour of the shortest specimens in Fig. 2 is thought to be associated with the relatively high shear stresses in the specimens (~ 60% of the shear strength for the 25 mm samples) and these have been excluded from the analysis below. The ratio of tensile strength to shear strength in these materials is about 30 and thus long thin specimens are necessary to eliminate possible effects of shear stresses.

Samples have also been tested in 4-point bending. For 10 and 20 mm inner spans and 60 and 70 outer spans, mean strengths were 1068 and 947 MPa.

Tensile tests were performed on samples with dimensions given in Table 2, which also summarises the strength results, after adjustment $V_f = 0.40$. The clear conclusion is that the strength measured in tension is not significantly dependent on specimen size.

4. DISCUSSION

The strength of ceramic fibres varies strongly with fibre test length although this was not measured here [5]. Generally the Weibull modulus estimated from tests at a single fibre length is in good agreement with that calculated from tests at different fibre lengths. One thus expects that the ratio of fibre strength at 2 specimen lengths is given by

$$(\sigma_1/\sigma_2)^m = (\ell_2/\ell_1) \tag{1}$$

A similar equation can be used for comparing strengths of composites, from a specific test type, where only the specimen length varies.

The Weibull modulus estimated from the scatter in bend strength of the composites for a individual specimen sizes is in the range 15-20. However, that estimated from equ. 1 for the 3-point specimens is only 3.1 and, using an appropriately modified equation, 2.7 for the 4-point bend specimens.

It is clear that the high scatter in strength (low Weibull modulus) with specimen span cannot simply be related to the low scatter in strength (high Weibull modulus) for samples tested at a specific span. This is because the strength variation with specimen span varies in a similar manner to the variation in the mean strength of fibres with variation in length. On the other hand the variation in strength for a particular span needs to be related to the variation in the strength of fibre bundles for a particular fibre length. Lamon [6] for example has shown that for individual silicon carbide fibres having a Weibull modulus of 5, the Weibull modulus associated with dry tows of fibres is 12 and for fibres impregnated by a polymer 15-20. This is because fracture of individual fibres is a single event whereas the fracture of a tow is a collaborative effect where individual fibres continue to break until the remainder cannot sustain the load.

For the tensile tests the lack of strength variation with sample size is anomalous but can be related to the failure mode. There was clearly a considerable end-tab contribution to the failure process, with much fibre pull-out from beneath the tabs, Fig. 3. The fracture appeared to have involved an extremely fibrous bundle-like failure, but with localisation of the bundle to a small region of the gauge length. This would account for the length insensitivity seen. The small increase in strength of the thicker specimens is again due to test technique influence. It is likely that the diamond machining process caused interruption of load-bearing plies which would be proportionally greater for thinner material. Better tensile testing techniques must be developed before tensile results can be compared with confidence.

When fibres are tested in a bundle, fibres fail successively until the stronger fibres fail collectively. The strength of the bundle

$(\bar{\sigma}_b)$ is less than that expected from the mean strength of the fibres $(\bar{\sigma})$ depending on the Weibull modulus according to [7]:

$$\frac{\bar{\sigma}_b}{\bar{\sigma}} = \frac{(em)^{-1/m}}{\Gamma(1 + 1/m)} \tag{2}$$

The ultimate strength of the composites can be related in principle to the fibre strengths. The strength of the composite should thus be the mean strength of the fibres multiplied by the relevant factor from equ. 2 (0.68) multiplied by the fibre volume fraction. From the fibre strength data in Fig. 1, and $V_f = 0.40$, the strength associated with the fibres in the composite is 583 MPa which is lower than all the strengths measured in bending. The simple theory [3] used to interpret the results here assumes that the fibres are relatively free slipping through the matrix between the matrix cracks. Although the fibre pull-out stress is relatively low (a few MPa) some fibre/matrix friction exists because this leads to the high work of fracture for these materials. This effectively reduces the active length of the fibres so that the strength is somewhat enhanced according to equ. 1. A quantitative explanation of this effect is obviously rather difficult. Lamon [6] finds for example that the strength of a fibre bundle impregnated with polymer is twice that for a dry fibre bundle. In the former case the fracture was relatively planar, indicating a relatively low effective fibre length.

5. CONCLUSIONS

- Ceramic composites do not display size effects in flexure which are consistent with strength distributions at each particular size tested. However, behaviour is consistent with the strength of fibre bundles and its variation with fibre length.

- Shear forces affect flexural strengths at low span to depth ratios

- Tensile strengths of unidirectional ceramic composites are difficult to measure, as large gripping contributions are present. Accurate data may require large and complex specimen geometry.

REFERENCES

1. R.W. Davidge, Fracture of Non-metallic Materials" edited by K. P. Herman and L.H. Larsson (ECSC, EEC, EAEC, Brussels and Luxembourg, 1987) pp 95-115.
2. D.C. Phillips and R.W. Davidge, Br. Ceram. Trans. J. 85 (1986) 123.
3. R.W. Davidge and A. Briggs, J Mat. Sci 24 (1989) 2815.
4. Composites Research Advisory Group, Royal Aerospace Establishment Technical Report 88012, February 1988, ed. P.T. Curtis.
5. C-H. Anderson and R. Warren, Composites, 15 (1984) 16-24.
6. J. Lamon, D Richon, E. Anderson and O. de Pous. Presented at Journées Nationales des Composites Paris October 1988 Ed. J.P. Favre and D. Valentin, AMAC Evry France.
7. B.D. Coleman, J. Mech. Phys. Solids, 7 (1958) 60-70.

TABLE 1: Standards for Tensile Testing

Tensile Parameter	CRAG Recommendation	Value Used
Specimen Gauge Length	100 - 150 mm	50 mm
Length of End-Tabs	> 50 mm	25 mm
Specimen Width	10 - 20 mm Constant	Uniform 10 mm
Thickness Profile	Uniform along Length	Nominally 2 mm

TABLE 2: Tensile Test Data

Specimen Size (mm)			Fibre Volume Fraction	Strength (MPa)	
Length	Thickness	Width		Actual V_f	$V_f = 0.40$
10	1	10	0.48	707 ± 26	589
50	1	10	0.48	693 ± 14	578
50	1	10	0.40	578 ± 35	578
50	2	10	0.40	615 ± 13	615

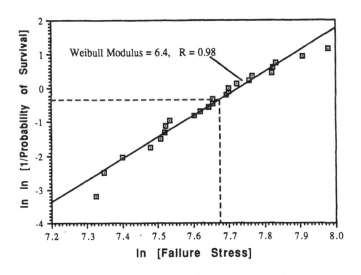

Fig. 1: Weibull Plot of Strenghts of Tyranno Fibres Extracted from the Glass Ceramic Matrix Composite by Dissolution in HF.

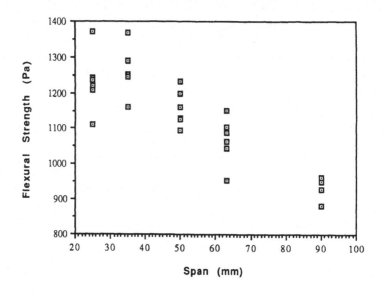

Figure: 2: Apparent Flexural Strength of Tyranno Fibre Composite in
3-Point Flexure

Figure 3: Tensile Test Specimen with 10 mm Gauge Length Length in
Tyranno Fibre Composite, Showing a Fibrous Fracture with a Large
End-Tab Contribution to the Failure.

CARBON FIBER REINFORCED SILICON CARBIDE COMPOSITES

K. NAKANO, A. HIROYUKI*, K. OGAWA

Government Industrial Research Institute Nagoya
1-1 Hirate-cho - Kita-ku - 462 Nagoya - Japan
*Nagoya Institute of Technology
Gokiso-cho - Showa-ku - Nagoya - Japan

ABSTRACT

Unidirectional carbon fiber reinforced silicon carbide composites have been fabricated by a process consisting of slurry impregnation followed by hot pressing. Characterization of the composites such as identification of crystal phase, observation of the metallographic structure, as well as characterization of flexural strength and fracture toughness at room and high temperatures (1400, 1600°C) in vacuo was carried out.

INTRODUCTION

Silicon carbide has excellent characteristics for use in high temperature structural materials and mechanical parts because of their heat-stability, superior strength at high temperature, and low density. However, it is brittle in the monolithic state. The major concern in utilizing silicon carbide in structural materials is in improving toughness.

Fiber reinforcement has recently been employed in toughing ceramics and is a promising method. Several processes have been employed for fabricating fiber reinforced ceramics. One example is the CVI/1-4/, and another example is the slurry method/1, 5-10/. The slurry method allows the formation of a ceramic matrix inside of the preform by slurry impregnation followed by sintering under ordinary pressure or hot pressing. This method has a short processing time and is inexpensive in practice compared to the CVI process. The authors have been trying to fabricate several kinds of long fiber reinforced ceramic matrix composites by using the slurry method which employs organosilicon resin as a slurry element/7-10/.

In the present work the carbon fiber reinforced silicon carbide composites were fabricated by the slurry process and the characterization of the composites such as identification of crystal phase, observation of metallographic structure, as well as characterization of flexural strength and fracture toughness at room and high temperatures was carried out.

I - EXPERIMENTAL PROCEDURE

1.1. Fabrication of composites

A high modulus type pitch carbon fiber (PETOCA HM-50: $\alpha = -1.1 \times 10^{-6}$/deg; $\rho = 2.16$ g/cm³; E = 500 GPa: Crystalline graphite X-ray diffraction pattern was observed) was used in this process. Fine powder of β-SiC [Ibiceram (0.3 µm) containing 0.4% SiO_2, 0.6% free carbon and 0.3% H_2O as impurities), and sintering additive of $A\ell B_2$ powder (B/Aℓ = 2.0) were used as filler. Unidirectional aligned prepregs were made by the filament winding method in which a slurry was used. The slurry consisted of the β-SiC and $A\ell B_2$ (7 wt % of total SiC content) powders, polysilastyrene (Nihon Soda PSS-400: average molecular weight 8000 - 12000; m.p. = 150 - 180°C), and a toluene solvent. The prepreg was cut into segment of equal size to form postforms. The postforms were isostatically pressed and then pyrolyzed in an Ar atmosphere. Composite ceramics were made by the hot pressing of pyrolyzed postforms at 1800 or 1850°C in an Ar atmosphere.

1.2. Characterization of composites

The identification of crystal phase was done by powder X-ray diffraction, and the metallographic structure was observed by SEM micrographs. The flexural strength [room temperature (RT), 1200, 1400, and 1600°C in vacuo], and fracture toughness (K_{1C}; RT, 1400, 1600°C in vacum) were measured by examining a sample, whose size was $3 \times 4 \times 40$ mm (the load was applied perpendicular to the fiber axis; the flexural strength by 3 points bending; K_{1C}: Single edge notched beam method by 4 points bending; Crosshead speed: 0.5 mm/min).

II - RESULTS AND DISCUSSION

2.1. Crystal phase and metallographic structure of composites

A trace of α-$A\ell_2O_3$ phase which could have been formed by the oxidation of $A\ell B_2$ in the fabrication process and also a trace of α-SiC phase were detected in addition to the main crystal phases of β-SiC and graphite by X-ray diffraction of the composites.

Fig. 1 shows SEM micrographs of the planes perpendicular (a) and parallel (b) to the fiber axis of a composite. The peripheries of the fibers were smooth, and only a few attack zones were observed around the fibers [Fig. 1-(a)]. This could be cause the bonding between the fibers and matrix were not very tight. A few cracks perpendicular to the fiber axis was observed in the plane parallel to the fiber axis, which could have resulted from the thermal expansion mismatch between the fiber and the matrix (Fig. 1-(b))/9-10/. The volum fraction (Vf) of the fiber of the composites in the present experiment was estimated to be 32.1 ± 3.3% by the SEM micrographs. The bulk density of the composites was estimated to be around 2.3 g/cm³ by the Archimedes method.

2.2. Mechanical properties of composites

Fig. 2 shows the flexural strength and open porosity of the composites. The room temperature flexural strength was clearly lower than those at high temperatures (1200, 1400 and 1600°C). At high temperatures, the average value of flexural strength corresponding to each temperature was nearly the same with each average falling within the error limits of the other two averages. Likewise the average value of the open porosity corresponding to each of 4 temperatures (RT, 1200, 1400, and 1600°C) was nearly the same with again each average falling within the error limits of the other averages. These suggest that the open porosity had little effect on the increase in the flexural strength at high temperature. Fig. 3 shows the fracture toughness and open porosity of the composites. The room temperature fracture toughness was evidently lower than those at high temperatures. In the high temperature region (1400, 1600°C), the difference of the average value of the fracture toughness between the two temperatures was nil within error limit. The average open porosity of these samples and of the samples at room temperature also had the same value within error limit, which suggest that the open porosity had little effect on the increase in the fracture toughness at high temperature. Although the fracture toughness values obtained in the present work do not have strict meaning because of the unideal brittle fracture of the composites (Fig. 4), they nonetheless display the increase of the fracture toughness with temperature increase. Fig. 4 shows examples of stress/displacement traces and micrographs of the failure samples in bending test (4 points) at room (a) and high temperatures [(b): 1400°C]. In the case of the high temperature test, the displacement at maximum stress tended to have higher value than that of room temperature test. The decrease of the stress, after the maximum stress point, as the displacement increase was rather slack at high temperature compared to decrease at room temperature. These observations suggest that the fracture energy of the composites at high temperature is higher than that at room temperature, which is consistent with the results of the facture toughness (Fig. 3). In the case of the room temperature bending test, the crack tended to propagate perpendicular to the fiber axis as shown in Fig. 4-(a), which represented the steep stress drop after the maximum stress point. However, in the case of the high temperature test, the crack tended to propagate parallel to the fiber axis (exfoliation fracture), which represented the slack stress decrease which occurred after a sharp drop from the maximum stress point (Fig. 4-(b)).

Fig. 5 shows SEM micrographs of the fiber/matrix interface along the fiber axis in the broken composites at room (a) and high [(b): 1600°C] temperatures. In the case of the broken composites at room temperature the matrix was peeled off-at the grain boundary between the fiber and matrix. Clear grain traces of the matrix were observed on the fiber (denoted by arrows in Fig. 5-(a)]. However, in the case of the broken composites at high temperature the matrix can be sheared off (by the shear stress parallel to the fiber axis) at the softening of the boundary layer between the fiber and matrix. Wavy slip lines which could be caused by the shearing of matrix were observed on the fiber parallel to the fiber axis [Fig. 5-(b): denoted by arrows]. This observation suggests that the crack propagation was interfered.with by the soft boundary layer. The crack propagation interference probably reflected the increase of the high temperature flexural strength and fracture toughness.

Fig. 6 shows SEM micrograph of a fractured surface of the high temperature broken composite (1400°C). Predominant fiber pull-out was observed as was true in the cases of the high toughness fiber-reinforced ceramics /5, 9, 10/.

CONCLUSION

Unidirectional carbon pitch fiber reinforced silicon carbide composites were fabricated by using the slurry process employing polysilastyrene, followed by hot pressing.

A few cracks were observed in the plane parallel to the fiber axis, which could have been the result of a thermal expansion mismatch between the fiber and matrix.

High temperature flexural strength and fracture toughness were higher than those of room temperature. This can be attributed to the fact that the crack propagation along the fiber axis was interfered with by the soft boundary between the fiber and matrix.

REFERENCES

1 - E. Fitzer and R. Gadow, Am. Cer. Soc. Bull., 65 (1986) 326 - 335
2 - A. J. Caputo et al., Am. Cer. Soc. Bull., 66 (1987) 368 - 372
3 - P. J. Lamiq et al., Am. Cer. Soc. Bull., 65 (1986) 336 - 338
4 - H. Hannache, J. M. Quenisset and R. Naslain, J. Matr. Sci., 19 (1984) 202 - 212
5 - K. M. Prewo, J. J. Brennan, and G. K. Layden, Am. Cer. Soc. Bull., 65 (1986) 305 - 322
6 - J. K. Guo et al., J. Mater. Sci., 17 (1982) 3611 - 3616
7 - K. Nakano et al., in "Sintering '87" (S. Somiya et al. ed.) (1988) 1350 - 1355, Elsevier Applied Science.
8 - K. Nakano et al., in "Development in the Science and Technology of Composite Materials" (A. R. Bunsell et al. ed.) (1989) 381-387, Elsevier Applied Science.
9 - K. Nakano et al., in "New Materials and Processes for the Future" (N. Igata et al..ed.) (1989) 1090 - 1095, The Nikkan Kogyo Shimbun, Ltd., Tokyo.
10 - K. Nakano, A. Kamiya, and H. Okuda, in "Metal and Ceramic Matrix Composites" (R. B. Bhagat et al. ed.) (1990) 185 - 192, The Minerals, Metals and Materials Society, Pennsylvania.

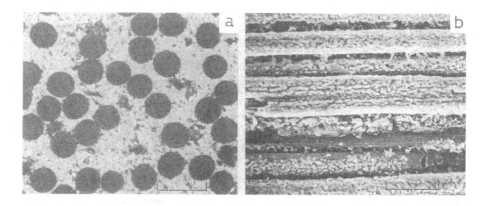

Fig. 1 - SEM micrographs of planes of composite.
a: perpendicular to fiber axis, b: parallel to fiber axis, scale: 20 μm.

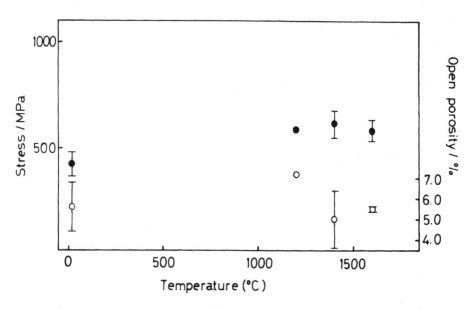

Fig. 2 - Flexural strength and open porosity of composites.
dark circle: flexural streugth, empty circle: open porosity.

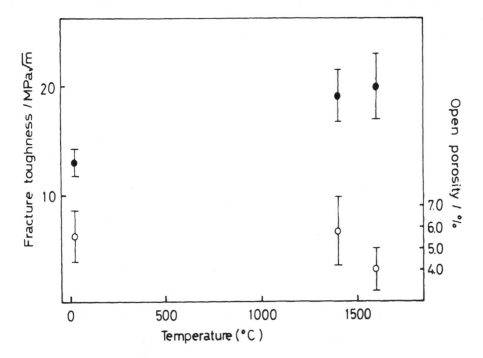

Fig. 3 - Fracture toughness and open porosity of composites.
dark circle: fracture toughness, empty circle: open porosity.

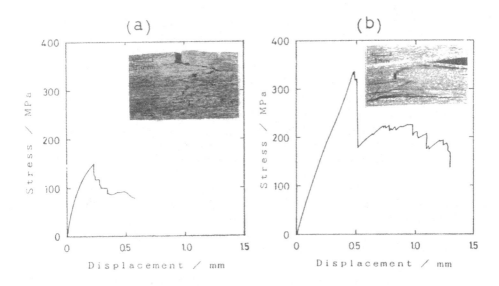

Fig. 4 - Stress/displacement traces and micrographs of the corresponding failure samples
in the bending test. (a): room temperature, (b): 1400°C.

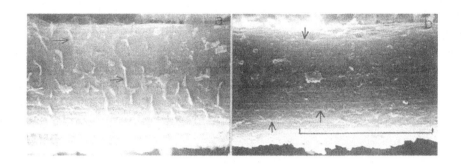

Fig. 5 - SEM micrographs of fiber/matrix interface along the fibers in the composites broken at room temperature (a) and 1600°C (b). Arrows show grain traces of matrix (a) and wavy slip lines which could be caused by the shearing of the matrix (b). Scale: 10 μm

Fig. 6 - SEM micrograph of a fractured surface of the high temperature breaking composite (1400°C).
Scale: 100 μm

OXIDE-BASED CERAMIC COMPOSITES

J. LEHMANN, G. ZIEGLER

University of Bayreuth
Inst. of Materials Research - PO Box 101251
8580 Bayreuth - West-Germany

ABSTRACT

Oxides and oxide-based materials with reinforcing se-condary-phase constituents have a potential, especially under the viewpoints oxidation resistance and reduction of processing temperature. Up to now, activities cover the area of particle or whisker introduction resulting in re-inforcing effects like crack shielding or crack deflec-tion. Only little work has been done in the field of con-tinuous- and short-fiber reinforcement of oxides.

The state-of-the-art of oxide-based ceramic composites is summarized. Novel processing techniques and new aspects of the traditional routes are described. Some essential aspects, such as the dispersion of reinforcements, the stability of coated carbon fibers in oxides and the impor-tance of the interfacial characteristics are pointed out.

1. INTRODUCTION

A promising development line to reduce the brittleness of ceramic materials and to achieve a damage tolerant be-haviour with controlled crack growth is the reinforcement by incorporating secondary-phase constituents. Different types of reinforcements are possible: the incorporation of whiskers and short fibers (randomly distributed or orien-tated) on one hand, and on the other hand unidirectional and cross-weaved bidirectional layers of continuous fibers or textile-like multi-dimensional pre-woven structures /1/. These composite materials can be processed via tradi-tional slip-casting, infiltration techniques or gas/liquid metal reactions. In the case of particulate components, additionally injection-moulding or pressing can be used.

In comparison to metal-matrix composites, where high-strength ceramic fibers are added to increase elastic modulus and strength at the expense of fracture toughness, fibers are added to ceramic matrices to increase fracture toughness, at least to the K_{IC}-values of brittle metals, sometimes at the expense of strength /2/. Nevertheless, in the case of whisker- or particulate-reinforced ceramics frequently toughness and strength could be increased simultaneously.

One important fact is that the energy dissipating effects are mainly depending on the interfacial characteristics. In this context the formation of interface layers through reactions during processing (especially if alkaline impurities are present /3/) and the control of the interface by fiber pre-coating are of importance.

2. DEVELOPMENT OF OXIDE-BASED REINFORCED COMPOSITES

Till now main activities cover the area of particle or whisker introduction (Table 1) leading to reinforcing effects like crack shielding or crack deflection and branching. Only little work has been done in the development of continuous- or short-fiber-reinforced oxides, like Al_2O_3 or mullite /4,5/.

2.1 ZIRCONIA-ALUMINA COMPOSITES

Zirconia-toughened alumina (ZTA) is considered as a promising group of ceramic composites. The toughness values reached are about up to a factor 4 higher compared with commercial alumina grades (Al_2O_3: K_{IC}-values about 3 to 4 $MNm^{-3/2}$). The mechanisms are based on the volume expansion of about 4 % and the shear strain of about 6 % connected with the tetragonal/monoclinic transformation during cooling down from sintering to room temperature. The effects can be described in the case of unstabilized zirconia as stress-induced transformation toughening leading to crack shielding and microcrack toughening by deflecting the propagating primary crack. In the case of a combination of ZrO_2-transformation toughening and whisker reinforcement very high toughness values of about 13.5 $MNm^{-3/2}$ are reported accompanied by strength values of ~ 700 MPa /6/. The addition of tetragonal ZrO_2-particles to alumina seems to be more effective in increasing strength, the whisker addition results in both the improvement of toughness and strength.

2.2 WHISKER REINFORCED OXIDES

According to theoretical predictions /7/ the degree of toughening increase is dependent upon the morphology of dispersoids (Fig. 1). Rod-shaped particles are predicted to be more effective in toughening than disk-shaped, which are again more effective than spheres. Based on these considerations, a lot of work has been done in the field of whisker reinforcement /8,9/ (Fig. 2). Whisker additions up to 50 vol.% to an alumina matrix increase the toughness values up to nearly 4 times to ~ 9 $MNm^{-3/2}$ starting from 2.5 $MNm^{-3/2}$ for the monolithic alumina. Strength is increased simultaneously up to a factor of 2.5 to a maximum strength of 900 MPa /10/. As toughening mechanisms crack bridging is most effective /11/ accompanied by crack deflecting and branching effects.

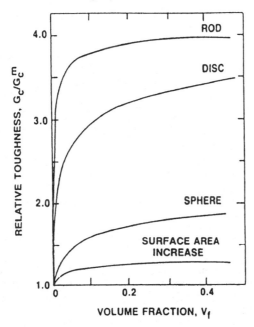

Fig. 1 - Toughness dependence upon dispersiod morphology /7/

The most important fact to initiate these toughening mechanisms is the formation of a relatively low bonded interface between whiskers and matrix. In the case of a strong bonding, for example based on glassy phases at the interface due to the surface-O_2-content of the whiskers or caused by the impurity level of the matrix material, a propagating crack will pass through the whiskers directly without any whisker-matrix friction. As a result, no essential reinforcing effect can take place. Glassy phases forming thin films or pockets in the interface whisker/

matrix were detected in some cases /12/, however, this point is still under controverse discussion. In any case, the interfacial bonding should mainly base on mechanical effects like thermal expansion mismatch and the surface morphology of the whiskers. Here, even small pull-out effects can take place.

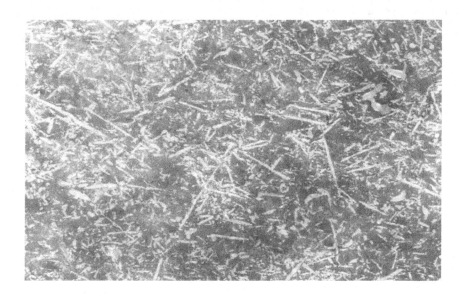

Fig. 2 - SiC-whisker-reinforced alumina (25 vol.% whisker, hot-pressed)

Using mullite as matrix material very promising toughness values were obtained by introduction of SiC-whiskers and ZrO_2-particles simultaneously. The highest values for fracture toughness were reported /13/ as 10.5 $MNm^{-3/2}$ for a mullite matrix containing 20 vol.% SiC-whiskers and 20 vol.% tetragonal ZrO_2.

Due to the carcinogenity of whiskers, platelets are thought to be a promising alternative. Introducing alumina platelets (5 vol.%) in partially Y_2O_3- and CeO_2-stabilized zirconia matrices /14/ the K_{Ic}-value was increased from 8.2 to 9.5 $MNm^{-3/2}$. However, the fracture strength decreased from 1430 to 735 MPa. This relationship between toughness and strength is observed in other systems, too, if the relatively large-sized platelets, only available up to now, are incorporated.

2.3 FIBER-REINFORCED OXIDES

As already mentioned above, little work has been done in the field of fiber-reinforced ceramic matrices /15/.

One promising example is the investigation of Singh and Gaddipati /5/. The improvement in crack propagation is caused by the BN-coating of the SiC-fibers (AVCO-SCS-6) (Fig. 3a-c). Maximum strength of 777 MPa and maximum strain of 1.2 % were achieved. The lower value of the interfacial shear stress for BN-coated SiC-fibers is expected to result in more fiber pull-out and only a gradual load drop during failure.

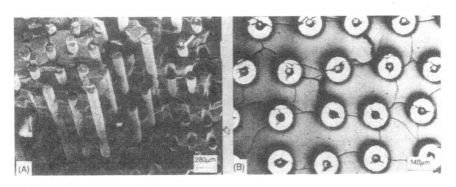

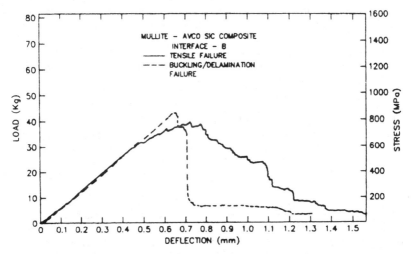

Fig. 3a-c - Failed mullite-silicon carbide composites showing
(a) filament pull-out, (b) matrix cracking and (c) load-deflection behaviour /5/

3. PROCESSING OF COMPOSITES

In the field of fiber-reinforced ceramic composites different serious problems have to be solved. Depending on the market situation, at present the thermal instability of the fibers at high temperatures during thermal compaction, is limiting the use of continuous or short fibers.

In addition, the chemical instability of reinforcing components will be enhanced by gaseous phases (e.g. nitrogen during sintering of Si_3N_4) or liquids (e.g. silicon during infiltration of C- or SiC-preforms). For oxide-based composites these effects are in general a little less critical than in nonoxide materials due to the lower compaction temperature and the processing environment usually not so aggresive. To solve these problems, thermally stable fibers, like C-fibers, have to be protected by suitable coatings, or processing techniques have to be developed which allow lower temperatures. By solving these problems one has to consider that the interface reinforcement/matrix has to be optimized with the aim of weak bonding.

In the case of short fiber-, whisker- or particle-reinforced composites one more problem is the proper dispersion of the reinforcements without causing damage.

3.1 POWDER ROUTE

Particulate-reinforced matrices, like ZrO_2-toughened Al_2O_3 or mullite, can be processed easily with the traditional forming methods (powder route) such as pressing, slip-casting or injection-moulding. In the case of whisker- and short-fiber-reinforcement mainly slip-casting is preferred to form compact bodies. Only this processing method is qualified for stress-free introduction of the reinforcing components.

Focusing on slip-casting and the resulting technological aspects, processing can be divided into two groups: whisker- and short-fiber-reinforced materials on the one side and continuous fiber-reinforced composites on the other side. In the case of whiskers and short fibers, the matrix material and the reinforcing components are the solid content in the slurry. To solve the problem of simultaneous dispersion of matrix and reinforcements, a specific combination of organic aids is necessary. For continuous-fiber reinforcement, fiber yarns or bundles have to be covered with slurry which contains the matrix material.

For thermal compaction uniaxial hot-pressing is the most common method. However, this technique leads only to simple-shaped parts like disks. For more complex-shaped components hot-isostatic-pressing (HIPing) is used to form composite materials with density values close to the theoretical density. Disadvantages of HIPing are the relatively high costs and the need of encapsulation. Pressureless sintering is the most convenient method for thermal compaction, however, it can only be used for particle- and whisker-reinforced materials with rather low content of the reinforcing component (nearly 10 vol.% whisker /16/.

following the powder route for processing continuous fiber-reinforced composites a pressure-assisted thermal compaction will be necessary depending on the matrix shrinkage.

3.2 REACTION-BASED PROCESSING

For oxide-based composites mainly two processing methods could be alternatives to the classical powder routes:
o   Sol-gel-infiltration
o   Directed oxidation (Lanxide).
In these cases, processing temperatures can be reduced by several hundred degrees compared to conventional techniques.

Sol-gel-infiltration:
This technique is advantageous in improving homogeneity of the single- or multi-phase matrix component. In addition, pressure-forming and injection-moulding techniques can be adapted. The infiltration should be performed step-wise in order to reduce the disadvantage of this method in comparison to the powder route/slurry techniques, the high shrinkage of the matrix material during pyrolysis.

Directed oxidation:
The Lanxide technique for processing composites bases on the reaction of a molten metal with a gaseous oxidant. An example for oxidic composites is the reaction of aluminium and adjacent inert fillers like particles and whiskers or two- or multi-dimensional fiber structures with oxygen. On the surface of the molten aluminium the reaction product Al_2O_3 is formed and will grow as a reaction front through the filler material. The reaction front will be supplied by molten metal due to capillar forces in the formed Al_2O_3-composite material. The residual metal content varies between 5 and 30 vol.%. For Lanxide processed Al_2O_3/SiC-fiber-composites strength values of up to 997 MPa and fracture toughness of up to 29 $MNm^{-3/2}$ were reported /16/. The disadvantage may be the residual metal content which limits the high-temperature properties.

4.   CONCLUSIONS

Only little work has been done in the development of continuous- or short-fiber-reinforced oxide-based composites compared to whisker reinforcement. This may depend on the technological problems during processing, and additionally, on the limited thermal stability of the fibers available at present. However, in our own investigations it was proved that even C-fibers can be kept stable in an Al_2O_3-matrix, if a suitable fiber coating, i.e. TiN, is developed in order to protect the C-fibers from oxidizing effects (Fig. 4). For high-temperature applications fiber-reinforced oxides seem to have a potential, particularly for applications in oxidic atmosphere.

431

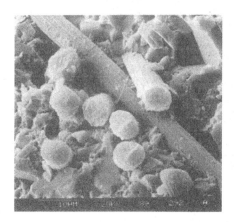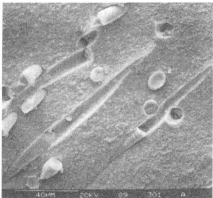

Fig. 4 - Short-fiber reinforced alumina
   a) TiN-coated C-fibers
   b) SiC-fibers (Tyranno)

In the case of whisker-reinforced composites the activities - at least in Germany - slowed down due to the carcinogenity, obviously more dangerous than asbestos. Under
technological viewpoints the development of organic aid-
combinations should become more important to guarantee the
stress-free introduction of fiber reinforcements to improve the so-called traditional powder routes like slip-
casting. Mechanical dispersion has to be replaced in general by chemical processing to avoid surface- or coating-
damages, or the reduction of the aspect ratio of the reinforcements. For particle- and platelet-reinforcement the
powder route is well suited.

In the scope of the development of ceramic composites
the impurity level of the reinforcing consituents seems to
have a decisive effect on the interface reinforcement/
matrix. At present, the impurity level of reinforcing constituents is significantly higher than that of the raw
materials used as matrix. The knowledge of the structure
and composition of the resulting interface is a preliminary condition to interpret and optimize the reinforcing
mechanisms.

Very promising techniques for processing composites,
like the sol-gel-infiltration and the Lanxide technique,
may solve the problem of matrix shrinkage during pressureless thermal densification, however, difficulties due to
residual closed porosity or residual metal content have to
be reduced. In this respect, the combination of the so-
called traditional and new processing techniques is an
interesting future aspect.

ACKNOWLEDGEMENT

Support regarding coatings by the Institut für Chemische Technik, Universität Karlsruhe, is kindly appreciated.

REFERENCES

1 - Rouby D., cfi/Ber.DKG 66, No. 5/6 (1989) 208-216
2 - Marshall D.B. and Ritter J.E., Am.Ceram.Soc. Bull., 66 [2] (1987) 309-317
3 - Prewo K.M., Am.Ceram.Soc.Bull., 68 [2] (1989) 395-400
4 - Lehmann J., Müller B. and Ziegler G., Euro-Ceramics, 1 (1989) 196-200
5 - Singh R.N. and Gaddipati A.R., J.Am.Ceram.Soc., 71 [2] (1988) 100-103
6 - Claussen N. and Petzow G., in: Proceedings of the Conference on Tailoring Multiphase and Composite Ceramics, edited by Tressler R.E. et al., New York: Plenum Press (1985)
7 - Faber K.T., Thesis, Univ. Berkeley (1982)
8 - Becher P.F. and Wei G.C., J.Am.Ceram.Soc., 67 [12] (1984) 267-269
9 - Tiegs T.N. and Becher P.F., Am.Ceram.Soc.Bull., 66 [2] (1987) 339-342
10 - Wagner R. and Ziegler G., J.Eur.Cer.Soc. (in preparation)
11 - Homeny J., Vaughn W.L. and Ferber M.K., Am.Ceram. Soc.Bull., 66 [2] (1987) 333-338
12 - Rühle M., Dalgleish B.J. and Evans A.G., Scripta Metallurgica, 21 (1987) 681-686
13 - Becher P.F. and Tiegs T.N., J.Am.Ceram.Soc., 70 [9] (1987) 651-654
14 - Heussner K.-H. and Claussen N., J.Eur.Cer.Soc., 5 (1989) 193-200
15 - Lehmann J., Müller B. and Ziegler G., J.Eur.Cer. Soc. (in preparation)
16 - Chiang Y.-M., Haggerty J.S., Messner R.P. and Demetry C., Am.Ceram.Soc.Bull., 68 [2] (1989) 420-428
17 - Wagner R., Thesis, Univ. Cologne (1989)
18 - Wei G.C. and Becher P.F., Am.Ceram.Soc.Bull., 62 [2] (1985) 298-304
19 - Fitzer E. and Gadow R., in: Proceedings of the Conference on Tailoring Multiphase and Composite Ceramics, edited by Tressler R.E. et al., New York: Plenum Press (1985)
20 - Wang J. and Stevens J., J.Mater.Sci., 24 (1989) 3421-3440
21 - Cales B., Mathieu P. and Torre J.P., Science of Ceramics, 14 (1987) 814-819
22 - Kim Y.-W. and Lee J.-G., J.Am.Ceram.Soc., 72 [8] (1989) 1333-1337

23 - Mukerji J. and Biswas S.K., J.Am.Ceram.Soc., 73 [1] (1990) 142-145

24 - Ruh R. and Mazdiyasni K.S., J.Am.Ceram.Soc., 71 [6] (1988) 503-512

25 - Okada K., Otsuka N., Brook R. and Moulson A.J., J.Am.Ceram.Soc., 72 [12] (1989) 2369-2372

26 - Panda P.C. and Seydel E.R., Am.Ceram.Soc.Bull., 65 [2] (1986) 338-341

Table 1 - Examples for oxide-based composites with various reinforcements

Matrix	Reinforcing Component Material	Shape	Strength [MPa]	Toughness [MNm$^{-3/2}$]
Al_2O_3	SiC	particle	530 /17/	
	SiC	whisker	800 / 8/	9 /18/
	SiC	fiber		10.5 /19/
	ZrO_2	particle	1080 / 6/	12 /20/
	ZrO_2/SiC	particle+		
		whisker	1000 /21/	13.5 / 6/
	TiC	particle	690 /22/	4.3 /22/
	TiN	particle	430 /23/	4.7 /23/
Mullite	SiC	whisker	450 / 6/	5.0 /18/
	SiC	fiber	650-	
			850 / 5/	
	ZrO_2/SiC	particle+	580-/ 6/	6.7-/ 6/
		whisker	720 /24/	11 /13/
ZrO_2	Mullite	whisker		15 /23/
	SiC	whisker	600 / 6/	11 / 6/
	Al_2O_3	platelet	735 /14/	9.5 /14/
Spinel	SiC	whisker	415 /26/	

SHEAR CHARACTERIZATION OF SILICON CARBIDE MATRIX COMPOSITES

P. PLUVINAGE, M. BOUQUET, B. HUMEZ, J.-M. QUENISSET*

Laboratoire des Composites Thermostructuraux
Europarc - 3 av. Léonard de Vinci - 33600 Pessac - France
*IUT - Université de Bordeaux I
351 cours de la Libération - 33405 Talence - France

ABSTRACT

Shear behavior of CMC composite laminates has been investigated through Iosipescu shear tests. The use of a finite element analysis for simulating these tests has allowed the actual stress field to be determined in the central part of the notched specimens. The complexity of the loading conditions has required the numerical assessment of the distribution of load transfer between specimen and device. As an example, a complete shear characterization of 1D and 2D SiC/SiC composites has been derived from the computer assisted tests. The results have proved that the damage and fracture mechanisms under shear loading are quite different depending on the fibrous structure.

1- INTRODUCTION

After establishing a constitutive law for representing a CMC behavior, it is necessary to identify all the related elastic and damage parameters by performing numerous numerical tests. Some of them are uneasy to carry on particularly the shear test, when the materials are semi-brittle and anisotropic as ceramic matrix composites (CMC). Thus, our need in shear characterization requires to make a choice among various methods by using the following main criteria :
  (i) ability of specimen manufacturing,
  (ii) cost of processing and machining,
  (iii) simplicity of the testing procedure,
  (iv) significance of the indications derived from the tests, particularly the
    possibility of obtaining the whole shear behavior.

Finally, a decision analysis technique based on these criteria has focussed our attention on the Iosipescu test /1,2/.

2- DESCRIPTION OF THE TEST METHOD

2-1- LOADING CONFIGURATION AND TEST GEOMETRY

The testing device, which has been designed, is derived from the studies reported by Adams and Walrath /3/. Its description and that of the specimen geometry are illustrated in figures 1 and 3. The main parameters defining the testing conditions are related to the application of load and the depth, angle and root radius of the notches /4/. The specimen is fixed in the device with the help of wedges, illustrated in figure 1, which enable specimens with some differences in dimension to be tested.

2-2 - MATERIALS

The composites which were tested in this study have been fabricated by chemical vapor infiltration of a silicon carbide matrix within a preform made of SiC (Nicalon) fibers. The various tested materials differ by their fibrous structure : unidirectional 1D composites $(0°)_n$ (0° corresponding to the direction 1), laminates $(90°_2, 0°_{10}, 90°_2)$ and cross-ply $(0°, 90°)_n$ made of unidirectional layers, and laminates made of bidirectional woven layers. In this type of CMC, the fiber volume fraction is about 40-45% and the more brittle and stiff componant is the matrix.

2-3- EXPERIMENTAL PROCEDURE

Two strain gauges were glued on each side of the specimen, in the central area at $\pm 45°$ to the specimen axis. Some of the specimens were also instrumented with gauges in the specimen axis in order to assess the tensile stress/shear stress ratio. In addition, specimen acoustic emission has been recorded during the tests to evaluate the significance of damaging mechanisms. Most of the specimens were tested at a cross-head speed of 0.02 mm/min and some of them were tested, with a strain gauges monitoring, at a strain speed of 0.02%/min .

3- NUMERICAL SIMULATION

3-1- FINITE ELEMENT METHOD

A modelling of the Iosipescu test has allowed a numerical simulation of the testing procedure. It has been shown that the boundary conditions generally used are not suitable to represent the actual specimen loading. In order to simulate the testing conditions, the numerical approach has concerned both specimen and loading parts of the whole set up /5/. The boundary conditions have been applied by monitoring the moving part of the device (figure 2).

A particular attention has been devoted to the load transfer between specimen and loading parts. Bonding elements have been used in the numerical simulation for representing compression and sliding at the loading surface of specimen. As a consequence, the numerical simulation has allowed the influence of the loading conditions and device stiffness to be pointed out.

3-2- NUMERICAL RESULTS

Experiments have shown the existence of undesirable tensile stresses in the central cross section. The tensile stress/shear stress ratio depends on material structure (i.e. orthotropy ratio, laminate sequences) and loading conditions (i.e.

clamping of specimen by wedges). For each type of composite, the boundary conditions used in the numerical simulation are modified, through the bonding elements, to adjust the resulting shear strain/tensile strain ratio to that measured with strain gauges.

The numerical simulation allows the distribution of shear stress in the central area of the specimen to be pointed out. It shows noticeable deviations from uniformity for the tested composites (figure 3). Despite the disagreement between the expected and actual stress fields, a small area undergoing nearly pure shear stress has been located in the vicinity of the specimen center where the strain gauges have been superposed. Whatever the conditions of testing, the numerical simulation has shown the need of a computer assistance as soon as the geometry or material nature of the specimen are changed /6/.

4- RESULTS OF THE COMPUTER ASSISTED TESTS

After validating the test procedure with various isotropic materials such as steel and aluminium alloy, which have shown deviations in the elastic constants lower than five per cent, the procedure has been used for evaluating the shear characteristics of the SiC/SiC composites previously presented. The numerical simulation has shown that the average shear stress $\sigma_{12} = F/S$, where F is the applied force and S the cross section between the two notches, is representative of the shear stress at the specimen center before the occurrence of damage phenomena. Despite early acoustic emission which can be related to crack propagation phenomena from the notch root as already reported /7/, the unidirectional composite shear behavior is quasi linear elastic, up to the degradation of the surface of load application.

The general feature of the stress-strain curves, for the other laminates, shows a significant linear domain (figure 4), which allows the calculation of a shear modulus. For the unidirectional laminate, the shear modulus value has been confirmed by oscillation tests in torsion performed on parallepipedic specimens. The elastic behavior under shearing of the $(90°_2, 0°_{10}, 90°_2)$ laminate is similar to that of the unidirectional laminate. The non linearity is caused by damage in the 0° plies protected by the external 90° plies. The order of magnitude of the shear modulus related to the cross-ply laminate is similar to those of the previous laminates. However, the failure shear strain and stress are more significant due to the laminate sequence. It is worthy to note that the shear modulus of the 2D woven laminate is higher and the shear failure strain is smaller, compared to the cross-ply $(0°, 90°)_n$.

All these curves show the great influence of laminate sequence and ply structure on the shear behavior.

5- CONCLUSIONS

Whatever the conditions used for performing shear characterization of CMC with the help of Iosipescu test, the numerical simulation has shown the need of a computer assistance as soon as the geometry and/or material nature of the specimens are changed.

The shear characterization of SiC/SiC composites processed by CVI reveals early damaging mechanisms leading to a decrease in the rigidity of laminated composites and a strong influence of the fibrous structure.

ACKNOWLEDGEMENTS

The authors wish to thank the SEP company (Etablissements de Bordeaux) for supplying the SiC/SiC composites and supporting this study.

REFERENCES

1. LEE S., MUNRO M.
'In plane shear properties of graphite fibre/epoxy composites for aerospace applications'
Aeronautical note NAE-AN22, 184, National Aeronautical Establishment, Ottawa, oct. 1984.
2. YEOW, BRINSON
'A comparison of simple shear characterization methods for composites laminates'
Composites, January 1978, pp 49-55.
3. ADAMS D.F., WALRATH D.E.
'Further development of the Iosipescu shear test method'
Experimental mechanics, june 1987, pp 113-119.
4. ADAMS D.F., WALRATH D.E.
'Current status of the Iosipescu shear test method'
Journal of composite materials, Vol. 21, june 1987, pp 494-507.
5. GAGNE C.R., HOOPER W.R., BECKWITH J.W.
'Shear response and failure in thick wall graphite/epoxy composites'
31st international SAMPE symposium, april 1986, pp 1604-1618.
6. BROUGHTON W., KUMOSA M., HULL D.
'An experimental - analytical investigation of intralaminar shear properties of unidirectional CFRP'
Proc., ECCM 3, Third European conference on composite materials, (A.R. BUNSELL et al.), Elsevier, 1989, pp 741-746.
7. BARNES J.A., KUMOSA M., HULL D.
'Theoretical and experimental evaluation of the Iosipescu shear test'
Composites science and technology, 28, 1987, pp 251-268.

Figure 1: Iosipescu shear test fixture.

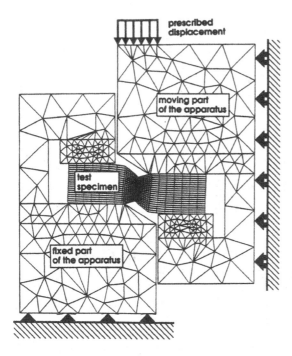

Figure 2: Numerical simulation of the Iosipescu test. Deformation of the mesh related to the applied boundary conditions.

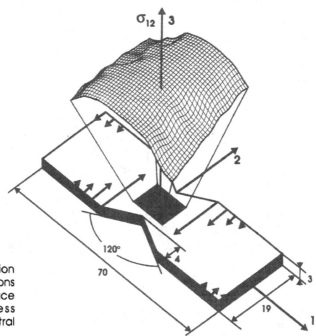

Figure 3 : Representation of the loading conditions at the specimen surface and shear stress distribution in the central area of the specimen.

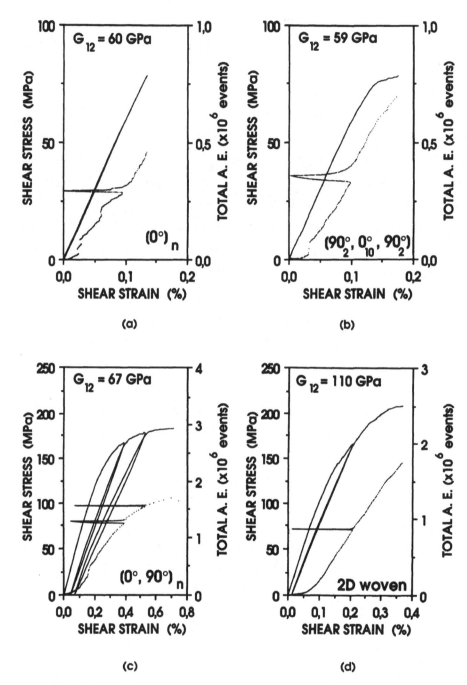

Figure 4 : Typical shear stress-strain curves and total acoustic emission signal for SiC/SiC composites laminates :

      (a) $(0°)_n$ (b) $(0°_2, 90°_{10}, 0°_2)$
      (c) $(0°, 90°)_n$ (d) 2D woven.

PULL-OUT MECHANISM IN BRITTLE CERAMIC MATRIX
REINFORCED WITH CONTINUOUS CERAMIC FIBRES

D. ROUBY, G. NAVARRE, M. BENOIT

Insa de Lyon - GEMPPM UA CNRS 341
20 Boulevard Albert Einstein - 69621 Villeurbanne Cedex - France

ABSTRACT

In ceramic-ceramic composites with continuous fibres, fracture surface morphologies are linked to toughening mechanisms and specifically to the pull-out process. The probability of fibre failure near a matrix crack for different stress transfer assumptions is analysed. The theoretical results are compared with experimental observations on real bidirectional SiC (Nicalon) - SiC composites.

INTRODUCTION

Monolithic ceramics and most ceramics reinforced with particles or whiskers generally fail in a more or less brittle manner. Ceramic matrix composites reinforced with fibres, however offer opportunities for greatly increasing toughness /1,2/. This improvement is mainly due to the bridging of matrix cracks by fibres. Globally, the resulting fracture energy depends on the bridging tractions times the matrix crack opening /3,4/. The fracture surfaces contain information about the stress transfer process between fibre and matrix where the interfacial properties play a role of great importance.

The proposal of this work is to analyse the influence of different microstructural parameters on the morphology of the fracture surfaces. In a first part, we describe some results obtained by calculations with different stress transfer hypotheses. In a second part, we compare these results with measurements performed on SiC (Nicalon)-SiC composites from the Société Européenne de Propulsion (SEP), exhibiting very large differences in pull-out lengths (from micrometer to millimeter).

I - MODELLING OF FRACTURE SURFACES

The main origin of toughening is that fibres fail not exactly at the matrix crack (Fig. 1a). As a matter of fact, due to the matrix crack, the fibre undergoes overloading which depends on the stress transfer; on the other hand, the fibre exhibits a certain distribution of local strength controlled by the flaw distribution along its axis. Therefore, under a given applied load, the fibre possesses a significant probability of breaking inside the matrix at a given distance from the matrix crack, defining the pull-out length l_p.

This problem was first analysed by Wells /5,6/. More recently, approaches based on constant interfacial shear stress τ^* were developed /3,7/, leading to comparable results: the average pull-out length, $<l_p>$, increases as the fibre Weibull modulus, m, and τ^* decrease and as the average fibre strength, $<\sigma^R>$, increases. Multiple matrix cracking tends to modify $<l_p>$ /7/. In the next sections, we will describe this analysis and particularly we study the effect of the shape of the stress transfer profiles (Fig. 1b, c, d and e) on the pull-out length distributions.

1.1. Fibre failure probability near a matrix crack.

The fibre is assumed consisting of a series of links (of length δx); the cumulative probability of failure of a link is given by the simplified weibull's statistics:

$$F_1 = (\sigma/\sigma_0)^m \, \delta x \tag{1}$$

where σ_0: scaling factor, linked to the average fibre strength and m: weibull's modulus. It is assumed that the entire flaw spectrum is largely repeated in the fibre overloading profile width. For the Nicalon fibre, values of m between 1,5 and 5 are reported /2,8,9/.

The stress σ acting on the link located at x depends on the peak stress, T (Fig. 1a), which depends on the applied load. Therefore, the probability density function of failure of the link at x when the peak stress attains T is given by /10/:

$$\phi\,(T,x)\,\delta T\,\delta x = \exp\left[-2\int_0^\Omega (\sigma/\sigma_0)^m\,dx\right]\,\{\partial(\sigma/\sigma_0)^m\,/\,\partial T\}\,\delta T\,\delta x \tag{2}$$

The exponential accounts for the survival probability of the other links (Ω is the stress profile half-width, see Fig. 1a); the partial derivative corresponds to the failure probability of the link at x. The cumulative fibre failure probability near the matrix crack is given by:

$$D = \int_0^\theta 2 \int_0^\Omega \phi(T,x)\,dx\,dT \tag{3}$$

If θ (the upper bound of T) tends to infinity, all the fibres fail and D = 1. In this analysis, the interaction between neighbouring fibres as well as stress singularities at matrix crack are neglected. The average pull-out length is given by:

$$<l_p> = D^{-1} \int_0^\theta 2 \int_0^\Omega x\,\phi(T,x)\,dx\,dT \tag{4}$$

and the pull-out length distribution by:

$$P(l_p) = 2\,D^{-1} \int_0^\theta \phi(T,x)\,dT \tag{5}$$

The average fibre strength near the matrix crack is given by:

$$<S> = \int_0^\infty 2 \int_0^\Omega T\,\phi(T,x)\,dx\,dT \tag{6}$$

and this term is generally not equivalent to $<\sigma^R>$. For fibres undergoing tensile stress, uniform in the gauge length L, Eq. 6 yields:

$$<\sigma^R> = \sigma_0\,L^{-1/m}\,\Gamma(1 + 1/m) \quad \text{with} \quad \Gamma(z) = \text{Gamma function} \tag{7}$$

1.2 Stress transfer profiles.

The fibre stress results on the addition of an overload profile (peak stress T) and of a nominal stress (at level B) corresponding to the fibre stress if there is no matrix cracking. For example, in the case of a tensile applied stress, σ_a, the two terms become:

$$T = \sigma_a / V_f \qquad \text{and} \qquad B = \sigma^{TH} + (\sigma_a E_f / E_c) \qquad (8)$$

where V_f: fibre volume fraction, σ^{TH}: residual thermal stress in the fibre, and E: Young's modulus (the indexes f, m and c refer respectively to fibres, matrix and composite).

In the case of a typical SiC (Nicalon)-SiC composite, σ^{TH} is compressive and then also B, so long as T is lower than approximately $<\sigma^R>/4$. In these conditions, the fibre failure probability outwarts of the overloading is very low. In the other hand, for the case of bending test on notched specimen, B decreases as x increases. In the next sections, the simplifying assumption that B = 0 is taken and the data used in the calculations are: Young's modulus: $E_f = 200$ GPa, $E_m = 350$ GPa; Poisson's ratios: $v_f = 0.25$, $v_m = 0.2$; coefficients of thermal expansion: $\alpha_f = 3 \times 10^{-6}\,°C^{-1}$, $\alpha_m = 4.6 \times 10^{-6}\,°C^{-1}$; temperature difference; $\Delta\Theta = -1000\,°C$; fibre strength : $<\sigma^R> = 2000$ MPa (with L = 2 cm); fibre radius: r = 7 µm

1.2.1. Bonded interface (Fig. 1b).

In the case of a strong interfacial bond, the overload profile is given by the following approximate expression /11,12/ (x > 0, B = 0):

$$\sigma = T \exp(-\beta x) \quad \text{where} \quad \beta^2 = -2 E_c / r^2 (1 + v_m) E_f (1 - V_f) \ln(V_f) \qquad (9)$$

From Eq. 4 (with $\theta = \Omega = \infty$, all the fibres are broken), the pull-out length becomes:

$$<l_p> = 1 / \beta m \qquad (10)$$

and the pull-out length distribution becomes:

$$P(l_p) = \beta m \exp(-\beta m l_p) \qquad (11)$$

which can be written in a normalised, non-dimensional, form as follows:

$$P(l_p) <l_p> = \exp(-l_p / <l_p>) \qquad (12)$$

It is interesting to notice that the non-dimensional distribution does not depend on the system parameters.

1.2.2. Interface with constant shear stress (Fig. 2c).

The profile is here triangular and the fibre stress is given by (x > 0, B = 0):

$$\sigma = T - (2 \tau^* x / r) \qquad \tau^*: \text{constant interfacial shear stress} \qquad (13)$$

The average pull-out length becomes (Eq. 4, with $\theta = \infty$ and $\Omega = T r / 2 \tau^* = d$) /3,7/:

$$<l_p> = r [\sigma_0^m (m+1) \tau^* / r]^{1/(m+1)} \Gamma\{1 + 1/(m+1)\} / 2 \tau^* (m + 1) \qquad (14)$$

For m > 1.5, $<l_p>$ can be obtained with the approximate expression (within 5% error):

$$<l_p> = (1/2) [r <\sigma^R> L^{1/m} / \tau^* (m + 1)]^{m/(m+1)} \qquad (15)$$

This expression illustrates the main trends of average pull-out length with the system parameters. As shown on Fig. 2a, the non-dimensionnal distribution is only few depending on the system parameters, as noticed by Sutcu /7/.

1.2.3. Coulombian friction (Fig. 2d).

In this case, the fibre undergoes radial compressive stress, due to the thermal strain misfit. In addition, in the zone of fibre overload, fibre and matrix elongations are not uniform and there are radius changes of the fibre and the matrix hole, due to Poisson's coefficient effect /6,13/. These Poisson's effects are of opposite sign if the fibre is in compression (during indentation test) or in tension (present study). A detailed analysis of the case of tension was done by Mc Cartney /14/. In a first approximation, the fibre overload can be given by /13/:

$$\sigma = A + (T - A) \exp (\gamma x) \quad \text{with} \quad \gamma = 2 \mu E^* v_f / r E_f \quad (16)$$

where $E^* = \{ [(1+v_m) / E_m] + [(1-v_f) / E_f] \}^{-1}$ and μ is the coefficient of friction.

The parameter $A = E_f (\alpha_f-\alpha_m) \Delta\Theta / v_f$ is an asymptotic limit. As shown on Fig. 1d, as the peak stress becomes closer to A, the width of the profile increases and this more and more quickly. If T = A, the overload stress is constant over the whole length, 2H, of the specimen (if it is assumed that fibres and matrix are perfectly loaded at the grips). The value of A is the higest fibre stress that it is possible to introduce by frictionnal stress transfer. Therefore, if $<\sigma^R>$ and A are close together, some fraction of the fibres cannot fail by the overloading due to a matrix crack.

The trends of the pull-out length distributions are shown on Fig. 2B, for m=5. If A and μ are large, $<l_p>$ is small and the distribution (cases a, b, c) is close to that given by a triangular profile (see Fig. 2A). The pull-out length decreases as μ increases (μ plays a similar role than τ^* in Eq. 14). Conversely, when A or μ decreases, the distributions (cases d and e) move rapidly to a limit (curve f) corresponding to the flat fibre stress shown on Fig. 1e, and $<l_p>$ tends to H/2, the quarter of the specimen length.

I - EXPERIMENTAL OBSERVATIONS

Observations of fracture surfaces are made on single edge notched beams broken in 3-point bending by SEM or optical microscopy. The material is a bidirectionnal SiC (Nicalon) - SiC composite, processed by SEP (France), which was submitted to some heat treatments leading to very different values of $<l_p>$. The measurements of the l_p's are performed ahead the notch front on the bundles normal matrix crack surface.

Fig. 3 shows two examples corresponding to very small $<l_p>$. The distribution, despite the scatter, can be well discribed by Eq. 12 refering to bonded interface (it is assumed that debonding occurs only after fibre failure, permitting pull-out). The values of $<l_p>$ give values of m comprising between 0.7 and 2.5 (for $0.3 < V_f < 0.6$) which are in agreement with the values reported for fibres damaged by the matrix densification /2,8/. The model of constant shear stress (Eq. 14) yields values of τ^* largely higher than 2000 MPa which are unrealistic. Short pull-out length account well for bonded interfaces. However, the present study does not take into account the stress singularities at the matrix crack surfaces and the debonding conditions.

Examples relative to longer pull-out lengths are shown in Fig. 4. There is some scattering (especially at low l_p's), but the distributions are always exponential, very different from the situations illustrated by the cases e, d and f on Fig. 2B. Similar results are obtained for $<l_p>$ on the order of 100 μm. This means that the asymptotic effect due to Coulombian friction and Poisson's effect seems not to play a critical role. However, it must be noticed that the numerical data give a value of A equal to 1280 MPa, somewhat lower than $<\sigma^R>$. The present results suggest that A is much higher than $<\sigma^R>$ and that the assumption of constant shear stress applies. A more detailed analysis should include the basis line at level B (see Fig. 1a) and its evolution before composite fracture. The influence of the more compliant carbon interfacial layer should be also taken into account.

CONCLUSION

Short pull-out lengths, illustrating a brittle behaviour, are better described with the bonded interface mode. Debonding before pull-out remains to be investigated. For longer pull-out lengths, giving higher fracture energies, the distributions are in agreement with the constant interfacial shear stress model. In the studied SiC (Nicalon) - SiC composite, the fretting thermal stresses are always sufficient to lead to fibre failure by interfacial stress transfer. Further work needs to analyse the effect of carbon coating on the fibres.

ACKNOWLEDGEMENTS

The authors wish to thank MM. Lamicq and Jouin for helpful discussions.

REFERENCES.

1 - Evans A.G. and Marshall D.B., *Acta Metall.*, 37 (1989) 2567-2583
2 - Thouless M.D., Sbaizero O., Sigl L.S. and Evans A.G., *J. Am. Ceram. Soc.*, 72 (1989) 525-532
3 - Thouless M.D. and Evans A.G., *Acta Metall.*, 36 (1988) 517-522
4 - Thouless M.D., *Acta Metall.*, 37 (1989) 2297-2304
5 - Wells J.K., PhD thesis 12497, Cambridge University, U.K. (1982)
6 - Wells J.K. and Beaumont P.W.R. , *J. Mater. Sci.*, 20 (1985) 1275-1284
7 - Sutcu M., *Acta Metall.*, 37 (1989) 651-661
8 - Simon G. and Bunsell A.R., *J. Mater. Sci.*, 19 (1984) 3649-3657
9 - Eckel A.J. and Bradt R.C., *J. Am. Ceram. Soc.*, 72 (1989) 455-458
10 - Oh H.L. and Finnie I., *Int. J. Fract. Mech.*, 6 (1970) 287-300
11 - Cox H.L. *Brit. J; Appl. Phys.*, 3 (1952) 72-79.
12 - Hsueh C.H., *J. Mater. Sci. Lett.*, 7 (1988) 497-500
13 - Arnault V., thèse de doctorat 89ISAL0098, INSA de Lyon, Fr. (1989)
14 - McCartney L.N., *Proc. R. Soc. Lond.*, A425 (1989) 215-244

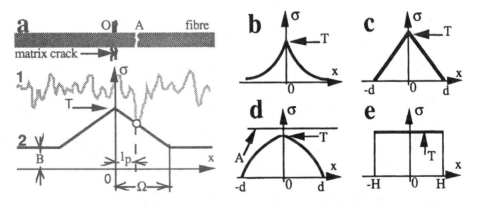

Fig. 1 - Stress transfer profiles. (a): Fibre failure (at A) near a matrix crack (at O); curve 1: local fibre strength; curve 2 : fibre stress. (b): perfectly bonded interface. (c): constant interfacial shear stress. (d): interface controlled by friction. (e): no stress transfer (for b, c d and e figures, it is assumed that B = 0).

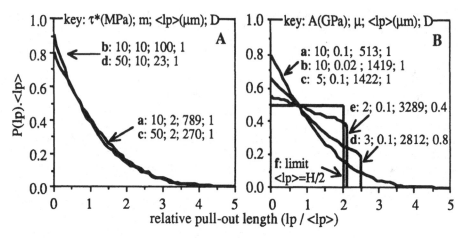

Fig. 2 - Theoretical pull-out length distributions (in non-dimensional scales)
(**A**) : Constant interfacial shear stress.(**B**) : Coulombian friction (H = 7 mm)
(the meaning of the number lists is given by the key on the top of each figure)

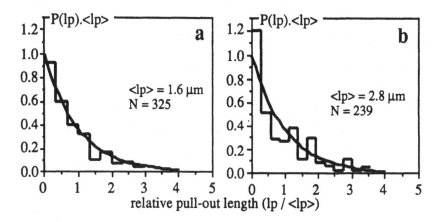

Fig. 3 - Non-dimensional pull-out length distributions (N : number of measurements)

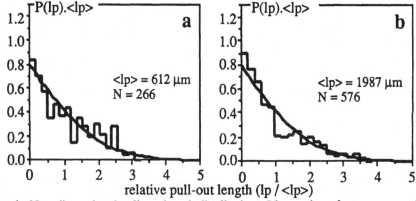

Fig. 4 - Non-dimensional pull-out length distributions (N : number of measurements)

A NUMERICAL SIMULATION OF THE MICROINDENTATION TEST FOR CHARACTERIZING THE FIBER MATRIX INTERFACE IN CERAMIC MATRIX COMPOSITES

N. PIQUENOT, J.-M. QUENISSET*

Laboratoire des Composites Thermostructuraux
1-3 Avenue Léonard de Vinci - 33600 Pessac - France
*Laboratoire de Génie Mécanique de l'IUT
Université de Bordeaux I - 33405 Talence - France

ABSTRACT

Various analytical approaches proposed for evaluating the conditions of load transfer at the fiber-matrix interface from microindentation tests have been compared to a numerical simulation of the testing method. The computer assistance of the test is shown to be usefull particularly to take into account important parameters such as the fiber volume fraction and the interphase characteristics which are not easily included in the analytical approaches.

1 - INTRODUCTION

The mechanical behavior of ceramic matrix composites is known to be strongly dependent on the conditions of load transfer at the fiber-matrix (F/M) interface. As a consequence, various testing methods have been developed for evaluating the interfacial shear strength [1] . The microindentation test based on the determination of the relationship between a load applied on a fiber cross section and its sliding displacement, offers several advantages such as the possibility of an "in-situ" measure and the apparent simplicity of indentation tests. In fact, this test deals with very precise measurement leading to specific devices and furthermore, the interpretation of the experimental results requires the use of models whose validity is often questionable. Thus, various theoretical approaches based on different assumptions and friction laws have been proposed to go with the technical improvement of the test [1].

Making use of a finite element analysis which enables the various types of frictional effects to be taken into account, a numerical simulation

447

of the microindentation test has been performed [2]. In addition to the definition of the limits of use of the analytical approaches, the simulation of the test allows the investigation of parameters which are difficult to include in theoretical analyses.

2 · COMPARISON BETWEEN ANALYTICAL AND NUMERICAL APPROACHES

The instrumented microindentation test originally developped by Marshall, enables the load F applied to a fiber flushing the specimen surface, to be simultaneously recorded with the indenter displacement δ given as follows :

$$\delta = h + u \qquad (1)$$

where u is the deviation in the fiber and matrix surfaces that is the fiber sliding and h the depth of impression at the fiber surface as expressed by the Meyer law :

$$h\,(F) = \alpha\,F^{\beta} \qquad \text{with } 0,5 \leq \beta \leq 1 \qquad (2)$$

2.1 - Differences in modelling

The various models which have been proposed for deriving interfacial characteristics from the test differ in the expression of u(F). It depends on both the friction law used for representing the load transfer at the F/M interface and the assumptions related to the matrix and fiber deformability. The main features of these models which have been reviewed elsewhere [1] can be summarized as follow:
(1) The Marshall's model only considers the longitudinal deformation of the fiber, assumes a constant frictionnal threshold τ_s at the F/M interface and leads to the following formula:

$$u(F) = \frac{1}{4\pi^2 R^3 E_f}\,\frac{F^2}{\tau_s} \qquad (3)$$

with R and E_f respectively the radius and rigidity of the fiber.
(2) All the other models take into account a linear friction law :

$$\tau = \tau_0 - \mu\,\sigma_i\,(z) \qquad (4)$$

in which μ is a coefficient of friction, $\sigma_i(z)$ is the compression stress related to the indentation load F along the F/M interface, and τ_0 is a friction threshold which includes the effect of the residual compression associated with the F/M thermal expansion mismatch. These models result in the following formula :

$$u(F) = \frac{C}{2\,E_f\,\mu k}\left[\frac{F}{\pi R} - \frac{\tau_0 R}{\mu k}\,\ln\left(1 + \frac{\mu k F}{\tau_0\,\pi R^2}\right)\right] \qquad (5)$$

where C and k are constants depending on the models as follows :

(a) in the model proposed by Shetty, the matrix stiffness is assumed negligeable compared to that of the fiber ($E_f/E_m \gg 1$) so that the constants are given by:

$$k = \frac{E_m \, v_f}{E_f \, (1 + v_m)} \quad \text{and} \quad C = 1 - 2v_f k \qquad (6)$$

with E_m, E_f, v_m, v_f the young Modulus and Poisson's ratio of the matrix and fiber,

(b) in the opposite, the model derived by Pérès from the studies of Chua and Piggott assumes the matrix compliance as negligeable which leads to $C = 1$, and an other expression of k:

$$k = \frac{v_f \, E_m}{E_f} \qquad (7)$$

(c) assuming the matrix compliance as non negligeable led Piquenot, Rouby and Arnault to modify the expression of k as follows:

$$C = 1 \qquad k = \frac{E_m}{E_f} \times \frac{v_f}{(1 + v_m) + (1 - v_f) \, E_m/E_f} \qquad (8)$$

2.2 - Numerical simulation

In order to get rid of any assumption required for analytical approaches, a program made up for computing frictional interaction between solids has been used for simulating the microindentation test [2]. The computer program based on a finite element analysis, is able to give the stress and strain fields within the fiber and matrix particularly at the F/M interface. It allows the numerical determination of the u(F) relationship without formulating any assumption excepted the elastic behavior of the components and composite environment.

2.3 - Comparison

The various analytical approaches and the numerical simulation have been applied to composite systems exhibiting either rigid fibers compared to the matrix as in the case of the SiC-LAS system or a rigid matrix compared to the fibers which corresponds to the SiC-SiC system whose matrix is obtained by chemical vapor infiltration [3]. The comparison between the related results gives rise to the following remarks :

(1) The Marshall's model leads to a good approximation of the u(F) relationship when the actual friction coefficient μ is very small (<0.01). As soon as μ becomes non negligeable, a significant deviation appears between the model and the numerical simulation as illustrated in fig. 1,2.

(2) The Shetty's model (2a) related to rigid fibers compared to the matrix give a reliable representation of a system such as the SiC-LAS composite while in this case, some deviation appears between the modified model (2c) and the numerical approach (n) ; (fig. 1).

(3) The model proposed by Pérès for a rigid matrix compared to the fibers, deviates significantly from the modified model (2c) and the numerical approach (n) (fig.2). As a consequence, systems such as SiC/SiC composites can be depicted by the modified model (2c) particularly when the Poisson's ratio is small.

(4) In general, the smaller the Poisson's ratio, the lower the deviation is between the numerical simulation and the modified model.

3 - USEFULNESS OF A COMPUTER ASSITED MICROINDENTATION TEST

The comparison between numerical and analytical approaches of the microindentation test shows the difficulties related to the interpretation of the related measures with the help of analytical approaches. Depending on the assumptions, the displacements u(F) can be strongly under or over estimated, although the last model which has been proposed (2c) gives an acceptable approximation of the conditions of load transfer at the F/M interface.

However, the use of a numerical simulation as an assistance of the experiments seems to be more reliable particularly when the studied composites sligthly deviate from the typical cases directly related to the model assumptions.

Furthermore, the analytical models are not able to take into account a change in the fiber volume fraction excepted through the value given to the friction threshold τ_0 which is not sufficient to represent the related change in the fiber environment. As an example the decrease in stiffness of the environment, corresponding to a change in the fiber volume fraction from 0.4 to 0.6 leads to a small increase in the fiber sliding of about 25 % of that related to a decrease in τ_0 (a decrease in the residual thermomechanical compression at the F/M interface) (fig. 3).

Finally, the presence of interphases at the F/M interface is not easily included in the analytical approaches while the extent of fiber sliding is significantly dependent on the interphase thickness and rigidity (fig. 4). The numerical simulation is able to take into account these interphases while they are well known to control most of the CMCs' mechanical behavior [4]

4 - CONCLUSION

Despite the suitability of various analytical approaches for interpreting the results of microindentation tests, the complexity of the composite systems presently studied, requires a computer assistance.

Provided such an aid, the conditions of load transfer at the F/M interface can be evaluated. However, we must keep in mind that these conditions are presumably strongly dependent on the loading conditions which unfortunately are quite different during microindentation tests and for instance tensile tests.

ACKNOWLEDGEMENTS

The authors are indebted to A. Degueil for stimulating discussions and the Société Européenne de Propulsion (SEP Etablissement de Bordeaux) for supporting this study.

REFERENCES

/1/ - N. PIQUENOT, V. ARNAULT, D. ROUBY in: Matériaux Composites pour applications à hautes températures, Proc. Edited by R. Naslain, J. Lamalle, and J.L Zulian. (AMAC/CODEMAC publications, Paris and Bordeaux) 81-92 (1990).

/2/ - A. DEGUEIL, J. LANCELLE Deuxièmes journées tendances actuelles en calculs des structures. Février 1983 Sophia-Antipolis.

/3/ - R. NASLAIN, J.Y ROSSIGNOL, J.M QUENISSET, F LANGLAIS in: Introduction aux Matériaux Composites Vol 2 (R Naslain, ed) CNRS/IMC Bordeaux (1985). pp 439-91

/4/ - R.A. LOWDEN, D.P. STINTON "The influence of the fiber-matrix bond on the mechanical behavior of Nicalon / SiC composites" Prepared by the Oak Ridge National Laboratory, Operated by Martin Marietta Energy Systems, Inc. for the U.S Departement of Energy under contract DE-AC05-84OR21400.

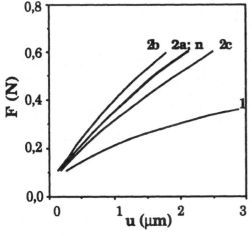

Figure 1: Load - F/M displacement curves related to the SiC-LAS system ($V_f = 0.4$, $\tau_0 = 1.7$ MPa, $\mu = 0.2$):
(1) Marshall's model;
(2a) Model of rigid fibers;
(2b) Model of rigid matrix;
(2c) Modified model;
(n) numerical simulation.

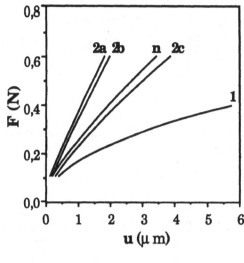

Figure 2: Load - F/M displacement curves related to the SiC-SiC system ($V_f = 0.4$, $\tau_0 = 10$ MPa, $\mu = 0.05$):
(1) Marshall's model;
(2a) Model of rigid fibers;
(2b) Model of rigid matrix;
(2c) Modified model;
(n) numerical simulation.

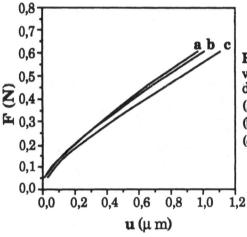

Figure 3: Influence of the fiber volume fraction on the sliding displacement of the fiber:
(a) $V_f = 0.4$; $\tau_0 = 30$ MPa;
(b) $V_f = 0.6$; $\tau_0 = 30$ MPa;
(c) $V_f = 0.6$; $\tau_0 = 20$ MPa.

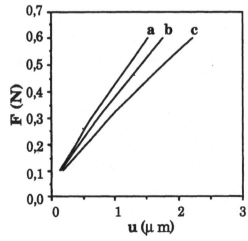

Figure 4: Load - F/M displacement curves for different thicknesses of interphase ($E_f = 200$ GPa, $E_m = 350$ GPa, $\upsilon_f = \upsilon_m = 0.3$): Interphase thickness:
(a) $e = 0$; **(b)** $e = e_1$; **(c)** $e = e_2 > e_1$.

CARBON-CARBON COMPOSITES

Chairman :   **Dr. W. HÜTTNER**
Schunk
Kohlenstofftechnik

PROPERTIES AND APPLICATION OF COMPOSITE MATERIALS OF CARBON-CARBON SYSTEM WITH ULTRADISPERSED FILLER

G. VOLKOV

Automechanical Institut
B. Semenovskaya 38 - 105839 Moscow - USSR

ABSTRACT

We considered a phisical model of carbon overmolecular structure forming deposition on crystallyzation surface during the process of graphite crystallites. Calculations done on the basis of physical model reveal quantitative relationships for the analysis of overmolecular structure. With the increase of the content of homogeneously crystallized particles the overmolecular structure changes from a crystal-oriented one — via transitive forms — to a quasi-isotopic one.

1 - INTRODUCTION

Overmolecular organization of macromolecules foresees physical, chemical and mechanical properties of modern carbon materials as a specific type of polymers. Some aspects of the mechanism of formation of overmolecular structure of carbon polymers need a more detailed treatment, that necessitates new theoretical investigations to be checked experimentally.

2 - THEORY

Graphite crystals are anisometric. This defines their interfitted stacking in the process of crystallization. Optical anisometry of graphite crystals make it possible to observe the forming structures received

455

the overmolecular name /1/, in polarized light. All the observed variety of overmolecular structures of graphitizing carbon is combination of two simplest kinds of overmolecular originations: primary and secondary (fig.1a). The secondary overmolecular origination is formed as a result of precipitation on the growth surface of the primary overmolecular origination of ultradispersed carbon particles.

Mathematical processing of the physical model defines the following functions:

for the primary overmolecular origination (fig.1b)

$$h_1 = \frac{2z\,tg^2\alpha/2}{1 - tg^2\alpha/2} \; ,$$

(1)

and for the secondary overmolecular origination (fig.1c)

$$tg\,\alpha/2 = \frac{\sqrt{R_1 h_2}\,(R_1 + 2z - h_2/2)}{(R_1 - h_2/2)\sqrt{2z(R_1 + z - h_2/2)}} \; .$$

(2)

The limiting case of precipitation strengthening of a carbon matrix with ultradispersed carbon inclusions conforms to their closest packing (fig.1d). When there was lesser content of dispersed particles the composite material forms crystal-oriented structure, and when the content is higher, it gets porousity retaining a quasi-isotropic structure.

3 - EXPERIMENT

The studied theoretical statements can be well simulated experimentally. The physical model with the closest packing of dispersed particles was selected for consideration from the point of view of technological treatment from a large scope of structures, i.e. from structures transient from crystal-oriented to quasi-isotropic ones. Composition materials of carbon- carbon system strengthened with maximal amount of ultradispersed particles of carbon obtained as a result of the proceedure shown in fig.1d were named uglesital on the analogy of the technological principles of structure formation and physical and mechanical performance. Maximal size of dispersed particles in the structure of uglesital reaches about 100 nm.

4 - PROPERTIES AND APPLICATION

Uglesital is marked by unique properties. These properties are practically isotropic due to quasi-isotropic structure that was formed by means of local orientation of graphite crystallites anisotropic structure in relation to a big number of crystallization centres.

Condensation-crystallization contact of dispersed particles ensures sufficient physical and mechanical qualities of uglesital. For example, the value of $\tilde{6}_{-8}$, $\tilde{6}_f$, $\tilde{6}_8$ of USB-15 uglesital is respectively 500, 200, 100 MPa, that exceeds the corresponding characteristics of carbon powder ceramics almost three times. With the increasing of temperature physicomechanical properties of uglesital increase. Working temperature of it is about 2000°C (fig.2). As to the specific strength uglesital exceeds by 5 times the tungsten. The constructional pro- perties of USB-15 uglesital that make it applicable in machine building are the following: modulus of elasticity, heat conductivity and thermal expansion coefficients equal respectively $2,1 \cdot 10^4$ MPa, 30 W/(m·°C), $6 \cdot 10^{-6}$ 1/°C.

Prevailing closed porosity of uglesital ensures their gas- and liquid proofness. USB-15 uglesital air gas permeability coefficient is about 10^{-10} m²/s, which is 10^8 less then carbon powder ceramic gas permeability coefficient.

Resistance to chemical and physical effects is deter- mined by orientation of graphite crystallites by a less reaction plane parallel to the external surface of car- bon dispersed particles.

Uglesital is practically chemical inert to all chemi- cal media. As far as chemical resistance in oxidizing me- dia is concerned, it exceeds carbon powder ceramics 300 times (Fig.3). Unlike the latter, uglesital are resistant in chromium, nitric and concentrated sulphuric acids with large oxidizing potential.

Chemical inertness insures biological compatibility with human tissue that allows to use uglesital as a mate- rial for endoprosthesis. Capacity for work of artifical heart valves made in Soviet Union provides closing elements from uglesital.

Uglesital is 15 times more resistant to cathode sputtering than any other known carbon material (fig.4). This fact determined application of uglesital for ther- monuclear reactor coating /2/.

Friction coefficient of uglesital in liquids is 5 times less than this coefficient of carbon powder cera- mics (fig.5). Under conditions of dry friction uglesital has a smaller value of intensive wear and starting tor- que of friction than antifriction graphites. It allows to apply uglesital as a constructional material for highly energetic fiction units of modern machinery.

We consider the most effective of uglesital use too:
- replacement of platinium electrodes in electroche- mical devices. The use of uglesitals reduces the coeffi- cient of variety of parallel definitions twice;
- working surface of gasodynamic bearings working in conditions of starting-stop.Uglesital practically has no wear traces in 2000 cycles at that the coefficient of friction at starting is minimum;

- electrodes of electrovacuum devices. Uglesital electric field strength at vacuum break-down is 140 kV/cm, treshold energy of sputtering by mercury ions is 110 eV:
- details working in chemical active media;
- details of high temperature devices and equipment;
- creating of new machines, mechanisms and equipment with a high level of consume properties permitting unique properties of uglesitals.

5 - REFERENCES

1. Volkov G.M., Chemistry of Solid Fuel, No.3 (1977),29-34.
2. Ivanov D. P., Pleshivtsev N. V., Garnov V. N., Zhukova L. M., Volkov G. M., Zakharova E. N. in "Questions of Atomic Sciences and Technology", series "Thermonuclear Synthesis", issue 4, (1986), 36-42, Moscow.

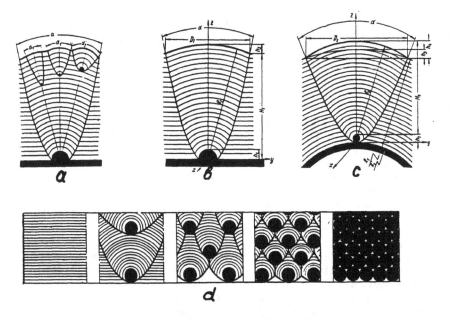

Fig.1 The formation of overmolecular structure of carbon: a - forming of primary and secondary overmolecular formation; b - primary overmolecular origination; c - secondary overmolecular origination; d - transition from crystal-oriented to quasi-isotropic structure.

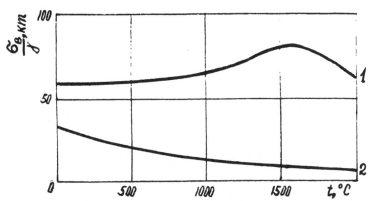

Fig.2   Specific strength of high-temperature materials:
1 - uglesital; 2 - tungsten.

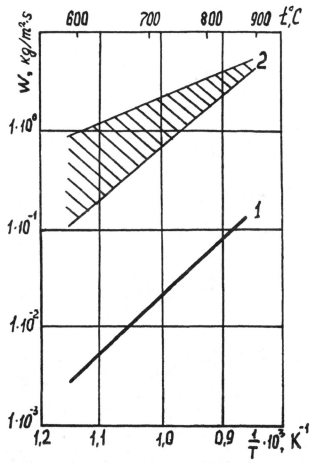

Fig.3   Carbon materials air oxidation rate:
1 - uglesital; 2 - carbon powder ceramic.

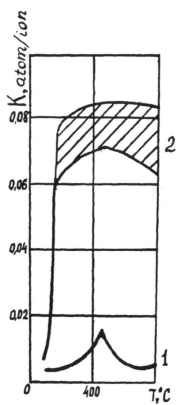

Fig.4 Cathode sputtering by 10 keV protons coefficient
       of carbon materials: 1 — uglesital; 2 — known
       carbon materials.

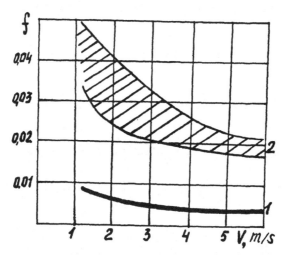

Fig.5 Steel friction coefficient of carbon materials
       in water at specific load 50 MPa on 03X13 steel:
       1 — uglesital; 2 — antifriction graphites.

460

THE ANISOTROPY OF THE CRACK GROWTH
RESISTANCE OF UNIDIRECTIONAL
CARBON-FIBRE/CARBON-MATRIX COMPOSITES

K. AHLBORN, T. CHOU*, Y. KAGAWA*, A. OKURA*

University of Tokyo
now c/o : Carl Freudenberg - PO Box 10 03 63
6940 Weinheim/Bergstr. - West-Germany
*University of Tokyo
7-22-1 Roppongi - Minato-ku - Tokyo 106 - Japan

ABSTRACT

The growth of cracks within an unidirectional C/C-composite and within the carbon-matrix is examined. In the first order the crack growth resistance depends on the orientation of the cracks towards the reinforcement. Parameters like fibre volume content and heat treatment temperature are of second order. The point of crack initiation at the tip of an artificial notch is dominated by thermal matrix pre-stresses. The R-curve concept is applied for the describtion of the fracture behaviour. The shape of the R-curve is closely correlated to the micro-failure processes acting during crack propagation.

INTRODUCTION

C/C-composites are one of the most promising materials for high temperature applications, because they have a good mechanical performance up to more than 2500 K in inert gas atmosphere[1]. Disadvantages are the sensitivity against oxidation, which limits the application in air to temperatures lower than 1200 K, and the weak shear strength of the graphite layers, which is due to the brittleness of the carbon-matrix. The properties of the composite and both components are anisotropic. The anisotropy of the matrix, as well as the size of the graphite layers and the void content depend on fabrication technique and temperature.

Because of the brittleness of the matrix the fracture resistance of a C/C-composite is an important design parameter. In an unidirectional composite the resistance against crack growth should depend significantly on the

461

orientation of the plane of a crack towards the direction of reinforcement. Three principal crack orientations can be distinguished (according to Fig. 5):

  a) The plane of the crack oriented parallel by a propagation perpendicular to the carbon-fibres (position para/perp),
  b) the crack oriented and propagating parallel to the fibres (position para/para), and
  c) the crack oriented and propagated perpendicular (position perp/perp).

1. THEORETICAL DESCRIPTION

The crack growth resistance of the brittle carbon-matrix is increased significantly by the addition of carbon-fibres. Only to a small amount the increment of crack growth resistance is caused by the high fracture surface energy of the carbon fibres. The crack growth resistance is much more effected by toughening processes like fibre bridging, fibre pullout, crack deflection and crack branching.

If a specimen with an artificial notch is loaded to the stress intensity $K_{i,c}$ a crack is initiated in the weaker component of the composite *(the first index defines the type of stress intensity factor: i=initial, c=critical, r=crack growth resistance; the second index indicates the material: m=matrix, f=fibre, c=composite)*. The stress intensity for crack initiation is easily described by the common equation:

$$K_{i,c} = \sigma_c \cdot \sqrt{a_i} \cdot Y \tag{1a}$$

Unlike in homogeneous materials this crack does not extend through the whole structure immediately - it is stopped as soon as it contacts a particle of the tougher component. In order to propagate the initial crack further, the applied stress has to be increased, because toughening processes create a damage zone in front of the notch tip and fibres bridge the crack. This behaviour is described by a fracture resistance curve (R-curve) in terms of a rising stress intensity $K_{r,c}$ as a function of crack extension Δa[2]. The common fracture mechanics do not apply any longer, though the basic fracture mechanisms may be linear elastic processes. The point of instability defined by the critical stress intensity factor $K_{c,c}$ in this case depends on specimen geometry, the stiffness of the load fixture and the strain rate[3,8]. As a result the common critical stress intensity is higher than the stress intensity at crack initiation:

$$K_{c,c} > K_{i,c} \tag{2}$$

If the size of the damage zone ahead of the notch becomes constant at a certain crack extension Δa, then the material inherent R-curve can be described by the function:

$$K_{r,c} = K_{i,c} + \Delta K_{r,c} \tag{3}$$

where $K_{r,c}$ and $K_{i,c}$ are measurable quantities and the increment $\Delta K_{r,c}$ is a function of Δa comprising the toughening mechanisms.

For the crack position perp/perp the stress intensity at crack initiation of a composite $K_{c,c}$ can be predicted from the fracture toughness of the matrix $K_{c,m}$, provided that :

i) the matrix is ideally brittle and has lower toughness than the fibre $(K_{i,m} = K_{c,m} < K_{c,f})$,

ii) the length of the artificial notch is longer than several fibre spacings,

iii) and that the composite can be modelled as a homogeneous, aniso-tropic material.

Then for a composite with $E_f = E_m = E_c$ and $\alpha_f = \alpha_m = \alpha_c$ it follows from eq. 1a:

$$K_{i,c} = K_{c,m} = \sigma_c \sqrt{a_i} \cdot Y \tag{1b}$$

For fibre composites with $E_f >> E_m$ the stress release in the matrix by the fibres has to be taken into account:

$$K_{i,c} = K_{c,m} \cdot E_c/E_m \tag{4}$$

where E_c is a function of the fibre volume content V_f according to the rule of mixture.

The prediction of the stress intensity at crack initiation $K_{i,c}$ of a fibre composite with a crack oriented parallel to the fibres is difficult. For the po-sition para/perp an additional strain- and stress-magnification occurs. Assuming a hexagonal fibre array and $K_{i,m} \approx K_{c,m}$ it holds[4]:

$$K_{i,c} = K_{c,m} \frac{1 - \left(\frac{V_f \, 4}{\pi}\right)^{0.5} \left(1 - \frac{E_m}{E_f}\right)}{1 - V_f \left(1 - \frac{E_m}{E_f}\right)} \tag{5}$$

In general the tensile thermal pre-stresses of the matrix $\sigma_{th,m}$ in a composite have to be taken into account, when $\sigma_f < \sigma_m$ and $E_f = E_m$. A convenient way of evaluation is to multiply eq´s. 4 and 5 with the thermal stress factor β[5]:

$$\beta = (1 + \sigma_{th,m}/\sigma_{u,m}) \tag{6}$$

where $\sigma_{u,m}$ is the ultimate tensile strength of the matrix.

2. EXPERIMENTS

Plates of the pure carbon-matrix and plates with unidirectional rein-forcement of different volume contents were prepared by Across Ltd. according to a recently developed method[6]. Preformed fibre bundles, con-taining the matrix powder and coated with a thin polyamid sheathing, were hotpressed at 873 K. Then the size of the graphite layers was grown by a secondary heat treatment at 1473 K and 2273 K. $K_{i,c}$ and $K_{r,c}$ for the position perp/perp were measured with single edge notched bendbars (SNBB) according to ref. 7. The crack growth behaviour of the composite for cracks oriented parallel to the fibres and that one of the carbon-matrix were measured with double cantilever beam specimen (TCB, DCB) with a new

463

loading technique, which allows the in-situ observation of the crack propagation with an polarized microscope[8].

3. RESULTS AND DISCUSSION

The crack growth resistance of the carbon-matrix is shown in Fig. 1. $K_{i,m}$ rises from 0.75 MPa $m^{0.5}$ at 873 K to about 1.15 MPa $m^{0.5}$ at 2273 K. The microscopical analysis shows that grain bridging and crack splitting cause a slight toughening effect and a R-curve of about 0.5 mm range[9].

For the crack position para/perp with the modulus ratio $E_{\perp,f}/E_m \approx 3$ it follows from eq. 5 that

$$K_{i,c} \approx K_{c,m} \cdot 0.7. \tag{7}$$

As shown in Fig. 2 the measured values of $K_{i,c}$ are much lower than predicted (about $K_{c,m} \cdot 0.2$), because the thermal pre-stresses cause a premature failure. Here a sharp rise of the R-curve and a range of about 2 mm is measured.

The R-curve of cracks in position para/para is presented in Fig. 3. A first increase of the crack growth resistance comparable to Fig. 2 is superimposed with a second rise. After about 4 mm crack extension crack bridges out of misaligned fibres start to carry load and to release the stress intensity at the crack tip.

The microscopical analysis shows that thermal pre-cracks oriented perpendicular to the fibres are located with constant spacing within the matrix. They are initiated during the cooling cycle of the first production step (873 K). Hence $K_{i,c}$ is extremely low for cracks in position perp/perp, because the matrix is thermally pre-stressed. After initiation the crack growth resistance rises sharply up to a plateau, due to the interaction of the main crack with the carbon-fibres: delamination processes occur which force the crack to propagate parallel to the carbon-fibres. With further crack extension extensive fibre bridging and fibre pullout become effective, causing a second progressive increase comparable to Fig. 3. The location of the measured plateau is equivalent to the conventional fracture toughness $K_{c,c}$. As shown in Fig. 4 $K_{c,c}$ depends on the fibre volume content or the Young's modulus ratio Ec/Em as predicted in eq. 4.

REFERENCES

1) K.J. Huettinger, Kohlenstoffwerkstoffe; Chemiker Zeitung, Vol. 112 (1988) 355.
2) D. Broek, Elementary Engineering Fracture Mechanics; Nijhoff Publ., Dortrecht (1986), Chap. 5.3.
3) M. Sakai, M. Inagaki; J. Am. Ceram. Soc.; 72 (1983) 388.
4) C. C. Chamis, in: Composite Materials Vol. 5, ed. L.J. Broutman (Academic Press, New York, 1974) pp.126-131.
5) Y. Kagawa and H. Hatta, Effects of Thermally Induced Stresses on the Interfacial Shear Strength and Failure Behavior of an SiC(CVD)/Glass Composite in contribution to: J. Japan Ceram. Soc.

6) T. Chou, A. Okura; in: The Materials Revolution through the 90's, Vol. 2, (7th Intern. Conf. BNF Metals Technology Center, Wantage, UK, 1989).
7) M. Jenkins et al., SAMPE Journal, May/June (1988) 32.
8) K. Ahlborn, Y. Kagawa and A. Okura, The in-situ observation of stable crack propagation in brittle solids, in contribution to Scripta Metallurgica.
9) K. Ahlborn, T. Chou, Y. Kagawa, and A. Okura; The influence of the heat treatment temperature on the fracture toughness of a fine grained carbon; to be contribtuted to Carbon.

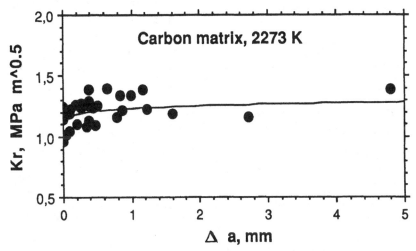

Fig. 1: Crack growth resistance curve of the carbon-matrix heat treated at 2273 K.

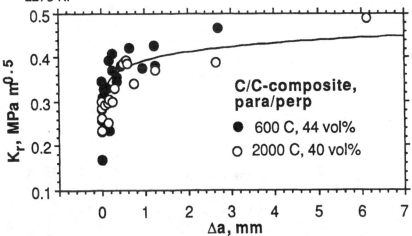

Fig. 2: Crack growth resistance curve of the unidirectional C/C-composite heat treated 873 K and at 22373 K for the crack position para/perp.

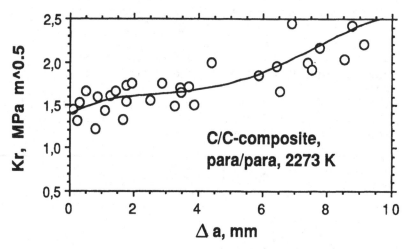

Fig. 3: Crack growth resistance curve of the unidirectional C/C-composite heat treated at 2273 K for crack position para/para.

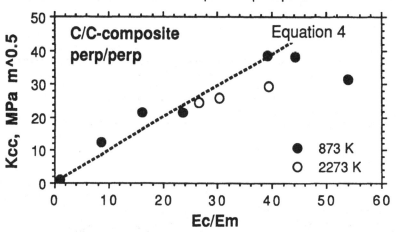

Fig. 4: Conventional fracture toughness of the unidirectional C/C-composite heat treated at 873 K and at 2273 K for the crack position perp/perp.

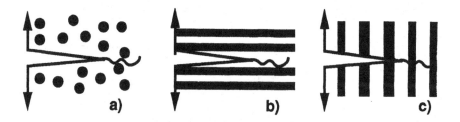

Fig. 5: Principal orientations of a notch towards the fibre reinforcement: a) para/perp, b) para/perp, and c) perp/perp.

TEM STUDIES OF INTERFACIAL STRUCTURES
IN CARBON/CARBON COMPOSITES

H.R. PLEGER, W. BRAUE, R. WEISS*, W. HÜTTNER*

DLR - Institute for Materials Research
PO Box 90 60 58 - 5000 Köln 90 - West-Germany
*Schunk Kohlenstofftechnik
Postfach 6420 - 63 Gießen - West-Germany

Abstract

Preliminary results of a TEM study on structural development of interfacial areas in carbon/carbon composites are reported. The investigation is based on well-defined microstructural standard conditions derived from a resin-based carbon/carbon model system.

Introduction

Carbon/carbon composites (C/C) exhibit high retained strength and stiffness, good thermal shock behavior and a remarkable damage tolerance which make them a very promising ceramic structural material for various aerospace applications. Treated with oxidation-resistant coatings C/C is expected to withstand even severe high-temperature service conditions /1/.

The combination of a brittle carbonaceous matrix with a brittle carbon fiber results in a complexity of different C/C microstructures which is unique compared to other ceramic composites with continuous fiber reinforcement. Properties of C/C composites strongly depend on the type of the carbon fiber, the fiber volume fraction, the fiber architecture and the nature of the matrix, e. g. highly anisotropic carbon matrices derived from pitch precursors or isotropic ones obtained from thermosetting resins.

The purpose of this study is to monitor the development of interfacial structures in a resin-based C/C model system by means of analytical transmission electron microscopy. Micromechanics and failure-related phenomena of C/C fiber/matrix interfaces are yet poorly understood thus requiring additional microstructural informations at a high scale of spatial resolution.

For reasons related to the core-skin heterogeneity of PAN-based carbon fibers and the development of the carbonaeous matrix complex interfaces of dynamic nature are expected for C/C composites. From previous TEM investigations of mainly pitch-derived materials /2/ it turned out that similar to C-fiber reinforced polymer matrices there is indeed no such "simple" interface in C/C which may account for the variations in thermal and mechanical properties observed at high temperatures.

Fiber/matrix interaction in C/C rather includes several interfacial structures and microcrack systems thus involving different types of bonding as weak and strong chemical bonding, mechanical interlocking and friction /3/. Despite the effects of raw materials and processing conditions at least four different interfacial structures contribute to the overall fiber/matrix interaction:

- between single filaments and the matrix ("intrabundle matrix", see fig. 4)

- between fiber bundles and the matrix ("interbundle matrix")

- between different matrix generations related to multiple impregnation

- between matrix and macroporosity (see fig. 4).

Materials and Methods

The investigation is based on well-defined microstructural standard conditons of a resin-based C/C material involving various HT and HM carbon fibers. Different sheets of the pressed and cured material were exposed to different heat-treatment procedures at 200° C intervals covering the complete processing range from low as 200° C (beginning of precursor pyrolysis) up to 2400° C (graphitization). Moreover, fully graphitized high-density commercial 2D C/C standard grades were included to the investigation. A Philips EM430 with the twin lens polepiece operating at 300 kV was used for the TEM work.

Evaluation of the correct preparation technique for thin C/C foils proofed to be a focal point. C/C composites are porous materials which are prone to deformation-induced artifacts like splitting of fiber bundles, fiber/matrix delamination, generation of microcracks and the like. Grinding – which works well for monolithic carbonaceous materials – of course fails when the integrity of more extended sample areas is mandatory. Ion-beam milling introduces thinning-artefacts too which however may be reduced under low beam voltage conditions (< 4 kV). Thin sections were dimpled prior to ion-beam-thinning. As revealed in the SEM micrograph (fig. 1) ion bombardment may yield an irregular rippled surface morphology. Surface striations correspond to different contrast in the TEM because of local thickness variations (fig. 2). According to /4/ the high amount fo neutrals in atom beam thinning probably offers a way out of this dilemma because they produce less-artifact bearing TEM foils compared to ion-beam-thinning. Optimization of TEM foil-preparation for this project is still under way and the potential of alternative methods like microtome-sectioning is currently investigated.

Results and Discussions

By comparing C/C materials which were heat-treated at low temperatures after curing, e. g. 200° C (fig. 3) with fully graphitized grades (fig. 4) the difficulties to prepare a dense, defect-free carbonaceous matrix become obvious. Macropores in the fully graphitized material are due to bubble formation during pyrolysis of the matrix precursor (fig. 3). Because of the low fracture strain of the matrix the failure of C/C is mostly initiated in the matrix and/or close to fiber/matrix-interfaces /5/ Therefore it is believed that defects like pores are favorable sites for crack nucleation and propagation under load. To emphasize this issue fractography after failure of C/C specimens are currently performed. Because of the high precursor shrinkage the 200° C to 600° C processing-temperature-range is most likely decisive with respect to fiber surface damage and development of interfacial structure. The type of fiber may be an important factor too.

A typical ensemble of microstructural features taken from an intrabundle matrix area of a fully graphitized 2D C/C composite is shown in the low-magnification TEM bright field image of fig. 4. Matrix graphitization is high with the graphite basal planes being aligned parallel to those of the fiber skin. Macropores typically are filled whith amorphous carbon (marked by black arrows in fig. 4).

As the amount of amorphous carbon is related to the oxidation resistance of the composite /3/ it should be reduced by prolonged exposure to graphitization conditions. Delamination zones between the outer fiber sheath and the matrix (marked by white arrows in fig. 4) constitute a characteristic microstructural feature of this C/C material which is strongly related to interfacial failure.

Conclusions

Several focal points of a TEM study on interfacial structures in C/C composites are put into perspective as they emerge during the early stages of the project. Preparation of artefact-free thin C/C TEM-foils is still a severe problem. A combination of TEM characterisation and detailed fractography of C/C microstructures after failure is probably the best aproach to meet the complexity of fiber/matrix interaction in C/C composites

References

1 J. D. Buckley, Am. Ceram. Soc. Bull. 67 (1988) 364

2 W. Kowbel, J. Don, J. Mater. Sci. 24 (1989) 133

3 L. H. Peebles, R. A. Meyer, J. Jortner, in: Interfaces in Polymer, Ceramic and Metal Matrix Composites (H. Ishida, ed.), Elsevier (1988) 1

4 C. P. Ju, J. Don, Metallography (1989), to be published

5 W. Hüttner, Z. Werkstofftech. 16 (1985) 430

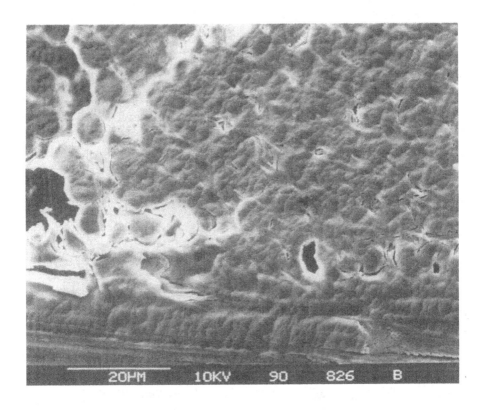

Figure 1: Surface morphology of thin 2D C/C TEM foil as imaged with secondary electrons in a scanning electron microscope. Bright areas correspond to thin parts ($< 5\mu$m) of the foil.

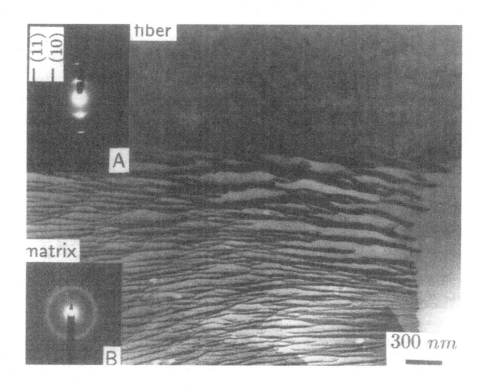

Figure 2: TEM bright field image of interfacial area from 2D C/C material prepared by heat-treatment at 200° C after pressing and curing. Striation effects in both fiber (A) and matrix (B) result from contrast variations because of rough and textured surface morphology

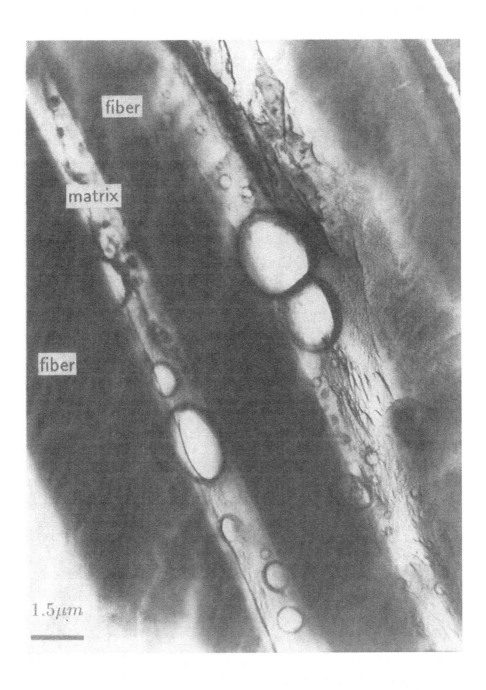

Figure 3: Interfacial area from 2D C/C material prepared by heat treatment at 200° C after pressing and curing (TEM bright field). Note bubble formation in matrix areas.

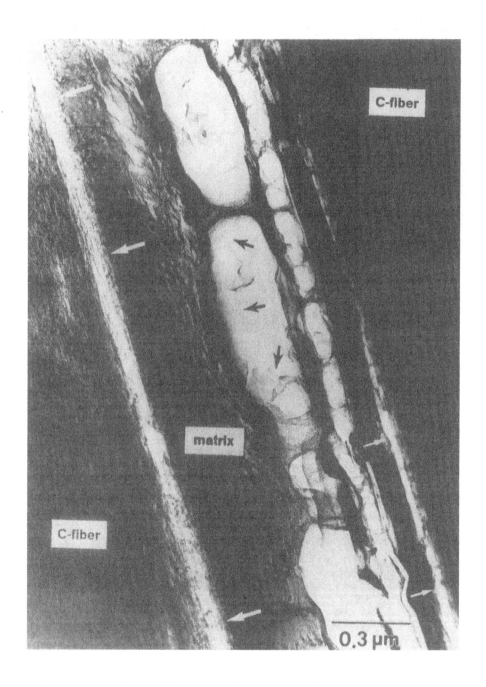

Figure 4: Intrabundle matrix area from fully graphitized 2D C/C material (TEM bright field image). White arrows mark fiber/matrix delamination zones, black arrows amorphous carbon bound to macropores respectively.

MICROSTRUCTURE, INTERFACES AND MECHANICAL PROPERTIES OF FIBRE REINFORCED GLASSES

B. MEIER, D. SPELMAN, G. GRATHWOHL

Institut für Werkstoffkunde II
Universität Karlsruhe (TH) - Kaiserstr. 12 -
7500 Karlsruhe 1 - West-Germany

Abstract

The mechanical properties, microstructure and interfaces of three glass matrix composites with different physical properties reinforced with SiC-fibres (NICALON) were investigated. The fracture behaviour of these composites was interpreted on basis of the calculated interfacial stresses and the chemical state of the interface fibre-matrix.

INTRODUCTION

The inherent brittleness of advanced ceramics and glasses is still a major problem, preventing the wider application of these materials for structural components. The addition of brittle reinforcing components, for example long ceramic fibres into these materials is a promising method to overcome this problem and results in drastically improved mechanical properties. The behaviour of such composites during loading is controlled by the stress transfer between the interacting brittle components. The reinforcing mechanisms such as crack deviation, multiple crack formation or pull-out are concentrated at and controlled by the interface between matrix and fibre. Therefore to interpret the mechanical behaviour, the interfacial properties must be understood. Furthermore the microstructure of the composites is of fundamental importance. In order to characterize any chemical changes during processing it is also important to analyze the reinforcing components in their initial state.

In this study the mechanical properties of three glassmatrix composites were measured. In order to investigate the influence of the thermal mismatch between fibres and matrix the glass matrices were chosen with differing thermal properties. The microstructure and the interface fibre-matrix were examined to interpret the mechanical properties of the composites.

EXPERIMENTAL METHODS

The mechanical behaviour of the composites was investigated by 3-point bending tests with a span suited to fibre reinforced materials. The interfacial shear strength between fibre and matrix was measured using the indentation method according to Marshall /2/. The interlaminar or macroscopic shear strength was also measured using different methods /3/. For the examination of the microstructure and the fibre - matrix interface optical microscopy, X-ray diffraction (Siemens D500), high resolution Auger electron spectroscopy (Perkin Elmer PHI600), scanning electron microscopy equipped with an energy dispersive analysis (Jeol JSM 840 equipped with a Kevex Quantum analyser) and element spectroscopy for chemical analysis (Perkin Elmer ESCA 5400) were used.

INVESTIGATED MATERIALS

The investigated components and their physical properties are listed in Table 1. As can be seen the thermal expansion coefficients of the matrices are lower (A-glass), slightly larger (Duran-glass) and much larger (Supremax-glass) than that of the SiC-fibres (NICALON). All glasses are delivered by Schott Glaswerke, West Germany. The composites were produced at the T.U. Berlin, West Germany /1/. The SiC-fibres were supplied by Nippon Carbon Ltd., Japan under tradename NICALON.

RESULTS

The axial, tangential and radial stresses on the matrix side of the interface were calculated according to Chawla /4/ for three glasses. For the A-glass the axial and tangential stresses are compressive while the radial stress is tensile, which tends to rupture the interface. In the case of the other two matrices where the thermal expansion coefficient of the matrix is higher than that of the fibres tensile stresses exist in the matrix in the axial and tangential direction, while the radial stresses are compressive in nature, as required for a purely frictional interfacial bond. More over, in the case of the A-glass tensile stresses exist in the fibre in contrary to the Duran- and the Supremax-glass where the stresses in the fibre are compressive. If the expansion coefficient of the matrix is too high the tensile forces can be larger than the strength of the matrix and cause microcracking. This was indeed observed in optical micrographs of Supremax composites in the as-received state.

Furthermore, the mechanical properties of the composites were investigated (Table 2). The bending tests on the three systems show Duran and Supremax materials to have similar strengths of slightly above 700 MPa. However, the variation of the bending strength was higher for the Supremax composites than for the Duran composites. Typical stress-strain diagrams are shown in Fig.1. The Supremax curves deviate from linear behaviour earlier than the Duran specimens. This is probably due to the higher axial residual stresses. By increasing the load on some Duran specimens in steps of 50 MPa and monitoring the accoustic emission the first microcracking on the tensile side of the specimens could be detected at about 250 MPa with series of

476

microcracks developing at between 400 MPa and 500 MPa. The matrix microcracking stress can be reasonably well predicted by the model from Marshall, Cox and Evans /5/. Only certain A-glass specimens show limited pullout, together with a controlled fracture.

The interfacial shear strength measurements show that Duran and Supremax composites have very similar properties, disproving the theory that the interface is held together purely by frictional forces. The measurements on A-glass show a large degree of scatter. Closer examination of the specimens reveals pores and precipitates in the matrix and also at the interface which explain this large scatter.

In order to facilitate a microanalytical investigation of the interface between fibre and matrix the samples were fractured in the HRAES stage to avoid any contamination. The micrograph of the polished sample and the secondary electron image of the in situ fractured composite with A-glass matrix is shown in Fig.2. As can be seen a porous region exists between the fibre and matrix. It follows that debonding will take place preferably in this porous region. The point analysis in the pores (Fig.2b point 3 and 4) reveals carbon. At the points 1 and 2 the matrix elements Si,O,B were detectable but also small amounts of C. The energy dispersive analysis reveals the elements Cu and Mn in the precipitates. Taking into account that the A-glass contains Cu and Mn it seems to be possible that these elements act as a foaming agent as it is a known fact from literature /6/. Concerning the other glass matrices it was found in case of optimized processing parameters that no foaming effect could be detected at the interface fiber-matrix. The HRAES investigation immediately after in situ fracture of the composite with Supremax matrix demonstrated the presence of a C-rich layer at the interface fibre-matrix. No diffusion of matrix elements into the fibre could be detected. Sometimes P was found to be present in the C-rich layer. Analyzing the imprints left by the fibre in the matrix after fracture a high concentration of C was also measured. It follows from this that the fibre is detached from the matrix in the C-rich layer. After sputtering away about 7nm with Ar-ions one finds an O-rich layer with a minor amount of C. The shift of the LMM transition from 92 eV to 76 eV shows that the Si is in an oxidized state. Further sputtering leads to the proportions of the fibres in their initial state. In comparison with the sputter rate of SiO_2 the layer thickness is at least 20-30 nm. Concerning the Duran matrix composite the situation at the interface fibre-matrix is rather different. The secondary electron image with the associated point analysis is illustrated in Fig.3. Two types of fibre surfaces could be seen; one type show a flow structure (indicated point 1) while the other type (point 2) is very smooth. The HRAES point analysis at point 1 reveals the elements of the glass matrix, especially O,B,Si,Al and a little C. The layer with the flow structure is estimated to be 0.5-1 μm in thickness. The HRAES point analysis immediately after in situ fracture at the smooth fibre surface displays a C-rich layer with minor amounts of B and O. After sputtering about 40 nm with Ar-ions carbon is still the dominating element, but a Si peak is now observable with a decreased O and B peak. At a sputter depth of about 200 nm the composition is comparable with that of the fibre bulk composition but B is still present. Hence follows that B had diffused into the fibre.

CONCLUSIONS

Generally speaking, carbon was found to be present in all investigated composites with SiC-fibres (NICALON) at the interface fibre-matrix. In comparison with the initial state of the fibre surface /7/ it became evident that the thickness of the C-rich layer increased during the sintering process. Interfacial cracks propagate through these C-rich layers. Therefore a considerable effect on the interfacial properties can be expected. Possibly a degradation of the fibre properties follows from this as it is described in /8/. Furthermore, a closer investigation concerning the diffusion of B into the fibre is planed. Taking into account that B is used as a sintering additive in SiC ceramics an influence on the microstructure in the outer surface seems to be possible. B could also promote the crystal growth of SiC as it was found when Al diffuses into the fibre /9/. The similarity between the stress-strain curves for the Duran and Supremax composites is due to the almost identical interfacial shear strengths. The departure of the Supremax curve from the linear behaviour at lower stresses is probably due to the higher axial residual stresses, which promote microcracking in the matrix. However, the results from the identations tests are almost identical for the two systems, although the residual stresses are quite different. It follows from this that the interfacial shear strength is controlled by other processes, i.e. the C-rich layer, which was found in both systems. In case of A-glass composites a large scatter in τ was found. When τ is high then no debonding occurs and the specimens fail brittely. When τ is low microcracking and pull-out can take place. This results in regions with considerable pull-out and other regions where the crack runs straight through the specimens. Possibly, a porous interface can lead to a tougher composite.

REFERENCES

1. Hegeler H., Brückner R., J. Mat. Sci. 24 (1989),1191-1194
2. Marshall D.B., J. Am. Cer. Soc. 67 (1984), C 259-260
3. Spelman-Kranich D., Jahresbericht 1986, Institut für Keramik im Maschinenbau, Universität Karlsruhe, 15-18
4. Chawla K.K., "Composite Materials, Science and Engineering", Springer Verlag, Berlin - Heidelberg - New-York (1987)
5. Marshall D.B., Cox B.N., Evans A.G., Acta. Met. 33 (1985), 2013-20
6. Köse S., Bayer G., Glastechnische Berichte 55 (1982), 151
7. Meier B., Hamminger R., Grathwohl G., Colloque de physique, Colloque C1, Supplement au n°1, Tome 51, janvier 1990, C1-885 - C1-889
8. Ponthieu C. et al., Colloque de physique C1, suplement au n°1, Tome 51, janvier 1990, C1-1021 - C1-1026
9. Bender B.A. et al., Ceram.Eng.Sci.Proc., 1984, p.513-525

Figures

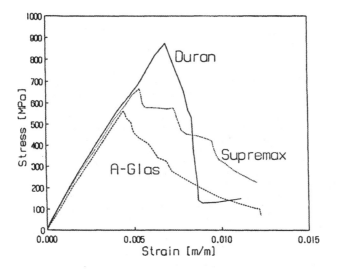

Fig.1: Typical flexural stress-strain diagramm of three different SiC-fibre (Nicalon) / glass matrix composites.

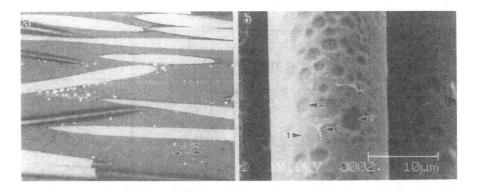

Fig.2: Micrograph (a) and secondary electron image (b) of the composite SiC-fibre (Nicalon)/A-glass matrix.

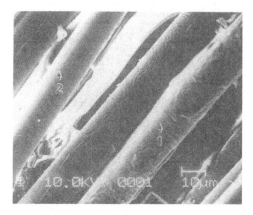

Fig.3:   Secondary electron image of the in situ fractured composite SiC-
         fibre (NICALON)/Duran matrix.

Tables

Table 1:   Thermal expansion coefficient ($\alpha_{(RT-Tg)}$), transformation-
           temperature (T_g) and Young's modulus (E) of the used
           components.

component	α [$10^{-6}K^{-1}$]	T_g [°C]	E [MPa]
SiC-fibre	2,70·	--	180
A-glass	2,30	620	73
Duran	4,37	530	63
Supremax	4,60	730	90

Table 2:   Results of the mechanical characterization of the three composites.

Matrix	σ_B [MPa]	E [GPa]	ILSS [MPa]	τ [MPa]
A-glass	556	128	--	29±16
Duran	731±128	125±3	48±5	13±2
Supremax	708±72	138±3	37±3	11±2

σ_B: bend strength   E : Young's modulus
ILSS: interlaminar shear strength τ : interfacial shear strength

SIC-FIBER REINFORCED CORDIERITE : INFLUENCE OF MATRIX MODIFICATIONS ON MECHANICAL PROPERTIES

C. REICH, R. BRÜCKNER, W. PANNHORST*, M. SPALLEK*

Institut für Nichtmetallische Werkstoffe
Techn. Universität Berlin - Englische Str. 20
1000 Berlin 12 - West-Germany
*Schott-Glaswerke - Mainz - West-Germany

ABSTRACT

The influence of minor additives of certain oxides (P_2O_5, Cr_2O_3, TeO_2, TiO_2, V_2O_5) to SiC-fiber reinforced MAS glass ceramic is studied with respect to crystallization during hot pressing and to mechanical properties. Some additives lead to comparable strength and toughness like those of pure cordierite composites, other lead to lower strength and show brittle fracture behaviour. The concentration of the additives is also of importance.

I-INTRODUCTION

Since carbon fiber reinforced polymers are accepted as reliable engineering materials by now, the need of matrix materials, which are not susceptible to environmental degradation due to moisture, oils, or fuels, has become obvious.

For important applications of composites, the mechanical properties must often be related to composite density. Therefore most metals are less suited for matrix materials. Among ceramic materials oxide-glasses have the lowest densities and they show a strong temperature dependence of their viscosity. This leads to composite fabrication-procedures in a manner similar to that used for resin matrix composites. The development of glasses reinforced with long ceramic fibers has been demonstrated by several research groups /1,2/.

With SiC-fiber reinforced glasses application temperatures of tough composites have reached the range of 800 -850°C, pure silicaglass-composites even overcome the 1000°C-limit, but show lower values of fracture toughness /3/. Even higher temperature-stability can be achieved with glass ceramic composites, which provide the unique capability to be densified in the low viscosity state and then be crystallized.

For this study we chose cordierite ($2MgO\ 2Al_2O_3\ 5SiO_2$) as matrix material for SiC-fiber reinforced composites. Cordierite is a low density glass-ceramic and has a thermal expansion coefficient well compatible with SiC-fibers.

Modifying the stoicheiometric cordierite matrix was thought to solve the following problems, known from recent experiments:

- Crystallization of the cordierite base glass starting at temperatures of about 850°C inhibits composite densification at low temperatures.

- The matrix-viscosity is too high at the preferred hot-pressing temperatures.

- The fiber/matrix wetting behaviour is insufficient.

According to Durville and coworkers /4/ additions of Cr_2O_3 should alter the crystallization of cordierite parent glasses. A crystallization-path via $MgCr_2O_4$ should result in small crystallites. Also P_2O_5 and TiO_2 additives are known to show pronounced effects on the crystallization of the matrix. TeO_2 and V_2O_5 are known to influence the surface tension in multicomponent glasses strongly. These oxides were introduced into a cordierite base glass to investigate possible influences of the wetting between fiber and matrix.

II-EXPERIMENTAL METHODS

In our investigation we used SiC fibers supplied under the tradename NICALON by Nippon Carbon, Japan. The cordierite matrix-material was modified with Cr_2O_3, P_2O_5, TiO_2, TeO_2 and V_2O_5 by melting the cordierite base glass with various amounts of one of these oxides.

A new process was used for composite prepreg fabrication: The glassy matrix powder was infiltrated into the fiber bundle with the help of an alkoxide binder based on partially hydrolyzed tetraethoxysilan. This binder-system avoids the binder burnout step necessary when using Polymeric binders /5/. The prepregs with 30-50

Vol % unidirectionally alligned fibers were hot pressed to nearly 100 % theoretical density.

The crystallization-behaviour of the modified base glasses was compared with that of stoicheiometric cordierite base glass by DTA-methods and X-ray diffraction. The mechanical properties of the composites were investigated by 3-point bending with a span to depth ratio larger than 20 to give predominantly tensile failure.

For the examination of the composite microstructure optical and scanning electron microscopy were used. In order to study the effects of the modified matrix chemistry on the fiber/matrix-interface high resolution Auger electron spectroscopy (HRAES) is best suited /6/.

III-SUMMARY OF THE RESULTS

The DTA measurements showed slight variations of the crystallization peak temperature for most additives; a pronounced shift was observed for the glass modified by P_2O_5. Investigations of composite materials by X-ray diffraction revealed a dependence of the crystalline phases formed on the modifying components.

TiO_2 modified composites show TiC and/or TiO diffraction peaks. Cr_2O_3-additions led to complex changes in crystal phases at low pressing temperatures (MgAl- and MgCr-spinel). At higher pressing temperatures these phases were dissolved and cordierite remained as the stable phase, partly with glass phase and mullite.

The bending strength of composites with modified cordierite matrix were 2 to 4 times larger and the toughness by more than 10 times larger than that of the corresponding unreinforced matrix. Composite toughness, qualitatively expressed by the amount of fiber pull-out and partly by stress-strain diagrams, may be influenced by choosing appropriate matrix modifications.

The influence of the minor additives can be summarized as follows

- P_2O_5, TeO_2 and Cr_2O_3 doped composites show comparable values in bending strength and toughness as compared to the stoicheiometric cordierite composites;

- additions of TiO_2 in small amounts lead to moderate values of strength and toughness;

- additions of V_2O_5 cause lower strength values and brittle fracture behaviour;

- larger amounts of TiO_2 and Cr_2O_3 reduce drastically strength and cause brittle fracture behaviour.

As an example for a low and high pull-out effect corresponding to a brittle and a tough fracture behaviour, respectively, Fig. 1 shows the fracture face of a brittle composite due to a high Cr_2O_3 content and Fig. 2 that of a tough composite due to a low Cr_2O_3 content.

In the case of brittle Cr_2O_3-modified composites HRAES-point analyses reveals Cr, C and O in porous structures at the fiber/matrix interface. This suggests that a reaction has occured between fiber and matrix resulting in a fiber degradation and/or a strong interfacial bond.

HRAES of composites containing less Cr_2O_3 show carbon-rich interfaces, responsible for the tough behaviour and the high work of fracture.

REFERECES

1. D.M. Dawson, R.F. Preston, A. Purser
   Ceram. Eng. Sci. Proc. 8 (1987) 815-821

2. K.M. Prewo, J. Mater. Sci. 17 (1982) 3549-3563

3. K.M. Prewo, J. Mater. Sci. 17 (1982) 1201

4. F. Durville, E. Champagnon, E. Duvalle, G. Boulon
   J. phys. chem. solids 46 (1985) 701-707

5. R. Brückner, H. Hegeler, Chr. Reich, W. Pannhorst,
   M. Spallek, Paper presented at the ACerS 91th annual
   meeting, Indianapolis, April 1989

6. B. Meier, G. Grathwohl
   Fresenius Z. Anal. Chem. 333 (1989) 388-393

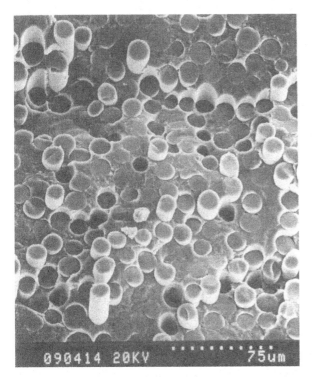

Scanning electron micrographs of the fracture face of SiC/Cordierit-composites showing various degrees of fiber-pullout

Fig. 1: Matrix with high Cr_2O_3-content

Fig. 2: Matrix with low Cr_2O_3-content

BOLT BEARING STRENGTH AND NOTCH
SENSITIVENESS OF CARBON-CARBON

A. THEUER, F.J. ARENDTS

Institut für Flugzeugbau - Univ. Stuttgart
Pfaffenwaldring 31 - 7000 Stuttgart - West-Germany

ABSTRACT

Carbon fibre reinforced carbon (C/C) as a structural composite material is well known for its good thermomechanical properties above 2000 $^{\circ}$C. Main applications in current aeronautical and space programmes are predicted for heatshields and propulsion systems. Many efforts have to be done to protect the material from oxidation above 500 $^{\circ}$C (SiC-matrix). Nevertheless the basic mechanical characteristics of this coated C/C depend mainly on the properties of the fibre skeleton. Some principal effects of stress-concentration and bolt bearing strength of C/C are shown in here.

1. INTRODUCTION

The strength of bolted or riveted joints in structures is of certain interest. First, because of the fitting of a thermally loaded part into the structure. Second aspect is the investigation of the failure mechanism in the area of high stress-concentration around a loaded or unloaded hole. Several theoretical and experimental investigations have been done, especially considering the phenomenons of fibrous composites /1/.

The various mechanisms of failure - tension, bearing, shearout and cleavage - depending on geometric parameters and stacking sequences are not easy to explain. For an exact theoretical description of the failure it is even necessary to concider a 3-dimensional stress analysis. Only few of them have been done for advanced composites. The present work shows the results of experimental tests on C/C-specimens with loaded and unloaded holes.

2. EXPERIMENTAL

2.1 Material

The C/C-material was produced and kindly made available by SIGRI GmbH Meitingen, West-Germany. The material is processed from 2-D woven and from unidirectional prepregs out of yarns of 3000 PAN based high modulus fibres of about 7 μm diameter. The laminate thickness varies from 1.1 mm to 3.7 mm, depending on the amount of layers for the different plates. The woven plates were made out of strips of 80 mm width. The warp direction of the strip gives the 0°-direction of the plate.

2.2 Testing

The size of the specimens with unloaded holes was defined as 170 mm x 30 mm. The hole diameters were 3, 5, 8, 10, 12 and 15 mm. For the definition of the geometrie, see fig.1.

Some pretesting of the bolt bearing behaviour showed that even big holes (high d/w) did not fail in the net-cross-section. For this reason, and to save material, the width of the specimens was limited to 20 mm, with bolt diameters of 5, 8 and 10 mm. The edge distance from the centre of the hole had the value of 20 mm. The bearing failure was defined as 2% of the bolt diameter /2/. The bearing deflection was measured with the device shown in fig.2. This device was developed in the institute, following examples in /2/ and /3/. It is possible to measure the direct bolt deflection without the strain in the specimen.

Both tests were static in room-temperature with unconditioned material. The loading was done path-controlled with a speed of 2 mm/min.

3. RESULTS

The evaluation of the results was done with the method of Hart-Smith /1/ (fig.3). The attained test loads P were related to the maximum ultimate tensile strength σ_{ut} wt of the cross section. This defines the structural efficiency η

$$\eta = \frac{P}{\sigma_{ut} w t} .$$

For fibrous composites η is found between the curve of ductile material and elastic isotropic material. Woven C/C is well known for its pseudo-plastic behaviour, pointed out in the curves of the unloaded case. They almost reached the theoretical ductile strength - the 45° specimens with factors η more than 1 (physically impossible) were made out of a different plate.

Exept cleavage, all failure modes could be attaint, whereas bearing failure occured the most often. The woven material (90 and 0,45,90) showed bearing failure with increasing load even for the higher d/w-ratios.

The unidirectional series have a lower level of testing loads. The change in strength at d/w of 0.4 is coherent with the change of plates. It shows the sensitivity of the C/C material for small variations in manufactoring process.

An evident result is the strengthening effect of 45° layers for the woven as well as for the unidirectional material. This was demonstrated by many authors for advanced composites with resin matrix, whereas only a few of them are mentioned here /1/,/3/. For C/C-material this effect is even greater, because of the mechanical toothing character of the carbon matrix.

4. CONCLUSIONS

The strength of a bolted joint in C/C-structure depends mainly on the stacking sequence, i. e. the amount of $+\Theta$ and $-\Theta$ layers. This influence is even more dominant than for CFRP, where the tendency was shown in the different investigations.

REFERENCES

1. Hart-Smith L. J., "Bolted Joints in Graphite-Epoxy Composites", Douglas Aircraft Company, NASA Langley Contract Report, NASA CR-144899, January 1977

2. Kraft H., Schelling H., "Statische Festigkeitsversuche an zweischnittigen gebolzten Fügungen aus CFK zur Ermittlung der Lochleibungsfestigkeit", DGLR-Bericht 84-02, Symposium Berlin, Novembre 1984

3. Collings T. A., Beauchamp M. J., "Bearing deflection behaviour of a loaded hole in CFRP", Composites, Vol. 15 No. 1, January 1984

	tensile strength σ_{ut}	[MPa]	
UD	$[0/90/0/90]_s$	493	×
	$[+45/-45/+45/-45]_s$	52	×
	$[0/+45/-45/90]_s$	234	×
weav	$[90/90/90/90/90/90]_s$	110	×
	$[45/45/45/45/45/45]_s$	39	×
	$[0/+45/-45/90]_s$	104	× ×

Table 1: Tensile strength datas from SIGRI (×) and IFB (××)

Fig. 1: Specimen geometrie

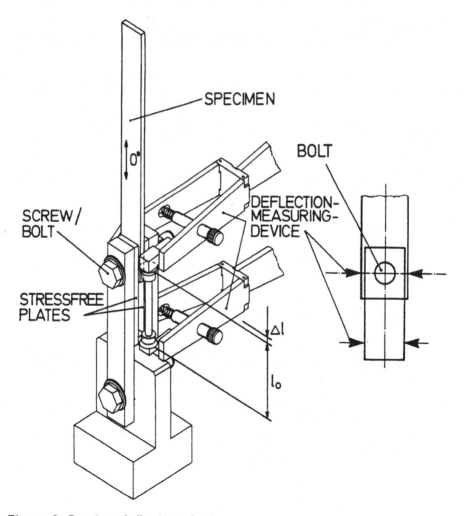

Figure 2: Bearing deflection device

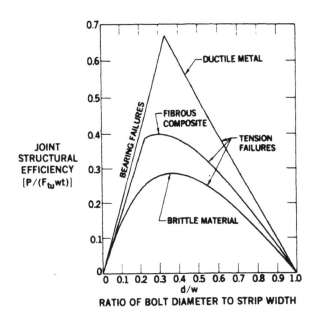

Figure 3: Strengths of bolted joints in ductile, fibrous composite and brittle materials /1/

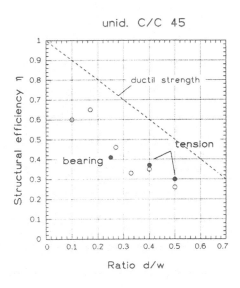

O   unloaded holes   ● loaded holes

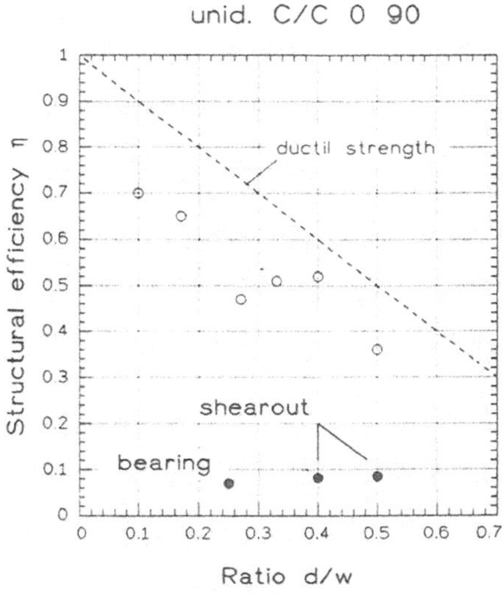

Fig. 4: Strength of the
       [0/90/0/90]ₛ-laminate

Fig. 5: Strength of the
       [+45/-45/+45/-45]ₛ-laminate

491

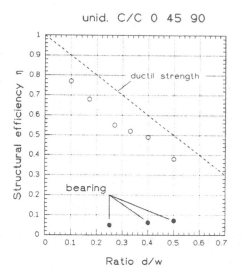

Fig. 6: Strength of the
[0/+45/-45/90]$_s$-laminate

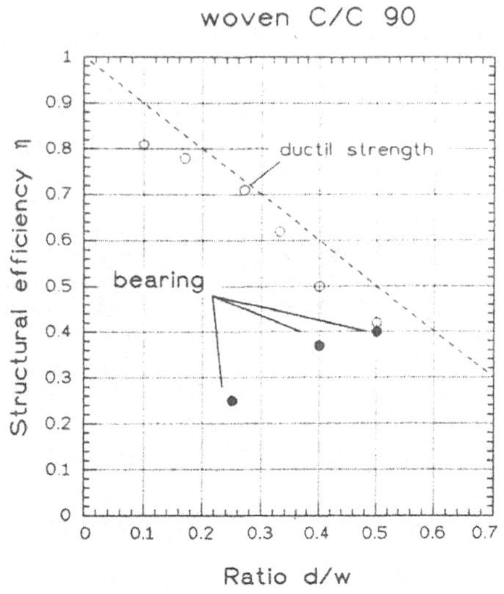

Fig. 7: Strength of the
[90/90/90/90/90/90]$_s$-lam.

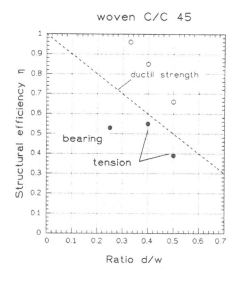

Fig. 8: Strength of the
[45/45/45/45/45/45]$_s$-lam.

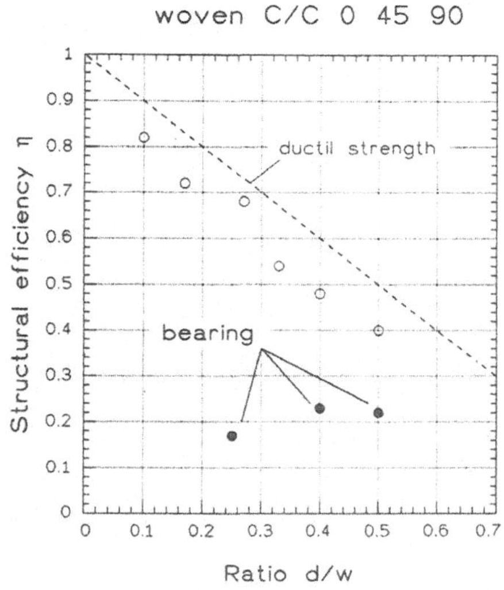

Fig. 9: Strength of the
[0/+45/-45/90]$_s$-laminate

DELAMINATION

Chairman : **Prof. J.L. LATAILLADE**
CERMUB - ENSAM

DELAMINATION BUCKLING OF LAMINATED PLATES

G. STEINMETZ, F.J. ARENDTS, R. NETHING

Universität Stuttgart - Institut für Flugzeugbau
Pfaffenwaldring 31 - 7000 Stuttgart 80 - West-Germany

ABSTRACT

The compressive strength of composite plates can be considerably reduced if they are damaged by delaminations. To study the failure process buckling loads of laminates that contain rectangular, circular and elliptical delaminations are calculated using two models based on the Rayleigh-Ritz and finite element method. The influence of several parameters such as stacking sequence, flaw size, flaw location and flaw shape is investigated and the results are plotted in diagrams. It can be distinguished between local buckling of the debonded sublaminate, buckling of the total plate and an interaction of both cases.

I - INTRODUCTION

Structural components made of composite materials usually contain defects, which might be caused during manufacture or service. They can be divided in three principal types: broken fibers, cracks in the matrix material and delaminations. Among them the interlaminar matrix delamination is of major importance since it leads to a considerable loss of the flexural stiffness. Especially if the structure is subjected to in-plane compressive loading, its load carrying capacity could be reduced considerably due to premature instability failure. Depending on the size and location of the delaminated area either local, global or combined (local and global) buckling can be observed. If the delaminated ply group is thin in comparison to the total thickness of the composite plate local instability will be very likely. In this case the overall behaviour of the plate is hardly influenced and the buckling can be described as a 'thin film model . If the delamination area is small in comparison to the total area of the laminate the overall buckling behaviour might hardly be affected. In this case only global buckling has to be considered. In all other cases there exists a coupling between local and global buckling, which will be called combined or mixed buckling.

The out-of-plane deformation of the sublaminate causes high interlaminar stresses at the edge of the buckled region and is often followed by a delamination growth that could be either stable or unstable. The latter case usually leads to the total failure of the structural component. Whether the delamination grows after buckling depends on many parameters including sublaminate size, sublaminate in-plane and flexural stiffness and the interlaminar fracture toughness of the material. In order to describe the damage process until final failure the post-buckling behaviour has to be considered. Then it would be possible to analyze the stress distribution at the edge of the debonded region and to apply some fracture criterion to predict whether the delamination will grow. This procedure is rather complex and would be beyond the scope of this paper. Therefore, just the instability problem will be considered. It should be kept in mind that the calculated buckling loads are generally not equivalent to the ultimate strength of the damaged plate but they represent a preliminary estimate of the damage propagation onset.

Buckling and post-buckling of delaminated layers has been addressed by several researchers analytically as well as experimentally. Theoretical models have been developed for strip /1, 2, 3, 4, 5/, circular /6/ and elliptical /7, 8, 9/ delamination shapes. The strip delaminations were treated in one-dimensional models applying the 'thin film' or the global model which takes into account the coupling between local and global buckling. The two-dimensional models, which were applied to circular and elliptical delamination shapes, were all based on the 'thin film model . This paper presents a two-dimensional analysis, which is able to consider the most general case of coupled buckling. Furthermore the analysis allows for bending-stretching coupling, that could lead to a drastic reduction of the buckling load.

II - ANALYSIS

Two analyses were used to calculate the load at which the instability point was reached: the finite element (FE) and the Rayleigh-Ritz (RR) method. For both types of analyses it was assumed that each part of the composite plate (the two sublaminates and the undamaged region) behave according to the classical laminate theory /10/ and that the linear buckling theory is appropriate.

2.1 The Rayleigh-Ritz Method

In the 'thin film' model it is assumed that only the delaminated plies buckle while the overall plate is not influenced. Therefore the sublaminate can be considered seperately.

The well known RR method was used to solve buckling problem. Based on a suited displacement function with unknown coefficients the total potential energy of the sublaminate was determined. The Trefftz stability criterion states that instability will occur when the second derivative of the total potential energy is equal to zero. This gives a set of simultaneous linear algebraic equations that represent the eigenvalue problem. A detailed description of the RR method is given in many references for example in /11/.

Generally the debonded ply group is not a symmetric laminate and the bending-stretching stiffnesses B_{ij} are not equal to zero. For a highly anisotropic laminate they can take considerable values and therefore dominate the energy equation. Neglecting them will lead to buckling loads that are too high. This analysis was based on the strain energy of a symmetric laminate. The bending-streching coupling was taken into account by substituting D_{ij}^* for the bending stiffnesses D_{ij}. D_{ij}^* is called the 'reduced bending stiffness' /11/ and is calculated as follows:

$$D^* = D - BA^{-1}B$$

2.1.1 Displacement Function

The transverse displacement function $w(x,y)$ has to represent the deformed shape of the sublaminate. Because of simplicity only rectangular delaminations have been considered. The boundary conditions $w = w_{,x} = w_{,y} = 0$ must be satisfied along all edges which is based on the assumption so that the sublaminate behaves like a clamped plate. The shape funktion

$$w(x,y) = \sum\sum\sum\sum C_{mnop} \sin(m\pi x) \sin(n\pi x) \sin(p\pi y) \sin(o\pi y)$$

was chosen. Each of its terms satisfies the above boundary conditions. Depending on the type of loading (compression, shear) and the materials properties different buckling shapes can exist. In this analysis up to 6 terms have been used. Thus a maximum of two halfwaves can be represented.

For the more general configurations, i.e. elliptical and circular shaped delaminations and/or coupled buckling, the RR method would not be adequate to cope with the complex behaviour. Therefore the finite element method has been used for those cases.

2.2 The Finite Element Analysis

In the present study the general purpose finite element program PERMAS /12/ has been employed to calculate the buckling loads. The composite plate was idealized using 4 noded rectangular and 3 noded triangular, layered plate elements /13/, which follow the classical laminate theory and may be composed of up to 25 layers with orthotropic material properties. The delamination seperates the panel into three parts: the two sublaminates, which have the size of the debonded area, and the undamaged region called the base laminate. Each part was modeled by the above mentioned plate elements. At the edge of the delamination the nodes of the elements were connected with rigid levers called offset vectors (see fig. 1) in order to allow for the coupling between the different portions of the plate. Thus the fact that the neutral plane of bending has different z-coordinates for each part of the plate has been taken into account.

A preprocessor was developed to generate different meshes of layered plates containing rectangular, circular and elliptical delaminations. The mesh is rather course in the undamaged region of the laminate and may optionally be refined in the debonded area. Two degrees of refinement are possible.

In several test-runs a comparison between the RR results and those obtained by the FE analysis was done considering an undamaged, square laminate with orthotropic material behaviour. By varying the orthotropic axis with respect to the axis of loading an influence on the accuracy of both methods could be shown. The precision of the FE method depends also on the degree of mesh refinement. A mesh of at least 8 x 8 elements was found to give results that are accurate enough. The RR method is even more precise except if the angle between orthotropic axis and loading axis is in the range of 40 to 60 degrees.

Because of material anisotropy generally no symmetry condition could be used to simplify the FE model. Therefore the entire plate was modeled in the analysis. A typical mesh of a panel containing an elliptical delamination can be seen in fig. 2. The deformed shape represents the first buckling mode.

The buckling eigenvalue problem was solved by simultaneous vector iteration, which is described in /14/.

III - RESULTS

By varying the delamination size 'a' (normalized to the total length of the plate) and its thickness 't' (normalized to the total thickness of the laminate) buckling loads could be determined. They were nondimensionalized with respect to the critical load of the undamaged plate and called load reduction factors (LRF).

Laminates with different stacking sequences and delamination shapes have been investigated. In order to keep the varied parameters limited only square plates under uniaxial compression loading were considered. For all calculations the material data of a typical carbon fiber reinforced epoxy known as 914C/T300 were used.

In figure 3 to 6 the load reduction factors are plotted versus the normalized delamination thickness 't'. Figure 3 shows the results that apply to a square plate with a square delamination in its center. The plate consists of a 16-ply laminate with the stacking sequence $[0_2/+45/0_2/-45/0/90]_s$. The plate is simply supported along all edges. The dashed lines represent the values obtained by the RR method. Up to a certain sublaminate thickness they coincide well with the FE solutions drawn by solid lines. The region where the results of both methods are identical corresponds to local buckling. The line where the LRF = 1 represents global buckling. In this case the delamination does not reduce the buckling load. In all other cases a combination of local and global buckling is encoutered. A principle sketch of the three different instability modes is shown in figure 7. It is noteworthy that the RR method represents a more effective way to analyze local buckling of rectangular delaminations since the large numerical effort of a FE analysis needs much more computing time.

As another example the quasi-isotropic laminate with the stacking sequence $[45/0/-45/90_2/-45/0/45]_s$ and with a circular delamination in the center of a simply supported plate was analyzed. The results are shown in figure 4. If the number of debonded plies is less or equal to three the plate seems to be rather susceptible to local buckling. The reason is that the extensional stiffness of the outer ply group is relatively large in comparison to the rest of the laminate whereas its flexural stiffness is low. Another reason for the small buckling loads could be the constrained transverse displacement caused by the adjacent 90°-layers which imposes a compression load rectangular to the applied load. This situation will be changed if the sublaminate includes the 90°-layers, which leads to a considerable increase of the LRF.

From figure 3 and 4 it is seen that in the case of coupled buckling the LRFs stay almost constant if the debonded area is small enough (a ≤ 0.4). This means that the thickness of the delaminated ply group does not have much influence on the overall stiffness. Therefore it would not make a difference if the sublaminate was just removed. In addition the instability load of the coupled buckling deviates at the most 10% from that of global buckling.

A similar behaviour was found for panels with elliptical delaminations. Some results are shown in figure 5.

In another example the influence of the angle between the axis of an elliptical delamination and the loading axis is demonstrated. The results can be seen in figure 6. Again in the region of combined buckling the instability load differs only slightly from the global buckling load, which means that this plate is only susceptible to local instability until the buckling load of the undamaged plate is reached.

It should be noted that some other laminates were treated. Also the in /9/ reported phenomenon of tensile buckling was observed. Due to the large number of parameters, that could be varied, it is almost impossible to extract conclusions of general validity. Nevertheless some major statements will be summarized in the following.

IV - CONCLUSIONS

1. The buckling load of laminated plates that contain delaminations is very sensitive to the size and location of the damage.

2. Among the buckling modes it can be distinguished between local, coupled and global instability.

3. The application of RR method is well suited for the analysis of rectangular sublaminates that are susceptible to local instability, since the computing effort is small in comparison to that of a FE analysis.

4. In the case of coupled buckling and for general delamination shapes the FE method seems adequate. It gives good results if the buckling sublaminate is modeled by a mesh of at least 8 x 8 elements with a bilinear displacement function.

5. The ply layup within the delaminated material has a strong effect on the buckling load. The introduction of high load bearing $0°$-plies into the sublaminate reduces the buckling stress. This effect is lowered if the bending stiffness of the debonded plies is relatively large.

6. If the normalized delamination size 'a' is less than 0.4, which covers the range of most practical problems, the instability load in the case of coupled buckling stays almost constant and never drops below 90% of the buckling load of a perfect plate. In practice this slight decrease might hardly be resolved since other effects such as geometrical imperfections will also reduce the buckling load. Therefore it seems justified to neglect coupled buckling if $a \leq 0.4$.

7. Nothing can be said about the ultimate compressive strength of the analyzed plates since damage propagation is not considered. For a complete understanding of the problem an investigation of the post buckling behaviour including delamination growth is already in process.

V - REFERENCES

1 - S. Sallam and G. J. Simitses, Delamination Buckling and Growth of Flat, Cross - Ply Laminates, Composite Structures 4, pp. 361, 1985

2 - G. J. Simitses, S. Sallam and W. L. Yin, Effect of Delamination of Axially Loaded Homogeneous Laminated Plates, AIAA Journal, Vol. 23, No. 9, Sept. 1985, pp. 1473

3 - W. L. Yin, S. N. Sallam and G. J. Simitses, Ultimate Axial Load Capacity of a Delaminated Beam-Plate, AIAA Journal, Vol. 24, No. 1, Jan. 1986, pp. 123

4 - W. L. Yin, The Effects of Laminated Structure on Delamination Buckling and Growth, Journal of Composite Materials, Vol. 22, June 1988, pp. 502

5 - I. Sheinman, M. Bass, O. Ishai, Effect of Delamination on Stability of Laminated Composite Strip, Composite Structures 11 (1989), pp. 227

6 - J. D. Webster, Flaw Criticality of Circular Disbond Defects in Compressive Laminates, NASA CR - 164830, March 1981

7 - Herzl Chai, Charles D. Babcock, Two-Dimensional Modelling of Compressive Failure in Delaminated Laminates, Journal of Composite Materials, Vol. 19, Jan. 1985, pp. 67

8 - Christos Kassapoglou, Buckling, Post-Buckling and Failure of Elliptical Delaminations in Laminates under Compression, Composite Structures 9 (1988), pp. 139 - 159

9 - K. N. Shivakumar and J. D. Whitcomb, Buckling of a Sublaminate in a Quasi-Isotropic Composite Laminate, Journal of Composite Materials, Vol. 19, Jan. 1985, pp. 2 - 18

10- Robert M. Jones, Mechanics of Composite Materials, Scripta Book Company 1975

11- J. M. Whitney, Structural Analysis of Laminated Anisotropic Plates, Technomic Publishing Company

12- Ernst Schrem, Handbook for Linear Static Analysis, INTES Publication UM 404, INTES GmbH Stuttgart

13- Horst Parisch, Ernst Schrem, QUAD4 and TRIA3 Shell Elements in PERMAS, INTES Publication UM 402

14- D. Nagy, PERMAS-III/2 Linear Buckling, INTES Publication No. 210

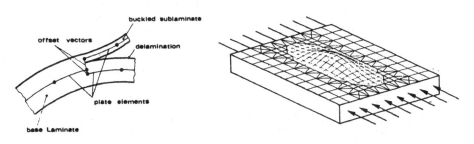

figure 1   figure 2

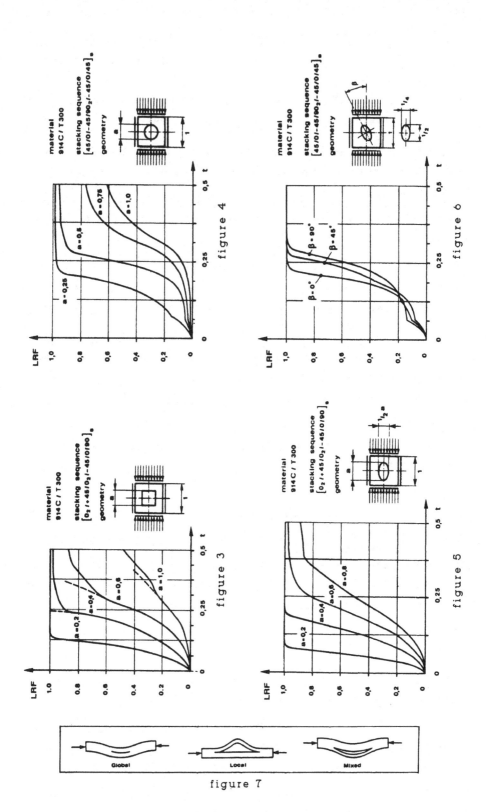

figure 4

figure 6

figure 3

figure 5

figure 7

502

NEW DATA REDUCTION SCHEMES FOR THE DCB AND ENF TESTS OF FRACTURE-RESISTANT COMPOSITES

W.-T. CHANG, I. KIMPARA*, K. KAGEYAMA*, I. OHSAWA*

Ishikawajima-Harima Heavy Industries co. Ltd
1 Shin-Nakahara-cho - Isogo-ku - 235 Yokohama - Japan
*University of Tokyo
7-3-1 Hongo - Bunkyo-ku - Tokyo 113 - Japan

ABSTRACT

Proposed in this study are new data reduction schemes for the DCB test and the ENF test. The former uses the relation between normalized crack length and the cubic root of the load-line compliance. The latter uses the CSD (Crack Shear Displacement) or the CSD compliance. R-curves both in the Mode I loading and in the Mode II loading were obtained to apply those schemes to Celion 6000/PMR-15 composite system. This system showed the interlaminar fracture resistant behaviour in both modes, which means the increasing toughness as the crack propagates.

INTRODUCTION

Advanced composite materials such as Carbon fiber/Polyimide system have been applying to the development of aerospace structures. In the laminated composite structures, the main problem might be the delamination failure. To investigate the interlaminar fracture behaviour, usually undertaken for Mode I and Mode II are the Double Cantilever Beam (DCB) test and the End Notched Flexure (ENF) test, respectively, /1/. In this study, new data reduction schemes for these tests are proposed to determine the R-curve, which is the relation between the fracture toughness and the crack length. Celion 6000/PMR-15 composite is investigated on the interlaminar fracture with the experimental consideration on the crack initiation.

I - EXPERIMENTS

1.1. Material and test specimens

The material employed in this study was a carbon fiber/polyimide composite supplied by FIBERITE, which consisted of continuous unidirectional carbon fibers of type Celion 6000 in a matrix of PMR-15. Composite panels were fabricated in an autoclave using the

..tional prepreg supplied by the manufacturer. A folded double layer of KAPTON
..t of thickness 50 μm was inserted between the two plies at the center plane to produce
a starter crack. After the fabrication of panals, DCB specimens (as shown in Fig. 1) and
ENF specimens (as shown in Fig. 2) were cut from each panel. Alminium end blocks
were bonded to the cracked end for the DCB specimen using the epoxy 2-component
paste adhesive ('Araldite', supplied by CIBA-GEIGY). Mode I precrack of about 8 to 10
mm was produced using a wedge for DCB and ENF specimens. After that, the folded
KAPTON film of each specimen was cut in two. The fiber volume fraction of specimens
was 56.5%.

1.2. Test procedure

1.2.1. DCB test

Tests were performed at a constant crosshead rate of 2mm/min. To moniter the
position of the crack tip, one edge of the specimen was coated with a white typewriter
correction fluid. Crack length was measured using an optical microscope. Loading and
unloading were repeated several times in the procedure, and a Load-Displacement curve
was obtained.

The effect of the crosshead rate on the fracture toughness was also examined by
changing the crosshead rate range of 0.1 to 50 mm/min.

1.2.2. ENF test

Tests were performed at a constant crosshead rate of 3mm/min (the span length 2L as
shown in Fig.3 was 100mm), and a load versus load line displacement curve was
obtained as same as the conventional ENF test procedure, /1/. In this study, the crack
shear displacement (CSD as shown in Fig.3) were additionally measured with a CSD
gauge, /2/, because the CSD is more sensitive to the crack growth than the load line
displacement. After the test was completed, the specimen was wedged in two in Mode I
crack propagation. Initial crack length was directly measured from the fracture surface of
specimen. Moreover to observe the crack initiation, test specimens were loaded up to and
then unloaded at several levels of loads before the maximum load. The crack extension
was also directly measured from the fracture surface of specimen in the same way as
mentioned.

II - DATA REDUCTION SCHEMES

In general, the energy release rate, G, can be given by the following equation. In
the equation, P is the applied load, B is the width of specimen, C is the load-line
compliance, which is defined as the ratio of the load-line displacement to the applied load,
and a is the crack length.

$$G=(P^2/2B)(dC/da) \tag{1}$$

2.1. DCB test

Proposed for the DCB test is the new data reduction scheme of the fracture toughness
using the relation between the normalized crack length, a/H, and the cubic root of the
load-line compliance, $C^{1/3}$, as given by Eq. (2). Then Mode I energy release rate, G_I, is
given by Eq. (3).

$$a/H = A_0 + A_1 C^{1/3} \tag{2}$$

$$G_I = 3P^2 C^{2/3}/2BA_1 H \tag{3}$$

Here, A_0 and A_1 are experimental constants and H is the semi-thickness of specimen.

Advantages of the proposed scheme compared with the conventional ones, /1,3/, are minimizing the effect of the deviation in specimen thicknesses by normalizing the crack length, and not using the crack length value for the data reduction of the fracture toughness. The crack length and the fracture toughness during the crack propagation can be calculated continuously from the load versus displacement curve.

2.2. ENF test

The load-line compliance, C^{BT}, and Mode II energy release rate, G_{II}^{BT}, for the ENF test can be obtained based on the elementary beam theory as follows, /4/, using characters as shown in Fig.3. E_L is the longitudinal elastic modulus of specimen.

$$C^{BT} = (2L^3 + 3a^3)/8E_L BH^3 \tag{4}$$

$$G_{II}^{BT} = 9a^2 P^2 C/2B(2L^3 + 3a^3) \tag{5}$$

In the conventional ENF test, usually accepted is the fracture toughness calculated from the maximum load, initial load-line compliance and initial crack length. The conventional characterization has a definitive ambiguity of the fracture toughness in case of the occurrence of subcritical crack growth.

In this study, the characterization of the fracture toughness at the crack initiation and during propagation was investigated, because such a subcritical crack growth was observed. The crack initiation was defined as the 2.5% crack growth relative to the initial crack length.

First proposed is the new data reduction scheme of the fracture toughness using the CSD or the CSD compliance, which is defined as the ratio of the CSD to the applied load. The CSD compliance, λ^{BT}, is given by the following equation based on the elementary beam theory, /2/.

$$\lambda^{BT} = \delta/P = 3a^2/2E_L BH^2 \tag{6}$$

From Eqs.(1), (4), and (6), Mode II energy release rate, G_{II}^{BT}, can be given in the form of including the CSD compliance or the CSD as Eq. (7) or Eq. (8).

$$G_{II}^{BT} = 3P^2 \lambda/8BH \tag{7}$$

$$= 3P\delta/8BH \tag{8}$$

These equations are not in the form of including the crack length.

Second proposed is the approximate procedure using the CSD compliance for determining the crack length during propagation. It is given by the following equation from Eq.(6). Here subscripts i and init represent a value during propagation and the value at initiation, respectively.

$$a_i = a_{init}(\lambda_i/\lambda_{init})^{1/2} \tag{9}$$

The crack length during propagation can be calculated from Eq.(9), and then Mode II energy release rate from Eq.(5) or Eq.(8).

III - EXPERIMENTAL RESULTS

3.1. Mode I test results

A typical Load-Displacement curve is shown in Fig.4. Experimental constants A_0 and A_1 in Eq.(2) were obtained as shown in Fig.5. Then the relation between Mode I energy release rate and crack extension can be obtained as shown in Fig.6. Figure.6 shows that the fracture toughness is low at the crack initiation and increases as the crack propagates. The average value of fracture toughness at the crack initiation, G_{ICi}, is $258 J/m^2$. Little effect of crosshead rate on the fracture toughness was observed in the range of 0.1mm/min to 50mm/min.

3.2. Mode II test results

A typical Load-CSD curve and Load-Load line displacement curve is shown in Fig.7. A subcritical crack growth can be noticed as the nonlinear region of Load-CSD curve in Fig.7. In this study, the crack initiation was defined as the 2.5% crack growth point, which corresponded to the intersection of the 12.6% less secant line and the Load-CSD curve as shown in Fig.7. From Eq.(6) based on the elastic beam theory, however, a 2.5% crack growth gives a 5% increase of CSD compliance. The theory indicates that the 2.5% crack growth point corresponds to the intersection of the 5% less secant line and the Load-CSD curve. This difference comes from the nonlinearity of material.

Figure 8 shows the relation between Mode II energy release rate and crack extension. The figure shows that the fracture toughness is low at the crack initiation and increases as the crack propagates. The average value of fracture toughness at the crack initiation, G_{IICi}, is $558.1 J/m^2$ for Eq.(5) and $630.1 J/m^2$ for Eq.(8). It is pointed out that the fracture toughness calculated from Eq.(8) gives the 12.9% higher value compared with that calculated from Eq.(5). This might come from the nonlinearity of material as mentioned.

IV - DISCUSSIONS AND CONCLUSIONS

Celion 6000/PMR-15 composite shows the interlaminar fracture resistant behaviour both in Mode I and in Mode II. The ratio of G_{IICi} to G_{ICi} is about 2.2 to 2.4.

The R-curve in the Mode I loading can be obtained from the load versus displacement curve to use the proposed data reduction scheme for the DCB test.

The R-curve in the Mode II loading can be obtained only from the load versus CSD curve to use the proposed data reduction scheme for the ENF test. The fracture toughness calculated from Eq.(8), however, gives the higher value compared with that calculated from Eq.(5).

REFERENCES

1 - Carlsson L.A. and Pipes R.B., Experimental Characterization of Advanced Composite Materials, Prentice-Hall, New Jersey (1987)

2 - Kageyama K., Kikuchi M. and Yanagisawa N., submitted to ASTM STP : Fatigue and Fracture, (1989)

3 - Whitney J.M., Browning C.E. and Hoogsteden W., J. Reinforced Plastics and Composites, Vol.1 (October 1982), pp.297

4 - Russell A.J. and Street K.N., Proc. ICCM-IV, (1982), pp.279

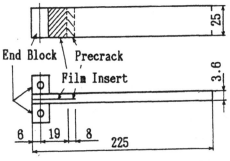

Fig.1 - Dimensions of DCB specimen

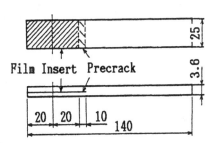

Fig.2 - Dimensions of ENF specimen

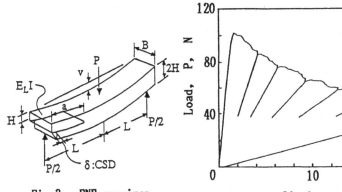

Fig.3 - ENF specimen

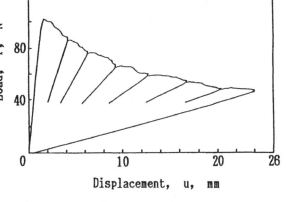

Fig.4 - Load-Displacement curve
in the Mode I loading

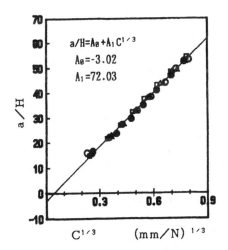

Fig.5 - Normalized crack length, a/H,
vs. cubic root of the load
line compliance, $C^{1/3}$

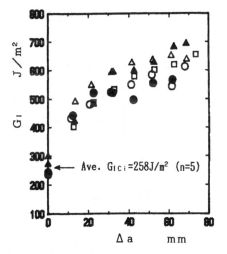

Fig.6 - The R-curve in the Mode I
loading

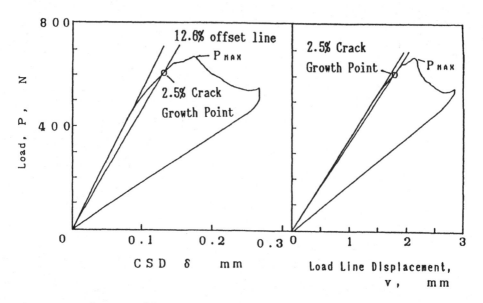

Fig.7 - Load-CSD curve and Load-Load line displacement curve
in the Mode II loading

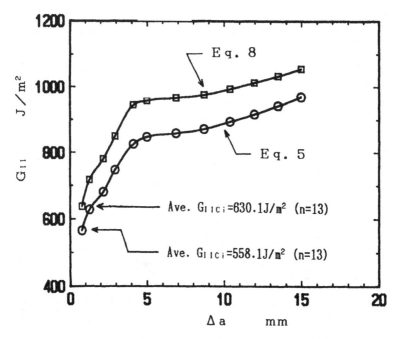

Fig.8 - The relation between Mode II energy release
rate, G_{II}, and crack extension, Δa

APPROACHES FOR IMPROVING THE DAMAGE TOLERANCE OF COMPOSITE STRUCTURES

J. BRANDT, K. DRECHSLER, F.J. ARENDTS*

MBB GmbH - PO Box 801109 - 8000 Munich - West-Germany
*University of Stuttgart - 7000 Stuttgart 80 - West-Germany

ABSTRACT

The damage tolerance of carbon-fibre-reinforced composites tends to be considerably lower than that of many metals or composites containing glass or organic fibres. One result is the susceptibility to impact, which may lead to damage such as delaminations.
Because of the serious nature of this problem, considerable efforts have been undertaken to develop solutions. The approaches presented in this paper are:

- 3-d Fibre Reinforcement
- Composite Design
- Matrix Toughness
- Interleafing

The different concepts are discussed, taking various matrix systems, interleafing techniques and reinforcing types into account. In detail, damage area, residual strength and elongation after impact as well as through-penetration energy were determined to compare the different approaches for monolithic and sandwich structures.

INTRODUCTION

The mechanical properties of composites are mainly determined by the fibre orientation. Optimal design for maximum stiffness and strength is possible when the reinforcing fibres are oriented absolutely straight in loading direction. On the other hand, this construction tends to be very susceptible to interlaminar loading occurring during

509

impact, lightning or machining due to the low interlaminar performance (see fig. 1). Several attempts were carried out to overcome this problem and to improve the damage tolerance of composites. One concept is the interleafing of conventional prepreg systems with tough layers to stop interlaminar failure propagation. Another one is based on the improvement of matrix toughness itself, using toughened thermosets or thermoplastic systems. The most effective improvement is expected when the layers are bonded together by reinforcing fibres. Such 3-d reinforcement is possible by the application of new textile technologies such as 3-d weaving, 3-d braiding, knitting or stitching, enabling the production of 3-d textile preforms which can be impregnated with thermoset or thermoplastic matrix systems.

In the following, various concepts are discussed comparing damage area and residual mechanical performance after impact as well as through penetration behaviour and fracture mechanical properties.

I   METHODS TO IMPROVE THE DAMAGE TOLERANCE OF COMPOSITES

1.   3-d Fibre Reinforcement

The most promising technologies for manufacturing 3-d textile structural composites are weaving and braiding. In both techniques, the mechanical properties of 3-d composites can be adjusted over a wide range according to the requirements by the choice of fibre arrangement and the fraction of fibres in thickness direction. An extensive overview concerning the mechanical properties and parameter studies with glass and carbon fibres (AS4) as well as thermoset (standard epoxy) and thermoplastic matrix systems (PEEK) is shown in /1/ and /2/.

To give an idea of the potential of 3-d fibre structures, some results with 3-d weavings are presented in figures 2 to 4. In figure 2 and 3, the potential of different basic 3-d weavings to improve the interlaminar properties is shown. The interlaminar shear strength is effected above all by the arrangement of z-directional fibres. The greatest potential for an improvement has the 3-X structure, because of the shear load carrying capability of the z-directional fibres laying in 45 degree direction. The greatest advantage of all 3-d weavings compared to 2-d is visible in peel tests, where the z-directional fibres are loaded in tension. Figure 3 shows the level of force during the test for 2-d laminates and a typical 3-d weaving with a 5 per cent z-fibre fraction.

The effect of this improvement in fracture-mechanical properties is demonstrated in compression-after-impact tests. Figure 4 shows the difference in compression strength after impact of various 2-d and 3-d woven composites. Besides higher strength, it is remarkable that 3-d carbon fibre composites reach a breaking elongation up to 0.9 per cent in compression-tests after impact of 6.7 J/mm. The reason for this superior behaviour is the local restriction of the damage area due to a prevention of extensive delaminations.

510

A further innovative approach is to produce whole structural parts using near net shape 3-d textile preforms being woven or braided in one step. While the weaving technology is suitable above all for stiffened flat panels, the braiding technique enables the production of complex shaped profiles. Figure 5 shows the fibre geometry of panels with integrally woven stiffeners.
This design leads to a high structural damage tolerance, because a debonding of panel and stiffeners is nearly impossible.

2. Matrix Toughness

Naturally textile structural composites show a decrease in in-plane stiffness and strength performance compared to laminates of unidirectional tapes caused by the fibre curvature or loss of fibres in loading direction. Therefore it is necessary to use 2-d reinforced laminated composites in structural applications where the maximum utilization of fibre stiffness and strength is required. To improve the damage tolerance of these composites, improved matrix toughness is a possible key. Several attempts have been made to modify conventional brittle epoxy resins by adding of rubber-base tougheners or thermoplastic polymers. On one hand, this modification improves toughness significantly, but on the other hand hot/wet performance is reduced, which limits the applicability of such modified epoxy systems.

Tough composites in combination with good hot/wet performance are possible using thermoplastic matrices such as PEEK, which yield an order of magnitude increase in toughness compared to conventional epoxy systems /3/.

In figure 6 the compression strength after impact of different toughened and standard composites is shown after impacts differing in magnitude. Figure 7 demonstrates the capability of thermoplastic matrix systems to improve the fracture toughness G_{Ic}.

3. Interleafing

To improve the damage tolerance of 2-d composites with standard matrix systems, interleafing with tough films between the laminate layers is a possible way to go.

In an experimental study, the influence of the type of interleafing material as well as the count and stacking sequence of interleafing layers were examined. Some results of CAI and fracture toughness tests are shown in figures 8 and 9. The materials tested are a tough epoxy-based adhesive film with a thickness of 0.4mm (Cyanamid FM377) and a thermoplastic film with a thickness of 0.025mm (Cyanamid HXT-441). The baseline material is a modified BMI system, Narmco 5245/T800. The fracture toughness tests, which were performed using one interleafing film in the mid-plane, showed a significant

511

improvement compared to the baseline material. The laminate with the epoxy film for example, reaches G_{2c} values nearly 10 times of those of the non-interleafed material. The best CAI results can be obtained using thin thermoplastic films. The pre-impact compression strength is not reduced significantly if only three layers are used. The position of interleafing layers is of great importance with regard to optimizing the benefit of interleafing. The best post-impact performance was reached when all three films are concentrated on the upper side between laminate layers 3/4, 4/5 and 5/6. This stacking sequence seems to best limit laminate damage.

Compared to toughened matrix systems, interleafing leads to a weight penalty because of the low stiffness and strength of the additional resin layers which proportionately reduce stiffness and strength of the laminate. This penalty can be minimized by a selective toughening of critical locations where three-dimensional interlaminar stresses occur.

CONCLUSIONS

The different concepts showed their potential for improving composite damage tolerance in an extensive experimental study comparing interlaminar shear strength, fracture toughness and compression after impact properties.
One possible way to improve damage tolerance is **toughening** of brittle epoxy resins. The penalty of this matrix systems is the poor hot/wet performance, limiting the applicability. As well good toughness and hot/wet performance is possible using **thermoplastic matrices**, which, on the other hand, require new manufacturing processes. A significant improvement of composite toughness is possible by **interleafing** standard epoxy system laminates with tough films. The weight penalty of these method can be limited by a selective interleafing of critical areas.
The greatest potential for improved damage tolerance is offered by a **3-d reinforcement** which prevents delaminations and restricts damage areas drastically. 3-d composites are suitable above all for structural components where impact properties are of greater importance than stiffness and strength. Of special interest is the use of near net shape 3-d textile preforms, which for example enable the manufacturing of panels with integrally woven stiffeners.

REFERENCES

1 - Brandt J., Drechsler K., Preller T., Manufacturing and Mechanical Properties of 3-D Fibre Reinforced Composites, SAMPE European Conference, Birmingham, 1989
2 - Arendts F.J., Brandt J., Drechsler K., The Application of 3-D Reinforced Fibre Preforms to Improve the Properties of Composites, SAMPE Symposium, Reno, 1989
3 - Brandt J., Richter H., Hochleistungsverbundwerkstoffe mit thermoplastischer Matrix - Möglichkeiten und Grenzen, Kunststoffe 77, 1987
4 - Brandt J., Warnecke J., Influence of Material Parameters on the Impact Performance of Carbon-Fibre-Reinforced Polymers, SAMPE European Conference, Munich, 1986

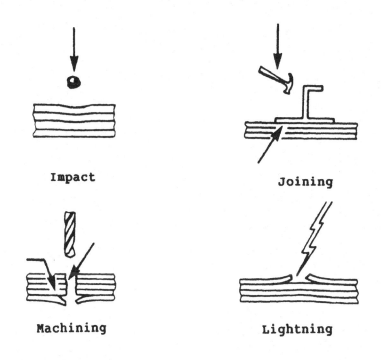

Fig. 1 - Interlaminar Loading of Composites

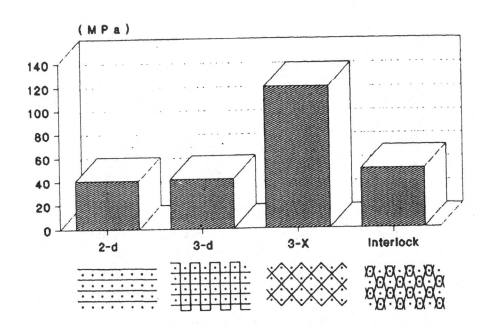

Fig. 2 - Interlaminar Shear Strength of Different 3-d Weavings

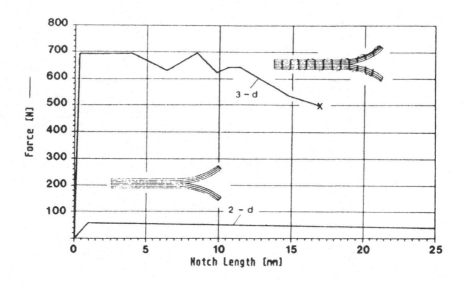

Fig. 3 - Peel Test with 2-d and 3-d Composites

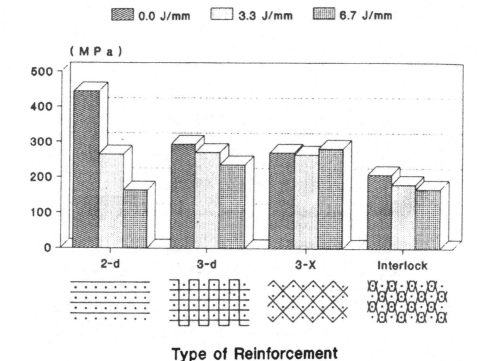

Type of Reinforcement

Fig. 4 - Compression Strength After Impact of Different 3-d
Weavings

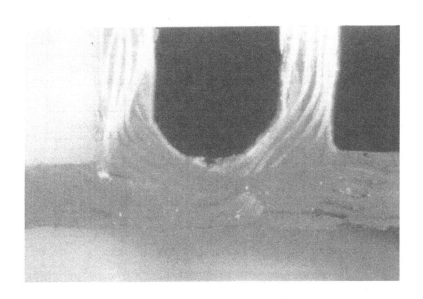

Fig. 5 – Composite Panel with Integrally Woven Stiffeners

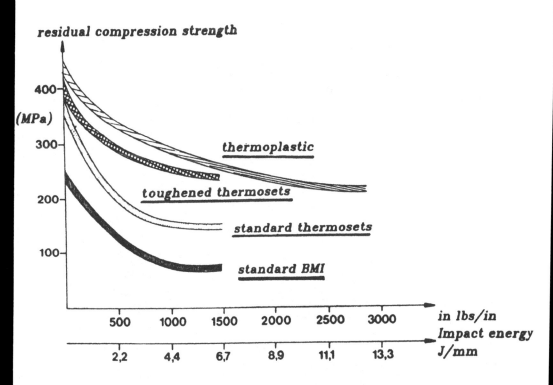

Fig. 6 – CAI-Performance of Toughened Matrix Systems

515

LAMINATE		G(IC) [J/m2]	G(IIC) [J/m2]
5208/T800	Standard Epoxy	87	154
5245/T800	Standard Epoxy	352	556
5250-4/IM7	Toughened BMI	280	620
APC2/IM6	Thermoplastic	1800	2200

Fig. 7 - Fracture Toughness of Different Matrix Systems

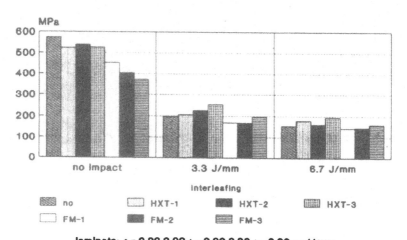

laminate: +,-,0,90,0,90,+,-,0,90,0,90,+,-,0,90,+,-//sym
Interl.: -1: 6/7,12/13; -2: 6/7,12/13,18/19; -3: 3/4,4/5,5/6

Fig. 8 - CAI-Performance of Interleafed Composites

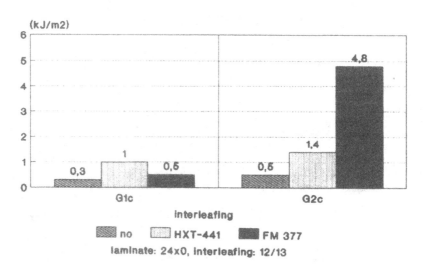

laminate: 24x0, interleafing: 12/13

Fig. 9 - Fracture Toughness of Interleafed Composites

STUDIES ON COMPRESSIVE FAILURE IN UNIDIRECTIONAL CFRP USING AN IMPROVED TEST METHOD

J. HAEBERLE, F.L. MATTHEWS

Imperial College - Dept of Aeronautics
Prince Consort Road - SW7 2BY London - Great Britain

ABSTRACT

A new test method for compression testing of unidirectional composites is proposed. It includes elements of established methods and, it is suggested, combines their advantages and avoids their shortcomings. A major postulate was to keep the rig and the specimen as simple as possible whilst at the same time reducing specimen constraints. Test results comparing different test methods are presented. The effect of some factors of specimen preparation on the measured strength is summarised. A specimen design which offers a more uniform stress distribution emerges from the observations made during testing. The failure mode of macrobuckling is then examined considering the nonlinear elastic behaviour of unidirectional CFRP.

INTRODUCTION

Following a literature review /1/ it was felt that an intensive study was necessary in order to establish the most meaningful test method for compression testing of composites. In short, the conclusions of the survey were that the compressive strength of a unidirectional CFRP system is difficult to determine, and to a large extent the experimental result is dependent on many factors such as method of load introduction, specimen design and preparation, and fixture alignment, to name a few. A series of experiments led to the development of the test method described below.

I - EXPERIMENTAL PROCEDURE

1.1 Test Method

The design of the rig is based on the fixture developed at Birmingham University /2/, but an end loaded tabbed specimen with a gauge length of 10 mm is used, rather than

517

a waisted untabbed RAE-type specimen as proposed by Barker and coworkers /2/. The specimen is only slightly clamped in end blocks and loaded on its ends. This resembles the loading configuration of the Boeing modified ASTM D695 method with the advantage that end failures of the test piece are avoided. By means of a high precision die set frictional effects are minimised and axial loading of the test piece is guaranteed. Figure 1 shows the main features of the rig and the specimen. The other methods used for the comparison are the CRAG (modified Celanese); IITRI; and the modified ASTM D695 method. A short description of these methods is found in /1/.

1.2 Materials and test piece preparation

The material used throughout the tests was Fibredux 914C/Courtaulds XA-S. The cured panels were postcured in the recommended cycle and stored in vacuum. After gritblasting, end tab plates were bonded onto the CFRP panels using room temperature curing adhesive. From these plates the individual test pieces were cut on a diamond-tipped saw. The tab surfaces were then ground flat (see Figure 1) and the test pieces instrumented with strain gauges on either side of the gauge section. All tests were carried out at room temperature.

1.3 Parameters examined

The parameters examined were: surface of the the gauge section by using different peel plies in the autoclave; fibre misalignment: material cut 2° off axis from a unidirectional panel and material layed up ±1°; thickness and length of the gauge section: 8, 12 and 16 plies and 5 to 20 mm respectively; and specimen width: 10 and 5 mm. Different end tab materials were included as well, the standard being 0/90° glass fibre/epoxy.

II - TEST RESULTS AND DISCUSSION

2.1 Comparison of different test methods

Results from a testing short course where different methods were compared are summarised in Figure 2. The mean strength result achieved with the proposed method is considerably higher than those obtained with CRAG and IITRI, a fact that can be attributed partly to lower stress concentrations introduced by the testing hardware, since the specimens have similar dimensions. It can be shown that, for instance, transverse forces as experienced by the test piece in the Celanese rig could cause additional curvature of surface fibres near the tab edge, resulting in premature failure of the specimen; see Figure 3. In practice the new (ICSTM) rig proved to be reliable and the easiest to use.

2.2 Influence of parameters of test piece production

None of the parameters listed above influenced the measured apparent strength significantly when identical slenderness ratios were compared. The only factor which seemed to influence the result was the end tab/CFRP interface as described below.
FE calculations predict high longitudinal stress peaks near the edge of the end tab. The stiffer the tab the higher are these predicted stress peaks. Surprisingly, specimens where the gauge section was machined out from thicker material showed comparatively high

test results, although these tabs resembled the stiffest in the present investigation. What seemed to be a contradiction of the FE analysis could be explained after a close examination of the specimens during the test. Figure 4 (a) shows a typical stress strain curve for these specimens. The local divergence of the strain gauge signals on opposite sides of the gauge section could be identified as the start of interlaminar shear failure starting at the corner of end tab and gauge section, i.e. the start of delamination of the end tab. Therefore stress concentrations were relieved in this critical area, allowing higher average failure stresses. This mechanism is thought to be a further reason for the higher results achieved with the ICSTM jig as the through-thickness pressure with the indirect (i.e. when the load is introduced by shear) compression test methods reduces the tendency of the tab tips to debond. Somewhat erratic delaminations account for a part of the scatter found in compression testing of composites.

2.3 Improved Specimen

It was found that, indeed, failure stresses up to 13% higher than these obtained with the standard specimen could be achieved when the delamination of the tab edges was controlled. This was realised by including a small strip of PTFE tape into the adhesive layer or by just leaving out a strip of some 5 mm beside the gauge section when grit-blasting the panels. Five specimens produced in that way gave a mean value of 1806 MPa with a coefficient of variation of 4.8 %; see Figure 2 (DELST).

2.4 Failure modes

The fracture surfaces of all the broken specimens looked similar macroscopically regardless of the test method or the parameter of manufacture. In almost all cases failure started near the end of the gauge section, adjacent to the tab edge. The fracture surfaces usually were inclined at an angle of about 75° to the fibre axis either in thickess or in width direction of the specimen. This indicated a fibre-kinking type of failure with a sharply defined band of kinked fibres, as sometimes observed in arrested failures /3/. The massive post-failure damage to the surfaces usually made it impossible to verify the kink bands microscopically.

The higher stresses achieved with the improved specimen presented a new problem. As can be seen from Figure 4 (d) even specimens of 10 mm gauge length now failed in a macro buckling mode, i.e. overall specimen instability, as indicated by the diverging signals of the opposing strain gauges. The Euler equations are known to overpredict the buckling stress for short composite bars by a large margin. If, however, the local tangent modulus of elasticity $E_t(\varepsilon)$, which decreases considerably with strain as pointed out in Figure 5, rather than the initial modulus is considered to be relevant, and the shear deformation is included in the differential equation of the buckled shape, the prediction of the critical stress becomes much more realistic. The critical stress for macro instability of the test piece can then be calculated using the following equation /4/:

$$\sigma_{cr} = \sigma_{Eu} \, 1/(1+n \, \sigma_{Eu}/G),$$

where the Euler buckling stress is taken here as:

$$\sigma_{Eu} = C \, \pi^2 E_t(\varepsilon*) / s^2.$$

n is a geometric factor subject to the cross section (n=1.2 for a rectangular cross section), G is the initial shear modulus of the material, C is a factor to characterise the end conditions of the compressed bar (C=1 for hinged and C=4 for built-in ends), s is the slenderness ratio of the bar and ε^* is the mean failure strain. The equation is solved iteratively and plotted for several clamping conditions in Figure 6. It can be seen how the shear deformation gains influence on the buckling stress for small slenderness ratios. The test results for specimens of different gauge lengths are included in Figure 6. The standard specimens with gauge lengths of 12.5 mm and more, which failed mainly in a macro buckling mode, and the improved specimen with 10 mm gauge length seem to fit the curve with a clamping constraint constant C of about 1.8 reasonably well. There is a tendency towards smaller values of C with decreasing slenderness ratio of the test piece. This might be explained by the higher stiffness of the shorter specimens compared to the test rig. It should be mentioned that macrobuckling of the specimen, which is elastic initially, initiates compressive failure on a microlevel, but at unknown local stress levels. Local compressive strains of elastically buckled specimens often exceeded 2%. The fracture surface and its location, however, gives no evidence of the initial type of failure.

CONCLUSIONS

High strength results achieved with the method presented indicate that stress concentrations which cause premature failure of coupon specimens are reduced compared to conventional test methods. The more so when the test piece is prepared so that the end tab edge can debond easily and symmetrically without sacrificing lateral support, as it would if the edges were tapered. It is, however, evident from the results that test coupons with standard dimensions might not be suitable to determine the compressive strength of high strength CFRP systems. The decreasing modulus of elasticity might lead to a buckling failure of the test piece even for smaller slenderness ratios than those presented.

ACKNOWLEDGEMENTS

The authors wish to acknowledge the support of the Ministry of Defence (Procurement Executive).

REFERENCES

1. Haeberle, J.G. and Matthews, F.L., "Compression Test Methods for Composites and Theories Predicting the Compressive Strength of Unidirectional CFRP." Proceedings of the conference "Applied Solid Mechanics-3", April 6/7 1989 at the University of Surrey, Guildford, Elsevier Applied Science, 1989.

2. Barker, A.J. and Balasundaram, V., "Compression Testing of Carbon Fibre-Reinforced Plastics Exposed to Humid Environments". Composites, Vol. 18, No. 3, July 1987.

3. Haeberle, J.G. and Matthews, F.L., "The Compressive Mechanical Properties of Fibre- Reinforced Plastics". Third Interim Report, Department of Aeronautics, Imperial College, London, May 1989, (unpublished).

4. Timoshenko, S.P. and Gere, J.M., "Theory of Elastic Stability", McGraw-Hill Kogakusha Ltd, Tokyo, 1961

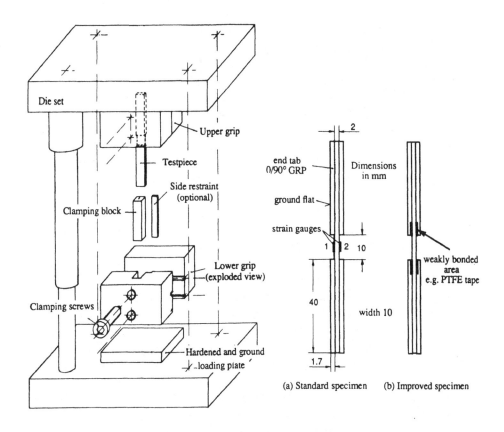

Fig. 1: Imperial College (ICSTM) compression test rig and specimen

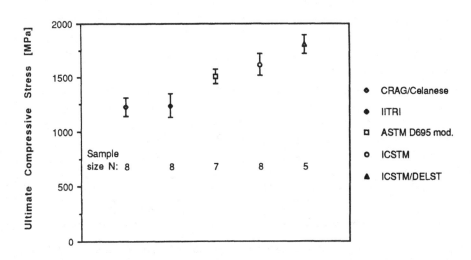

Fig. 2: Results for the compressive strength of 914C/XA-S using different test methods

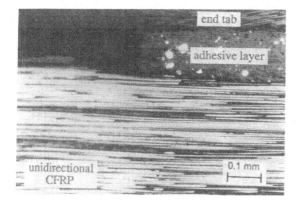

Fig. 3:

Illustration of the effect of through-thickness loading on the curvature of fibre bundles.

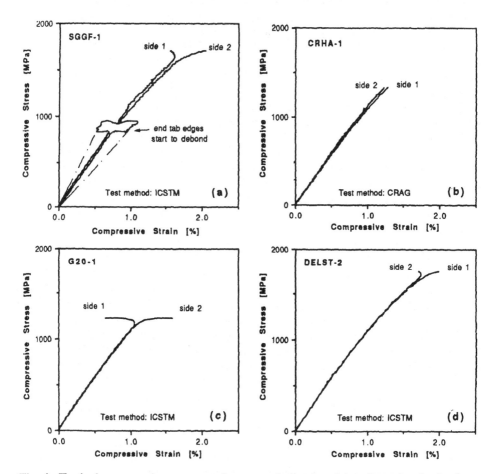

Fig. 4: Typical compressive stress strain curves, indicating debonding of end tab edges (a); failure without macro buckling (b); macro buckling: of a specimen with 20 mm gauge length (c) and of an improved specimen with 10 mm gauge length (d)

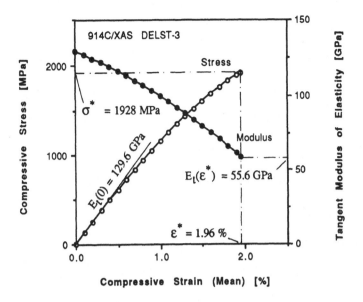

Fig. 5: Tangent longitudinal modulus of elasticity for a typical stress strain curve

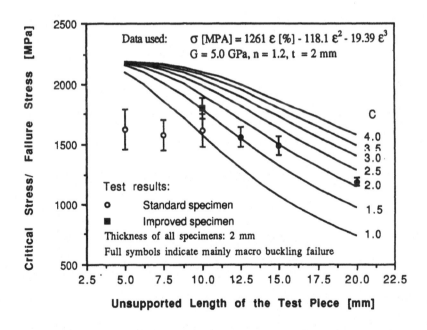

Fig. 6: Variation of the critical compressive stress for macro buckling with gauge length of the testpiece including test results of specimens with different gauge lengths

A STUDY OF CARBON FIBRES UNDER AXIAL COMPRESSION

C. GALIOTIS, P.C. NICHOLAS, T.D. MELANITIS

Materials Dept. - Queen Mary & Westfield College
Mile End Road - E1 4NS London - Great Britain

ABSTRACT

A new method for determining the critical compressive strain and compressive modulus of single PAN-based carbon fibres is presented. This technique combines a cantilever beam to load the fibres in tension or compression and Raman Spectroscopy to map the strain along the fibres. The mechanical properties of the fibres , and their mode of failure, in compression, were found to be dependent on the Young's modulus and the crystallinity of the material.

I - INTRODUCTION

The axial compressional behaviour of carbon fibres has been studied in the past employing a variety of methods such as, the elastica-loop test /1/ and compression tests on embedded, in a polymer matrix, single fibres /2/. Recently, a method was developed to apply well defined compressive strains to single filaments by bonding them to the compressible side of an elastic polymer beam /3,4/ . The technique, although successfull for polymeric fibres, failed for carbon fibres, due to difficulties in assessing the compressive damage in opaque systems . This method, has been modified by the authors /5/ , so that a version of a Cantilever Beam (CB) configuration was used to load the fibres and a Laser Raman Probe was used as the strain monitoring system /6,7/ .

II - EXPERIMENTAL PROCEDURE

2.1 Samples and preparation.

Eight PAN-based (Courtaulds Grafil) carbon fibres of different moduli and graphitisation procedures were examined in this programme . Their properties are listed in Table 1. Some of them , produced particularly for this programme, are denoted as Lab Fibres (LF). The test specimens were prepared by aligning single carbon fibres on the top surface of PMMA bars (5.5x10x50 mm^3).The fibres were bonded to the bars using a thin acrylic film.

2.2 Loading Configuration - Cantilever Beam (CB)

The Cantilever Beam (CB) is shown in Fig.1 . The beam can be flexed up or down, subjecting the fibre to compressive or tensile load, respectively. The strain e' at any point along the fibre is calculated from the distance x from the fixed end of the CB and the deflection y_{max} at the free end following the formula :

$$e(x) = \frac{3.t.y_{max}}{2.L^2}.(1 - x/L) \qquad (1)$$

where t is the thickness and L is the free length of the CB. The Raman microscope probe, then, can be scanned along the fibre, enabling us to obtain the Raman Frequency of the fibre as a function of the applied compressive strain.

2.3 Raman Spectroscopy

Raman Spectra were taken with a 514.5 nm Ar-ion laser. The laser beam was focused on a $2\mu m$ spot on the fibre and the backscatterd light was collected by a microscope objective and directed to a SPEX 1877 triple monochromator. A Wright Instruments CCD camera was employed as the photon counting system for recording the Raman Spectra.

III - RESULTS

The Raman Spectra of the first order region for most of the fibres tested are presented in Fig.2. Two prominent frequency bands are observed; one at about 1360 cm^{-1} (D) which is related to the disorder in the graphitic structure /8/ and another at about 1580 cm^{-1} (G), assigned to the E_{2g} vibrational mode of graphite /8/. The ratio of the intensities of the two bands is associated with the graphite crystal size /8/ . The strain dependence of the Raman Frequencies of carbon fibres in tension has been found to be linear /6,7/ : $\Delta v = v - v_0 = a . e_f$ (2)

where Δv is the frequency shift under e_f fibre strain. Table 1 contains the results for the tensile strain dependence of all the tested fibres. The tensile load has been applied to the fibre using the CB configuration and independently verified by loading the fibres on a conventional microextensometer.

The results for compression can be classified into two categories; the first includes the fibres that broke in compression by transverse cracking as the HMS fibre of Fig.3; the second category includes the fibres that failed in compression by bulging , as the IM fibre of the Fig. 4 . The slopes of the Raman Frequencies in compression (and tension) represent, in fact, the Raman - Frequency - Gauge-Factor (RFGF or the term a of eq.2) in compression (and tension ,respectively) [Table 1.].

IV - DISCUSSION

4.1 Critical Compressive Strain to failure
The CB/Raman technique is an excellent method for measuring the critical compressive strain to failure of an opaque material. Even in cases of non-catastrophic failure, when the optical observation is insufficient, the frequency plateau formation of Fig.4 is a strong indication that the fibre cannot take up any further load. It can be seen that the critical compressive strain to failure decreases with increasing modulus (Table 1). Small discrepancies can be explained in terms of different manufacturing routes followed for some of these fibres.

4.2 Modulus in Compression
A noticeable result of the Table 1 is the consistent difference in the value of RFGF in compression, compared with its corresponding value in tension, for all types of fibres. As Raman Spectroscopy is ONLY sensitive in bond deformations, this difference indicates that in compression the graphitic bonds are not contracted by the same amount as they are extended in tension. That event is a strong indication of modulus softening in compression. Additionally, this difference in RFGF values in compression and tension, becomes more significant as the fibre modulus decreases (ratio RFGF(C)/RFGF(T) of Table 1). According to this observation, the modulus softening in compression for low modulus fibres is more pronounced.

4.3 Mode of Compressive Failure.
In general, high modulus fibres fail in compression by transverse cracking (Fig. 5a). After the first failure, the strain in the fibre recovers till it reaches its critical value, where, the fibre cracks again (Fig.3). A different type of compressive failure was observed in the second category of fibres, represented by the IM (Fig.4). This failure here is by bulging, beyond which, the fibre still supports an important amount of residual compressive load. The LFO-378 fibre exhibited both modes of failure.

V - CONCLUSIONS

The variety of manufacturing procedures results in fibre products,for which the mechanical properties do not follow strictly the traditional trend indicated by their Young's Modulus. Therefore, a "crystallinity" classification of the experimental data, as is expressed by the ratio I_D/I_G /6,8/ provides a better insight to the problem. To summarise, it seems that the Raman features of carbon fibres can reveal very interesting aspects of the fibre structure, which control the compressional behaviour of the material. As displayed in Fig.6 , highly crystallined fibres, with I_D/I_G $\leq$ 0.50 , are highly sensitive to external compressive loading and exhibit RFGFs $\geq$ 7 cm^1 / % in compression. Fibres of this category fail in compression by transverse cracking. Fibres with I_D/I_G >0.50 and RFGF(C) < 7.0 cm^{-1}/ % fail in compression by a bulging/yielding mechanism.

ACKNOWLEDGEMENTS

This programme of research is funded by grants from RAE-MoD, DTI, Courtaulds Graphil plc, SERC and QMW College, whose contribution is gratefully appreciated.

REFERENCES

1. D.Sinclair,J.Appl.Phys.,21 (1950) 380
2. H.M.Hawthorne & E.Teghtsoonian, J.Mat.Sci, 10 (1975) 41
3. S.J.Deteresa et al, ibid. 23 (1988) 1886
4. S.J.Deteresa et al, ibid. 19 (1984) 57
5. N.Melanitis & C.Galiotis, to be published
6. C.Galiotis & D.N.Batchelder, ibid. Letters, 7 (1988) 545
7. C.Galiotis et al, ibid. 19 (1984) 3640
8. F.Tuinstra & J.L.Koenig, J.Comp.Mater., 4 (1970) 492

TABLE - 1

Fibre	HMS	HMS4	LF0	LF1	LF2	LF3	IM	XAS
Modulus GPa	390	345	378	405	370	335	305	230
Diameter μm	6.5	7.0	5.0	5.0	5.0	5.0	5.0	7.0
I_D/I_G	.25	.30	.51	.60	.75	.85	.75	.85
RFGF/T cm^{-1}/%	-11.7	-10.5	-10	-9	-8.5	-7.5	-7	-8
RFGF/C cm^{-1}/%	10.5	10.1	7.0	6.8	6.3	4.7	4.7	5.1
I(RFGF) C/T	.90	.95	.70	.75	.73	.63	.68	.68
e_f(T) %	.95	1.0	1.3	1.15	1.27	1.45	1.7	1.5
e_f(C) %	.45	.50	.55	.57	.67	.72	.7	>.9
Mode of fail.	CR	CR	CR/BL	BL	BL	BL	BL	BL

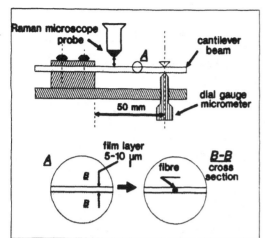

FIG.1 The Cantilever Beam (CB)

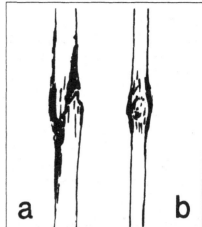

FIG.5 Compressional failure modes

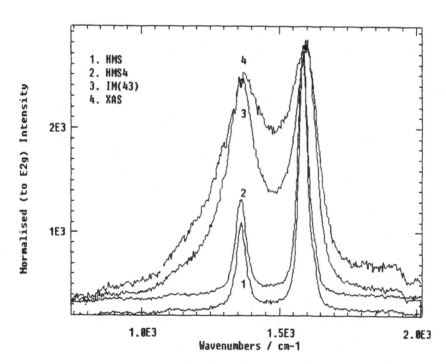

FIG.2 Raman Spectra of the Commercial fibres

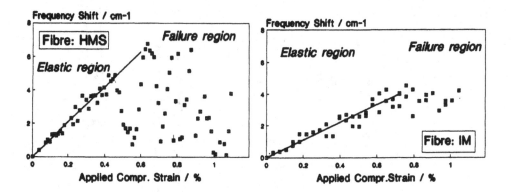

FIG.3 HMS fibre in compression

FIG.4 IM fibre in compression

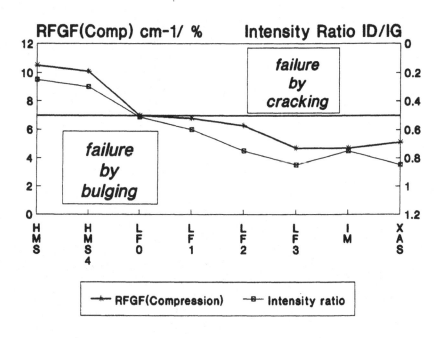

FIG.6 The Raman features of carbon fibres with reference to the compressional mode of failure

FIBRES

EFFECT OF CARBON FIBER PROPERTIES ON CARBON FIBER REINFORCED PLASTIC STRENGTH

G. GUNYAEV

All-Union Institute of Aviation Materials
Moscow - Radio Street 17 - 107005 Moscow - USSR

ABSTRACT

MECHANICAL PRODERTIES OF CARBON FIBER REINFORCED PLAS-
TICS (CFRP) ARE CONSIDERED IN THE FORM OF FUNCTIONAL DEPEN-
DECIES OF CARBON FIBERS PROPERTIES. STRUCTURAL PROPERTIES
OF CFRP AND THEIR FRACTURE CHARACTER ARE IN TIGHT CONNEC-
TION WITH FIBER PROPERTIES, THEIR STABILITY AND STRUCTURE
OF REINFORCEMENT. THE REQUIREMENTS TO CARBON FIBER PRO-
PERTIES SHOULD BE FORMULATED WITH TAKING INTO ACCOUNT WHO-
LE BODY OF FACTORS DETERMINATIVE THE PERFORMANCE OF COMPO-
SITE MATERIAL IN STRUCTURE UNDER APPROPRIATE KINDS OF LOA-
DING.

High strength in structural high-modulus carbon fiber
reinforced plastics is due to the realization of carbon
fiber strength which is defined not only by elastic-strength properties of fiber, but also by their surface state
and geometric parameters.

Filament properties are characterized by a number of
elastic-strength (strength σ_f , modulus E_f), geometric
(diameter d_f , distortion φ , specific surface S_f and
roughness factor K_f) and energetic (surface energy V_f)
parameters; scattering of these properties is defined by
variation factor with corresponding index (V_σ , V_E ,etc).
S_f and U_f parameters determine the adhesion strength of
carbon fibers with polymeric matrices along the surface
and can be replaced with τ_{fm} compex parameter.

Basing on the experimental and theoretical data, carbon fiber reinforced plastic properties can be represented

as a function of carbon fiber properties:
- tensile strength, $\sigma_x^+ = f(\sigma_f, d_f, \psi, \tau_{fm})$
- compressive strength - $\sigma_x^- = f(E_f, d_f, \psi, \tau_{fm})$
- fatique strength - $\sigma_x^N = f(\sigma_f, E_f, \psi, \tau_{fm})$
- fracture toughness factor - $K_x^c = f(\sigma_f, \tau_{fm}, d_f)$
  at $\tau_m > \tau_{fm}$

Consideration of functional dependences (without composite fracture mechanism analysis) leads to the conclusion about predominant effect of reinforcing fiber strength (σ_f) on composite strength properties under loading along reinforcement orientation: static and fatigue tensile strength, fracture toughness, etc. But the experimental data /1,2/ testify, that average strength increase of carbon fibers does not always result in the proportional increase of carbon fiber reinforced plastic tensile strength. If the last one fails as a result of initiation of macrocracks out of the most critical defect through the sound material, then to increase these properties of carbon fiber reinforced plastics not only the increase of carbon fiber average strength is necessary but also scattering degree reduction of their strength and modulus.

The compressive strength of modern carbon fiber reinforced plastics along fiber orientation does not depend on carbon fiber strength; compressive fracture of carbon fiber reinforced plastic is due to the local stability loss of filaments long before the loading exceeds fiber ultimate strength.

Compressive strength increase of carbon fiber reinforced plastics should be expected with the carbon fiber modulus (E_f) increase, as in this case the $E_f \cdot d_f$ parameter defining fiber stability is increased /3/. Proceeding from theoretical study the composite dynamic fatigue /4/ resistance should be increased with the carbon fiber modulus increase in consequence of the proportional stress reduction, appearing in polymeric matrix and along the interface. But in practice for carbon fiber reinforced plastics, compressive and fatigue strength decrease with the use of high modulus carbon fibers is found, that can be explained by surface activity decrease of these fibers and, as a result, by considerable decrease of their adhesion strength (τ_{fm}) with polymeric matrix.

Tensile strength of carbon fiber reinforced plastics is increased in consequence of scale dependence of carbon fiber strength from cross-section area (d_f diameter) and fiber length, in particular, due to the decrease of fiber effective length in composite. But with carbon fiber thinning the sensitivity of carbon fiber reinforced plastics to pore formation /1/ and stability loss (fiber distortion) The thinner are carbon fibers, the lower is compressive strength of carbon fiber reinforced plastics and their crack resistance (K_x^c parameter decrease); cross-section geometry of carbon fibers also effects compressive strength.

Considerable effect on strength properties of carbon fiber reinforced plastics is produced by texture shape of reinforcing fiber, used for making composites in the shape of twisted and untwisted tows, consisting of filaments, cord tapes, where twisted threads tows are bonded by flimsy weft, and also fabrics. The indicator of fiber buckling degree during their twisting and interweaving are misorientation angles γ (deviation from straight-line direction) the values of which are calculated basing on the equation, given in the work /1/: taken into account are filler texture characteristics, filament diameter, their, number and package density in the tow or thread, twisting degree, the "diameter" ratio of interweaving tows of warp and weft and also interlaying spacings in fabrics and tapes. It was proved experimentally /2/, that the decrease of fiber misorientation angles down to 2-3 degrees, due to lower twisting degree, the increase of tow metric number, and also the decrease fo weft threads density in tapes and fabtics results in the increase of carbon fiber reinforced plastic strength.

The use of twisted tows and threads for carbon fiber reinforced plastics defines the spesific tensile fracture mechanism of carbon fiber reinforced plastics, when fracture occurs due to damage accumulation, not as a result of separate filaments failure, but as a result of bonded bundles (tows, threads) failure. Thus, the tow or thread is an independent reinforcing element with its own values of elastic-strength properties, geometric dimensions, effective length, etc. Similar model was considered earlier, when describing elastic-strength properties of whiskerized carbon fiber reinforced plastics. In this case not the strength of filaments but the strength of bonded tow (bundle) of filaments is realized in the carbon fiber reinforced plastic, and hence its tensile strength is correlated to the strength of bonded filament tow.

Adhesion of carbon fibers with polymeric matrix along the interface (τ_{fm}) produces the decisive effect on carbon fiber reinforced plastics shear and compressive strength and defines, to a great extent, their dynamic fatigue and crack resistance. Positive effect of adhesive strength along the interface on carbon fiber reinforced plastics strength properties manifests itself in case, when adhesive strength is lower, compared to matrix cohesive strength τ_m inside and near the interface layer. If $\tau_m < \tau_{fm}$, then the further increase of adgesive strength without the matrix strength increase results in the decrease of carbon fiber reinforced plastics fracture toughness and shear and compressive strength stabilization. In accordance with the data of the work /5/ the ratio $\tau_{fm} \sim 0.64\,\tau_m$ should be maintained between the adhessive strength along the interface and matrix strength. In this case interlayer shear failure of carbon fiber reinforced plastics has combined cohesion-adhesion mode. In order to increase adhesion strength along the interface, carbon fibers are activation treated, as a result of which the value of carbon fibers specific surface S_p, roughness coefficient K_p and surface energy U_p are increased /2/. The concentration of chemically active centres and functional groups on carbon fibers surface is also increased. Surface layer etching results not only in the change of S_p, but also microrelief of carbon fiber surface due to the opening of various-shaped pores. Pore shape and their sizes define, to a great extent, the possibility of their filling with the binder and thus the total contact degree between the phases in carbon fiber reinforced plastic.

To develop carbon fiber reinforced plastics, it is necessary to have wide enough range of reinforcing fillers, differring both in their texture parameters, and elastic-strength properties and surface character of carbon fibers.

REFERENCES

1. Tumanov A.T., Gunyaev G.M., Lutsau V.G., Stepanichev E.I. "Composite Mechanics". 2, (1975), 248-257.

2. Gunyaev G.M. Structure and Properties of Polymeric Fibre Composites. (1981), 216. "Khimiya", Moscow.

3. Tumanov A.T., Gunyaev G.M., Yartsev V.A. In "Fibre and Dispersion-Strengthened Composites". (1976), 156-159. "Nauka", Moscow.

4. Gunyaev G.M. In "Fibre and Dispersion - Strengthened Composites". (1976), 144-148. "Nauka", Moscow.

5. Gunyaev G.M. "Composite Mechanics", 6, (1977), 981-986.

THE TENSILE STRENGTH OF SINGLE CARBON FIBRES OF HIGH-STRAIN TYPE

C. BROKOPF

Germanischer LLoyd
Vorsetzen 32 - 2000 Hamburg 11 - West-Germany

ABSTRACT

An analytical and experimental investigation on the ultimate tensil strength of single carbon fibres of 'high-strain' type is carried out. By a special testing mechanism it was possible to measure the strength of fibres up to minimum fibre length of 0.6 mm and to analyse the fracture surfaces by SEM afterwards. The mean values of fibre strength showed an increased strength for decreasing fibre length. It could be veryfied that two different failure modes are responsible for fibre breaks: the first one is caused by fibre surface flaws and dominating for low failure stress and the second one dominates for high failure stress. It was possible to describe the probability of failure by a strength distribution with two terms of Weiball type, one for each failure mode.

EINLEITUNG

Das Versagensverhalten von Kohlenstoff-Einzelfasern wurde bisher recht ausführlich an den Fasertypen 'High Stength' und 'High Modulus' untersucht /1,2/. Die Bedeutung solcher Untersuchungen ist darin zu sehen, daß erst eine gesicherte Erkentnis des Festigkeitsverhaltens von Einzelfasern vorliegen muß, um eine einigermaßen aussagefähige theoretische Bestimmung von Bauteilfestigkeiten durchführen zu können.

THEORIE

Die bisher durchgeführten Untersuchungen über die Zugfestigkeit von Einzelfasern führten auf die Modellvorstellung, eine Faser als Kette aneinandergereihter Segmente mit zufällig verteilter Festigkeit zu betrachten (weakest-link -theory). Die unterschiedlichen Festigkeiten beruhen auf Fehlstellen in der Faser selbst. Eindeutigstes Anzeichen für das Zutreffen dieser Modellvorstellung ist die in allen Untersuchungen festgestellte signifikante Zunahme der Faserzugfestigkeit mit abnehmender Messlänge der Faser. Die kürzesten Faserlängen in den bisherigen Untersuchungen lagen bei 1-2 mm. Die Wahrscheinlichkeit des Faserversagens ließ sich dabei durch eine zwei-parametrige Weibull-Verteilung /3/ beschreiben:

$$Pf = 1 - exp[-\frac{l}{l_0}\left(\frac{\sigma}{\sigma_0}\right)^w]$$

Dies bedeutet auch, daß eine einzige Fehlstellenversteilung über die gesamte Faserlänge deren Versagen bestimmt. Nur in einzelnen Arbeiten /2/ fanden sich Hinweise auf ein Abweichen von diesem Verhalten bei kurzen Faserlängen.

Der neue Fasertyp 'High Strain' basiert nach Angaben der Hersteller auf dem Fasertyp 'High Strength', wobei es jedoch durch modifizierte Herstellungsverfahren möglich wurde, die Fehlstellen in der Faser zu verringern und somit deren Festigkeit zu vergrößern. Diese Verbesserung konnte jedoch nicht allgemein an unidirektionalen Zugproben von Composites bestätigt werden, sondern es ergab sich überraschenderweise eine große Abhängigkeit der Festigkeit vom Matrixwerkstoff. In der folgenden Untersuchung soll geklärt werden, inwieweit ein verändertes Versagensverhalten der Einzelfaser für dieses Phänomen verantwortlich sein kann.

EXPERIMENTELLES

Eine Prüftechnik für die Zugprüfung an Einzelfasern wurde entwickelt, damit auch der Bereich sehr kleiner Faserlängen (< 1mm) berücksichtigt werden kann. Dazu wird eine Einzelfaser in eine hochtransparente Harzmatrix eingegossen, und die Zugprüfung erfolgt an Harzproben mit einer in Belastungsrichtung liegenden Faser. Die Belastung der Faser findet durch Schubspannungen über die Harzmatrix statt. Dies erlaubt auch, schon einmal versagte Fasersegmente noch höher zu belasten und damit weitere Versagensereignisse festzustellen. Damit ist auch eine Reduzierung des sonst notwendigen Probenumfangs verbunden. Die für eine vollständige Krafteinleitung notwendige Faserlänge wird durch die Qualität der Faser/Matrixhaftung festgelegt und ergibt die minimale Prüflänge der Fasersegmente. Für die Bestimmung der effektiven Faserbelastung bzw. -länge müssen Krafteinleitungseffekte sowie eventuelle Vorbelastungen der Faser (Schrumpfung des Harzes u.a.) berücksichtigt werden.

Während des Versuches sind die bei unterschiedlichen Belastungen eintretenden Bruchstellen eines Fasersegmentes unter polarisiertem Licht gut zu erkennen, und sowohl die Belastung als auch die Fehlstellen werden kontinuierlich registriert. Die Ausgangslänge der Einzelfasern betrug 30 mm und für die Versuche wurde eine maximale Harzdehnung von 2,3 % festgelegt. Die Höhe der (gewählten) Belastungrenze war für die Untersuchung des Bruchverhaltens der Faser ausreichend und bei höheren Belastungen wären zusätzliche Effekte, z.B. Kriechen, zu berücksichtigen gewesen. Im Anschluß an den Belastungsversuch wurde in einer speziell entwickelten Apparatur das Harz chemisch verascht, wobei die Lage der Fasersegmente zueinander nicht verändert wurde. Zur Bestimmung der Versagensursache wurden die Bruchflächen der Fasersegmente danach elektronenmikroskopisch analysiert.

ERGEBNISSE UND AUSWERTUNG

In der Auswertung wird die anfänglich 30 mm lange Einzelfaser theoretisch in gleiche Segmente einer beliebig festzulegenden Länge unterteilt. Diese werden als eigenständige, belastete Fasern betrachet. Die jeweilige Versagensspannung dieser Segmente wird aus der aufgezeichneten Matrixbelastung berechnet. Dabei zeigte sich, daß die Mittelwerte der Versagensspannung mit abnehmender Segmentlänge ansteigen. Dies entspricht den bisherigen Untersuchungen und bestätigt die Annahme, daß das Versagen aufgrund von zufällig über die Faserlänge verteilten Fehlstellen eintritt.

In diesem Fall sollte die Fehlstellenverteilung ebenfalls durch eine Weibull-Verteilung beschreibbar sein. Trifft dieses zu, so müßte in einem doppelt-logarithmischen Diagramm der Verlauf der mittleren Versagensspannung über der Faserlänge eine Gerade ergeben. Wie die Werte in Bild 1 zeigen, lassen sich die ermittelten Werte für den Bereich bis zu 2 mm Faserlänge mit hoher Genauigkeit durch eine Gerade annähern, jedoch zu kleineren Faserlängen hin ist eine deutliche Abweichung von dieser Linearität erkennbar. Jedoch scheinen auch diese Werte durch eine Gerade, nur mit unterschiedlicher Steigung, anzunähern zu sein. Dies würde bedeuten, daß die Fasern in diesem Bereich aufgrund einer differierenden Fehlstellenart versagen, deren Verteilung aber auch durch eine Weibull-Verteilung zu beschreiben ist. Unter Berücksichtigung des Zusammenhangs zwischen der Steigung der Geraden und dem Parameter w der Weibull-Verteilung stellt sich das Versagensverhalten folgendermaßen dar: Für Faserlängen > 2 mm dominiert eine Fehlstellenart, die weniger häufig auftritt, jedoch die Faserfestigkeit stark herabsetzt. Im Bereich kleinerer Faserlängen dominiert eine Fehlstellenart, die häufig vorkommt jedoch die Faserfestigkeit weniger herabsetzt. Der Übergang zwischen beiden Bereichen ist dabei kontinuierlich. Vergleicht man die Parameter der Weibull-Verteilung für Faserlängen > 2 mm mit den Ergebnissen bisheriger Untersuchungen /1,2/, so läßt sich feststellen, daß die Häufigkeit dieser Fehlstellenverteilung abgenommen hat. Dies bestätigt die Angaben der Faserhersteller über die Steigerung der Faserfestigkeit durch Fehlstellenminimierung mit verbesserten Produktionsmethoden. Eine mathematische Beschreibung dieser doppelten Fehlstellenverteilung in der Faser ist möglich, wenn vorausgesetzt wird, daß die beiden Fehlstellenverteilungen voneinander unabhängig sind. Allgemein gilt für eine Versagenswahrscheinlichkeit Pf aufgrund mehrerer unabhängiger Ereignisse i :

$$Pf = 1 - \prod (1 - Pf_i)$$

Damit ergibt sich für eine bikausale Versagensursache, d.h. einem Versagen aufgrund zweier Fehlstellenarten, die beide durch jeweils eine Weibull-Verteilung beschreibbar sind:

$$Pf = 1 - exp[-\left(\frac{l}{l_0}\right)\left(\left(\frac{\sigma}{\sigma_{01}}\right)^{w_1} + \left(\frac{\sigma}{\sigma_{02}}\right)^{w_2}\right)]$$

Aus den in den Versuchen ermittelten Versagensspannungen der Fasersegmente wurden die Parameter der Verteilung bestimmt. Die Werte der Parameter sowie die daraus berechneten und die experimentell ermittelten Versagenswahrscheinlichkeiten sind in Bild 2 dargestellt. Unter Berücksichtigung einer allgemein tolerierten Messungenauigkeit von ca. 5 % ist die Übereinstimmung zwischen den theoretisch und experimentell ermittelten Wahrscheinlichkeiten sehr gut. Auch die aus der theoretischen Verteilung bestimmten mittleren Faserfestigkeiten für entsprechende Faserlängen zeigen eine hervorragende Übereinstimmung (Abweichung < 2,5%) zu den empirischen Werten. Die größten Abweichungen finden sich, wie auch schon Bild 2 andeutet, für den Bereich, für den nicht eine einzige Fehlstellenverteilung das Versagen dominiert.

Die bisher angesprochenen zwei Arten von Fehlstellen sollen im folgenden spezifiziert werden. Da es möglich war, jeder Bruchstelle ihre Versagensspannung zuzuordnen, wurde eine Untersuchung der Bruchoberflächen mittels eines Rasterelektronenmikroskops durchgeführt. Aus Veröffentlichungen /4,5/ war bekannt,

539

daß Fasern von Typ 'High Strength' und 'High Modulus' vorzugsweise an Stellen versagen, an denen in der Faser ein größerer Hohlraum (Lunker) vorliegt oder an denen die Faseroberfläche Anomalien aufweist. Dazu zählen Fehlstellen direkt unter der Oberfläche als auch anscheinend mechanisch entstandene Beschädigungen der Faseroberfläche, die eventuell bei Herstellung oder Verarbeitung der Faser entstanden sind. Die Untersuchung der Versagensquerschnitte zeigte, daß praktisch ausschließlich zwei unterschiedliche Oberflächenstrukturen auftreten. Die eine zeigt eine von einem Punkt der Faseroberfläche ausgehende fächerartige Struktur der Oberfläche (Bild 3). Dies ist ein eindeutiger Hinweis auf das Vorliegen eines Oberflächenfehlers in diesem Punkt. Die Betrachtung der zugehörigen Bruchspannung zeigte, daß diese Fehlstellenart in dem Bereich des Versagens mit geringeren Festigkeiten dominiert (Bild 4) und entspricht damit der ersten Fehlstellenverteilung der bikausalen Versagenswahrscheinlichkeit. Die zweite Art der Bruchoberflächen zeigt eine über den gesamten Querschnitt einheitlich starke Rauhigkeit. Der Bruch verläuft normal zur Faserlängsachse, und es ist keine Versagensursache visuell feststellbar. Dies, und da das Versagen bei hohen Belastungen eintritt, führt zu der Vermutung, daß hier Fehlstellen der Faserstruktur selbst für das Versagen verantwortlich sind. Damit wäre auch die zweite Fehlstellenart der Faser bestimmt. Auffällig ist das völlige Fehlen des Versagens aufgrund von Lunkern. Jedoch stimmt dies wieder mit den Herstellerangaben über die Verbesserungen bei der Faserherstellung überein, welche eine Minimierung der bisher dominierenden Versagensursache vorsieht und zu den neuen Fasern von Typ 'High Strain' führte.

REFERENZEN

/1/ P.W. Barry
Experimental data for the longitudinal Strength of UD Fibreous Composites
Fib. Sci. Tech. 11 (1979)

/2/ S. Chwastiak; J.B. Barr; R. Didchenko
High Strength Carbon Fibres from Mesophase Pitch
Carbon Vol. 17 (1979)

/3/ W. Weibull
A statistical Distribution Funktion of Wide Applicability
J. Appl. Mech. 18 (1951)

/4/ J. Breedon Jones; J.B. Barr; R.E. Smith
Analysis of Flaws in High-Strength Carbon Fibres
J. Mat. Sci. 15 (1980)

/5/ S.G. Burnay; J.V. Sharp
Defect Strukture of PAN-based Carbon Fibres
J.Micr. Vol. 97 (1973)

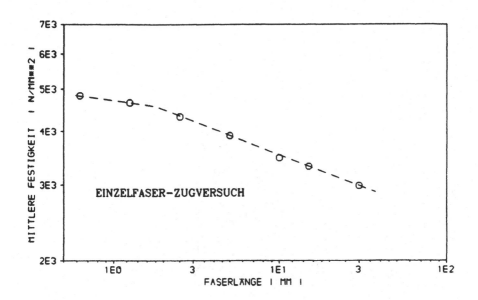

Bild 1 - Abhängigkeit der mittleren Zugfestigkeit von Einzelfasern von Ihrer Prüflänge

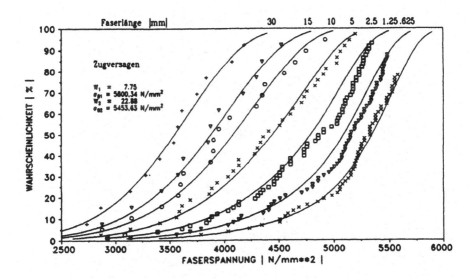

Bild 2 - Vergleich der theoretisch berechneten Wahrscheinlichkeit des Einzelfaserversagens mit den empirisch ermittelten Wahrscheinlichkeiten des Versagens

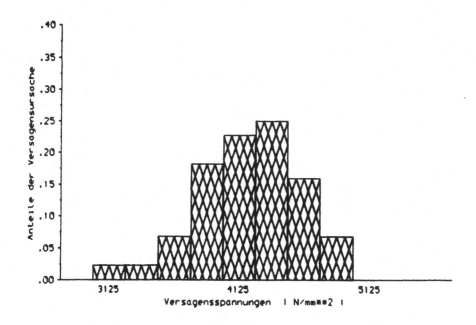

Bild 4 - Verteilung der Häufigkeit des Versagens aufgrund von Oberflächen-
fehlern von zugbelasteten Einzelfasern über ihrer Versagensspannung

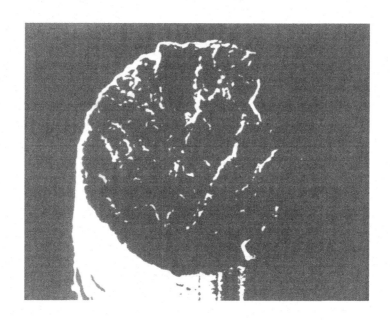

Bild 3 - Charakteristische Struktur der Bruchoberfläche einer Einzelfaser für
ein Versagen aufgrund eines Oberflächenfehlers
(Vergrößerungsfaktor: 15000)

STATISTICAL EVALUATION OF THE STRENGTH BEHAVIOR OF THIN FIBRE MONOFILAMENTS FOR THE CHARACTERIZATION OF FIBRE REINFORCED COMPOSITES

J. GÖRING, F. FLUCHT, G. ZIEGLER*

DLR - Institute of Materials Research - 5000 Köln 90 - West-Germany
*University of Bayreuth - West-Germany

1. Abstract

In this paper, an equipment for testing the tensile strength of thin single fibres ($d > 4$ μm) is presented. Different evaluation methods are pointed out to describe the strength distribution of carbon fibres with a diameter of about 5 μm based on the Weibull-statistics. The significance of these evaluation methods - implying variations in fibre diameter, volume effects and real structural effects of the fibres - for the mechanical characterization of fibres and composites is discussed. The strength distribution measured by the single-fibre test equipment is compared with the results of testing fibre bundles of the same material. After considering the statistical effects, a good correlation of the strength values and Weibull modulus calculated from bundle and monofilament test results is found.

2. Introduction

Fibre reinforcement is used more and more to improve the mechanical properties of polymer, metallic and ceramic materials. Especially in the case of ceramic matrix materials many different failure mechanisms are discussed to improve toughness, and additionally properties such as strength and creep behavior.

The properties of the composites are controlled by a number of parameters, one essential factor is the strength distribution of the fibres incorporated in the composite /1/. In the scope of the development of the composites, moreover, the change in mechanical properties of the fibres during coating and/or particulary during processing has to be known.

In principle, there are two methods to measure the tensile strength of fibres. One is the testing of fibre bundles, and the other one is to test single fibres. In composites with low strength and/or low Young's modulus of the matrix materials a great number of fibres are participating in the damage process. In this case, bundle testing seems to be more suitable to describe the failure mechanism in the composite.

In ceramic matrix composites, different damage mechanisms are discussed where the decisive interaction between matrix and fibre is controlled by a single crack (bridging effects in crack waves or pull out)/2,3/. In this case, only a small fibre volume is participating in the reinforcing process, thus the use of strength values from single-fibre tests seems to be more advantageous for describing the damage behavior of the material. Furthermore, using monofilament testing, it is possible to describe the strength distribuiton of the fibres by different evaluation methods. This allows a more detailed description of the strength distribution of the fibres and the separation of statistical and structural effects.

3. Methods and Materials

For single-fibre strength measurements the fibres are fixed on a special sample holder shown in figure 1 /2/. The fibre is glued on a paper bridging the center of a circular hole (diameter about 5 mm). On both sides of the hole the paper is separated and bridged by a polymer fibre. The whole arrangement is installed in a tensile testing equipment with a special load cell for very small loads (maximum load 1 N). After fixing the holder, the polymer fibres are melt without contacting and loading the test fibre. By using this procedure, fibre damage during handling is nearly avoided. This is very important for correct measurements, because in the case of damage the fibres with the low strength values will be separated and neglected in the strength distribution.

The tensile test is carried out under a constant crosshead speed of 0.2 mm/min. The diameter of the fibres is measured after testing by using SEM.

The procedure of characterizing the strength values and distribution is demonstrated by the example of carbon fibres (T800,TORAY, mean diameter about $5\mu m$) and a silicon carbide fibre, (NICALON, Nippon Carbon, mean diameter about $14\mu m$). The change of the strength values during processing of a RBSN composite is demonstrated with carbon fibres (M40B, TORAY, mean diameter about $6.5\mu m$).

4. Results and Discussion

The two parameter Weibull distribution is used to describe the strength values (eq.1). Here F is the probability of failure, &si. the applied stress, V the effective tested volume, V_o the unit volume, σ_o the normalizing factor and m the Weibull modulus /5/.

$$F = 1 - \exp[-(V/V_o)(\sigma/\sigma_u)^m] \qquad (1)$$

The maximum likelihood method was used to calculate the Weibull parameters from the strength values.

Using the first evaluation method in characterizing the strength distribution, the statistical volume effect is neglected (see eq.1 $V/V_o = 1$). Based on 32 single fibre tests the mean strength was calculated to 7144 ± 1855 MPa and the Weibull parameter to $\sigma_o = 7839$ MPa and m = 4.4. Figure 2 shows the strength as a function of diameter, where a great scatter of the diameter from 4.0 to 6.4 μm is observed (the mean diameter of the fibres is $5.11\mu m$). Furthermore figure 2 shows that the strength values decrease with increasing diameters. This effect may be explained by two facts. One is a statistical effect where the stress is decreasing with increasing tested volume of the fibres. The second one may be a change of the fibre structure with different diameters caused for example by the variation of the fabrication conditions. In order to separate the statistical effect, the test volume has to be taken into consideration. This means that the volume ratio V/V_o in eq .1 has to be considered. To calculate the unit volume V_o any diameter could be taken, in this case the mean diameter of the fibre $d = 5.11$ μm is used. The Weibull parameters are calculated to $\sigma_o = 7709$ MPa and $m = 5$. This Weibull modulus m is the correct value in eq. 1. Additionally, figure 2 gives the strength values after statistical correction. The decrease of the strength values with increasing diameter is still observed. This result proves that the strength reduction with larger diameters is a real material effect. This tendency is observed in a great number of other carbon and silicon carbide fibres, too.

To compare two different fibre types with various diameters, it is sometimes more reasonable to compare the strength values related to the same unit volume. In table 1 the strength values of two different fibre types (T800 with $d_m = 5.11 \mu m$ and SiC-NICALON with $d_m = 14 \mu m$) are determined from the measured load and diameter of each fibre. Moreover, the data related to a unit volume based on the mean diameter of each fibre grade as well as on a common diameter (in this case $10 \mu m$ are chosen), are given. In the case of the directly measured values, the strength of the NICALON fibre is much smaller than the strength of T800 ($\sigma_{NICALON} = 0.45\sigma_{T800}$). With the same unit volume calculated from a diameter of 10 μm, the difference is getting smaller ($\sigma_{NICALON} = 0.75\sigma_{T800}$).

When describing the damage mechanism in a composite material, it is often not possible to relate the strength values and the diameters. In this case, it could be more reasonable to describe the strength distribution using strength values calculated from the load and the mean diameter, without considering the real diameter of the single fibres. This means, the variation in diameter, the influence of the diameter or the volume on the strength values and the scatter of the strength data caused by the defects or microstructural effects are summarized in one strength distribution. These assumptions are also made by calculating the Weibull parameter from a bundle test /6,7/. Under the assumption of a constant diameter of $5.11 \mu m$, the Weibull parameters are calculated to $\sigma_o = 8216$ MPa and $m = 7.2$. At MPI in Stuttgart /7/ the strength distribution of the same material was investigated in a bundle test with 12000 single fibres and a test length of 200mm. In order to compare the results from the bundle and single-fibre tests it was necessary to eliminate the statistical effect caused by the different test lengths of the fibres. With the Weibull modulus $m = 5$ the strength values are calculated from eq.2

$$\sigma_b/\sigma_s = (V_s/V_b)^{1/m} = (l_s/l_b)^{1/m} \tag{2}$$

where σ_b is the single-fibre strength related to the test length of the bundle $l_b = 200$mm and σ_s is the strength of the single-fibre with $l_s = 5.4$mm. The results of the bundle and the single fibre measurements are summarized in table 2. A relatively good agreement between the results of the single-fibre- and bundle test results can be observed. This result demonstrates that in the case of the fibre T800 the whole failure population of a 12000 filament bundle is found in 32 single fibres with a small test length. However, in general it may not be expected that other fibre materials show the same homogeneity as the T800 fibre, resulting in differences in the values from single fibre and bundle tests. As a consequence, the results of both testing methods - the bundle and the single-fibre strength values - are necessary to characterize a fibre material completely.

One essential example which demonstrates the significance of single-fibre testing is to determine the changes in mechanical properties of the fibres during processing of the composites. During fabrication of ceramic matrix composites, high process temperatures and aggressive atmospheres are applied in some cases resulting in microstructural changes in the fibre material and in strength degradation. Sometimes it is possible to remove single fibres from the compound which are long enough for measuring their strength values. Figure 2 shows the strength distributions of a carbon fibre (M40B) as received, after coating with TiN, after annealing under nitridation conditions and removed from an RBSN composite /8/. Especially in the field of ceramic matrix composites, the change in mechanical properties of the fibres during processing is one of the most critical aspects in producing this kind of materials. In this connection, the examination of fibre properties is an important task.

5.Conclusions

Different evaluation methods are presented to describe the strength distribution of fibre materials using single-fibre tension tests. Based on the results of the various evaluation methods, it is possible to characterize the fibres regarding the microstructural effects and to compare various fibre materials. Furthermore, it allows a better description of the crack extension mechanism, especially in ceramic matrix composites. With carbon fibres (T800), a good correlation between the results in single-fibre and bundle testing is found. In the case of RBSN composites, it was possible to remove single fibres from the compound. This example demonstrates the significance of single-fibre testing to describe the change in fibre properties during processing.

6.References

[1] D.Rouby, cfi/Ber. DKG 66(1988) 208-246

[2] H.Stang, S.P.Shah, J.Mater. Sci. 21 (1986) 953-957

[3] J.K.Wells, P.W.R.Beaumont, J.Mater. Sci. 20 (1985) 1275-1284

[4] B.Kanka, Patent, Offenlegungsschrift DE 3815425 A 1

[5] W.Weibull, Ingvetenskaps Handl. Nr.151, Stockholm 1939

[6] Z.Chi, T.W.Chou, G.Shen, J.Mater. Sci. 19 (1984) 3319-3324

[7] Th.Helmer, A.Wanner, K.Kromp, vorgetragen beim Festigkeitsseminar über keramische (Verbund-) Werkstoffe 1989, MPI Stuttgart

[8] B.Kanka, G.Ziegler Proceedings of the 7th-World Ceramics Congress & Satellite Symposia Montecanti Terme, Italy, June 24-30 1990

Material	Unit Volume	Strength in MPa	Weibull modulus m	σ_O in MPa
C-Fibre T800	$V_O/V=1$	7144 ± 1855	4.4	7839
	$V_O(d_m=5.11\mu m)$	7078 ± 1650	5.0	7709
	$V_O(d=10\mu m)$	5411 ± 1261	5.0	5893
SiC-Fibre NICALON	$V_O/V=1$	3244 ± 1015	3.7	3597
	$V_O(d_m=14\mu m)$	3234 ± 1226	3.0	3634
	$V_O(d=10\mu m)$	4047 ± 1534	3.0	4548

Table 1. Strength values and Weibull parameters of T800- and NICALON- fibres as measured and considering the unit volumes V_o calculated from the mean diameter of each fibre and from a constant diameter of 10 μm

	Single Fibre $l_O = 5.4mm$	Single Fibre $l_O = 200mm$	Bundle Test $l_O = 200mm$
Number of Fibres	32	32	12000
Weibull modulus m	7.2	7.2	8.8
Weibull parameter σ_O in MPa	8216	3615	3590

Table 2. Weibull parameters calculated from single-fibre and bundle testing

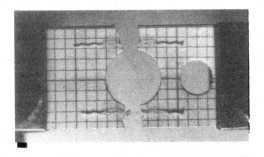

Figure 1. Fixture for tensile testing of thin single fibres (NICALON- fibre, $d_m=14$ μm)

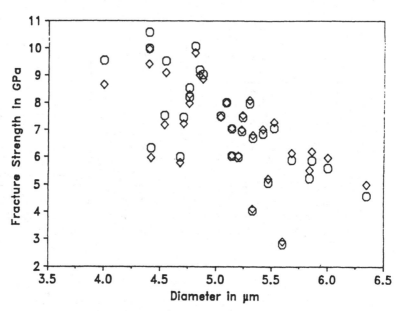

Figure 2. Strength values of the T800- fibre as a function of the diameter as measured (o) and based on a unit volume calculated from the mean diameter $d_m = 5.11 \mu m$ (◊)

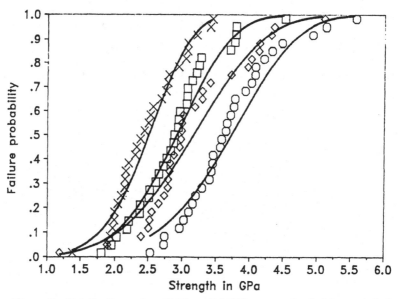

Figure 3. Strength distribution of a C-fibre (M40B) as-received (o), coated (◊ annealed under nitridation conditions outside the compound (□)an after removing the fibres from the RBSN (reaction bonded Si_3N_4) con posite (×)/8/

A STUDY OF THE DISTRIBUTION OF FIBER FRAGMENT LENGTHS BY A CONTINUOUSLY MONITORED SINGLE FILAMENT COMPOSITE TEST

H.D. WAGNER, A. EITAN

Materials Research Dept - The Weizmann Institute of Science
76100 Rehovot - Israel

ABSTRACT

A continuously monitored single-filament composite (CM-SFC) test was conducted to measure the stress at which successive fiber breaks occur in the single fiber fragmentation process. The purpose of this exercise was to explore the possibility of using this test as a simple alternative means of (i) measuring the size effect in single fibers, (ii) calculating the Weibull shape and scale parameters for fiber strength, (iii) calculating the fiber/matrix interfacial shear strength from the extrapolated value of fiber strength using the loading history of a single fragmentation test, rather than from the value of fiber strength extrapolated from extensive testing of single fibers at various gauge lengths, as is usually done. These are aspects of the SFC test which have largely been ignored so far. The results presented here confirm the possibility of using the CM-SFC test for such purposes. A possible effect of fiber pre-tensionning on fragmentation results is also briefly discussed.

I-INTRODUCTION

The fragmentation phenomenon, described first by Kelly and Tyson /1/, is currently among the most intensively researched aspects of composite micromechanics (see references in /2, 3/). This phenomenon is, in fact, a rich source of micromechanical information and we believe that some of the potential of this test has so far not been taken advantage of. Most researchers use the data obtained in the final stage of the test, namely at the saturation limit, to calculate the interfacial shear strength via the Kelly-Tyson expression /1/ or one of its variants /4/. We believe that if continuous monitoring of the stress at which the fiber breaks occur is available, then (i) accurate estimations of the interfacial strength and interfacial energy may probably be calculated from data available at the very early stages of the fragmentation test, that is, at relatively low applied loads, when the first fragments are created, (this aspect of the phenomenon will be dealt with in a forthcoming article); (ii) a new and simple alternative way to calculate the Weibull shape and scale parameters of the single fiber, and thus the size effect, might be available. So far these aspects of the fragmentation test, and their potential usefullness, have been neglected in the literature, and the interest was directed more towards data produced at the saturation limit.

Here, we further explore the possibility of utilizing the entire loading history of a single-fiber fragmentation test to calculate the length effect on fiber strength, an aspect of the fragmentation test which thus far has been ignored altogether. The possibility of predicting the values of both the Weibull shape and scale parameters (and not only the shape parameter as assumed in reference 2), as well as the interfacial shear strength, from a continuously monitored single-filament composite test will also be assessed.

II-THEORY

We first review a few key features /2/. In the past various statistical models for the distribution of fragment lengths at saturation have been assumed on the basis of goodness of fit. Examples are the Weibull, the lognormal and the Gaussian distributions. The fit of experimental data to any of these models was always found to be acceptable but clearly this goodness-of-fit in no way justifies the use of such models as physically valid solutions. In reference 2 we argued that (i) if the flaws are assumed to be distributed along the fiber length according to a spatial Poisson process, as is commonly assumed /5/, then far from the saturation limit the resulting fragment lengths follow a shifted exponential distribution, and (ii) close to the saturation limit the shifted exponential distribution is still the most physically plausible model, even though it is only approximately correct. According to (i) above, the number of defect occurences in any fiber interval of length s is a Poisson random variable with parameter λs, where λ is a positive real number. In other words, the probability of finding exactly k defects in the length interval s is $(1/k!)(\lambda s)^k e^{-\lambda s}$. The parameter λ is the mean number of defect occurences per unit length, or the intensity of the process. Of interest is the random variable S, the distance between neighbouring defects along the fiber length. The probability that S is larger than an arbitrary value s, $\Pr\{S \geq s\}$ is equal to the probability that no defect occurs on the interval length s. Since the underlying process is a Poisson process, this is $\Pr\{S \geq s\} = e^{-\lambda s}$ and therefore the cumulative distribution for S, the distance between neighbouring defects (or the fragment length), is the exponential distribution $\Pr\{S \leq s\} = 1 - e^{-\lambda s}$. However, it is necessary for the tensile stress in the fiber to build up along a small but finite length $l_c/2$ on both sides of a break location, up to the value of the peak fiber stress. Within this 'forbidden' zone, of total length l_c, the applied stress will almost always be less than the stress needed to fail the defect. Thus a shift parameter has to be included in the fragment length cumulative distribution function:

$$Pr\{S \leq s\} = 1 - e^{-\left(\frac{s-\mu}{\zeta}\right)} \tag{1}$$

where $\zeta = 1/\lambda$ is the average spacing between breaks (or average fragment length), and μ is approximately equal to $(3/4)l_c$ at the saturation limit (rather than $l_c/2$ as assumed previously /2/). This picture is exact as long as the 'forbidden zone' is much smaller than the average spacing between two breaks, because the probability of the next weakest defect being within a 'forbidden zone' is negligibly small. Close to the saturation limit, however, this probability is not negligible anymore, and the process cannot be exactly modeled as a random (Poisson) process anymore: the data will probably deviate from the exponential distribution.

We now deal with the following problem: in equation 2, how does the average fragment length ζ depend on the applied stress? Assuming that the strength of a fiber, σ_f, obeys the Weibull/weakest link model, and if the fragmentation test is viewed as a tensile test in which independent fiber samples (of varying lengths) in a series arrangement are submitted simultaneously to an (average) applied stress, we obtain the desired relationship between the average fragment length ζ and the fiber stress:

$$\zeta = \alpha^\beta \sigma_f^{-\beta} \left\{ \Gamma\left(1 + \frac{1}{\beta}\right) \right\}^\beta \tag{2}$$

where α and β are the Weibull scale and shape parameters for strength, respectively, and $\Gamma(.)$ is the Gamma function. The reverse form, given in equation 2, is preferred because a continuously monitored experiment yields a set of fragment lengths as a function of the applied stress. Closer to the saturation limit we expect equation 2 to cease to be applicable because as the fragment lengths reach their saturation limit they cannot divide any further, whatever the stress increase, and therefore the average fragment length will become insensitive to the applied stress. The link between the applied stress σ_a and the fiber stress σ_f is assumed here to be given by $\sigma_f = (E_f/E_m)\sigma_a$ (where E_f and E_m are the fiber and matrix moduli, respectively).

This result is of importance because it provides a much simpler method to measure the effect of fiber length on strength (the size effect), and the Weibull shape and scale parameters for strength, β

and α, from a single fragmentation test rather than from the conventional, extensive, testing of single fibers at various gauge lengths /5/. Using equation 2 a plot of $\ln(\zeta)$ against $\ln(\sigma_f)$ yields a straight line with slope equal to the Weibull shape parameter β. The Weibull scale parameter α can then be calculated from the value of the intercept, which is equal to $\beta[\ln\alpha + \ln\Gamma(1+\frac{1}{\beta})]$.

III-EXPERIMENTAL RESULTS AND DISCUSSION

An enhanced version of our custom-made mini-tensile testing machine /6/ was used for the CM-SFC tests, which includes an electronic signal mixing system. This enables us to simultaneously record fragmentation events (from a video camera fitted to a stereozoom microscope) and real-time stress and strain data from the screen of the microcomputer which drives the experiment. See Figure 1 in reference 2. CM-SFC test results were obtained with various types of fibers, as described in detail in /3/. The matrix was DER 331 cured with TEPA. The sample preparation procedure and testing is described in /3/. Results from the CM-SFC tests were compared with data for single fiber statistics at various gauge lengths from literature sources, when these were available.

As an example, strength statistics at various gauge lengths for single intermediate modulus carbon fibers /3, 7/ was used to calculate a Weibull scale parameter at unit gauge length (1 mm) of 8960.6 MPa and a Weibull shape parameter of 6.1. A CM-SFC sample containing this fiber was tested up to the saturation limit, yielding a total number of 41 fragments at a saturation stress of 45.7 MPa. The mean fragment length obtained at a given fiber stress level was plotted in ln-ln coordinates using the procedure described in the theoretical section. We found that the data fit reasonably well a straight line. Some deviation from linearity occured below a fragment length of about 1 mm. A linear regression analysis was performed through the linear data points only (ranging from the sample original gauge length down to a lower limit which was obtained arbitrarily by multiplying the average length of the fragments at saturation by two). The slope and intercept obtained yielded (using the ln-ln version of equation 2) a Weibull shape parameter $\beta = 8.64$ and a scale parameter at unit length (1 mm) of $\alpha = 9345.2$ MPa. These results compare quite well with those obtained above from single fiber strength against length. Possible sources for the differences are discussed in /3/. Taking into account the fact that the fragmentation data presented here involve the testing of one sample only (instead of many more in conventional single fiber strength versus length tests), the fit between the results from both tests is certainly very good. This observation is further reinforced if one uses both sets of results above to calculate the interfacial shear strength τ using the Kelly-Tyson formula $\tau = K(d_f\sigma_f(L))/(2L)$, where $K = 0.75$, d_f is the fiber diameter, $\sigma_f(L)$ is the strength of a fiber fragment at the saturation length L, which may be calculated if the Weibull shape and scale parameters are available. The results for τ using both experimental procedures for the obtention of α and β are 30.1 and 30.9 MPa, and, as seen, the differences are negligible. Results with other types of fibers (PRD-172 low modulus carbon, Kevlar 49, E- and S-Glass) are equally satisfying, as demonstrated elsewhere /3/. Figure 1, for PRD-172/epoxy, is presented as a typical example. This confirms the validity of the approach proposed here for the measurement of the size effect (or for the obtention of the Weibull parameters for fiber strength) based on the CM-SFC test, rather than from extensive testing of single fibers of various lengths.

An interesting effect on the fragmentation results was observed as follows. Strong differences in fragmentation results appear between seemingly identical samples, apparently due to single fiber pre-tensionning (which helps keep the fiber straight during matrix polymerization). There are two possible explanations for the effect of fiber pre-tension on the interfacial shear strength (or on the number of fragments at the saturation limit): (i) a de facto bias in the statistics of fiber strength because more fibers break prior to being embedded in the resin. Thus only the stronger fibers survive and they yield less fragments at saturation; (ii) higher pre-tension weights induce more strain energy in the fiber. When the fiber breaks, the released energy is dissipated in various ways, including the formation of interfacial debonding zones, which are larger if more energy is released, that is, if more pre-tension is applied. This means that the _active_ fragment lengths are shorter than the observed break-to-break distance, and therefore a lower number of breaks occur subsequently, yielding less fragments at saturation. This second explanation may not be relevant in the case where interfacial debonding does not occur, such as for the PRD-172 fiber. This fiber pre-tensionning effect, if proved correct, is of importance since it may represent one cause for the discrepancies often observed between the results from SFC test data produced by different laboratories using identical materials and preparation procedures. More work is needed to clarify this point.

IV-CONCLUSIONS

We have presented selected results of an exercise conducted with various single fiber composite systems tested by means of a continuously monitored fragmentation procedure. Detailed results are available in a separate article /3/. We have shown that the proposed experiment may be used with a reasonable degree of accuracy as a simple alternative means to the extensive testing of single fibers at various gauge lengths, as usually done, for the following purposes: (i) measuring the size effect in single fibers, (ii) predicting the Weibull shape and scale parameters for single fiber strength, (iii) calculating the fiber/matrix interfacial shear strength from the extrapolated value of fiber strength. The accuracy of the results obtained for the scale parameter with the present method depends strongly on the values of the fiber and matrix Young's moduli, which should be accurately determined for each fiber. A possible effect of fiber pre-tensionning on fragmentation results (including the value of the interfacial shear strength) was observed and briefly discussed, but this aspect definitely deserves more research.

This work was supported by the U. S. - Israel Binational Science Foundation. H. D. Wagner is the Incumbent of the J. and A. Laniado Career Development Chair.

REFERENCES

1. Kelly, A. Tyson, W. R., Journal of the Mechanics and Physics of Solids, 13 (1965) 329.

2. Wagner, H. D., Eitan, A., Applied Physics Letters, (1990) in press.

3. Yavin, B., Gallis, H. E., Scherf, J., Eitan, A., Wagner, H. D., (1990) submitted.

4. Netravali, A. N., Schwartz, P., Phoenix, S. L., Polymer Composites, 10 (1989) 385.

5. Wagner, H. D., Journal of Polymer Science - Polymer Physics, 27 (1989) 115.

6. Wagner, H. D., Steenbakkers, L. W., Journal of Materials Science, 24 (1989), 3956.

7. Verpoest, I., Desaeger, M., private communication.

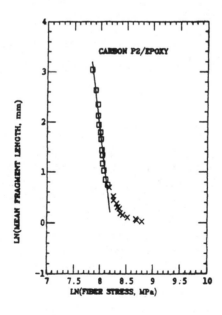

Figure 1: Ln-ln plot of the mean fragment length against fiber stress for carbon LM/epoxy. The least-square regression line through the data described by an open square symbol only gives r^2 = 0.98.

PREPARATION AND MECHANICAL PROPERTIES OF ARAMID FIBRES FROM BLOCK COPOLYMERS

B. HELGEE, C.-H. ANDERSSON*

Dept of Polymer Technology - Chalmers University of Technology
41296 Göteborg - Sweden
Swedish Institute for Textile Research
PO Box 5402 - 40229 Göteborg - Sweden

ABSTRACT

Aramid block copolymers containing stiff and flexible segments have been synthesized, spun into fibres and subjected to tensile testing. Both the Young's modulus and the tensile strength of the fibres decreases with increasing length of the flexible segment while the elongation at break increases. As spun fibres show a well defined yield point. Heat treatment of the fibres up to 500 °C results in a 2-3 times increase in Young's modulus while the elongation at break decreases. The fibre containing the longest flexible blocks becomes very brittle upon heat treatment at 500 °C.

INTRODUCTION

In the present paper a study of stiff chain aromatic polyamides containing flexible blocks will be reported. Block copolymers with stiff and flexible segments were synthesized and the length of the flexible segments was varied. The polymers were spun into fibres according to the dry jet wet-spinning technique. As spun as well as heat treated fibres have been evaluated with tensile tests and fractographic analysis.

The properties of stiff chain aramid fibres have their origin in the chemical structure of the polymers and the morphology of the fibres. When aramid fibres are spun from liquid crystalline solutions the resulting fibre develops a high degree of orientation and exceptional mechanical properties /1-3/. Morphological studies of this type of aramid fibres revealed a pleated lamellar structure with crystallites oriented along the fibre axis /4-6/. The interactions between fibrils and lamellas are of hydrogen bond character /7/. The

555

properties of stiff chain aramid fibres include high modulus/high strength tensile behaviour /1-9/, low creep and stress relaxation rates and some strain hardening beyond a well defined yield point /7,8/. In axial compression the fibre deforms plastically due to microfibrillar buckling /1-3/ while in transverse compression they behave like conventional syntetic textile fibres /9/.

EXPERIMENTAL

All polymers used in this report for the spinning of fibres were synthesized by a solution polymerization technique. Stoichiometric amounts of terephthaloyl chloride (TPC) and the diamines 4,4'-diaminobenzanilide (DABA), for the ridgid blocks, and 4,4'-diaminodiphenylsulfone (DDS) for the flexible blocks were used. All stiff polymers contained only TPC and DABA, a system earlier studied by Preston et.al. /10,11/.

Polymers with flexible blocks were prepared as block copolymers in a two step procedure. For polymer data see Table 1.

Fibre spinning

The spinning of fibres from the polymers was performed with a small scale wet spinning unit from Bradford University Research Ltd. The spinning solutions were 14 % polymer in concentrated sulfuric acid. All these solutions showed the stir opalescence typical of liquid crystalline solutions. The spinning unit was equiped with a single hole, 50 μm diameter, spinneret. The spinning was performed in the dry-jet wet spinning mode. After drying some fibres were heat treated at either 370 °C or 500 °C for approximately 3 s and under very low tension. Fibre diameters were in the range 25 - 50 μm with little variation in each individual run.

Tensile testing

The strength and Young's modulus of individual fibres were measured using a model 1122 Instron testing machine equipped with either a 5 N or a 20 N load cell and pneumatic grips. The fibres were mounted for testing with drops of epoxy adhesive to two L-shaped pieces of card held together by a spring clip, Figure 1, /12/. After testing, the fibre diameter was measured using a Watson split image micrometer.

Fibre samples were tested with test lengths: 10, 50, and 100 mm. Each sample consisted of 10 individual fibre tests. A crosshead speed of 5 mm/min was used.

For the evaluation of the breaking strain, the yield point and the Young's modulus, the machine and fibre mounting contributions to the measured strain have to be eliminated. However, these are not depending on the gauge length, thus:

556

$$\delta_m(L,F) = \delta_f(L,F) + \delta_o(F) \qquad (1)$$

where $\delta_m(L,F)$ is the recorded elongation, $\delta_f(L,F)$ is the elongation of the fibre specimen and $\delta_o(F)$ is the contribution from other sources to the measured elongation.

For the evaluation of Young's modulus one of the following matematically equivalent expressions can be used.

$$\delta_m(L,F) = \frac{\sigma}{E_f} L + \delta_o(F) \qquad (2)$$

or

$$\frac{1}{E_m} = \frac{1}{E_f} + \frac{\delta_o(F)}{\sigma\ L} \qquad (3)$$

Where σ is the applied stress, E_f the Young's modulus of the fibre, E_m the appearent Young's modulus and L the specimen length.

RESULTS AND DISCUSSION

The results from tensile tests are summarized in Tables 2 and 3 and typical fracture surfaces are shown in Figures 2 - 4. In the as spun state all fibres exibited a distinct yield point and two regions with differing Young's moduli termed E_1 and E_2.

For as spun fibres it is evident from Table 2 that the incorporation of flexible segments into the polymer chains results in a decrease in both the Young's modulus and the strength.

The influence of heat treatment of the fibres is shown in Table 3. Heat treatment of all stiff fibres results in an increase in the modulus with increasing temperature followed by a subsequent decrease in elongation at break. The strength of the fibres is unaffected by this treatment.

In Table 3 it is also shown that the introduction of flexible blocks in the polymers results in a decrease in both the modulus and the strength of the fibres. The decrease in modulus is dependent upon the length and the amount of flexible blocks. This is in contrast to the as spun fibres.

Heat treatment of fibre IV at 500 °C produced very brittle fibres that could not be collected on the take up reel without breaking. The elongation at break of this fibre is also comparatively low.

For fibres heat treated at 370 °C a well defined yield point could be detemined.

557

The mechanical properties of the fibres can be discussed in the terms of a combination of the microcomposite model /13/ and the chain-end model /14/. In this case however, these models have to be modified for microparacrystallinity /15,16/.

ACKNOWLEDGEMENTS

The authors gratefully acknowledge Mrs Amri Larsson for skilled assistence with electron microscopy. Financial support from The Swedish National Board for Technical Development (to B.H.) is gratefully acknowledged.

REFERENCES

1. DeTeresa S.J., Porter R.S., Farris R.J., J. Mater. Sci., 23 (1988) 1886-1894
2. Allen S.R., J. Mater. Sci., 22 (1987) 853-859
3. Dobb M.G, Johnson D.J. and Saville B.P., Polymer, 22 (1981) 960-965
4. Li L.-S., Allard L.F. and Bigelow W.C., J. Macromol. Sci.-Phys., B22 (1983) 269-290
5. Morgan R.J., Pruneda C.O. amd Steele W.J., J. Polym. Sci., Polym. Phys. Ed., 21 (1983) 1757-1783
6. Wagner H.D., J. Mater. Sci. Lett., 5 (1986) 439-440
7. Ericksen R.H., Polymer, 26 (1985) 733-746
8. Northolt M.G. and v. d. Hout R., Polymer, 26 (1985) 310-316
9. Skelton J., J. Textile Inst., 56 (1965) T454-T464
10. Preston J., Smith R.W. Black W.B. and Tolbert T.L., J. Polym. Sci. Part C, 22 (1969) 855-865
11. Preston J, Kriegbaum W.R. and Kotek R., J. Polym. Sci. Polym. Chem. Ed., 20 (1982) 3241-3249
12. Andersson C.-H. and Warren R., in "Advances in Composite Materials", (Bunsell A. et. al., eds.) (1980) 1129-1130,AMAC - Pergamon, Oxford.
13. Knoff W.F., J. Mater. Sci. Lett., 5 (1987) 1392-1394
14. Smith P. and Termonia Y., Polymer Commun., 30 (1989) 66-68
15. Hindeleh A.M. and Hosemann R., J. Phys. C, 21 (1988) 4155-4170
16. Hindeleh A.M. and Abdo Sh.M., Polymer,30 (1989) 218-224

Table 1. Polymer data.

Fibre nr	Inherent viscosity /dLg	Amt. flex. block /%	Block length[a] stiff	flex.
I	9.3	0		
II	3.26	22	49	27
III	4.16	28	49	41
IV	3.80	50	49	99

[a] Theoretical degree of polymerization based on stoichiometry.

Table 2. Tensile properties of as spun fibres.

Fibre nr	Flex. block /%	E_1 /GPa	E_2 /GPa	Yield point σ /GPa	/%	σ_B /GPa	ε_B /%
I	0	52	40	0.43	0.62	1.87	4.2
II	22	24	6	0.26	0.9	0.61	5.6
III	28	28	13	0.29	0.9	0.86	4.1
IV	50	24	11	0.34	1.8	0.69	5.5

Table 3. Tensile properties of heat treated fibres.

Fibre nr /temp/°C	Flex. block/%	E_1 /GPa	E_2 /GPa	Yield point σ /GPa	ε /%	σ_B /GPa	ε_B /%
I/370	0	65	–	–	–	1.82	2.5
I/500	0	105	–	–	–	1.88	0.7
II/500	22	65	–	–	–	0.64	1.9
III/370	28	40	32	0.30	0.5	0.93	2.1
IV/500	50	54	–	–	–	0.75	0.5

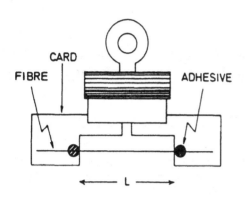

Figure 1. Fibre mounted for tensile testing.

Figure 2. Fracture surface of as spun fibre I.

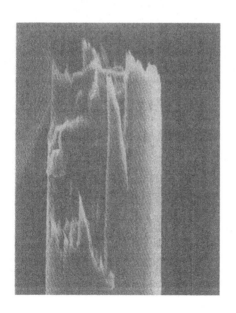

Figure 3. Fracture surface of as spun fibre III.

Figure 4. Fracture surface of as spun fibre IV.

STRENGTH AND MICROSTRUCTURAL ANALYSIS
OF PRD 166 FIBRE

V. LAVASTE, M.H. BERGER, A.R. BUNSELL

Ecole des Mines - Centre des Matériaux
BP 87 - 91003 Evry - France

ABSTRACT

Fibre PRD 166 (Du Pont de Nemours, U.S.A.) is a new continuous ceramic fibre candidate for high temperature reinforced composite materials. The fibre is polycrystalline consisting of α-alumina and zirconia. A transmission electron microscopy study has revealed three grains sizes and the results of X-ray diffraction show that the bigger α-alumina grains have a tendancy to be oriented perpendicular to the axis of the fibre. An intergranular phase has also been identified which can attain a thickness of 10 Å.

INTRODUCTION

One of the first ceramic fibres proposed as reinforcement for metal matrix composites was the fibre FP developed by Du Pont de Nemours (U.S.A.). Fibre FP is a continuous polycristalline filament with a greater than 99% purity of α-alumina /1/. This results in a very high Young's Modulus, 380 GPa, approaching that of bulk α-alumina. The polycristalline grains of the fibre have an average size of about 0,5 µm which gives it a rough surface. These are the causes of the low tensile strength (1,6 GPa with a 0,64 cm gauge length) and strain (0,4 % with a 25 cm gauge length).

There is now another fibre produced by the Du Pont Company called PRD 166. This is also a continuous polycristalline filament but consisting of 20 % of zirconia and 80 % of α-alumina. This fibre has been developed to overcome the disadvantages of fibre FP due to its brittleness. The manufacturer has announced an increase in tensile strength of 50 % and therefore the strain could reach 0,6 % /2/. The PRD 166 fibre is an interesting candidate for reinforcing high temperature materials.

This paper deals with the microstructural caracterisation of the fibre PRD 166 and with the implications of grain size and orientation on its tensile behaviour.

I - EXPERIMENTAL PROCEDURE

A PHILIPS EM 501 scanning electron microscope has been used to observe the general aspect and dimensions of the fibre. Before being observed fibres are cut into very small parts (< 5 mm length) and metallized with Au-Pd to avoid charging.

The identification of different phases has been performed by X-ray diffraction on a computerised SIEMENS D 500 equipped with a Co $K\alpha_1 K\alpha_2$ (40 kV, 20 mA, iron filter) and a ELPHYSE linear proportional counter filter detector. The specimens have been examined in both the reflection and transmission modes. The samples studied are fibres reduced to a powder by grinding and aligned fibres coated in an epoxy resine and polished until 150 μm thickness for transmission.

Transmission electron microscopy allows the crystallinity to be studied, the nature of phases identified, as well as the study of grain size distribution. A PHILIPS EM 430 300 kV electron microscope equipped with a TRACOR EDX X-ray selective energy analyser was used. To prepare thin foils, fibres were coated in an epoxy resin and mechanically thinned down to a thickness of 50 μm. Electron transparency is reached by ion milling at an angle of bombardement of 15° for ten hours then 10° for one hour.

II - RESULTS

2.1. General aspect of the fibre

The PRD 166 fibres supplied by Du Pont were very homogeneous. They presented a regular external shape and had an apparently circular cross section of about 17 μm diameter (fig. 1). The fibre surfaces were extremely rough typical of a strongly crystallised sample. The grain size seems to have been about 0,5 μm. These grains presented no special orientation on the surface and had a prismatic form.

2.2. Microstructure of the fibre

2.2.1. Crystalline composition

No glassy phase was detected by X-ray diffraction ; the fibre PRD 166 was highly crystalline and the only phases detected were those of alumina in the alpha phase and tetragonal zirconia. A closer study between 31° and 38° (2θ) (the field where the ($\bar{1}$11) and (111) plans of monoclinic zirconia diffract) confirmed the absence of the monoclinic phase at this detection scale /3/. Using the X-ray analyser coupled with the TEM the presence of yttrium has been detected within the zirconia grains. Yttria-oxide has evidently been introduced in very small quantities (< 2 % mol) in order to stabilize the zirconia in the tetragonal phase.

2.2.2. Crystalline organisation

Transmission electron microscopy has revealed that the structure of the PRD 166 fibre was more complex and very different from most frequently studied ceramic samples. Indeed fig.2 and fig.3 show the superposition of grains in a region having a thickness of about 1000 Å and a large dispersion of the grain size. However, the fibre consists of large prismatic grains of α-alumina in the range 0,5 μm to 1 μm, smaller prismatic zirconia grains (0,1 μm to 0,5 μm) which appear very dark, and these are bathed in a multitude of very small (~ 20 nm) α-alumina cristallites.

In the very thin areas of the thin foil an intergranular phase of about 10 Å can be seen (fig.4a). Fig.4b exhibits a very thin grain boundary. The white fringe is the image of a potential discontinuity between the two neighbouring grains. The thickness is about 5 Å and it is not realistic to consider the boundary as an amorphous phase at this scale.

Generally, zirconia grains are not twinned because of yttria-oxide which stabilizes these grains in the tetragonal phase. Nevertheless, a few twinned grains have been performed. The fig.5 shows twinning observed on a small zirconia grain of about 25 nm.

X-ray diffraction gives results about the texture of the fibre. The comparison between the relative intensities of rays from the (110) and (116) diffraction planes of α-alumina in reflection and in transmission shows that the [001] axes of α-alumina grains have a tendancy to be oriented perpendicular to the axis of the fibre. The orientation affects first the big alumina grains in the range 0.5 μm to 1 μm because these grains grow inside the fibre during the thermal treatment during the elaboration stage. The small crystallites of about 20 nm should not be concerned.

III - DISCUSSION

The strength of strong ceramics depends primarily on the microstructure developed during firing - i.e. on the sizes and shapes of the grains formed from the powder particules and the strength and continuity of the bond between these grain /4/.

The big grains included in a multitude of small crystallites are a source of flaws. The interface between their large facets and surrounding materials can be considered as a flaw according to the Griffith theory /5/. An important stress concentration takes place at this interface due to the elastic anisotropy and depends on the grain size and the shape of the crystal boundary. For a grain size d and radius of curvature ρ the concentrated stress σ produced when a stress σ_a is applied will be :

$$\sigma \approx \sigma_a (\Delta E/E) (d/\rho)^{1/2}$$

ΔE is the difference in the elastic stiffness E of neighbouring crystals in different direction. Now, the big α-alumina grains are oriented (the C axis of the hexagonal unit cell has a tendency to be perpendicular to the axis of the fibre) ; therefore, they exhibit a large surface in the plane which is perpendicular to the axis of the fibre and this can limit the strength of the fibre in this direction.

Another cause of the reduction of tensile strength of the fibre is the intergranular phase. One reason is that the atoms in the boundary region are far from the equilibrium positions they have in the grain and so are unable to form bonds as strong as those they form in a perfect crystal or in a grain boundary.

CONCLUSION

The fibre PRD 166 consists of α-alumina and tetragonal partially stabilized (with Y_2O_3) zirconia. Three grains sizes have been observed : large prismatic grains (0,5 μm to 1 μm) and very small crystallites (about 20 nm) of α-alumina and zirconia grains (in the range 0,1 μm to 0,5 μm). Bigger alumina grains have a tendency to be oriented : the [001] axis of the hexagonal unit cell of α-alumina was found to be perpendicular to the axis of the fibre. An intergranular phase has been seen between the small grains and it was observed to be very thin (about 10 Å).

The presence of the large oriented α-alumina grains and a fine intergranular phase can be expected to have a major influence on the tensile properties and in particular the strength of the fibre.

REFERENCES

1 - Romine J.C., New high temperature ceramic fibers. Ceramic Engineering and Science Proceedings, Vol.8, n°7-8, (July-August 1987).
2 - Dhingra A.K., Alumina fiber FP. Phil. Trans. R. Soc. Lond. A294, 411-417, (1980).
3 - Thorel A., Influence de la température sur les propriétés d'une zircone partiellement stabilisée. Mémoire C.N.A.M. soutenu le 2 juillet 1984.
4 - Kelly A., Macmillan N.H., Strong solids. Oxford Science Publications, Clarendon Press, Oxford (1986).
5 - Griffith A.A., Phil. Trans. R. Soc. A221, 163, (1920).

Fig. 1 :
S.E.M. photograph of the aspect of the PRD 166 fibre.
(regular form, circular cross-section, rough surface).

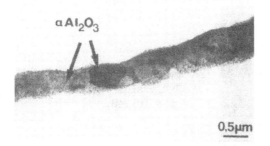

Fig. 2 :
T.E.M. photograph of general aspect of a thin section of the fibre showing large grains and microcrystallites of α-Al_2O_3.

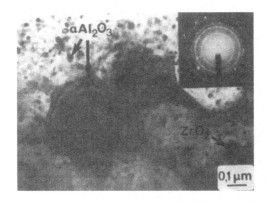

Fig. 3 :
T.E.M. photograph of the microstructure of the fibre.

Fig. 4 :
T.E.M. photograph of grains separated by an intergranular phase
of about 10 Å.

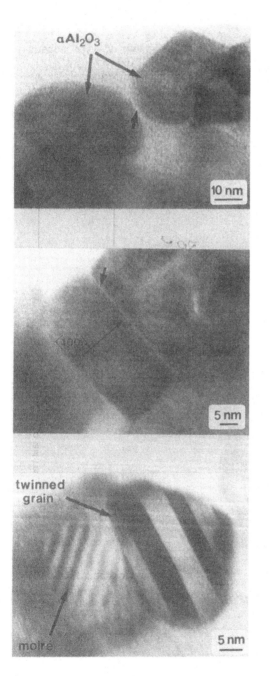

Fig.4a : α-Al_2O_3 grains

Fig. 4b :
ZrO_2 grains : {100} planes
are imaged.

Fig. 5 : ZrO_2 twinned grain.

THE CHARACTERISATION OF POLYSILANES AND POLYCARBOSILANES FOR THE PRODUCTION OF SILICON CARBIDE FIBRES

R.J.P. EMSLEY, J.H. SHARP

University of Sheffield - School of Materials
Northumberland Road - S10 2TZ Sheffield - Great Britain

ABSTRACT

A range of polysilane and polycarbosilane copolymers has been synthesised as potential precursors for the fabrication of silicon carbide fibres for use in high temperature composites. The molecular weight distribution of the polymers has been determined by gel permeation chromotagraphy and shown to depend on the molecular structure in the copolymer. Thermogravimetry gave information about the temperatures at which loss in mass occurred and the ultimate ceramic yield, which was variable. Some of the polymers have been spun into fibres and factors affecting their ease of spinning are discussed.

INTRODUCTION

Improved ceramic fibres are urgently required for the reinforcement of both metal matrices and ceramic matrices. The present generation of continuous ceramic fibres do not meet the target properties set by engineers for the most demanding applications, partciularly in the aerospace industry. A new fibre retaining good mechanical properties at temperatures above 1000°C would have a large potential market and would be an important commercial development.

The currently available ceramic fibres for the reinforcement of advanced composites have been reviewed recently /1/ and found wanting compared with the properties required. It seems likely that fibres with improved thermal and mechanical properties will be produced from the group of covalently-bonded, light-weight non-oxide ceramic systems. Of these the most developed system is currently silicon carbide, although other carbides and also nitrides and borides are potential systems.

Silicon carbide fibres can be made either by chemical vapour deposition or by the pyrolysis of polymeric precursors. The former route inevitably leads to rather thick fibres, usually in excess of 100 μm, which are difficult to incorporate into ceramic matrices. Finer fibres of 10-20 μm thickness can be made by the latter route, which was pioneered by Yajima and co-workers /2,3/. Their approach has led to two commercial products, Nicalon and Tyranno, which are exploited in both metal matrix and ceramic matrix composites.

These fibres do, however, have disadvantages as high performance fibres at elevated temperatures. To improve upon them requires (a) reduction in the oxygen content, (b) reduction in the carbon content, and (c) maintenance of hot strength. Nicalon fibres are non-stoichiometric with a C:Si ratio greater than one and a significant oxygen content. They exhibit pronounced grain growth and a corresponding drop in strength at temperatures above 1100°C. These disadvantages stem from the processing of the polymeric intermediates. It is highly desirable to eliminate the oxidative cross-linking process and to synthesise a polymer with a 1:1 Si:C ratio. Maintenance of hot strength will then depend on production of a fibre which is either amorphous or microcrystalline. In the latter case small grain size must be maintained, perhaps by incorporation of a grain growth inhibitor.

It is believed /4/ that the ceramic yield of preceramic polymers can be improved by the incorporation of either unsaturated groups or ring species, such as vinyl and phenyl species. Yield should also be increased if the polymer contains branched structures and also if the metal atom is incorporated into the polymer backbone rather than in the side groups.

RESULTS AND DISCUSSION

To examine this hypothesis, a range of polysilanes and polycarbosilanes has been synthesised by the alkali metal dechlorination of a mixture of chlorinated silane monomers in an inert solvent. Two systems have been used, sodium in o-xylene and potassium in tetrahydrofuran (THF). The effect of changing side groups and increased networking in the polymer has been examined and their influence on molecular distribution, ceramic yield and spinnability observed. Some of these properties have been seen to be inter-related.

The polymers have been characterised by gel permeation chromatography (GPC) to determine the molecular weight distribution. Polymeric structure was ascertained by infra-red (IR), ultra-violet (UV) and nuclear magnetic resonance (NMR) spectroscopy and ceramic yield was determined by thermogravimetry (TG) up to 1500°C.

The use of difunctional monomers such as dichloromethylphenyl-silane, $(C_6H_5)(CH_3)SiCl_2$ (I), or a mixture of I with dichloro-dimethylsilane, $(CH_3)_2SiCl_2$ (II), results in a linear polysilane product in which silicon atoms are bonded together in the polymer backbone. As discussed below these polymers have been shown to give poor ceramic yields, whereas polymers having branched structures, particularly those in which silicon is bonded directly to carbon in the main chain of the polmer have greatly improved ceramic yields.

The degree of branching which takes place is dependent on the relative proportions and functionalities of the various monomers used, and may be expressed in terms of the molar functionality, F, of the copolymer /4/ according to the formula:

$$F = \frac{x_1 f_1 + x_2 f_2 \cdots + x_n f_n}{x_1 + x_2 \cdots + x_n}$$

where x_n is the number of moles of the n th monmer and f_n is the functionality of the n th monomer. A mixture of monomers each having f=2 leads to the formation of a linear copolymer. The incorporation of monomers with f>2 results in branching of the polymer chain. A mixture of monomers having a value of F≥3 leads to a networked polymer which will be insoluble and infusible. A mixture with F=2.0 should give on average a linear structure. Values of F between 2 and 3 lead to a mixture of three types of polymerisation products:
  (a) solid products which are fully cross-linked and are, therefore, insoluble and not thermoformable
  (b) solid products which are soluble in toluene, xylene and THF, and have molecular weights > 1000
  (c) liquid products with molecular weight < 1000.
Only those products in group (b) are suitable for spinning into fibres. The data reported below relate only to this fraction unless otherwise specified.

GPC measurements, such as those shown in Fig. 1, were used to determine molecular weight distributions, which varied from 1000 up to a maximum of 335,000. The distribution sometimes showed only a single peak, but frequently bimodal or even trimodal distributions were observed (Fig. 1). No definite dependence of the molecular weight distribution on the value of F was seen, both linear (curves A and C) and networked systems (curves B and D) gave a wide range of products. The molecular weight distribution was, however, found to be more sensitive to the type of substituent groups present in the monomers. A 1:1 copolymer of I and II gave high molecular weight material in large quantities (Fig. 1A), whereas another 1:1 copolymer of II with methyl-2-(3-cyclohexenyl)ethyldichlorosilane, $C_6H_8.CH_2.CH_2.Si(CH_3)Cl_2$ (III), gave mostly oligomers, that is group (c) products. The GPC curve of the group (b) product from this polymerisation is shown in Fig. 1C, which indicates that although some high molecular weight material was obtained, most of it had a M.W. of around 3000. This change in molecular weight distribution must be entirely due to the replacement of the phenyl group in polymer A by the ethylcyclohexenyl (ECH) group in polymer C. It seems that the large steric bulk of the phenyl group promotes linearity in the polymer, whereas the larger but less hindering ECH group allows the formation of small cyclic silane species in preference to the promotion of high molecular weight linear polymer.

The GPC curves of two networked polymers are shown in Fig. 1, curves B and D. When the appropriate amount of dichloromethyldimethylchlorosilane, $(CH_3)_2ClSiCHCl_2$ (IV), was added to a mixture of monomers II and III to give a copolymer with F = 2.5 (system B), an appreciable yield of products with high molecular weights in excess of 50,000 was obtained (Fig. 1B). Another copolymer with an F value of 2.5 was prepared from a mixture of monomers I, II and IV and gave rather more low molecular weight product (Fig. 1D). Although the reasons for this are not as clear as in the case of the linear polymers it may be that the phenyl group in some way hinders branching and so restricts the size to which polymer molecules can grow, but again in the case of the ECH group there is less steric hindrance so polymer growth can take place relatively unhindered.

Thermogravimetry of polymers A and E, both linear polysilanes, showed a dramatic loss in mass at temperatures above $350^{o}C$ (e.g. Fig. 2E, which relates to the homopolymer of II). This is ascribed to the loss of organic substituents accompanied by the loss of volatile silicon species from the backbone of the polymer. The evolution of the latter species is due to the relative ease with which the Si-Si bond is cleaved, compared with the energy required to break either the Si-C or C-C bonds. The volatile silanes formed by this process are suggested by Sinclair /5/ to be in the form of $Si(CH_3)_4$ or similar molecules or even in the form of 6-membered silane rings. Such reversion reactions are particularly bad in linear polysilanes where chain scission and recombination to volatile species is favoured.

Networked polymers have been synthesised in an attempt to overcome this problem by pinning the silicon atoms securely into the structure by utilising 3 or even 4 bonds of an individual silicon atom, instead of the usual 2 bonds in a linear polymer. The TG curves of two such polycarbosilanes with F = 2.5 are shown in Fig. 2, curves D and F. The incorporation of Si-C bonds increased the thermal stability of the polymer and retention of the silicon content. In copolymer D, the addition of monomer IV, which is trifunctional, yielded a networked polymer, as well as introducing Si-C bonds, hence its relatively high thermal stability. However, the degree of pinning of the silicon is not great and consequently gives rise to a ceramic yield of only 30%. In polymer F there is the same degree of networking (F = 2.5) and increased pinning due to the use of a vinylchlorosilane. The ceramic yield increased to 55% (Fig. 2F). An even higher ceramic yield was obtained from a commercial Nicalon precursor copolymer (Fig 2N), in accordance with Yajima's proposed structure /2/, which is highly networked and contains a large number of Si-C bonds.

The spinning properties of the polymer systems investigated have differed widely; their behaviour has been shown to be closely related to the molecular weight distribution and the structure of the polymer. The linear polysilane A may be easily melt spun into fine diameter fibres (<30 μm). However, the corresponding linear polysilane C cannot be spun into fibres as it is a viscous liquid at ambient temperature. The difference in the behaviour of these apparently similar polymers can be seen to be due to the wide molecular weight distribution of polymer A which includes a large quantity of high molecular weight material. By comparison, polymer C lacks the large quantity of high and intermediate molecular weight material which would make it suitable for spinning.

The fibres produced from polymer A are delicate by comparison with polypropylene, but less so than those from the Nicalon precursor or from vinyl-containing monomers which are highly networked. In both cases, these fibres are so brittle that they have to be classed as very delicate and handled with great care. Some polymers produced from vinyl monomers have such a high molecular weight and are so highly cross-linked as to be infusible, but are still soluble in solvents such as THF. For these materials it may be that dry spinning from a suitable solvent is one way that fibres can be made. Dry spinning from solution is, however, a complicated process that can be hazardous because of the use of inflammable solvents.

CONCLUSIONS

The synthesis and evaluation of several possible preceramic polymers has been undertaken. They have been characterised with respect to molecular weight distribution, polymer structure, ceramic yield and spinnability. While the ceramic yield increased, the ease of spinning into fibres decreased with increased extent of networking in the poymer. The molecular weight distribution showed no such obvious trend, but was more sensitive to the type of substituent side group incorpoated into the copolymer by judicious selection of monomers. Polymers with F = 2.5 had a high degree of networking and usually formed soluble solid products in good yield. Their thermo-formability was, however, closely related to their molecular weight distribution; the higher the molecular weight the more likely it was that a particular polymer was found to be non-thermoformable and so decomposed before melting. Those networked polymers which could be melt spun produced extremely brittle fibres making further handling difficult. Linear polymers, on the other hand, yielded fibres which could be easily spun but had extremely low ceramic yields, too low to be able to maintain fibre integrity during pyrolysis.

ACKNOWLEDGEMENTS

We are grateful to Mr.P. Rowlands for assistance with some of the experimental work, Dr. H. Rudd of Harwell for supplying the Nicalon precursor polymer, and SERC for the award of a studentship to RJPE.

REFERENCES

1 - Emsley, R.J.P., Sharp, J.H. and Bailey, J.E., Brit. Ceram. Proc., 45 (1990) 139-151
2 - Yajima, S., Phil. Trans. Roy. Soc. London A, 294 (1980) 419-426
3 - Yajima, S., Hasegawa, Y., Hayashi, J. and Iimura, I., J. Materials Sci., 13 (1978) 2569-2576
4 - Wynne, K.J. and Rice, R.W., Ann. Rev. Mat. Sci., 14 (1984) 297
5 - Sinclair, R.A., in "Ultrastructure Processing of Ceramics, Glasses and Composites" (L.L. Hench and D.R. Ulrich, ed.) Wiley-Interscience, New York (1984) 256-264

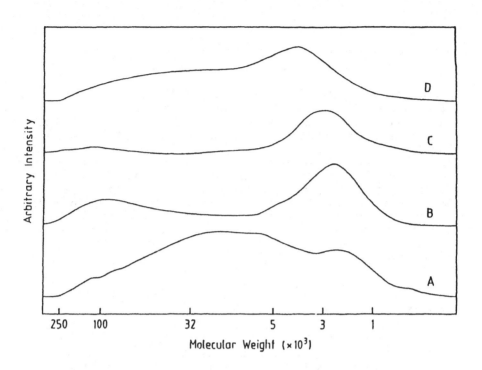

Fig.1 - GPC curves for copolymers A-D (see text)

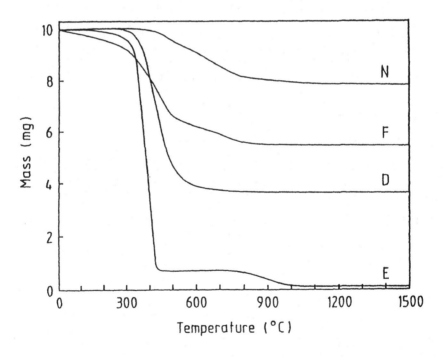

Fig. 2 - TG curves obtained on 10 mg samples of copolymers D, E, F and N (see text), heated at 10°C/min. in nitrogen

FIBERAMIC [R] : A NEW SILICON CARBONITRIDE CERAMIC FIBER WITH HIGH THERMAL STABILITY

G. PEREZ, O. CAIX

Rhône Poulenc Recherches
Centre de Recherches des Carrières - 85 Av. des Frères Perret
BP 62 - 69192 Saint-Fons Cedex - France

Abstrac

RHONE-POULENC is developing a new silicon carbonitride fiber named FIBERAMIC ®.
The main feature of this fiber is its good thermomechanical behavior up to 1300°C - 1400°C.
With a homogeneous amorphous structure based on a continuum of silicon-centered tetrahedrons, this fiber can be used at elevated temperatures because of its structural and chemical stability.
Mechanical and thermomechanical investigations made with FIBERAMIC have demonstrated good high temperature durability such as a retention of 80% of the room temperature strength at 1400°C as well as a high temperature stability after ageing in oxidizing atmosphere. Current efforts are devoted to the production of FIBERAMIC ® at the pilot plant scale and the evaluation of thermomechanical properties of FIBERAMIC ® based composites.

1. Introduction

In recent years, high temperature materials such as ceramic and metal matrix composites are gaining considerable attention for specific components used in space, aeronautics and automotive industries. One of the keys of this development is the avaibility of a stable ceramic fiber, even in oxidizing atmosphere at high temperature. A wide range of ceramic reinforcements have been commercially available such as alumina-silica fiber or silicon oxycarbide fiber. However, they lack oxydative stability at elevated temperatures (above 1000°C - 1200°C) mainly because of their chemical and structural evolutions during heating.

Pure silicon carbide monofilaments can also be prepared by chemical vapor deposition on a carbon substrate but their very high diameters (140μm) don't allow the fiber to be woven into fabrics commonly used. RHONE-POULENC is developing a silicon carbonitride fiber named FIBERAMIC ® which is a continuous textile yarn of 250 or 500 filaments obtained through pyrolysis of a polysilazane precursor. This concept can be readily understood with reference to the synthesis of carbon fibers which are industrially used today. This route permits the production of continuous ceramic filaments with small diameters. The aim of this Paper is to give information on the preparation of FIBERAMIC ® and to describe its structure and mechanical properties.

2. Experimental

Room temperature mechanical properties are measured by single filament test methods using a gauge lengh of 10 mm and an extension rate of 2 mm/min. Strength and diameter correspond to the average from about 30 individual tests. Elastic moduli are determinated by the compliance method. Thermomechanical properties are mesured also by single filament tests method using a gauge length of 25 mm. This equipment was developed in collaboration with Société Européenne de Propulsion. Temperature is controlled along the filament at +/-5°C. Tests are performed under air after a stabilization of the filament in the hot furnace for 3 min.

3. Preparation of FIBERAMIC

Figure 1 shows the four steps necessary to produce FIBERAMIC ®. In the first one, polysilazane precursors are obtained by reacting ammonia or amine with chlorosilanes :

$$2 \equiv Si-Cl + 3NH_3 \longrightarrow \equiv Si-NH-Si \equiv + 2NH_4Cl$$

With this type of polymer we can adjust the average molecular weight in the range of 1,000 to 20,000 in order to select a polymer with good spinnability. Green fibers can be prepared by melt or dry spinning. A continuous green yarn with 250 or 500 filaments is produced. In order to avoid the melting of green fiber during pyrolysis, RHONE-POULENC has patented thermal and chemical curing processes suitable for polysilazanes such as :

$$\equiv Si-H + \equiv Si-CH=CH_2 \xrightarrow{\text{Pt, h}\nu} \equiv Si-CH_2-CH_2-Si \equiv$$

The development of the new curing processes leads to the preparation of ceramic fibers with a low oxygen content as compared to that obtained through oxydization or hydrolysis.

The cured fiber is then converted into ceramic silicon
carbonitride fiber by pyrolyzing in an inert atmosphere in
the range of 1200°C to 1400°C; this produces flexible,
black fibers. Their optimum properties and stability are
related to the densification of the structure (see fig.2).

4. Structure and properties of FIBERAMIC ®

Structural Description

The typical composition of FIBERAMIC ® is Si:57wt%,
N:22wt%, C:13wt% and O:8wt%, in phases it is equivalent to
about only 5 to 10% of free carbon and 90 to 95% of
$SiN_xC_yO_z$. Figure 3 shows X-RAY diffraction patterns of
FIBERAMIC ® after a 5 hour'heat treatment : it remains
amorphous up to 1400°C; therefore no defect appears by
local excess grain growth. In fact, ^{29}Si-NMR measurements
(see fig.4) show that FIBERAMIC ® can be described as a
continuum of silicon-centered tetrahedrons with nitrogen,
carbon and oxygen as first neighbours. The structure is
homogeneous and the free Carbon is organized in small clusters as shown by bright field transmission electronic
microscopy (see fig.5).

Mechanical and thermomechanical properties

Results of mechanical testing on FIBERAMIC ^{R} are as
follows : Figure 6 shows the distribution of diameters; the
average value is about 14 μm and the distribution is narrow. FIBERAMIC ® exhibits typical brittle failure with an
average tensile strength in the range of 1500 - 2200 MPa
and Weibull parameter of about 7 (see fig.7). The failure
is due to small surface defects as shown on figure 8.
Measurement of FIBERAMIC ® toughness, based on Griffith
equation, gives an average k_{1c} value of 2.5 MPa.m$^{-1/2}$ despite an amorphous structure, as compared to monolithic
silicon carbide (4.5 MPa.m$^{-1/2}$). It is established that the
control of the shape and size of surface flaws is the way
to reach high failure strength. Elastic moduli of FIBERAMIC
is about 200 GPa, related to the density and composition of
silicon carbonitride. As expected from its thermal stability, FIBERAMIC ® has an excellent thermomechanical behavior in an oxidizing atmosphere as shown on figure 9,
silicon carbonitride retains 80% of its room temperature
strength when tests are performed at 1400°C. After ageing
in oxidizing atmosphere, FIBERAMIC ® tensile strength decreases slowly with time as shown on figure 10, the fiber
retains more than 85% of the initial stress after 100 hours
at 1000°C or after 10 hours at 1200°C. Those results show
that the chemical and structural stabilities of the amorphous silicon carbonitride fiber up to 1400°C provide interesting thermomechanical properties in oxidizing atmosphere

5. Conclusions

The new ceramic fiber FIBERAMIC ® described in the present paper offers good mechanical properties at room temperature in comparison to the other ceramic reinforcements. Amorphous silicon carbonitride presents very interesting intrinsic properties in terms of thermomechanical behavior up to 1400°C in oxidative atmosphere. The material and the product line present a large potential for further improvement and optimization and should find an interesting place among ceramic, metal or glass composites.
Various grades of FIBERAMIC ® are under development in order to optimize the thermomechanical properties of the fibers and composites.

6. Acknowledgements

The authors wish to thank the team of Mrs OBERLIN (laboratoire M. Mathieu - PAU - France) for its intensive work and microscopic studies as well as Mr TAUTELLE (laboratoire de spectroscopie du solide, Jussieu - PARIS VI France) for the NMR analysis and interesting discussions.

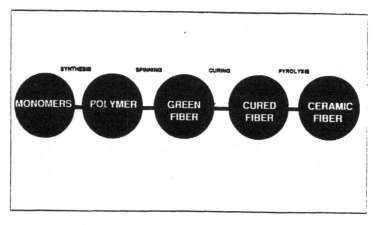

Figure 1 **Preparation of SiNC Fiber**

576

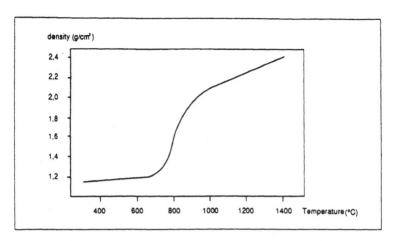

Figure 2 **Density versus temperature**

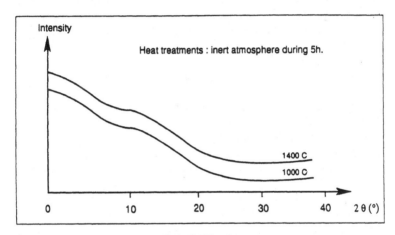

Figure 3 **X-RAY diffraction patterns**

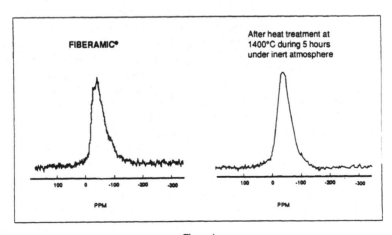

Figure 4
^{29}Si – NMR measurements of FIBERAMIC®

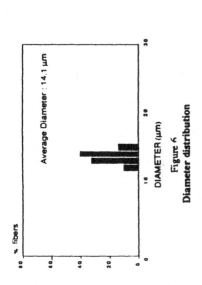

Average Diameter : 14.1 µm

DIAMETER (µm)

Figure 6
Diameter distribution

% fibers

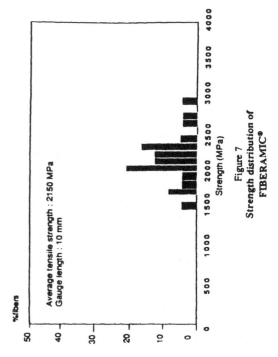

Average tensile strength : 2150 MPa
Gauge length : 10 mm

Strength (MPa)

Figure 7
Strength distribution of
FIBERAMIC®

%fibers

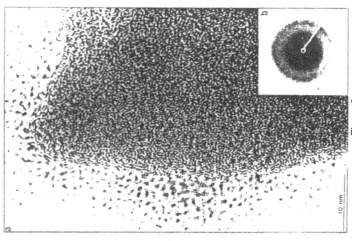

Figure 5
Bright field Transmission Electronic
Microscopy of FIBERAMIC®

10 nm

578

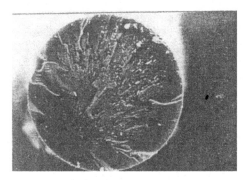

Figure 8
SEM analysis of failure surface

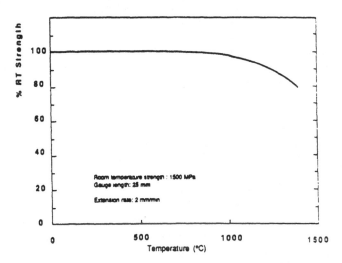

Figure 9
Strength retention versus temperature in air
(after 3 min. of stabilization time)

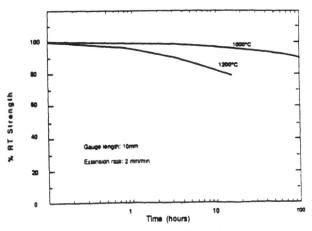

Figure 10
Strength retention after ageing in air
versus time

CONSTITUENTS
KOMPONENTEN

Chairman : **Dr. A.R. BUNSELL**
Ecole Nationale Supérieure
des Mines de Paris

LIMITS TO THE MECHANICAL PROPERTIES OF FIBRES

A. KELLY, N.H. MACMILLAN*

University of Surrey - Guildford GU2 5XH Surrey - Great Britain
*Oregon State University - Dept of Mechanical Engineering
Corvallis - Oregon 97331-6001 - USA

ABSTRACT

The latest theoretical estimates of the possible stiffness and strength of fibres attainable at room temperature are compared with experiment. Some remarks are made on strength at very high temperatures.

Figure 1 compares experimental values (top) with theoretical values for the elastic modulus divided by the density at room temperature for a number of materials. The theory for the elastic modulus of a perfect specimen represents the latest values reviewed by Macmillan 1., and Macmillan and Kelly, 2. and the density has been taken as the defect free X-ray density at room temperature. The experimental values are the largest known to the present authors and the values of the density are experimental ones. The values of density in both cases have been converted to specific gravity taking the value for water to be exactly 1. It can be seen that there is some room for improvement in the stiffness of carbon fibres. The absence of an accurate theoretical value for thin layer platelets of metals, eg Cu:Ni is noteworthy.

Figure 2 compares experimental (top) and theoretical (bottom) values of strength to density (specific gravity). The room for improvement in the strength of silicon carbide is very noteworthy: the more certain value (no query mark) for polyethylene of 19 GPa exceeds the highest experimental value (not shown) of 6 GPa by a factor of about 3.

Of great practical import at present is the question of the possible values of strength of fibres at elevated temperatures up to 2000°C. Figure 3 shows the experimental values for the highest strengths of specimens of alumina Al_2O_3. Strengths of greater than 2 GPa at temperatures in excess of 1600°C are noteworthy. We are carrying out calculations of the temperature dependence of the ideal strength. In all cases the dependence of this quantity on temperature is too slow to account for the experimental values.

REFERENCES

1. N. H. Macmillan, The maximum strength theoretically obtainable from a fibre in a high temperature composite in "Mechanical and Physical Behaviour of Metallic and Ceramic Composites", (S. I. Anderson et al. ed.), (1988) Risø National Laboratories, Denmark.

2. N. H. Macmillan and A. Kelly, Ann. Rev. Mater. Sci. (1990) in press.

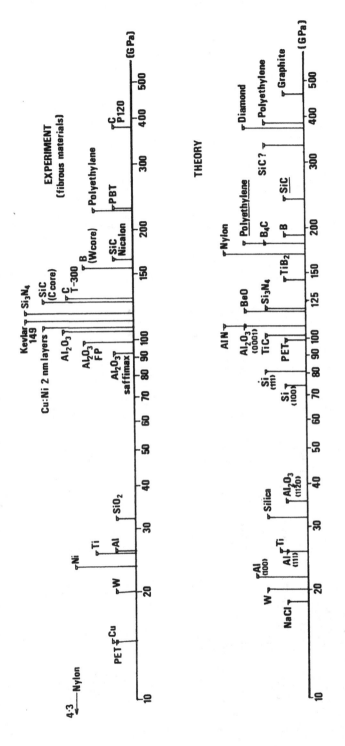

Figure 1. Experimental and theoretical values of elastic modulus divided by specific gravity.

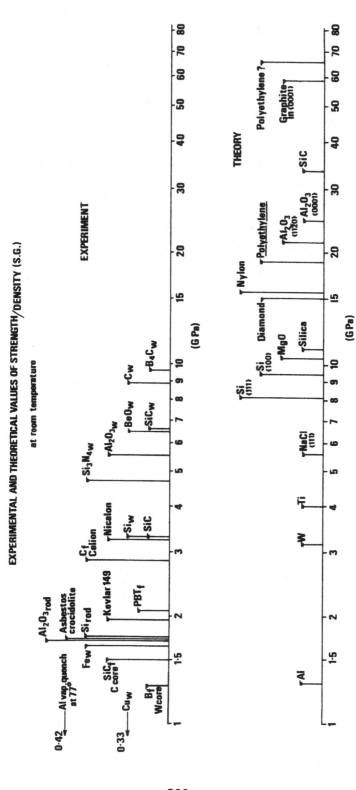

Figure 2. Experimental and theoretical values of breaking strength at room temperature divided by the (experimental) specific gravity

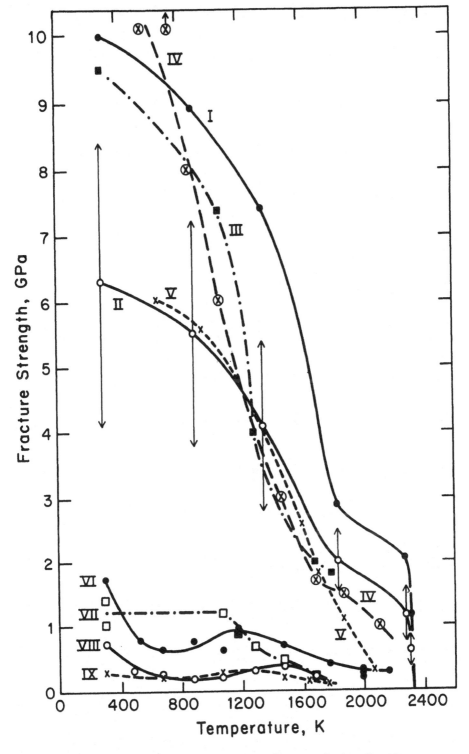

Figure 3. Temperature dependence of the fracture
of various forms of alumina (Al₃0₂)

ALLYLPHENOXYIMIDE MODIFIED BISMALEIMIDES
A NEW SERIES OF TOUGH HIGH TEMPERATURE IMIDE RESINS
FOR ADVANCED COMPOSITES

P. KÖNIG, H. STENZENBERGER, W. RÖMER

Technochemie GmbH Verfahrenstechnik
Gutenbersgstrasse 2 · Postfach 1265 · 6915 Dossenheim · West-Germany

ABSTRACT

New bis[3(2-allylphenoxy)phthalimides] are presented, which can be synthesized by a nucleophilic displacement reaction from bis(3-nitrophthalimides) and the sodium salt of o-allylphenol. These new allylterminated comonomers can easily be melt blended with bismaleimides to provide high temperature resins with excellent toughness and stiffness properties. Investigations are made to improve fracture toughness by modifying BMI / bis(o-propenylphenoxy)benzophenone - and BMI / bis [3(2-allylphenoxy)phthalimide] resins with Poly(etherimide) (Ultem 1000)

INTRODUCTION

Advanced composites based upon thermosetting resins, such as epoxies and bismaleimides, are finding increased use in the aerospace and electrical industries because of their outstanding temperature properties and advantageous processing characteristics. On account of their high cross link density, these matrix resins are brittle leading to a low impact tolerance in composites. To improve the fracture toughness of these resins, it is necessary to reduce the cross link density, however care has to be taken that the bismaleimides' inherent good properties, like the high glass transition temperature, are not adversely affected. Several chemical concepts for improving the fracture toughness of bismaleimides have already been described in literature /1/, such as I. modification of the backbone structure of the BMI-monomers, II. addition of reactive elastomers

/2/, III. linear chain extension of the bismaleimides
via Michael addition /3/, IV. copolymerisation with
allyl- or propenylterminated comonomers /4,5,6/ and a
relatively new approach, V. modification with thermoplastics /7,8/.Excellent results for improving toughness
whilst maintaining the good high temperature properties
are achieved by copolymerisation of the bismaleimides
with allyl- or propenylterminated comonomers and by
modification with nonreactive thermoplastics /9/.
It is believed that the effectiveness of a thermoplastic
in a thermoset resin is dependent in various factors
such as type of thermoplastic, heterophase distribution, interfacial adhesion between the heterophase and
the matrix, the BMI-resin and BMI-resin comonomers, cure
conditions and molecular weight between crosslinks in
the matrix resin. Changes in the backbone chemistry of
the BMI and the BMI-comonomers should make it possible
to influence the morphology/compatability of the cured
system.
In this paper new allylterminated phenoxyphthalimides
are presented, which in combination with bismaleimides
lead to tough high temperature resins.
A comparison is made between BMI/o-propenylphenoxybenzo-
phenone and BMI/bis(o-allylphenoxyphthalimide) resins,
both systems modified with polyaryleneether (Ultem 1000)
for enhanced toughness. Morphological and toughness
differences should provide insight into morphology/
toughness-correlation of thermoplastic modified bismaleimides.

I   EXPERIMENTAL

1.1 Synthesis of Bis[3-(2-allylphenoxy)phthalimides]

The synthesis of bis[3-(2-allylphenoxy)phthalimides] is
in principle a two step reaction. In the first step
bis(3-nitrophthalimides) are synthesized according to
the procedure reported by White et al /10/ and in the
second step the bis(3-nitrophthalimides) are reacted
with the sodium salt of o-allylphenol to the
corresponding bis [3- (2-allylphenoxy) phthalimides]
(Fig.1) /11/.

1.2 General Procedure for the Copolymerization of Bis-
     maleimide with Bis[3-(2-allylphenoxy)phthalimide]
     and Preparation of Copolymer Castings

Bismaleimide resin (BMI), a low melting blend of bismaleimide building blocks, and the
bis(allylphenoxy)phthalimide are melt blended at a
maximum temperature of 150°C by use of a rotary evapora-
tor. The homogeneous clear resin melt is degassed under

vacuum and cast into a parallel epipedic steel mould and cured under a pressure of 5 bars for 1 h at 160°C plus 3 h at 180°C plus 4 h at 210°C. After demoulding the resin casting was postcured for 5 h at 240°C.

1.3 Neat resin specimen preparation of thermoplastic modified Bismaleimide and thermoplastic modified BMI/Allylphenoxy-phthalimide copolymers.

Neat resin test specimens were obtained from plaques prepared by compression moulding of resin powders. Compression moulded specimens were obtained by first combining a 60 % by weight solution of the bismaleimide resin or the BMI/allylphenoxyphthalimide copolymer in methylene chloride with a solution of the thermoplastic in the same solvent. Subsequently methylene chloride was removed to give a residue which was B-staged, ground to a fine powder and then dried to remove last traces of solvent. This material was compression moulded into plaques.
Neat resin flexural properties were determined according to the DIN 53452 specification. Fracture toughness and fracture energies were measured by use of compact tension specimens. Specimen dimensions are: 3.18cm x 3.18cm x 0.65cm. The test conditions were those described in ASTM E399 to determine plain strain fracture toughness.

1.4 Laminate fabrication

Unidirectional prepregs were manufactured via a drum winding process. The roving impregnation bath contained the methylene chloride solution of the respective resin. The fabricated tape was dried on the mandrel to strip off solvent. The dried prepregs were compacted in a vacuum bag at 70°C before use. The prepregs were moulded in a low pressure autoclave moulding process at 190°C and 5 bars.

II RESULTS AND DISCUSSION

Aromatic nucleophilic displacement of activated bishalo and bisnitro compounds with bisphenols has been used extensively to synthesize high molecular weight polyetherketones /12,13/ and polyethersulfones /14/. Polyetherimides like Ultem 1000 were synthesized from bis(nitrophthalimides) /10,15,16/ with bisphenol salts in polar solvents. The nucleophilic displacement of nitro groups in 3-nitrophthalimides by phenoxide anion is a fast and complete reaction at low temperatures.
This reaction type offered the possibility of synthesizing a series of new bis(allylphenoxy)imides by simply reacting bis(3-nitrophthalimides) with the sodium salt

of o-allylphenol to yield bis[3-(2-allylphenoxy)phthal-imides] (Table 1).

However, because of the alkaline reaction conditions, there is the possibility for allyl groups to isomerize to propenyl groups and all of the bis[3-(2-allyl-phenoxy)phthalimides]synthesized in this work therefore contain 4-5 % of bis[3-(2-propenylphenoxy)phthalimides] as evidenced by H-NMR spectroscopy /11/.

The chemical structures and properties of all bis[3-(2-allylphenoxy)phthalimides] synthesized in this work are shown in Table 1. 4,4'bis[3-(2-allylphenoxy)phthalimido] diphenylmethane and 2,4- bis[3-(2-allylphenoxy)phthal-imido]toluene could be synthesized in 75 % yield and high purity. All other compounds required additional purification steps to achieve >95 % HPLC purity.

2.1 Bismaleimide / Bis[3-(2-allyl-phenoxy)phthalimide]-copolymers

Bis(allylphenyl)-compounds are of particular interest because these can be copolymerized with bismaleimides (BMI) in a wide range of molar ratios to provide excellent tough high temperature thermoset resins.

2,4-bis [3-(2-allylphenoxy) phthalimido] toluene (BAPPT) and 4,4'- bis [3-(2-allylphenoxy) phthalimido]diphenyl-methane (BAPPD) were selected for preliminary copoly-merization studies. A eutectic bismaleimide-mixture was melt blended with the respective allylphenoxyphthalimide with different molar ratios and cured.Typical DSC-traces are given for two BMI/BAPPT-mixtures in Figure 2. The resulting plaques were characterized by a series of mechanical tests (Table 2). The BMI/BAPPT systems show very high Tg's for all weight ratios investigated (around 280°C). Also the elastic modulus at 250°C is very high (2.95-3.23 GPa) with no decrease with increasing BAPPT concentration. A dramatic change, however, is seen for the fracture energies (G_{Ic}). BMI/BAPPT ratios around 1:1 by weight, which represent a 2:1 molar BMI/BAPPT ratio, provide fracture energies of around 200 J/m². However, only a 10 % by weight reduction of BAPPT results in dramatic loss of the fracture energy.

The BMI/BAPPD systems show similar interrelations. The 50:50 mixture provides a fracture energy of 180 J/m², whereas for the 60:40 blend the G_{Ic} decreases to around 100 J/m². It is obvious that as with other bis(allyl-phenyl) comonomers the BMI/comonomer-stoichiometry has to be close to 2:1 to maximize toughness.

2.2 Laminate properties

Due to processing reasons the BMI/BAPPT system was

selected for prepreg and laminate fabrication because BAPPT has the lowest melting transition of all the bis (allylphenoxy) phthalimides synthesized. The visco-sity-temperature correlation of this prepreg resin is shown in Figure 3. In Table 3 the mechanical properties of a T800/BMI/BAPPT-laminate are compared to laminates consisting of a BMI/COMPIMIDE TM123-system which is the basis for tough BMI-prepreg formulations and T 800 fib-res with standard epoxy size respectively bismaleimide size. The BMI/BAPPT-system shows excellent matrix dominant properties like 90⁰ flexural properties, 0⁰ shear strength and G_{Ic} fracture toughness (double cantilever beam test), which are even better than the properties of the control system on BMI-sized T800-fibres.

2.3 Modification of Bismaleimide with high molecular polyaryleneetherimide

Within an ongoing in-house research and development programme engineering thermoplastics are screened and incorporated into bismaleimides for enhanced toughness. It was observed that the outstanding toughness of the thermoplastics employed translates into BMI/TP blend systems (Fig.4). It is obvious that the toughening effect at the concentration investigated (0 – 30 %) is below that which would be expected from the rule of mixture. Poly (etherimide) (Ultem 1000) was found to be attractive with Compimide 796/Compimide TM 123 as the host thermoset for two reasons:

△ The toughness improvements at low TP-concentrations seem to be sufficient, i.e. doubling of the neat resin G_{Ic} at a 13 % by weight concentration.
△ Ultem 1000 is soluble in methylene chloride and therefore provides a means of blending it with the Bismaleimide/Compimide TM 123 blend via solution techniques.

The only disadvantage of Ultem is its relatively low glass transition temperature of ca.220⁰C. Glass tran-sition temperatures of > 250⁰C are attractive for TP's in BMI thermosets.
Our commercial system Compimide 796/Compimide TM 123 was modified with various levels of Poly(etherimide) (Ultem 1000) and used as matrices for carbon fibre composites based on Torayca's T 800 (epoxy sized) fibres. A series of matrix dominant properties have been generated for laminates with varying amounts of Ultem 1000. It is obvious that the toughness related properties such as the critical strain energy release rate (G_{Ic}), the end notched flexural strength (G_{IIc}) and to a minor extent the edgewise delamination resistance (EDL) increase with

increasing Ultem concentration. The G_{Ic} for the all Ultem composite is as high as 1500 Joules/m² while the equivalent thermoset value is 319 Joules/m². Ultem concentrations around 10 % have to be employed to move the blend system significantly into the direction of the all thermoplastic composite.

Scanning electron microscopy studies show that the C 796/TM 123/Ultem system at low (ca. 10 %) Ultem concentration provides a polyphase morphology in which the thermoset is the continous phase with large clusters of thermoplastic. These large TP-clusters themselves carry thermoset nodular inclusions.

Work is now in progress to incorporate Poly(etherimide) (Ultem 1000) in Bismaleimide/bis[3-(2-allylphenoxy) phthalimide (BAPPT) system (see para 2.1 this report). It is expected that the Ultem - Poly(etherimide) combined with BAPPT should provide a different morphology in comparison to the BMI/Compimide TM 123 for the blend system.

Neat resin and composite results will be available for the verbal presentation.

CONCLUSION

The nucleophilic nitrodisplacement reaction of 3-nitrophthalimides and the sodium salt of o-allylphenol provides a new family of bis[3-(2-allylphenoxy)phthalimides] in high yield. These products can be copolymerized with low melting bismaleimides to provide tough copolymer networks. Due to their strong chain rigidity the allylphenoxyphthalimides contribute to the good high temperature properties of the new BMI/comonomer-resins. They show glass transition temperatures in the range of 280°C. The mechanical properties, however, are dependent on the molar ratio between the resin ingredients and are optimum for a 2:1 stoichiometry of BMI and bis [3-(2-allylphenoxy)phthalimide].Composite investigations show that the good neat resin mechanical properties of a BMI / BAPPT-system translate directly into excellent laminate properties. Modification of BMI/bis (o-propenylphenoxy)benzophenone (TM 123) and BMI / bis [3-(2-allylphenoxy) phthalimide] resins with the engineering thermoplastic Poly (etherimide) (Ultem 1000) offers the possibility of increasing the toughness potential of these basis thermoset resins to a great extent.

REFERENCES

1. H.D.Stenzenberger, P.König, M.Herzog, W.Römer, Internat. SAMPE, Techn.Conf. 19, 372 (1987)
2. A.J.Kinloch, S.J.Shaw ACS, Polym. Mater. Sci. Engn. 49, 307 (1983)

3. H.D.Stenzenberger et al. 30 th Nat. SAMPE Symp. 30, 1568 (1985)

4. H.D. Stenzenberger, W. Römer, M. Herzog, S. Pierce, M.S.Canning, K.Fear, 31st. Internat. SAMPE Symp.31, 920 (1986)

5. J.J. King, M. Chaudhari, S. Zahir, 29 th Nat. SAMPE Symp. 29, 392 (1984)

6. H.D. Stenzenberger, P. König, M. Herzog, W. Römer, S.Pierce, M.Canning, 32nd. Internat. SAMPE Symp.32, 44 (1987)

7. H.D.Stenzenberger, W.Römer, M.Herzog, P.König,33rd. Internat. SAMPE Symp. 33, 1546 (1988)

8. H.D. Stenzenberger, W. Römer, M. Herzog, P .König, Internat. SAMPE Symp.34, closed session proc.(1989)

9. P.König, H.D.Stenzenberger, M.Herzog, W.Römer,Proc. of 3rd.Europ.Conf.on Composite Materials, 43 (1989)

10. D.M.White, T. Takekoshi, F.J. Williams, H.M.Relles, P.E.Donahue, H.J.Klopfer, G.R.Loucks, J.S. Manello, R.W. Schluenz, J. Polym. Sci. Polym. Chem. Ed. , 19 (1981) 1635

11. H.D.Stenzenberger, P.König,High Performance Polym., 1 (2), 133 (1989)

12. T.A.Attwood, P.C. Dawson, J.L. Freeman, L.R.J. Hoy, J.B.Rose and P.A.Staniland, Polymer,22, 1096 (1981)

13. E. Radlmann, W. Schmidt and G.E. Niksch, Makromol. Chem., 130, 45 (1969)

14. R.N.Johnson, A.G.Farnham, R.A.Clendinning, F.W.Hale and C.N.Merriam, J.Polym.Sci.,A15, 2375 (1976)

15. J.G.Wirth and D.R.Heath,US Patent 3, 838,097 (1974)

16 T.Takekoshi, J.G.Wirth, D.R.Heath, J.E.Kochanowski, J.S.Manello and M.J.Webber, J.Poly.Sci., Polm.Chem. Ed.,18,3069 (1980)

R	Mw (g/mole)	Mp [1] (°C)	Purity [2] (Area %)
-(CH₂)₆	640,2	106–108	95,2
⟨⟩-CH₃	646,4	85–91	95,4
⟨⟩-CH₂-⟨⟩	722	174–178	98,2
⟨⟩-SO₂-⟨⟩	772	171–178	95,1
⟨⟩-O-⟨⟩-C(CH₃)₂-⟨⟩-O-⟨⟩	934,7	91–99	86,0

(1) Capillary tube
(2) HPLC purity, UV detection 220 mm

Table 1: Properties of bis[3-(2-allylphenoxy)phthal-imides]

BMI = eutectic bismaleimide blend
Modifier: BAPPT = 2,4 - Bis [3(2-allylphenoxy)phthalimide] toluene
BAPPD = 4,4'- Bis [3(2-allylphenoxy)phthalimide] diphenylmethan

PROPERTY	BAPPD (% Wt) 40	50	BAPPT (% Wt) 38	48	49	52
Tg (tan) °C	--	--	286	278.5	284	280.5
Flexural Properties (RT/Dry)						
Strength, MPa	110	125	94	89	114	115
Modulus, GPa	4.10	3.99	4.24	4.00	4.04	4.04
Elongation, %	2.70	3.06	2.23	2.31	2.89	2.81
Flexural Properties (177°C/Dry)						
Strength, MPA	92	103	63	89	94	94
Modulus, GPa	3.34	3.33	3.74	3.80	3.53	3.31
Elongation, %	2.80	3.23	1.70	2.48	2.65	2.72
Flexural Properties (250°C/Dry)						
Strength, MPa	89	103	61	86	95	92
Modulus, GPa	2.69	3.00	2.95	2.95	3.23	3.18
Elongation, %	4.77	3.99	2.08	3.16	3.07	3.17
Fracture Energy G_{IC}, J/m^2	104	178	88	146	165	207
Moisture Gain [2], %	2.47	2.34	2.73	2.42	3.19	2.41

(1) cure : 1 h 160°C, · 3 h 180°C, · 4 h 210°C, · 5 h 240°C
(2) 1000 hours, 94 % RH, 70°C

Table 2:   Mechanical properties of cured BMI/BAPPT (BAPPD) - blends

C 796 = Compimide 796
BAPPT = 2,4 - Bis (o-allylphenoxyphthalimide) toluene
TM 123 = Bis (o-propenylphenoxy) benzophenone
Fibre = T 800, epoxy sized

PROPERTY		C 796 / BAPPT T 800	C 796 / TM 123 T 800	C 796 / TM 123 T 800 (PI)
0° Flex. Strength	23°C	2110	1833	1747
(MPa)	250°C	1437	1243	1268
0° Flex. Modulus	23°C	149	153	155
(GPa)	250°C	150	146	152 (?)
90° Flex. Strength	23°C	136	92	99
(MPa)	250°C	85	69	75
90° Flex. Modulus	23°C	10.3	8.6	8.7
(GPa)	250°C	7.12	9.2	7.3
0° ILS	23°C	117	103	103
(MPa)	250°C	69	51	48
0 +/- 45 ILS	23°C	77	62	81
(MPa)	250°C	63	52	43
G_{IC} (J/m^2)	23°C	468	319	319
EDL first fail.	23°C	234	126	202
Ultimate	23°C	462	522	571
CAI (MPa)	23°C	--	--	135

Table 3:   Mechanical properties of T800/BMI/BAPPT-laminate.

BMI: COMPIMIDE 796/TM 123/Ultem 1000   Fibre: Torayca T800

%Ultem		0	4,76	9,0	13,04	26	100
Neat resin G_{IC} (J/m^2)		85	–	–	171	312	1050
Fibre Size		EP	EP	EP	EP	EP	EP
90° Flex.Strength	23°C	92	84	95	95	–	–
(MPa)	250°C	69	55	42	29	–	–
90° Flex. Modulus	23°C	8,7	8,5	9,8	9,7	–	–
(GPa)	250°C	9,2	7,9	7,0	4,9	–	–
0° Short Beam Shear	23°C	103	97	94	93	–	–
(MPa)	250°C	51	56	43	22	–	–
0±45° Short Beam Shear	23°C	62	62	76	72	–	–
(MPa)	250°C	51	44	30	14	–	–
G_{IC}–DCB (J/m^2)	23°C	319	369	352	585	770	1500
G_{IIC}–ENF (J/m^2)	23°C	637	879	916	899	–	–
$EDL\left[(\pm25)_2 90\right]_s$	23°C						
(MPa)	first failure	126	111	134	142	–	–
	ultimate	522	517	554	595	–	–

Table 4:   Mechanical properties of T800/BMI/Ultem
laminates.

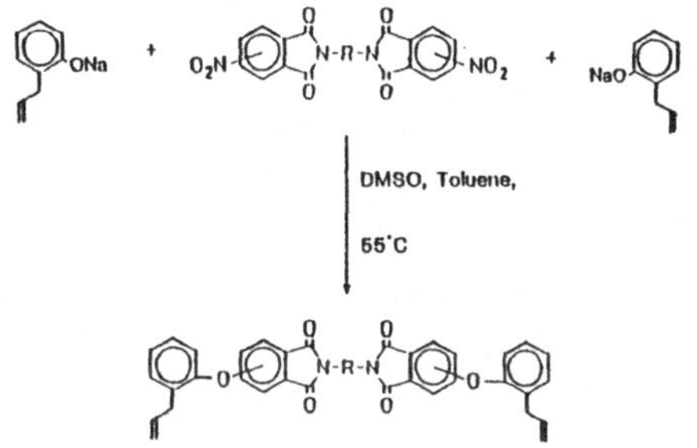

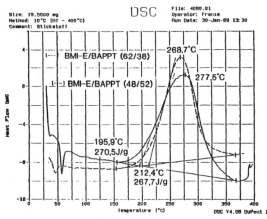

Fig.1: Synthesis of
bis (allyl-
phenoxy)
phthalimides

Fig.2: DSC-traces
of BMI/BAPPT
-blends

BMI = Eutectic bismaleimide blend
BAPPT = 2,4 - Bis [3 (2 allylphenoxy) phthalimide] toluene

597

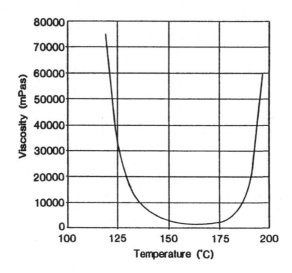

Fig.3: Viscosity-temperature-correlation of a BMI/BAPPT
(50/50) blend

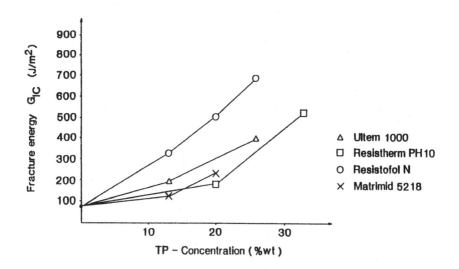

Fig.4: TP-concentration/toughness-correlation of TP-
modified BMI

THE INVESTIGATION AND UNDERSTANDING OF IMPROVED CFRP COMPOSITE MATERIALS

P. SIGETY, H. BOOKHOLT*, C. BROKOPF**, P. CURTIS***
W. 'T HART****, I. KRÖBE*****

ONERA - 29 Av. de la Division Leclerc - BP 72 - 92322 Chatillon - France
*FOKKER - PO Box 7600 1117 Schiphol - The Netherlands
**DLR - Pfaffenwaldring 38-40 - 7000 Stuttgart 80 - West-Germany
***RAE - R 50 Building - Farnborough/Hants GU14 6TD - Great Britain
****NLR - PO Box 153 - 8300 Ad Emmeloord - The Netherlands
*****MBB - Postfach 10 78 45 - 2800 Bremen 1 - West-Germany

ABSTRACT

Evolution of constituents (matrix, fibre and interface) of "improved" composite materials is examined and improvement of tensile fibre strength and matrix toughness is displayed. These changes induce an increase of directly related performances of composites (tensile strength, transverse tensile behaviour, G_{IC}). On the contrary, disappointing results are induced by the connected negative changes of constituents (decrease of fibres diameter, lowering of matrix flow stress) and results of compressive strength – plain or impacted specimens – make the estimation of global improvement more questionable.

INTRODUCTION

After about ten years of increasing utilization of composites in aeronautical structures – for more and more critical parts – a new generation of carbon fibre composite materials gradually emerged in the middle 1980's. This evolution was intended to answer the aeronautical users request for higher static performances (strength, modulus, design strain) and better damage tolerance (to delamination or impact for instance). In the framework of the GARTEUR organization (Group for Aeronautical Research and Technology in EURope), an Action Group joining representatives of European research establishments and aeronautical companies was constituted with intend:

– to correlate measurements achieved with components and simplified composites to performances evaluated with structural laminates;

– to identify and quantify these materials parameters which differentiate the properties of improved composite materials from those of first generation carbon fibre composites;

– to collect more general data on the failure of improved composite materials to give a wider perspective on their potential use in aircraft structures.

The complete set of experimental results of this work is available in the form of a technical publication from ONERA. It seems nevertheless interesting to present the most striking points for which "new composites" represent a real improvement when compared with the previous generation, and also the points for which the evolution is more disappointing.

I-MATERIALS AND FABRICATION

Three types of "new generation" composites have been selected as representative of the so-called "improved composites":

– FIBREDUX T400H – 6376C from CIBA-GEIGY is constituted of a CIBA toughened flow controlled epoxy reinforced by TORAYCA high failure strain fibres. This composite is representative of new systems for civil aircraft (higher failure strain, reasonable price);

– Rigidite T800 – 5245C from BASF is constituted of a NARMCO modified bismaleimide matrix reinforced by TORAYCA intermediate modulus fibres. This composite is representative of new systems for primary structure and military aircraft (impact strength, toughness);

– APC2 from ICI is constituted of a polyetheretherketone (PEEK from ICI) semi-crystalline thermoplastic matrix reinforced by AS4 fibres from HERCULES (intermediate between standard high strength fibres and new high failure strain fibres). This composite is representative of the new generation of high toughness systems used for their high level of damage tolerance.

Laminates were fabricated following the suppliers recommendations. Even when post-curing is a possible option proposed by the suppliers (6376 and 5245 based systems), the shortest curing cycles (final temperature of 180°C during 2 hours) were chosen to follow the most common practice of industry. Conditioning at 84% RH was applied until weight equilibrium – as representative of the world's worst average climate /1/. When possible, results obtained with new composites were compared with results issued from the previous generation and, when available, results from T300-5208 (NARMCO) and T300-914 (CIBA) systems were used.

II-COMPONENTS AND INTERFACE ASSESSMENT

2.1 Tests with neat resin specimens

As they correspond to the main problem encountered with organic matrices, results of toughness measurements (classical K_{IC} tests with CT10 specimens) are presented in the bar chart of Fig. 1. New resins display an appreciable improvement when compared with the older ones (with the exception of post-cured 914). Results obtained with PEEK are noteworthy as its K_{IC} is seven times K_{IC} of thermoset resins. 5245, on the contrary, is the most brittle of the new materials.

2.2 Measurement of fibres performances

Results of tensile tests of single filaments carried out at ONERA are presented in Fig. 2. Despite the slight increase of modulus observed during tension, results presented here are plotted assuming a strict linear elastic behaviour of the fibres (the constant modulus is in fact the secant modulus at fracture). The increase of strength observed with new fibre is clear when compared with the reference T300 fibre. This improvement is obtained either by increasing the ultimate strain with no change of moduli (T400 and AS4) or by increasing the modulus with no change of ultimated strain (T800). Single filament tensile tests were also carried out at RAE and the results were compared with tow tests results (Fig. 3). Tow tests were realized either with a classical high strain bifunctional epoxy (MY750) and, when possible, with the base resin used with composite. The trend observed with single filament tests are confirmed by tow tests, but it must be noticed that (except with 914) impregnation with the high strain resin always entails the highest results. Taking these results as a reference, it can be also observed that T800-5245 is an efficient system whereas efficiency of T400-6376 system is more questionable.

2.3 Measurement of interfacial adhesion

Direct measurement of fibre-matrix adhesion is a hard task as it involves micro-mechanical tests. It is generally replaced by macro-mechanical tests (for instance ILSS test) which, in best cases, constitute only an indirect assessment of interfacial behaviour. Corresponding to the requirement of a direct assessment of interfacial adhesion, pull-out and fragmentation tests (/2/ and /3/ respectively) were carried out at ONERA. We must precise that the first type of experiment corresponds to a measurement of ultimate interfacial shear stress, and the second to a measurement of the efficiency of interfacial load transfer. We can notice here that the transfer shear stress (τ_m) of new systems is lower than in the case of the older T300-5208 system. These low results of τ_m correspond either to a relatively low ultimate shear stress τ_u (T800-5245) or to a plastic flow of the resin near the interface (T400-6376 for which extraction of the fibre did not occur at the interface). Due to experimental difficulties (high viscosity, high melt point) the pull-out test could not be applied to AS4-PEEK, but it can be assumed that again extraction of the fibre would occur by plastic flow (high observed adhesion between fibre and matrix low shear strength of the matrix).

III-TESTS WITH COMPOSITE MATERIALS

3.1 Introduction

From the complete set of results achieved in the framework of this cooperation (starting from simple UD data until impact and fatigue behaviour of structural laminate) we have selected only a few results corresponding either to noticeable differences with older systems or to difference with what could be inferred from values obtained from tests with single components.

3.2 Results

From tensile tests carried out at RAE with strongly orientated multidirectional tensile specimens (Fig. 4) it appears that high performances of T800 fibres indeed correspond to high values of composite tensile strength. Improvement realized with other new systems is more questionable when compared with the older XAS-914 system whose efficiency was already noticed (Fig. 3). Compressive strength of new systems is more disappointing as the older XAS-914 system remains the best system, whereas the low value obtained with T800-5245 is noteworthy. Despite the lack of complete interpretation of these results, it must be emphasized that improvement realized with the new fibres was obtained by decreasing the fibres diameter and that microbuckling of fibres under compression is then promoted. Furthermore, the relatively low fibre-matrix adhesion (T800-5245) or the low shear flow stress (T400-6376 and APC2) can also be evoked as lowering the compressive strength. Tests related to the toughness of matrices – transverse ultimate strain and G_{Ic} measurements – were also realized (Fig. 5 and 6). It must be noticed that in both cases the high toughness of PEEK matrix corresponds to a real improvement with regard to other systems. As noticed in Fig. 1, 6376 and 5245 resins realized an improvement only when compared to an older brittle resin (5208) but the results are comparable with 914. As a consequence, G_{Ic} values of T400-6376, T800-5245, T300-914 and XAS-914 are comparable (and higher than for T300-5208), but T300-914 remains the best thermoset resin based system, probably due to the high level of adhesion between fibre and matrix. Finally, results of compression after impact (CAI) tests are presented in Fig. 7. High velocity impact (v = 50 m/s) was applied to compression pre-strained quasi-isotropic coupons. It can be observed that in comparison with the conventional HEXCEL T300-F593 system the new systems show improved CAI behaviour that is most pronounced for the zero pre-strain condition. However, when the pre-strain at impact increases the CAI decreases. This was most noticeable for T800/5245. Further, the relatively poor compression strength for APC2 cannot be related to its high fracture toughness. This is because of the rate-dependent material behaviour of APC2, i.e. the damage growth under dynamic conditions (high v impact) cannot be described by statically determined toughness properties.

CONCLUSIONS

Changes touching constituents of "improved" composite materials mainly improve on the one hand tensile performances (improvement of fibre tensile strength) and on the other hand behaviour related to the matrix toughness (transverse tensile behaviour, G_{Ic}). These improvements are nevertheless obtained by a decrease of fibre diameters and of matrix plastic flow stress (6376 and PEEK). In the case of T800-5245 system the weakness of the interface must also be noticed. A strong sensitivity to compressive loading, either for static behaviour or during impact tests, is then induced. The global evolution is consequently questionable with regard to the requested performances.

ACKNOWLEDGEMENTS

This cooperative work of six European research institutes and aeronautical companies was carried out in the framework of GARTEUR organization – Structures and Materials Division.

REFERENCES

1. Environmental effects in the testing of composite structures for aircraft
   BAE Report MSM-R-GEN-0603 (1985)

2. M. Sanchez and G.D. Désarmot, in "Proceeding of the 5th National Symposium on Composite Materials" (Pluralis ed.) 9-11 September 1986, JNC5, Paris

3. W.A. Fraser, F.H. Ancker, A.T. Di Benedetto, in "Proceedings of the 30th Ann. Tech. Conf." (1975) Reinf. Plastics/Comp. Inst., The Soc. of Plastics Ind., Inc., section 22-A

	Pull-out tests τ_u (MPa)	Fragmentation tests τ_m (MPa)
T400/6376	> 125	
T800/5245	103	36
AS4/PEEK		28
T300/5208	150	45

Interfacial shear stresses (ONERA)

Table 1

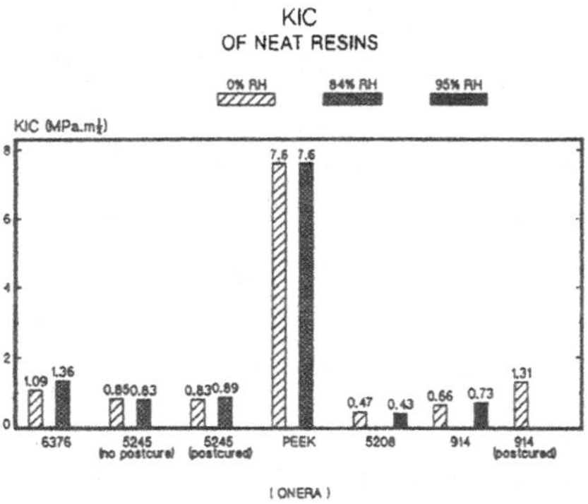

KIC OF NEAT RESINS

Figure 1

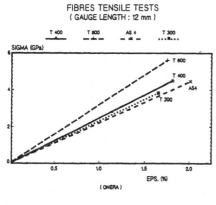

FIBRES TENSILE TESTS
(GAUGE LENGTH : 12 mm)

Figure 2

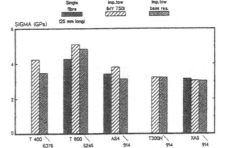

SINGLE FIBRE AND TOW TEST DATA

Figure 3

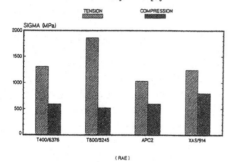

Figure 4

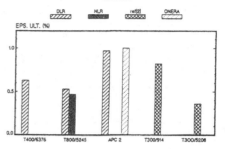

Figure 5

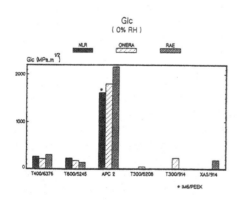

Figure 6

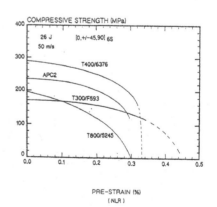

Figure 7

RIGID STRUCTURAL ELEMENTS MADE FROM POLYIMIDE FIBERS

C. BARKER

Lenzing AG - 4860 Lenzing - Austria

ABSTRACT

Using solvent-spun amorphous polyimide fibers, rigid structural elements for use in the aerospace, automotive and electrical industries can be made. The forming process uses only heat and pressure and therefore avoids the need for binder materials and long curing cycles. The excellent properties with respect to: thermostability, non-flamability, low smoke emission and low generation of toxic gases at decomposition and thermal insulation, make these products well suited for use in aircraft interiors and mass transit vehicles.

INTRODUCTION

With an increase in the requirements placed upon the thermal characteristics regarding heat release, smoke density and toxic gases in the transportation industries, manufacturers are looking twards new polymeric materials to help meet these requirements. Rigidified polyimide felt structures offer an interesting and cost effective solution to these problems. Based upon a newly developed polyimide fiber, complex components made from thermally rigidified polyimide felt can be economically fabricated into interior structures such as sidewall panels and airducts.

I-PRODUCTION TECHNIQUES

The production of rigid structures based upon polyimide fibers can be broken down into three different stages: fiber production, needled felt fabrication and rigidification.

1.1 Fiber Production

The solvent-spun fibers are based upon a recently developed amorphous fully imidized polyimide /1,2/. The polymer consists of a polycondensate of benzophenon 3,3',4,4'-tetracarboxylicdianhydrid and aromatic diisocyanates.

During the fiber processing, the fibers undergo a drawing process. In addition to increasing the strength of the fibers, this treatment causes the fibers to shrink when exposed to temperatures around the Tg, 330°C. This shrinkage property is the key element in the production of rigid parts /3/.

1.2 Needle Felt Fabrication

Needle felt fabrication begins with crimped polyimide staple fibers. The fibers are carded to form a batt with a prefered fiber orientation. The batts are then combined and densified by a needle loom to form needle felts. The needle felt making process is very well known and highly automated, which makes it a very economical fabrication process.

A further advantage of needled felts is that the physical properties of the final felt product can be varied through simple adjustments at various stages in the felt fabrication. This allows the density, strength and the shrinkage characteristics of the final rigidified product to be optimized for each application.

1.3 Rigidification

The rigidification process consists primarily of exposing the polyimide felt to temperatures around the Tg (330°C) with or without external pressure. Although this polyimide has no real melting point, SEM-studies have shown that cohesive bonding between individual fibers begins to take place during the shrinkage process, Figure 1. Since the shrinkage process takes place almost immediately and without additional binders, the total time needed is dependant only upon how long it takes to heat all of the fiber mass up to the given temperature.

For making flat or contoured panels with densities ranging between 0.2 g/cm³ and 0.8 g/cm the felt is constrained in a frame or simple matched mold while in the oven. This allows for the felt to shrink in the thickness direction but not in the surface plane. The finished panel will permanently retain the shape of the constraining frame or tool.

A variation of this production method consists of reheating rigidified felt to the Tg and then forming it in a cold matched die tool under pressure. Once cooled the formed part will retain the shape of the mold. Since this process takes only a few minutes, thermally formed 100 % polyimide parts can be produced in large quantities very economically.

Seamless hollow tubes of various shapes can be made by allowing polyimide felt tubes to shrink around metal mandrils. In this case the heat capacity of the mandril usually determines how long it will take for the rigidification to be completed. This process is excellent for making high-temperature resistant air ducts and conduits with a variety of cross-sections, Figure 2.

II-PROPERTIES AND APPLICATIONS

Depending upon the manufacturing process used, the physical properties of the rigidified parts will vary. The excellent thermal properties of the original polyimide fiber are retained by the final products.

2.1 Physical Properties

Since the density and physical properties vary depending upon the fabrication process used, a listing of typical property ranges is given below:

Density	$0.19 - 0.8$ g/cm^3
Tensile Strength	$6.2 - 30$ MPa
Tensile Modulus	$0.1 - 0.21$ GPa
Percent Elongation	$10 - 25\%$
Thermal Conductivity	$0.04 - 0.05$ W/m°K
Cont. use Temp.	270°C

2.2 Thermal Properties

Aircraft sidewall panels were chosen as a typical application of the rigidified polyimide felt to demonstrate the noteworthy thermal properties of these structures.

Interior sidewall panels made from the rigidified polyimide felt tested in an OSU test chamber were shown to have very low heat release values, Figure 3. In addition, the limiting oxygen index (L.O.I.) value for these products has been measured to be 44% and Figure 4 shows the flammability properties acc. to FAR 25.853(a). Figure 5 shows the low values of smoke emission and toxic gas generation at decomposition acc. to ATS 1000.001.

2.3 Applications

Because of the excellent thermostability, non-flamability, low smoke emission and low generation of toxic gases at decomposition and good thermal insulation of the rigidified structures, applications in aircraft interiors have become very promising. Interior sidewall panels made from rigidified polyimide felts are presently being examined because of the increased restrictions on the interior materials and because of the potential cost savings presented by the economical production methods.

Aircraft airducts made from rigidified polyimide felts pose an economical alternative to present systems. Because of their built-in insulating properties and reduced fabrication costs polyimide felt ducts fulfill the weight and other requirements at lower costs.

607

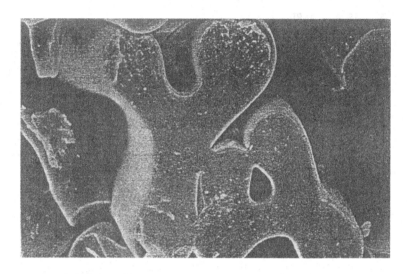

Polyimide fibers after shrinkage process
in a needled felt.
Magnification: 2000 x

Figure 1.   Interfiber Bonding

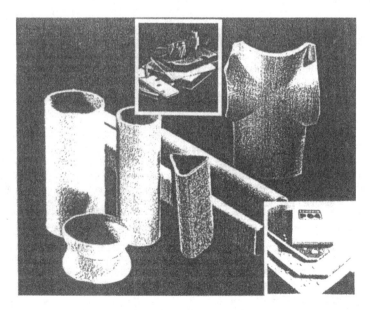

Figure 2.   Seamless Tubular Shapes

```
┌─────────────────────────────────────────────────────────────┐
│                                                             │
│        SMOKE & TOXIC GAS GENERATION                         │
│           (Max Ds & ppm at 4 Min.)                          │
│                                                             │
│                                          Crush Core         │
│                            PYROPEL       Honeycomb          │
│                                                             │
│              HCN             6               0              │
│              NO2             2               1              │
│              SO2             0               1              │
│              HF              0               0              │
│              HCL            0               0               │
│              CO           <100            <100             │
│                                                             │
│              Ds            .58             .21              │
│                                                             │
│                                                             │
│  • Test Method ATS - 1000.001                               │
└─────────────────────────────────────────────────────────────┘
```

Figure 5. Smoke and Toxic Gas Generation

REFERENCES

1.   K. Weinrotter, H. Grieβer, Chemiefasern/Textilindustrie, 35/37 (1985) 409 - 410

2.   K. Weinrotter, Chemiefasern/Textilindustrie, 37./89.(1987)

3.   K. Weinrotter, R. Vodunig, in "Polyimides: Materials, Chemistry and Characterization"; Elesvier Science Publishers B.V. (C. Feger, M.M. Khojasteh and J.E.McGrath, ed.) (1989) pp. 769 - 778

TYPICAL AIRCRAFT SIDEWALL
OSU HEAT RELEASE

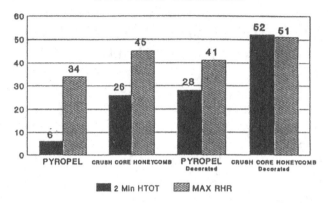

Figure 3. Heat Release Values

PYROPEL WALLPANEL
60 SEC. VERTICAL FLAMMABILITY
FAA 25.853(a)

	Flame Time	Burn Length	Flaming Time of drippings
Sample #1	0 sec.	1.4 in.	0 sec.
Sample #2	0 sec.	1.5 in.	0 sec.
Sample #3	0 sec.	1.3 in.	0 sec.

Figure 4. Flammability Properties

QUALITY ASSURANCE AND QUALITY CONTROL ENHANCEMENT IN COMPOSITE ENTERPRISES

I. CRIVELLI-VISCONTI, G.C. CAPRINO, L. NELE

University of Naples - Piazzale Tecchio - 80125 Naples - Italy

ABSTRACT

The SPRINT Program covers research financed by E.E.C.. The objective of this Program is the improvement of knowledge transfer between Research Centers and firms.

Within the framework of the SPRINT Program, the RA10 project has been designed for development of "QUALITY" within Small and Medium Enterprises.

In this paper the early results of the research are related and some informations about future actions are given.

INTRODUCTION

The SPRINT Program covers researches financed by the Commission DG XIII of the E.E.C.. The purpose of SPRINT Program is to encourage the knowledge transfer within Technical Centers and University Laboratories on the one hand, and between these facilities and Small and Medium Size Enterprises (SMES) on the other hand, in order to increase the competitivity.

Within the framework of the SPRINT Program, the RA10 project has been designed for the development of "QUALITY" within COMPOSITE MATERIALS PROCESSING INDUSTRIES.

The working group in charge of the development of the SPRINT Program consists of seven partners from the E.C. countries, namely:

- CENIM (Spain)
- CETIM (France)
- CRIF (Belgium)
- DIMP (Italy)
- INEGI (Portugal)
- PERA (England)
- SINTEF (Norway)

CETIM: Technical Center of Mechanical Industries
This is a professional technical center financed by the Ministry for Industry and by the Federation of Mechanical Industries.

CENIM: Centro   Nacional   des   Investigaciones Metalurgicas.

CRIF: Center for Scientific and Technical Research for the Metal Production Industry.
A professional technical center working in Belgium.

DIMP: Department of Engineering for Materials and Production
A department of the University of Naples specialized in materials study and application.

INEGI: Institute of Mechanical Engineering and Industrial Management.
A technical center of the University of Porto in charge of technological transfer from the University to firms.

PERA: Production Engineering Research Association.
A professional technical center essentially privately financed for assistance to firms.

SINTEF: The Foundation for Scientific and Industrial Research.
A Norvegian Institute for assistance to firms.

1. OBJECTIVES AND WORKING ARRANGEMENTS

The objective initially set by the seven partners for the SPRINT RA 10 program was that of analyzing how the question of "QUALITY" is tackled within SMES working in the field of composite materials.
For this purpose, a questionnaire has been drafted and circularized among European firms.
The informations collected evidences a certain number of gaps, especially with respect to the knoweledge of control methods and how to manage "QUALITY". A considerable number of firms overstimates the resources - organizational and human - dedicated to "QUALITY", whilst only few firms demonstrate that they really use Quality Assurance procedures.

After the analysis of such responses, the working Group has decided to develope a classification criterion for levels of quality for the product and firms themselves, for each type of production technology.

2. CRITERION OF CLASSIFICATION

The manufactured parts, and, consequently, the companies, are to be classified into three categories. As far manufactured parts are concerned, the categories are defined with respect to their technological application, dependent upon levels of utilization.

Inspection procedures and test methods should be related to three main stages in the manufacture of composite parts and to three categories of components.

The three main manufacturing stages are:
- incoming raw materials;
- processing and process condition;
- final part inspection.

The three categories of components, referring to their use, are as follow:
- severe or critical;
- process or semi-critical;
- service or non-critical.

The definition of each category is given in Tab.1. It is not necessary for a component to fulfill all the requirements of a particular category for it to be contained within that category.

In Tab.2 is shown a classification criterion proposal for HAND LAY-UP TECHNOLOGY.

3. DEVELOPMENT OF RESEARCH: EUROPEAN NETWORK

Within the framework of the SPRINT RA10 program, information has been collected related to the question of "QUALITY" from SMES working in the field of composite materials.

Such information has highlighted gaps, especially concerning the concept of "QUALITY" and the knowledge of inspection and monitoring methods.

This evaluation gave birth to the idea of proposing a criterion of classification for products and for companies working in the field of composite materials processing.

The implementaion of the full program provides for the drafting of a QUALITY ASSURANCE manual in support of QUALITY CONTROL recommendations for all technologies.

Once all the information on the controls to be implemented according to the criterion proposed had been collected, the working group of the SPRINT RA10 program decided to proceed to the setting-up of an European Network for Quality.

Such a network will be designed to aid and support all those companies which cannot or do not want to invest directly in quality preferring to avail themselves of the collaboration of outside facilities for quality inspection and control.

The Network will consist of proven Companies and Laboratories in the E.C. who are ready and willing to make their own facilities, equipment and know-how available to outside Companies.

CETIM: static and dynamic test set.

CRIF: acoustic emission test set.

INEGI: 4 points bending test set.

DIMP: ultrasonic test set.

616

CATEGORY	DEFINITION
1- Severe or Critical	Those components which operate in extreme conditions and where continuity in operation is regarded as critical. Where failure will cause danger to life and/or property and/or cause a very expensive shut down and/or necessitate extensive and expensive repairs.
2- Process or Semi-Critical	Those components which may or may not operate in extreme environments and conditions of service. Where failure can cause lengthy and expensive shut downs, repair is costly and there is limited danger to personnel, property or the environment.
3	Those components which operate in non-extreme environments and conditions of service, e.g. ambient temperature, atmospheric pressure, and are not subjected to possible chemical or solvent attack. Where they are subjected to low stresses and failure of the component will not necessitate expensive repairs and will cause minimal damage to other equipment and little or no loss of production. There is no danger to personnel and environmental damage is not a consideration.

TABLE 1 - DEFINITION OF COMPONENT CATEGORIES

Q.C. PROCEDURES	TEST METHOD (Where applicable)
General Recommendations	
All tests and monitoring equipment to be calibrated at frequent intervals.	
Each stage of manufacturing certified by qualified Q.A. inspector.	
Quality control of personnel to be independent of production personnel.	
Material In	
Determine binder solubility of CSM*	Time to break for weighted CSM. Supported in styrene.
Verify construction of woven roving	Determine warp and weft ends per unit
Resin	Determine S.G. Viscosity and gell time
Processing	
Check each ply, as it is layed, against laminate sequence.	
Record change of operator or shift change.	
Maintain constant recommended humidity	
Final Product Inspection	
Fault detection to be performed on each component.	N.D.T. ultrasonic X Ray
Mechanical testing to be performed on samples produced at the same time as the components preferably where possible on sample off cutts from each component.	As for Class II

*Chop Strand Mat

TABLE 2.I - CLASS I COMPONENTS

617

Q.C. PROCEDURES	TEST METHOD (Where applicable)
General Recommendations	
All tests and monitoring equipment to be calibrated at regular interval. Provide process record sheets to record all details from materials in through progression to final part inspection. Ensure component is being manufactured to the current drawing/specification.	
Materials in	
Check weight of reinforcement	Accurate balance
Check width of reinforcement	Steel rule
Ensure certificate of conformance received	
Resin	Viscosity and gell time
Use material in correct rotation	
i.e. oldest first	
State at recommended condition of temperature and humidity	
Inspect glass material for damage or inclusions	Bark lighting
Processing	
Ensure laminate lay up sequence and procedures are correct and are followed	
Check expiry date of materials	
Allow sufficient time to research	
Correct processing	Visual
Operators to use clean gloves	
Prepare glass cloth in room separate form other operations	
Glass bo be protected from contamination during cutting handling in storage	
Maintain high standard of cleanliness	
Maintain correct ambient temperature	
Manufacture test panel to same specification as component	Thermometer
Final Part Inspection	
Materials testing to be conducted on random components	Tensile, flexural (routine)
	Chemical, impact, H.D.T., electrical (if required)
Fibre volume fraction	Burn off
State of cure	Barcol hardness
Check dimensions	Metrology

TABLE 2.N - CLASSE II COMPONENTS

618

Q.C. PROCEDURES	TEST METHOD
Materials In	
Check for correct material	Visual inspection
Record batch numbers	Visual inspection
Inspect for damage	Visual inspection
Processing	
Maintain correct ambient	Thermometer
Temperature (refer to suppliers recommendations)	
Ensure correct resin mix	Accurate weighing
Ensure adequate mixing	Visual inspection
Maintain reasonable standard of cleanliness	
Adequate wet-out	Visual inspection
Lay-up one layer of glass at time	
Record name operator	
Final Part Inspection	
Inspect for faults voids-delaminations inclusion air bubbles-surface	Visual inspection
Defects colour variation striations	
Check dimensions	

TABLE 2.III - CLASSE III COMPONENTS

619

A NEW TYPE OF KNITTED REINFORCING FABRICS AND ITS APPLICATION

J. SARLIN, P. PENTTI, T. JÄRVELA

TUT - Institute of Plastics Technology
PO Box 527 - 33101 Tamperre - Finland

ABSTRACT

This work deals with a new type of knitted fabric, which acts as a unidirectional reinforcement. A knitted fabric provides good processibility, which is useful in various composite production methods. The production method of the fabric itself is rather versatile, which makes it easy to use reinforcement fibres of different kinds. In the light of the preliminary results it seems that using this reinforcement it is possible to obtain properties comparable to conventional unidirectional reinforcing materials.

INTRODUCTION

The use of knitted reinforcing materials has been rather limited compared to woven reinforcements. The main reason for this has been the undisputed advantages of the weaving technique and the highly developed methods. Applying the conventional knitting technique does not necessarily yield a fabric with the best possible reinforcing effect. This work presents some preliminary results obtained with a new type of knitted reinforcement, whose brand name is "FOXFIBER". This fabric acts as a unidirectional reinforcement, where the perpendicular knitted structure offers good processing strength. The processibility of the fabric in the manufacture of laminates is excellent. Using this method, reinforcements have been produced, for example, of glass, carbon and aramid fibres.

I – REINFORCING MATERIAL

1.1. Structure of the reinforcement

The structure of the FOXFIBER fabric consists of at least two fiber types, which can be called primary and secondary fiber. The secondary fiber forms the knitted structure while the primary fiber acts as the actual reinforcement.

The secondary fiber must be one with good knitting properties, and usually the fibers used have been various PET fibers. The knitted fabric is rather loose, and the actual reinforcing primary fiber is located inside the knitted structure. Although it is loose, the fabric has good processing properties.

The primary fiber, which is surrounded by the knitted fabric, is completely unidirectional, and the knitted structure does not cause any wave-like changes in the direction of the fiber, which can occur in woven fabric because of the binding method. The production method does not set any practical limitations to the primary fiber material, and here we tested fabrics reinforced with such materials as glass, aramid, carbon and UHMWPE fibers.

One typical feature of the FOXFIBER production method it is versatility. The method permits the use of several different primary fibre materials at the same time, and it has been tested by manufacturing hybrid reinforcements such as glass-aramid and carbon-aramid. The versatility of the production method also allows profitable production of even short production lots.

1.2. Properties of the fabric

A fabric produced by the knitting technique is somewhat thicker than a woven reinforcement of the same basis weight. The thickness of the fabric is usually 0.6...1.2 mm. This thickness mainly consists of the knitted structure, which sets a minimum for the thickness. Using different binding structures the thickness can be varied to some extent. On the other hand, increasing the basis weight of the primary fiber also increases the thickness of the fabric. Due to the binding structure the compressibility of the fabric is generally 10...20%, when the used pressures are of the same magnitude as used in the manufacture of laminates.

The basis weight of the primary fiber has generally been about 70...90 g/m^2 and the total basis weight of the fabric has varied in the range 500...900 g/m^2, which leaves 350..830 g/m^2 for the basis weight of the reinforcement. So the basis weights are rather high.

The tensile strength and modulus of the fabric in the direction of the primary fibre is based on the corresponding properties of the reinforcement fiber itself, in other words, the fabric is rather strong and tenable in this direction. In the perpendicular direction the mechanical properties are based on the knitted structure of the secondary fibre. In this direction the tensile strength is sufficiently high to prevent damages to the fabric during production. Owing to the knitted structure the fabric is rather flexible in this direction, which can be useful in the production of objects of different shapes.

II – COMPOSITES

2.1. The production of laminates

Here the epoxy laminates were produced by manual lay-up, and immediately after the lay-up they were pressed using one of the mould surfaces in a hydraulic press. This enabled us to vary the thickness of the laminate and the amount of reinforcement in the specimens. Furthermore, both sides of the laminate could be made rather even. In the lamination the good wetting properties of the fabric could be observed. The laminates were postcured at elevated temperatures, after which specimens were produced according to the standard.

2.2 Properties of the laminate

The properties of the laminates were examined mainly on the basis of tensile, flexural and impact tests and delamination properties. The primary fiber content of the laminates was generally 37...42 percent by volume. The measured properties reflected primarily the reinforcement content and the properties of the reinforcing fiber. The measured properties were comparable to those of conventional unidirectionally reinforced composites. The knitted secondary fiber structure seemed to have no negative effects on the properties of the specimens.

Acknowledgements

The authors wish to acknowledge Managing Director J. Saarikettu of Foxwear Ltd. for the opportunity to publish this work.

PROPERTIES OF ALUMINA TRIHYDRATE FILLED
EPOXY GLASS LAMINATES

G. PRITCHARD, R. WAINWRIGHT

Kingston Polytechnic
Penrhyn Road - Kingston-up-Thames - Surrey KT1 2EE - Great Britain

ABSTRACT

A non-halogenated fire retardant matrix system for glass laminates is evaluated. It consists of a dispersion of alumina trihydrate (ATH) in an epoxy resin. The matrix shows improved critical oxygen index and tensile modulus, but a lower tensile strength and ultimate elongation at break than the epoxy resin itself. The elongation is improved by the addition of an elastomeric additive to produce a three-component system. Glass laminates without any added elastomer showed an early kink in the stress-strain curve, but with added elastomer, the kink was much delayed.

INTRODUCTION

The high flammability of epoxy and polyester resin matrix materials and their dense smoke evolution on burning are major disadvantages in structural composites. Improvements can be achieved by using halogen modified resin formulations, but undesirable halogen-rich fumes are evolved during combustion. Consequently, halogen-free fire retardant additives are of interest. Alumina trihydrate (ATH) is a halogen-free additive capable of reducing the cost, flammability and smoke density of thermoplastics and thermosets, but high filler loadings are required /1/ which considerably alter the mechanical properties.

This paper reports some properties of epoxy resins containing ATH, and discusses the effects of the ATH filler on the mechanical properties of the matrix and the corresponding glass fibre laminates.

I - MATRIX PROPERTIES

1.1 Materials and Testing

The epoxy resin used in this study (Epikote 816) was a diglycidyl ether of bisphenol A modified with a reactive diluent. The curing agent (Epikure T) was a mixture of primary, secondary and tertiary aliphatic amines. The resin and curing agent were supplied by Shell UK Ltd. Several grades of ATH were obtained from BA Chemicals Ltd and from Croxton and Garry Ltd. An amine-terminated butadiene-acrylonitrile rubber, Hycar ATBN, was obtained from B.F.Goodrich Chemical (UK) Ltd.

Tensile tests were carried out using an Instron testing machine model 1114. Fracture energy tests using tapered double cantilever beam specimens were performed on the above instrument. Viscosity measurements were performed using a Ferranti concentric cylinder viscometer. Flammability tests were performed using a Stanton Redcroft FTA oxygen index instrument.

1.2 Viscosity

The addition of a dispersion of ATH to uncatalysed epoxy resin results in an excessive increase in its viscosity (Fig.1). Two methods were used to minimise the viscosity. It has been shown /2/ that the minimum viscosity of a dispersion containing two sizes of solid particles at a given filler loading is achieved by combining fine and coarse particles in a ratio of 1:3 v/v. This was done by using an appropriate grade of ATH (ON 920V, ex. Croxton and Garry) supplied already in the required form, i.e. a 1:3 mixture. Secondly, the viscosity decreased at constant filler loading if a surface treatment was added to the ATH before mixing.

1.3 Flammability

The critical oxygen index of ATH-filled epoxy resin increased with increasing filler loading (Fig. 1). This is attributed to three main factors:

(1) Replacement of epoxy resin with ATH results in a reduction in the amount of flammable material.

(2) The chemical decomposition of ATH is an endothermic process. At 200°C, ATH decomposes as follows:

$$Al_2O_3.3H_2O \text{ -------> } Al_2O_3 + 3H_2O \qquad (+ 300 \text{ KJ/mol})$$

(3) The production of water dilutes any combustible gases evolved, and thus reduces the temperature in the vicinity of the combustion site.

1.4 Mechanical Properties

The addition of an ON-920V dispersion to the epoxy resin affected the tensile properties (see the curves in Fig.2). The tensile modulus increased as the filler loading was increased, as

626

expected for a dispersion of inorganic particles in a relatively soft matrix /3/. However, the tensile strength and especially the ultimate elongation decreased as the filler loading was increased. This behaviour could be explained by assuming that the ATH particles constitute crack-like defects; the practical strength is then determined by the critical flaw size, which can be calculated from the Griffith equations for an elastic material under plane strain conditions. Neglecting plasticity, and assuming that the ATH particles have an aspect ratio of unity,

$$\text{Stress at failure} = \sqrt{\frac{EG_{IC}}{\pi a(1 - \mu^2)}} \qquad - [1]$$

$$\text{Strain at failure} = \sqrt{\frac{G_{IC}}{E\pi a(1 - \mu^2)}} \qquad - [2]$$

where
G_{IC} is the fracture energy,
E is the tensile modulus,
μ is Poisson's ratio,
2a is the critical flaw size.

The addition of ATH usually results in a large increase in the critical flaw size. We assume that critical flaws consist of agglomerations of particles (Fig. 3) or are formed during testing and occur because of microcracking between ATH particles in the matrix at low stresses (ATH particles are relatively weak compared with some other mineral fillers /4/). Also, ATH varieties such as ON-920V tend to form large, weakly bonded clusters of particles, which constitute large flaws and accentuate the problem. Scanning electron microscopy showed that some particles are easily cleaved by cracks (Fig. 4), depending on the angle between the crack and the crystallographic planes. An improvement was achieved by using the FRF 20 grade (ex BA Chemicals) for the later work on laminates, as this does not form clusters so easily, although it imparts a slightly higher viscosity.

The main focus of attention in this work was therefore how to improve the elongation of FRF 20-filled epoxy to produce a tougher matrix for glass fibre laminates. This was approached by considering how G_{IC} in equation [2] might be increased. G_{IC} values for ATH/epoxy compositions increased as the filler loading was increased (Table 1). However, the addition of a small amount of liquid rubber has previously been shown to increase G_{IC} values in epoxy resins by energy dissipating mechanisms /5/, /6/ and /7/. Therefore, in order to increase G_{IC} further, 10 parts by weight per hundred of resin (pphr) of an amine terminated butadiene acrylonitrile (ATBN) rubber was added to ATH/epoxy mixtures. This also lowered the modulus of the matrix.

II - LAMINATE PROPERTIES

Glass fibre laminates were prepared using the matrix resin system described above, without added rubber. The glass was woven roving with a surface density of $2200g/m^2$. The tensile strength of the laminates decreased as the filler loading was increased. However, a higher apparent elongation at break was observed for the laminates containing ATH than for those which contained no particulate dispersion.

In ATH-filled laminates, the stress-strain graph showed a very early kink (Fig. 5), suggestive of substantial matrix cracking before ultimate failure. The relief of strain energy in the laminate during the tensile tests enabled a higher elongation to break to be achieved, but the material was damaged at low stresses. As the filler loading of ATH in the laminates was increased, matrix cracking initiated at a progressively earlier stage. On adding ATBN rubber to the matrix, the onset of matrix cracking was much delayed (see asterisk, Fig. 2). Rubber modification (10pphr) lowered the tensile modulus of the laminates by up to 20%.

III - CONCLUSION

A rubber-modified, ATH-filled epoxy glass laminate system is reported which possesses low flammability. Careful consideration of the nature of ATH enabled processing difficulties to be lessened and the mechanical properties of the laminates to be improved. Further work is in progress to establish optimum formulations and properties.

Acknowledgement
This work has been carried out with the support of Procurement Executive, Ministry of Defence.

REFERENCES

1 - K. Evans and R. Shaw, Reinf. Plast., (1980) 273.
2 - H. Hsieh, Polym. Eng. Sci., 18 (1978) 928.
3 - A. Moloney, H. Kausch, T. Kaiser and H. Beer, J. Mater. Sci., 22 (1987) 381.
4 - H. Katz and J. Milewski, "Handbook of fillers for plastics", Von Nostrand Reinhold, New York, (1987) P.292.
5 - W. Bascom, R. Cottington, R. Jones and P. Peyser, J. Appl. Polym. Sci., 19 (1975) 2545.
6 - C. Bucknall and T. Yoshii, Br. Polym. J., 10 (1978) 53.
7 - S. Kunz-Douglass, P. Beaumont, and R. Ashby, J. Mater. Sci., 15 (1980) 1109.

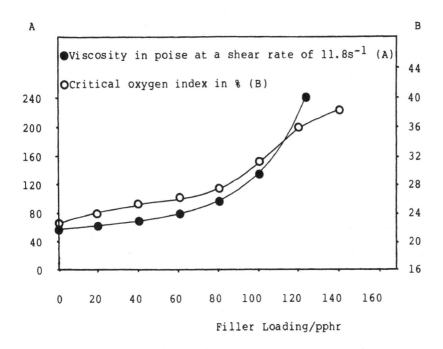

Fig. 1 - Viscosity (at 25°C) and critical oxygen index (at 20°C) of an ATH - filled epoxy as a function of filler loading

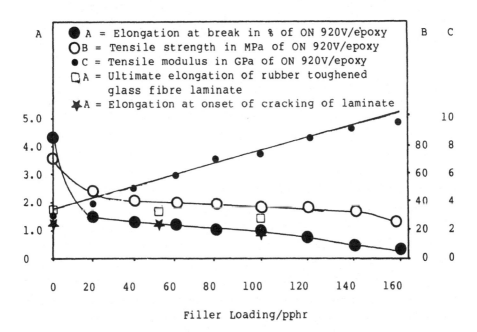

Fig. 2 - Tensile properties of an ATH - filled epoxy as a function of filler loading for castings and laminates at 20°C

Table 1 - Fracture energies and critical flaw sizes of FRF 20 - filled epoxy

Filler Loading (pphr)	Fracture Energy* (KJ m^{-2})	Calculated[+] critical flaw size/um
0	0.06	21
20	0.32	223
40	0.27	252
60	0.21	261
80	0.20	321
100	0.20	443

* Using tapered double cantilever beam specimens
+ Using the Griffith equation, assuming plane strain conditions with an estimated Poisson's ratio of 0.40 for unfilled and 0.32 for filled specimens. Experimental flaw size found to be 260um at 20pphr.

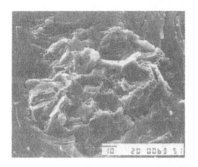

Fig. 3 - SEM micrograph showing a large critical flaw

Fig. 4 - SEM micrograph showing trans-particle failure of an ATH particle at a fracture surface

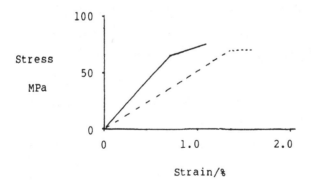

Fig. 5 - Stress-strain curves for ATH (100pphr)/epoxy glass laminates with (···) and without (—) rubber modification

IMPACT AND ENERGY ABSORPTION
SCHLAGVERHALTEN UND ENERGIE ABSORPTION

Chairman : **Prof. I. VERPOEST**
Catholic University of
Leuven

AN EXPERIMENTAL STUDY OF IMPACT-DAMAGED PANELS UNDER COMPRESSION FATIGUE LOADING

R. AOKI, J. HEYDUCK

DLR - Institute of Structures and Design
PO Box 80 03 20 - 7000 Stuttgart - West-Germany

ABSTRACT

The damage development under compressive fatigue loading of impact-damaged CFRP blade stiffened panels were investigated experimentally with help of the ultrasonic-in situ method (USIS). The quantification of the US-rear wall echo shows an experimental correlation with the degree of degradation of the laminate.

INTRODUCTION

The investigation of the capability of a fibre reinforced composite structure to tolerate a reasonable level of damage or defects that may be encountered during manufacture or in service is a very important topic that can influence the design criteria of a structural part. Even with impacts that show little indications of damage at the surface, the matrix damage may be significant, and therefore its ability to stabilize the fibres in compression may be seriously degraded. Thus, impact damage is a critical design consideration and compression is a critical loading mode.

This paper investigates the problem of damage development in blade-stiffened CFRP-panels subjected to impact damage and subsequent compressive fatigue loading.

The damage development was monitored with the ultrasonic-in situ method (USIS) which allows a quantification of the damage progress in the panel. The experiments show that the damage progress and the degradation of the laminate influence the buckling mode of the specimen. An experimental correlation between quantification of US-rear wall echo at different locations and change of strain at a specific load level is shown.

1. MATERIALS AND EXPERIMENTAL METHOD

The materials used to fabricate the blade stiffened panels were T300/Code 69 unidirectional tape prepreg, and A002/Code 69, fabric prepreg, both graphite expoxy 180°C cure systems. The lay-up of the specimens is shown in Tab. 1.

The panels were cured following the manufacturer's recommended cure cycle after the one-shot method - which allows to cure the panel and the stringers in one autoclave cycle - developed in our institute /1/. The specimens were cut from big panels using a diamond saw and the loading surfaces were ground flat and parallel to assure an uniform axial compressive load introduction. The configuration and the dimensions of the tested specimens are shown in Fig. 1.

The barely visible immpact damage (BVID) was introduced in a drop weight test facility /2/ with following parameters: hemispherical steel impactor with a diameter of 10 mm, a mass of 0.625 kg and a drop height of 0.28 m. Visible impact damages in the laminates were introduced by a bullet having a diameter of 5.6 mm and a velocity of 280 m/s (E=107 J). The location of the impact damage was nearly in all cases in the centre of the midbay between the 2 stringers.

The compression fatigue tests (R=10) were performed on a Schenk servo-controlled hydraulic test machine at constant amplitude, load controlled, with harmonic axial loading at a frequency of 5 Hz. The specimens were simply supported at the loaded ends, and the lateral edges were unsupported, as shown in Fig. 2.

The ultrasonic-in situ (USIS) method developed in our institute /3/ was used to evaluate the degree of damage imparted to the panel through the impact and the damage progression due to the compression fatigue loading. USIS allows the US inspection of specimens without removing them from the loading frame. The US-equipment consists of a Krautkrämer KB6000, a selfmade scanning mechanism, Fig. 2, an IBM AT personal computer and an IBM 4381 for data management. The results of the US puls-echo measurement are the standard C-scan and the

H-scan. The H-scan shows the isohypsi, lines connecting areas of equal rear-wall echo (RE) amplitude, offering the possibility to quantify the damage development in the specimen.

The strain was measured at accessible parts of the panel with an extensometer from Schenk with a gage length of 10 mm. The out-of plane deformation of the panel surface could be observed with shadow-Moiré fringe patterns.

2. TEST RESULTS AND DISCUSSION

Laminates 1 and 2 show no big difference in the projected damaged area under BVID conditions, but the difference between both laminates is considerable in the case of the bullet impact, Fig. 3. The visible damage in laminate 1 is a hole of almost the same size as the impacting particle and in laminate 2 the damage extends over the whole length of the specimen, Fig. 9.

The damage development in the scanned area of a composite specimen can be quantified calculating the mean value of all measured US-RE amplitudes, i.e. U_i. Results of the laminate 1 are presented in Fig. 4. The mean value of the measured RE amplitude (U_i) at different load cycles were normalized with their respective value at the beginning of the test, without impact damage (U_1). The ratio U_i/U_1 shows that the specimens tend to pass through distinct phases or stages during their life. The BVID damaged specimens show a decrease of U_i/U_1 after the impact, a moderate decrease of U_i/U_1 over a larger number of load cycles and an acceleration of degradation at the last stage of loading until the specimen eventually fails. All specimens show a similar behaviour, including the not-damaged specimens F201 and F204. This means that initial imperfections and material deterioration are here the predominant factors affecting the damage development and not the imparted impact damage. The high decrease of U_i/U_1 for specimen F100 as loading starts is due to the damage caused by 3 different impacts. The steep decrease of U_i/U_1 for specimens F202 and F203 was caused by a load increase. All the specimens were loaded at about 5000 N.

The ratio of the enlarged impact damaged area (D_i) at the corresponding load cycle to the impact damaged area before fatigue loading (D_i) for BVID specimens are plotted versus the load cycles in Fig. 4. The tendency is similar to those found in the literature /4/, a gradually growing damage area up to a steep increase at the last loading stage before failure.

The behaviour of laminate 2, after the impact damage, is alike to laminate 1, but the second stage is more flat, see Fig. 5. An analogous behaviour of the stiffness degradation was also observed with flat CFRP specimens by different authors /5, 6/. The impact damaged area of the BVID specimens grows faster than in the case of laminate 1, due to the faster damage progression with UD layers, compared to specimens made of fabric layers. Specimen U4 shows a faster growth of damaged area due to 10% higher loading than specimen U2.

A comparison of the change of U_i/U_1 over the load cycles for laminate 1 (F6b) respectively laminate 2 (U5b), which were damaged by a bullet is depicted in Fig. 6. As already shown in Fig. 3 the damaged are of laminate 2 is larger than that of laminate 1, consequently, the decrase of U_i/U_1 is much steeper, meaning that the relative degradation of the whole specimen is higher than in the case of laminate 1. The enlargement of the damaged area at higher load cycles is similar for both laminates, as shown in Fig. 6. A better overview of the damage development is given in Fig. 7 where H-scans at different load cycles are shown for both laminates.

The degradation of laminate 1 under compression fatigue loading causes a change in the buckling mode. At the beginning, there was a symmetric buckling mode shape pattern with 1 halfwave at each bay. During compression fatigue loading the buckling mode shape pattern changed to 2 halfwaves, as shown by the shadow-Moiré pictures of the skin-side of two impact damaged blade stiffened panels in Fig. 8. The length of the halfwave changes, depending on the location of the impact damage in the midbay. The change of the buckling mode was also stated by the change of the slope of the load-strain curves. Strain data measured mainly in the locations RM and LM as well as in the stiffeners (see Fig. 1) for laminate 1 are presented in Fig. 9. The strain level at which the slope change was observed, decreased as the number of cycles was increased. The linear regression line fit is not intended to suggest that the ϵ-N data fit a straight line but, rather, to illustrate an observed trend. For laminate 2 the load was too low for buckling.

A correlation between the ratio U_i/U_1 measured at the limited areas RM and LM and the strain ratio ϵ_1/ϵ_i (ϵ_1 strain at the first loading at a specific load) measured at these locations after different number of load cycles was observed in laminate 1, as shown in Fig. 10. The change of the strain ratio is also a measure of stiffness change. This means, that through the quantification of the US-RE measurement it is possible to find the degree of degradation of a laminate.

3. CONCLUSIONS

The behaviour of impact damaged composite panels under compression fatigue loading is controlled by a number of interacting factors including material properties, load level, initial imperfections, intensity of stress, damage modes, distribution of damage, and intensity of damage.

The damage induced into a laminate through an impact depends among other factors on the material properties and the energy of the impactor. Laminates made of fabric CFRP are more damage tolerant than those made of UD CFRP tapes.

USIS is an appropriate method for the quantification of the US measurement. It opens the possibility to correlate the NDI inspection results with the degree of degradation of the structure. It seems that impact damage in stiffened panels is so important that its effects on buckling should be studied in greater detail.

An understanding of the factors that affect the long-term behaviour of laminates must be achieved to satisfy damage requirements for structural performance.

4. REFERENCES

/1/ D. Wurzel, S. Dehm, in Proc. ICAS 86-4.62
    London (1986)

/2/ K. K. Stellbrink, R. M. Aoki, Effect of Defect on
    the Behaviour of Composites, ECCM IV,
    October 25-28, Tokyo/Japan, 1982

/3/ R. M. Aoki, J. Heyduck, Damage Development in CFRP
    and its Detection, ECCM 3, March 20-23,
    Bordeaux/France, 1989

/4/ R. E. Horton, J. E. McCarty, Damage Tolerance of
    Composites, Engineering Materials Handbook,
    Volume 1, Composites, ASM International 1987

/5/ H. T. Hahn, Fatigue Behaviour and Life Prediction
    of Composite Laminates, Composite Materials:
    Testing and Design, ASTM STP 674, 1979

/6/ R. A. Simonds, et al., Effect of Matrix Toughness
    on Fatigue Response of Graphite Fiber Composite
    Laminates, Composite Materials: Fatigue and
    Fracture, Second Volume, ASTM STP 1012,
    Philadelphia, 1989

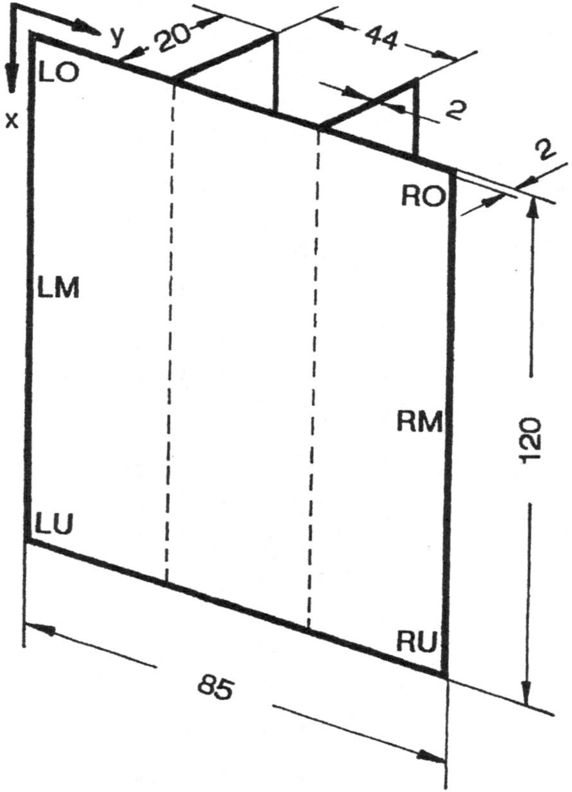

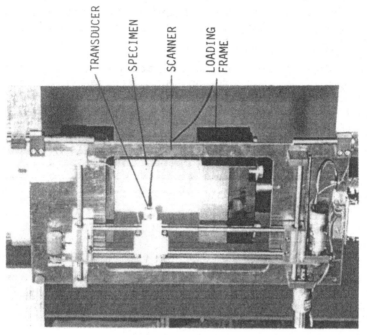

Fig.1 - Specimen geometry

Fig.2 - Compression fatigue test set-up

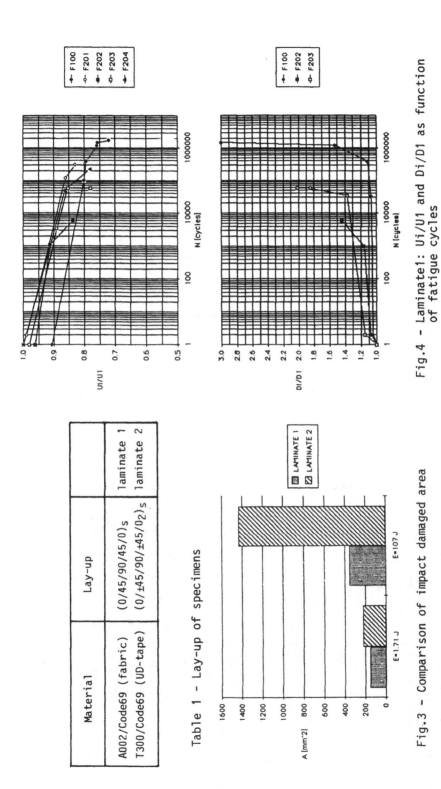

Table 1 - Lay-up of specimens

Material	Lay-up	
A002/Code69 (fabric)	$(0/45/90/45/0)_s$	laminate 1
T300/Code69 (UD-tape)	$(0/\pm45/90/\pm45/0_2)_s$	laminate 2

Fig.4 - Laminate1: Ui/U1 and Di/D1 as function of fatigue cycles

Fig.3 - Comparison of impact damaged area

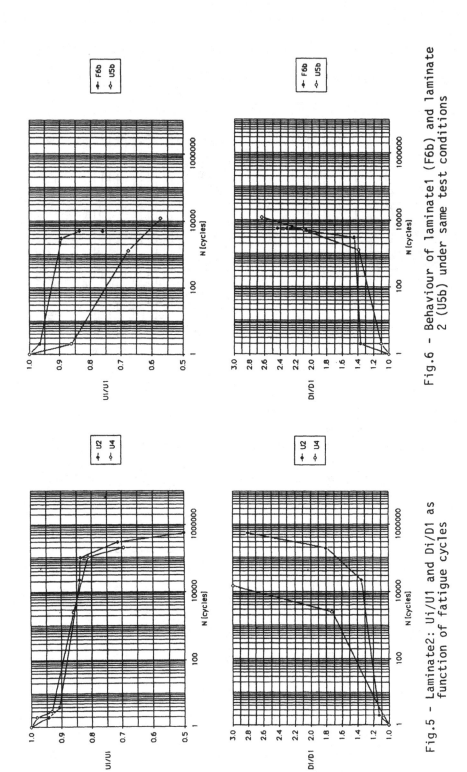

Fig.5 - Laminate2: Ui/U1 and Di/D1 as function of fatigue cycles

Fig.6 - Behaviour of laminate1 (F6b) and laminate 2 (U5b) under same test conditions

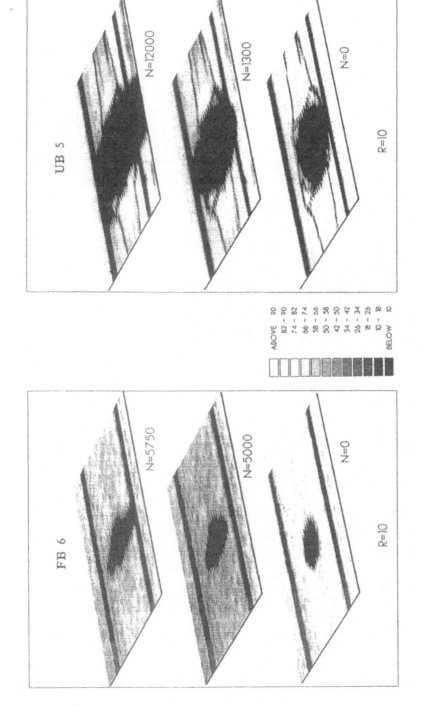

Fig.7 – H-scan pictures after different fatigue load cycles of impact damaged blade stiffened panels laminate1 (left) and laminate2 (right). Lower RE-values mean higher degradation of laminate.

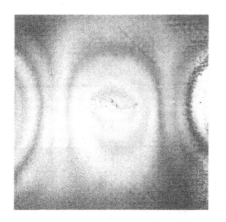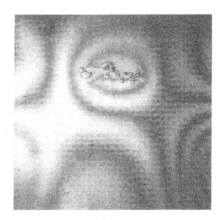

Fig.8 - Shadow-Moiré pictures, laminate1

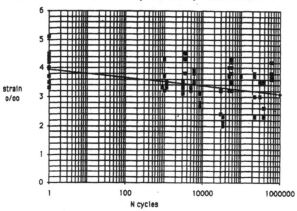

Fig.9 - Strain at onset of buckling mode change, laminate1

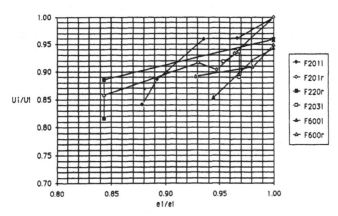

Fig.10 - Correlation between Ui/U1 and $\epsilon 1/\epsilon i$ measured at limited areas LM and RM of laminate1

ENERGY ABSORPTION OF COMPOSITES AS AN ASPECT OF AIRCRAFT STRUCTURAL CRASH-RESISTANCE

C.M. KINDERVATER

DLR - Institute of Structures and Design
Pfaffenwaldring 38-40 - 7000 Stuttgart - West-Germany

ABSTRACT

The paper presents investigations on the energy absorption behaviour of composite materials and generic structural elements with the aim to improve the occupant's survivability in crashing vehicles. Emphasis is given to the energy dissipating mechanisms of crushing composites depending on the constituent materials, crush initiation, laminate architecture, and structural configuration. Results from static and dynamic crushing of aircraft subfloor intersection elements and keel beam sections are presented. First attempts are made to predict the force-deflection curves of those elements.

1. INTRODUCTION

Aircraft structural crashworthiness requires the maintenance of a protective shell around the occupants in addition to absorbing vehicle's kinetic energy. The design goal is a well tuned energy absorption management which includes the landing gear, the fuselage structure and the seat system. Besides the perspective of reduced weight, design flexibility, and low fabrication costs composite materials offer a considerable potential for lightweight energy absorbing structures. This fact also attracts the attention of the automotive industry. Occupant crash protection in a car is a major design issue and mandatory requirements are set. Also for military helicopters crashworthiness regulations exist /1/, whereas for all other aircraft categories up to now only comparable low emergency landing conditions are required.

A number of research groups conduct investigations on the energy absorption behaviour of composites /2-6/. However, the data base is still small and the understanding of how composites absorb energy in a crash is rather limited caused by the numerous influencing parameters. Up to now, no applicable analytical tools are developed to predict those complex fracture processes. Most researchers still rely on experimental studies. In Fig.1 the energy absorption performance of metal and composite elements under compression loading is compared in terms of specific energy and crushforce efficiency. Both design parameters are derived from the force-deflection curve of a crush test. The value of the specific energy relates the absorbed energy to the crushed mass of the structure. High values indicate the lightweight absorber. The crushforce efficiency is the ratio of the average- to peak-crushforce level expressed in percent. The rectangular shaped curve of the ideal-plastic absorber has a crushforce efficiency of 100 percent. As can be seen in Fig.1, many structural composite elements have higher specific energies in combination with high crushforce efficiencies than comparable metal parts. This will now be overviewed in more detail for tubes and generic aircraft subfloor elements such as intersections of beams, and bulkheads, and keel beam sections, respectively.

2. COMPOSITE ENERGY ABSORBING MECHANISMS UNDER COMPRESSION

2.1 Fundamental Crushing Processes

Basically, composites are considered as low energy absorbing materials due to their low strain to failure behaviour. Metals, however, have high failure strains up to 60 percent and their structures can absorb energy by folding mechanisms with plastic hinge formation and material plastification. Plastification in composites is almost not apparent - only to a very small amount in some matrix materials and some synthetic reinforcing fibres. Instability dominated failures caused by global or local buckling generally lead to low energy absorption. The key mechanism of high composite energy absorption under compression loading is a strength controlled formation of a microcrack pattern which spreads out locally in the laminate. The formation of such a propagating crushfront has to be initiated by the action of a so-called "trigger" or crush initiator. The morphology of the crushfront which is generated by the trigger determines the level of the crushforce. Furtheron, this complex sequence of local failures is controlled by the laminate's architecture (fibre and matrix properties, fibre orientation, stacking sequence etc.), and by the geometrical specimen-/or structural configuration. On the macroscopic level several investigators have tried to catalog typical laminate crush modes such as transverse shearing, brittle fracturing, lamina bending, or local buckling /2/. Mostly, a mixture of several modes occurs. Anyhow, the tremendous amount of intra-/and interlaminar failures such as fibre-, matrix-, and interface fractures, delaminations plus friction between broken parts in the crushfront generate a "quasi- plastification" energy dissipating process in the laminate.

2.2 Energy Absorption of Tubular Elements

Cylindrical tube geometry has proven to be one of the most favourable shapes for energy absorption. Also composite cones show high energy absorption performance with the advantage of a self-triggering capability /3/. Tube and cone structures can directly be used as absorbers in seats, mounting devices, landing gears, or car bumpers. In terms of specific energy square and rectangular cross-sections are less effective, Fig.2. The specific energy of composite tubes can exceed aluminium tubes by the factor of two, and the force-deflection curves can have an almost rectangular shape. The scope of specific energies of carbon/epoxy (CFC), glass/epoxy (GFC), and aramid-fibre/ epoxy (AFC) angle-ply tubes are shown in Fig.3. The values range between 15 and 100 kJ/kg depending on the material and fibre orientation. Combinations of angle-ply layers and unidirectional layers in the crush direction can improve the energy absorption, especially with GFC-tubes. Also, unidirectional layers supported in-/and outside by circumferential layers have resulted in high specific energies of 110 kJ/kg /7/.

Slightly increasing cross-sections at one tube end such as bevel or tulip type triggers work very well for crush initiation, Fig.4. Also, crushing over metal dies or cones is commonly used as trigger mechanism, however less efficient. As an example, for a CFC-tube (T300/LY556) with +/- 60° fibre orientation with a bevel trigger 1.3-times higher crushforce level compared to a cone-triggering could be achieved. Other researchers have reported on even more severe influence of triggering, especially for fracture dominated failure modes which are observed with CFC-and GFC- tubes. Triggering is less important with AFC-tubes because they fail primarily by local instabilities (folding) comparable to aluminium tube failures.

Tube failure modes basically do not change from static to dynamic crushing, however, some differences in energy absorption can be observed. CFC-tubes show up to 20 percent degradation under impact loading, whereas material systems such as the new high performance polyethylene fibre Dyneema SK60/epoxy or carbon fibres embedded in a thermoplastic polyamid matrix (PA) show ratios of dynamic to static crushforce levels up to 1.5, Fig.5. Also under elevated temperature (up to 120 ^{o}C) the energy absorption performance decreases mainly caused by the degradation of the matrix, Fig.6. Although thermoplastic matrices (PEEK,PA, PEI) with carbon-, glass- or aramid-fibre reinforcement offer considerable potential from the manufacturing point of view, they do not show better energy absorption performance than epoxy systems, Fig.7.

3. CRASH-RESISTANT AIRCRAFT SUBFLOOR STRUCTURES

3.1 Design Philosophy

The subfloor structure is a crucial element in the energy absorption system of a crashing aircraft, especially when the landing gear is retracted, and for helicopters where the vertical impact velocity is often very high. To minimize cost and weight penalties a dual function structural concept with load carrying capability and crash-resistance should be realized in the subfloor structure whenever possible. Crash simulation studies showed that subfloor element crush-characteristics with moderate initial stiffness and then slightly increasing crushforce levels resulted in tolerable overall crash response. Fig.8 overviews the generic structural components that have to be considered for crash-resistance optimization of the fuselage.

3.2 Energy Absorbing Subfloor Beam Designs

Sine wave web beams are the most efficient subfloor design concepts yet evaluated: they combine high load carrying capability, high energy absorption in the web direction, and post-crush structural integrity by using hybrid lamination techniques (mixture of carbon- and aramid-fibre laminates). Specific crushing stresses of various sine wave beam sections determined under quasi-static and dynamic crushing are shown in Fig.9. The specific crush stress is determined by dividing the average crush stress (avg. crushforce/cross-section) by the laminates' density. The specific crushing stresses in Fig.9 are plotted against the CFC-volume fraction of the web laminate. Pure AFC- and CFC-beam webs and various AFC/CFC- hybrid configurations were investigated. The dynamic specific crushing stresses were not consistently higher than the static stresses or vice versa. Depending on the laminate`s configuration specific crushing stresses between 20 and 60 kJ/kg were measured from the tests. High values were obtained for web laminates having a percentage between 30 and 40 percent AFC. These web laminates also have good post crush integrity and acceptable shear buckling stiffness.

For HSIN-hybrid sine wave webs a procedure suggested in /8/ was verified to predict the specific crush stresses of a structural element, i.e. the sine wave web, by the summation of area-weighted crush stresses of characteristic elements, i.e. ring segments with an opening angle of 145°. In Fig. 9 also the crushing results of HSIN-laminate tubes and 145°-segments are included. It was found that for sine wave webs having an opening angle of the characteristic element less than 180° (half-tube segment) the suggested procedure provides an over-estimation of the web`s crush response. For proper predictions of the specific crush stresses a correction factor depending on the opening angle has to be introduced.

Crush initiators play an important role in the energy absorption of sine wave beams. Caused by the upper and lower caps bevel- or tulip-triggers can not be used. Trigger slots at the bottom of the web reduce the peak loads essentially without affecting the absorption performance, Fig.10. However, notch-type triggers also reduce essentially the shear load carrying capability of the web and therefore the bending strength and stiffness of the fuselage.

3.3 Subfloor Intersections

Structural cruciforms, i.e. intersections of beams and bulkheads represent typical floor structure sub-elements and their crush-characteristics contribute essentially to the overall crash response of a subfloor assemblage. Under vertical crash loads cruciforms are hard-point stiff columns which do create high decelerative loads at the cabin floor level. Starting from an aluminium baseline cruciform taken out from a commuter type aircraft various aluminium and composite cruciform designs having single and multiple notched edge joints, corrugated edge joints, and increasing AFC-share of the laminate at the mid-section were investigated. Fig.11 shows some selected crush-characteristics, and in Fig.12 the energy absorption performance of the elements is overviewed. An improved hybrid design variant (HTP-element) with a column-like mid-section, integrated bevel triggger at the bottom, and tapered edge joints showed 2.8-times higher absolute energy absorption than the aluminium baseline design. Composite elements showed weight savings between 15-30 percent. The highest weight savings compared to aluminium baseline elements were achieved by pure CFC-cruciforms. This elements, however, disintegrated completely under crushing but absorbed comparable high levels of energy.

A first attempt of practice-directed crush prediction of cruciform elements was made /9/. A comparison of predicted and measured load-deflection curves showed good agreement for aluminium as well as composite cruciforms, Fig.13. Based on the fact of qualitatively similar failure curves of composite and aluminium elements for both acceptable approximations of the peak failure- and mean- crushforce were found. As test reproducibility is possible concerning failure loads within 10 percent, and 30 percent concerning energy absorption, approximate calculation methods are adequate rendering results in the same error limits.

4. CONCLUSIONS

- Fracture processes on the microscopic level lead to "quasi-plasti-fication" energy absorption capability of composites which can be equal or even higher than for metals.

- Composite tubes and cones are very efficient light weight absorbers. Their energy absorption level is influenced by geometrical parameters, material selection, and laminate architecture. Triggering plays a crucial role to achieve high and stable crushing.

- For a composite subfloor adequate absolute energy absorption compared to an aluminium structure can be achieved without additional energy absorbing elements.

- Sine wave subfloor beams are very efficient multifunction design concepts. Emphasis should be directed to improve integrated triggers.

- Composite cruciforms compare quite favourable to aluminium elements. Good crush-characteristics in combination with structural integrity by hybridization can be achieved. Weight savings between 15-30 percent were realized. For the composite elements somewhat higher manufacturing efforts have to be taken into account.

- When attention is paid to structure specific failure modes and configuration peculiarities an acceptable prediction accuracy of force-deflection behaviour is possible.

- Applicable analytical tools and appropriate material models still need to be developed to predict the crushing behaviour of composites.

5. REFERENCES

1. MIL-STD-1290 A (AV), Military Standard Light Fixed and Rotary-Wing Aircraft Crash Resistance, 26 Sept. 1988.

2. G.L. Farley, 43. Annual Forum of the American Helicopter Society, 1987.

3. J.N. Price and D. Hull, Composite Science Technology, 28, (1987), 211-230.

4. P.H. Thornton, Composite Science Technology, 40, (1986), 199-223.

5. J.K. Sen, Journal of the American Helicopter Society, Vol. 32-No.2, April 1987.

6. R.L. Boitnott and C.M. Kindervater, 14. European Rotorcraft Forum, 12.-15. Sept. 1989, Amsterdam.

7. C.M. Kindervater and K.F.M. Scholle, in New Generation Materials and Processes, Saporiti/Merati/Peroni, Milano, 1988.

8. G.L. Farley, Journal of the American Helicopter Society, Technical Note, April 1989.

9. C.M. Kindervater and H. Georgi and U. Körber, AGARD-CP-443, December 1988.

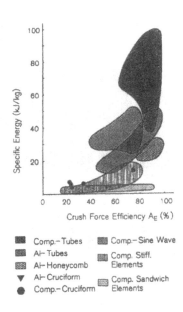

Comp.– Tubes Comp.– Sine Wave
Al– Tubes   Comp. Stiff.
Al– Honeycomb Elements
Al– Cruciform Comp. Sandwich
Comp.– Cruciform Elements

Fig.1 Energy absorption performance of metal and composite elements

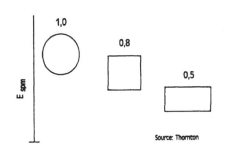

Fig.2 Cross-section influence on the specific energy

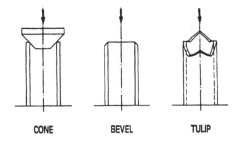

CONE   BEVEL   TULIP

Fig.4 Trigger mechanisms for tubular elements

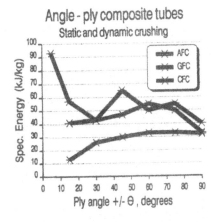

Fig.3 Specific energy of angle-ply composite tubes

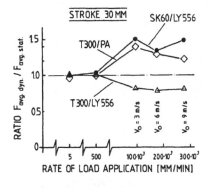

Fig.5 Dynamic to static crushforce levels for various composite tubes

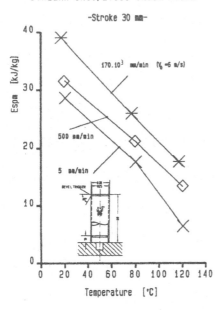

DYNEEMA SK60/LY556 CRUSH TUBES

-Stroke 30 mm-

Fig.6 Temperature influence on specific energy

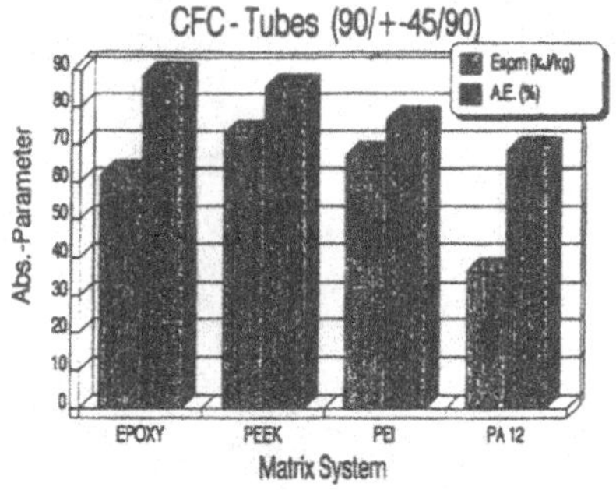

CFC-Tubes (90/+-45/90)

Fig.7 Energy absorption of epoxy versus thermoplastic matrices

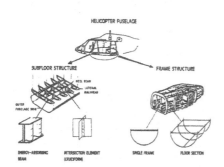

Fig.8 Generic composite structures for crash-resistance optimization

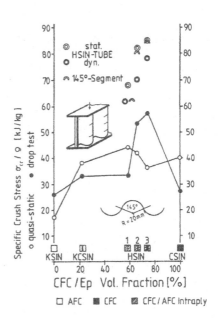

Fig.9 Sine wave beam energy absorption performance

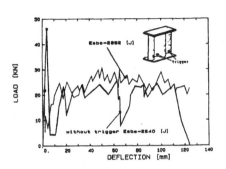

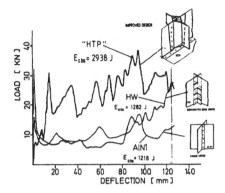

Fig.10 Influence of trigger mechanism on crushing of sine wave beam webs

Fig.11 Cruciform crush-characteristics

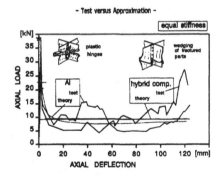

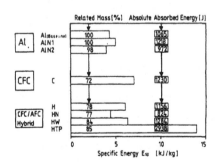

Fig.13 Prediction of cruciform load-deflection curves

Fig.12 Cruciform energy absorption performance

651

TESTING OF COMPOSITES AT HIGH STRAIN RATES USING ± 45° SHEAR TEST

C. WOLFF, A.R. BUNSELL

Ecole des Mines de Paris - Centre des Matériaux
RN 7 - BP 87 - 91003 Evry - France

ABSTRACT:
The aim of this study is to show the use of a high speed hydraulic testing machine (12m/s,50kN) for impact studies on composite materials Experimentaly strain rate effects on the energy absorption in shear using ±45° tensile test on glass-epoxy laminates have been established.

INTRODUCTION

The optimisation of the use of composite materials in vehicles,in particular for structural load bearing components in cars,requires an understanding of the energy absorption mechanisms occuring during a crash.This is necessary so as to optimise the material and the structure in terms of mass,volume and cost saving,and to guaranty better survival chances for passengers in high speed collisions.

When we consider the behaviour of structures submitted to impact,for exemple car components in crash conditions at speed between 20 and 50 km/h,an estimation of strain rates is /1/:

-100 to 1000 /s at the point of impact

-0,1 to 10 /s for the other parts of the structure

In the case of composite structures designed for energy absorption purposes (an exemple is given/by the progressive crushing of tubes /2/),dissipation of energy takes place in a fractured zone near the point of impact.Therefore the highest strain rates also have to be considered .

Depending upon the strain rates the testing methods will be different.For strain rates over 100/s,we are in the field of dynamic effects,which means that the testing speeds are near the wave propagation speed in the material,so that strains can be non uniform.The Hopkinson pressure bar technique /3/,based on stress wave propagation,can be used in this case either in tension or in compression to induce a uniform stress state until fracture of the specimen.

For middle range strain rates (1 to 100/s) little work has been done yet.Rotem and Lifschitz /4,5/ used a falling mass testing arrangement to impact specimen of various ply orientations at speed and strain rate up to 4m/s and 30/s respectively.Others /6/ reached rates up to 500/s with explosives instead of a mass.Internal burst test has been used too with different means of pressurizing /7/,using drop hammer up to 10/s or explosives up to 150/s.

The main problem when studying strain rate effects is the control of testing conditions,especially strain rate,so as to have good reliability of results. Moreover impact testing requires new methodology to be useful in industrial laboratories.

I.M. DANIEL /8/ was the first to use an hydraulic machine for testing in tension at speeds up to 5m/s. More recent work /9/ has shown the possibilities of this technique in controlling strain rate for testing polymers and for fracture mechanics applications.

This work aims to show the possibilities in the control of testing conditions and the problems to be solved. We will use such a device for applying ±45° tensile test ,to study strain rate effect on energy absorption in shear.

The interest of shear testing is to highlight the behaviour of both matrix and fibre-matrix interface with strain rate. Their resistance will have a determinant influence on the cohesion of the composite and therefore on impact properties.

EXPERIMENTAL PROCEDURE
Testing method in shear at high strain rates
High speed hydraulic testing machine
This machine is identical to a static machine but can generate speeds up to 12m/s for 50 kN . The complete displacement is 100 mm,however acceleration and deceleration of the piston needs about 10 to 20 mm each,depending upon the load on the specimen. A slack adaptator is necessary to allow the piston to reach the nominal speed before loading the specimen (fig.1).

The position of the piston is vertical and a high-frequency response (60 kHz) piezo-electric load-cell is put beyond the lower grip. We used self gripping jaws to avoid slipping in the grips.

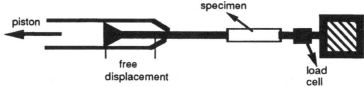

figure 1: Princip of high speed testing.

Shear testing
Application of shear testing methods to high speeds implied that we used fixtures able to be submitted to high speeds up to 10 m/s and therefore they must have a simple design and to be as light as possible to avoid ringing,which will perturb load measurement. Among the different shear test methods /10/ ,off-axis test (unidirectional specimen with fibers oriented at 10 to 15 degrees to tensile direction ,recording of strains in three directions) and ±45° test (ASTM standard D3518-76) agree with the requirements above. However changes of shape of off-axis specimen will introduce bending moments at the ends and also load components transverse to load direction,which are harmful to impact tests.

For ±45° test specimen,analysis· of in-ply stresses and strains from recording of longitudinal ε_L and transverse ε_T strains,and tension stress σ ,gives inplane shear response of the material :

-Shear stress is : $\qquad \tau = \sigma /2$

-Shear deformation is : $\qquad \gamma = \varepsilon_L - \varepsilon_T$

Strain gauge measurement

Test on ±45° laminates induce large strain and we had strain of about 13% in the longitudinal direction and 9% in compression in transverse direction. We used high-strain strain gauges Kyowa KLM6=A9 with EC30 bonding.

Wheastone bridge which are adequate for very small strain measurement with high precision,becomes non linear for large deformation (for 10% strain ,non linearity is 10%),and we prefered to build a constant intensity powersupply with a 100 kHZ bandwidth for measuring large variations in resistance (about 25Ω for a initial 120Ω strain gage). For large deformations we have to use real strain depending upon resistance of the strain gage by:

$$eps=(1/f).Log(1+\Delta R/Ro)$$

f:gage factor,ΔR:resistance variation,Ro:initial resistance.

Material and specimen

The material is glass-epoxy composite made by filament winding and cured in a vacuum air bag to produce a laminated plate. (The matrix material was epoxy Dow Chemical DER331 with anhydride curing agent and Vetrotex P122 E-glass fibers). The nominative dimensions of the test section were 100 x 25 x 2 (mm3),aluminium tabs 1mm thick were bonded to the specimen.

Data recording

We used a four channel digital oscilloscope to store and then to transfer and to process the results on a PC computer.

We recorded the response of the longitudinal and transverse strain gauges,the load and the displacement of the moving grip.

RESULTS

Strain gauge measurement has shown good reliability. For 58 strain gauges used,6 failed during test,2 were unstuck after impact.

Analysis of high speed test

Testing in impact conditions may be subjected to various perturbations,thus to perform a good analysis of test results for studying strain rate effects,we must determine the best test conditions.

Several facts may cause inadequate test conditions for our experiments:

-load transmission to the specimen:the movement of the piston must be transmitted in a time as short as possible to the specimen,so as to have loading at constant speed Contact problems in the slack-adaptator may be important here.

-several causes may change strain rate during experiment:

+misalignement of grips and specimen

+slipping in the grips

+homogeneity of the material (it may change the response of strain gauge with position on specimen)

+ringing of the fixtures which will perturb load cell response.

From this analysis we can infer three criteria of good test conditions:
-records of strains versus time show that constant strain rate is achieved after 1.5% of deformation. Thus strain rate effect analysis may only have meaning beyond 3% shear strain.
-comparison between deformation from the strain gauge (local strain) and deformation from displacement of the upper grip (overall strain) shows when transmission of speed to specimen was good (no problems of misalignement or slipping),and when overall and local strains are in agreement (homogeneity of strain).
-ringing will become significant with increasing impact speed, due to shock waves becoming much more important. The actual grips,designed to avoid the slipping of the specimen,have a ringing frequency of 2.3 kHz. The perturbation of load cell response becomes too much important above 1m/s to carry out analysis on rough curves.

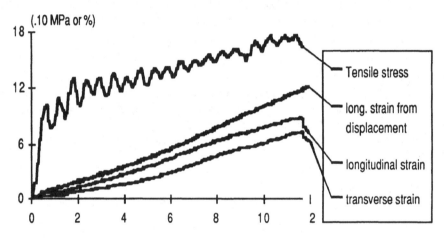

figure 2 : stress and strain records from test at 1m/s.

Discussion
Following the analysis,we selected tests in agreement with criteria and drew shear stress strain curves for speeds between 2mm/mn and 1 m/s (fig.3) from ±45° test relations.
Scatter of the experiments was then smaller than 15% of measured stress values,and usually smaller than 5%. Therefore we can conclude a good reliability of test results in controlled testing conditions.
Trends are towards an increase of flow stresses and ultimate stresses with strain rate,and decreasing of ultimate strain. Moreover change is much more important near yield point (from 40 MPa to 63 MPa) than near fracture point (from 70 MPa to 85 MPa).
Deformation are very large showing an important ductility in shear and strains decreasing from 19% to 15%.
Similar behaviour of glass-epoxy composite has been shown by /5/ on ±45° tensile specimen with axial strain of 5%,and /7/ on burst tests on ±45° tubes with hoop strain of 14% for strain rates near 10/s and even up to 80/s.

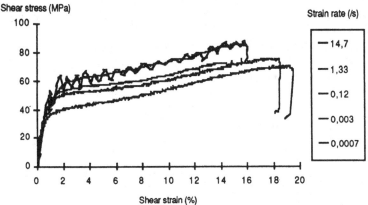

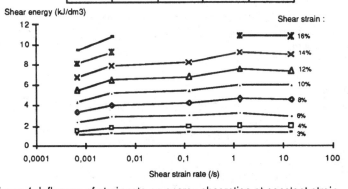

Shear stress (MPa)

Strain rate (/s)

— 14,7
— 1,33
— 0,12
— 0,003
— 0,0007

Shear strain (%)

figure 3:Influence of strain rate on shear stress-strain curves

A good analysis of strain rate effect can be achieved by considering the evolution of stress or,much better in our case,of absorbed energy from 0 to a given strain level. This is shown figure 4 for strains between 3 and 16% and strain rate up to 10/s (test at 1m/s).
Effect of strain rate show two intervals of strain rate sensitivity:around 0.001 and between 0.1 and 1/s. This variations are significant compared to experimental scatter.
A good representation can be given by a Eyring type law used for polymers:

$$W = A \cdot \gamma \cdot \log(\dot{\gamma}) + B \cdot \gamma \qquad \text{(W energy in kJ/dm}^3\text{)}$$

The coefficients for the four strain rate intervals are:

	1	2	3	4
A	0.1	0.02	0.09	0.006
B	0.9	0.65	0.7	0.8

Shear energy (kJ/dm3)

Shear strain :

16%
14%
12%
10%
8%
6%
4%
3%

Shear strain rate (/s)

figure 4 :Influence of strain rate on energy absorption at constant strain.

The main limitation to the ±45° tensile test is due to the stress component normal to the fibres which will lead to premature failure under biaxial stress state. Its value depends on anisotropy and Poisson ratio and represents 20% of axial stress for glass-epoxy laminate./11/ has shown the possibility of optimising fibre angle at ±41° in laminates to reduce the normal stress component.For this angle,the stress redistribution due to shear non-linearity leads to a small compressive stress.
However a comparison with the static rail-shear test /12/ shows a good agreement of shear response for glass-epoxy composites.

For the previous analysis we assumed that the specimens did not show strain non-uniformity due to shock wave effect [6].

If we consider an homogenous elastic material,a better idea can be given by a comparison of time of wave propagation through the specimen (t_p) and time to failure (t_f) :

$t_p = l / c$ l : length of specimen   v : testing speed

$t_f = l . \varepsilon / v$ ε : failure strain   c : elastic wave propagation velocity

Failure will take place after propagation of wave if : $t_p < t_f$.

So the critical corresponding velocity will be : $v_{\sigma 1} = \varepsilon . c$.

If we assume that tolerable strain variations due to shock wave is $\Delta\varepsilon$,the corresponding velocity is $v_{\sigma\Delta} = \Delta\varepsilon . c$.

The values of critical speeds for a glass-epoxy (modulus 8GPa,density 2,c=2000 m/s) are :

ε (%)	$v_{\sigma 1}$ (m/s)	$v_{\sigma\Delta}$ (m/s) for $\Delta\varepsilon/\varepsilon$=10%
1	20	2
10	200	20

In our case (ε =10%,v=1m/s) we are beneath these critical velocities,which moreover do not consider damping in the material. However when testing materials with small failure strains,for example pure resins,shock wave propagation may become of significance.

CONCLUSION

High speed testing with a hydraulic machine applied to shear testing has shown good reliability of results when testing conditions are controlled.

Extension to higher speeds over 1m/s will need:

-lightening of grips to increase ringing frequency,but with sufficient clamping force to avoid slipping of specimen.

-Fourier transform analysis in the case of ringing.

-Evaluation of stress wave effects and strain non uniformity on specimen.

The analysis of the ±45° shear test needs a better knowledge of stress biaxiality on shear response and failure.

Acknowledgement:

High speed testing has been performed at the Impact laboratory,Mechanical Engineering Department,University of Liverpool.

This study was supported by RENAULT Research Group.

REFERENCES

1.STELLY M., Revue Française de Mécanique n°73,1980.

2.Wolff Ch.,Bunsell A.R., 5ème jour. nationale DYMAT,Bordeaux 14/12/89.

3.Harding J.,Sc. and Eng.of Comp. Mat.,Vol1,n°2,(April-June 1989).

4.Rotem A.,Lifshitz J.M., 26th Annual SPI Conf.,10-G(1971)

5.Lifshitz J.M., J.Comp.Materials,10,p.92(1976)

6.Armenakas A.E.,Sciarammella C.A., Exp. Mechanics,october,p.433,(1973).

7.Al-Salehi F.A.R.,Al-Hassani S.T.S.,Hinton M.J., J.of Comp. Materials,Vol 23,p.288,(March 1989).

8.Daniel I.M., Int.Conf. of Comp. Materials,p.1003.

9.Béguelin Ph.,Barbezat M., 5ème jour. nationale DYMAT,Bordeaux 14/12/89.

10.De Charentenay F.X.,Kamimura K., Annales des composites AMAC,Vol.2,(Décembre 1982).

11.Lifshitz J.M.,Gilat A., P.444,(December 1979).

12.Sims D.F., J.Comp.Materials,7,p.124(1973).

IMPACT RESISTANCE AND COMPRESSIONAL PROPERTIES OF THREE-DIMENSIONAL WOVEN CARBON/EPOXY COMPOSITES

L. DICKINSON, M.H. MOHAMMED, E. KLANG

North Carolina State University
Box 8301 - 27695 Raleigh NC - USA

ABSTRACT

This paper evaluates 3–D orthogonal woven carbon/epoxy composites. Preforms were manufactured on an automatic 3–D weaving machine developed at N. C. State University, USA. Three different orthogonal structures were produced. Preforms were consolidated with two types of epoxy using vacuum infiltration molding. Consistent quality of consolidation proved to be very difficult. The mechanical properties were compared with traditional tape and fabric quasi–isotropic laminates. While the in–plane properties were inferior, the 3–D orthogonal woven composites offer improved damage tolerance.

I. INTRODUCTION

The properties of advanced composite do not only depend on the reinforcement and matrix materials, but also on the arrangement of the fibers and their interface with the matrix. By proper choice of fiber orientations, a wide range of mechanical and physical properties may be designed in laminated composites. However, the properties of laminates in the thickness direction are typically very poor with delamination being a very common failure mode of these materials. Several processes and structures have been developed to produce composites with sufficient reinforcement in the third (thickness) direction. These include stitching multiple layers of 2–D fabrics, 3–D braiding and 3–D weaving. Stitching layers together requires additional processing and labor. The direct manufacture of a three–dimensional fiber preform offers economical advantages as well as an opportunity for improved mechanical properties. Improved damage tolerance and the ability to form specific net or near net shapes are of particular importance.

II. PREFORM MANUFACTURE

A method of manufacturing 3–D multi–layer structures has been developed at the Mars Mission Research Center of North Carolina State University. Mohamed et

hydraulic grips at a constant extension rate of .254 cm/min (0.1 in/min). SBC and CAI testing was done on a 534 kN (120 Kip) Baldwing machine at a constant rate of loading of 44.4 kN/min (10 Kip/min). Impacting was done with a low velocity air gun (developed at NASA Langley) firing small aluminum spheres . The CAI panels were compressed in a test fixture which held the panel on the top and bottom surface at the edges in a simply supported fashion while applying the compressive load. A more complete description of the test apparatus and procedures is given by Dexter et. al., /3/.

For the 828 composites: three tensile and SBC tests were conducted for each structure; two panels of each structure were impacted one at 27.1 J (20 ft-lbs) energy and one at 40.7 (30 ft-lbs) for the CAI test. For the 3501-6 composites: two tensile and three SBC tests were conducted for each structure; one panel of each structure was impacted at 27.1 J (20 ft-lbs) for the CAI. Pulse echo C-scans were performed on the CAI panels with a Testech C-scan Bridge and tank with an Automation Industries ultrasonic instrument. Image analysis software was used on the C-scan data to determine the damage area for the impacted specimens.

5.2 Results and Discussion

Estimates of the fiber volume fractions of the O1, O2, and O3 preforms were calculated to be 46, 45 and 44 % respectively, /2/. The averages (little variation) of tensile and compressive data are presented in table I. Table II is a summary of the CAI test data. Impact energy is presented in energy per unit thickness of the panel to account for variations in the panel thickness.

The 3501-6 panels exhibited significantly better tensile and in particular, compressive properties than did the 828 panels. The 828 is a much softer resin as was seen in the modes of failure in compression. The 3501-6 panels failed in a typical brittle manner while the 828 composites resulted in a more ductile type failure. The 828 matrix yielded allowing the fibers to buckle. The void content in the resin pockets of the 828 panels was noticeably higher than that of the 3501-6 panels. The observed variations of tensile and compressional properties among the structures follow no consistent pattern and may be explained by the poor consolidation quality of some of the panels. The tensile strengths may be misleading due to the fact that nearly all the coupons failed near the grips.

Fowser et. al reported on a 3-D orthogonal T300/3501-6 composite material, /4/. While noting that the structure and volume fractions were different, he stated that the 3-D fabric was weaker and less stiff than 0/90 laminates. Fowser also found microcracks in the resin rich pockets caused by the thermal stresses induced during curing. He put forth the theory that the resin pockets "act as initiation sites for failure." Similar cracks can be seen in the 3501-6 panels used in this study, figures 3-4. The 828, while having more voids, developed fewer cracks due to being cured at a lower temperature.

As with the compression/tension data, there are no conclusive patterns relating the CAI data to the O1, O2 and O3 structures (table II). Two of the impacted panels [828/O1 13.7 J (10 ft-lb) and 3501-6/O1] did not fail through the impacted areas. Variations in the quality of resin impregnation and slight differences in thickness may account for the inconsistent variation in properties. The least amount of Z yarns (O3) may also be more than is necessary to inhibit impact damage. Most all of the CAI panels failed in a buckling or combination of buckling and compressive modes. Buckling is not typical in the CAI test for composites. Typical CAI tests are conducted on panels with the same or greater thickness but with a higher modulus. The 3501-6 panels were compressed somewhat during the molding process and being thinner thus received a higher impact energy per unit thickness. The 3501-6 is also a more brittle resin. These two facts may explain the greater reduction of compressive strength in the 3501-6 panels.

al. described the process and the automatic 3–D weaving machine, /1,2/. The structures are formed in a way similar to that of traditional weaving. Yarns (tows) are placed axially (X, warp direction), transversely (Y, filling direction) and in the thickness (Z) direction. As in ordinary weaving, warp yarns are separated into layers to allow for filling insertion. In this case there is more than one layer, and thus more than one shed (area between separated layers). Multiple needles are used to insert doubled filling yarns between the layers. Z yarns are fed into the machine parallel to the warp yarns and separated into two layers controlled by harnesses. When the Z yarns are moved by crossing the harnesses, a vertical component of yarn is laid into the structure. Figure 1 shows a photograph of the automatic 3–D weaving machine.

III. PREFORM STRUCTURE

Preforms can be manufactured with different cross sections (e.g. T, I shapes) and different internal structures. 3–D orthogonal woven panels with a rectangular cross section are described in this paper. A schematic of the 3–D orthogonal structure is shown in figure 2. Figures 3 and 4 are micrographs showing actual structure. Three variations of the orthogonal structure with three different amounts of Z direction reinforcement were made. The first orthogonal (O1) structure has one Z yarn for every warp (axial) yarn (figure 2). The second (O2) structure has one Z yarn for every two warp yarns while the third (O3) has one Z yarn for every three warp yarns. The number of warp yarns was kept constant in all three structures. These structures differ from woven fabrics not only because of the Z reinforcement, but also because there is no internal in plane yarn crimping (i.e. the yarns remain essentially straight). Interlacing occurs only at the surfaces of the preform where the filling and Z yarns change direction. Typically untwisted yarns are used and the tows assume a rectangular cross section (figures 3,4).

IV. CONSOLIDATION (RTM)

Preforms were made of GY30–500 Celion® carbon fibers. The warp and Z yarns were 12k tows while the filling yarns were doubled 6K resulting in 12k tows. Preforms were consolidated with two different resins. A set of panels were consolidated with EPON 828, a room temperature curing epoxy selected because it was easy to work with. A second set of panels were consolidated with 3501–6 resin (Hercules) at NASA Langley Research Center, Hampton VA. A similar resin transfer molding (or more specifically, vacuum infiltration molding) process was used in each case. Cured panels measured approximately 27.9 X 15.2 X 0.635 cm (11 X 6 X 1/4 in.).

All the fibers in the bundles seemed to be well wetted. However, voids were found in the resin rich pockets unique to this structure. Some of these voids may be associated with the shrinkage of the epoxy. The 3501–6 panels had fewer voids but more microcracks. These cracks were mainly in the resin pockets, but sometimes extended into the fiber bundles. Typical voids and microcracks are shown in figures 3 and 4.

V. PHYSICAL TESTING

5.1 Test Procedures

Tensile coupons measured 2.54 X 25.4 or 2.54 X 22.9 cm (1 X 10 or 1 X 9 in.), Short Block Compression (SBC) coupons 4.44 X 3.81 cm (1.75 X 1.5 in.) and Compression After Impact (CAI) panels 12.7 X 25.4 cm(5 X 10 in.). Tensile testing was done on untabbed specimens with a MTS 222 kN (50 Kip) machine with

Dexter et. al. reported the reduction of compressive strength after impact for quasi–isotropic panels made from AS4/3502 tape and T300/934 8–harness satin woven fabric was 75% and 66% respectively. Figures 5 and 6 compare Dexter's damage area and ultimate strain with O2 and O3 data (averaged). Table II and figures 5 and 6 show that the 3–D orthogonal structure, with ordinary epoxies, offers damage tolerance better than those of conventional tape and 8–harness fabric composites.

While the 3–D orthogonal material may not compare favorably with laminates in in–plane properties, it does offer potential for better damage tolerance. The 3–D orthogonal woven composite is not a quasi–isotropic material. However, Hatta et. al. report that the in–plane shear properties are better than those of a woven laminate, /5/. If future work produces 3–D orthogonal fabrics with more in–plane fibers and an optimum number of Z fibers, and the problems of consolidation can be solved, the 3–D orthogonal material offers potential for an impact resistant composite. This material may prove to be economical as well since the fabric can be made on an automatic machine in one processing step eliminating the labor intensive laminate layup processing step.

VI. CONCLUSIONS

1. 3–D orthogonal fabrics can be produced on an automatic machine in one processing step.

2. Consolidation of the 3–D orthogonal material is very difficult and is a subject of further work by the authors.

3. Given improvements in consolidation and optimum material design, the 3D material may be able to compete with traditional laminates for design purposes.

4. Even without quality consolidation the 3–D orthogonal material offers better damage tolerance than traditional laminates.

5. Although 3–D orthogonal material can not be made to simulate an isotropic or quasi–isotropic material, its potential in–plane shear weakness may not be as severely limiting a factor for certain design applications as is generally believed.

VII. ACKNOWLEDGEMENTS

The authors would like to express their appreciation for the assistance of Mr.s H. Benson Dexter, Donald Smith, Gregory Hasko and Richard Chattin at the NASA Langley Research Center, Hampton Va. USA.

This work is supported by NASA under Grant No. NAGW–1331. The authors are grateful for this valuable support.

VIII. REFERENCES

1. Mohamed, M. H., Zhang, Z., and Dickinson, L. C., "Manufacture of Multi-layer Woven Preforms," ASME, Advanced Composites and Processing Technology, MD–Vol. 5, Book No. 00484, pp. 81–89, (November 1988).

2. Mohamed, M. H., Zhang, Z., and Dickinson, L. C., "Weaving of Net Shapes," First Japan International SAMPE Symposium, Chiba Japan, (Nov. 28–Dec. 1, 1989).

3. Dexter, B. H. and Smith, D. L., "Woven Fabric Composites with Improved Fracture Toughness and Damage Tolerance," NASA Conference Publication 3038, Fibertex, (1988).

4. Fowser, S., Wilson, D., Chou, T. and Pipes, R., "Influence of Constituent Properties and Geometric Form on the Behavior of Woven Fabric Reinforced Composites," Progress Report for NASA Grant NAG–1–378, (June 1986).

5. Hatta, Hiroshi, and Yamashita, Shu, "Compressive Strength of Tri–Axial 3D Fabric Composites," First Japan International SAMPE Symposium, Chiba Japan, (Nov. 28 – Dec. 1, 1989).

Table I. Tensile and Compressional Data.

	STRUCTURE/ RESIN	STRENGTH (Ksi)	STRENGTH (MPa)	MODULUS (Msi)	MODULUS (GPa)	ULTIMATE STRAIN (%)	POISONS RATIO
T	01/828	46.5	321	5.58	38.5	**	0.081
E	02/828	35.0	241	5.53	38.1	0.72	0.049
N	03/828	41.9	289	5.92	40.8	0.85	0.052
S							
I	01/3501-6	66.5	459	6.76	46.6	1.18	0.027
L	02/3501-6	80.7	557	7.74	53.4	1.27	0.039
E	03/3501-6	82.0	565	7.42	51.2	1.30	0.031
	01/828	70.4	486	5.26	36.2	**	0.084
C	02/828	64.9	447	5.21	35.9	**	0.059
O	03/828	73.7	508	5.42	37.4	**	0.062
M							
P	01/3501-6	82.4	568	6.69	46.1	**	0.041
	02/3501-6	92.7	639	7.69	53.0	**	0.041
	03/3501-6	86.3	595	6.98	48.1	**	0.027

NOTE:
** STRAIN GAGES FAILED BEFORE ULTIMATE LOAD.

Table II. Compression After Impact Data.

STRUCTURE/ RESIN	IMPACT ENERGY/ UNIT THICKNESS (in-lb/in)	IMPACT ENERGY/ UNIT THICKNESS (J/cm)	STRENGTH (ksi)	STRENGTH (MPA)	ULT. STRAIN (%)	DAMAGE AREA (in²)	DAMAGE AREA (cm²)	STRENGTH LOSS ## (%)
01/828**	423	18.8	30.9	213	0.640	0.36	2.3	34
02/828	972	43.2	24.8	171	0.526	1.25	8.1	29
03/828	852	37.9	28.8	199	0.595	1.38	8.9	31
01/828	1277	56.8	28.4	196	0.580	0.98	6.3	39
02/828	1234	54.9	22.7	157	0.490	3.33	21.5	35
03/828	1332	59.3	26.8	185	0.558	0.84	5.4	36
01/3501-6	1092	48.6	34.0	234	0.631	++	++	49
02/3501-6	1128	50.2	31.7	219	0.515	1.85	11.9	61
03/3501-6	1294	57.6	29.4	203	0.552	2.45	15.8	64

NOTES:
** THIS SPECIMEN WAS HIT WITH 2 BALLS, FASTEST ONE AT 13.6 J (10 FT-LBS).
++ THE DAMAGE AREA WAS UNCLEAR IN THE C-SCAN.
THE PERCENTAGE LOSS IN COMPRESSIVE STRENGTH DUE TO IMPACT.

Figure 1.   Automatic 3-D
Weaving Machine.

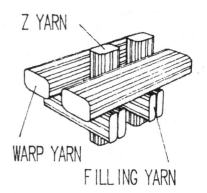

Figure 2. 3-D Orthogonal (O1)
Structure.

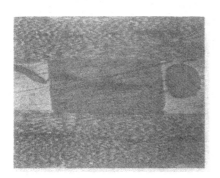

Figure   3.   Micrograph
(49.2M) of O1 Structure,
Showing a Void in Resin
Pocket, and Crack Through
Warp Yarn.

Figure 3.   Micrograph (49.2M)
of O1 Structure, Showing
Microcracks   in   Resin
Pockets.

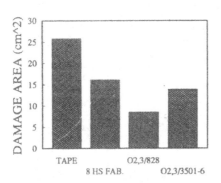

Figure   5.   Impact   Damage
Area of Compared Fiber
Structures.

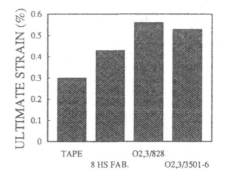

Figure   6.   Ultimate   Strain
After Impact for Compared
Fiber Structures.

NON DESTRUCTIVE TESTING
ZERSTÖRUNGSFREIE PRÜFUNG

Chairman : **Mr. C. LE FLOCH**
Aérospatiale

INELASTIC MODELLING OF ACOUSTIC EMISSION SIGNATURE IN COMPOSITE MATERIALS

R. PYRZ

University of Aalborg
Pontoppidanstraede 101 - 9220 Aalborg - Denmark

ABSTRACT

Experimental and analytical techniques are employed in the present study to identify a damage parameter with the acoustic emission signature in woven glass fiber reinforced composite under tensile loading. The damage parameter is estimated by the use of stereological methods and its relation to acoustic emission monitoring is enlighted.

INTRODUCTION

Due to large variety of micro-damage modes influenced by the composite build up, the constituent properties as well as by interactions among constituents the identification procedure still faces significant difficulties particulary in situations when one wants to collect the information about an internal pattern changes in real time. We can not underestimate the need for the microstructure descriptors which are monitored preferably with the utilization of non-destructive techniques and at the same time are uniquely related to microstructure rearrangements. The inclusion of such descriptors characterizing the generic material property and microstructural information is essential in constructing effective constitutive theory on macrolevel. In the present study the microstructure parameter which reflects the development of particular microcracks mode is determined with utilization of stereological principles and its relation to acoustic emission signature is shown.

I - EXPERIMENTAL PROCEDURE

The material used in the present study was woven glass fabric-polyimide composite with plane weave fabric. In order to limit the modes of

667

damages only one layer of reinforced fabric with thickness 0.16 mm was used, the warp and weft yarns being oriented at 45° to the loading direction. The glass fiber content in the warp yarns was about 20% larger than in the weft yarns. During all tensile experiments performed with constant strain rate $5 \cdot 10^{-5}$ [1/s] the pattern of microcracks was monitored by taking microphotographs at predetermined time intervals while the acoustic emission was detected with wide-band transducer supplemented with guard censors in order to eliminate extraordinary signals coming from gripping area.

II - RESULTS AND DISCUSSION

The predominant mode of damage in the form of matrix cracking along the warp direction was observed at moderate strain levels, Fig. 1. The average length of individual microcrack corresponded to the dimension of the plane weave cell. At larger strains microcracks tend to join along the preferred direction and a number of small secondary microcracks appears along the weft direction. The microstructure parameter D is defined as the primary microcracks length density and calculated from microphotographs with the use of stereological concepts /1-2/. The result of calculations is shown in Fig. 2 where separation of the microstructure parameter evolution into two distinctive regions is clearly seen. This finding is supported by the acoustic emission, Fig. 3. The saturation strain ϵ_s divides the development of AE hits into two stages suggesting different mechanisms operative in the first and second region as observed on the microphotographs. A direct correspondence between the AE hits and the microstructure parameter D is seen in Fig. 4, where curves were normalized with respective values achieved at saturation point. Such correspondence does not exist for second region because the calculations of the parameter D do not include the secondary microcracks on the one hand, and the number of AE hits is enlarged by these microcracks and their interaction with primary microcracks on the other. AE parameters exhibit the Kaiser effect while unloading the specimen in the first region, Fig. 5. This behaviour may be analytically resembled by the microstructure parameter if we relate D to the strain in the following way

$$D = ||D||_\infty - a \, ||\epsilon||^n_\infty, \qquad \epsilon \leq \epsilon_s, \qquad (1)$$

with $a = 1.199 \cdot 10^6$ and $n = 3.69$. The symbol $|| \ ||_\infty$ designates the Lebesque norm with the property that it takes a present value of the argument when the argument is monotically increasing and it is constant, keeping its previous maximum value otherwise /3-5/. On the physical grounds such behaviour is fully acceptable since the microcracks nucleation cannot take place under unloading path and the pattern of microcracks and their changes terminates at the point of unloading switching on again whenever the previous strain is exceeded. It is worth to mention that formula (1) describes the parameter D from Fig. 4 as well. For higher strain amplitudes $\epsilon > \epsilon_s$ applied in the loading-unloading-reloading cycle a significant acoustic emission is monitored at strain levels below those previously attained, leading to the breakdown of the Kaiser effect. This occurs because of the redistribution of residual stresses during unloading period. The result of the redistribution is that additional microfailure will occur during reloading. In such a case the acoustic emission can be described by the norm $||\ ||_p$, /3/, while the parameter D must be redefined in order to identify the additional emission. Attempting to mathematically model the microstructural behaviour as observed by a non-destructive measurement it appears that key variables so happened to have exactly the same mathematical representation as the Lebesque norms. It may therefore in some way physically justify the use of norms in constitutive theory.

REFERENCES

1 - Pyrz R., in "Inelastic Deformation of Composite Materials", (G. J. Dvorak, ed.) (1990), RPI Troy
2 - Weibel E.R., "Stereological Methods", vol. 2, (1980), Academic Press, N.Y.
3 - Pyrz R., Special Report No. 2, (1989) Inst. of Mech. Engng., Univ. of Aalborg
4 - Schapery R.A., in "Advances in Aerospace Structures and Materials" (S.S. Wang, W.K. Renton, eds.)(1981) 5-20, ASME AD-01
5 - Fitzgerald J.E., Vakili J., Exp. Mech., 13 (1973) 504-510

Fig. 1. - Damage pattern at strain $8 \cdot 10^3$ μs and $32 \cdot 10^3$ μs, respectively.

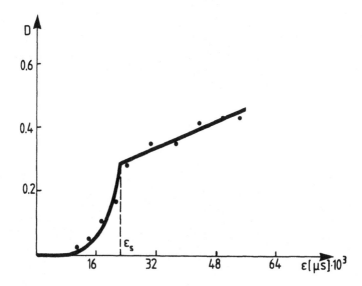

Fig. 2 - Cumulative distribution of microstructure parameter D.

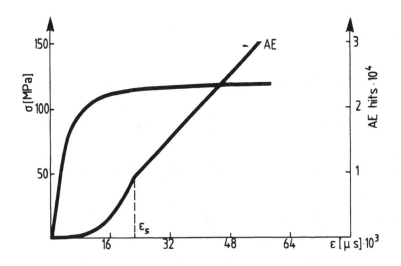

Fig. 3 - Stress-strain diagram and AE hits.

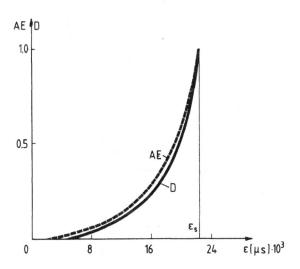

Fig. 4 - Normalized AE hits and microstructure parameter D.

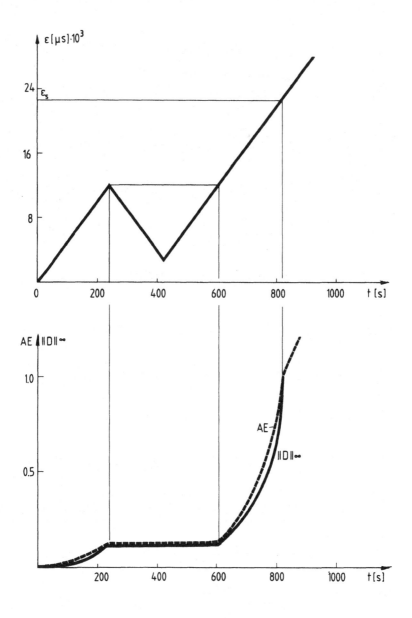

Fig. 5 - Kaiser effect.

SPREADING OF LIQUID DROPLETS ON CYLINDRICAL FIBERS : ACCURATE DETERMINATION OF CONTACT ANGLE

H.D. WAGNER, H.E. GALLIS, E. WIESEL

Materials Research Dept - The Weizmann Institute of Science
76100 Rehovot - Israel

ABSTRACT

The characterization of the physicochemical nature of interfaces is a key problem in the field of advanced fibrous composites. In the case of thin cylindrical monofilaments wetted by liquid polymers, the macroscopic regime contact angle, which reflects the energetics of wetting at the solid-liquid interface, is difficult to measure by usual methods. A numerical method is proposed in this article for the calculation of macroscopic regime contact angles from the shape of a liquid droplet spread onto a cylindrical monofilament. This method, which builds on earlier theoretical treatments by Yamaki and Katayama /1/, and Carroll /2/, very much improve the accuracy of the contact angle obtained. Experimental results with high-strength carbon, para-aramid, and glass fibers, are presented to demonstrate the high degree of accuracy of the method proposed.

I-INTRODUCTION

The macroscopic regime contact angle, which reflects the energetics of wetting at the solid-liquid interface, is difficult to measure by usual methods in the case of very thin cylindrical fibers. On the other hand, the characterization of the physicochemical nature of interfaces is a key problem in the field of advanced fibrous composites. A method for determining the contact angle of liquid droplets on cylindrical surfaces was developed simultaneously and independently by Yamaki and Katayama /1/, and Carroll /2/. Unfortunately, measurements of the contact angle based on this method are unable to provide an accuracy of better than about 5 degree. A simple extension of the method of Yamaki and Katayama, and Carroll, is presented here, from which highly accurate values of the contact angle may be obtained. This is demonstrated experimentally from the spreading of glycerol and/or epoxy droplets on carbon, aramid, and glass fibers.

II-PROFILE OF A LIQUID DROPLET ON A MONOFILAMENT

Yamaki and Katayama /1/ and Carroll /2/ have derived an analytical expression for the reduced droplet length L as a function of the reduced maximum droplet thickness T and of the macroscopic contact angle θ (see Figure 1), as follows:

$$L = 2[a\, F(\phi,\ k) + T\, E(\phi,\ k)] \tag{1}$$

where L=(droplet length/cylinder radius), T=(droplet half-thickness/cylinder radius), and a is given by

$$a = \frac{T\cos\theta - 1}{T - \cos\theta} \tag{2}$$

F and E are Legendre's standard incomplete elliptic integrals of the first and second kind, respectively, which are tabulated. The arguments ϕ and k are calculated by using the expressions:

$$\sin\phi = \left(\frac{1}{k^2}\left(1 - \frac{1}{T^2}\right)\right)^{\frac{1}{2}} \tag{3}$$

and

$$k^2 = 1 - \frac{a^2}{T^2} \tag{4}$$

(In the above analysis, the effect of gravity is neglected and the droplet is assumed to be in thermodynamic equilibrium). Since the dependence on the contact angle in Equation 1 is complex this expression cannot be inverted and the contact angle must be obtained by reading from a plot of L against T where lines corresponding to various values of θ are drawn. This involves a great deal of reading error, as observed in our laboratory as well as by other researchers /3, 4/. This will be illustrated later.

We have recently developed a simple numerical algorithm by which the contact angle may be determined with very high accuracy /5/. This algorithm, now described, helps exploit to its full extent the method proposed by Yamaki and Katayama, and Carroll.

First Equation 1 is modified as follows. Rather than Legendre's form, the symmetrised variants /6/ of the incomplete elliptic integrals of the first and second kind are preferred for computational purposes:

$$R_F(x,y,z) = \frac{1}{2} \int_0^\infty \frac{d\xi}{\sqrt{(\xi+x)(\xi+y)(\xi+z)}} \tag{5a}$$

$$R_D(x,y,z) = \frac{3}{2} \int_0^\infty \frac{d\xi}{\sqrt{(\xi+x)(\xi+y)(\xi+z)^3}} \tag{5b}$$

Then it can be shown that Legendre's standard integrals can be expressed in terms of R_F and R_D as follows:

$$F(\phi,k) = (\sin\phi)\, R_F(\cos^2\phi,\ 1-k^2\sin^2\phi,\ 1) \tag{6a}$$

$$\begin{aligned} E(\phi,k) = (\sin\phi)\, R_F(\cos^2\phi,\ 1-k^2\sin^2\phi,\ 1) \\ - \tfrac{1}{3}k^2\sin^3\phi\, R_D(\cos^2\phi,\ 1-k^2\sin^2\phi,\ 1) \end{aligned} \tag{6b}$$

In view of equation 3, useful simplifications arise as follows: (i) The first argument (i.e. $\cos^2\phi$) of R_F and R_D in equations 6a and 6b is calculated in a straightforward manner; (ii) The second argument (i.e. $1-k^2\sin^2\phi$) of R_F and R_D in equations 6a and 6b may be immediately replaced by $1/T^2$.

Equation 1 is now rewritten in terms of R_F and R_D by means of equations 6a and 6b:

$$L = 2[\sin\phi\,(a+T)\,R_F - \tfrac{1}{3}Tk^2\sin^3\phi\,R_D] \tag{7}$$

Based on equation 7 the following scheme was adopted for accurate determination of θ (the values L^* and T^* are measured experimental readings):

675

(1) A first guess θ_1 of the contact angle is chosen (using Carroll's plot), and a value T_1 is calculated using Equations 1 to 4, and T^*.

(2) A second guess θ_2 is made (close to the first one), and a second value L_2 is calculated as in (1);

(3) A value θ_3 is obtained by the interpolation (with i=3)

$$\theta_i = \frac{\theta_{i-2}-\theta_{i-1}}{L_{i-2}-L_{i-1}} L^* + \frac{L_{i-2}\theta_{i-1}-L_{i-1}\theta_{i-2}}{L_{i-2}-L_{i-1}} \tag{8}$$

(4) The value of the contact angle obtained is then compared to the previous one and if the difference between these is larger than a predetermined threshold value the process is continued (with i=4, 5,...) until a satisfactorily accurate value is obtained.

A Fortran 77 program was prepared based on the scheme proposed above, which proved to be simple, accurate and reliable. In /5/ we plotted the contact angle (in the range $0°\leq\theta\leq90°$) as calculated from the above method, against the reduced droplet length for reduced droplet thicknesses ranging between 1.5 and 5. The accuracy of the contact angle obtained from this plot was found to be better than that obtained from Figure 2 in Carroll's paper /2/. However the use of the Fortran program in an interactive mode yields much more accurate values of θ than those read from the plot in /5/, and this is the preferred method.

We now present a few results recently produced in our laboratory, to illustrate the method. Data were obtained at room temperature with (i) droplets of glycerol spread on ex-polyacrylonitrile extra high strength carbon fibers (ACIF-XHT, diameter of 6.8 μm, from Afikim Carbon Fibers, Israel), (ii) droplets of epoxy resin (CY 223 from Ciba Geigy, no hardener used) spread on single para-aramid filaments (Kevlar 49, diameter of 11.9 μm, from du Pont de Nemours, Inc.), and (iii) droplets of epoxy resin (DER 331 from Dow Chemical, no hardener used) spread on glass fibers (E-glass 5139, diameter of 18 μm, from Vetrotex International). For the carbon/glycerol system the average contact angle from 9 droplets is 55.8° with a coefficient of variation of 14.7 pc. This result is very close to the one obtained by Chang et al. /7/ for the same system by using the Wilhelmy (electrobalance) technique, namely 58.9°. For the aramid/epoxy system the average contact angle from 14 droplets is 15.8° with a coefficient of variation of 19.3 pc. For the glass/epoxy system the average contact angle from 27 droplets is 19.7° with a coefficient of variation of 8.5 pc.

The scheme presented here for the measurement of the contact angle of each droplet gives much more accurate results for the contact angle than the evaluations from Carroll's plot (Figure 2 in Reference 2). For example, a range of 45-50° is found from Carroll's plot for a glycerol droplet on carbon fiber, with reduced droplet length equal to 10 and reduced thickness equal to 4, whereas an accurate value of 48.42° is found with our method.

With very thin fibers (such as carbon) the droplet-to-droplet variability in contact angle is relatively large, whatever the method, and that the improvement in accuracy on the contact angle is very clear on a per droplet basis only. The fibers tested here had an epoxy-based surface sizing, and it is probable that inhomogeneities on the fiber surface due to the application of the sizing are partly responsible for this variability. Contact angle results currently obtained in our laboratory with unsized carbon fibers show a much lower droplet-to-droplet variability (coefficients of variation of about 8 percent with 15 fibers tested). Possible inaccuracies in the experimental measurements (mainly of droplet length) may constitute another source of variability in the contact angle results, but this is substantially reduced by using the largest droplets only. With thicker fibers, such as E-glass, the method is very efficient as the droplet-to-droplet variability in contact angle is much reduced, as illustrated above.

III-CONCLUDING REMARKS

Based on an earlier scheme by Yamaki and Katayama /1/, and Carroll /2/, a simple method was proposed to determine with high accuracy the apparent contact angle of a liquid droplet formed on a thin cylindrical monofilament. A plot may be constructed /5/ which gives the apparent contact angle against the reduced droplet length for a variety of values of the reduced thickness of the droplet. The highest accuracy results, however, are obtained by means of an interactive Fortran program based on the iteration scheme proposed. Because of its simplicity and high accuracy, the method proposed can be used with great advantage in studies of wettability of reinforcing filaments by liquid resins in advanced composites.

This work was supported in part by a grant from the National Council for Research and Development (Israel) and the K. F. A. Juelich (Germany). H. D. Wagner is the incumbent of the Jacob and Alphonse Laniado Career Development Chair.

REFERENCES

1. Yamaki, J., Katayama, Y., Journal of Applied Polymer Science, 19, (1975) 2897.

2. Carroll, B. J., Journal of Colloid and Interface Science, 57 (1976) 488.

3. Gilbert A. (private communication).

4. Nardin, M., Ward, I. M., Materials Science and Technology, 3 (1987) 814.

5. Wagner, H. D., Journal of Applied Physics, 67 (1990) 1352.

6. Carlson, B. C., "Special Functions of Applied Mathematics", (New York, Academic Press, 1977).

7. Chang, H. W., Smith, R. P., Li, S. K., Neumann, A. W., in "Molecular Characterization of Composite Interfaces" (Polymer Sci. & Technol. Ser.: Vol. 27), H. Ishida and J. Kumar, Eds., (New Jersey, Plenum Press, 1985), 413.

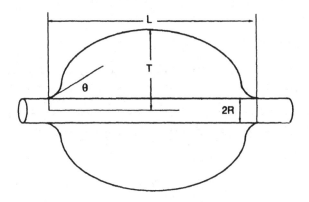

Fig. 1 - Liquid droplet in equilibrium on a cylindrical fiber.

MONITORING INTERFACIAL PHENOMENA IN SINGLE FIBRE MODEL COMPOSITES

C. GALIOTIS, H. JAHANKHANI

Materials Dept - Queen Mary & Westfield College
Mile End Road - E1 4NS LONDON - Great Britain

ABSTRACT

The technique of Laser Raman Spectroscopy was applied in the study of aramid fibres and aramid/epoxy interfaces. A linear relationship was found between Raman frequencies and tensile strain for a single kevlar 49 fibre in the air. Model single fibre composites subjected to various tensile loading and transfer lengths for reinforcement were measured. Finally, an analytical model was used to derive the interfacial shear stress distribution along the interface.

I-INTRODUCTION

Interfacial Shear Strength measurements are normally conducted either on composite specimens by means of a standard short beam shear, four point shear and flexural strength tests or model single fibre composites by means of the, pull-out, fragmentation, microbond or microdebond types of test. As far as tests on composite specimens are concerned, the state of stress during testing is rarely in-plane shear so the results might not be representative of the true interfacial shear strength. On the other hand, it can be argued that tests on single fibre model composites cannot be considered truly composite tests, since they are conducted on a single fibre /1,2/.

In this paper a new technique for interfacial measurements is presented which can be applied equally well to a single fibre surrounded by an infinite matrix and to a single fibre in a full composite, thus, enabling the examination of truly interfacial phenomena, as well as, more complex phenomena encountered in full composites. This technique employs a Laser Raman microprobe in order to measure the strain in individual fibres at the microscopic level and can be applied to all composites provided that the matrix is relatively transparent and the fibre possesses a relatively high degree of crystallinity.

Interfacial studies have been conducted on short Kevlar 49 fibres embedded in epoxy resin matrices. The use of short fibres was preferred since it enabled the strain and interfacial shear stress mapping along the whole length of the fibre, at any level of applied load.

II-EXPERIMENTAL

Raman spectra were taken by employing the 632.8 nm line of a He-Ne laser. A modified Nikon microscope was used to focus the incident beam to a 8 m spot on the fibre. The 180° backscattered light was collected by the microscope objective and focussed on the entrance slit af a SPEX 1877 triple monochromator. A cooled CCD imaging detector was employed for recording the Raman spectra.

To obtain the relationship between Raman frequency and tensile strain, single Kevlar 49 filaments were loaded/unloaded in the air on a specially made micrometer to 2% strain. For the interfacial studies in tension Kevlar 49 filaments of approximately 2.8mm in length were chopped from a continous yarn using a pair of ceramic scissors. The chopped filaments were embedded in a dogboned PTFE moulds containing Ciba-Geigy HY/LY 1927 two-part solvent free epoxy resin, using 100 parts (by weight) of resin to 36 parts of hardener. The dogboned specimens were cured for 7 days at room temperature to avoid thermal shrinkages. The composite sample was strained in the fibre direction by means of a purpose made loading device up to first fibre fracture. At each level of applied load the fibre was scanned with the Laser Raman microscope probe to obtain the distribution of Raman frequencies along its length. The strain on the specimen was also measured independently using foil strain gauges mounted on the epoxy bars before testing.

III-RESULTS/DISCUSSIONS

In Figure (1) a typical Raman spectrum within the 1580-1640 cm^{-1} frequency range at 0% and 2% tensile strain is shown. The high intensity peak at 1615 cm^{-1} has been found to be most sensitive to mechanical load /3/ and was, therefore, used for strain measurements in this work.

As shown in Figure (2), a linear relationship with a slope of (-4.37±0.10) cm^{-1}/% is obtained between Raman Frequency and strain upon loading the fibre in tension. The value of the above least-square fitted slope is used to convert the Raman frequency into fibre strain in all subsequent experiments. On unloading, a deviation from linearity, and a curve of higher slope were obtained, while a permanent tensile fibre strain of 0.2% was measured at the end of the experiment. It has been reported elsewhere /4/ that there are two mechanisms contributing to the elastic extension of aramid fibres; a crystallite stretching parallel to the molecular chains and a crystallite rotation involving certain degree of plastic deformation.

The variation of the fibre strain along the whole length of the embedded Kevlar 49 fibre before and after fibre fracture is shown in Figure (3) for selective matrix strains of 0.5%, 1.0%, 1.5% and Figure (4) respectively. The fibre broke approximately at the middle, indicating (a) good axial alignment of the fibre in the loading direction and (b) maximum tensile strain at that point. The fibre length over which the load is built up from each end of the fibre is defined here as the transfer length, l_t. Figure (5) shows l_t as a function of matrix strain for both ends of the fibre. The data show a gradual increase of l_t with strain (region I) followed by a dramatic increase over approximately 1.5% of strain (region II).

The strain-transfer profiles correspond to the Cox or shear-lag model /5/ whereby the load is built up from the fibre ends and becomes maximum at the middle of the fibre. Since, in our model composites, there is no influence of neigbouring fibres and/or fibre ends, then it is reasonable to assume that the applied tensile load is transferred to the fibre solely, by a shearing mechanism between fibre and matrix. If we now consider an infinitesinal length dx at a distance x from the fibre end then the force equilibrium for this length illustrated in Figure (6) requires that :

$$F - (F + dF) + 2\pi r \ dx \ \tau_{(rx)} = 0 \qquad (1)$$

where F is the tensile force acting on the fibre, τ, is the shear stress at the fibre/matrix interface at a distance x, and r is the fibre radius. By solving (1) as for the fibre tensile strain, e_f, we finally have :

$$\tau_{(rx)} = E_f \ \frac{r}{2} \left[\frac{de_f}{dx} \right] \qquad (2)$$

where E_f is the tangent fibre modulus at the fibre strain e_f. Equation (2) is also valid for interfacial regions where stress concentration effects are thought to be present.

The derivatives de_f/dx at each position along the fibre length can be derived, each time, from the slope of the strain-transfer profiles. The modulus, E_f, and the diameter, d, have been determined independently /6/ for the same bundle of Kevlar 49 fibres. Hence from equation (2), the interfacial shear stress distribution, τ, along the fibre can be obtained. This is shown in Figure (7), where the strain-transfer profile at 1.0% strain is converted, point to point, to the interfacial stress profile along the whole length of the fibre. As expected the interfacial shear stress takes up its maximum value at the fibre tips and sharply decays to zero at a distance equal to the transfer length.

To speed up ISS Data collection, fifth degree polynomial functions were fitted to the strain vs length data enabling fast determination of the de_f/dx at any position along the fibre. The two ends of the fibre are examined separately and the ISS values obtained from both ends of the fibre for five different levels of applied strain can be taken from Table (1). Typical ISS profiles for the left hand fibre end is shown in Figure (8a,b). The overall shape of the ISS profiles up to 1.3% of applied strain indicates good fibre/matrix bonding. The high interfacial shear stress (62 MPa, Table 1) developed at the LH fibre tip is bound to lead to local matrix yielding and/or debonding with any futher increase of the applied load. The ISS profiles at 1.5%, 1.8% and 2.5%, Figure (8b), provide evidence for that. At 2.5% of applied strain, three distinct regions noted (a) plateau corresponding to complete fibre debonding and to which the applied load is transferred purely by friction; (b) gradual take up of ISS corresponding to an area of partial bonding and (c) good bonding is envisaged, as concluded from the shape of the ISS profile at that point.

In Figure (9), maximum ISS obtained from both ends of the fibre is plotted as a function of fibre strain at the middle of its length. It can be seen that up to about 1.2% of applied strain the maximum ISS increases proportionally to the applied strain and then decreases, at 2.5% strain, to values of approximately 27 MPa and 30 MPa for the LH and RH fibre ends, respectively. This value coincides with the yield strength of the bulk matrix which has been measured to be approximately 26 MPa /6/.

The results presented here show clearly that the maximum shear stress that our model system can accommodate prior to matrix yielding/fibre debonding is approximately 62 MPa. This is considered to be a close approximation to the actual interfacial strength of Kevlar 49/1927 Ciba Geigy RT-cured epoxy system.

IV - CONCLUSIONS

The Non-Destructive Technique presented here can be applied successfully to:
(a) determine the interfacial properties of a Kevlar/epoxy system by obtaining the strain profile of a single embedded fibre under various degrees of matrix tension.
(b) map the interfacial shear stress distribution along individual fibres.

V-ACKNOWLEDGEMENTS

We would like to thank E.I. du pont de Nemours and Co. for financial support and for supplying the Kevlar 49 fibres and, also, Ciba Geigy (UK) for supplying the epoxy resin.

VI-REFERENCES

1. J.M.Whitney, I.M.Daniel, B.R.Pipes, "Experimental Mechanics of Fibre Reinforced Composite Materials", (Society for Exp. Mech.) (1984).
2. R.M.Jones, "Mechanics of Composite Mateials", (Hemisphere Pub. Co.).
3. I.M.Robinson, R.J.Young, C.Galiotis, D.N.Batchelder, J. Mat. Sci., 22 (1987) 3642.
4. S.Van der Zwaag, M.G.Northot, R.J.Young, I.M.Robinson, C.Galiotis, D.N.Batchelder, Polymer Comm., 28 (1987) 276.
5. H.L.Cox, Brit. J. Appl. Phys., 3 (1952) 72.
6. H.Jahankhani, C.Galiotis, J. of Comp. Mat., to be published.

Table 1. Maximum ISS values and ISS values at the fibre tip for both left-hand (LH) and right-hand (RH) ends of the fibre at different levels of applied strain.

Matrix strain / %	Max.fibre strain / %	ISS at fibre tip (LH) / MPa	Max. ISS (LH) / MPa	ISS at fibre tip (RH) / MPa	Max. ISS (RH) / MPa
0.5	0.36	14.1	15.2	18.7	at tip
1.0	0.94	48.4	at tip	51.1	at tip
1.3	1.27	62.0	at tip	18.1	33.0
1.5	1.53	2.4	34.1	7.1	34.1
1.8	1.84	6.7	26.1	7.2	30.0
2.5*	2.54	11.1	25.6	9.5	31.8

*Onset of fibre fracture. Results from one fragment only (LH)

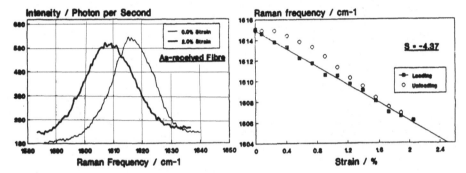

Fig.1 - A typical CCD Raman spectrum for Kevlar 49 fibre at 0% and 2.0% strain.

Fig.2 - Raman frequency as a function tensile strain.

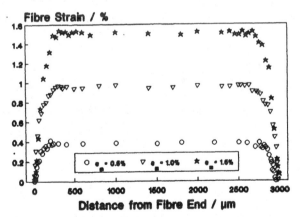

Fig.3 - Axial tensile strain in the fibre as a function of fibre length at 0.5%, 1.0%, 1.5% matrix strain.

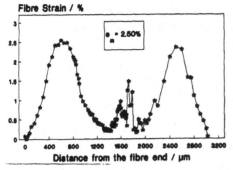

Fig.4 - Axial tensile strain in the
fibre as a function of fibre
length at 2.5% matrix strain.

Fig.5 - Fibre transfer
length as a function
of matrix strain.

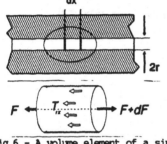

Fig.6 - A volume element of a single
fibre surrounded by an infinite
matrix showing the force balance.

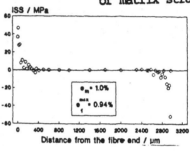

Fig.7 - ISS profile along the length of
the fibre.

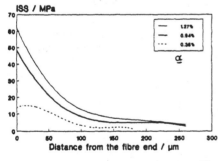

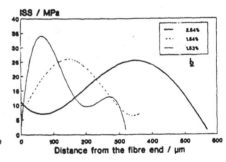

Fig.8 - ISS profiles for (a) 0.5%,1.0%,1.3%
levels of matrix strain and (b) 1.5%
1.8%,2.5% levels of strain for LH
fibre end.

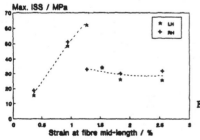

Fig.9 - Maximum ISS as a function of strain
fibre mid-length for both LH and RH
ends of fibre.

STRAIN MEASUREMENT IN FIBRES AND COMPOSITES USING RAMAN SPECTROSCOPY

R.J. YOUNG, R. YOUNG, C. ANG

Manchester Materials Science Centre - UMIST
M60 1QD Manchester - Great Britain

ABSTRACT

It has been found that well–defined Raman spectra can be obtained from a wide range of high–performance polymer fibres. It is also found that on the application of tensile stress or strain the bands in the Raman spectra shift generally to lower frequency. This gives information upon the molecular strain in the materials. In addition, it has been found that when the fibres are incorporated in an epoxy resin matrix, the Raman technique can be used to determine the point–to–point variation of strain within the fibre. It is shown that the strain in high modulus rigid rod fibres can be evaluated in an epoxy composite subject to both axial tension and compression. Hence, the compressive stress–strain behaviour of the fibre can be measured and the onset of compressive failure through kink–band formation can be determined.

I–INTRODUCTION

Over recent years there has been considerable interest and excitement concerning the use of Raman spectroscopy to follow the deformation of high performance polymer fibres. It is found that well–defined Raman spectra can be obtained from individual fibres using a Raman microscope system and that the Raman bands are sensitive to the level of applied stress or strain. This behaviour has been demonstrated for a wide variety of materials such as polydiacetylenes /1,2/, aromatic polyamides /3,4/, rigid–rod polymer fibres /5,6/, polyester fibres /7/, gel–spun polyethylene fibres /8,9/, carbon fibres /10/ and even ceramic fibres /11/. In general, it is found that the bands in the Raman spectrum shift to lower frequency on the application of a tensile stress or strain indicating that the covalent bonds in the structure undergo direct molecular deformation which causes a reduction in the interatomic force constant for the bond. Analysis of this phenomenon has given a unique insight into the way in which macroscopic deformation in these materials is transferred to the molecules in the structure. Additionally, since spectra can also be obtained from the fibres within a polymer matrix it has been possible to use the Raman technique to evaluate the deformation of fibres within a composite and obtain detailed information concerning composite micromechanics /2/.

This present report is concerned with the deformation of fibres of the rigid–rod molecule poly(p–phenylene benzobisthiazole) (PBT) /5/ subjected to both tension and compression. It is shown that although PBT fibres have impressive tensile properties /5/ their behaviour in compression is relatively poor. They fail at low strains in compression through a process of kink band formation.

II–EXPERIMENTAL

The fibres used in this present study were as–spun (AS) and heat–treated (HT) fibres of PBT /5/. They were supplied by the USAF Materials Laboratory, Dayton, Ohio in the form of filaments of the order of $14\mu m$ diameter and the HT fibres had a modulus of 270GPa. Their properties have been described in detail elsewhere /5/.

The single–fibre composites were prepared using Ciba–Geigy XD927 which is a two–part solvent–free cold–setting epoxy resin. A "dog-bone" shaped PTFE mould was filled with half of the resin/hardener mixture and this was allowed to set partially before the fibre and rest of the resin were added. In this way a specimen was produced with a single fibre aligned along the length of the specimen. After setting at room temperature for 7 days a thin film resistance strain gauge was attached to the specimen near the centre of the specimen to allow overall strain to be measured. All the specimens were polished and holes were drilled at both ends. They were then deformed in a "Minimat" materials tester.

The specimens used for compression testing were essentially the same as the tensile ones. They were obtained by cutting the two ends of the "dog–bone" shaped composites leaving a small rectangular block. A strain–gauge was also attached and it was deformed in the compression cage of the "Minimat". The compression specimens were observed during deformation using an optical transmission microscope in order to view the formation of kink bands.

Raman spectra were obtained from the fibres and films during deformation using a Raman microscope system as described elsewhere /5,6/. The spectra were collected using a highly sensitive Wright Instruments Charge coupled device (CCD) camera. The beam was focussed to a $2\mu m$ spot and was polarised parallel to the fibre axis for all measurements. The beam could be focussed on to the fibre through several mm of resin within the single fibre composite and this enabled in situ deformation of fibres to be monitored directly within the composite.

III–FIBRE DEFORMATION

The formation of kink bands occurs in fibres of both AS and HT PBT during bending or compression. It can be readily observed using both optical and electron microscopy. In fact both types of fibres were found to contain some kink bands in the as–received state which must have formed either during manufacture or handling. Figure 1 shows a series of micrographs taken at different levels of applied strain showing the development of the kink bands in AS PBT. It can be seen that there is a rapid increase in the number of kink bands above 0.76% strain. The development of the kink bands is shown in more detail in Figure 2 where the number of kink bands along a given length is plotted as a function of compressive strain. The rapid increase in the number of kink bands above about 0.8% strain can be clearly seen for the AS PBT fibres which have a modulus of 195GPa. It should also be noted that the higher modulus HT PBT fibre (E = 270GPa) fails by kink band formation in compression at a significantly lower strain of about 0.3% (figure 2).

It has been found /5/ that the position of the $1475cm^{-1}$ Raman band in PBT is sensitive to the level of applied stress or strain and for free–standing fibres shifts to lower frequency on the application of a tensile strain /5/. This behaviour is also found for the fibres within the epoxy resin bars and the shift for a HT PBT fibre in a resin subjected to tension is shown in figure 3. The rate of shift (ie slope of the line) is similar to that of the free–standing fibre /5/ and similar behaviour is also found for the AS PBT. The shift in Raman band peak position with applied compressive strain is shown in figure 4 for HT PBT and it can be seen that the behaviour is quite different from that obtained in tension. There is a shift to higher frequency and after an initial shift the data points become scattered. This corresponds exactly with the onset of kink band formation which relieves the applied compression but causes local stress concentrations in the kink band boundaries. The critical strain for kink band formation of about 0.2% for HT PBT (figure 4) is consistent with that found from direct observation (figure 2).

Another aspect of the behaviour is shown in figure 5 where the data showing the change in Raman frequency from figures 3 and 4 are replotted as function of composite strain (both tensile and compressive). This can be thought of as a full stress–strain curve for the fibre in both tension and compression. The first point to note is that slope of the curve in compression is lower than that in tension. This is a direct indication that the compressive Youngs modulus of the fibre is lower than the tensile Youngs modulus. This is a highly–significant finding since it is not normally possible to measure the stress–strain behaviour of a fibre in compression because of buckling.

IV–CONCLUSIONS

It has been demonstrated that Raman spectroscopy can be used to measure molecular strain in rigid–rod polymer fibres of PBT subjected to both tension and compression. The compressive failure of PBT fibres through the formation of kink bands has been examined in detail and it has been shown that Raman microscopy is a unique method of following the compressive stress–strain behaviour of individual fibres. Although this paper has been concerned with a specific application of the technique to the PBT system, the method is of general application to a wide variety of polymer, fibre and composite systems [1–11].

ACKNOWLEDGEMENTS

The work presented above has been supported by research grants from the USAF European Research Office and the SERC.

REFERENCES

1. Galiotis, C., Young, R.J. and Batchelder D.N., Journal of Polymer Science, Polymer Physics Edition 16, (1983), 2483.
2. Young, R.J., Polymer Single Crystal Fibres, In Developments in Oriented Polymers–2, ed I.M. Ward, Applied Science, London, (1987) p.1.

3. Galiotis, C., Robinson, I.M., Young, R.J., Smith, B.E.J. and Batchelder D.N., Polymer Communications 26, (1985), 354.
4. Van der Zwaag, S., Northolt, M.G., Young, R.J., Robinson, I.M., Galiotis, C. and Batchelder, D.N. Polymer Communications, 28, (1987), 276.
5. Day, R.J. Robinson, I.M., Zakikhani, M. and Young, R.J., Polymer 28, (1987), 1833.
6. Young, R.J. Day, R.J. and Zakikhani, M., Journal of Materials Science, 25, (1990), 127.
7. Fina, L.J., Bower, D.I. and Ward, I.M. Polymer, 29, (1988), 2146.
8. Prasad, K. and Grubb, D.T., Journal of Polymer Science, Polymer Physics Edition 27, (1989), 381.
9. Kip, B.J., van Eijk, M.C.P., Leblans P.J.R. and Meier, R.J., Molecular strain in high-modulus polyethylene fibres during stress relaxation studied by Raman microscopy, Paper presented at the Rolduc Polymer Meeting-5, (1990).
10. Robinson, I.M., Zakikhani, M., Day, R.J., Young, R.J. and Galiotis, C., Journal of Materials Science Letters, 6, (1987)1212.
11. Day, R.J., Piddock, V., Taylor, R., Young, R.J. and Zakikhani, M., Journal of Materials Science, 24, (1989), 2898.

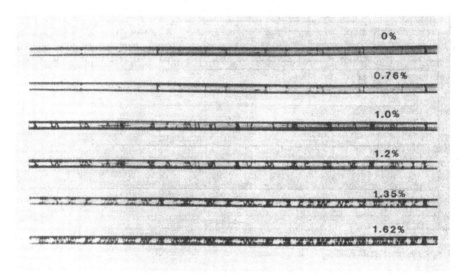

Figure 1.   A series of optical micrographs of the formation of kink bands ... a single PBT fibre at different levels of applied compressive strain.

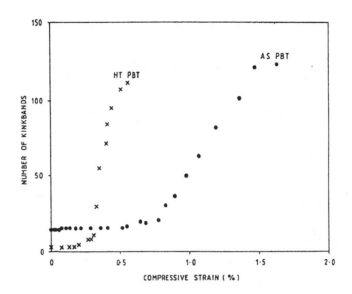

Figure 2.   Increase in the number of kink bands over a length of fibre with compressive strain for both the AS and HT PBT fibres.

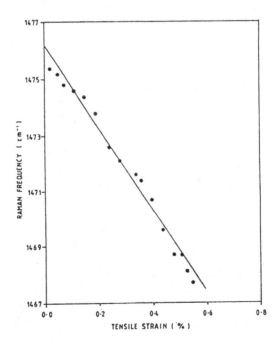

Figure 3.   Variation of the position of the peak of the 1475cm^{-1} Raman band with tensile strain for a HT PBT fibre in the single fibre composite.

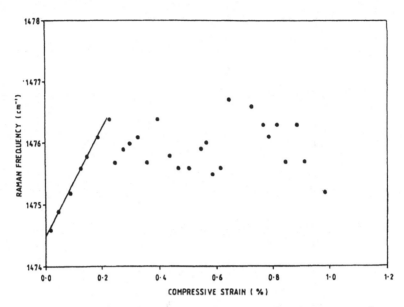

Figure 4.   Variation of the position of the peak of the 1475cm^{-1} Raman band with compressive strain for a HT PBT fibre in the single fibre composite.

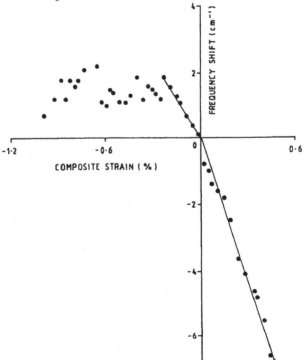

Figure 5.   Change in the frequency of the 1475cm^{-1} Raman band peak with strain (both tensile and compressive) for the HT PBT single fibre composite.

IDENTIFICATION OF SUBSURFACE DEFECTS BY A THERMAL
METHOD USING A SENSIBILITY ANALYSIS

A. DEGIOVANNI, A.-S. LAMINE*, A.-S. HOULBERT, D. MAILLET

LEMTA - CNRS 875 - Ecole des Mines
Parc de Saurupt - 54000 Nancy - France
*LSGC-ENSIC - Rue Déglin - 54000 Nancy - France

ABSTRACT

A subsurface defect in a composite material can be characterized in term of depth and thermal resistance after a heat pulse - "flash" - perturbation on its front side. A sensitivity analysis of the rear-side contrast thermogram is done and various methods of identification of these two parameters are reviewed. The use of experimental Laplace transforms seems to bring better results than the classical one-point evaluation.

INTRODUCTION

In carbon-epoxy composite materials built of parallel layers of fibers - held together by a resin matrix - the defect often stems from a lack of bonding between two layers and thus consist in an air layer of small thickness (a few micrometers).

So far, in industry, nondéstructive control of such materials has been mostly carried out by means of ultrasonic or X-ray methods. These methods have proved their efficiency but present the disadvantage of both high cost and difficulty of implementation.

Transient thermal methods can also be applied thanks to the development of quantitative infrared thermography. Our work takes place within the framework of such techniques and is sponsored by the Dassault Company.

I - THERMAL NON DESTRUCTIVE EVALUATION

1.1. Principle of the method

Let's consider a defect - a disbond for example -located in a slab of infinite radial r extent (Fig.1). This slab can be a composite built of sequences of homogeneous isotropic or ansisotropic materials if the number of layers is large enough to allow the homogenization with the use of equivalent thermal characteristics, /1/. The case of one-dimensional (1D) transient heat transfer, created for example by a homogeneous pulse of energy by unit area Q on the front side of this slab, is considered in this article.

In the following we will use notation λ for axial conductivities - direction y - a for axial diffusivities and ρc for thermal volumetric capacities. If the defect (d) is formed of an air layer of conductivity λ_d, of thickness e_d small compared to the total slab thickness e ($= e_1 + e_2$) , the defective slab can be modelled by a contact resistance $R_c = e_d/\lambda_d$ with no thickness, surrounded by the two external layers (1 and 2). For usual air disbonds the capacity $(\rho ce)_d$ of the defect can be neglected, /2/.

1.2. Direct 1D - model

Using the quadripole method wich can be applied in any 1D - transient thermal problem, /2/, /3/, it can be shown that the Laplace transform $\theta(p)$ of the rear side thermogram-temperature T versus time t - can be expressed analytically :

$$\theta(p) = QR\left\{\alpha \sinh(\alpha) + R_c^* \ \alpha^2 \sinh(\alpha x^*)\sinh[\alpha(1 - x^*)]\right\}^{-1} \qquad (1)$$

with : $R = e/\lambda$ $R_c^* = R_c/R$ $x^* = e_1/e$ $\alpha = e\sqrt{p/a}$

This equation is valid, if heat losses can be neglected, for a heat pulse excitation, the initial temperature (t=0) of the slab being equal to zero. We only deal in this article with rear-side detection ; an expression similar to equation (1) can be obtained for the front-side temperature, /2/. Temperature T can be made dimensionless if it is divided by its adiabatic final value :

$$T^* = \frac{T}{Q/(\rho ce)}$$

If θ^* is the Laplace transform of T^* and if subscript 0 denotes a thermogram on a defectless slab $\left(R_c^* = 0\right)$ after a pulse of same energy Q, it is possible to define a reduced thermal contrast $\left(T^* - T_0^*\right)$ whose Laplace transform is :

$$\Delta\theta^*(p) = -\frac{e^2}{a} \ \frac{R_c^*}{\sinh(\alpha)} \ \frac{\sinh(\alpha x^*)\sinh[\alpha(1-x^*)]}{\sinh(\alpha) + R_c^*\alpha \ \sinh(\alpha x^*)\sinh[\alpha(1-x^*)]} \qquad (2)$$

The originals T^*, T_0^* and ΔT^* can be calculated as a function of time t (or of its reduced value $t^* = at/e^2$) thanks to an algorithm developed by Stehfest /4/. Their typical shapes are given on Fig. 2.

II - IDENTIFICATION PROBLEM

Inversion of equation (2) shows that contrast ΔT^* is a function of Fourier number t^* and of the defect parameters : reduced depth x^* and contact resistance R_c^* :

$$\Delta T^* = \Delta T^* \left(t^*, \; x^*, \; R_c^* \right) \tag{3}$$

2.1. One point evaluation

The contrast curve (Fig.2) shows a maximum T_{max}^* for a reduced time τ_m. These two characteristics of the maximum can be calculated for various values of parameters x^* and R_c^* and a plot can be drawn (Fig.3) showing iso-resistance and iso-depth curves in a $\left(\tau_m, \; T_{max}^* \right)$ graph. Let's note that ΔT^* remains unchanged if x^* is replaced by $(1 - x^*)$ in equation (2) which means that rear-side evaluation cannot discriminate between two symmetrical locations of the defect. Experimental time variation of temperature on the rear-side of a real slab can be recorded at the defect level - $T(t)$ - and outside it - $T_o(t)$ - after a pulse using an infrared analizer. The difference of the two curves $\Delta T(t)$ can be plotted versus time and experimental values of t_{max}, ΔT_{max} and T_{omax} are extracted from it (same shape as Fig.2). Experimental characteristics $\tau_m \, (= at_{max}/e^2)$ and $\Delta T_{max}^* \, (= \Delta T_{max}/T_{omax})$ are calculated and used as entries of diagram 3. Parameters of the defect R_c^* and x^* are identified on the same diagram.

This technique is very simple to implement but, since it uses only one point of the contrast thermogram, it yields the whole measurement error linked to that point.

2.2. Sensitivity analysis

Beck has shown /5/ that introduction of techniques connected to "inverse methods" in the thermosciences allows a more pertinent approach to parameter identification problems. Equation (3) allows the introduction of the two sensitivity coefficients :

$$s_x = x^* \; \frac{\partial \Delta T^*}{\partial x^*} \qquad s_R = R_c^* \; \frac{\partial \Delta T^*}{\partial R_c^*}$$

Their calculation is very simple because their Laplace transforms can be derived from equation (2). This equation being linear in R_c^*, for low values of this parameter, s_R and ΔT^* are equal.

Both coefficients s_x and s_R are plotted versus reduced time on Fig.4 for $R_c^* = 0.065$ and $x^* = 0.4$ and 0.5. It can be easily seen that they are not proportional which means that x^* and R_c^* are independant making their identification possible. s_R being higher than s_x, rear-side evaluation gives better values for R_c^* than for x^*. For a defect in the plane of symmetry of the slab, sensitivity to x^* is equal to zero.

2.3. Weight functions and experimental Laplace transforms

The most appropriate entities to be used for this type of identification must integrate the signal and balance each point by the sensitivity to the desired parameter at one and the same time :

$$M_\gamma = \int_0^\infty s_\gamma \, \Delta T \, dt \qquad \text{with } \gamma = x \text{ or } R$$

This ideal approach stumbles over the fact thas s_x and s_R depend on the a-priori unknown parameters x^* and R_c^* .

The sensitivity coefficients must be therefore approximated by another weight function which can be a negative exponential function :

$$M_\gamma' = \int_0^\infty \Delta T \, \exp(- p_\gamma t) \, dt \qquad \text{with } \gamma = x \text{ or } R_c \qquad (4)$$

The main advantage for this last approach is that an analytical expression exists for M_x' and M_R' since they are the Laplace transforms of ΔT for $p = p_x$ and $p = p_R$. These coefficients p_x and p_R must be optimized : one way to achieve this is to plot functions $s_\gamma \Delta T^*$ and $\exp(- p_\gamma^* t^*)\Delta T^*$ versus reduced time t^* for different values of parameters x^* and R_c^* and to adjust the values of p_γ^* ($= e^2 p_\gamma/a$) in order to get synchronous maxima for the two types of functions. This technique is valid for p_R^* whose optimized value is equal to 1. Yet optimized values of p_x^* must be positive which does not fit with experimental integration.

Another way to look at this problem is to notice that the following ratio :

$$E = \frac{\Delta\theta_x^*}{\Delta\theta_R^*} \, \frac{1 + (e^2/a) \, \alpha_R \, \sinh(\alpha_R) \, \Delta\theta_R^*}{1 + (e^2/a) \, \alpha_x \, \sinh(\alpha_x) \, \Delta\theta_x^*}$$

with : $\qquad \Delta\theta_\gamma^* = \Delta\theta^*(p_\gamma) \qquad \alpha_\gamma = e\sqrt{p_\gamma/a} \qquad \gamma = x \text{ or } R$

does not depend on R_c^*. A network of curves of E versus x^* has been plotted for different values of p_x with the preceeding value of p_R. It shows that for any fixed value of x^*, the slope of each curve increases with p_x : there is therefore a higher dependance of E on x^* for high values of p_x . For practical reasons - finite time of the first measurement and necessity of a minimum weighted signal in integral (4) - an optimal value $p_x^* = 5$ has been chosen. Experimental evaluation is quite direct : an identification function of E versus x^* can be plotted ; experimental values of the $\Delta\theta_\gamma^*$'s allow the identification of x^* and R_c^* is calculated afterwards using equation (2) written for $\Delta\theta_R^*$.

III - EXPERIMENTAL RESULTS AND DISCUSSION

Experiments have been done on 6 cm x 6 cm slabs (e = 2 mm) of a carbon-epoxy composite material, presenting an inclusion - two 1 cm x 1 cm sheets of teflon of 25 micrometers thickness each - simulating an air disbond ofthe order of 5 micrometers in the slab plane of symmetry

$(x^* = 0.5)$. The heat pulse was produced by a flash-tube lamp whereas temperature was monitored by an infrared camera connected to a micro-computer.

One point evaluation was implemented to identify R_c^* and x^* starting from three experiments on the same sample. Experimental points 1, 2 and 3 - Fig.3 - give quite scattered identified depths which is quite natural if one reminds that sensitivity s_x is equal to zero for a symmetrical defect. Identified resistances stay quite close for the three experiments but they turn out to be three times higher than their nominal value of 0.065 (point 2L). This can be explained by air layers trapped between teflon films and between composite material -increased resistance - and by the volumetric capacity $(\rho c)_d$ of teflon which cannot be neglected. A three layer model developed by the quadripole method yields a couple $(\tau_m , \Delta T_{max})$ that can be plotted on the 2 layer diagram of Fig.3 (point 3L). Points 1, 2 and 3 are closer to this last point.

Implementation of the experimental Laplace transform method on the same thermograms - points A_1 and A_3 - first yields quite scattered reduced depths. Sensitivity to x^* being there equal to zero this parameter must be removed from the model and its nominal value of 0.5 can be taken for x^* , giving 3 close identified resistances - points B_1 , B_2 and B_3 on Fig.4. These resistances get closer to their nominal value (2L).

CONCLUSION

A method of thermal non destructive evaluation of slabs of defective composite material has been presented. Different techniques of parameter identification have been reviewed with respect to this particular problem. A sensitivity analysis has confirmed the difficulty of defect depth evaluation for rear-side detection. Preliminary study of the experimental Laplace transform technique compared to one-point evaluation shows that the first technique yields a good reproductivity in term of identified contact resistances.

REFERENCES

1. A.DEGIOVANNI, Revue Générale de Thermique, 339 (1990) 117-128.
2. A.S.LAMINE, A.DEGIOVANNI, D.MAILLET, in "Advancing with composites", Int. Congress on composite materials (May 10-12, 1988) 313-320, Milano.
3. A.DEGIOVANNI, Int.J. Heat Mass Transfer, 31 (1988) 553-557.
4. H.STEHFEST, Com. A. C. M., 13 (1970) 624.
5. J.V.BECK, B.BLACKWELL and C.J.SAINT-CLAIR, in "Inverse heat conduction - ill-posed problems" (J.WILEY and Sons ed.) (1985), New-York.

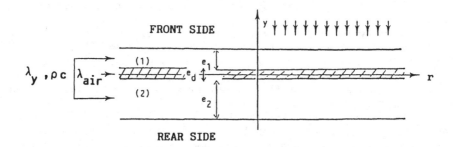

Fig 1: Defect geometry and modelling

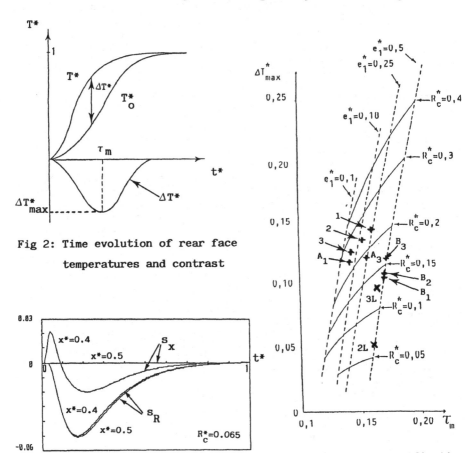

Fig 2: Time evolution of rear face
temperatures and contrast

Fig 4: Sensitivity coefficients

Fig 3: One point identification
diagram and comparison of 2
identification techniques

REAL-TIME INSPECTION OF CARBONFIBER COMPOSITE STRUCTURES UNDER THERMAL AND MECHANICAL LOAD BY ACOUSTIC EMISSION

J. BLOCK

DLR - Institute for Structural Mechanics
Flughafen - 3300 Braunschweig - West-Germany

ABSTRACT

The weight-specific strength and stiffness values of CFRP laminates are significantly superior to those of metals and alloys. However, this advantage has to be paid with a very complicated fracture behaviour under load, which makes real-time inspection necessary. Crack phenomena can be reliably detected and pursued by means of acoustic emission if it is taken into consideration, that the inhomogeneity and the anisotropy of carbonfiber composites require special test procedures. Such procedures have been developed in order to investigate the growth of cracks both under mechanically and thermally induced stresses between fibers and matrix.

INTRODUCTION

The complexity of local stress and strain distributions in advanced fiber composites is tremendously higher than in homogeneous and isotropic materials, since the mechanical and thermal properties of the constituent components (fibers and matrix) are significantly different. This inevitably leads to a manifold of defects such as broken misaligned fibers, delaminations, and cracks, when the structure is subjected to external and/or internal stresses. The respective onset level is rather low. If it is being accepted as a limiting design criterion, the characteristic advantages of the material are mostly wasted. This comes particularly true for CFRP laminates in aerospace applications, where mass reduction is essential. The unique combination of the low specific density and the high strength and stiffness of carbonfibers can only be fully utilized when structures are designed "damage-tolerant". Tolerating a certain amount of damage, however, may only be accepted under the supposition that all occurring defects can be monitored and then classified with respect to their severity, particularly when the security demands are such high as for aircraft structures.

Thus it is clear that - apart from stationary NDE techniques - suited real-time inspection methods are needed. In the following it is shown that acoustic emission (AE) analysis can be successfully used for this purpose, if special adaptations to some inhomogeneity effects take place.

I - DAMAGE MECHANISMS IN FIBER COMPOSITES AND ACOUSTIC EMISSION

Any release of elastic energy by fracture mechanisms leads to both longitudinal and transversal acoustic waves running through the material. That portion that reaches the surface will propagate in form of Rayleigh and/or Lamb waves and may finally be detected by sensitive piezoelectric transducers. As long as cracking occurs only in the matrix material (or in the fiber-matrix-interface, respectively), the related energies remain little. The onset of fiber fracture correlates with much higher energies per unit of area ($\sim$ several orders of magnitude, corresponding to the difference in Young's modulus, see /1/).

The measurable AE energy is - at least approximately - linked with the originally released elastic energy. Typical AE records measured on CFRP laminates therefore consist of a large number of low-energy *events* (from matrix-dominated mechanisms) and a much smaller portion of fiber-controlled high-energy *events*. The energy content of each event is coupled with its peak amplitude. There is a number of well-known classification models which rely on the AE event energy or peak amplitude, e.g., A. Rotem's model /2/, or A. Pollock's famous (and often up-dated) amplitude distribution model /3/. Other AE event parameters (i.e. rise time, counts, and duration) are only of minor importance in common evaluation models. This can become a critical omission if the carbonfiber composite under inspection is subjected to uninterrupted mechanical loading.

1.1. Emission from laminates under mechanically applied stress

The common AE analysis technique is based upon the assumption of discrete, burst-like events (see Fig. 1). Any sequence $v' = 1, 2, ..., n', ..., N'$ of emission mechanisms, which are well separable in space and/or time, shall be registered as a sequence $v = 1, 2, ..., n, ..., N$ of event data sets, which subsequently become the basis of evaluation. The above-mentioned evaluation models altogether depend on the supposition that $N = N'$.

It is obvious that $N = N'$ is a strong idealization. Different kinds of correlation errors may occur in real tests, such as remote signals from external noise sources, a wrong selection of the registration threshold, signal splitting due to reflection or dispersion inside the composite structure, and signal overlapping in the time domain. The latter effect (overlapping) is the most severe and cannot be compensated by technical means alone. It is a material effect typical of fiber composites, where the inhomogeneity leads to emission rates that are vastly higher than in comparable metallic structures.

Now if a sequence of damage mechanisms occurs in such a quick succession that the durations D of the corresponding acoustic signals overlap, then the respective event-based data (including the event energy and amplitude) become wrong. In practice, each duration D has even to be elongated by a time interval θ which includes all system "dead times" and "rearm times" and may be much longer than the typical event durations themselves.

A mathematical model has been developed /4/ which comprises D and θ (using a "generalized" event duration $\tilde{D} = D + \theta$, see Fig. 1) and which does not depend upon hardware specifications. It is based on the following assumptions:

1. The emission process is *stochastic* during small intervals T of time (i.e. small compared with the total test duration), and

698

2. the time distance between successive emissions in the fiber composite during *T* is governed by a *Poisson distribution*.

A number of tests have been made in order to check these assumptions. Evidence was found that they describe the reality in a satisfactory approximation. The model predicts the probability ϕ for any arbitrary acoustic burst signal n' (out of N' in total during T) to become registered as one -and only one- event n (out of N in total) as

$$ \phi_{n'} = \left(1 - \frac{1}{T} \sum_{v=1}^{N} \tilde{D}_v \right) \exp \left(-\frac{N' \tilde{D}_{n'}}{T} \right) $$

The expression under the summation sign is easily measurable with all standard AE equipment. It is the "cumulative duration" plus the product of θ (which is a characteristic constant for the specific test system and test mode) and N (which is the number of events recorded during T). If no overlapping takes place, ϕ is equal to 1. On the other hand, $\phi \rightarrow 0$ stands for the worst case. Further developments of the model /5/ going out from the above equation led to estimated values for the ratio between N and N' and allowed thus an estimated compensation.

The model was used for the tracing of damage propagation in CFRP laminates up to imminent failure, where the emission rate is such high that reliable AE inspection results are otherwise not obtainable (Fig. 3).

1.2. Emission from laminates under thermally induced stress

CFRP laminates excel not only by their mechanical properties and their low specific weight, but also by their peculiar coefficient of thermal expansion, α, in fiber direction. This coefficient is a superposition of (1) the "normal" positive α of the matrix, and (2) a fiber-governed component which may become negative because of the graphite structure of the carbonfibers. The superposition is of course only applicable to the macroscopic laminate. In microscopic dimensions shear stresses between fibers and matrix are always being generated when the temperature changes. If the gradient in temperature becomes large enough, then cracks will develop and emit acoustic waves.

The N' emissions generated in a certain time T, during which the laminate undergoes such thermal loading, ought to be recorded as N events in analogy to the above model. In this case, however, generally *more* events are monitored than "audible" mechanisms in the material occur, i.e. $N' \leq N$. Instead of the overlapping problem there is the problem of "noisy" heating and cooling equipment, fasteners, and other auxiliary devices. A lot of technical precautions were taken in DLR's test program (e.g., specimen clamps made of teflon) In order to minimize all kinds of emissions coming from the environment of the composite structure.

II - EXPERIMENTAL RESULTS

2.1. Mechanical Tests

In order to thoroughly understand the "damage mechanics" of CFRP laminates, DLR's Institute for Structural Mechanics has investigated these materials in a comprehensive test program /6,7/. A number of mechanical tests were accompanied by AE measurements in the frequency range about 150 kHz in order to identify 2 characteristic damage levels: (1) the onset of crack growth, and (2) a dramatic increase in damage progression initiating the "final" phase in the specimen's life.

Fig. 2 summarizes the results. Mean values from a large number of tests are displayed in a stress-strain diagram containing average σ-ε-curves for 4 different laminates. The model described above has been applied for this with success. If necessary, the individual test results were rectified point-wise, as shown in Fig. 3.

2.2. Thermal tests

If the CFRP laminate is part of a spacecraft and being exposed to the large temperature gradients occurring in a low earth orbit (LEO), the stresses in the interface will inevitably lead to cracks, as discussed above. This phenomenon will become particularly important for future projects such as large light-weight truss structures, e.g. for space stations or antennas. The DLR, therefore, has initiated an experimental program focussed on this specific topic /8/.

Different CFRP specimens have been cyclically exposed to temperatures between -160 °C ($\sim$ shadow) and $+100$ °C or more ($\sim$ sunlight). Consequently, a large number of microcracks has been detected in the laminates under investigation. Their growth has successfully been pursued in real time by AE measurements with high sensitivity.

Fig. 4 shows the damage development for different CFRP laminates. Maximum and minimum temperature have been selected similar to the conditions in LEO, but each of the 400 thermal cycles has been shortened in time. (Those time intervals, where the real structure would be more or less in a thermal equilibrium with solar irradiation, have been cut off). It is obvious that crack growth starts quickly and becomes slower after a number of cycles, which may easily be interpreted as a saturation phenomenon. The release of AE energy when displayed as a function of temperature during *one* cycle leads to "threshold" temperatures characterizing the damage initiation in certain laminates.

These results, which were obtained in real time by means of AE inspection, could be conclusively verified by non-destructive and destructive *post-test*-inspection. The investigations are being continued.

CONCLUSIONS

Acoustic emission is a helpful tool for real-time investigations on CFRP laminates. The irreversible process of damage progression can be resolved in the time domain, i.e. as a function of stress, strain, or temperature. Emphasis was placed upon the specific problems due to the inhomogeneity of the material, which lead (1) to high emission rates and overlapping problems under *mechanical* load and (2) to a lively crack growth when the structure is exposed to *thermal* gradients.

REFERENCES

1. E. Fitzer (Ed.), Carbon Fibres and Their Composites. Springer Verlag, Berlin/Heidelberg/New York/Tokyo (1985)
2. A. Rotem, Composites Technology Review, 6 (1984) 145-158
3. A. Pollock, Non-Destructive Testing, 6 (1973) 264-269
4. J. Block and G.W. Ehrenstein, Materialprüfung, 29 (1987) 67-70
5. J. Block, Detektion von Schädigungsgrenzen in kohlenstoffaserverstärkten Kunststoffen mittels Schallemissionsanalyse. Theses, Univ. Kassel (1988)
6. H.W. Bergmann et al., DFVLR-Forschungsbericht 85-45 (1985)
7. H.W. Bergmann et al., DFVLR-Forschungsbericht 88-41 (1988)
8. W. Hartung, in /7/, 262-293

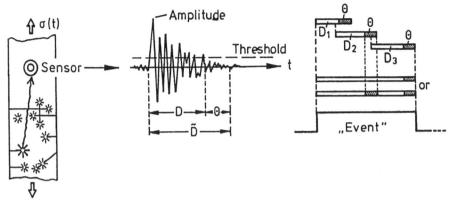

Fig.1 - The inhomogeneity of fiber composites leads to high emission rates so that the individual "event" signals may overlap

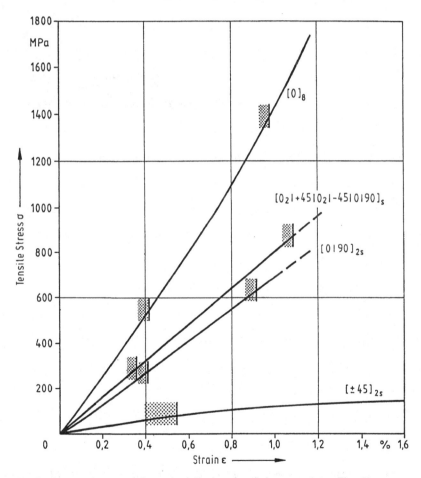

Fig.2 - Lower and upper characteristic levels of damage (see Fig. 3), summarized for 4 different stacking orders and displayed on average σ-ϵ-curves

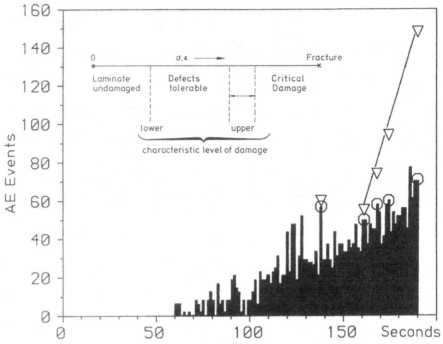

Fig.3 - The onset of cracks and an increase in crack growth rate (~ emission rate) short before failure, detected by means of the proposed model

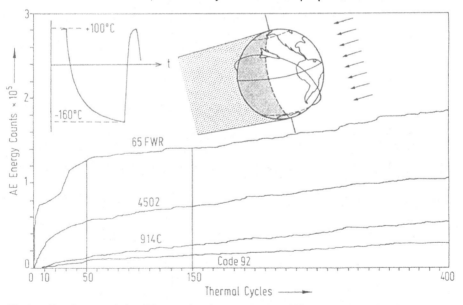

Fig.4 - Crack growth in different laminates during 400 thermal cycles (~ temperatures as in low earth orbits) monitored with AE

X-RAY DETERMINATION OF CRISTALLINITY OF THERMOPLASTIC SEMICRYSTALLINE SYSTEMS BY RULAND'S METHOD

G. MARRA, A. ADEMBRI, P. ACCOMAZZI, F. GARBASSI

Enimont - c/o Istituto Donegani - Via Fauser 4 - 28100 Novara - Italy

ABSTRACT

A new application of Ruland's method has been developed for semicrystalline matrix composites. The experimental profile has been obtained by simulating the cristalline peaks of the polymer matrix with Cauchy functions and performing a linear combination of the amorphous polymer and the pure fiber spectra. The Ruland's method takes into account also that cristalline defects reduce the observed intensity of the peaks by introducing an appropriate correction term for the crystalline total intensity. The values of cristallinity obtained from several samples are widely discussed in order to correlate the intrinsic meanings of the assumption we made on the cristallographic disorder.

INTRODUCTION

The correlation between degree of crystallinity and mechanical properties of semicrystalline systems is well established [1] [2]. Since the structure and morphology of semicrystalline resin matrix can impact the final properties of composites, it is essential to determine, in an absolute way, the degree of crystallinity in order to optimise the mechanical properties of the end product. Although there are other methods for such purpose we used WAXS (Wide Angle X-ray Scattering) because it is insensitive to fibers, fillers and all the other components which can make data unreliable. The determination of the degree of crystallinity is not a trivial task because common methods such as DSC (Differential Scanning Calorimetry) and density measurements need the exact value of the polymer weight fraction that is difficult to determine if carbon fibres are present as reinforcement. Moreover the presence of voids at the fibre/matrix interface and the interactive crystallization phenomena during the scanning, make X-ray an interesting tecnique because insensitive to such phenomena. The goal of this work is to perform a method to determine the absolute value of crystallinity of polymers and composites obtained using the same matrix. We have chosen as test material, the PEEK/CF composite, because it is one of the most promising material for its outstanding properties and furthermore it can be processed by several methods including the new F.I.T. technology (Fibers Impregnated with Thermoplastic).

I- BACKGROUND

1.1 Theoretical basis

The Ruland's method has strong theoretical bases /3/ /4/, it is the only one which takes into account the reduction of peaks intensities due to thermic disorder and structural defects; the degree of crystallinity x_{cr} is calculated introducing a "ad hoc" correction term and so we have:

$$x_{cr} = \frac{K(s_p)}{R(s_p)} \qquad (1)$$

where

$$K(s_p) = \frac{\int_{s_0}^{s_p} I_{cr}(s) s^2 ds}{\int_{s_0}^{s_p} I_{tot}(s) s^2 ds} \qquad (2)$$

and

$$R(s_p) = \frac{\int_{s_0}^{s_p} \overline{f^2}(s) D(s) s^2 ds}{\int_{s_0}^{s_p} \overline{f^2}(s) s^2 ds} \qquad (3)$$

where: $s = \frac{2\sin\theta}{\lambda}$ with λ the wavelenght in $\mathring{A}$, $I_{cr}(s)$, $I_{am}(s)$ and $I_{tot}(s)$ are the coherent intensity of the crystalline phase, of the amorphous phase and of the whole sample respectively; s_0 and s_p are related to the starting and ending angle of the measure; they must be chosen so that the above integrals are indipendent from x_{cr}. $\overline{f^2}(s)$ is the mean square of atomic scattering factors of the atoms in the polymer and $D(s)$ is a disorder function having two different expressions depending on first kind defects (thermal vibrations, vacancies and mechanical stresses) or second kind defects (loss of long range periodicity, also known as paracrystallinity) /5//6/.

$$D(s) = \begin{cases} \exp(-ks^2), & \text{for first kind defects;} \\ \frac{2\exp(-as^2)}{1+\exp(-as^2)}, & \text{for second kind defects.} \end{cases}$$

where k describes the amplitude of atomic displacement due to thermal vibrations and a describes the paracrystallinity distorsions. The Ruland's method has been modified and computerized by Vonk /7/ who has shown that, for most polymers, for values of s_p less than 1 Å^{-1}, K is nearly linear in s^2, that is:

$$K(s_p) = 1 + bs_p^2 \qquad (4)$$

moreover, when $D(s) = \exp(-ks^2)$ a relationship between k and the angular coefficient b has been found, and precisely $b = k/2$. Combining the equation (1) and (4) we have:

$$R(s_p) = \frac{1}{x_{cr}} + \frac{bs_p^2}{x_{cr}} \qquad (5)$$

The degree of crystallinity and disorder factor are obtained from the intercept and slope of K versus $s_p{}^2$ plot.

1.2 Theoretical development

The integrated total intensity, Q_{tot}, of the X-rays scattered by polymer can be expressed as the sum of the integrated intensities scattered by the amorphous and crystalline phases, Q_{am} and Q_{cr}. For a composite system, there is an additional scattering due to the reinforcing phase, Q_{fil}, so we have:

$$Q_{tot} = Q_{cr} + Q_{am} + Q_{fil}$$

after correcting the raw data from incoherent scattering subtraction and Lorentz polarization factor, the last and perhaps most important step to do is to separate the composite diffraction pattern into its proper components. While it is rather easy to obtain the completely amorphous matrix and pure fibers spectra, it is absolutely impossible to have a 100 % crystalline sample. In order to reproduce such pattern, we simulated it adding together Cauchy functions which have the following expressions:

$$I(2\theta) = \frac{A}{1 + \frac{(2\theta - B)^2}{C^2}}$$

where A, B, and C are the peak parameters. To reproduce the experimental spectra, we have to calculate a linear combination of the different contributions with the proper scaling parameters. The final intensity will be:

$$I_{teo}(2\theta) = \alpha I_{am}(2\theta) + \beta I_{fil}(2\theta) + \sum_{i=1}^{n} \frac{A_i}{1 + \frac{(2\theta - B_i)^2}{C_i^2}}$$

where n is the number of peaks to be fitted while α and β are the scaling parameters for the amorphous matrix and pure reinforcing phase. All these parameters are to be determined by minimising the residual function:

$$F = \int_{s_0}^{s_p} |I_{exp}(s) - I_{teo}(s)|^2 ds$$

where I_{exp} is the composite experimental spectra.

II-EXPERIMENTAL

All spectra have been performed by using a semi-automated Philips 1050/70 diffractometer, equipped with a solid-state scintillation counter and reflection geometry of scattering was used; copper radiation was used. Data analysis on Unisys 1100 computer and fit parameters are obtained by using a minimisation routine included in G.M.P. (Graphic Mathematical Package). The crystallinity of Filkar CF/PEEK grade 150 produced with F.I.T. technology, and APC 2 (AS 4 CF/PEEK) laminates were tested. The neat amorphous resin, PEEK grade 150, was obtained as described in literature /8/, while the reinforcing phases were extracted by dissolving the matrix in concentrated sulphuric acid.

III-RESULTS AND DISCUSSION

The spectra of the three samples are shown in fig 1, 2 and 3, where the contribution of each phase profile are reported as well. In the case of neat PEEK the fitting procedure has given very reasonable parameters and a good linear behaviour has been observed in $R(s_p^2)$, while concerning the composite samples some problems have been encountered. In fact the peak arising from the (002) reflection of the reinforcing phase is not easily to be fitted using the spectra of the pure fiber; any attempt for reproducing the experimental pattern in this way did not give us reasonable coefficients α and β and a strong non linearity in $R(s_p^2)$ was observed. So we decided to add in corrispondence of the (002) reflection a new narrow peak which describes the decreasing in crystallinity of the fibers due to the extraction procedure.

The results relative to the tested samples are shown in table 1. The representation of the values of R versus s_p^2 are plotted in Fig. 4 and the extrapolation of s_p^2 versus R gives $1/x_{cr}$. A very good linearity of R has been found for all the samples and concerning the thermal parametar k we found a significative reduction passing from the neat matrix to the PEEK/CF (F.I.T) composite which means an increasing of the crystallographic order.

With regards to APC 2 composite the value of crystallinity has been evaluated subtracting from the crystalline pattern a high contribution of the

(002) peak, and a high value of the slope of R has been found. This can be an indication of the not simple composition of such composite.

REFERENCES

1 - M.F. Talbott et al., *J. Composite Materials*, **21**, (1987) 1056
2 - J.N. Chu, J.M. Schults, *J. Material Science*, **24**, (1989) 4538
3 - W. Ruland, *Acta Crystallographica* **14**, (1961) 1180
4 - W. Ruland, *Polymer*, **5**, (1964) 89
5 - R. Hosemann, *J. Applied Physics*, **34**, (1963) 25
6 - R. Hosemann, *Z. Physik* **128**, (1960) 1 and 465
7 - C.K. Vonk, *J. Applied Crystallography*, **6**, (1973) 148
8 - D.J. Blundell, B.N. Osborne, *Polymer*, **24**, (1983) 953.

Table 1		
samples	$x_{cr}(\%)$	k
PEEK 150G	22.76	4.07
PEEK/CF (F.I.T.)	23.9	0.
PEEK/CF (APC 2)	16.3	5.0

Table 1 Crystallinity values and calculated thermal parameters.

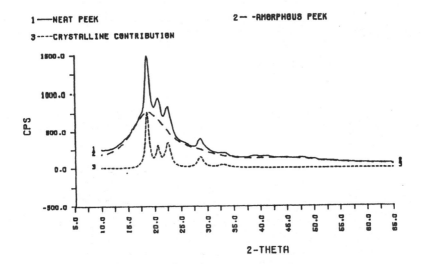

Fig. 1 Experimental diffraction pattern of neat PEEK and relative deconvolution in its amorphous and crystalline contributions.

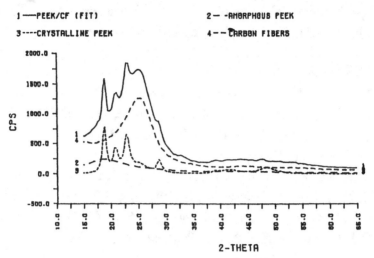

Fig.2 The same of fig.1 for PEEK/CF (F.I.T.)

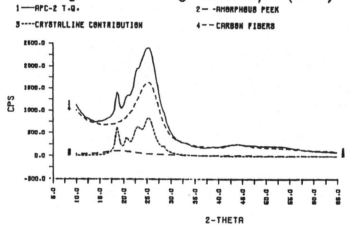

Fig.3 The same of fig.1 for APC-2

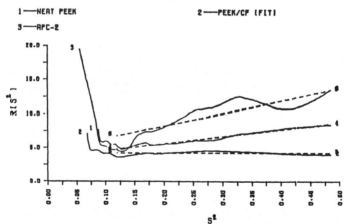

Fig.4 Vonk plots of the three samples

QUALITY ASSESSMENT OF GFRP TUBES WITH SHORT TERM ACOUSTIC EMISSION TEST

B. MELVE

Sintef Production Engineering - 7034 Trondheim - Norway

ABSTRACT

Filament wound tubes of glassfibre/epoxy, glassfibre/polyester and glassfibre/polyester with a surface mat were tested in three point bending with acoustic emission monitoring. In order to distinguish better between the layups, the tubes were subjected to a short term cyclic bend loading. There was a large difference in the level of acoustic emission activity between the epoxy matrix and polyester matrix tubes having the same layup. A surface mat reduced the acoustic emission of the polyester tubes. The "noisiest" material was the polyester matrix thus indicating a more brittle material than the epoxy material. The short term test gave a good distinction between the behaviour of the different materials and layups.

I - INTRODUCTION

Acoustic emission (AE) is a well established method for quality control of composite material products /1/, and procedures exists for testing of vessels and tanks /2-4/, glassfibre booms for bucket trucks /5/ and pipelines of fibre reinforced composites /6/. For other products similar test procedures have been used and developed. For instance have fan blades /7/, balsawood core sandwich vessels /8/ and many other products been tested with acoustic emission. In the present case GRP tubes for whip antennas have been investigated. The main application area for these antennas have been where corrosion resistance is needed i.e. for ships and offshore platforms. The conducting wires for the antennas can be inserted in the middle or braided onto the tube between some of the reinforcing layers. The antennas are subjected mainly to bending loads and therefore have to be tested in bending during an acoustic emission test. The usual method for qualifying such antennas have been to use the ultimate fracture load in three point bending of a 1 m long tube. But when monitoring the acoustic emission behaviour in such a fracture test, the correlation between onset of acoustic emission and fracture load is not always good /9-10/. Some other criterion than the maximum load may provide a better prediction of the long term performance of the tube. One alternative is to use a cyclic loading instead of a fracture test. By selecting a suitable load range for

709

the loads, a good characterisation can be obtained during a short period of time. The maximum bending load was below the expected load for total failure within the realistic loading range.

II - EXPERIMENTAL

2.1 Tube specimens

The antenna tubes were produced by the socalled "Howald process" which is a combination of filement winding and a pultrusion process. The process has been used for fishing rods and konical structures. The inner reinforcement layer was 1.2 mm thick and oriented in the hoop direction and the outer layer was 1.3 mm thick and oriented axially as shown in figure 1. The tubes had a diameter of 33 mm and a wall thickness of 2.5 mm. Three different material combinations were tested:

- a) Glassfibre/polyester
- b) Glassfibre/polyester with a fine surface mat
- c) Glassfibre/epoxy

2.2 Acoustic emission equipment

A Physical Acoustic Corporation (PAC) 3000/3104 system was used for the tests. Two PAC µ30 sensors from (frequency range 100-500 kHz) were mounted on the specimens approximately 10 cm from the load point. The pre-amplifiers amplified the signals 40 dB. The dynamic amplification of the equipment was 40 dB and the threshold was 0.4 V. The load cell output voltage was also given as a parametric input to the acoustic emission equipment. The AE sensors were coupled to the tubes with silicone grease and held tight by elastic adhesive tape.

2.3 Loading

The loading was done in a screw-driven universal testing machine (Instron). The loading rate was 10 mm/min. The span between the supports was 70 cm. The load supports were covered with 3 mm thick rubber pads in order to decrease contact stresses and stop noises coming from the loading machine and surroundings. The cycling was done in load control from a minum of 15 kg and a maximum of 85 kg. All tests were done at roomtemperature. The test setup is shown in figure 2.

2.4 Bend analysis of the tubes

The deflection and outer fibre strain of thinwalled and linear-elastic laminate products can be analysed by bending theory with the Youngs modulus E exchanged with the E^o of the laminate. E^o is the engineering stiffness of the midplane of the laminate /11/. The MIC-MAC TUBE program developed by Tsai /11/ was used for analysing the tubes. The program uses laminate theory for the calculations. According to these calculations did the outer fibre strain vary from to 0.07% to 0.36 % during a load cycle from 15-85 kg.

III - RESULTS

The tests lasted 15-20 minutes. This time period allowed for an appropriate number of cycles to be conducted. The rate of acoustic emission events was also acceptable and did not exceed the capacity of the instrumentation.

The load for acoustic emission events is plotted versus time (or cycles) in figure 3 for the three layups. It is clear that the unmodified polyester tube have the highest number of events all through the load cycle. The most important events are belived to be those appearing at the highest load: "primary events". The other events might be called "secondary events". The secondary events can come from friction sources when crack surfaces with fibres open or close. The polyester matrix layups probably have more microcracks giving rise to the frictional acoustic emission. A surface mat seems to reduce the number of microcracks as shown in figure 3 b). The epoxy matrix is the most quiet material because of fewer microcracks.

The amplitude of the acoustic emission events versus load is shown in figure 4 a)-c). It is seen that the unmodified polyester matrix system have the largest number of high amplitude events. These events are indicating that for instance fibres are broken or that large matrix cracks developes. The surface layer on the polyester tube does reduce the number significantely.

Polyester matrices usually have a smaller elongation before failure than epoxy matrices /12/. For the same loading a larger number of microcracks will develop in the polyester layups and then give more acoustic emission events. If fatigue is combined with agressive media, the microcracks can increase the rate of uptake of the medium because diffusion rates in cracks is much higher than in uncrack bulk material.

IV - CONCLUSIONS

- The epoxy resin layups give less noise than the polyester resin. This can indicate a better long term capability for the epoxy layups.
- A fine surface layer decreased the number of microcracks giving rise to acoustic emission.
-A short term cyclic loading gives a good possibility to separate between materials and layups, and obtain more information than a single cycle to fracture.

AKNOWLEGDEMENTS

Mr. Jostein Espedalen, Comrod A/S, Tau is thanked for the supply of tubes. The work was sponsored by the Nordic Fund for Technology and Industrial Development through contract P87148.

REFERENCES

1 - T. L. Swanson, Development of codes, standards and practises for acoustic emission examination of composites in the USA, Second International Symposium on Acoustic Emission from Reinforced Composites, Society for the Plastics Industry (SPI) Montreal, 1986

2 - Recommended Practice for Acoustic Emission Testing of Fiberglass Tanks/Vessels, 37th Annual Conf., Reinforced Plastics/Composites Institute, Society for the Plastics Industry (SPI), 1982

3 - ASTM E1067-85, Standard Practice for Acoustic Emission Examination of Fiberglass Reinforced Plastic Resin (FRP) Tanks/Vessels, American Society for Testing and Materials, Philadelphia

4 - Acoustic Emission Monitoring of Fiber Reinforced Plastics Vessels, American Society of Mechanical Engineers, (ASME) Boiler and Pressure Vessel Code, Section V, Article 11, 1985

5 - ASTM F914-85, Standard Test Method Acoustic Emission for Insulated Aerial Personnel Devices, American Society for Testing and Materials, Philadelphia

6 - E1118-86, Standard Practice for Acoustic Emission Examination of Reinforced Thermosetting Resin Pipe (RTRP), American Society for Testing and Materials, Philadephia

7 - T.N. Crump and J.J. English, Acoustic emission testing of RP fan blades used for air cooled heat exchangers and cooling towers, Second International Symposium on Acoustic Emission from Reinforced Composites, AECM-2, Society for the Plastics Industry (SPI), Montreal, 1986

8 - P. Oulette, S.V. Hoa and L. Li, A Procedure for Acceptance Testing of FRP Balsawood Core Pressure Vessels, Third Int. Symp. on Acoustic Emission from Composite Materials, AECM-3, Paris, The American Society for Nondestructive Testing, Columbus, 1989

9 - M. Hval and B. Melve, SINTEF report STFA87114, Trondheim, 1987

10 - B. Melve and M. Hval, Quality Control of CFRP Ski-poles using Acoustic Emission, 2nd. Japan Sweden Symposium on Composite Materials, March 1988, Tsukuba, Japan, 1988

11 - S. W. Tsai, Composite Design, Think Composites, Dayton, 1988

12 - Kompositboken, Swedish Plastics and Rubber Institute, Stockholm, 1988

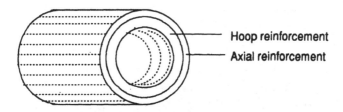

Hoop reinforcement
Axial reinforcement

Figure 1 The layup of the glassfibre reinforcement used in the tubes.

712

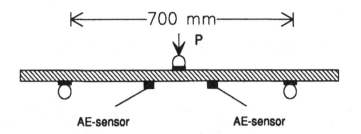

Figure 2 Experimental setup of specimens with sensors mounted.

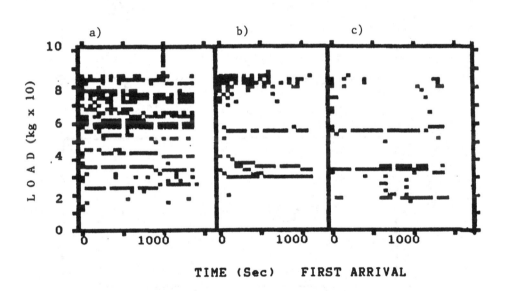

TIME (Sec)   FIRST ARRIVAL

Figure 3 Load at acoustic emisson events versus time (cycles) for a) glassfibre/polyester,
b) glassfibre/polyester with surface mat and c) glassfibre/epoxy.

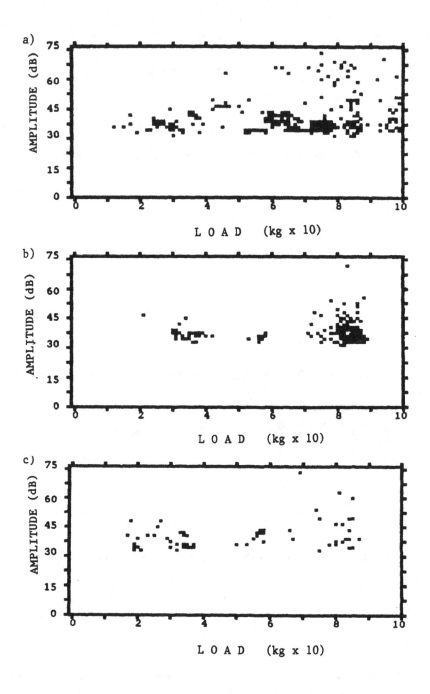

Figure 4 Amplitude of the acoustic emission events versus load for a) glassfibre/polyester
b) glassfibre/polyester with surface mat and c) glassfibre/epoxy.

ANALYSIS
ANALYTISCHE METHODEN

Chairmen :   **Prof. M. NEITZEL**
BASF, now University
of Kaiserslautern
Prof. S.W. TSAI
University of Stanford

ON THE THERMOMECHANICAL ELASTIC-PLASTIC
RESPONSE OF A CLASS OF FIBROUS COMPOSITES
A UNIFIED APPROACH

K.P. HERRMANN, I.M. MIHOVSKY*

Laboratorium für Technische Mechanik
Paderborn University - Pohlweg 47-49 - 4790 Paderborn - West-Germany
*Dept. Math. and Informatics - Sofia University
Anton Ivanov Str. 5 - 1126 Sofia - Bulgaria

ABSTRACT

An approach to the investigation of the elastic-plastic thermomechanical response of a class of fibrous composites is proposed. The approach is unified in the sense that it reduces an entire class of problems to particular cases of a certain general problem of the plasticity theory. The predictions of the approach are shown to agree with available experimental data.

INTRODUCTION

A series of recent authors' works has been devoted to certain aspects of the thermomechanical response of a class of low-fibre volume fraction composites, /1-4/. Unidirectionally reinforced composites with continuous elastic fibres perfectly bonded to an elastic-perfectly plastic matrix have been considered. The cited works treat the axisymmetric model problems of thermal (matrix cooling) and mechanical (longitudinal extension) loading of an initially stress-free composite unit cell consisting of a single circular cylindrical fibre with a coaxial matrix coating. They deal basically with the process of redistribution of stresses due to progressive matrix plastification.

A series of both qualitative conclusions and quantitative results concerning basic features of this process have been derived and then used for the prediction of the overall response of the composites. The predicted response proves to be quite realistic but it still needs a comparison with the behaviour of real composites (with fabricationally in-duced internal stresses) under real operational conditions.

The analysis of the results obtained in /1-4/ shows that from a mechanical point of view the two model problems are approached in the same manner. It appears at the same time that different regimes of matrix plastification develop depending upon the loading type and its intensity. From the view point of the mathematical plasticity theory the two problems prove to be governed by the same set of partial differential equations. Changes in the combinations of constituents properties, loading status, and/or load intensity imply changes in its type (hyperbolic, elliptic, parabolic). Thus, the solutions of this set and, respectively, the regimes of plastification may really appear to be quite different in their properties.

I - BASIC CONSIDERATIONS

The present work follows the idea that the considered approach should be first of all based upon the analysis of the interactions between the principal effects of the fibre reinforcement (stiffening, strenghthening, shrinkage, and stress concentration) and matrix ductility (limited elastic response).

As is known from observations on real composites of the class considered their behaviour "in the fibre direction" is rather elastic-like than elastic-plastic. This seems to be due to the stiffening effect. The stiff elastic fibres prevent the development of irreversible plastic parts of the axial (along the fibres) matrix strain ε_z which are large with respect to the elastic ones $\varepsilon_z{}^{el}$. Thus, when considering the elastic-plastic response of such composites one should rather take into account the current $\varepsilon_z{}^{el}$-strains instead of neglecting them as is the case in the common plasticity approaches.

To realize such an account is quite a difficult task. It is clear at the same time that with progressive matrix plastification the increase in the $\varepsilon_z{}^{el}$-strain is restricted due to its limited elastic response. One may then assume that an average value $\varepsilon_z{}^*$ of the latter strain exists upon which the $\varepsilon_z{}^{el}$-increments become negligible with respect to the plastic ones and the plastification develops further under the condition $\varepsilon_z{}^{el} = \varepsilon_z{}^*$. Note that in thermal problems the $\varepsilon_z{}^{el}$-strains should not involve the thermal component. The $\varepsilon^*{}_z$-strain is specific for a given combination of a composite structure and the associated loading status. Its identification is a part of the analysis of the composite response.

Due to the stress concentration effect the matrix should start yielding at the interface. Upon certain transitional phases of plastification with increasing $\varepsilon_z{}^{el}$-strains the condition $\varepsilon_z{}^{el} = \varepsilon_z{}^*$ should be first achieved at the interface. The latter phase should not be expected to change considerably the initial linear elastic response.

The shrinkage effect will be considered in the next section. Due to the restriction of axial symmetry the plastic zone should keep the form of an annulus. Along with the latter restriction the assumptions of perfect bonding and fibre continuity allow to consider the plane cross-section hypothesis to apply to the cell under typical technological and operational loadings. Thus, when the cell is referred to cylindrical coordinates (r, θ, z) (the z-axis being the axis of the fibre) the normal stresses σ_i, $i = r, \theta, z$ should be the principal ones in both the fibre and the matrix, respectively.

The fibre and matrix cross sections occupy the regions $0 \leq r \leq r_f$ and $r_f \leq r \leq r_m$, respectively. The Young's moduli, the Poisson's ratios, and the thermal expansion coefficients of the fibre and the matrix are denoted by E_f and E_m, v_f and v_m, α_f and α_m, respectively. The matrix material obeys the von Mises' yield condition and the associated flow rule. Its tensile yield stress is σ_y. The quantities E_i, v_i, α_i, σ_y are temperature independent.

II - THE UNIFIED APPROACH

2.1 General Plasticity Problem

Since the elastic parts of the strains in a plastically deformed material are related to the stresses through Hooke's law (cf. for example /6/) then, in accordance with the previous considerations, the following relation holds true within the plastified annulus $r_f \leq r \leq R_c$ of the matrix phase

$$\sigma_z = E_m \, \varepsilon_z{}^* + v_m \, (\sigma_r + \sigma_\theta) \tag{1}$$

This relation allows the reduction of the von Mises yield condition to the equation of an ellipse in the $(\sigma_r, \sigma_\theta)$-plane

$$(\sigma_r - \sigma_\theta)^2 + (\sigma_r + \sigma_\theta - 2E_m\varepsilon_z^* \ \text{cotan}\ \phi/\sqrt{3})^2 \ \tan^2 \phi - 4\sigma_y^2/3 = 0 \qquad (2)$$

The stresses σ_r and σ_q being coordinates of the points of the yield ellipse (2) are representable in the form

$$\left.\begin{array}{c}\sigma_r \\ \\ \sigma_\theta\end{array}\right\} = \frac{E_m \ \varepsilon_z^*}{1-2v_m} + \frac{\sigma_y}{\sqrt{3}\ \sin\phi}\ \cos(\omega \pm \phi), \qquad (3)$$

where

$$\sin \omega = (\sigma_\theta - \sigma_r)\ \sqrt{3}/2\sigma_y; \quad \tan \phi = (1-2v_m)/\sqrt{3}, \qquad (4)$$

and the angle $\omega = \omega(r)$ defines the position of the stress states given by (3) along the yield ellipse.

These stresses satisfy the equilibrium equation

$$\frac{d\sigma_r}{dr} + \frac{1}{r}\ (\sigma_r - \sigma_\theta) = 0 \qquad (5)$$

Now the set of equations (1)-(5) specifies a plane stress-like perfect plasticity problem in the sense that the latter is analogous to the known classical plane stress problem of the perfect plasticity theory, cf. /6/. The cases $|\omega| < \varphi$ (or $|\omega-\pi| < \varphi$), $\varphi < \omega < \pi-\phi$ (or $\varphi < \omega-\pi < \pi-\varphi$), and $|\omega| = \phi$ (or $|\omega-\pi| = \phi$) are easily shown to correspond to elliptic, hyperbolic and parabolic types of the set, or equivalently of the phases of plastic redistribution of the stresses.

2.2 General Scheme

Along with the approximate boundary condition

$$\omega_{Rc} = \text{arc cos}\ [-E_m\ \varepsilon_z^*/\sigma_y\ (1+v_m)] \qquad (6)$$

the integration of equation (5) implies the implicit $\omega(r)$-dependence

$$R_c^2 \sin \omega_{Rc} = r^2 \sin \omega \ \text{exp}\ [(\omega-\omega_{Rc})\ \text{cotan}\ \phi],\ r_f \le r \le R_c \qquad (7)$$

where $\omega_{Rc} = \omega(R_c)$. Equation (6) reflects the standard assumption of plastic incompressibility.

The continuity conditions at the interface $r = r_f$ (for the radial displacement) and the elastic-plastic boundary $r = R_c$ (for the radial stress) imply through equation (7) the $R_c\ (\varepsilon_z)$-dependence and specify thus the stresses within the unit cell as functions of ε_z. Note that for the considered class of problems the general form of the elastic stress distributions is known from the general axisymmetric elasticity problem. The desired strain versus load intensity dependence, i.e. the overall composite response, follows then from the equilibrium condition of the axial stresses.

2.3 Particular Problems

When a given combination of a composite and a loading type is under consideration then the specific elastic response for this combination defines a specific $\varepsilon_z^{el,pl}$-value of the ε_z^{el}-strain at which the matrix starts yielding as well as a specific point of initial yielding on the yield ellipse. Depending upon the position of this point and the way in which it will further move along the ellipse with progressive loading different phases of plastification (cf. Section 2.1) and expansion of the plastic zone will occur and develop in the matrix.

As is shown in /1-4/ matrix cooling implies a hyperbolic regime with a monotonically expanding plastic zone. A similar state develops under longitudinal extension provided $(v_m - v_f) > (1+v_m)(1-2v_m)/3$ holds true. In the opposite case an elliptic state occurs and a sudden entire matrix plastification takes place.

It is easy to observe that any plastic stress redistribution, i.e. any motion of the representative point along the yield ellipse corresponds either to an increase or to a decrease in the shrinkage effect, i.e. in the value of $\sigma_r(\omega_{rf})$, $\omega_{rf} = \omega(r_f)$, cf. equation (7). Since the general scheme from Section 2.2 involves as a particular step the determination of the $\omega_{rf}(\varepsilon_z)$-dependence then shrinkage control should be achievable through either appropriate load intensities and programmes (for a given composite) or appropriate sets of constituents properties (for given loading programmes).

As it follows from equations (3) maximum possible shrinkage is achievable at $\omega_{rf} = \pi - \phi$. The corresponding maximum plastic zone radius R_c^* is given by equation (7) with $r = r_f$, $\omega_{rf} = \pi - \phi$. The point $\omega_{rf} = \pi - \phi$ is a parabolic point of the yield ellipse. As is shown in /1-3/ plastic instability may occur at this point and cause certain modes of failure of the composite (debonding, fibre breaking and fibre pull-out). Thus the parabolic stress states are critical and their occurence should be avoided. The process of matrix cooling is easily seen to be a process of monotonically increasing shrinkage.

Note that as it follows from the general scheme a series of initially assumed ε_z^*-values implies a series of predicted deformation curves. The identification of the actual ε_z^*-strain is a matter of a comparison between these curves and the real one, cf. /3/. In any case the actual ε_z^*-value should be close to the value $\varepsilon_z^{el,pl}$ specified above.

The presence of internal axisymmetric stress fields in the composites does not imply principal changes in the general approach. In such cases the scheme should be simply coupled with the standard analysis of residual stresses of the plasticity theory (cf. for example, /6/).

III - APPLICATIONS

3.1 Matrix Cooling in Absence of Internal Stresses

When applied to this particular case the unified approach implies the following implicit form of the $\varepsilon_z(T_m)$-curve, T_m being the matrix temperature with $T_m \leq 0$:

$$\Delta \varepsilon_z = \alpha_m \Delta T_m \left[1 - \frac{E_c}{1 + E_c - R_c^{*2}(1 - A + A \Delta \varepsilon_z / \Delta \varepsilon_z^*)/r_m^2} \right] \tag{8}$$

where $E_c = E_f r_f^2 / E_m r_m^2$, $A = 1 - r_f^2/R_c^{*2}$, $\Delta \varepsilon_z = \varepsilon_z - \varepsilon_z^{el,pl}$, $\Delta T_m = T_m - T_m^{pl}$, $\varepsilon_z^{el,pl} = -\varepsilon_z^*/E_c$, $T_m^{pl} = -\varepsilon_z^*(1+E_c)/\alpha_m E_c$, $\Delta \varepsilon_z^* = \Delta \varepsilon_z (T_m^*)$, T_m^* being the temperature of the critical state, cf. Section 2.3. At the point of initial matrix yielding ($\varepsilon_z^{el,pl}$, T_m^{pl}) the deformation curve (8) smoothly deviates form the straight line $\varepsilon_z = \alpha_m T_m (1+E_c)$. The latter describes the linear elastic response of the composite.

The $R_c(T_m)$-dependence follows from equation (8) and the equation

$$R_c^2(\Delta \varepsilon_z) / R_c^{*2} = 1 - A \left[1 + \Delta \varepsilon_z \frac{2v_f (1-2v_m) E_f \cos \phi}{(\pi - \phi - \omega_{Rc})^2 \sigma_y (1+v_f)(1-2v_f)} \right] \tag{9}$$

Figure 1 shows according to equation (8) the predicted response of an Al/C fibrous composite with $E_m = 6 \cdot 10^4 \text{N/mm}^2$, $E_f = 24 \cdot 10^4 \text{N/mm}^2$, $v_m = 0.31$, $v_f = 0.27$, $r_f/r_m = 1/10$, $\sigma_y = 135 \text{N/mm}^2$, $\alpha_m = 18.2 \cdot 10^{-6} \text{K}^{-1}$ (data taken from /8/). The lines L_i, $i = 0, 1, 2, 3$

correspond to initial $\varepsilon_{z,i}^{*}$-values defined as $\varepsilon_{z,i}^{*} = (\varepsilon_z^{el,pl} - \alpha_m T_m^{pl})(0.8 + 0.2i)$. These lines deviate smoothly at the points P_i from the straight line L_e. The latter represents the linear elastic response. Here and below the star signs "*" correspond to critical states of the composites.

3.2 Heating of a Composite with Internal Stresses

The line LE in Fig. 2 reflects the experimental data obtained by Larsson /5/ in heating a SS/W-2% ThO$_2$ (SS stays for a stainless steel, type 304) composite from its as -fabricated state, i.e. from a state with internal stresses induced by a hot isostatic pressing. As it should be expected during this heating the composite should pass through a stress free state. This state is reported to be achieved at approximately 430°C. Accordingly, if the predictions of the present approach were correct then equation (8) along with the linear elastic $\varepsilon_z(T_m)$-dependence should describe an arc, say BC, of the real LE-curve. The response of the stress-free composite defined by equation (8) is presented by the curve HM. Now it is a matter of a translation of the latter curve to prove that the arc OD simply laps over the arc BC with point B corresponding to the stress-free temperature.

The $R_C(T_m)$-dependences for this case and for the case of matrix cooling (Section 3.1) with i = 1 are derived by the equations (8), (9) and shown in Fig. 3 as L-and HM-curve, respectively.

3.3 Further Implications

As is shown in /3,9/ the approach allows to justify the applicability of the known Dugdale crack model /10/ to the class of radial matrix cracks. Fig. 3, curve S, shows the temperature dependence of the length s of the thin plastic zone at the tip of an radial edge matrix crack of length l = 0.3 r_m in the case of matrix cooling (Section 3.1, i=1).

Details concerning the influence of the fibre volume fraction and constituents properties on the $R_C(T_m)$ and $s(T_m)$-dependences are given in /11/. Note that these dependences are relevant to the important class of fibre-matrix cracks interactions.

Further implications of the approach concerning the model problems of matrix cooling and longitudinal extension can be found in /1-4,7,9,11/.

REFERENCES

1. Herrmann K., Mihovsky I., Rozprawy Inzynierskie-Engn. Trans., 31 (1983) 165-177
2. Herrmann K., Mihovsky I., Mechanika Teoretyczna i Stosowana, 22 (1984) 25-39
3. Herrmann K., Mihovsky I., DFG-Research Report, LTM, Paderborn University (1987)
4. Herrmann K., Mihovsky I., ZAMM, 68 (1988) T193-T194
5. Larsson L.O.K., In "ICCM/2-Proceedings" (B. Noton et al, eds.) AIME, Warrendale, (1978) 805-821
6. Kachanov L.M., "Foundations of the Theory of Plasticity", North-Holland, Amsterdam-London (1971)
7. Herrmann K., Mihovsky I., to appear.
8. Herrmann K., In "SM Study No 12" (I. Prowan, ed.) University of Waterloo, (1978) 313-338
9. Herrmann K., Mihovsky I., ZAMM, 67 (1987) T196-T197
10. Dugdale D.S., J. Mech. Phys. Solids, 8 (1960) 100-104
11. Herrmann K., Mihovsky I., In "Proc. Euromech Colloquium 255", Paderborn University (1989), to appear.

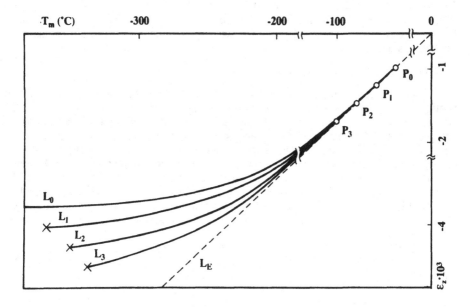

Fig. 1. Predicted deformation curves (matrix cooling).

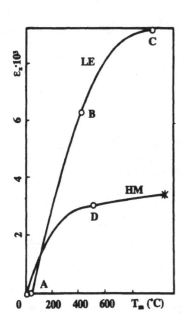

Fig. 2. Experimental (LE) and predicted (HM) deformation curves.

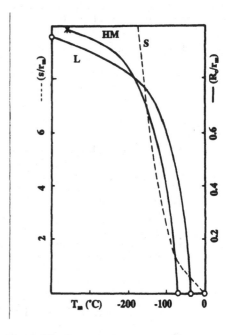

Fig. 3. Plastic zone sizes versus temperature dependences.

ANALYSIS OF ADHESIVE BONDED GENERALLY ORTHOTROPIC CIRCULAR CYLINDRICAL SHELLS

O.T. THOMSEN, A. KILDEGAARD

University of Aalborg - Institute of Mech. Eng.
Pontoppidanstraede 101 - 9220 Aalborg East - Denmark

ABSTRACT

The problem of an adhesive bonded lap joint between two dissimilar orthotropic circular cylindrical laminated shells is considered. The principal directions of orthotropy do not have to coinside with the principal directions of curvature, and the external loads are allowed to be of non-axisymmetric type. The adhesive layer is modelled in two ways. The first approach assumes the adhesive layer to behave as a linear elastic material. The second more realistic approach takes into account the predominantly inelastic behavior of many polymeric adhesives, and it is shown, that the non-linear behavior affects the adhesive stress distribution even at low load levels. The developed numerical solution procedures have been used to conduct a parametric study, and a few general design recommandations are given.

I - INTRODUCTION

The importance of adhesive bonding in technology has been generally recognized, and a considerable amount of analytical, finite element and experimental work has been carried out on the subject of adhesive joints /1-6/. Most of this theoretical and experimental work has been dealing with the investigation of adhesive joints between plate elements /1-3/. However, adhesive bonding techniques are also being employed in joining multilayer composite systems made of thin, anisotropic, circular shell elements, and the papers regarding the investigation of such tubular joint are very few /4-6/.

In general, the reviewed references (except Hart-Smith /2/ and Adams and Peppiatt /3/) assume the adhesive layer as well as the shell elements to behave as linear elastic materials. This assumption does provide useful information about the intensities of stress concentrations and their location, but the results do not reflect the true stress distribution or behavior at appreciable levels of loading, because of the predominantly inelastic nature of many polymeric structural adhesives.

The main objectives of the present paper is to consider the general problem of tubular lap joints between two dissimilar circular cylindrical shells (easily extended to other adhesive joint configurations such as scarf or stepped lap joints). The investigation is more general than the reviewed references regarding tubular adhesive lap joints /4-6/ in the sense, that the principal directions of orthotropy do not have to coinside with the principal directions of curvature, and the loads are allowed to be of non-axisymmetric type. The present study also extends the solution based on linear elastic assumptions for the adhesive layer to include adhesive non-linearity by the use of an empirical effective stress/strain concept suggested by Gali, Dolev and Ishai /7/.

II - ELASTO-STATIC FORMULATION

The tubular lap joint configuration composed of two dissimilar orthotropic laminated shell elements subjected to axisymmetric as well as non-axisymmetric external loadings is shown in Fig.1.

The analysis presented in this paper is based on a general theory of orthotropic laminated shells developed by Kalnins /8/, which includes the effect of the transverse shear deformations.

The equilibrium equations for the individual laminated shell elements (i=1,2) can be written as (differentiation with respect to the space coordinates x and θ is indicated by a comma):

$$N^i_{x,x} + \frac{1}{R^i} N^i_{\theta x,\theta} = -p^i_x \quad ; \quad N^i_{x\theta,x} + \frac{1}{R^i} N^i_{\theta,\theta} = -\frac{Q^i_\theta}{R^i} - p^i_\theta \quad ;$$

(II.1) $$Q^i_{x,x} + \frac{1}{R^i} Q^i_{\theta,\theta} = \frac{N^i_\theta}{R^i} - p^i_z \quad ; \quad M^i_{x,x} + \frac{1}{R^i} M^i_{\theta x,\theta} = Q^i_x - m^i_x \quad ;$$

$$M^i_{x\theta,x} + \frac{1}{R^i} M^i_{\theta,\theta} = Q^i_\theta - m^i_\theta \quad ;$$

where N^i_x, N^i_θ, $N^i_{x\theta}$, $N^i_{\theta x}$ = membrane stress resultants;

M^i_x, M^i_θ, $M^i_{x\theta}$, $M^i_{\theta x}$ = bending and twisting moment resultants;

Q^i_x, Q^i_θ = transverse shear stress resultants;

P^i_x, P^i_θ, P^i_z = load terms corresponding to summation of external surface loads and the adhesive layer stresses σ_z, τ_{zx}, $\tau_{z\theta}$;

m^i_x, m^i_θ = moments of the surface loads and adhesive stresses with respect to the reference surface.

The generalized strain-displacement relations for each laminated shell element (i=1,2) are:

(II.2) $$u^i_{,x} = \epsilon^i_{x0} \quad ; \quad v^i_{,x} = \gamma^i_x \quad ; \quad w^i_{,x} = \gamma^i_{xz} - \beta^i_x \quad ; \quad \beta^i_{x,x} = \kappa^i_x \quad ; \quad \beta^i_{\theta,x} = \delta^i_x \quad ;$$

(II.3) $$\begin{cases} \epsilon^i_{\theta 0} = \frac{1}{R^i} v^i_{,\theta} \quad ; \quad \kappa^i_\theta = \frac{1}{R^i} \beta^i_{\theta,\theta} \quad ; \quad \gamma^i_\theta = \frac{1}{R^i} u^i_{,\theta} \quad ; \quad \delta^i_\theta = \frac{1}{R^i} \beta^i_{x,\theta} \quad ; \\ \gamma^i_{\theta z} = \beta^i_\theta - \frac{1}{R^i}(v^i + w^i_{,\theta}) \quad ; \end{cases}$$

where u^i, v^i, w^i = components of displacement of the reference surface in the x-, θ- and z-directions respectively;

724

ϵ^i_{x0}, $\epsilon^i_{\theta 0}$ = normal strains of the reference surface;

κ^i_x, κ^i_θ = curvature change components in the x-, θ-directions;

γ^i_x, γ^i_θ = components of the in-plane shearing strain $\gamma^i_{x\theta}$;

δ^i_x, δ^i_θ = components of twist change in the x-, θ-directions;

β^i_x, β^i_θ = rotations of the normal to the reference surface.

The stiffness relationships for the i'th shell element (i=1,2) can be expressed by the following relation:

$$(II.4) \quad \begin{aligned} &(N^i_x, \ N^i_\theta, \ N^i_{x\theta}, \ N^i_{\theta x}, \ M^i_x, \ M^i_\theta, \ M^i_{x\theta}, \ M^i_{\theta x}, \ Q^i_\theta, \ Q^i_x) \\ &= [D^i] \ (\epsilon^i_{x0}, \ \epsilon^i_{\theta 0}, \ \kappa^i_x, \ \kappa^i_\theta, \ \gamma^i_x, \ \gamma^i_\theta, \ \delta^i_x, \ \delta^i_\theta, \ \gamma^i_{\theta z}, \ \gamma^i_{xz}) \end{aligned}$$

where $[D^i]$ is a (10,10)-matrix of shell stiffness coefficients, which are given in ref. /9/.

Equations (II.1-4) complete the system of equations which represents the mathematical model used for the laminated shell elements. The coupling between the shell elements is established through the constitutive relations for the adhesive layer, which as a first approximation is assumed to be homogeneous, isotropic and linear elastic. The constitutive relations are proposed in accordance with ref. /6/:

$$(II.5) \quad \begin{aligned} \tau_{zx}(x,\theta) &= \frac{G}{t} \{u^2 - \beta^2_x \frac{h}{2}2 - u^1 - \beta^1_x \frac{h}{2}1\} \ ; \\ \tau_{z\theta}(x,\theta) &= \frac{G}{t} \{v^2 - \beta^2_\theta \frac{h}{2}2 - v^1 - \beta^1_\theta \frac{h}{2}1\} \ ; \\ \sigma_z(x,\theta) &= \frac{E^*}{t} \{w^2 - w^1\} \ ; \end{aligned}$$

where G, E^* are elastic constants (E^* is somewhat larger than the E-modulus of the adhesive) and t is the thickness of the adhesive layer. Equations (II.5) express the compatibility of strains across the interfaces between the adherends and the adhesive layer.

The set of equations (II.1) through (II.5) represents a 10. order system of coupled partial differential equations containing 25 unknowns for each shell element. However, the number of unknowns can be reduced to 10 dependent variables /8/,/10/, henceforth called the fundamental variables, by simple rearrangement of the set of governing equations. Thus, if a column matrix is defined such that

$$(II.6) \quad \{y^i\} = \{u^i, \ v^i, \ w^i, \ \beta^i_x, \ \beta^i_\theta, \ M^i_{x\theta}, \ N^i_{x\theta}, \ M^i_x, \ N^i_x, \ Q^i_x\} \ ;$$

is the matrix of fundamental variables of the i'th shell (i=1,2), the governing equations can be reduced to a set of 1. order partial differential equations.

By expanding the variables of the problem in Fourier series, the dependency of the circumferential coordinate θ is eliminated, and the governing equations are reduced to a set of 1. order ordinary differential equations, which have to be solved subjected to boundary conditions specified at $x=-L_1$, $x=0$, $x=L$, $x=L+L_2$ (see Fig.1) and specified external surface loads.

The solution of the elasto-static problem is achieved numerically by application of the multisegment-method of integration /10/, which is based on a transformation of the original multiple-point boundary value problem into a series of initial value problems. The tubular lap joint configuration is divided into a finite number of segments, and the solution within each segment can be accomplished by means of any of the standard methods of direct integration (Adam's predictor-corrector method in the present analysis). Continuity of the fundamental variables at the separation points between the segments is ensured by formulating and solving a set of linear algebraic equations.

III - NON-LINEAR FORMULATION

The assumption of linear elasticity of the adhesive is not realistic since the response of most polymeric structural adhesives is predominantly inelastic in the sense, that plastic residual strains are induced even at low levels of external loading. The present non-linear formulation, which is applicable only to problems subject to axisymmetric boundary conditions and surface loadings, is based on an effective stress/strain relationship derived empirically from tests on bulk specimens, and it is assumed that the bulk and "in-situ" mechanical properties of the adhesives are closely correlated. All non-linear time and temperature dependent effects are still ignored.
It is widely accepted that the yield behavior of many polymeric adhesives is dependent on both deviatoric and hydrostatic stress components, and a consequence of this phenomenon is a difference between the yield stresses in uniaxial tension and compression /7/. This behavior has been incorporated into the analysis by applying a modified von Mises criterion suggested by Gali, Dolev and Ishai /7/:

$$s = C_s (J_{2D})^{1/2} + C_v J_1 \; ;$$

(III 1)

$$C_s = \frac{\sqrt{3}(\lambda+1)}{2\lambda} \; ; \; C_v = \frac{\lambda-1}{2\lambda} \; ; \; \lambda = \sigma_c/\sigma_t \; ;$$

where s = effective stress; J_{2D} = 2. invariant of the deviatoric stress tensor; J_1 = 1. invariant of the general stress tensor and λ = ratio between compressive and tensile yield stress. In the case of $\lambda=1$ Eqs. (III.1) reduces to the ordinary von Mises criterion.
The effective strain e is given by:

(III.2)

$$e = C_s \frac{1}{1+\nu} (I_{2D})^{1/2} + C_v \frac{1}{1-2\nu} I_1 \; ;$$

where ν = Poisson's ratio; I_{2D} = 2. invariant of the deviatoric strain tensor; I_1 = 1. invariant of the general strain tensor.
It is further assumed, that λ (see Eqs.(III.1)) and ν both can be regarded as constants throughout the strain range.
The developed approximate non-linear solution procedure is based on an iterative use of the elasto-static solution procedure. Initially a uniform E-modulus is assumed for each point of the adhesive layer, and the elasto-static solution procedure is applied. If the calculated effective stress, at a specific adhesive point, is above the proportional limit of the adhesive, then the E-modulus, is replaced by the secant-modulus obtained from the "experimental" stress-strain curve corresponding to the calculated effective strain. This is repeated for all the adhesive points, and the elasto-static solution procedure is rerun with the elastic moduli modified as described.
The described procedure is repeated until the difference between the calculated effective stresses and the stresses as obtained from the "experimental" stress-strain curve drops below a specified fraction (2%) of the calculated stress values. Convergence is usually achieved

within 5-10 iterations.

IV - RESULTS AND CONCLUSIONS

In order to show the applicability of the developed ·elasto-static and non-linear solution procedures, the case of two adhesive bonded laminated GRP-tubes subjected to external tension loading is considered. The GRP-tubes are assumed to be composed of identical 4-layer E-glas/epoxy laminates stacked in the sequence $[0^o,90^o]_s$. The adhesive choosen, AY103 from Ciba-Geigy, is a two-component plasticized epoxy. For the numerical results, the adhesive stresses are normalized with respect to the nominal stress in the outer shell element away from the joint region. The nominal stress for the axial tension load case is defined as:

(IV.1) $$\sigma_N = N_0/(2\pi R^2 h_2) \; ;$$

where N_0 is the prescribed value of the external tension load.
The geometry and the material properties used in the present study are given by:

geometry: R^1= 60.0 [mm]; R^2= 63.1 [mm]; $h_1 = h_2 =$ 3.0 [mm];

$L_1 = L_2 =$ 40.0 [mm]; $L =$ 20.0 [mm]; $t =$ 0.1 [mm];

adhesive: $E =$ 2.8 [GPa]; $\nu =$ 0.4; $\lambda \approx$ 1.3 ($\lambda = \sigma_c/\sigma_t$);
(AY103,/3/) $s_{ult} =$ 71.0 [MPa] (ultimate tensile stress);
$e_{ult} =$ 0.049 (ultimate tensile strain).

adherends: $E^i_{11} =$ 45.0 [GPa]; $E^i_{22} =$ 8.3 [GPa]; $\nu^i_{12} =$ 0.25;
(i=1,2) $G^i_{12} = G^i_{31} =$ 4.2 [GPa]; $G^i_{23} =$ 2.1 [GPa].

Fig.2 shows the adhesive stress distributions at the ultimate load level, obtained by applying the elasto-static and non-linear solution procedures respectively. The ultimate load bearing capability is predicted to $N_0 \approx$ 83.5 [kN] by the use of a maximum stress criterion.
The differences between the non-linear and linear solutions are of course strongly dependent on the level of loading, and Fig.3 demonstrates this feature for the analysed example by presenting the values of τ_z, σ_z at x=0 as functions of the nominal stress σ_N.
From Fig.3 it is seen that the adhesive non-linearity starts to affect the adhesive stress distribution at very low levels of loading, and the effect becomes increasingly important as the external load is increased. The general tendency is, that the severe stress concentrations smoothends out as the load is increased (as shown in Fig.2), and that the loads redistributes along the overlap. It should be noticed, that a certain degree of plastic yielding near the ends of the overlap zone is unavoidable, unless a brittle adhesive is used. Furthermore, it should be recognized, that the occurrence of these plastic zones is necessary in order to avoid a premature brittle failure.
The developed solution procedures have been used for conducting a parametric study, and on the basis of this a few general design recommendations are given in order to maximize the static and fatigue load carrying capability. The general design recommendations are:

1. Use adherends with high values of those of the adherend stiffnesses, which corresponds to the actual loading mode.

2. The bending- stretching coupling stiffnesses of the laminates should be zero, i.e. use symmetric laminates.

3. Whenever possible, join identical or nearly identical adherends.

4. Use an overlap length of at least ten times the minimum adherend thickness.

5. Use an adhesive with relatively low values of the elastic shear and tensile moduli. The adhesive should also possess relatively high ultimate strength values (tensile and shear).

6. Use an adhesive capable of at least a certain degree of plastic yielding, since the joint strength is governed by the ductility of the adhesive rather than the strength.

REFERENCES

1. M.Goland and E.Reissner, Journal of Applied Mechanics, 1 (1944) A17-A27.
2. L.J.Hart-Smith,"Adhesive-bonded single lap joints", NASA CR 112236, 1973.
3. R.D.Adams,J.Coppendale and N.A.Peppiatt, Adhesion, 2 (1978) 105-119.
4. R.D.Adams and N.A.Peppiatt, Journal of Adhesion, 9 (1977) 1-18.
5. J.L.Lubkin and E.Reissner, Journal of Applied Mechanics, 78 (1958) 1213-1221.
6. U.Yuceoglu and D.P.Updike, in "Emerging Technologies in Aerospace Structures, Design, Structural Dynamics and Materials" (ASME-publication), (1980) 53-65.
7. S.Gali, G.Dolev and O.Ishai, International Journal of Adhesion and Adhesives, (january 1981) 135-140.
8. A.Kalnins, Indian Journal of Mathematics, vol.9 no..2 (1967) 381-425
9. O.T.Thomsen, Ph.D. Thesis, (1989), University of Aalborg, Denmark.
10. A.Kalnins, Journal of Applied Mechanics, 31 (1964) 467-476.

FIGURES

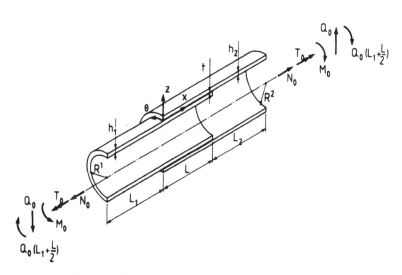

Fig.1 Tubular lap joint configuration composed of two dissimilar laminated shell elements subjected to external tension, torsion, bending moment and shear loading.

728

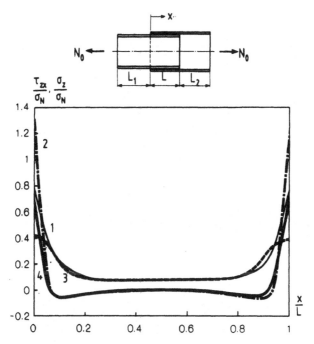

1) τ_{zx}, linear solution ;   2) σ_z, linear solution ;
3) τ_{zx}, non-linear solution ;   4) σ_z, non-linear solution ;

Fig.2 Adhesive stress distribution at ultimate tension loading;
$N_{0,ult} \simeq 83.5$ [kN] corresponding to $\sigma_N \simeq 70.2$ [MPa].

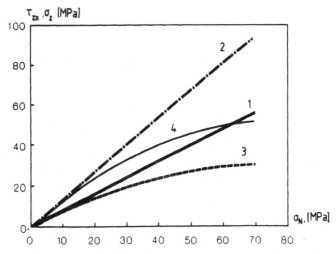

1) τ_{zx}, linear solution ;   2) σ_z, linear solution ;
3) τ_{zx}, non-linear solution ;   4) σ_z, non-linear solution ;

Fig.3 Relation between maximum adhesive stresses (x=0) and nominal
stress σ_N, obtained by application of the elasto-static and non-
linear solution procedures respectively ($\sigma_{N,ult} \simeq 70.2$ [MPa]).

CRITICAL BUCKLING LOAD FOR WOUND
CARBON EPOXY COMPOSITE CYLINDERS

J. CASTAGNOS, T. MASSARD*

CEA - CESTA - 33114 Le Barp - France
*CEA - CEB. III - Bruyères le Chatel - France

ABSTRACT

The aim of the theoretical and experimental study presented here is
to evaluate the critical buckling load of filament wounded composite
cylinders under external pressure . The first part discribes the experimental device . The second part deals with the hypothesis of the
finite elements study . The correlation between the experimental and
theoretical results and the validity of finite elements method in view
of this analysis are clearly discussed .

1-INTRODUCTION - OBJECTIVES

The design of structures combining lightweight and effectiveness
in the field of deterrent weaponry requires the use of composite
materials , and a thorough knowledge of their properties . In the case
of a cylinder subjected to external pressure , there normally are two
types of failure : strength failure and buckling .

On the critical buckling load of wound composite cylinders
subjected to external pressure , the results of the analytical
models dealing with this type of problem are far from reliable when it
comes to designing anisotropic composite structures.

On the contrary ,the critical load and the mode of buckling can be
correctly assessed using the finite elements method ,provided the validity of the model , the proper representation of the boundary conditions and of the behavioural laws of the material have been ensured .

2 - DESCRIPTION OF THE STRUCTURES

The proposed structures are cylinders made of graphite-epoxy and graphite-epoxy-honeycomb composite materials ; they are fitted with lightweight aluminium alloy stiffeners to minimise edge effects(Fig.1).
NOTE : Thickness and orientation of plies were designed to provide the best laminate strength (Table 1).

3 - DESIGN CONDITIONS

The load applied is a uniform external lateral pressure of 3MPa . The structure is considered to be under single support (Fig.1).

4 - EXPERIMENTAL DESIGN

4 - 1 Test facilities

The external pressure was applied uniformly over the tested specimen by means of an elastomer bag holding pressurized oil (Fig.5); this assembly was placed on a single support in a safety chamber. The pressure was increased in stages to measure deformations. Three tests are run for each type of material .

4 - 2 Instrumentation - measurements

Deformations were recorded over the entire circumference, at three different levels, by revolving displacement sensors (Fig.5):
- at H/2 for the measurement of the mean deflection (C2)
- at + X and at - X on each side of the centre plane (C1,C3).
The recording of the deformations was performed during the pressure increments to determine the mode of buckling : the number of lobes within the deformed circumference of the centre plane (Fig.6).
An acoustic emission data acquisition unit makes it possible, by means of a sensor (CA) installed on the internal face of the cylinder, to record, after amplification, the various noises emmited by the structure (disbonding, break of the layers).
The acoustic recording is maintained throughout the test to capture the initiation of the failure by buckling of the structure to deter-mine the critical buckling pressure.

5 - THEORETICAL DESIGN

5 - 1 Description of the models

As far as the thin shells of revolution are concerned, two types of models can be investigated :
- the use of a large number of solid elements (a quadrangle

with 4 or 8 nodes) leading to a very large number of nodes (Fig.2) :
- the use of specialized elements (shell with 2 nodes); this
type of element makes it possible to obtain a limited number of nodes,
however does not include the transverse stresses .

We have studied three possible models by combining the solid
elements for the stiffeners and the shell elements for the standard
part of the cylinder. We have compared each of these three models to
a reference solution to make the best possible choice for all the
design calculations performed on the composite cylinders with faithful
restoration of the behaviour of the cylinders, plus a minimum number
of nodes (size of the rididity matrix).

"COQ2" mesh

The composite cylinder is modeled by the shell elements representing the neutral layer ; bonding of the stiffeners with the composite
cylinder is simulated by using the relationships between the degrees
of freedom of the nodes of the cylinder shell elements and the corresponding degrees of freedom of the solid elements of the stiffeners
(Fig.3) .

"Q4-COQ2" mesh

The composite cylinder is modeled by the shell elements representing the neutral layer ; the bond of the stiffeners with the composite
cylinder is provided by combining the nodes of the stiffener elements
with the corresponding nodes of one half-thickness of the composite
cylinder modeled by the solid elements , and the external nodes of
this half-thickness with those of the corresponding shell elements.

"Q4+COQ2" mesh

Only the standard part of the cylinder between the two stiffeners
is modeled by the shell elements.

The two parts of the cylinder in contact with the stiffeners are
entirely modeled by the solid elements.

The connection between the "solid" part of the cylinder and the two
"shell" parts is provided through the relationships between the degrees of freedom of the "extreme" nodes of the "shell" part and the
corresponding degrees of freedom of the "solid" parts (Fig.3).

"Q4" and "Q8" meshes

The entire structure is modeled by the quadrilateral solid elements
with either 8 nodes (Q8) or with 4 nodes (Q4) (Fig.3).

5 - 2 Choice of the reference modelil

NOTE : For the preliminary studies , we took an isotropic material
with an elasticity modulus $E = 5.E10$ PA and a Poisson's coefficient

733

NU = 0.3 ; these characteristics correspond to those of an isotropic carbon-epoxy composite material (orientation of plies : 0°,45°,90°) .

The 8 node quadrilateral element is obviously more accurate than the 4-node quadrilateral element. However, for the preparation of meshes of composite cylinders by layer of material, and thus for the use of large number of elements and nodes, we evaluated the performance of a model in 4-node quadrilateral elements, as a feasible compromise between accuracy and the number of nodes.

The 8-node model is better adapted for the calculation of radial stresses in the area of the cylinder close to the stiffeners ; the evolution of the radial stresses is in fact homogenous over the entire standard part of the cylinder :
  - P on the external face
  - P/2 on the central fibre
  - 0 on the internal face.
The critical buckling pressure, PC, is the maximum pressure compatible with the stability of the structure. The buckling coefficient , defined by A = PC/PO , has the same value for the two models .

The mesh in quadrilateral solid elements with 8 nodes is the best reference for the rest of the study. However, for the model of a composite materal wound layer by layer, we can, without any loss of precision in the calculation of the critical load, use a model with 4-node elements (optimalization of the number of nodes).

5 - 3 Choice of the study model

Analysis of the radial displacements fails to reveal any significant difference between the three models and the reference, the closest being the "Q4-COQ2" model .
For axial displacement, the "Q4 + COQ2" model reveals a relatively high disparity with respect to the other models .
In the standard part of the cylinder, evolution of the stresses is homogenous for the three models, and differs slightly from that recorded for the reference. This is due to the number of elements in heigh (25 for the three models, 250 for the reference).The critical buckling pressure has the same value for the 4 models .

In view of the preliminary study, the "Q4-COQ2" model appears to offer a good compromise between a minimum number of nodes and a good analysis of the behaviour of the cylinder.

5 - 4 Choice of the number of elements

By doubling the number of elements in the standard part of

the cylinder and by associating a shell element of the cylinder with each external element of the stiffeners, a better reproduction of the behaviour of the cylinder under external lateral pressure is possible.

5 - 5 Study of the behaviour law of the material

To enter the mechanical characteristics of a wound orthotropic material into the design code used, two options were available :

"Wound"
This option, that can be used only with the shell and plate elements, makes it possible to describe a wound material. The characteristics are given in the local index of the shell (Fig.4).

"Orthomat"
This option, that can be used only for the solid elements, is limited to the description of a single layer of laminated material at a time. The characteristics are given in the overall index of the structure (Fig.4).

5 - 6 Choice of the option for the input of the characteristics

NOTE : The proposed models use both shell elements and solid elements for the description of the composite material .Two groups of characteristics are therefore necessary :
- in the local index of the shell elements
- in the overall index of the structure

The material used for this study corresponds to cylinder "A" .

From the results recorded from all the preliminary studies , the model "Q4-COQ2" and the wound option were retained , as making it possible to obtain directly the values of stresses in the local index of each winding .

6 - RESULTS - COMPARISON

Comparison between the results recorded theoretically and experimentally shows a relatively reliable reproduction of the phenomena of buckling in a cylindrical structure in composite material by the finite elements method (Table 2).

This study shows the value of the use of composite materials in the mechanical design , where weight is one of the major criteria in the choice of materials .For a cylinder in aluminium alloy of the same weight (M=1.312kg RE=174.284mm) , the acceptable lateral pressure would be limited to 0.5868 MPa.

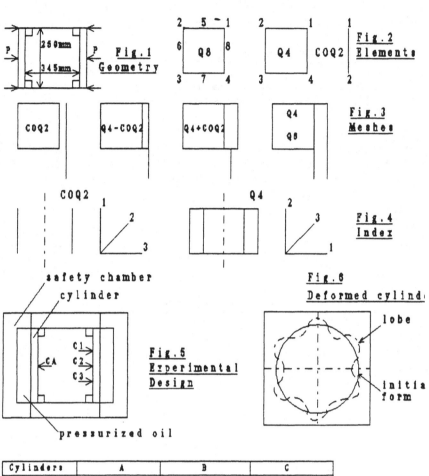

Fig.1 Geometry

Fig.2 Elements

Fig.3 Meshes

Fig.4 Index

Fig.5 Experimental Design

Fig.6 Deformed cylinder

Cylinders	A	B	C	
Fibers	T800	M40	T300	
Core	/	/	Nomex	
Thickness	3 mm	3 mm	1 + 5 + 1 mm	Table 1
Number plies	26	24	8+8	
Thickness of 1 ply	0.115 mm	0.125 mm	0.125 mm	Structures
Ply Orientation	4x90/25/-25/ 3x90/25/-25/ 2x90 /S	75/-75/35/-35/ 75/-75/35/-35/ 75/-75/35/-35 /S	2x90/50/-50/ 2x90/50/-50/ nida NOMEX /S	

Critical buckling pressure -MPa- (Mode)

Cylinders	A T800	B M40	C T300	
Experimental	1.64 (5-6)	2.35 (6)	1.83 (2)	Table 2
Theoretical	2.19 (5)	2.32 (6)	2.42 (2)	Results

THE UTILIZATION OF COMPOSITE LAMINATE THEORY IN THE DESIGN OF FILAMENT WOUND SYNTHETIC SOFT TISSUES FOR BIOMEDICAL PROSTHESES

G. MAROM, B. GERSHON, D. COHN

The Hebrew University of Jerusalem
Graduate School of Applied Science & Technology - 91904 Jerusalem - Israel

ABSTRACT

This paper draws attention to the advantages of using composite design considerations for the preparation of biomedical soft tissues. By the utilization of composite laminate design a wide selection range for the compliance is provided without a concomitant variation of other properties. The trend of the compliance as a function of the reinforcement angle is discussed for an angle-ply composite of low compliance constituents, as well as the implications on the stress-strain behaviour. Experimental examples pertinent to prosthetic arterial design are presented.

I - THEORETICAL BACKGROUND TO SOFT COMPOSITE DESIGN

The primary advantage of using composite design considerations in the preparation of soft tissue prostheses is that they provide a wide selection range for the compliance, and eventually the ability to tightly control the chosen value. In general, the compliance along the x axis, $1/E_x$, of an orthotropic lamina with its principal axes oriented at angle θ with respect to the coordinate axes is given by

$$\frac{1}{E_x} = \frac{\cos^4 \theta}{E_L} + \frac{\sin^4 \theta}{E_T} + \frac{1}{4} \left(\frac{1}{G_{LT}} - \frac{2\nu_{LT}}{E_L} \right) \sin^2 2\theta \qquad (1)$$

where E_x is the Young modulus in the x direction, E_L, E_T, G_{LT} and ν_{LT} are the Young moduli, the shear modulus and the Poisson ratio, respectively, of the composite lamina, and L and T denote the principal material axes (longitudinal and transverse). Clearly, compared with a homogeneous isotropic material, a composite structure

737

offers a much higher level of design sophistication through a choice of material combinations with a wide selection range of constituent volume proportions. Whereas a choice of properties and relative constituent concentrations determine the basic lamina properties, the orientation angle θ adds yet another dimension to the design options as expressed by Equation 1.

Examples of implications of Equation 1 with advanced composite materials comprising high modulus fibres are commonly given in composite material textbooks, for example [1]. It is obvious that at $\theta = 0°$, E_x is equal to E_L, and at $\theta = 90°$, E_x is equal to E_T. In the range $0° < \theta < 90°$ the trend of E_x depends on the elastic properties of the lamina, and can be determined by examining the first and second derivatives of Equation 1. It can be shown that depending on the value combination of E_L, E_T, G_{LT} and v_{LT}, the principal elastic constants, E_x can either be greater than E_L or smaller than E_T at some intermediate values of θ, as determined by the conditions expressed respectively by Equations 2 and 3.

$$G_{LT} > \frac{E_L}{2(1 + v_{LT})} \qquad (2)$$

$$G_{LT} < \frac{E_L}{2(E_L/E_T + v_{LT})} \qquad (3)$$

Equations 2 and 3, in fact, express the conditions for a maximum and minimum, respectively, for the function $E_x = f(\theta)$. The possibility of obtaining a minimum for E_x (a maximum for the compliance) bears significant implications on the design options for soft composite materials beyond the above mentioned inherent merits of composite design. To demonstrate this point let us consider a $\pm\theta$ angle-ply soft composite material whose main elastic constants are E_L = 11.0 MPa, E_T = 8.5 MPa, G_{LT} = 2.0 MPa and v_{LT} =0.4, which are expected to produce a minimum for E_x according to the condition set by Equation 3. Fig. 1 presents a plot of $E_x = f(\theta)$ for this material as expressed by Equation 1. It can be calculated that the minimum for E_x occurs at θ_{min} = 48.5°.

If under the action of stress on the compliant composite θ decreases, the combined effect of this change and of the presence of a minimum in the $E_x = f(\theta)$ function will govern the elastic response of the composite depending on the initial value of θ. Accordingly, three separate cases are considered due to the value of the initial winding angle, θ_{init}, with respect to the location of the minimum, namely, the initial θ is below, above or close to θ_{min}. Consequently, the shape of the stress-strain curve for the soft angle-ply lamina is determined, as shown in Fig. 1.

II - EXPERIMENTAL

Soft angle-ply composites were manufactured by filament winding according to the technique developed previously for arterial prostheses as described in [2]. (For additional examples of application of filament winding in arterial prosthesis manufacture and

with elastomeric resin matrices see Ref. 3 and 4, respectively.)   In
the present project a relatively large diameter mandrel (70 mm) was
chosen. This resulted, after cutting the filament wound tube
longitudinally and spreading it out, in a flat rectangular laminate,
from which specimens for tensile mechanical tests were taken.

Two material systems were employed, both comprising Lycra
polyether urethane fibres (segmented block copolymers of 4,4'-
diphenylmethane diisocyanate (MDI), polytetramethylene oxide (PTMO)
and ethyléne diamine) from Du Pont. These were Lycra T-336-C
monofilaments of 40 denier.   The first system, which consisted of
laminates with a range of winding angles, θ_{init} and a constant fibre
volume content of 49%, had a Pellethane 2363-80A matrix (block
copolymer of MDI, PTMO and 1,4-butanediol) from Dow Chemicals.   The
second system, which consisted of laminates with a constant winding
angle of θ_{init} = 73° and a range of fibre volume contents, had a
Hytrel matrix (polybutylene terephthalate - PTMO block copolymers)
from Du Pont.

The first material system (Lycra-Pellethane) was also employed to
manufacture filament wound tubes of 16 mm internal diameter, a
constant fibre volume content(49 %) and a range of winding angles of
8-71°

Tensile mechanical testing was carried out on flat, 5 mm wide, 70
mm long composite strips at a loading gauge of 50 mm. The loading rate
was 5 mm/min producing a nominal strain rate of 0.17 %/s. The initial
value of the laminate modulus, E_x, was determined at low strains (up
to 10 %), while the angle change as a function of stress was measured
up to 30 % srain.   The strain and the angle θ were measured at regular
stress intervals with a coordinate cathetometer (Gaertner), accurate
at the 1 μm level, attached to a digital readout processor
(Quadra-Check II, Metronics Inc.).

The filament wound tubes were tested by inflation with nitrogen
gas. The internal pressure was measured by a mercury manometer
(accurate to 0.5 mm Hg) and the hoop strain was measured with the
coordinate cathetometer. In order to induce a pure hoop stress
situation with a zero axial stress or strain, one end of the tube was
connected to a syringe whose piston was supported by a fixed end
spring. Under pressure, the piston could retract so that no
longitudinal stress was transfered to the tube.

III - RESULTS AND DISCUSSION

3.1. Effect of fibre volume fraction.

Flat samples of a Lycra fibre reinforced Hytrel composite at a
constant winding angle of 73° were checked. Fig. 2 presents the
results of E_x (θ_{init} = 73°) and E_y (90 - θ_{init} = 17°) against the fibre
volume fraction, Φ_f. Since the Hytrel matrix is stiffer than the
Lycra fibre, both E_x and E_y are *decreasing* functions of the fibre
volume fraction as predicted by Eq. (1). It is also observed that
$E_x < E_y$, because E_L is always greater than E_T (even for $E_m > E_f$). Due to
the anisotropy of the Lycra fibre, each curve should extrapolate to a
different E_f value at Φ_f = 100%. Because only the axial value of the
fibre could be obtained, the extrapolations in Fig.2 of both E_x and E_y
approximate an isotropic fibre modulus, E_f.

3.2. Effect of winding angle

The effect of the winding angle, θ_{init}, on the elastic properties of soft angle-ply laminae was studied with the Lycra fibre reinforced Pellethane composite by tensile testing of flat samples, and by pressure testing of filament wound tubes. The experimental results are presented in Fig. 3 for tensile testing and in Fig. 4 for internal pressure testing. The expected trend as predicted by the theoretical discussion and by Fig. 1 is evident. In Fig. 3 the experimental minimum is slightly lower than the calculated one, however, the fit would improve if a smaller G_{LT} value was assumed (for example, $G_{LT} = 1.5$ MPa).

3.3. Effect of angle reduction with stress

The influence of angle reduction on the shape of the stress-strain curve of the soft composite lamina is demonstrated with $\theta_{init} = 33.5°$ (were E_x corresponds to $\theta_{init} = 33.5°$, while E_y corresponds to $\theta_{init} = 56.5°$). Fig. 5 presents typical stress-strain curves for this material in the x and y directions up to a 30 % strain level. According to the results and analysis in Fig. 3, E_x and E_y are expected to be of the same magnitude, with the former slightly higher than the latter. This indeed is the case in Fig. 5 up to approximately 7 % strain level. Thereafter, it is apparent that E_x is an increasing function, while E_y is a decreasing function of strain, corresponding to the respective convex and concave stress-strain curves. Considering that the stress field acting on the natural artery reflects an approximately pure hoop loading, a requirement for a "J"-shaped hoop stress-strain trend can be met with an angle ply design of $\theta_{init} > \theta_{min}$.

VI - CONCLUSIONS

This paper draws attention to the advantages of using composite laminate design considerations for the preparation of biomedical soft tissues, which provides a wide selection range for compliance, with a given material combination. The compliance selection range is determined a priori by an available range of relative volume contents of the composite constituents. Moreover, a $\pm\theta$ angle-ply design provides an additional compliance magnitude control as well as a choice of compliance-strain dependencies.

REFERENCES

1   Mallick, P. K. Fiber-Reinforced Composites (Marcel Dekker, USA, 1988)
2   Gershon, B., Marom, G. and Cohn, D. 'New arterial prostheses by filament winding' Clinical Materials 5 (1990) 13-37
3   Leidner, J., Wong, E. W. C., MacGregor, D. C. and Wilson, G. J. 'A novel process for the manufacturing of porous grafts' J Biomed Mat Res 17 (1983) 229-249
4   Philpot, R. J., Buckmiller, D. K. and Barber, R. T. 'Filament winding of thermoplastic fibres with an elastomeric resin matrix' SAMPE J 25 (1989) 9-13

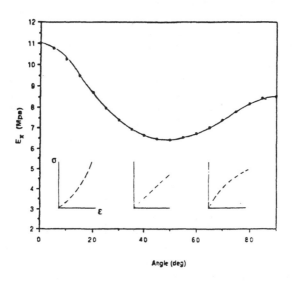

Fig.1:E_x as a function of winding angle
(E_L=11.0 MPa,E_T=8.5 MPa,G_{LT}=2.0 MPa,
ν_{LT}=0.4).

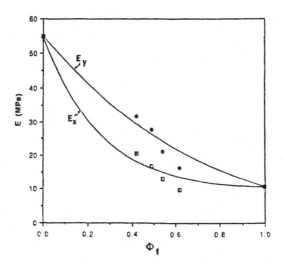

Fig.2: E_x and E_y as functions of fibre
volume fraction.

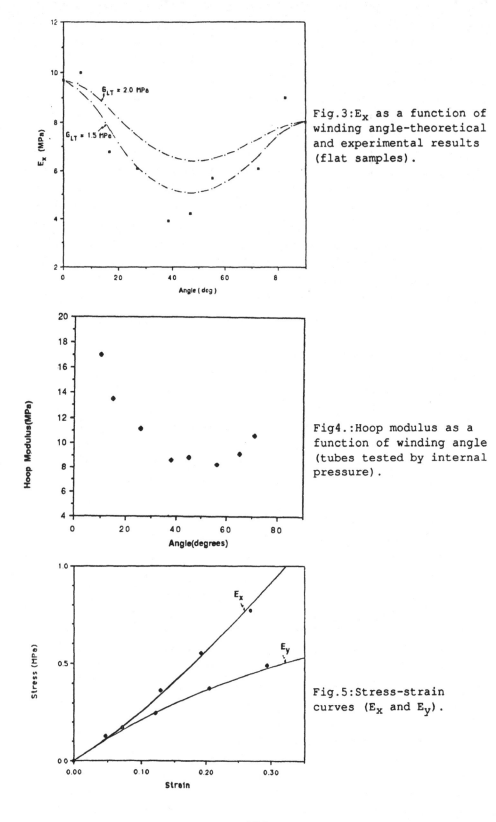

Fig.3:E_x as a function of winding angle-theoretical and experimental results (flat samples).

Fig4.:Hoop modulus as a function of winding angle (tubes tested by internal pressure).

Fig.5:Stress-strain curves (E_x and E_y).

742

EFFICIENT THERMAL ANALYSIS OF ANISOTROPIC COMPOSITE PLATES USING NEW FINITE ELEMENTS

R. ROLFES

DLR - Institute for Structural Mechanics
Flughafen - 3300 Braunschweig - West-Germany

ABSTRACT

The thermal behaviour of laminated composites is quite complicated due to their anisotropy and inhomogeneity. Therefore even thin 2D-structures have to be calculated as three-dimensional bodies, this leading to an enormous numerical effort. A thermal lamination theory (TLT) assuming a linear temperature distribution in thickness direction is established to reduce the problem to two dimensions. Basing on this theory a general quadrilateral finite element (QUADTL) for thin plates is developed. Numerical comparisons with three-dimensional calculations show the favourable behaviour of the new element.

INTRODUCTION

In most aerospace applications the thermal loads decisevely influence the design of the structures. For that the exact and effective calculation of temperature distributions in layered composites is of high importance. Assuming that the thermal conductivities of fibres and matrix are quite different (e.g. in CFRP high modulus fibres conduct the heat up to 900 times better than the matrix) the problem becomes inhomogeneous and anisotropic.

In the field of heat conduction in composites several papers appeared, beginning in the 1970's, that treated the problem analytically. Most of them were restricted to one or two dimensional heat flow in isotropic laminates, e.g. /1/, /2/. Anisotropy was taken into account by Padovan /3/, /4/ Chang /5/ and Tsou et al. /6/. While Chang dealt with homogeneous anisotropy only, Padovan handled combined inhomogeneity and anisotropy. In /3/ slabs and cylinders were investigated utilizing Fourier series. Solutions were given for an infinite slab with periodic temperature and heat flux fields in the plane of the laminate. The thermal loading acting in-plane was analysed in /4/ via piecewise continuous eigenfunctions. Tsou et al. /6/ developed a lamination theory for arbitrarily stacked laminates by assuming a constant temperature gradient in thickness direction throughout the whole laminate. They apply their theory to a periodically stacked bilamina cylinder with uniform temperature boundary conditions. In comparison with the exact solution they get a favourable correspondence. However, their theory will not be able to treat problems with non-uniform thermal loadings.

In fact all analytical approaches are restricted to special loadings, geometries or boundary conditions. Especially radiative boundary conditions or temperature dependence of the thermal properties, both causing nonlinearity, are nearly impossible to incorporate. These shortcommings were tried to overcome by Padovan already in 1974. In /7/ he carried out a three dimensional nonlinear calculation by use of the Finite Element Method (FEM). However, according to the inhomogeneity of the laminates each layer has to be discretised seperately, this leading to very high numerical effort. The semi-analytical approach in /8/ can reduce this for axi-symmetric bodies only. More recently Tamma and Yurko /9/ solve the two-dimensional problem with finite elements. They simply integrate the conductivity matrix over a couple of layers, thus being unable to take into account the stacking sequence.

The approach of Tsou et al. /6/ seems to be the most promising despite its inherent limitations already mentioned above. Although not established for that purpose, it can serve as basis for the development of a special two-dimensional finite element, which would enable us to treat general geometries, loadings and boundary conditions with rather low numerical effort.

Within the present paper an improved thermal lamination theory (TLT) for thin laminated composites is developed in continuation of the works of Tsou et al. Subsequently the formulation of a finite element basing on this theory and its numerical testing will be described.

I. THE NEW THERMAL LAMINATION THEORY (TLT)

Looking at a single layer with unidirectional reinforcement, the axis perpendicular to the plane is detected as principal direction of the heat flux. Thus the conductivity in that direction doesn't depend on the fibre orientation. Therefore the conductivity in the thickness direction of a multilayered composite is constant, if its constituents (fibre and matrix) remain unchanged from layer to layer. Furthermore the thickness h of the laminate is assumed to be small in comparison with a representative length. These two characteristics cause a temperature distribution in thickness direction, that is nearly linear. This was confirmed by several three-dimensional finite element calculations. It encourages to introduce the linearity in the new theory a priori. Then the temperature field can be expressed by the temperature T_0 of a reference surface and the gradient $T_{0,z}$ in thickness direction (z-direction).

$$T(x,y,z) = T_0(x,y) + T_{0,z}(x,y) \cdot z. \qquad (1.1)$$

It should be mentioned that the gradient $T_{0,z}$ can vary within the laminate's plane. Therefore no further assumption except the linearity of $T(z)$ is made. This is in contrast to the theory of Tsou et al., who predict a constant gradient throughout the whole structure.

Utilizing (1.1) a law of heat conduction of a laminate of the form

$$Q = -K_p \, \rho \qquad (1.2)$$

shall be derived, where ρ denotes the vector of functional degress of freedom (FDOF's) with components independent from z, $\mathbf{K}_\rho$ is the heat conductivity tensor of the whole laminate and $\mathbf{Q}$ the vector of thermally "equivalent forces". This can be achieved by integrating Fourier's law

$$\mathbf{q} = -\mathbf{K} \text{ grad } T \tag{1.3}$$

over the laminate's thickness. T, $\mathbf{K}$ and $\mathbf{q}$ are temperature, conductivity tensor and heat flux respectively. The gradient of the temperature field (1.1) can be expressed by five FDOF's

$$\rho := \begin{pmatrix} T_{o,x} \\ T_{o,y} \\ T_{o,z} \\ T_{o,z,x} \\ T_{o,z,y} \end{pmatrix} \tag{1.4}$$

and the matrix

$$\mathbf{S} := \begin{bmatrix} 1 & 0 & 0 & z & 0 \\ 0 & 1 & 0 & 0 & z \\ 0 & 0 & 1 & 0 & 0 \end{bmatrix}, \tag{1.5}$$

in the form

$$\text{grad } T = \mathbf{S} \cdot \rho. \tag{1.6}$$

The law (1.2) should connect the "equivalent forces" and the FDOF's uniquely. That's why five "equivalent forces" are needed. A natural choice of those is

$$\mathbf{Q} = \begin{bmatrix} Q_x \\ Q_y \\ Q_z \\ M_x \\ M_y \end{bmatrix} = \int_{(h)} \begin{bmatrix} q_x \\ q_y \\ q_z \\ q_x \cdot z \\ q_y \cdot z \end{bmatrix} dz. \tag{1.7}$$

Q_x, Q_y and Q_z are called heat flux resultants in x, y and z-direction respectively, M_x and M_y "moments" of heat flux in x- and y-direction respectively. $\mathbf{Q}$ can be expressed by $\mathbf{q}$ and the matrix $\mathbf{S}$

$$\mathbf{Q} = \int_{(h)} \mathbf{S}^T \mathbf{q} \, dz. \tag{1.8}$$

Inserting Fourier's law (1.3) and (1.6) into (1.8) gives

$$\mathbf{Q} = -\int_{(h)} \mathbf{S}^T \mathbf{K} \mathbf{S} \, dz \cdot \rho. \tag{1.9}$$

The heat conductivity tensor $\mathbf{K}_\rho$ of the whole laminate can thus be identified as

$$\mathbf{K}_\rho = \int_{(h)} \mathbf{S}^T \mathbf{K} \mathbf{S} \, dz. \tag{1.10}$$

The components of $\mathbf{K}$ are generally different for all layers. Thus the integral has to be evaluated layerwise. It occurs

$$K_p = \begin{bmatrix} A_{xx} & A_{xy} & 0 & B_{xx} & B_{xy} \\ & A_{yy} & 0 & B_{xy} & B_{yy} \\ & & A_{zz} & 0 & 0 \\ \text{sym.} & & & C_{xx} & C_{xy} \\ & & & & C_{yy} \end{bmatrix} \tag{1.11}$$

with

$$A_{lm} = \sum_{i=1}^{n} k_{lm}^i \, h^i \tag{1.12}$$

$$B_{lm} = \frac{1}{2} \sum_{i=1}^{n} k_{lm}^i \left((z^{i+1})^2 - (z^i)^2 \right) \tag{1.13}$$

$$C_{lm} = \frac{1}{3} \sum_{i=1}^{n} k_{lm}^i \left((z^{i+1})^3 - (z^i)^3 \right). \tag{1.14}$$

A_{lm} are the conductivities of the laminate, C_{lm} can be regarded as conductivity eccentricities and B_{lm} denotes coupling conductivities.

From (1.13) can be seen, that all B_{lm} vanish for symmetric laminates. In that case the heat flux resultants only depend on the temperature gradient of the reference surface. Furthermore, if we have a symmetric cross-ply laminate, A_{xy} and C_{xy} vanish too. Then the heat flux resultants in one laminate direction don't depend on the temperature gradient in the other direction. The laminate is completely thermally decoupled.

II. THE NEW PLATE ELEMENT QUADTL

The weak formulation of the general heat conduction equation for steady state is considered

$$\int_{\Omega} (\text{grad } v)^T \, K \, \text{grad } T \, dx + \int_{\Gamma_2} q^T \, n \, v \, ds = 0 \quad \forall \, v, \, T \in H^1 (\Omega) \tag{2.1}$$

$$T = \bar{T} \text{ on } \Gamma_1, \quad q^T n = \bar{q} \text{ on } \Gamma_2, \tag{2.2, 2.3}$$

where v are test functions for the temperature field T. Constant boundary conditions are chosen for simplicity. Other like e.g. convection make no difficulties and have been implemented in the code.

The temperature field and its gradient have to be expressed by the FDOF's to introduce the TLT into the functional (2.1). In addition to the five degrees of freedom, chosen in (1.6) to describe the gradient, T_o is needed to describe the temperature field

$$T = R \bar{\rho} \tag{2.4}$$

$$R = [\, 1, 0, 0, z, 0, 0 \,] \quad \bar{\rho} = \{\, T_o, T_{o,x}, T_{o,y}, T_{o,z}, T_{o,z,x}, T_{o,z,y} \,\}. \tag{2.5, 2.6}$$

For the discretisation with finite elements shape functions have to be introduced. Unless the total number of degrees of freedom is six, only two sets of shape functions are necessary, one for T_0, the other for the gradient in thickness direction $T_{0,z}$. All other FDOF's are derivatives of them.

The finite element QUADTL (**Quadrilateral, TLT**) is a general quadrilateral element for thin plates. The same bilinear shape functions N_i are used for both variables (T_0 and $T_{0,z}$) as well as for the approximation of the geometry. This concept is called double isoparametric and leads to an element with eight degress of freedom. Introducing the shape functions

$$\bar{\rho} = N \, \vartheta \qquad (2.7)$$

$$N = [\, N^1, N^2, N^3, N^4 \,] \quad \vartheta = \{\, T_0^1, T_{0,z}^1, T_0^2, T_{0,z}^2, T_0^3, T_{0,z}^3, T_0^4, T_{0,z}^4, \,\} \quad (2.8, 2.9)$$

$$N^I = \begin{bmatrix} N^I & 0 \\ N^I_{,x} & 0 \\ N^I_{,y} & 0 \\ 0 & N^I \\ 0 & N^I_{,x} \\ 0 & N^I_{,y} \end{bmatrix}, \qquad (2.10)$$

results in the equation

$$\sum_{I=1}^{N} \left(\int_A N^T K_p N \, da \, \vartheta + \int_{\Gamma_2} \bar{q} N^T R^T \, d\gamma \right) = 0, \qquad (2.11)$$

where N is the total number of elements. Denoting the first integral by I_p and the second by r gives the final equation

$$\sum_{I=1}^{N} (I_p \vartheta + r) = 0. \qquad (2.12)$$

A quadratic plate subjected to a constant heat flux acting on a small part of the surface with stacking sequence $[0_2 | + 45 | 0_2 | - 45 | 0 | 90]_{sym.}$ was selected as the model problem. The temperature distributions along lines A and B on the upper surface of the plate were calculated on a 10 x 10 grid. Figure 1 shows the very good agreement with a fully three dimensional calculation. The computing time on comparable machines was reduced by two orders of magnitude (250 sec. on a VAX 8530 versus 3 sec on an IBM 4381-R14). Further calculations demonstrated a significant influence of the stacking sequence on the temperature field. The element QUADTL reflected this effect properly.

It should be mentioned that the double isoparametric QUADTL represents only one possibility of transposing the TLT into the FEM. Other elements with different shape functions for T_0 and $T_{0,z}$ and/or of higher order might be interesting.

REFERENCES

1 - H. Murakami; A. Maewal; G.A. Hegemier: "A Mixture Theory for Heat Conduction in Laminated Composites", ZAMM 61, (1981) 305-314

2 - K.D. Hagen: "A Solution to Unsteady Conduction in Periodically Layered, Composite Media Using a Perturbation Method", J. of Heat Transfer 109, (1987) 1021-1023

3 - J. Padovan: "Conduction in Anisotropic Composite Slabs and Cylinders", Proc. 5th Int. Heat Transfer Conf., Japan Soc. of Mechanical Engineering, 1, (1974) 147-151

4 - J. Padovan: "Conduction in Multiply Connected Thermally Anisotropic Domains", Letters in Heat and Mass Transfer 2, (1975) 371-380

5 - Y.-P. Chang: "Analytical Solution for Heat Conduction in Anisotropic Media in Infinite, Semi-Infinite, and Two-Plane-Bounded Regions", Int. J. Heat and Mass Transfer 20, (1977) 1019-1028

6 - F.K. Tsou; P.C. Chou; I. Singh: "Apparent Tensorial Conductivity of Layered Composites", AIAA Journal 12, (1974) 1693-1698

7 - J. Padovan: "Steady Conduction of Heat in Linear and Nonlinear Fully Anisotropic Media by Finite Elements", J. of Heat Transfer 96, (1974) 313-318

8 - J. Padovan: "Semi-Analytical Finite Element Procedure for Conduction in Anisotropic Axisymmetric Solids", Int. J. Numerical Methods in Engineering 8, (1974) 295-310

9 - K.K. Tamma; A.A. Yurko: "A Unified Finite Element Modeling/Analysis Approach for Thermal-Structural Response in Layered Composites", Computers and Structures 29, (1988) 743-754

FIGURE

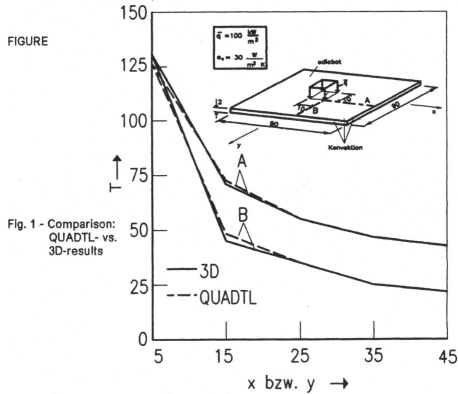

Fig. 1 - Comparison: QUADTL- vs. 3D-results

GEOMETRICALLY AND PHYSICALLY NON-LINEAR INTERFACE ELEMENTS IN FINITE ELEMENT ANALYSIS OF LAYERED COMPOSITE STRUCTURES

H. SCHELLEKENS, R. DE BORST*

Delft University of Technology
PO Box 5048 - 2600 GA Delft - The Netherlands
*Delft University of Technology
TNO Institute for Building Materials and Structures
Postbus 49 - 2600 AA Delft - The Netherlands

ABSTRACT

Delamination of laminated composite structures is a failure mode in which the location and the direction of the failure are known a priori. This allows the use of interface elements as an intermediate between the different layers of a composite. A formulation for an interface element is proposed which is applicable in physically and geometrically non-linear analyses of shell structures. A Coulomb friction model including gap-formation accounts for the physically non-linear delamination phenomenon.

INTRODUCTION

Laminated composites are known to develop large stresses at free edges which may result in an important failure mode for this kind of materials, namely delamination. The bond between the composite layers is broken and large discrete cracks appear. Although attention is usually focused on edge-delamination, delamination may also occur in the interior of the material, especially at locations where cracks in the matrix reach the adjacent plies.

An interface element consists of an upper and a lower plane. In contrast to continuum elements where stress-strain relations are used, the interface elements are governed by relations between tractions and relative displacements of the interface surfaces.

Within the elastic region the interface elements have a large dummy stiffness as to accomplish that no additional deformations are introduced. From the moment that the elastic limit is exceeded the traction-relative displacement relation becomes non-linear. In the paper the non-linear behaviour is governed by a Coulomb friction law in combination with a tension cut-off limit.

In order to allow large rotations and translations a Total-Lagrange formulation for updating the displacement field is used.

1 - FINITE ELEMENT FORMULATION

The finite element formulation has been developed for an m-noded shell interface in /2//4/. This element as is shown in Figure 1 has $3m$ translational degrees of freedom and $2m$ rotational degrees of freedom. The relations between tractions and relative displacements of the lower surface of the upper shell and upper surface of the lower shell are evaluated at integration points.

We introduce the nodal degrees-of-freedom vector a which reads

$$a = [\, u_n{}^1, u_n{}^2, .., u_n{}^m, u_s{}^1, u_s{}^2, .., u_s{}^m, u_t{}^1, u_t{}^2, .., u_t{}^m, \alpha^1, \alpha^2, .., \alpha^m, \beta^1, \beta^2, .., \beta^m \,]^T. \tag{1}$$

with $u_n^1 \dots u_t^m$ nodal translations and $\alpha^1 \dots \beta^m$ nodal rotations. For the continuous displacement field we have

$$u = [\, u_n{}^u, u_n{}^l, u_s{}^u, u_s{}^l, u_t{}^u, u_t{}^l \,]^T. \tag{2}$$

where the continuous displacement field of the lower side of the interface u^l equals

$$u^l = \sum_{i=1}^{m/2} N_i\, u^i + \sum_{i=1}^{m/2} N_i \left[\, v_{2,i}, -v_{1,i} \,\right] \begin{bmatrix} \alpha^i \\ \beta^i \end{bmatrix} \tag{3}$$

with N_i the interpolation polynomials and $v_{1,i} = [v_{1x}, v_{1y}, v_{1z}]^T$ and $v_{2,i} = [v_{2x}, v_{2y}, v_{2z}]^T$ the rotation vectors of node i. The rotation vectors are obtained according to

$$v_1 = \frac{t_i}{2} \frac{v_\xi}{|v_\xi|} \quad \text{and} \quad v_2 = \frac{t_i}{2} \frac{((v_\xi \times v_\eta) \times v_\xi)}{|(v_\xi \times v_\eta) \times v_\xi|} \tag{4}$$

in which v_ξ is $\left[\dfrac{\partial x}{\partial \xi}, \dfrac{\partial y}{\partial \xi}, \dfrac{\partial z}{\partial \xi}\right]^T$ and v_η is $\left[\dfrac{\partial x}{\partial \eta}, \dfrac{\partial y}{\partial \eta}, \dfrac{\partial z}{\partial \eta}\right]^T$, t_i is the thickness of the upper or lower shell element in node i. The displacement field of the upper side of the interface is given by

$$u^u = \sum_{i=m/2}^{m} N_i\, u^i + \sum_{i=m/2}^{m} N_i \left[\, -v_{2,i}, v_{1,i} \,\right] \begin{bmatrix} \alpha^i \\ \beta^i \end{bmatrix} \tag{5}$$

With the relative displacements Δu we arrive at the nodal displacement-relative displacement relation in the global coordinate system

$$\Delta u = B_g a \tag{9}$$

where B_g equals

$$B_g = \begin{bmatrix} -n & n & 0 & 0 & 0 & 0 & -n \cdot v_{2z} & -n \cdot v_{2z} & n \cdot v_{1z} & n \cdot v_{1z} \\ 0 & 0 & -n & n & 0 & 0 & -n \cdot v_{2y} & -n \cdot v_{2y} & n \cdot v_{1y} & n \cdot v_{1y} \\ 0 & 0 & 0 & 0 & -n & n & -n \cdot v_{2x} & -n \cdot v_{2x} & n \cdot v_{1x} & n \cdot v_{1x} \end{bmatrix} \tag{10}$$

The rotation matrix which is necessary to transform the B_g matrix to the local coordinate system in the integration points equals

$$R = \left[\, (v_\xi \times v_\eta),\ ((v_\xi \times v_\eta) \times v_\xi),\ v_\xi \,\right] \tag{11}$$

so the B_g-matrix transforms to the local coordinate system according to $B_l = B_g R^T$. If the constitutive relations for the interface are given by the matrix D

$$
D = \begin{bmatrix} d_n & 0 & 0 \\ 0 & d_s & 0 \\ 0 & 0 & d_t \end{bmatrix} \tag{12}
$$

the traction-relative displacement relation becomes

$$
t = D\Delta u \tag{13}
$$

in which $t = [t_n, t_s, t_t]^T$ is the vector containing the tractions. The element stiffness matrix K can now be derived following standard finite element procedures. /1/

In a geometrically non-linear analysis the directions of the local coordinate systems have to be updated within each successive loading step. To accomplish this the coordinates in the new situation are calculated according to

$$
x_i^t = x_i^{t-1} + u_i^t
$$

The new local directions v_ξ^t, v_η^t and $v_\zeta^t = v_\xi^t \times v_\eta^t$ can be determined. Because of space limitations the reader is referred to /5/ or /6/ for the calculation of the updated directions of the rotation thickness vectors.

2 - COULOMB FRICTION MODEL

In the sequel we assume that the delamination initiation is governed by Coulomb friction law with a tension cut-off limit. Brittle cracking occurs once the normal traction t_n exceeds the cut-off limit f_{ct} (see Figure 2). As this takes place the apex of the friction contour shifts to the origin and no adhesion is left. Sliding occurs when the value of the friction criterion

$$
\Phi = \sqrt{(t_s^2 + t_t^2)} + t_n \tan\phi - c \tag{14}
$$

equals zero. Here c denotes the adhesion and ϕ the interlaminar friction angle, t_n is the normal traction and t_s and t_t are the shear tractions.

2.1. Derivation of the incremental plastic relative displacements.

We define a plastic potential g with a dilatancy angle ψ as

$$
g = \sqrt{(t_s^2 + t_t^2)} + t_n \tan\psi - c \tag{15}
$$

by means of which the incremental plastic relative displacements in a finite loading step can be written as

$$
\Delta u^P = \int_t^{t+\Delta t} \dot{\lambda} \frac{\partial g}{\partial t} \tag{16}
$$

in which $\dot{\lambda}$ is a non-negative multiplier. Integrating this by a single point rule yields

$$
\Delta u^P = \Delta\lambda \frac{\partial g}{\partial t} \tag{17}
$$

with $\Delta\lambda$ the final plastic multiplier. The traction at the end of a loading step is written as

$$
t_e = t_t - D\Delta u^P \tag{18}
$$

where t_t is the elastic trial traction. Substitution of eq. (17) in (18) gives

$$t_e = t_t - \Delta\lambda D \frac{\partial g}{\partial t} \qquad (19)$$

Linearising the yield condition $\Phi(t,\kappa)$ at $t = t_e$ results in

$$\Phi(t,\kappa) - \Delta\lambda \frac{\partial \Phi}{\partial t}^T D \frac{\partial g}{\partial t} + \Delta\kappa \frac{\partial \Phi}{\partial \kappa} = 0 \qquad (20)$$

and for the plastic multiplier

$$\Delta\lambda = \frac{\Phi(t,\kappa)}{H + \frac{\partial \Phi}{\partial t}^T D \frac{\partial g}{\partial t}} \qquad (21)$$

κ is a measure for the effective plastic relative displacements and H is a hardening parameter equal to $-\frac{1}{\Delta\lambda} \frac{\partial \Phi}{\partial \kappa}^T \Delta\kappa$.

3 - EXAMPLE

An orthotropic composite plate

As an example the clamped square AS4/PEEK laminate plate of Figure 3 is considered. The plate is clamped at two edges and consists of four orthotropic plies connected by interface elements. The orientation of the plies is in the direction of the clamped edges. The material properties are summarised in Table 1. The structure is subjected to a point load in the centre. Due to symmetry only a quarter of the structure is modelled.

Table 1. Material properties

E_1	$134.0 \ 10^9$	G_{12}	$5.1 \ 10^9$	v_{12}	0.28
E_2	$8.9 \ 10^9$	G_{13}	$5.1 \ 10^9$	v_{13}	0.2
E_3	$8.9 \ 10^9$	G_{23}	$5.1 \ 10^9$	v_{23}	0.2

In the friction model we have taken for the adhesion $10.0E6$ N/mm^2. The values of the friction and the dilatancy angle are taken equal to 0.0025.

In figure 4 the load displacement curve for the mid-node is presented. Due to membrane action the plate stiffens with increasing load. Figures 5 to 7 present the delaminated areas for respectively the upper, middle and lower interface layer at a load of 5000 N. It is observed that delamination starts in the middle interface layers where maximum shear stresses appear. After a further increase of the load the delamination extends to the outer layers and starts to appear also near the supports.

ACKNOWLEDGEMENTS

The calculations presented in this paper have been performed with the finite element package DIANA of the T.N.O. Institute for Building Materials and Structures.

REFERENCES

1. Bathe, K.J., Finite Element Procedures in Engineering Analysis, Prentice Hall, New-Jersey (1982).

2. Beer, G., An Isoparametric Joint/Interface Element for Finite Element Analysis, Int. j. numer. methods eng., Vol. 21, 585-600 (1985).

3. de Borst, R., Vermeer, P., Non-Associated Plasticity for Soils, Concrete and Rock, Heron, Vol. 29(3) (1984).

4. Rots, J.G., Computational Modeling of Concrete Fracture, Dissertation, Delft University of Technology (1988).

5. Schellekens, J.C.J., Interface Elements in Finite Element Analysis, Report, Delft Univ. of Techn., to appear, (1990).

6. Surana, K.S., Geometrically Nonlinear Formulation for the Curved Shell Elements, Int. j. numer. methods eng., Vol. 19, 581-615 (1983).

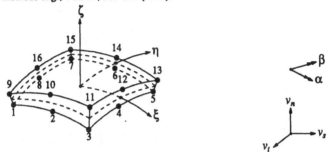

Figure 1. Quadratic shell interface element with degrees of freedom.

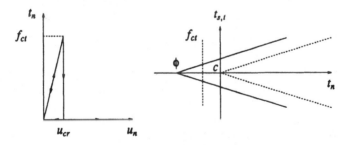

Figure 2. Coulomb friction model.

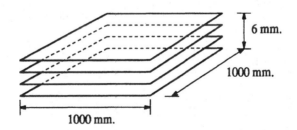

Figure 3. Four-ply AS4/PEEK laminate.

753

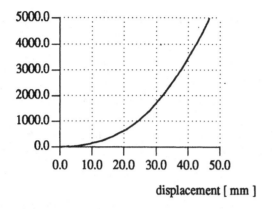

Figure 4. Load-displacement curve at centre.

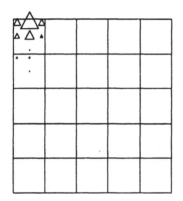

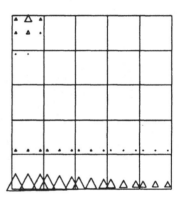

Figure 5. Delamination upper layer.   *Figure 6. Delamination middle layer.*

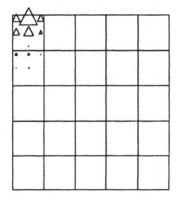

Figure 7. Delamination upper layer.

EVALUATION OF THE PROPERTIES OF SPATIALLY REINFORCED COMPOSITES

I. ZHIGUN

Institute of Polymer Mechanics
226006 Riga - USSR

ABSTRACT

Four groups of composites with spatial arrangement of reinforcement have been distinguished. Models for calculation of the elastic properties elaborated and their acceptability has been experimentally verified. Specific features of experimental determination of composite properties are indicated and recommendation for proper selection of loading type, specimen shape and dimensions given.

In the last years - in the USSR and abroad - there is a revival of interest in the development of composites with spatial reinforcement schemes due to their prospective application in many branches of technology /I-3/. The analysis of the structural schemes of reinforcement shows that four approaches can be distinguished in their development. The first approach is based on the formation of spatial bonds by curved fibers in one and the same direction (Fig. I,a). The second approach involves the formation of spatial bonds by the fibers in the third direction, by forming a system of three filaments (Fig. I,b).
The formation of the spatial bonds by whiskers or other discrete elements lies in the basis of the third approach (Fig. I,c). The fourth approach concerns the formation of the spatial bonds by introducing reinforcement in several directions (Fig. I,d).
To assess the advantages and disadvantages of

755

these composites, it is necessary that their models of calculation be known. One of the possible approaches to the development of design models for composites is considered here. It is based on the structure of composites as consisting of repeating structural elements - flat laminae (Fig. 2). The calculation of the elastic characteristics of a structural element with straight fibers placed in its plane is a generalized method, based on the conditions of a plane problem /4/ and the method of partial smoothing of reinforcement in one of the reinforcement directions /5/. Further approach to elaboration of the models of calculation for composites with different schemes of reinforcement is connected with taking into account the principle of summing up of separate repeating structural elements. This principle is based on the equality

$$\sigma_{k3}^{(\ell)} = \langle \sigma_{k3} \rangle ; \qquad \varepsilon_{ij}^{(\ell)} = \langle \varepsilon_{ij} \rangle , \qquad (1)$$

where $\ell = 1,2,\ldots, n$; $k = 1,2,3$; $i, j = 1,2$; n is the number of structural elements; the averaged stresses and strains in composite are put into French quotes. Employing the principle of summation of the structural elements with straight fibers in their plane and taking the structure of each group of composites into account, we can obtain the expressions for calculation of the elastic constants. Thus for composites of the first group they are given in a simplified form as follows:

$$\tilde{\tilde{E}}_x = \frac{(1-\psi^2)^2 + \dfrac{4(\alpha + 1 + 2\,\nu_{zx})}{\beta}\,\psi^2}{1 + 2\nu_{zx}\psi^2 + 4\alpha/\beta\,\psi^2 + \alpha\,\psi^4}\,E_x ; \qquad (2)$$

$$\tilde{\tilde{E}}_z = \frac{(1-\psi^2)^2 - \dfrac{4\alpha(1+2\nu_{xz}+1/\alpha)}{\beta}\,\psi^2}{1 + 2\nu_{xz}\psi^2 + 4/\beta\,\psi^2 + 1/\alpha\,\psi^4}\,E_z ; \qquad (3)$$

$$\tilde{\tilde{E}}_y = E_y ; \qquad (4)$$

$$\tilde{\tilde{\nu}}_{xz} = \frac{\nu_{xz}(1+\psi^4) + \left(1 + \dfrac{1}{\alpha} - \dfrac{4}{\beta}\right)\psi^2}{1 + 2\nu_{xz}\psi^2 + \dfrac{4}{\beta}\psi^2 + \dfrac{1}{\alpha}\psi^2} ; \qquad (5)$$

$$\tilde{\tilde{\nu}}_{yz} = \frac{\nu_{yz} + \psi^2\nu_{yz}}{1+\psi^2} \qquad (6)$$

$$\tilde{\tilde{G}}_{xz} = G_{xz}\left[1 + \frac{\beta\left(1 + \frac{1}{\alpha} - 2\nu_{xz} - \frac{4}{\beta}\right)\psi^2}{(1+\psi^2)}\right]; \qquad (7)$$

$$\tilde{\tilde{G}}_{xy} = \frac{(1+\psi^2)^2 G_{xy} G_{yz}}{G_{xy}\psi^2 + G_{yz}}, \qquad (8)$$

$$\tilde{\tilde{G}}_{yz} = \frac{(1+\psi^2)^2 G_{xy} G_{yz}}{G_{xy} + \psi^2 G_{yz}}, \qquad (9)$$

where $\alpha = E_x/E_z$; $\beta = E_x/G_{xz}$; ψ is the degree of fiber curvature, $\psi = tg\,\theta$, θ is the slope of fibers towards longitudinal axis of the element.

The elastic constants, entering the right-hand parts of relations (2)-(9), reflect the properties of materials, reinforced by straight fibers in two directions. They are calculated by the formulas, obtained for structural elements. Table I gives a comparison of theoretical and experimental values of the elastic constants of composites with certain schemes of reinforcement. As it is shown in Table I, the relations obtained describe the elastic constants of these composites very well. The shear strength of these composites exceeds 2 to 2.6 times and transverse tension strength 5 to 8 times that of laminated composites.

For composites, the structural schemes of reinforcement of which are formed of the system of three filaments (second group), the simplified relations, obtained in the case of neglecting the terms of the order ν_a^2, E_c/E_a, compared to a unity, take the form:

$$E_i = \mu_i E_a + \frac{(1+\mu_k)\left[(1-\mu_i-\mu_j)^2\mu_i + (1+\mu_i+\mu_j)\mu_j\right]E_c}{(1-\mu_k)(1-\mu_i-\mu_j)(\mu_i+\mu_j)}, \qquad (10)$$

$$G_{ij} = \frac{1+\mu_i+\mu_j}{(1-\mu_i-\mu_j)(1-\mu_k)}G_c; \quad i,j,k = 1,2,3; \ i \neq j \neq k. \qquad (11)$$

The indices "f" and "m" refer to reinforcement and matrix, respectively. The values of elastic constants calculated according to these relations, are in a good agreement with experimental data (Table 2), obtained for composites with different reinforcement schemes.

The composites, in which the spatial bonds are formed by discrete fibers or whiskers (third group) have, as a rule, random distribution of discrete elements throughout the volume. The relations for calculation of the elastic characteristics are as follows

$$E_y = E_z = \frac{[1+(n_x-1)\mu_\beta]E_\beta}{[\mu_\beta+n_z(1-\mu_\beta)][1+(n_x-1)\mu_\beta]-(n_x\nu_{zx}^{*2}-\nu_\beta^2)(1-\mu_\beta)\mu_\beta} ; \quad (12)$$

$$\nu_{zx} = \nu_{yx} = \nu_{zx}^{*}(1-\mu_\beta)+\nu_\beta\mu_\beta; \quad (13)$$

$$G_{xz} = G_{xy} = \frac{m_x(1+\mu_\beta)+1-\mu_\beta}{m_x(1-\mu_\beta)+1+\mu_\beta} G_{xz}^{*} \quad (14)$$

Here, $n_x = E_\beta/E_x^{*}$, $n_z = E_\beta/E_z^{*}$, $m_x = G_a/G_{xz}^{*}$; μ_β is the reinforcement coefficient by fibers, without whiskers; E_β, G_β are the elastic characteristics of fibers.

The characteristics of a modified matrix (desig nated by asterisk) are expressed in terms of coefficients of stiffness matrix, while the latter - coefficients of stiffness matrix of a unidirectional lamina, reinforced by whiskers. A s it is shown in Table 3, there is a satisfactory correlation between the theoretical and experimental data for this group of materials.

The studies show that the use of whiskerized reinforcement leads to the increase in transverse tension and interlaminar shear strengths by I.5 to 2 times as against the characteristics of conventional reinforcement-based materials.

The construction of models for calculation of composites with spatial arrangement of reinforcement is a more complicated task. One of the approaches to calculation of the elastic characteristics of such structures is treated in /6/.

One of the most efficient ways of assessment of the accuracy of theoretical models for determination of composite properties is the comparison of theory with experiment. Therefore, selection of the loading type on specimen, specimen shape and dimensions of its gage section and also loading section is of great importance for obtaining of reliable data about mechanical properties of composites. The multitude of experiments show that, even in tensile tests, obtaining of reliable strength data on specimens of limited length is very complicated due to slipping or fracture at the sites of load application. To find out the reasons of this phenomenon, the problem was solved, taking into account the actual loading type on specimen in tension /7/. It follows from the analysis of the solution to this problem that fracture of a standard-shape spec imen is due to the overload of the outer fibers (Fig. 3). The overload is removed in two ways - by changing the standard shape of specimen and the length

758

of the loaded sections. The test specimen is a strip
with reusable tabs, mounted on the sections of fast-
ening. Optimal thickness of tabs is 0.20 to 0.25 of
the thickness of the strip specimen (Table 4). Thus
the length of the loaded section is also of significance.

The $\overline{\sigma}_x$ -values are calculated for the section,
corresponding to the end of load application zone,
at $E_x / E_z = 20$, $E_x / G_{xz} - \nu_{zx} = 150$.

The complexity of compression tests is due to
splitting of specimen end faces, longitudinal delamination or specimen failure outside gage section. In
compression tests, one of the main problems is correct selection of loading type with external forces.
The compression load can be applied in three different ways: normal forces over the end faces (I); normal forces over end faces and tangential forces over
side surfaces (2); tangential forces over side surfaces (3). The first loading type does not allow to
obtain correct values of strength characteristics due
to high stress concentrations at the sites of load
application, i.e. due to splitting of end faces /8/.
Two other loading types permit to uniformly transmit
external forces to the specimen, without visible damages in the loaded sections right up to specimen failure and to obtain stable and reproducible values of
characteristics (Table 5).

REFERENCES

I - Zhigun I. and Polyakov V. Behavior of Spatially
    Reinforced Composites. Riga, Zinatne (1978)
2 - Tarnopol'skii Yu., Zhigun I. and Polyakov V.
    Spatially Reinforced Composite Materials. Moscow.
    Mashinostroyeniye, (1987).
3 - Lamicq P. Proc. of AIAA/SAE I3th Propulsion Conf.,
    (1977) I-4.
4 - Bolotin V. Strength Analysis, I2 (1966) 3-3I.
5 - Malmeister A., Tamuzs V. and Teters G. Resistance
    of Rigid Polymer Materials. Riga, Zinatne (1972).
6 - Kregers A. and Melbardis J. Polymer Mechanics, I
    (1978) 3-8.
7 - Polyakov V. and Zhigun I. Polymer Mechanics, I
    (1977) 63-74.
8 - Zhigun I., Polyakov V. and V. Mikhailov. Polymer
    Mechanics, 6 (1979), IIII-III8.

Table I
Theoretical and Experimental Values of Moduli of Elas
ticity and Shear of GFRP, Formed of the System of Two
Filaments

Material[*]	Characteristics[**]				
	E_1/E_x	E_2/E_y	E_3/E_z	G_{12}/G_{xy}	G_{13}/G_{xz}
G-I-I0-65	I.03	I.II	0.9I	0.84	0.87
G-II-32-50	I.I3	I.07	I.09	-	-
G-III-I5-48	0.96	0.97	I.05	0.9I	0.98
G-IV-I4-49	0.92	0.97	-	0.86	-
G-V-I7 -52	I.07	I.I2	-	0.90	I.0I

G is GFRP, Roman figure designates reinforcement
scheme, Arab figure designates the angle of fiber
curvature and fiber volume fraction, respectively.

Table 2

Characteristics	Material[*]			
	G-I-59	G-II-63	G-III-45$_q$	C-I4-43
E_1/E_x	I.I3	I.20	I.2I	I.I8
E_2/E_y	I.02	I.03	I.I4	I.I8
E_3/E_z	0.92	0.89	I.34	0.90
G_{12}/G_{xy}	I.I5	0.98	I.0I	I.I0
G_{13}/G_{xz}	0.86	0.9I	I.00	I.II
G_{23}/G_{yz}	0.96	0.88	0.96	I.II

G is GFRP, C is CFRP, I, II, II designate reinforcement schemes, Arab numbers - fiber volume fraction,
q are quarts fibers. The experimentally obtained
values of characteristics (IO MN/m) are marked by
superscript.

Table 3

Reinforcement	$\mu_{кр}.\%$	$\mu_a.\%$	G_{13}/G_{xz}	E_1/E_x
Carbon fiber tape; whiskerized	2.0	46	0.77	I.099
Carbon fibers, whiskerized	4.5	64	I.03	0.99
Carbon fiber braids, whiskerized	7.6	46	0.8I	0.94

Table 4

Dependence of $\bar{\sigma}_x = \sigma_x/\rho$ on Relative Length of Loaded Section (a/h) and Specimen Thickness Ratio (H/h)

a/h	H/h					
	1.0	1.2	1.4	1.6	1.8	2.0
	$\bar{\sigma}_x$					
3	2.55	2.47	2.42	2.49	2.52	2.53
5	1.92	1.73	1.71	1.74	1.76	1.78
7	1.64	1.32	1.36	1.39	1.42	1.44
9	1.46	1.13	1.14	1.19	1.23	1.26
11	1.37	0.97	1.00	0.98	1.10	1.14

Notes: a is length of loaded section; H is specimen thickness at the sites of loading (specimen with tabs), h is thickness of strip. σ_x is normal stress in gage section, ρ is nominal normal stress

Table 5

Dependence of Relative Strength $\bar{R}_x$ in Tension of Composites on Loading Type on Specimen

Material	Characteristics	Loading Types		
		1	2	3
3 D with carbon matrix	$\bar{R}_x$	0.54	0.89	1.00
	v	16.8	7.5	4.0
Based on polymer matrix reinforced by system of two filaments	$\bar{R}_x$	0.83	0.96	1.00
	v	7.2	4.1	2.8
Unidirectional CFRP	$\bar{R}_x$	0.59	0.92	1.00
	v	9.00	8.2	7.9

Notes: $\bar{R}_x$ are given in relation to ultimate strength of composites, obtained according to the third loading type; v , % is coefficient of variation of strength values.

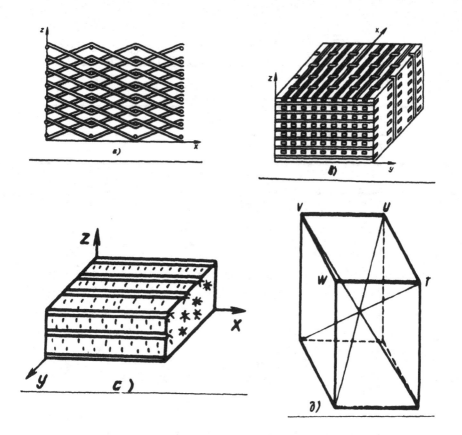

Fig. I   Schemes of reinforcement: a – with wavy fibers in one of the directions; b – with straight fibers in three directions; c – with filamentary crystals between fibers; d – reinforcement in n direction.

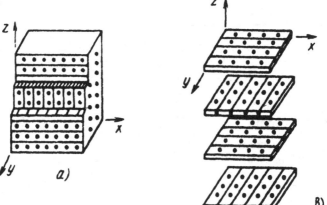

Fig. 2 Division of a Composite into Laminae: a – fiber array in material; b – layer layup in zy plane.

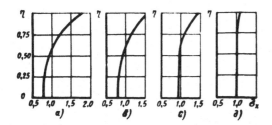

Fig. 3 Stress diagrams $\bar{\sigma}_x = \sigma_x/\rho$ through specimen
section, calculated at $E_x \angle G_{xz} = 150; c/h=1; a/h=5$:
a) $\Delta \ell = 0;$ b) $\Delta \ell = 2.5h$; c) $\Delta \ell = 5: h;$
d) $\Delta \ell = 8.5h$. c , h is specimen height in
the loaded zone and gage length; a is the loaded
section, $\Delta \ell$ is the distance from the end of loaded section to reference section.

NON-HOMOGENEOUS BARS UNDER TENSION, PURE BENDING AND TEMPERATURE LOADS

H. RAPP

Messerschmitt-Bölkow-Blohm GmbH
Helicopter Division - PO Box 801140 - 8000 Munich 80 - West-Germany

ABSTRACT

Non-homogeneous bars may show discrepancies in stiffness and thermal expansion between the engineering theory of bending and experiment. The reason for this are additional stresses perpendicular to the bar axis due to different Poisson's ratios of the used materials. Especially, if fiber reinforced materials are used, these stresses have to be taken into account. A theory for evaluating the stiffness of those non-homogeneous bars is shown. Other methods of calculation, based on the engineering theory of bending and the classical laminate theory are also given. Test results are in good correlation with the theory.

INTRODUCTION

In modern aerospace design and construction fiber reinforced materials are very often used. With this new materials effects become important which could be neglected for isotropic materials. One of these effects is the influence of different Poisson's ratios in non-homogeneous bar-like structures on the deformation under tension, pure bending and temperature loads.

When loading such non-homogeneous bars, the different materials want to deform independent from each other. But as they are bonded together, their deformations are restrained by the neighbouring materials (fig. 1). The result are additional stresses perpendicular to the bar axis which increase the resulting bar stiffness compared to a beam where all materials have the same Poisson's ratio.

In /1/ Musschelischwili shows a way to calculate the stiffness of such non-homogeneous bars by using complex stress functions. Some authors follow his way and calculate stresses in non-homogeneous circular tubes /2/-/5/, but no one explicitly determines the influence of different Poisson's ratios on the bar stiffnesses and coefficients of thermal expansion.

In the following a way to calculate the stiffness and the coefficients of thermal deformation of non-homogeneous bars is given. The theory is based on the finite element method and therefore there are no restrictions on the bar cross-section geometry. Special tests show a good correlation between this theory and experiments.

I - ENGINEERING THEORY OF BENDING

In the engineering theory of bending (ETB), equation (1) describes the relation between the strain ϵ_x in axis direction ($\epsilon_x = a + by + cz$) and the resulting cross section loads F_X, M_y and M_z shown in fig. 2.

$$\begin{pmatrix} EA & ES_z & ES_y \\ ES_z & EI_z & EI_{yz} \\ ES_y & EI_{yz} & EI_y \end{pmatrix} \cdot \begin{pmatrix} a \\ b \\ c \end{pmatrix} = \begin{pmatrix} F_z + F_{th} \\ -M_z - M_{thz} \\ M_y + M_{thy} \end{pmatrix} \tag{1}$$

This equation results from the one dimensional stress state (in which the stress σ_x is the only non-zero stress) and the linear strain distribution over the bar cross section. Furthermore besides the usual assumptions (infinite small deformations, linear-elastic, orthotropic material) the stresses and strains shall not vary along the bar axis.

The coefficients of the left matrix in (1) describe the stiffnesses of the bar:

EA   : tension stiffness, EI_y : bending stiff. about y-axis,
ES_y : elastic moment about y-axis, EI_z : bending stiff. about z-axis,
ES_z : elastic moment about z-axis, EI_{yz}: bending deviation stiffness.

With these coefficients, the position of the elastic and thermal center and the orientation of the principle axis can be calculated. The thermal loads in the right vector allow us to determine the coefficient of thermal expansion α_t and of thermal curvature κ_{ty}, κ_{tz} (/6/).

In the case of homogeneous bars or of non-homogeneous bars with materials, which have all the same Poisson's ratio, the stiffness coefficients can be evaluated directly from the material and geometry data:

$$EA = \int_A E\,dA, \qquad ES_y = \int_A Ez\,dA, \qquad ES_z = \int_A Ey\,dA,$$
$$EI_y = \int_A Ez^2\,dA, \qquad EI_z = \int_A Ey^2\,dA, \qquad EI_{yz} = \int_A Eyz\,dA, \tag{2}$$

These stiffnesses can be used as approximate solutions for non-homogeneous bars made from materials with different Poisson's ratios. But when these differences are quite large, significant errors can occur. Therefore a more exact theory is necessary.

II - CLASSICAL LAMINATE THEORY

The classical laminate theory (CLT) can also be the basis of determining bar stiffnesses and thermal deformations. For simple laminated
rectangular bars (see fig. 1, left) with all layers in one plane
(either the x-y- or the x-z-plane) the bar stiffnesses and the CTE can
be determined directly from the stiffness matrix of the laminate with
$n_y=n_{xy}=m_y=m_{xy}=0$. In the CLT different Poisson's ratios in each layer
are always taken into account, but as the basis of the CLT is the
infinite plate, edge influences can not be considered. For bars however, where the width to length ratio is very small and the width to
height ratio is near 1.0 (fig. 1), edge influences are usually important. For this reason these stiffnesses in general are too high.

III - THREE DIMENSIONAL THEORY

From equation (1) it can be seen, that it is possible to calculate the
stiffness coefficients from a known strain distribution and the
resulting cross section loads. This shows the way to calculate the bar
stiffnesses and the coefficients of thermal deformation:

1. Assume a linear strain distribution over the bar cross section.
2. Determine the resulting loads due to this strain distribution.
3. Evaluate the stiffnesses and thermal coefficients from
   equation (1).

3.1 Calculating the resulting cross section loads

For a circular tube, the stresses due to a linear strain distribution
ϵ_x can be determined directly. The three dimensional equations can be
reduced to a two dimensional problem of a state of generalized plain
strain by introducing the assumption of non-variing stresses and
strains along the bar axis. The resulting equations can be solved
analytically /6/.

For other cross sections than the circular tube analytical solutions
are no longer possible. So approximation methods have to be used. One
of the best known is the method of finite elements because of their
general applicability. With the same assumptions as above, again a two
dimensional problem of a state of generalized plain strain has to be
solved. The state of generalized plain strain is necessary as a
non-constant strain distribution has to be taken into account.

From the calculated stress distribution the resulting cross section
loads can be derived by integrating the stress σ_x over the bar cross
section. The axial force and the bending moments result from (3):

$$F_x = \int_A \sigma_x \, dA, \qquad M_y = \int_A \sigma_x z \, dA, \qquad M_z = -\int_A \sigma_x y \, dA. \qquad (3)$$

3.2 Evaluation of the stiffness and thermal coefficients

To calculate all unkown stiffness and thermal coefficients from equation (1), four independent load cases are necessary, e.g.:

1. Pure tension, a≠0, b=c=0, $\Delta T = 0$.
2. Pure bending about z-axis, a=0, b≠0, c=0, $\Delta T = 0$.
3. Pure bending about y-axis, a=b=0, c≠0, $\Delta T = 0$.
4. Temperature load, a=b=c=0, $\Delta T \neq 0$.

The stiffnesses and the resulting thermal loads are a direct result from equation (1). The other cross section data (such as position of the elastic center, orientation of principle axis, etc.) and the coefficients of thermal deformation have to be calculated from these data /6/.

IV - EXAMPLES AND EXPERIMENTAL VERIFICATION

To prove this way of calculating the bar stiffness and the coefficients of thermal deformation, special tests are performed. A bar with quadratic cross section is investigated, the geometry is shown in fig. 2. All used materials were CFRP laminates with the fiber orientation +45° and 90°.

In fig. 3 to fig. 5 the stiffnesses and the CTE are plotted versus the portion of +45° laminate of the whole cross section. It can be seen that there are very large discrepancies between experiment and the stiffness calculated from the engineering theory of bending. The reason for this is the very large Poisson's ratio of the +45° laminate (about 0.78) which is nearly totally restraint by the high stiffness of the 90° laminate in the transverse direction. This results in a very high additional stiffness of the +45° laminate.

V - CONCLUSIONS

Non-homogeneous bars made from materials with different Poisson's ratios may show discrepancies in stiffness and thermal deformation between experiment and the engineering theory of bending. Especially, if fiber reinforced materials are used, these effects become important. For the stiffness and thermal coefficients evaluation the following recommendations can be given:

- The above described procedure (following the three dimensional theory) for evaluating the bar properties should be preferred to the engineering theory of bending and the classical laminate theory.

- Thin-walled bars can be treated with sufficient accuracy by a combination of the CLT and the engineering theory of bending.

- Stiffnesses, calculated with the engineering theory of bending, are only reliable, if the differences in the Poisson's ratios of the used materials are small.

- Stiffnesses, calculated with the engineering theory of bending, are a lower bound, those, calculated with the classical laminate as a rule are an upper bound of the real stiffnesses.

- Thermal coefficients should always be calculated with the theory of chapter II.

VI - REFERENCES

1 - Musschelischwili N.I., Einige Grundaufgaben zur mathematischen Elastizitätstheorie, Carl Hanser Verlag, München, 1971

2 - Lo K.H., Bending of Laminated Beams, Thesis; New York, 1973

3 - Lo K.H., Conway H.D., Effect of Change of Poisson's Ratio on the Bending of a Multi-Layered Circular Cylinder, Int. J. of Mech. Sciences, 16 (1974) 757-767

4 - Lo K.H., Conway H.D., Plane Stress and Plane Strain Assumptions in the Stress Analysis of Laminated Bodies, Int. J. of Mech. Sciences, 18 (1976) 1-4

5 - Schile R.D., A Non-homogeneous Beam in Plane Stress, Journal of Applied Mechanics, Sept. 1962

6 - Rapp H., Inhomogene Balken unter Temperatur-, Zug-, und reiner Biegebeanspruchung, Thesis, München, 1988

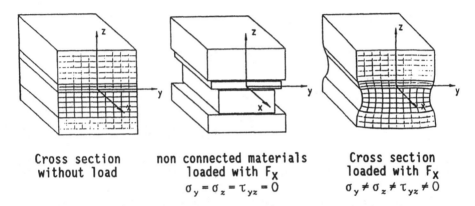

| Cross section without load | non connected materials loaded with F_x $\sigma_y = \sigma_z = \tau_{yz} = 0$ | Cross section loaded with F_x $\sigma_y \neq \sigma_z \neq \tau_{yz} \neq 0$ |

Fig. 1 - Non-homogeneous bar under load in axis direction

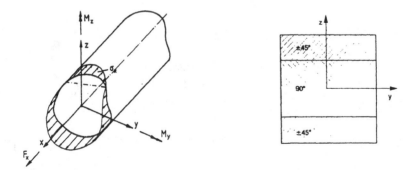

Fig. 2 - Definition of the resulting cross section loads, cross section of test bar

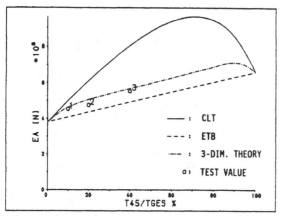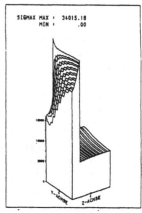

Fig. 3 - Tension stiffness of the test bar, distribution of stress σ_x versus one quarter of the cross section

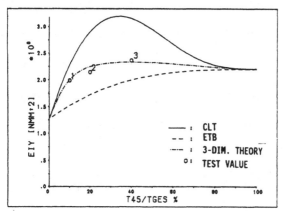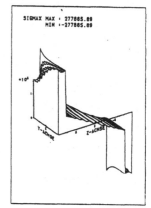

Fig. 4 - Bending stiffness about y-axis, distribution of stress σ_x versus one half of the cross section

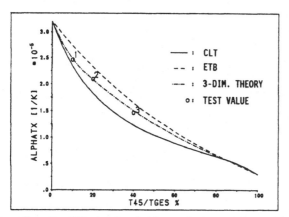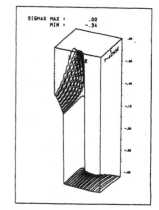

Fig. 5 - Coefficient of thermal expansion, distribution of stress σ_x versus one quarter of the cross section

EVENT : AN EXPERT SYSTEM
FOR COMPOSITE'S DESIGN

H. KERN, J. JANCZAK, A. NITSCHE

University of Dortmund - Lehrstuhl für Werkstofftechnologie
PO 500500 - 4600 Dortmund 50 - West-Germany

ABSTRACT

Composites give the possibility to generate requested properties by combining of single materials. But the design of composites is a task of high complexity which has to consider many constraints. To gain this challenge computer based systems, e.g. expert systems, finite element methods, materials data bases etc. are essentially necessary.
In this paper we present the first prototype of the expert system EVENT which supports the engineers by solving the problem of composites' design.

EINFÜHRUNG

Die Verbundwerkstoffe haben unser tägliches Leben revolutioniert. Vom Tennisschläger bis zu Flugzeugen und Space Shuttle werden sie mit großem Erfolg eingesetzt. Sie weisen öfters excellente Eigenschaften auf, die homogene Werkstoffe weit übertreffen. Sie bieten dabei die Möglichkeit, ein Material entsprechend den Anforderungen zu entwickeln. Diese Freiheit des Designs wird aber heutzutage nicht vollständig genutzt. Es haben sich bis jetzt keinerlei Methoden als Standards durchgesetzt, die den Entwurf von optimalen Verbundwerkstoffen in die Konstruktion des Bauteils integrieren /1/. Da die Konstruktion des an die gestellten Anforderungen angepaßten Materials einen sehr komplexen und nicht vollständig erforschten Vorgang darstellt /2/, kommen in diesem Bereich immer noch die "trial and error" Methoden

zur Anwendung. Der Experte versucht dabei, dieses Verfahren zu verbessern, indem er seine Auswahl gemäß seinem Erfahrungsschatz nach den Ergebnissen der früheren Materialuntersuchungen anderer Probewerkstoffe und nach den Ergebnissen der einzelnen Standardberechnungen trifft. Die Modelle zur Vorhersage der Eigenschaften von Verbundwerkstoffen sind zur Zeit noch nicht genügend ausgereift /3/, doch dienen sie ebenso wie das Erfahrungswissen des Experten zur Verringerung der Zahl der nötigen, aber aufwendig herzustellenden Materialproben mit anschließenden Materialprüfungen.
In diesem Artikel wird der erste Prototyp des Expertensystems "EVENT" (Expertensystem zum Verbundwerkstoff ENTwurf) vorgestellt. Das Erfahrungswissen der Experten findet in diesem System ebenso Eingang wie bewährte mathematische Verfahren und Kennwerte bekannter Werkstoffe.

I-DIE ARCHITEKTUR DES SYSTEMS

Die Zentrale von EVENT bildet das Kern-System. Es übernimmt die Anforderungen und steuert den Lösungsprozeß. Daneben werden externe Module eingesetzt. Dies sind ein Datenbanksystem, ein Finite-Elemente-Programm und ein Modul zur Bearbeitung der mathematischen Modelle für Verbundwerkstoffe.

1.1 KERN-SYSTEM

Das Kern-System steuert den gesamten Lösungsfindungsprozeß. Dabei bedient es sich der anderen Module und des in ihm gespeicherten Expertenwissens. Diese Systemkomponente nimmt die Anforderungen entgegen und schränkt anschließend die Menge aller Verbundwerkstoffe schrittweise auf diejenigen ein, die dem Anforderungsprofil genügen. Dabei setzt es das heuristische Wissen der Experten ein. Dieses interne Wissen liegt als regelbasiertes Wissen deklarativ vor. Die Regeln des Expertensystems beschreiben Zusammenhänge zwischen den Beanspruchungen, den Eigenschaften, der Verbundart und der Herstellungstechnologie, die sich durch mathematische Modelle nicht darstellen lassen. Das durch sie repräsentierte Wissen gibt die Zusammenhänge auf einem abstrakteren und allgemeineren Niveau als die übrigen externen Module wieder. Als Beispiel können folgende Regeln genannt werden:
"wenn
    Beanspruchung = starke mechanische Belastung
    und
    Anforderung = Leichtbau
dann
    Verbundart = Faserverbund"

"wenn

   mechanische Belastung = Querkräfte

dann

   Herstellungstechnologie = kein thermisches Spritzen"

Das durch Regeln programmierte Wissen des Experten beschleunigt die Suche nach geeigneten Verbundwerkstoffen wesentlich. Hierin spiegelt sich die Auswahl der praxiserfahrenen Experten wieder. Das Kern-System ist in der Art der Blackboard-Architektur aufgebaut /4/. Die vorliegenden Informationen werden durch eine Reihe von sich unabhängig arbeitenden Experten (wie mechanische, thermische Experte, Korrosions-Experte) bearbeitet. Diese Struktur bietet einen wichtigen Vorteil an: die Möglichkeit einer beliebigen Erweiterung des Systems und der durch ihn verarbeiteten Aspekten /4/.

1.2 DATENBANK-MODUL

Exakte Daten über das physikalische, mechanische, chemische und thermische Verhalten der homogenen Werkstoffe, aus denen die Verbundwerkstoffe konstruiert werden, findet sich nicht in der Wissensbasis, sondern in einem externen Datenbanksystem wieder. Es stehen beispielsweise folgende Größen für jeden Werkstoff zur Verfügung: E-Modul, Streckgrenze, Zugfestigkeit, Dichte, Poissonsche Zahl, Schmelztemperatur, spezifische Wärme usw. Es sind drei Tabellen dieses Typs vorhanden, eine für die Werkstoffe des Volumen-Materials (Matrix), eine für die Fasern und eine für die Beschichtungen. Die Angaben über die chemische Verträglichkeit zweier Werkstoffe sind in Form einer Verträglichkeitsmatrix abgespeichert. Eine weitere umfangreiche Tabelle stellt die Korrosionstabelle dar.

1.3 FEM-MODUL

Das FEM-Programmsystem ermöglicht die Lösung der mechanischen und thermischen Probleme, die mit geschlossenen Modellen nicht immer lösbar sind. Da aber Berechnungen mit einem FEM-System sehr zeitaufwendig sind, wird angestrebt, die notwendige Anzahl der Aufrufe dieses Moduls möglichst zu minimieren. Für dieses Ziel werden die für einen Vergleichwerkstoff durchgeführten Berechnungen sehr genau analysiert und auf deren Basis die Ergebnisse für andere Werkstoffe simuliert.

1.4 MODELLE-MODUL

Mit diesem Modul erfolgen die Transformationen der Belastungen in die Eigenschaften der homogenen Werkstoffe sowie in die Summen- und im weiteren in die Komponenteneigenschaften der Verbundwerkstoffe.

Die ersten werden durch physikalische Zusammenhänge unterstützt, die zweiten unter Anwendung von Gefüge-Eigenschafts-Gleichungen /5/.

II-DER ENTWURF DER VERBUNDWERKSTOFFE MIT DEM SYSTEM EVENT

Der erste Schritt bei der Arbeit mit dem System ist die Definition der Anforderungen an den Verbundwerkstoff. Informationen, wie Geometrie, Lagerung und Belastung des Werkstücks, teilt der Benutzer dem Expertensystem im Dialog mit. Die anschließende Analyse dieser Daten führt zu einem Eigenschaftsprofil des zu konstruierenden Werkstoffs, das durch die Kenngrößen charakterisiert wird, die in der externen Datenbank für die Komponentenwerkstoffe gespeichert sind. So werden beispielsweise Benutzereingaben über die Einsatztemperatur zu einem Kriterium für die Schmelztemperatur transformiert. Im Bereich mechanischer Belastungen wird diese Transformationsaufgabe durch das Finite-Elemente-Programm für einen Vergleichswerkstoff unterstützt. Aus diesen Ergebnissen werden z.B. Werte für den E-Modul, die Bruchdehnung und die Festigkeit abgeleitet.

Da sich diese Werte auf den Verbundwerkstoff beziehen, die Datenbank aber Werte für die Komponenten enthält, müssen die möglichen Kombinationen einzelner Werkstoffe betrachtet werden. Da die Anzahl dieser Paare mit der Anzahl der betrachteten Komponenten wächst, ist hier eine Vorselektion nötig. Dazu wird versucht, die Eigenschaften in Komponenteneigenschaften zu transformieren, um die Werkstoffe einzeln betrachten zu können. Es wird dabei mit negativen Selektionskriterien gearbeitet, um die Anzahl der in Frage kommenden Komponenten zu reduzieren. Werkstoffe, die nicht zur Erstellung der gewünschten Eigenschaften führen können, werden ausgesondert.

Die übrigen Werkstoffe werden miteinander kombiniert. Dabei werden auf die vorselektierten Werkstoffe alle noch nicht verwendeten Kriterien, wie z.B. Verträglichkeit der Komponenten miteinander, angewendet. Es werden Kombinationen zum Faser-, Schicht- und Teilchenverbundwerkstoff berücksichtigt. Die Entscheidung für eine Verbundwerkstoffart wird durch Regeln unterstützt. Als Ergebnis bekommt der Benutzer eine Liste der Verbundwerkstoffe, die die gestellten Anforderungen erfüllen. Es werden die Art des Verbundes, seine Zusammensetzung sowie auch die Herstellungstechnologie genannt.

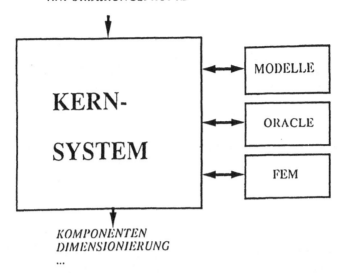

ANFORDERUNGSPROFIL

KERN-SYSTEM

MODELLE

ORACLE

FEM

KOMPONENTEN DIMENSIONIERUNG

...

Bild 1. Die Architektur des Systems EVENT

LITERATUR

1. F. Lockett, The Provision of Adequate Materials Property Data, in Proceedings of the 6th International Conference on Composite Materials and Second European Conference on Composite Materials ICCM & ECCM, Elsevier Applied Science, London and New York, 1987

2. U. Behrenbeck, M. Effing M., Konstruieren mit Faserverbundwerkstoffen — von den Materialkennwerten bis zu den Maschinensteuerdaten, Werkstoffe & Konstruktion, 3 (1988)

3. A. Levesque and G. Taylor, Advanced Industrial Materials, ASM-News 18 (1987)

4. H. Nii, Blackboard Systems: The Blackboard Model of Problem Solving and the Evaluation of Blackboard Architectures (Part One), in AI Magazine 8(1986)

5. G. Ondracek, The Quantitative Microstructure-Field Property Correlation of Multiphase and Porous Materials, Reviews on Powder Metallurgy and Physical Ceramics, 3 (1987)

OPTIMUM DESIGN OF LAMINATES
UNDER ASPECTS OF STRENGTH

M. WEGENER, W. MICHAELI

Institut für Kunststoffverarbeitung - Pontstrasse 49 - 5100 Aachen - West-Germany

ABSTRACT

Continuous fibre reinforced composites have excellent mechanical properties compared with classical design materials. Because of the anisotropy of the material and also the laminate properties it is difficult to get full use of the composite material potential. This paper offers an easy method for optimum strength design of multi-layer laminates by diagrams and presents a strategy for more complex optimization problems.

INTRODUCTION

It is widely known that the mechanical properties of fibre reinforced plastics are excellent. Comparing them to classical design materials, they reveal some striking superiority, most of all with respect of weight. The properties of these materials, however, are submitted to the direction; stiffness and strength only reach an optimum in the direction of the fibres. These values are usually employed in comparisons.

It is essential for the designer to provide for optimum utilization of the material's anisotropic properties, considering the individual aim of the design (strength, stiffness, stability etc.). There is little advice available on this topic. In fact, quasi isotropic or almost quasi isotropic laminates (0°/90°/±45° laminates) are very frequently investigated. However, they almost fail to utilize the full potential of the material, most of all in respect to strength (fig. 1). The module of a quasi isotropic laminate is only one third as high as that of a merely unidirectional laminate; strength falls down to 1/8 to 1/10 of the unidirectional value!

I - OPTIMIZATION - A TYPICAL DESIGN TASK

Taking these things into account, the designer does not only have the "possibility", but rather the indispensable "duty" to try and achieve the optimum of the anisotropic material properties.

The design of laminates suited to the individual material and load conditions, revealed to be more difficult than expected. Design strategies such as /1 - 3/ are not satisfying, since they handle only restricted layer angles or only layer thicknesses. However, it is both of the two parameters that have a steep effect on strength. Computing programs indeed make it possible for the designer to examine various lay-outs with various angles and layer thicknesses within a short time. For the purpose of finding a laminate that completely use the potential of the material, the designer usually must examine all the variants systematically. In view of the high number of design variables with multi-layer laminates (one angle and one value of thickness for each layer, leaving apart the design variables of stacking sequence and the fibre volume content), it is impossible to effect searching systematically. Starting from a 4-layer laminate with only 5 different angles and layer thicknesses, the number of possible designs comes up to a rough 20,000; if a fine searching mesh with 10 different angles and layer thicknesses is used, then 7 million laminate designs will have to be checked.

For the purpose of obtaining an "exact" solution of this task, numerical optimization methods must be employed, which was shown by Wissmann and Eschenauer /4, 5/. Generally speaking, with these deterministic gradient methods, every problem can be solved, no matter how complicated. Nevertheless, it is not very easy to employ, and therefore is not recommended for some (all) cases. Moreover, Wissmann says /5/ that the user must avail of far-reaching experience.

II - OPTIMUM STRENGTH DESIGN BY VARIANT STRATEGY

In the framework of some investigations carried out by the IKV /6/, an easy variant strategy of optimum laminate lay-out was developed, which uses set values adequate to the material to reduce the number of the variables down to the two most essential ones.

The strategy is based on the idea of locating one of the direction of the maximum principal stress and, through easy variation calculations, finding out the optimum of difference angles and layer thickness ratio. The optimum value is found at the point where the requirement for laminate thickness reaches its minimum. This strategy's findings for a material can be summarized in two diagrams (fig. 2 and 3). The diagrams present the optimum angle and layer thickness, subject to the principal stress or normal force ratio (N_I/N_{II}).

So for example the optimum laminate design for a pressure vessel (the ratio of the principal stresses is two) has a difference angle of ±42,5° and an infinite thickness ratio which means that the "0°-ply" in maximum principal stress direction (N_I) is not necessary.

A comparison with other laminates shows a significant weigth saving by using this simple diagrams (fig. 4). It is possible to reduce the laminate thickness about 70% in comparison to a quasi isotropic laminate and about 30% in comparison to an optimum laminate design given by netting analysis (±54,7°)

These diagrams are developed for a glasfiber/epoxy-laminate. But they are also usefull for the design of other continous fiber reinforced composites.

Detailled analyses show that the optimum normally provides the profit of a minimum in the sensitive transverse loading on the different layers. Only to a very low extent is the strength in fibre direction being utilized; the transverse loading on the individual layers essentially determines the strength of a multi-layer laminate.

In fact, this strategy leads to a pronounced improvement in the use of the laminate materials' potential of strength. The strategy is however restricted to a single load case and simple laminates.

III - MORE COMPLEX PROBLEMS - ANOTHER STRATEGY

For more complex problems, the strategy of evolution is worth some consideration. It is particularly suited if many variables are to be taken into account /7/ during optimization and has been successfully employed several times. For instance, in the optimization of the geometry of a superheated steam nozzle it leads to such solutions that do not fit physical model conceptions. Yet, experiment have shown that geometries generated by the strategy of evolution is indeed superior to all the other known geometries /8/. In addition, in this strategy of "intelligent control of coincidence", the numerical problems that are involved in deterministic methods (e. g. during derivation of the target function) are omitted. Neither is its convergence rate necessarily lower than that of the classical deterministic methods. The strategy of evolution distinguishes itself by:

- easy mathematical basis,

- high resistivity to discontinuities of the target function and
- high probability in finding a global optimum, which is of advantage when the "topology" of the target function is unknown /8/ (as is the case with).

The strategy was successfully applied to the problem of optimum laminate design. A comparison to the variant strategy mentioned also shows very good coincidence with the findings even there is no reduction of the variables with the evolution strategy.

Because of type of the algorithm the computing time e.g. the number of iterations is comparatively high. But if the optimum is found it is possible to adjust the parameters of the strategy (number of parents, recombination, ...) to the type of the optimize problem. This leads to a significant reduction of iterations and computing time for the next similar optimize tasks.

The evolution strategy seems to be an efficient tool to handle new types of optimize problems which is very commom for engineers work.

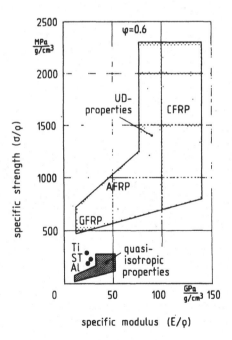

Fig. 1 - Specific properties of FRP-laminates

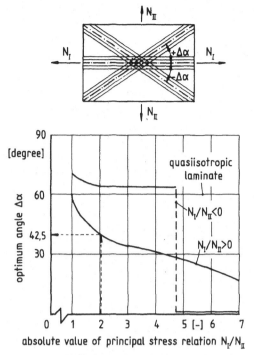

Fig. 2 - Optimum angle depending on the principal stress
ratio

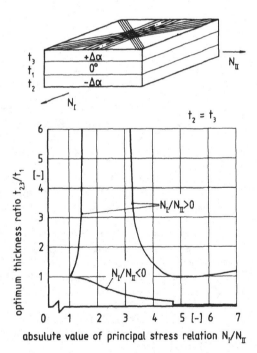

Fig. 3 - Optimum thickness ratio depending on the principal
stress ratio

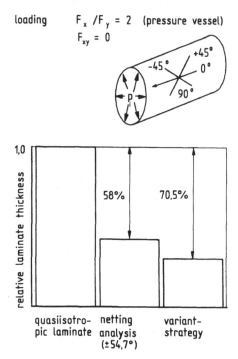

loading $F_x/F_y = 2$ (pressure vessel)

$F_{xy} = 0$

Fig. 4 - Relative laminate thickness of a pressure vessel

REFERENCES

1 - Tsai, S. W. Hahn, H. T. Composite Materials Workbook, Air Force Wright Aeronautical Laboratories, Wright-Patterson AFB, Ohio 45433, March 1988

2 - Chamis, C. C. Design Procedure for Fibre Composite Structural Components: Panels Subjected to Combined In-Plane Loads, 40th Annual Conference, Reinforced Plastics/Composites Institute, SPI January 1985

3 - Tsai, S. W. Composites Design, Think Composites, Dayton, Ohio, 1988

4 - Eschenauer, H. Optimierung ebener Flächentragwerke aus Verbundwerkstoff, Zeitschrift für Flugwissenschaften und Weltraumforschung, annual volume 8, volume 6

5 - Wissmann, W. J. Entwurf und Optimierung von faserverstärkten Strukturen unter mehrfachen Belastungen, Final report on the research project, DFG-SPP: Konstruktionsforschung, TH Darmstadt 1976

6 - Wiartalla, A. Entwicklung einer Strategie zur Auslegung von flächigen Faserverbundlaminaten, unpublished diploma thesis, IKV Aachen 1988

7 - Schwefel, H. P. Numerische Optimierung von Computermodellen mittels der Evolutionsstrategie, Birkhäuser Verlag, Basel, 1977

8 - Schwefel, H. P. Optimierungsstrategien, Lecture held at IKV Aachen on 4th of September 1989.

APPLICATION
ANWENDUNG

Chairmen : **Prof. S.W. TSAI**
University of Stanford
Prof. I. CRIVELLI-VISCONTI
University of Naples

ADVANCED COMPOSITES WITH POLYMER MATRIX SELECTION AND APPLICATION CONDITIONS

B. KNAUER

Technical University - 8027 Dresden - East-Germany

Die werkstoffgerechte Konstruktion wird durch ein Polymerverbundgesamtsystem gefördert. Die Auswahl-kriterien für Verbundkomponenten werden vorgestellt. Durch gleichzeitige Entwicklung der inneren Struktur und der äußeren Gestalt werden Anwendungsbedingungen bestmöglichst berücksichtigt.

1. Ein Polymerverbundgesamtsystem

Die werkstoffgerechte Konstruktion stellt in der Gegenwart einen Engpaß dar. Die volle Ausschöpfung der Werkstoffvorzüge und die Umgehung von Nachteilen wird unter Beachtung der Verbundbildung mit Eigenschaften, die nicht summarisch eingehen, immer weniger beherrscht. Es hat sich bei der Hochschulausbildung bewährt, mit einem System, z. B. der polymeren Matrix, die Kombination mit Metall, organischen und anorganisch-nichtmetallischen Verstärkungsmaterial als Beispiel exemplarisch zu erfassen, um danach beliebig neue Verbunde zu entwickeln (Abb. 1). Polymerverbunde werden nicht vordergründig als standardisiertes Halbzeug genutzt, sondern entsprechend der Anforderungen im konstruktiven Entwicklungsprozeß geschaffen. In Wechselwirkung von Matrix- und Verstärkungsmaterial sowie den Verarbeitungs- und Anwendungsbedingungen werden dabei gleichzeitig die innere Struktur als Werkstoff und die äußere Gestalt als Konstruktion festgelegt. Das Entwerfen von inhomogenen und teilweise stark anisotropen Bauteilen ist damit der Weg zur bestmöglichen Anpassung von Konstruktionen an gegebene Beanspruchungen (Abb. 2). Die innovativen Anwendungen werden in Einheit von Werkstoff- und Konstruktionstechnik umgesetzt. Der Trend liegt bei neuen Gebrauchswerten durch mehrphasige, heterogene Werkstoffe mit stark gestaffelten Größen der Gefügebestandteile. Den Oberflächen und Grenzschichten kommt dabei wachsende Bedeutung zu. Der Weg ist nach der Steuerung der chemischen Zusammensetzung, den Elementarverteilungen und Bindungszuständen als mikrostrukturelles Konstruieren zu kennzeichnen. Es wird in zwei Ebenen, d. h. der Erzeugnis- sowie Verbundentwicklung angegangen und bedeutet, daß zur Funktionserfüllung der Gesamtkonstruktion die Verbunde nicht nur als Werkstoff, sondern als funktionell gestaltete Konstruktionselemente angesehen werden müssen.

Das Polymerverbundgesamtsystem (PVGS) stellt deshalb geometrische Elemente in den Vordergrund (Abb. 3). Alle bekannten Verbunde lassen sich einordnen, wenn in jedem Feld bis zu 6 Unterteilungen als UD-, MD- und BD-Einfachverbunde sowie als Mischprinzipe der Verstärkungsmaterialien vorgenommen werden. Die mechanisch-physikalischen Kennwerte der Verbunde werden auf rechnerisch-experimentellen Weg bestimmt, so daß das System beliebig viele Verbunde darstellen kann. Unter Berücksichtigung der morphologischen Struktur der Verstärkungselemente, gestaffelter Orientierungen und Volumenteile lassen sich daraus eine nahezu unbegrenzte Zahl von Verbunden ableiten. Die Zeit-, Temperatur- und Medieneinflüsse werden über Berechnungs- und teilweise auch Struktur-Eigenschaftsbeziehungen im Auswahl- und Entwurfsprozeß berücksichtigt /1/.

2. Verbundauswahl

Mit Ein- und Ausgangsgrößen sowie Randbedingungen läßt sich das Anforderungsbild jeder Entwicklungsaufgabe beschreiben. Eine Präzisierung der Aufgaben führt zu Kennfunktionsanforderungen:

- mechanische Kennwerte (Zug, Druck, Biegung, Schub, Torsion, statisch dynamisch)

- thermische, elektrische, korrosive, optische Bedingungen, Lebensdauervorgaben

- Wünsche zum Feuchtigkeitsverhalten, zur Flammwidrigkeit, Dichte u. a.

Nachfolgend sind Festforderungen eindeutig zu kennzeichnen und nach dem k. o.-Prinzip zu bewerten.

Auf der Grundlage von Problemanalysen wurde ein Fondssystem entwickelt, das für Matrixmaterialien und ausgewählte Verbunde mit Rechnerunterstützung zur Kennwertsuche und Werkstoffauswahl eingesetzt wird/2/:

Fonds A: Werkstoffauswahl

Fonds B: statisch-thermisches Langzeitverhalten

Fonds C: Kennfunktionen dynamisches Verhalten

Fonds D: tribologische Kennwerte

Fonds E: verarbeitungstechnische Werte

Die Bestimmung der Werte erfolgt auf der Basis von internationalen Normen. Alle Hersteller von Werkstoffen sind verpflichtet, ihr Material ausreichend charakterisiert der Datenbank zur Kenntnis zu geben. Es ist dadurch möglich, komplexe Informationen zusammenzustellen, um über Betriebsgrenzen hinweg, die Werkstoffauswahl durchzuführen /3/. Als Ablauf haben sich 4 Schritte bewährt:

1. Werkstoffvergleiche zwischen allen Werkstoffklassen mit einem Minimum an Informationen und Entscheidung zur einzusetzenden Klasse

2. Mit Primärdaten zu den Gebrauchs- und Funktionseigenschaften erfolgt die Vorentscheidung zur Werkstoffgruppe Verbunde

3. Grob- und Feinbemessung und damit die Festlegung des genutzten Feldes im PVGS mit den Angaben zur Matrix und dem Verstärkungsmaterial

4. Parallel zur Detailgestaltung (auch der Krafteinleitungsbereiche) kann der Verbund mit seinen Makro- und Mikroelementen, der Anisotropie, den Schichten den Fasern und/oder Teilchen umfassend zeichnerisch und verbal dargestellt werden

Rückgriffe sind erforderlich, wenn der Materialwechsel bei einer Komponente bzw. auch die konstruktive Gestaltung des Verstärkungsmaterials ansteht, so als Abstandsgewebe, -gewirke oder Kraftliniengerechter Fadenlegung (Abb. 4). Eine automatisierte Werkstoffauswahl wird deshalb bei Verbundkonstruktionen nicht befürwortet.

3. Anwendungen

Für das PVGS wurden in größerer Zahl Verbundanwendungen analysiert und letztlich 25 mit unterschiedlichem Aufbau, geordnet nach Einfach- und Mischverbunden, innerhalb jedes der 8 Felder charakterisiert. Hochleistungsverbunde sind dabei die für ausgewählte Eigenschaftskomplexe verstärkten Werkstoffe, die ein über den klassischen Konstruktionswerkstoffen liegendes und durch Normwerte gesichertes Beanspruchungsniveau aufweisen. Sie führen zu innovativen Anwendungen mit konstruktiver und funktioneller Beanspruchung, so bei der Steife und der Reibungsenergiedichte. Hochleistungsverbunde stellen für die moderne Volkswirtschaft eine unverzichtbare Basis dar, deren Bedeutung für Maschinen und Anlagen, Bauelemente und Geräte ständig zunimmt. Man muß jedoch auch erkennen, daß von jedem Anwendungsfall wieder Rückschlüsse auf die Werkstoffauswahlstrategie ausgehen. Werden Lebensdauerakten geführt oder Schädigungsmessungen im Betriebszustand geführt, dann sind neue Verknüpfungen von Werkstoff- und Fertigungsparametern gegeben. Gleiches gilt für die Kosten. Der Fertigungsentschluß setzt einen nachgewiesenen Anwendernutzen voraus, der in der Regel sich aus einem berechenbaren Gebrauchswertzuwachs ergibt.

Literaturverzeichnis

/1/ Knauer, B.: Grundlagen zur konstruktiven Erzeugnisentwicklung auf der Basis eines Polymerverbundgesamtsystems, Promotion B, TU Dresden 1989

/2/ Informationssystem Werkstoffe-Grundwerkstoffspeicher Plaste-Hrg.: Zentralinstitut für ökonomischen Metalleinsatz, Dresden 1988

/3/ Dupp, G.: CAMPUS die Datenbank der Kunststoffhersteller zur Werkstoffvorauswahl in: Die Zukunft des Konstruierens, SKZ Fachtagung Mannheim Mai 1990

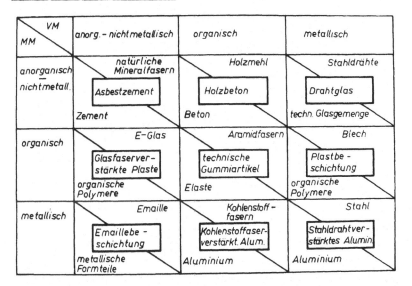

VM \ MM	anorg. - nichtmetallisch	organisch	metallisch
anorganisch nichtmetall.	natürliche Mineralfasern / Asbestzement / Zement	Holzmehl / Holzbeton / Beton	Stahldrähte / Drahtglas / techn. Glasgemenge
organisch	E-Glas / Glasfaserverstärkte Plaste / organische Polymere	Aramidfasern / technische Gummiartikel / Elaste	Blech / Plastbeschichtung / organische Polymere
metallisch	Emaille / Emaillebeschichtung / metallische Formteile	Kohlenstofffasern / Kohlenstofffaserverstärkt. Alum. / Aluminium	Stahl / Stahldrahtverstärktes Alumin. / Aluminium

Abb. 1 Verbunde als Beispiel unbegrenzter Werkstoffkombination

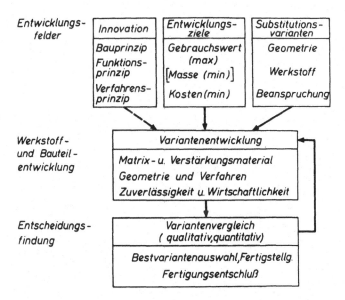

Entwicklungsfelder

Innovation	Entwicklungsziele	Substitutionsvarianten
Bauprinzip	Gebrauchswert (max)	Geometrie
Funktionsprinzip	[Masse (min)]	Werkstoff
Verfahrensprinzip	Kosten (min)	Beanspruchung

Werkstoff- und Bauteilentwicklung

Variantenentwicklung
Matrix- u. Verstärkungsmaterial
Geometrie und Verfahren
Zuverlässigkeit u. Wirtschaftlichkeit

Entscheidungsfindung

Variantenvergleich
(qualitativ, quantitativ)
Bestvariantenauswahl, Fertigstellg.
Fertigungsentschluß

Abb. 2 Werkstoffgerechte Konstruktion

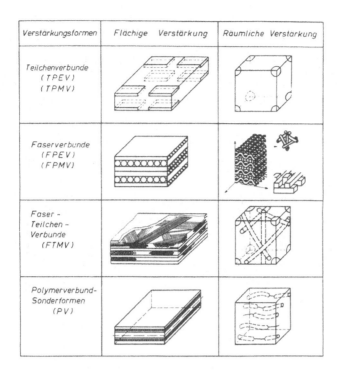

Verstärkungsformen	Flächige Verstärkung	Räumliche Verstärkung
Teilchenverbunde (TPEV) (TPMV)		
Faserverbunde (FPEV) (FPMV)		
Faser- Teilchen- Verbunde (FTMV)		
Polymerverbund- Sonderformen (PV)		

Abb. 3 Polymer-Verbundwerkstoff-Gesamtsystem

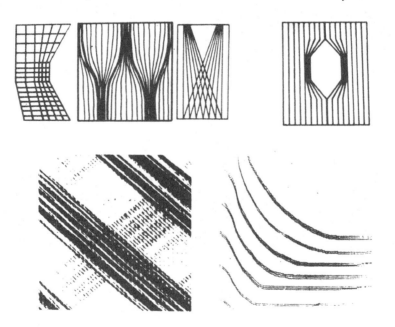

Abb. 4 Fadengelege

EFFICIENT PROCESS OF REINFORCING
THE COMPOSITE TURBOMACHINERY BLADES

L. GORSHKOV

Central Institute of Aviation Motors
Ciam 2 Aviamotornay St. - 111250 Moscow - USSR

In the analysis of the stress-strain state in static, quasistatic, and dynamic problems the composite material of blades is most often simulated as an anisotropic continuous medium with physicomechanical properties which happen to be certain integral characteristics of its reinforcement layers [1 -3].

An analysis of naturally twisted beams simulating a blade in such a quasihomogeneous model of a composite material will show that, owing to the coupling between tensile and torsional strains in twisted bodies, the distribution of normal and shearing stresses over the foil part of a blade as well as the untwisting angle of its peripheral section depend strongly on the anisotropy index of its elastic properties (ratio of longitudinal modulus of elastisity to shear modulus).

At certain values of the parameters characterizing the anisotropy of the material and the initial geometry there may appear high compressive stresses at the edges of a naturally twisted beam or blade under tension, and these stresses together with the shearing stresses between layers can cause delamination of the edges and a local loss of stability by the reinforcement layers.

For selecting the reinforcement scheme for a blade made of a composite material and for evaluating how a rational layout of the reinforcement layers depends on the properties of the fibers and the matrix, we will use the modified engineering theory of naturally twisted beams [4]. According to this theory, the stess-strain state of a blade will be determined with the laminate struoture of the material taken into account.

A rational design of the reinforcement for a blade profile is evaluated starting with the allowable untwisting angle of the peripheral section and the stress state of each reinforcement layer.

The angle through which the peripheral section of a blade untwists during operation of the machine is

$$\varphi = \int\limits_{R_o}^{R_p} \frac{d\varphi}{dr} \, dr \, , \qquad (1)$$

where

$$\frac{d\varphi}{dr} = \frac{1}{\theta^2 \, I_p^{*2}} \left[\frac{P \, \theta \, I_p^*}{F^*} + M_\xi \, \theta \, \frac{I_{p\xi}^*}{I_\xi^*} - M_\eta \, \theta \, \frac{I_{p\eta}^*}{I_\eta^*} - M_{tw}o \right]$$
$$q_1 - \frac{}{F^*}$$

The normal stresses in the reinforcement layers are

$$\sigma = E \left[\varepsilon_o + \frac{M_\xi}{I_\xi^*} \, \eta - \frac{M_\eta}{I_\eta^*} \, \xi - \frac{d\varphi}{dr} \, \theta \left(r^2 - \frac{I_{p\eta}^*}{I_\eta^*} \, \xi - \frac{I_{p\xi}^*}{I_\xi^*} \, \eta \right) - \alpha \, T \right]$$
$$(2)$$

The strain ε_o of the blade axis in this equality can be expressed as

$$\varepsilon_o = \frac{q_1 \dfrac{P}{\theta^2} + M_\xi \dfrac{I_p^* I_{p\xi}^*}{I_\xi^*} - M_\eta \dfrac{I_p^* I_{p\eta}^*}{I_\eta^*} - M_{tw}o \, \dfrac{I_p^*}{\theta}}{q_1 \, \dfrac{F^*}{\theta^2} - I_p^{*2}} \qquad (3)$$

Supplementary temperature factors are introduced to evaluate the influence of changing the linear expansion ratio of the material along blade sections into power factors (1) - (3).

They are as follows

$$P^o = \int\limits_F E \, \alpha \, T \, dF \quad ; \quad M_\xi^o = \int\limits_F E \, \alpha \, \eta \, T \, dF \quad ;$$
$$(4)$$
$$M_\eta^o = -\int\limits_F E \, \alpha \, \xi \, T \, dF \quad ; \quad M_{tw}^o = \theta \int\limits_F E \, \alpha \, r_1^4 \, T \, dF \, .$$

The relation for determining the shearing stresses which appear at the edges of the blade profile, between

layers of the composite material, is
$$\tau = \sigma \, \theta \, r_1 . \qquad (5)$$

The following symbols are used in expressions (1) – (3).

P, centrifugal force on a blade at a given section; M_ξ, M_η, bending moments on centrifugal and gas forces about the referred principal axes of a cross section; M_{tw}o, twisting moment of centrifugal forces; φ, relative untwisting angle of a cross section; $\theta = d\alpha/dr$, relative angle of natural (initial) twist of a cross section; ε_o, strain of the blade axis; σ, normal stresses in the reinforcement layers at points of a cross section; τ, shearing stresses between reinforcement layers at points of a cross section; E, modulus of elastisity of reinforcement layers at points of a cross section; G, shear modulus of reinforcement layers at points of a cross section; ξ, η, coordinates of points on the principal axes of a cross section ($r_1^2 = \xi^2 + \eta^2$); F, area of a blade cross section; T^o, geometrical rigidity in twist of reinforsement layers relative to the center of twist;

$$q_1 = T^* - \frac{I_{p\xi}^{*2}\, \theta^2}{I_\xi^*} - \frac{I_{p\eta}^{*2}\, \theta^2}{I_\eta^*} + \theta^2 I_{4p}^* \; ;$$

$$I_p^* = \int_F E\, r_1^2 \, dF \; ; \qquad\qquad I_\eta^* = \int_F E\, \xi^2 \, dF \; ;$$

$$I_\xi^* = \int_F E\, \eta^2 \, dF \; ; \qquad\qquad I_{p\xi}^* = \int_F E\, \eta\, r_1^2 \, dF \; ;$$

$$I_{p\eta}^* = \int_F E\, \xi\, r_1^2 \, dF \; ; \qquad\qquad I_{4p}^* = \int_F E\, r_1^4 \, dF \; ;$$

$$T^* = \int_F G\, T^o \, dF \; ,$$

referred physicogeometrical characteristics of a cross section.

The physicogeometrical characteristics of a section in equalities (1) – (4) evidently depend on the elastic properties of each reinforcement layer, its configuration, and its position relative to the referred center of gravity of a given cross section.

For evaluating the dependence of the scheme choice on the properties of the fibers and the matrix, we examined three types of composite material with an aluminum matrix and characterized by different values of longitudinal and transverse parameters. The first two types of composite material obviously differ essentially in the kind of reinforcement and thus in the characteristics along fibers. The transverse and shearing parameters of these materials have low values. The transeverse and shearing parameters of the third material have rather high values. The third type of material has elastic properties with the least anisotropy.

For analyzing the dependence of the choice of rational reinforcement scheme on the basis of relations (1) - (3), we have calculated the stress-strain state for the -model blade made of either the first or the second type of material with various volume fractions of reinforcement layers laid at ±15°, ± 30°, or ±45° angles to the radial axis of the blade profile.

The graph in Fig.1 indicates that for a blade of the composite material of the first type the opening angle of its peripheral section is smallest with 70% of the reinforcement layers laid at ±30°angles to the profile axis. The angle of elastic blade untwisting becomes larger with the layers laid at ±45° angle to blade axis, although the material has then also a higher modulus of elasticity.

In the blade made of the second type of composite material the angle of elastic untwisting of its peripheral section is smallest with the ·reinforcement layers laid at a zero angle, i.e., all laid parallel to the blade axis. In this case the angle of blade untwisting depends so much more on the modulus of elasticity of the high-modulus reinforced aluminum in tension along the fibers than on the shear modulus that the angle of blade untwisting will increase upon any deviation of the reinforcement layers from the longitudinal profile axis. In the material of the second type, the sharp decrease of the modulus of elasticity in tension, upon some disorientation of layers, is not compensated by some increase of the shear modulus, wich is characteristic of composite materials with a high degree of anisotropy.

The feasibility of laying the reinforcement layers in the composite material of the first type at an angle to the blade axis so as to minimize the untwisting angle of the blade cross sections has been established by calculation of the maximum normal sresses in all reinforcement layers. The graph in Fig.2 depicts the dependence of the maximum normal stresses and of the strength margin under breaking stresses on the volume fraction of crossing layers for three different reinforcement angles (±15°, ±30°, ±45°) in longitudinal and crossing layers.

The number reinforcement layers laid at ±45°,±30° or ±15° angles is determined by the allowable strength margins in specific products.

When the composite material of the third type is used for producing blades, its transverse and shear parameters having rather high values and its properties varying most softly in tension; then it becomes possible to lay many more reinforcement layers at an angle to the radial axis of the blade profile and to lay individual layers at an angle near ±45° so that the shear modulus of the composite material will be increased as much as possible to make the effective torsional rigidity of the profile part of the blade higher. In this case the increase of the shear modulus overrides the decrease of the tensile modulus of elasticity.

Additional evaluations of the influence of changing the linear expansion ratios of the composite material along layers showed that even at constant temperature along blade section in the composite material layers there can occur additional temperature stress beside stresses induced by centrifugal and aerodynamic forces. At nonuniform temperature field the stress distribution disposition can be changed, besides, in consequense of connection of deformation of stretching and torsion in blades, the origin of temperature stresses can influence the character of elastic deformations in blade.

Based on the preceding analysis of the stress-strain state for a model compressors blade made of composite materials, one can conclude that the choice of a rational reinforcement scheme for a statically loaded blade is determined principally by the way in which the properties of the unidirectionally oriented composite material vary with the fiber lay angle to the load line.

LITERATURE CITED

1. I.A.Birger, Design of Blades for Strength [in Russian],.Moscow (1956).
2. I.A.Birger, B.F.Shorr, G.B.Iosilevich, Design of Mashine Parts for Strength [in Russian], 3rd ed.,Moscow (1979).
3. B.F.Shorr,"Design of naturally twisted blades for strength," in: Trans. Central Institute of Aircraft Engine Design [in Russian], No.256, Moscow (1954), pp. 2-18.
4. L.A Gorshkov, "Particularities in the calculation of the state of stress and strain for naturally twisted blades made of composite material," in: Trans. Cetral Institute of Aircraft Engine Design [in Russian], No. 887, Moscow (1980), pp. 77-89.

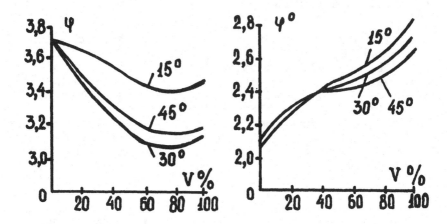

Fig.1 – Untwisting angle of the peripheral blade
section for a blade made of the first (a)
and the second (b) type of composite
material.

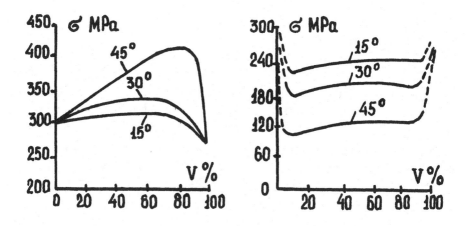

Fig.2 – Maximum normal stresses (a, b) in
longitudinal and in crossing layers.

ADVANCED COMPOSITES IN VEHICLES - CURRENT STATE AND FUTURE PROSPECTS FOR APPLICATION

I. KUCH

Daimler-Benz AG - Postfach 80 02 30 - 7000 Stuttgart 80 - West-Germany

ABSTRACT

The current state of application of fibre-reinforced plastic (FRP)-components in the automobile is demonstrated by the development of hybrid connecting rods and FRP rear axle components.
Examples are given of low volume production for FRP components from which future applications are deduced!

I. INTRODUCTION

Experience gained in the past decade in the field of advanced composites for automobile application proved the industrial applicability of these materials on prototypes.

For the new decade, too, it is a long-term objective to save energy resources and reduce environmental pollution resulting from production and waste disposal, while at the same time improving safety standards, reliability and comfort. For volume production in the automotive industry additional costs involved in the application of lightweight components can be justified only, if all the above-mentioned objectives are taken into account. The characteristic properties of new materials mean that they may fulfill these requirements better than conventional materials.

II HYBRID CONNECTING RODS

Reasons for the development of a FRP connecting rod are possible weight reductions of the rotating and oscillating masses of the connecting rod itself as well as the crankshaft. This leads to increasing engine speed and lower noise emissions.

To solve this task, suitable FRP materials and designs for a highly mechanically stressed part in chemically aggressive service fluids (used oil) at high temperatures up to 170 °C were to be developed.

In order to gain experience under conditions close to practical operation, a 2-litre 4-cylinder standard production engine was selected for testing purposes. It was required to split the connecting rod, similarly to the standard production component.

At the very beginning of the project it had become obvious that, due to a higher parts variety of the split fibre-reinforced composite connecting rod as compared to the steel version, subsequent development into a standard production version would not be possible. However building and testing the components allowed clear conclusions to be drawn about the noise characteristics, the possible fuel savings and the service lives with a large number of load cycles. The experience gained from this can also be applied to other highly stressed components, in particular to engine components.

The split version of the connecting rod, which is provided for in-engine test, consists of tension loop, compression shank made of CFRP and a split crankshaft bearing, small connecting rod eye made of titanium (Figure 1). The weight reduction achieved is approx. 30 % compared with the standard steel version. Tests in hot motor oil have been carried out, the in-engine test will start in summer 1990.

Unsplit versions of connecting rods offer considerable advantages for the design because the material is better utilized and the number of parts needed can be reduced. This results in better weight savings and easier production.

However, modified and more expensive engine concepts are required for this purpose.

Unsplit connecting rods can be produced by directly winding the tension loop onto the crankshaft journals, or in combination with a built-up crankshaft.

They were tested in a Mercedes-Benz M 102 engine. The test had to be finished after 1236 operating hours due to a secondary damage.

III DEVELOPMENT OF REAR AXLE COMPONENTS

The FRP rear axle was selected as a functional prototype with the objective of obtaining advantages through reductions in unsprung masses, making use of the composites special spring-loaded properties and damping qualities, in order to improve handling even further.

Figure 2 shows the development concept towards the realisation of a FRP rear axle. A first step was the design of a composite spring link, which represents the most extremely loaded axle component. It is designed as a rigid part and forms the joining element between the axle support and the wheel carrier.

We are aware of the fact that mere substitution of one metallic part does not allow to make optimum use of the composites positive characteristics, and that a great deal of effort is required to satisfy the geometrical boundary conditions. The most important task, however, which is to prove the suitability of fibre composites for automotive applications, is fully attainable.

Now that proof was furnished by experience gained in manufacturing and operational trials in the first phase, greater modifications to the vehicle are justified in further stages, requiring the substitution of several metallic parts by one composite component. The leaf spring link represents the second step combining the load bearing function of the spring link with the spring-loaded properties of a helical spring. The hinge points of the axle are retained, and axle kinematics are only slightly modified.

However, the objective of developing an overall composite concept as the third step can only be achieved by a complete departure from the metallic design, and by including the axle carrier. It is only after the realisation of this concept that a reasonable comparison of the cost/benefit ratios of steel and composite materials can be made.

Within this development concept a hand-laminated CFRP spring link was manufactured and tested.
After simulation test runs, two components were fitted to a Mercedes 190 E 2.3 and tested over 113 000 km within 2 years. Caused by a strong impact, damage occured and the axle was removed.

The German Aerospace Research Establishment, DLR Stuttgart, developed and manufactured thermoplastic spring links made of carbon fibre structures pre-impregnated with polyetherimide (PEI) and with comingled fabrics of carbon and PEEK fibres.

These spring links were statically and dynamically tested by being subjected to staged operating stresses. No damage occured.

By substituting the metallic spring link and the helical spring by an FRP component glassfibres are used due to their high energy storage capacity and the low material price.
Static and dynamic tests were carried out. Installation and test in a car are provided in summer 1990.

IV FURTHER APPLICATIONS AND FUTURE PROSPECTS

Although at present advanced composites are not yet used in volume production, concepts are available from small and medium production which are suitable for further development.

One example for the volume production of aligned continuous fibre composites is the leaf spring of the Corvette (General Motors Company) which has been in use for several years. Since in Europe there are no leaf spring axle concepts for middle-range and luxury-class cars, this is not a case for substitution. However, this concept is very interesting for vans as well as light and heavy trucks. The consequence of this is that almost all manufacturers of metal leaf springs are cooperating with manufacturers of FRP springs.

In 1985 Ford Motor Company introduced a hybrid propshaft for the Econoline model. Due to capacity and quality problems the entire FRP propshaft production was discontinued in 1986, although it had been possible to confirm unproblematic operation by parallel European productions (Ciba-Geigy). In spite of this, automotive industry is progressing in the development of FRP prophafts in for small volume production. Here, the low tool costs have a favourable effect on the unit price. In addition, the crash characteristics of the overall vehicle can be improved.

For several years, fibre composites have been successfully applied in the bodies of ralley and sports cars.
For this purpose either bodies in white are panelled by
gluing or bolting of advanced fibre composite components
(Audi, Porsche), or primary structures, such as floor
groups (BMW Z1) are additionally made of solid fibre
composite or sandwich structures.

The examples given show that - provided that advanced
fibre composites are applied in accordance with their
load bearing properties - volume production of some
individual components, such as leaf springs, is ahead of
us. However, body structures now as before can only be
produced in small series. In addition, there is a potential for application wherever highest weight-specific
mechanical characteristics are required to fulfill
overall technical concepts, for example in sports and
racing car constructions.

References:

/1/ I. Kuch, P. Kümmerlen, Potential for Lightweight
    Design of a Hybrid Connecting Rod for Automobile
    Application will be published May 1990, SAMPE
    Conference 1990

/2/ I. Kuch, Development of FRP Rear Axle Components
    Materials and Processing, ELSEVIER; Amsterdam 1989

/3/ J. Kretschmer, G. Zoll, Faserverbundwerkstoffe
    im Fahrzeugbau, Fortbildungsseminar der DGM,
    05.-09.03.90, Bad Nauheim

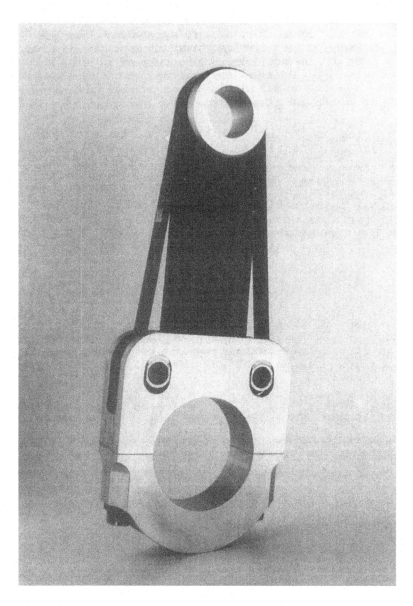

figure 1: Hybrid Connecting Rod

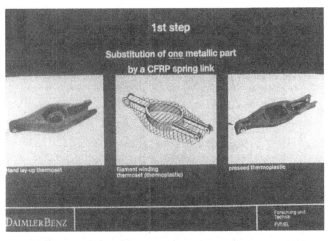

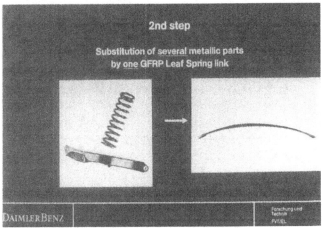

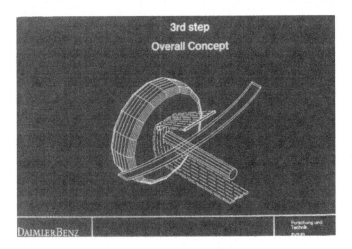

fig. 2: Development Concept for a FRP Rear Axle

THERMAL INSULATION OF CRYOGENIC TANK FOR A SPACE TELESCOPE BY USING A PRETENSIONED SUSPENSION OF FIBER REINFORCED COMPOSITES (FRC)

B. BONGERS, O. HAIDER, W. TAUBER

MBB Unternehmensbereich Hubschrauer Abt HE224
Postfach 80 11 40 - 8000 Munich 80 - West-Germany

ABSTRACT

For the thermal insulation of cryogenic tanks in satellite applications fiber reinforced composite (FRC) materials are preferable because of their low thermal conductivity and high tensile strength compared to metallic materials. At the ISO-satellite the main liquid helium (LHe) tank is suspended by one spatial framework and eight pretensioned chain strands at each side. Frameworks and chain strands are acting as a thermal barrier and therefore made of FRC. To meet the various and, in parts contradictive requirements, sophisticated design approaches are chosen for the structural parts.

INTRODUCTION

In 1988 the European Space Agency (ESA) made the final decision to built the earth orbiting Infrared Space Observatory (ISO).

An essential part of scientific satellites equipped with infrared telescopes are cryogenic tanks providing a supply of LHe. The LHe is used for cooling down the scientific equipment and the infrared telescope to a temperature of approximately 3 K. This avoids screening of the scientific instruments by the characteristic infrared radiation of the satellite itself.

The LHe-tank with its content represents a considerable mass and has to be connected with the main structure of the satellite by an adequate designed suspension.

The main requirements are:
- Transfer of forces due to launch vibration, while specified system-eigenfrequencies are met.
- Thermal insulation of the LHe tank from the satellite structure to minimize cooling losses.

I - SELECTION OF MATERIALS

For a thermal insulation application FRC-materials are preferable because they show a very low thermal conductivity and high tensile strength. A further outstanding property of some types of carbon fibers is the very low or even negative coefficient of thermal expansion parallel to the fiber orientation. Thus it is possible to cover a wide range of coefficients of thermal expansion with adequate tuned laminates. Fig. 2 shows the ratio of stiffness and strength to thermal conductivity at LHe-temperature (4.2 K) and at room temperature. At 4 K both, the ratio of strength and stiffness to thermal conductivity show significant higher values for high-tensile (HT) FRC in comparison with glass fiber composites. At room temperature S-glass- and E-glass-, KEVLAR- and carbon fiber composites have almost the same ratio for the stiffness. For the strength ratio, however, S-glass shows the best values. With respect to these material properties FRC-materials are used for structural parts of the ISO satellite system subjected to wide temperature ranges. First the low transmission of heat from the outer structure to the LHe-tank should give a long mission duration; second the additional thermal stresses should remain small when filling the tank with helium. As the structural parts of FRC are attached directly to the LHe-tank, the thermal and mechanical properties of the materials, especially at cryogenic temperatures have to be investigated.

II- FUNDAMENTAL REQUIREMENTS AND DESIGN OF THE TANK SUSPENSION

2.1 Main Design

The LHe-tank of the ISO is designed as an annular tank with toroidal shaped bottoms at each side. At the transition between the inner cylindrical surface and the bottoms, ring shaped planes provide mounting flanges for a spatial framework used as a final heat barrier (Fig. 3). The framework consists of a quadratic belt which is connected to the LHe-tank by eight corner struts and eight center struts. On each of the four corners of the belt two high loaded chain strands of FRC transfer the loads generated by launch vibration and initial pretension forces onto the outer support structure.

2.2 Fundamental Requirements

For the design of the framework components and the chain strands several requirements, partly contradictive, are governing:
- Transfer of high loads together with high stiffness in order to reach the required eigenfrequencies of the tank suspension.
- Minimize the cross sectional areas keeping in mind the above mentioned requirements. Selection of materials with low heat conductivity giving a low heat transfer via the suspension to the LHe-tank.
- The different coefficients of thermal expansion of the connected materials (LHe-tank made of aluminium, framework and chain strands made of FRC, brackets made of titanium) require a sophisticated design to avoid structural damage by thermal loads.

These requirements are met for the framework and chain strand by selection of optimal combinations of the following parameters:
- Selection of reinforcing fibers
- Selection of suitable semiproducts and manufacturing processes
- Definition of the laminate lay-up for load transfer and load introduction into the structural parts
- Design of the load introduction areas

2.3 Design of structural parts

2.3.1 Brackets

Titanium brackets at the corners of the belt as well as at the connection points and foot points of the struts join the tank to the framework and the framework to the chain strands (Fig. 3).

For the connection brackets titanium is chosen for the following reasons:
- Load transferring elements from various directions meet at these points. Consequently, an isotropic material in this case gives more simple design and manufacturing.
- Among the metals titanium shows low thermal conductivity and high stiffness and strength.

Especially the brackets at the corners of the belt are transferring high tension forces from the belt into the chain strands and have to sustain the compression loads of the corner struts. Six rods have to be joined at each corner of the belt. To minimize bending moments resulting from elastic deformations the bracket design provides one single intersection of the centerline of all joining rods. To prevent losses in stiffness the bolted connections between brackets and FRC-parts were designed to low stress levels (Fig. 1).

2.3.2 Belt

The belt shows a quadratic shape with a solid cross section of 30 mm in square for each belt leg (Fig. 1). Tension loads from the belt legs are transferred into the corner brackets via shear bolts. For the belt this leads to a fiber lay-up with a great portion (83%) of unidirectional fibers oriented along each belt leg and a small fiber portion (17%) of ± 45° fiber orientation. The unidirectional fibers are of a ultra-high-modulus-type (UHM) providing the required tensile stiffness. The fibers with ± 45° orientation are of a high-tensile-type (HT) giving sufficient bearing strength for the bolted connections. To prevent backlash between belt and corner bracket even at cryogenic temperature the bracket flanges are pressed to the belt by two high tensioned bolts at the inner side of the belt corner.

2.3.3 Corner Struts

Each corner strut shows a rectangular solid cross section of 53 mm x 21 mm. At the strut ends spherical bearings with shear bolts are used for mounting the struts to the corner bracket of the belt and to the bracket at the LHe-tank. The spherical bearings avoid any bending moments in the corner strut resulting from relative distortions between belt and LHe-tank due to different thermal expansion during cool-down of the satellite.
The fiber lay-up of the strut is similar to that of the belt. The fiber portion with ± 45° orientation is increased at the ends of the strut providing sufficient bearing strength for the spherical bearings. In the homogeneous part of the strut the major fiber portion is oriented along the longitudinal direction giving the required compression stiffness of the strut. The tapered fiber layers are covered by continuous layers at the outer strut surface avoiding any peel stresses.

2.3.4 Center Struts

The center struts are stiffening the spatial framework in the lateral direction (y-z-plane) showing a rectangular solid cross section and a bolted connection to the belt brackets and to the LHe-tank brackets. The requirements in stiffness and strength are low compared to that of the corner struts. The center struts therefore are manufactured by machining a raw FRC-plate with a lay-up of 80% unidirectional fibers and 20% fibers in ± 45° direction referring to the strut longitudinal direction.

The relative distortions between the strut ends are
compensated by elastic deformation of the strut itself
saving the spherical bearings.

2.3.5   Chain Strands

Each chain strand consists of two double links and
two single links. The chain links are acting as a loop
showing an unidirectional fiber lay-up. A low amount of
woven fabric at each vertex provids sufficient transverse
strength in this area. For the inner chain link HT-carbon
fibers are used, all other links are made of HT-glass
fibers (S-glass).

III   Testing

To verify the selected design, development testing
was done for each part on required overall stiffness and
strength and on local strength respectively.
Component testing was done on a complete spatial
framework verifying required stiffness and strength.
Strength was tested with design load including a safety
factor of 1.5.

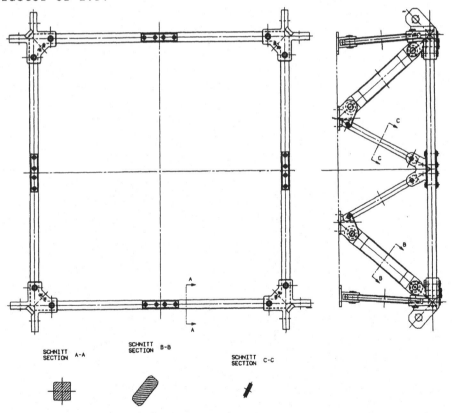

SCHNITT
SECTION A-A

SCHNITT
SECTION B-B

SCHNITT
SECTION C-C

Fig. 1 - Spatial Framework

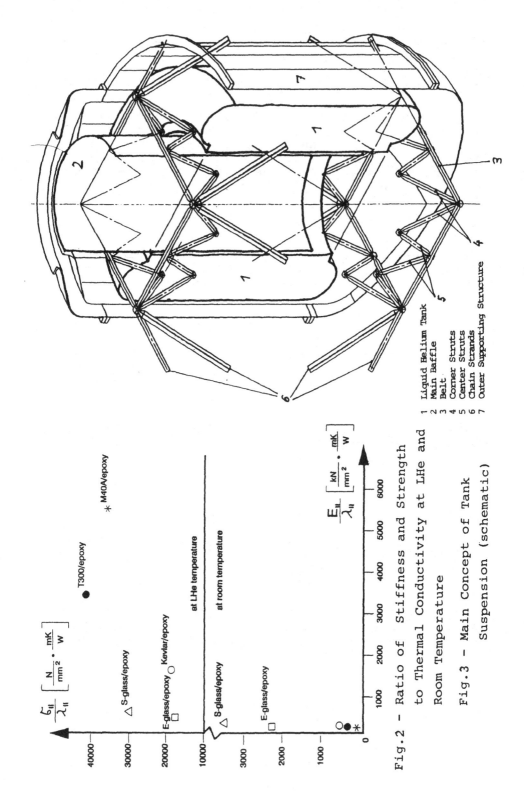

Fig.2 – Ratio of Stiffness and Strength
to Thermal Conductivity at LHe and
Room Temperature

Fig.3 – Main Concept of Tank
Suspension (schematic)

1 Liquid Helium Tank
2 Main Baffle
3 Belt
4 Corner Struts
5 Center Struts
6 Chain Strands
7 Outer Supporting Structure

810

COMPOSITES FOR LIGHT-WEIGHT ROBOT STRUCTURES

H. HALD, F. KOCIAN, P. SCHANZ, R. STEINHEBER

DLR - Institute for Structures and Design
Pfaffenwaldring 38-40 - 7000 Stuttgart 80 - West-Germany

ABSTRACT

As a first composite application the aluminum main arm of an industrial robot was substituted by a GFRP arm made by a one-shot filament winding process. Significant gains in weight, acceleration and load carrying capacity could be achieved.

A second application for a light-weight manipulator system is based on the combination of a new type of joint drives and ultra-light carbon fiber frame structures, which are made by filament winding technique on a lost core. Structural design and results of a FEM-analysis were verified by a static loading test.

1. Introduction

Present robots and manipulator systems mostly operate in a series production line with repetitive moving paths and equivalent load conditions. Although the structural design normally looks more or less like a "dino-saur", end point accuracy can only be realized by calibration schemes, because kinematic errors due to fabrication and backlash in the joints exist [1]. This provides an increasing need for sensor control related to the environment and further progress will lead to relative pointing precision. In fact, this opens the floor for light-weight and more flexible structural design concepts for future1 advanced robot systems leading to improved dynamic properties, higher load carrying capacity and reduced weight.

Cast aluminum arm structures of present robots mostly indicate stress and strain levels of negligible order. Coming to light-weight structural concepts, of whatever kind, stress and strain levels will increase due to a more economical use of material.

In this sense, the use of composite materials primarily because of their high specific stiffness properties is very attractive. To come really into broader use, composite related manufacturing problems, integration of driving units, bearings and electronic equipment first have to be solved by adequate design concepts [2] and [3]. To keep costs at a competitive level, the expensive material has to be used without waste and a light-weight structural design is necessary. Generally, there are two main fabrication routes for composite components:

1. Fabric plies, impregnated either "by hand", injection under pressure, preimpregnated (prepregs) with different kinds of resins, are stacked in molds and cured under certain temperature and pressure conditions leading to "shell-type" structures.

2. Filament winding of endless rovings onto a core offers an automization potential using modern NC-machines. Closed shells as well as frameworks principally can be made. Problems mainly arise from the necessary non-concave surface design, the removability of the enclosed core and the integration of components into the structure. Embedding of prefabricated inserts as rigid attachment points is one way to make an integral one-shot manufacturing process without any post machining.

2. Integral GFRP Robot Arm Structure

As a first application we substituted the aluminum main arm of a MANU-TEC r2 robot with a composite structure [3]. The aluminum arm is an open shell with an attached heavy cap. Considering the structural concepts mentioned before, we chose the filament winding technique. Figure 1 shows the r2 robot with the final GFRP composite main arm.

2.1 Structural design

Calculations of stresses within the aluminum structure, which are due to external loads, structural mass, geometry and system dynamics, indicated a very low stress/strain level combined with a high structural stiffness, i.e., there was no need for use of stiff carbon fibers. Looking at structure costs and the fabricating process we chose glass fibers and an epoxy resin (Bakelite Rütapox L20) as matrix.

Shaping the arm as a closed shell principally allows a drastic increase in natural torsion stiffness, and a filament winding angle of $\sim45°$ combined with additional UD-tapes in the 0°-direction provide sufficient overall stiffness.

Thermal loads may become a problem for a closed composite structure, if heat sources like motors are located inside, due to the lower thermal conductivity compared to aluminum. In the case of the r2 arm the external attachment of the motors prevents heating-up problems.

In contrast to the aluminum arm, internal components are mounted sequentially in the closed shell from the motor side. It was one development goal to avoid post-machining of the composite structure. Thus, we tried to mold the beds for the bearings directly during the manufacturing process. Pretests indicated, that due to the curing process a shrinkage of the bearing bed diameter (150 mm) of less than 0.04 mm (0.026%) occurs, that easily can be compensated by the original mold tolerance; the quality of contour reproduction is very high.

2.2 Core and manufacturing process

To fulfil tolerance requirements for attachment planes and to realize bearing beds in situ a precise segmented steel core was used. The smooth outer surface contour was achieved by means of prefabricated foam pieces which were put onto the steel core and coated afterwards with wax.

The manufacturing process starts with local fabric reinforcements where aluminum inserts for the attachment of the gear boxes (Figure 2) are fixed to the core by screws. Then the filament winding begins and after completion and curing the steel core can be removed.

2.3 Data and test results

Compared to the aluminum structure a mass reduction from 12 kg down to 4 kg could be achieved. The complete integrated arm was attached and tested on an original r2 robot. Neither component integration nor the mechanical tests caused any problems. Although only a single component of the robot was substituted, an increase in acceleration of 13% or an increase in load carrying capacity of 34% was obtained [3].

3. Ultra-light Carbon Fiber Frame Structures for a Manipulator with ROTEX-Geometry

Derived from the German RObot Technology EXperiment ROTEX [4], [5], which is aimed to be flown with the next Spacelab mission in 1991, the development of a terrestrial simulation and training model for the astronauts is going on, because the original flight robot can not operate in a 1-g environment.

3.1 General Structural Design

This model is based on an overall modular design concept with ROTEX geometry and kinematics (6 axes). Combining advanced joint drives with a new gear box design, developed by our colleagues from the "Institute for Flight Systems Dynamics" [1], and ultra-light CFRP frame structures, a lightweight manipulator with high performance characteristics can be achieved. The overall length in stretched configuration is ~1.23 m without end-gripper.

Referring to sections 1 and 2, preliminary calculations indicated that a frame structure is expected to be the most light-weight design concept with specific advantages over shells:

- A certain potential for an automated one-shot manufacturing process exists.
- Load introduction can be realized easily with low additional mass [6].
- Internal heat sources like motors and electronic equipment do not cause any heat accumulation problems.
- As the frame structure consists of unidirectinal reinforced CFRP, it is expected to give a structure without any thermal strain effects, thus, positional accuracy can be assured even in a changing thermal environment, e.g. in space applications.

A first hand-made prototype of such a typical frame structure was made for feasibility demonstration and testing purposes and is shown in Figure 3. The structure has a length of 336 mm between the axes and a mass of only 212 gr. Prefabricated CFRP-shells are imbedded into the continuously wound frame structure in situ. The fabrication was done with HT carbon fibers and the Rütapox L20 resin. First, it was intended to fit the joint drives into these shells with a circumferential friction bearing between gear box housing and CFRP-shell. The load introduction from the gear box driving flange into the CFRP-shell and the frame structure, respectively, should have been achieved by six single pins as shown in Figure 3.

During the development process the complete joint bearing was included in the gear box. Additionally, the risk of thermal induced clamping within the friction bearing, probable heat accumulation problems in the encapsulated joints, the results of a static structure loading test and FE-calculations led to a redesign of the structural attachment to the gear box as well as the structural topology. Figure 7 gives an impression of this redesign compared with the first development step (Figure 3). Now, the load introduction is done by fiber loops which are directly located at nodal points of the framework. The loops are fixed to the aluminum housings by means of prestressed inserts. The structure no longer contains any embedded components and is wound without interruption.

3.2 Manufacturing Technology

The basic component for the fabrication of such a frame structure using filament winding is a "lost core", which is provided with guiding grooves to contain the carbon fiber rovings during the fabrication process. A prefabricated, flexible master form is inserted into a rigid mold (Figure 4). To assure geometrical accuracy of the frame structure, the gear box housing dummies at the joint positions are fixed together by a rig, assembled with the mold and positioned on a precise jig. Now, the core can be poured in layers from a specific gypsum material. Then, the cured core can be attached to a filament winding machine. After curing the CFRP-structure, the rig and the housings can be removed and the plaster core can be crumbled.

3.3 Finite Element Analysis

Our present frame structure is not a framework in the classical sense. Due to the continuously winding process, the single frame struts are connected rigidly, thus, bending moments can be transferred over the knots and a combined longitudinal/bending loading can result for the struts. Only convex or flat surfaces can be realized and slightly curved struts may be induced by geometrical constraints. Thus, only a finite element analysis can quantify the influence on structural stiffness, buckling behaviour and local damage sensitivity.

Calculations were performed with the FE-system PERMAS using the beam elements BECOS and BECOC, whereby composite orthotropy and the Tsai/Wu failure criteria were taken into account. The curvature of each frame strut was simulated by a polygon of four beam elements resulting in 648 elements with 3246 degrees of freedom. The FE-analysis was carried out in [7] for cylindrical frame tubes and for the optimized frame structure as shown in Figure 7.

Obviously, structural stiffness decreases with increasing frame strut curvature because of additionally induced bending moments and deflections within the struts. In the case of the present prototype structure with a typical ratio of slenderness of the struts of ~12 a decrease in stiffness of ~25% is apparent and a more facetted core to give straight struts is recommended.

Comparing the buckling behaviour of a shell tube versus a frame tube, a shell needs a certain amount of circumferential stiffness (90°-winding) to prevent buckling which does not increase the overall structural stiffness. Due to the much higher lateral stiffness of the frame struts, there is no need for an extra anti-buckling reinforcement.

Using composite materials in a frame structure, calculations indicate a significant effect. The higher the fiber stiffness is, the greater is the difference between shear and Young's modulus and the relation E/G, respectively. Now, comparing a frame structure of either a composite or an isotropic material with equal Young's modulus and strength, the same displacements under bending loads will arise when shear deformation effects are ignored. Thus, at a certain load and/or deflection level a lower shear stress level or a greater shear strain will occur in the composite material in contrast to the isotropic one. This provides a different state of stress for the two types of materials and leads to a more uniaxial stress condition in the composite with a better material efficiency. Additionally, a change in load distribution within the single struts in the composite frame structure may occur with increasing loads and displacements, respectively.

Summarizing, a composite frame structure even with very stiff fibers is higher loadable than an equal structure made of an isotropic material. Figure 5 quantifies this effect for the optimized structure of Figure 7 as a relative structural safety factor compared to isotropic materials. The shear loads mentioned above are mainly induced by a non-regular structure topology, that is reflected by the different slope of the curves for the three axes in Figure 5.

If shear yield strength or strain of the matrix is reached first, matrix cracking may occur, leading to a pseudo-reduction of the shear modulus, changes in load path and a reduced overall structural stiffness.

Calculated values for the displacements and the torsion angle of the optimized frame structure are outlined in Figure 6 for the safe load level.

3.4 Static Loading Test

To verify the basic design principles and the FE-calculations, a path-controlled static loading test was carried out with the prototype frame structure. It was fixed rigidly to a flange at one of its ends and was loaded at a certain angle by means of an attached extension rod, simulating the most severe combination of shear, torsion and bending loads.

Figure 6 shows a linear torsion and displacement behaviour up to the safe load level. Further load increase was accompanied by cracking noises and leads to an untypical non-linear deformation behaviour for a CFRP-structure. This may be caused by local matrix shear failure, leading to load path changes within the frame structure, as predicted by FEM; in any case, a safety margin of j = 1.54 was achieved. The final failure was due to to a shear-break of a diagonal strut near a knot, nevertheless the structure was able to withstand a further 74% of the yield load without a catastrophic failure, thus indicating a certain damage tolerance behaviour.

Strain gauge measurements on struts generally indicated a nearly linear stress-strain behaviour up to the yield load; the ultimate strain of corresponding longitudinal struts of the upper and lower side of the structure were

-0.6% and +0.3%, thus, indicating that the compression face struts are deformed additionally by bending.

Looking at the calculated stiffnesses (FEM) in Figure 6, one has to take into account, that they are carried out for the already optimized topology with the new type of load introduction (Figure 7). Due to only a minor change of diagonal framework topology, there exists a good correlation with the measured torsion; bending stiffness will increase significantly compared to the prototype due to longitudinal reinforcements.

Summary

The developments demonstrated and confirmed the feasibility and the basic design goals of two different structural concepts for very light-weight and one-shot fabricated robot arm structures, which are both based on the filament winding technique. Calculations indicated, that composite anisotropy properties are well suited to a frame structure design and stiff, highly loadable and thermally neutral structures can be realized by use of carbon fibers. To assure high and reproducable quality at competitive costs, the next step will be a completely automated, NC production process.

Acknowledgement

The authors want to thank Mr. Lenz and Mr. Nikolopoulos for their engagement in the development process and for their support in the laboratory.

4. Bibliography

[1] Dietrich, J.; Hirzinger, G.; Gombert, B.; Schott, J., *On a Unified Concept for a New Generation of Light-Weight Robots,* First Int. Symposium on Experimental Robotics, Montreal/Kanada, 1989

[2] *Optimierung der Hüllkonstruktion eines Robotergelenkarmes,* "Der Konstrukteur", 9/88

[3] Kocian, F., *Entwicklung, Bau und Test eines Industrieroboterarmes in Faserverbundbauweise (FVW),* DLR, Institut für Bauweisen- und Konstruktionsforschung, Studienarbeit, 1988

[4] Hirzinger, G., *The Telerobotic Concepts of ROTEX - Germany's first Step into Space Robotics* 39th IAF Congress, Bangalore/Indien, 1988

[5] *ROTEX Preliminary Design Review,* Cologne, Feb. 1989

[6] Grüninger, G., *Möglichkeiten der Krafteinleitung in faserverstärkte Bauteile,* Düsseldorf, VDI-Verlag 1977, (VDI-Ges. Kunststofftechnik)

[7] Kocian, F., *FEM-Analyse und -Optimierung eines ultraleichten Roboterarmes in integraler CFK-Gitterbauweise,* DLR, Institut für Bauweisen- und Konstruktionsforschung, Diplomarbeit, 1990

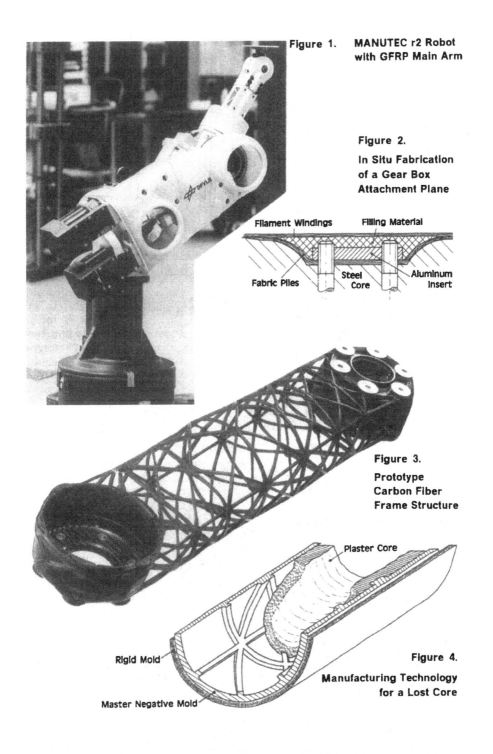

Figure 1.   MANUTEC r2 Robot with GFRP Main Arm

Figure 2.

In Situ Fabrication of a Gear Box Attachment Plane

Filament Windings   Filling Material

Fabric Plies   Steel Core   Aluminum Insert

Figure 3.

Prototype Carbon Fiber Frame Structure

Plaster Core

Rigid Mold

Master Negative Mold

Figure 4.

Manufacturing Technology for a Lost Core

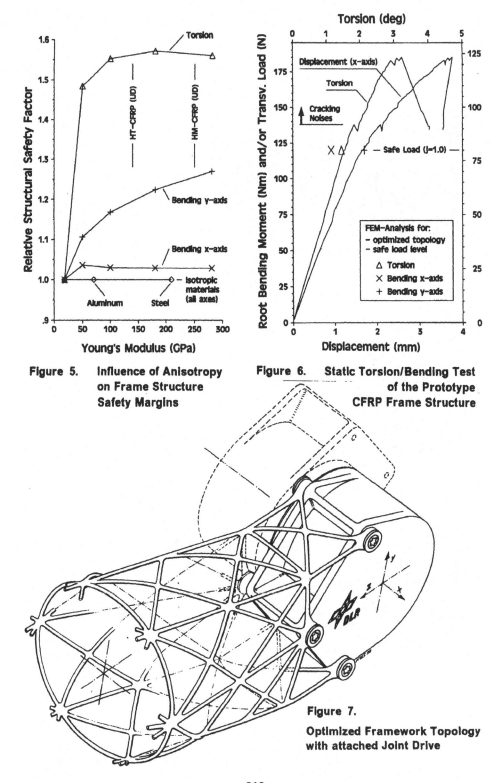

Figure 5. Influence of Anisotropy on Frame Structure Safety Margins

Figure 6. Static Torsion/Bending Test of the Prototype CFRP Frame Structure

Figure 7.

Optimized Framework Topology with attached Joint Drive

MANUFACTURING ENGINEERING COMPONENTS WITH CARBON FIBRE REINFORCED PEEK

P.-Y. JAR, P. DAVIES, W. CANTWELL, C. SCHWARTZ,
J. GRENESTEDT, H.-H. KAUSCH

Ecole Polytechnique Fédérale de Lausanne
32 Chemin de Bellerive - 1007 Lausanne - Switzerland

ABSTRACT

The use of thermoplastic matrix composites allows a wide range of manufacturing methods to be considered for the production of engineering components. However, potential problems are associated with the high forming temperatures and with the use of semi-crystalline matrix materials whose structures may be sensitive to processing history. This paper summarises work to investigate these aspects of thermoplastic composite forming, and presents results showing the influence of forming parameters on short and long term properties of carbon fibre / PEEK laminates.

INTRODUCTION

The potential of thermoplastic matrix composites was recognized in the early 1970's and considerable development was carried out on polysulfone composites [1,2]. However it was not until the commercial introduction of carbon fibre reinforced poly(ether-ether-ketone) (PEEK) in the 1980's that adequate solvent resistance was achieved to make these materials serious contenders for primary aircraft structure. More recently a number of other semi-crystalline thermoplastics have also been proposed as matrix materials and these constitute a very active area of research. The general properties of these materials, such as their good impact resistance, are now well established but less work has been published on processing-property relationships. For a semi-crystalline matrix such as PEEK the important forming parameters are :
a) Forming temperature
b) Forming pressure
c) Hold time at temperature
d) Cooling rate.
Variations in all of these may occur when complex parts are being produced and the forming equipment available may also impose limitations. A study showing the

influence of these parameters on component properties is therefore essential if parts are to be moulded with confidence. This paper summarizes such a study, with particular emphasis on the influence of forming parameters on fracture behaviour. Over 100 interlaminar fracture specimens and around 100 fatigue specimens have been tested in this part of the project. In the limited space available it is not possible to discuss all the experimental procedures, but details may be found in the references.

I   FORMING TEMPERATURE

Previous work has shown the effect of forming temperature on the microstructure and mechanical behaviour of carbon fibre/PEEK formed in an autoclave /3/. Those results suggest that 370°C is the lowest temperature admissible to obtain a well-consolidated flat panel, which is consistent with the manufacturer's recommendation /4/. Work on diaphragm forming has indicated that lower temperatures may be used but the parts were first heated to 385 °C in that work and then cooled to lower temperatures to be formed /5/. In the present work, in order to assess the variations in forming temperature which may be recorded across complex parts, V-shaped panels were moulded using a matched die technique. Three different 16 ply lay-ups were used: (0), (0/90) and (0/90/+45/-45). Typical temperature distributions across a part are shown in Figure 1, while Figure 2 shows a polished cross section of the point of the "V" indicating that consolidation is satisfactory. While it may be possible to compensate to some extent for mould temperature variations, in complicated components it is preferable to know how far the forming window can be extended. A series of flat panels was therefore formed at temperatures of 360, 380, 400, 420 and 440°C.

Differential Scanning Calorimetry (DSC) showed no significant differences in the melting behaviour of samples from these panels. All had degrees of crystallinity around 30% except that formed at 440°C for which 27% was measured. However, DSC is not very sensitive to differences in morphology. Examination in transmitted light of thin sections taken from all panels showed a clear transition in matrix structure from 380 to 400°C, which has been reported previously, e.g./6/. Larger spherulitic structure is observed in specimens formed at the higher temperatures.

Short term interlaminar fracture behaviour was then examined. Mode I (DCB) results are shown in Figure 3. For forming temperatures above 400°C significantly higher initiation values were measured, while at 360°C slightly lower values were obtained. Initiation (from a 13 micron thick starter film) was defined by the onset of non-linearity on the load-displacement plot /7/. Propagation values were independent of temperature. SEM photos from the initiation regions, just in front of the starter film, Figure 4, show clear differences in the matrix behaviour for the panel moulded at 400°C and similar fracture surfaces were observed for specimens moulded at higher temperatures.

An indication of the influence of forming temperature on long term behaviour is given by tension-tension fatigue tests on +/-30° specimens, Figure 5. No effect is apparent in this case.

II   FORMING PRESSURE

Low forming pressure may result in panels with incomplete consolidation. Tests were undertaken to examine the influence of this parameter on properties. For example, panels were moulded applying different pressures during the hold time at

380°C, from 0.8 to 2 MPa, (the recommended pressure is 1.4 MPa), followed by cooling under the recommended 2 MPa pressure, and no change in interlaminar fracture behaviour was observed. In Figure 5 fatigue data for two pressures at the 380°C forming temperature also show no influence of pressure in this step of the process.

III   HOLD TIME AT FORMING TEMPERATURE

Panels have been produced using the standard forming cycle but holding at temperature for 30 minutes instead of 5 minutes. Previous work has shown that this longer hold time results in fewer nucleation sites, more nucleation on fibres and a stronger fibre-matrix interface /8/. No effect was found on the fracture behaviour here ; a mean mode I initiation value of 1300 J/m^2 was measured for the longer hold time.

IV   COOLING RATE

The influence of cooling rate on both structure and on short and long term properties has been described in detail elsewhere /9/, and will not be discussed at length here. This parameter was initially thought to be critical in determining the fracture resistance of the material due to its influence on the degree of crystalline perfection. Indeed slower cooling rates do promote higher degrees of crystallinity, but the range of cooling rates likely to be encountered in forming processes do not appear to be detrimental to the fracture resistance of current commercial material.

More work is needed however to quantify the effect of fast cooling rates on engineering properties of laminates as high internal stresses can develop in these materials /9,10/.

ACKNOWLEDGEMENTS

This project was supported by the Swiss Fonds National project NFP19. The technical assistance of Mr. Serge Fleury of the Polymers Laboratory,EPFL, Brian Senior of the Institut Interdépartmental de Microscopie, EPFL, and of Mrs. G. Keser of ABB, Baden, is gratefully acknowledged.

REFERENCES

1 - Hoggatt JT, Proc. 20th Nat. SAMPE Symp., May (1975), 606.
2 - Maximovich MG, Proc. 19th SAMPE Symp., (1974), 262.
3 - O'Bradaigh CM, Mallon PJ, Comp. Sci. & Tech., 35 (1989) 235.
4 - ICI Fiberite data sheets
5 - Manson J-A E, Seferis JC, J. Thermoplastic Comp. Mats., 2 Jan. (1989) 34.
6 - Blundell DJ, Crick RA, Fife B, Peacock J, Keller A, Waddon A, J. Mat. Sci. 24, (1989) 2057.
7 - Davies P, Cantwell W, Kausch HH, Comp. Sci. & Tech., 35 (1989) 301.
8 - Lee Y, Porter RS, Poly. Eng. & Sci., 26, 9, (1986) 633.
9 - Davies P, Cantwell W, Jar P-Y, Richard H, Neville DJ, Kausch HH, "Cooling Rate Effects in Carbon Fibre/PEEK composites", presented at 3rd ASTM Symp. on Fatigue & Fracture, Orlando Nov. (1989).
10 - Jeronimidis G, Parkyn AT, J. Comp. Mats. 22, May (1988) 404.

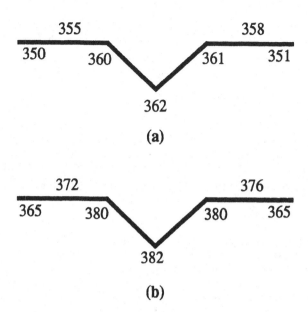

Figure 1. Temperature profiles measured at the centre of 120mm wide, 200 mm long mould for V-shaped panels. Set temperatures : a) 360 °C, b) 380 °C

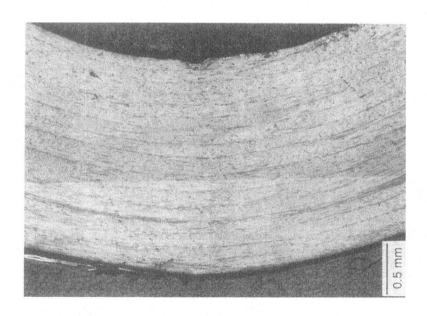

Figure 2. Section through the "V" point of a unidirectional panel formed at 380 °C, showing consolidation.

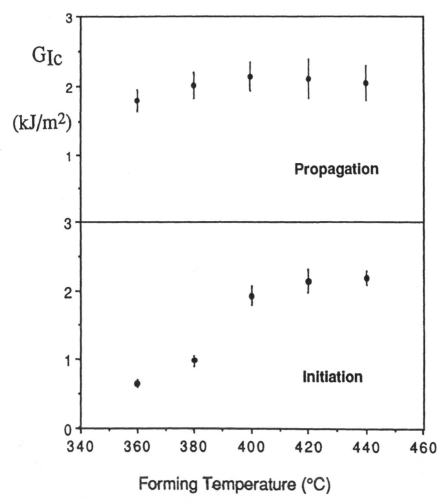

Figure 3. Mode I DCB results for flat 24-ply panels moulded at different temperatures. Initiation from 13 micron films. Propagation values correspond to R-curve plateau. Mean, maximum and minimum values are shown.

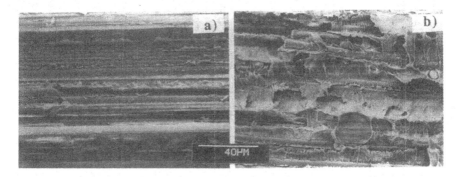

Figure 4. Scanning electron micrographs taken from initiation regions of DCB specimen fracture surfaces. a) 380 $^{\circ}$C (360 $^{\circ}$C specimens similar) b) 400 $^{\circ}$C (420 and 440 $^{\circ}$C similar).

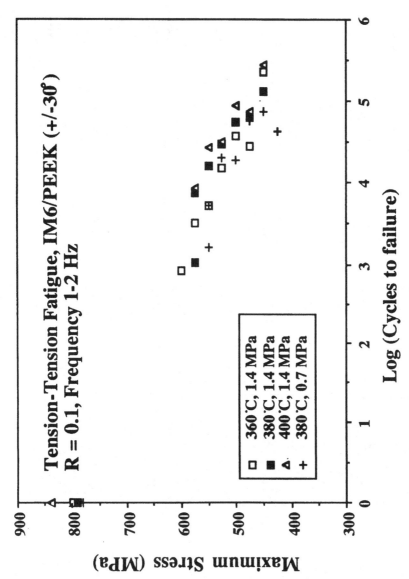

Figure 5. Examples of fatigue results showing the influence of forming temperature and forming pressure. Frequency selected to keep surface temperature below 35°C.

DETERMINATION OF OPERATIONAL FAILSAFE CHARACTERISTICS
OF THE SAILPLANE OF COMPOSITE MATERIALS

V. PAULAUSKAS, J. BAREISIS*, A. USHAKOV**

Construction Designing Bureau "sports Aviation"
Spalio Revoliucijos 11-18 - 234340 Prienai - Lithuania
*Kaunas Polytechnical Institute
**Moscow CAHI

ABSTRACT

The paper presents operational survivability characteristics investigation methods for the LAK - 12 "Lietuva" sailplane parts made of composite materials in certification stage. Rated conditions, controllability and maintainability characteristics have been determined. The paper includes the full-scale sailplane units survivability test results. Operational survivability characteristics of the composite materials (CM) parts determined by way of theoretical and experimental investigation provide the specified in-service safety level of the LAK-12 "Lietuva" sailplane during its entire life time.

INTRODUCTION

We chose for investigation object the most critical and highly loaded parts of sailplane LAK-12 "Lietuva" made of CM - such as wing and stabilizer integrity of which substantially determine safe life of the sailplane.

The wing panel and stabilizer enclosure represent a three-layer structure covered by glass plastic and epoxy resin skin. The spar caps are made of unidirectional carbon plastics.

The main criterion for providing the safe life of the sailplane parts made of CM is "an operational survivability" which, in the stage of designing, is predicted by design technological means, further is substantiated by calculation and experimental analysis in certification stage and is maintained in use by periodical integrity checking and recovery of the construction strength (repairs).

I - RATED CONDITIONS

The desing conditions in certification stage determine amount and type of the load applied, also climatic factors spectrum influence upon properties of the CM parts during the full-scale structures tests for failure of the CM parts.

They are formed on the basis of the technical requirements according to the wide-scale production data and analogue constructions service, also sample test data.

The desing conditions include :

• rated conditions related to climatic factors spectrum influence upon crack resistance of CM parts,

• rated conditions related to damage ability,

• safety factors related to residual strength of the CM parts.

1.1. Rated conditions related to climatic factors spectrum influence upon crack resistance

The rated conditions related to climatic factors influence are formed with reference to basing area data and flight mission in intended service and comprise :

a/ Climatic factors spectrum variation extreme values :

- ambient temperature t_{min} = - 10°C, t_{max} = 54°C
- ambient air humidity P_{min} = 20 %, P_{max} = 98 %
- atmospheric pressure p_o = 760 Hg

b/ Heating and moisture content of the CM parts corresponding to the extreme climatic conditions

Taking into account good thermoinsulating properties of the foam plastic used as a filler in the enclosures of the wing panels and the stabilizer, also availability of water balast in the wing torsion boxes one may conclude that only external glass plastic skin of the wing panel and the stabilizer enclosure will be subjected to intense heating due to sun radiation. At the specified maximum ambient temperature value we assume t_r = 104°C as rated heating temperature. For the rest parts of the wing and the stabilizer made of CM including carbon plastic materials we assume t_r = 22°C as rated heating temperature value. Equilibrium moisture content level in glass plastics and carbon plastics on the base of epoxy matrices is assumed to be equal to G = 0.84 % in service which corresponds to a wide range of climatic factors spectrum extreme values variation.

1.2. Rated conditions related to damage - ability

The desing conditions related to damage ability in certification stage include :

• dimensions and recommendations on model building of theoretical technological faults (appearing in all the stages of part manufacturing) that exist in the part at the moment of its putting into service and would not be revealed during its life,

• dimensions and recommendations for model building on in - service damages due to middle and low speed impacts on the parts.

We assume $2L_{adm}$ as a rated dimension of the technological fault beginning with which we can reliably detect the fault in production process by instrumental checking, also checking by such nondestructive methods as acoustic, XR, impedance. All intermediate products (elements, parts) are subjected to postoperational inspection. It is assumed that defects of $2L>2L_{adm}$ dimension are eliminated by repair or rejection. The $2L_{adm}$ value has been set on the base of the data characterizing efficiency of the test methods used in wide scale production, also considering the production faults list for the LAK-12 sailplanes made by the results of technical inspections during the sailplane manufacture stage, by inspecting repair requiring sailplanes, also by the comments received from the clubs during the period of 1983-1986.

1.3. Safety factors related to residual strength of the CM parts

With reference to the methods given in /1/ the required safety level of the CM parts according to "operational survivability" conditions is provided by inserting safety factors related to residual strength f_{rs}. At certification trial of full-scale constructions life time and residual strength of the CM parts should be confirmed for the main cases of loading :

a/ when design technological defects $2L_{adm}$ and operational damages not revealed during its life time are available. For this CM part damage case the recommended value in (1) is $f^a_{rs} = 1.5$. Consequently, the full-scale structures of the wing panels and stabilizer with the abovesaid damages after fatigue tests in time corresponding to its life time and taking into consideration the specified safety factors, should maintain the allowed residual strength

$$(P_s{}^{adm})^a = 1.5 * P_o{}^{max} * \frac{f_{cr}}{K_t * K_h} \tag{1}$$

Were : $P_o{}^{max}$ - maximum operational load,

f_{cr} - factor evaluating reduction of crack resistance characteristics for the CM parts at a long-term complex influence of climatic factors in entire service life,

K_t, K_h - factors taking account of short-term reduction of crack resistance characteristics due to influence of temperature and humidity.

b/ When operational damage of the CM parts is obvious after general inspection and the moment of damage appearance is fixed by the glider pilot. According to recommendations given in (1) for this CM part damage case $f_{rs} = 0.67$

Among the abovesaid damages there is the damage caused by bird encountering. Consequently, trial of the full-scale construction, particularly, wing panels and stabilizer, with leading edge, operational damage of $2L°_{adm} = 100$ mm has shown that allowed residual strength :

$$(P_s{}^{adm})^o = 0.67 * P_o max * \frac{f_{cr}}{K_t * K_h} \tag{2}$$

is sufficient.

If during the full-scale constructions tests influence of the climatic factors is simulated then $f_{cr} = K_t = K_h = 1$.

While testing the wing panels and stabilizer climatic factors impact is not simulated, therefore according to rated conditions related to climatic factors spectrum impact on the CM parts crack resistance we assume for compressed zones of the construction $f^c{}_{cr} = 1.00$, $K^c{}_t = 1.00$, $K^c{}_h = 0.83$, and for extended zones of the construction $f^e{}_{cr} = 1.00$, $K^e{}_t = 1.00$, $K^e{}_h = 1.00$. After inserting the above said values into equations (1,2) we obtain for compressed construction zones :

$(P_s{}^{adm})^a = 1.8 * P_o{}^{max} = 120 \% P_r$

$(P_s{}^{adm})^o = 0.8 * P_o{}^{max} = 53 \% P_r$

for extended structures :

$(P_s{}^{adm})^a = 1.5 * P_o{}^{max} = 100 \% P_r$

$(P_s{}^{adm})^o = 0.67 * P_o{}^{max} = 44 \% P_r$

where P_r - rated load.

II - SPECIFIED CHARACTERISTICS OF CONTROLLABILITY

Generally, the interval between inspections of the CM parts is determined by intensity of operational damage apperance, inspection reliability, in-service stresses, CM parts crack resistance. Taking into consideration the sailplane service conditions (wide dispersion of the clubs, lack of instruments and skilled staff for carrying out inspections) dimensions of rated technological and in-service damage are chosen so that relatively great damage of the CM parts does not result in safety level drop in comparison with he specified value.

Fig. 1 shows requirements for controllability of wing panel and stabilizer structure sections and areas in terms of detection probability dependence p(2L) upon damage dimension at instrumental test and visual inspection.

The repaired CM parts after fatigue test carried out in time corresponding to service life and taking into account the specified reliability factors related to longevity should maintain admissible $(P_s^{adm})^a$, determined from equaltion (1). This condition sets requirements for efficiency of the damaged CM parts repair methods.

III - EXPERIMENTAL DETERMINATION OF THE RESIDUAL STRENGTH CHARACTERISTICS FOR DAMAGED CM PARTS

Two wing panels and one stabilizer of the LAK-12 sail-plane have been to tested.

The wing test program included loading in sine cycle of variable sign by applying force simulating varying load of + 5.3 and - 3.0 in flight.

Rated technological faults and in-service damage were applied to wings. After running the specified number of loading cycles both wing panels endured 125 % of rated static load without obvious residual damage or strain.

Artificially applied damages in high load spots of the structures did not expand in loading process.

Rigidity and acting stresses variation was within measurement dispersion.

It allows to conclude that operational survivability characteristics of the CM parts determined by way of theoretical and experimental investigation provide the specified in-service safety level of the sailplane LAK - 12 "Lietuva" during its entire life time.

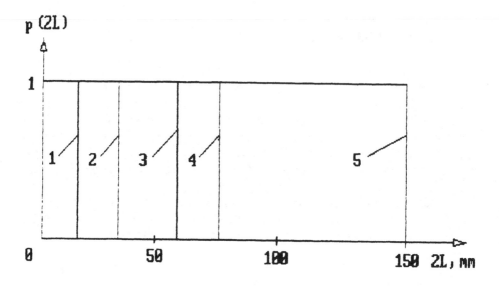

Fig. 1 Requirements for controllability of the CM parts :

1/ Special instrumental checking of the wing spar caps (crack and delamination detection),

2/ Special visual inspection of wing ribs (crack detection),

3/ Special visual inspection of the wing ribs (delamination detection),

4/ General complex (visual and instrumental inspection) of the three-
layer wing and stabilizer enclosures in junction area,

5/ General complex inspection of the three-layer wing and stabilizer enclosures in regular area.

REFERENCES

1/ SELIKHOV A.F. and USHAKOV A.E. (1988) : The probability failure model of damaged composite structures. Presented at Inter. Conf. on Composite Materials and
structures. FRP Research Center, Indian Inst. Technol., Madras, India.

2/ Study of specified life renewal of the sailplane LAK-12 "Lietuva". Scientific-techni-
cal report N° 197, Construction-designing bureau "Sports Aviation", Prienai, Lithuania, 1988

STRUCTURAL COMPONENTS OF FIBRE REINFORCED THERMOPLASTICS

W. WERNER

Dornier Luftfahrt GmbH - PO Box 1303
7990 Friedrichshafen - West-Germany

ABSTRACT

A new advanced production process for aircraft structural components of carbon fiber reinforced thermoplastics (CFRT) has been developed in several studies. For the first time, the new carbon fiber reinforced plastics (CFRP) composite has successfully met the far-reaching requirements of the German aviation authority as a horizontal stabilizer leading-edge profile and as an air brake. The components are now being tested in regular flight operation.

I - THERMOPLASTIC SEMI-FINISHED PRODUCTS FOR STRUCTURAL COMPONENTS

In conventional CFRP processing, carbon fibers are positioned into random or directed patterns to form webs which are impregnated with synthetic resin components. These impregnated woven or non-woven materials are called prepregs or semi-finished products. The resin component (matrix) used in prepreg manufacturing usually is an epoxy resin, i.e. a thermoset plastic. Depending on the desired characteristics, different resins can be mixed to obtain a matrix system. For the new CFRT composite, a matrix system has been used that does not consist of thermoset resins but of modern thermoplastic resins. Thermoplastic matrices differ from thermoset resin matrices, for example, by their better impact resistance, higher elongation at rupture, improved erosion resistance and, sometimes, higher temperature resistance and unlimited shelf life.

Unlike thermosetting resins, thermoplastics can be repeatedly molten and, in principle, molded, fused or formed much like metals. As the thermoplastic prepregs available to date are not sticky and, therefore, cannot be draped in molds, new production and processing methods have been developed. Both the autoclave and the molding technique have been successfully tested. Combined with infrared radiators and hot-gas systems, the production cycles can be reduced when compared with the processing of fibre reinforced thermosets, depending on the component dimensions.

At Dornier, the thermoplastic resins polyether ether ketone (PEEK) and polyether imide (PEI) have been studied for structural applications and have been found suitable. Experience from several technology programs is available for these materials. PEEK is used in the unidirectional carbon fiber prepreg of the Fiberite/ICI Company, offered under the tradename of APC-2. In cooperation with the material manufacturer and the official certification authority, this prepreg has been qualified for flight certification of two of the Alpha Jet's structural elements: the air brake and the horizontal stabilizer leading-edge box. PEI is offered as a woven prepreg, for example by Ten Cate.

II - ALPHA JET HORIZONTAL STABILIZER LEADING-EDGE BOX AND AIR BRAKE

In one study, a horizontal stabilizer leading-edge box and an air brake for the Alpha Jet were made of carbon fiber reinforced thermoplastics and certified by the official German authority. The study comprised production development of the thermoplastic leading-edge box and the air brake, manufacturing of the components, demonstration tests, and flight certification by the official authority. After the successful opening of the flight envelope, both components are now under endurance testing in normal troop service. The Alpha Jet is an aircraft which offers long-term experience with the use of fiber reinforced plastic components, for example in production air brakes, endurance testing of the rudder and flight testing of the CFRP horizontal stabilizer.

With the testing of the horizontal stabilizer leading-edge box over a prolonged period, problems with regard to impact and erosion behavior of comparable components such as the wing and the fin are covered as well. The dimensions of the leading-edge skin were taken from the sheet-metal skin.

The Alpha Jet's air brakes can be considered to be a typical structural component with the characteristics of: double-curved surfaces, high load density, concentrated load applications, and joining by rivets and bonding. Air brake production can be subdivided into the three main sectors of spar production, shell production, and joining of the spar, shell, and metal parts. For the design of the thermoplastic air brake, the existing dimensions of the production air brake with epoxy matrix were basically maintained. The individual parts of the air brake were manufactured with the production toolings.

III - USE OF FIBER REINFORCED THERMOPLASTICS IN CIVIL AIRCRAFT PROGRAMS

The use of fiber reinforced thermoplastics for components up to medium size seems to be most promising when the molding technique is applied. Prepressed fiber reinforced thermoplastic plates are heated with radiators and formed in the cold mold. Production cycles in the minute range can thus be obtained. This production process can be automated in principle. Prepressed thermoplastic plates can be bought from the manufacturer and stored indefinitely. In the first processing station, the plates are cut to the final dimensions of the component. The next stations are: heating, molding, and removal of the finished components. Several molding processes in sequence can also be imagined, much like in metal processing.

After discussion with the departments for design, work planning, and value analysis, ribs were selected as suitable components for the use of fiber reinforced thermoplastics. Here, the existing know-how gained in former studies of fiber reinforced thermoplastic molding could be put to use. Therefore, the ribs of the Dornier 328 landing flap will be made of fiber reinforced thermoplastics. This would be the first application of thermoplastics for a structural part in a civil aircraft. Another possible application would be the landing flap ribs of the Airbus A 321.

IV - OUTLOOK

Fiber reinforced thermoplastics will be of increasing importance. As fiber reinforced thermoplastics can be repeatedly molten and processed much like metals, economic production processes are feasible. This includes, for example, a filament winding process, where the thermoplastic tape directly molten onto the component under production so that no subsequent curing is required. In the long term, this may lead to a significant reduction of the production costs of structural components. Dornier intends to keep a leading role in the development and provision of new production technologies for fiber reinforced thermoplastics in Europe.

Leading-edge box of the Alpha Jet's horizontal stabilizer made of
carbon fiber reinforced thermoplastics
(Luftbildfreigabe-Nr. Regierung OBB.GS-300/8506/80)

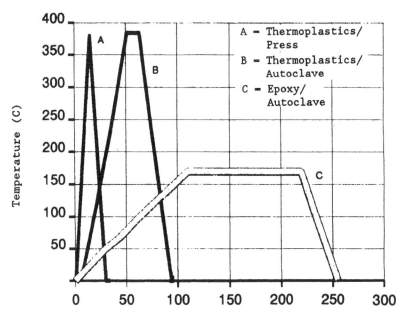

Comparison of the production cycles of thermoplastics and epoxy
resins

PROPERTIES
EIGENSCHAFTEN

Chairmen :　　**Prof. J.F. ARENDTS**
University of Stuttgart
Dr. A. SAVADORI
Enichem

IMPROVEMENT ON COMPOSITE CONNECTING-ROD DESIGN

I. CRIVELLI-VISCONTI, A. LANGELLA, L. NELE, G. DELLA VALLE

University of Naples - Piazzale Tecchio - 80125 Naples - Italy

ABSTRACT

The use of composite connecting rod in alternative engines offers noticeable advantages such as inertial forces reductions and high fatigue resistance.
Several studies have shown the possibility to manufacture composite connecting rods satisfying the static load resistance requirement. The past research had been carried out without limitations concerning geometry and dimensions of the conrod.
In this study the Authors use a finite element method on several models of carbo-resin conrod for a two strokes one-cylinder engine which show geometry and dimensions according to the production engine.
Several prototypes have been realized and the theoretical results have been substantially confirmed by the experimental analysis.

1. INTRODUCTION

Polymeric matrix composite materials are more and more used in engine applications to realize mechanical structural components.
It's a long time since several researchers proposed a study on composite materials to obtain propeller shaft, connecting rods, pistons, etc..
In this framework this study can be inserted, as a continuation of previous works [1], aiming to define optimum shape and fabrication technology of conrod for a single-cylinder engine, to permit its immediate use on previously existing engines.

In fact, at the present moment of automobile industry it is easier to provide for the only substitution of a steel component with a composite material, rather than redesign the whole engine with composite material. A composite material conrod must be designed by considering some fundamental aspects such as the dimension, the stiffness and the strength, and no last the high temperature operation for a composite (~ 200°C). In this regard, studies are reported (2,3) on carbon fiber composites with polyimide resin which have given excellent results in oil immersion tests for a period of 1000 h at 180°. It has been noticed neither a notable loss of mechanical properties nor a significant oil corrosive etch the composite.

In this work, however, we have been mainly concerned with the shape of the conrod, while aiming at a technological cycle to obtain composite material conrod using more traditional epoxy resins rather than polyimide ones.

2. ACTING FORCES ON CONNECTING ROD

This work refers to a CAGIVA two-cycle engine with the following data:

S	piston surface	660.5 mm^2
r	radius of crank	25 mm
l	conrod length	48 mm
δ	r/l	0.52
m_a	mass of organs in alternate motion	255 g
m_b	mass of conrod	150 g
W	angular speed	$1{,}047 \text{ rads}^{-1}$
P	maximum pressure	81.5 daN/mm^2

tab. 1

Acting forces on conrod are due to gases pressure in the cylinder and to inertial forces, while friction forces can be neglected for strength design.

The connecting part of two rod eyes is fundamentally stressed to traction and compression.

Maximum tensile and compression stresses, on which it is normally carried out the conrods design, are achieved in practice by considering the beginning of intake phase and the blast and they are:

$$P_{traz} = m_a r w^2 (1+\delta) + m_b r w^2 (1+\delta/2) = 1{,}362 \text{ daN}$$

$$P_{comp} = P_{max} S = 2{,}008 \text{ daN}$$

Since slenderness ratio δ is smaller than the minimum value of 60, the Eulerian load determination is not necessary [4].

Therefore, the design of conrod is bound to traction and compression alternating stresses.

3. DESIGN CONDITIONS AND DIMENSIONS

The design of a composite material element, such to substitute a steel one, develops in two phases. In the first place it is necessary that the two components have the same longitudinal or flexional stiffness. Therefore, from the stiffness of the steel element material's characteristics, piece geometry and stacking sequence can be calculated.

In the next phase the resistance of the conceived structure is verified in the most dangerous points.

However, in our case it was not possible to design for an equal stiffness structure since it is necessary that the conrod has to be inserted in CAGIVA engine with no essential changes.

In the present study, we were forced to fix element shape and geometry based on the available space, and then determine the best stacking sequence for the strength and stiffness conditions.

Figure 1 shows a detail of the engine section, with the steel conrod at limit positions with respect to the cylinder internal walls and the shape of composite rod, obtained imposing the lowest distance between rod lateral surface and cylinder walls.

4. CONROD STRUCTURES

Two possible shapes of conrods have been selected.

The conrod with external loop consists of two indipendent elements: an external loop to absorb traction and flexure loads, with inside a connecting part of two rod eyes to support compression loads (fig. 2a). The external loop of the loop conrod was made with a unidirectional composite ribbon, while the central body was made with a laminated composite plate.

The design of the one piece-conrod (fig. 2b) was carried out assuming a single plate, whose geometry depends on the maximum dimensions showed in Fig. 1.

In both conrods, two antifriction axle boxes are inserted in the two eyes. The thickness of two conrods is bound on the shape and dimensions of the engine shaft.

We have employed, in this study, a commercial material, TORAYCA T400#3620, unidirectional pre-preg whose characteristics are in the following table 2:

orientation	elasticity modulus (daN/mm^2)			strength (daN/mm^2)		
	tensile	comp.	shear	tensile	comp.	shear
0°	14000	13000	500	240	160	11
90°	900	900	500	55	30	11

Tab. 2 - Material characteristics

5. FINITE ELEMENT ANALYSIS

The stress and strain states rising in the loaded structure were calculated with finite element method.

Four nodes quadrilateral plane elements were used for conrod modelling. The piston pin and the pin were modelled with beam elements. These elements have been assumed an of infinite rigidity and are convergent in the center of the piston pin and of the pin.

These beam elements were used to distribute the compressive and tensile loads on the conrod eyes, so that, in this way, we cold simulate the real loads distribution on the eyes. The mesh used for the one piece conrod for the tensile and compressive load are respectively reported in fig. 3 and fig. 4.

The finite element analysis showed that, for both conrod, the normal stress had very low values.

Considering this results the shear stresses are preeminent in the failure criterion. Some exemples of the shear stress distribution in the loop conrod and one piece conrod are reported in fig. 5 and fig. 6.

In tab. 3 and tab. 4 the stacking sequence and thicknesses and the normal and maximum shear stresses respectively for the loop conrod and for one piece conrod are reported.

The stacking sequence and thicknesses have been choosen in order to minimize the shear stresses.

These shear stress values should assure a good behaviour both for dynamic load and for resistance reduction due to technology processes.

Laminae	thickness (mm)	σ_1	σ_2 (daN/mm^2)	τ_{12}
0°	3.0	7.9	2.45	2.16
+45°	1.25	54.4	0.51	1.19
-45°	1.25	-5.4	3.01	-1.19
90°	0.5	41.0	1.07	-2.16

Tab. 3 - Loop conrod

Laminae	thickness (mm)	σ_1	σ_2 (daN/mm^2)	τ_{12}
0°	3.0	16.3	0.5	-2.1
+45°	1.0	-19.6	2.0	-0.5
-45°	1.0	37.5	-0.5	0.5
90°	1.0	1.64	1.07	2.1

Tab. 4 - One piece conrod

6. MANUFACTURING TECHNOLOGY OF PROTOTYPES

The employed technology to realize the conrod weighs heavily on the final cost and this incidence is valued about 70% of total cost.

This incidence is due to manufacturing technology being too slow for mass production, while automation costs could be easily recovered at the high production rate of automobile industry.

In this framework we have cosidered the possibility of automated technologies during:
a) vacuum moulding
b) ribbon winding

However our prototypes have been manually made with no particular equipment.

6.1 LOOP CONNECTING ROD

This conrod consists of two parts: a central body and an external loop.

The central part is made with vacuum stamping by the following operations:
- laminae cutting
- laminate deposition
- polymerization

Each lamina was manually cut using simple devices.

The laminate deposition was carried out compacting each lamina with vacuum permanence for about 1'

On the central body, so obtained, two metal axle boxes were placed into the conrod.

In this phase a mold was used , fig. 7, which has two parallel pins necessary for a correct positioning of the axle boxes. A unidirectional composite ribbon is then wound around the central body and axel boxes.

The loop, so obtained, reaches the final shape, leaning it self on the central body, by the action of two lateral wedges.

The loop conrod presents a weight reduction of 53% respect to the metal one.

6.2 ONE PIECE CONROD

This conrod is made with vacuum moulding by the following operations:
- laminae cutting
- laminate deposition
- polymerization
- drilling and inserting axle boxes.

The latter operation was very important and tools and cutting parameters had to be carefully choosen to avoid surface damage.

6.3 TECHNOLOGICAL CYCLE

The numerical definition of the cycle has been performed by considering the actual dimension of the pre-preg layers, and the final thickness of the parts.

Assuming, as it was the case in the present study, that no excess resin was present in the pre-preg, the applied pressure was calculated by considering the total number of layers, and the shrinkage necessary to reach the final thickness. The pressure applied by closing the mold, shold have produced a Vf around 60%.

The tension on the pre-preg ribbon around the central body was of the order of a few daN/cm.

7. CONCLUSIONS

The two types of conrod have been subjected to static traction and compression test showing the feasibility of the design.

We are now on the way of performing actual testing mounting the conrods on real engines in order to verify the dynamic behaviour.

It can be useful to note that while the one piece conrod can be used in monocylinder engines, the loop conrod is more easily able to modified is use on a multicylinder engine is to be considered.

REFERENCES

1 - I. Crivelli Visconti, A. Langella and C. Fernandes "Study of a connecting rod in composite materials", Advancing with composites, Milan 1987.

2 - R. Kruger and H.D. Beckmann, "Gasoline engine components made of CFRP", Elsevier Science Pubblishers, Amsterdam 1985.

3 - J. Kretschmer, "Load bearing parts in ground car vehicles out of fiber reinforced plastics", Elsevier Science Pubblishers, Amsterdam 1986.

4 - R. Giovannozzi, Costruzioni di macchine, Patron ed.

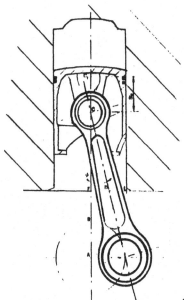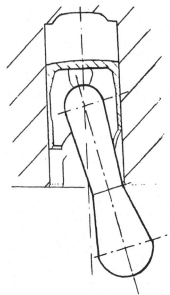

fig. 1

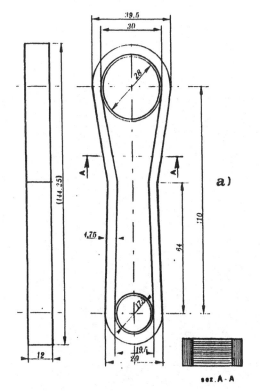

a)

sez. A - A

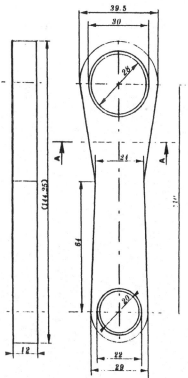

b)

sez. A /

fig. 2

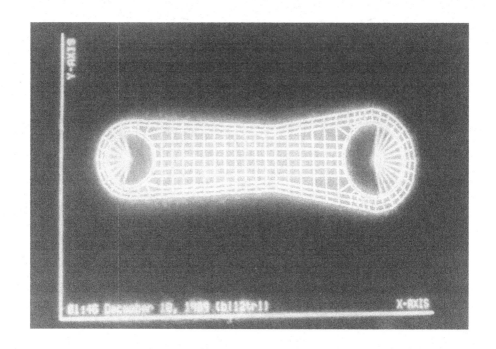

fig. 3 - Mesh for tensile load, one piece conrod

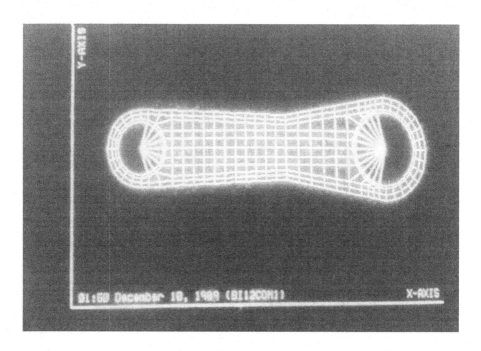

fig. 4 - Mesh for compression load, one piece conrod

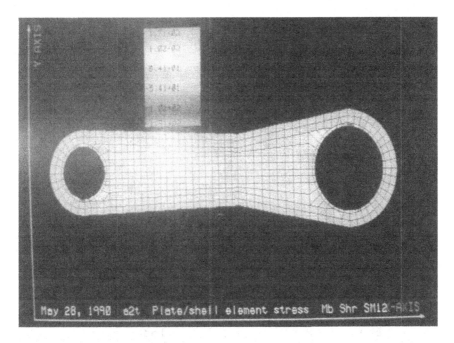

fig. 5 - Shear distribution for traction loads - loop conrod

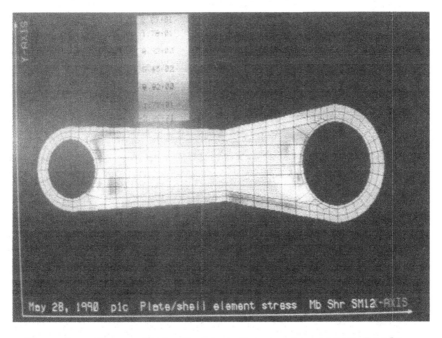

fig. 6 - Shear distribution for compression loads - one piece conrod

fig. 7 - Mould used for the loop conrod

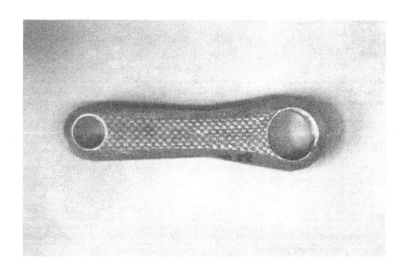

fig. 8 - Loop conrod

3-D REINFORCING FABRICS FOR MONOLITHIE AND SANDWICH-COMPOSITES

W. BÖTTGER

Vorwerk & Co. Möbelstoffwerke GmbH & Co. KG
Vorwerkstr. 4 - 8650 Kulmbach - West-Germany

ABSTRACT

New developments of textile structures made of high efficiency reinforcement materials as glass, aramide, carbon and ceramic offer new possibilities for manufacturing composites.

Introduction

The rationalization at manufacturing composites makes it necessary to use made-to-measure reinforcement textiles. In addition to this, it is essential to lay out the structures according to the requested mechanically strains, i.e. to design application orientated. In a close co-operation with MBB, the Institut für Flugzeugbau, the University Stuttgart and Vorwerk new reinforcement structures have been developed within the scope of a BMFT-research programme, which are making allowance for these requests.

PRODUCTION

Distance Fabrics

consist of 2 fabric layers bonded together by integral woven vertical arranged threads. The thread length between the fabric layers specified the overall height of the sandwich structure.

Distance fabrics are impregnated by hand or automatic arrangements as conventional single layer fabrics. After squeezing out the surplus resins the distance fabrics erect to the requested height between the 2 and 16 mm getting extreme, solid and stiff sandwich structures.

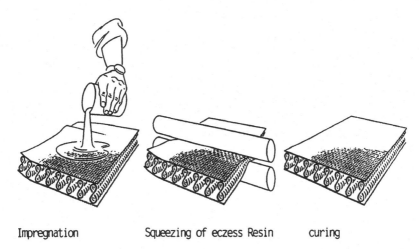

Impregnation   Squeezing of eczess Resin   curing

Working with Distance Fabrics

Multi Layer Fabrics

consist of 2 and more same or different fabric layers, of same or different materials. The thickness could be between 0,10 mm and several mm. The integral fabric construction is specified by the requests of the end product and the optimized manufacturing methods. The excellent characteristics concerning the mouldability makes multi layer fabrics especially usable for preform manufacturing.

3-D-Fabrics

High requests to special characteristics profiles have led to the development of fabrics showing additionally to the x/y-axis reinforcement threads in the z-axis. Modern weaving technologies enable to manufacture pre-forms of different forms and measures.

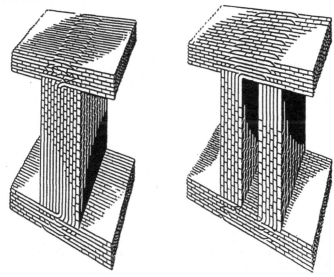

3 - D - J - Beam

Multi Axial Fabrics

offer the possibility to manufacture quasi-isotropical rein-forcement textiles. Reached by it, that in addition to the running threads of 0/90 degrees such with an angle of, i.e. +/- 45 degrees are arranged.

By variation of angles between 30 and 60 degrees special requests could be fulfilled.

The typical lay out for multi axial fabrics of the reinforcing threads without the crimp-effects of weaving fabrics leads to the full utilization of the fibre characteristics.

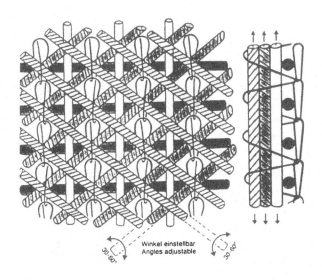

Winkel einstellbar
Angles adjustable

30 60°   30 60°

Weaving of Multiaxial-Fabric, 0,90, $\pm$ 45°

Summary

By incorporation the manufacturers of reinforcement textiles before beginning of the development phase for manufacturing composites it is multiple possible to get economic solutions by made-to-measure structures. A new generation of reinforcement materials shows new possibilities and chances for manufacturing composites.

850

LOAD TRANSMITTING ELEMENTS FOR FIBER REINFORCED RODS LOADED BY LONGITUDINAL FORCES

R. SCHUETZE

DLR - Institute for Structural Mechanics
3300 Braunschweig - West-Germany

ABSTRACT

It is particularly difficult to introduce axial loads to fiber reinforced rods. By use of an optimal design, the load transmitting sections of the rods have low masses and are nevertheless able to carry high axial loads. First, examples of form-locking and force-locking load transmissions will be described. Afterwards, two load transmitting elements are considered, which are especially suited for fiber reinforced rods. One of them is an integrated form-locking version and the other one a load transmitting element with the possibility of a later assembly on the rod. The minimum load carrying capacity of both under compression loading is the compressive strength of the rod itself, and under tension loading a value of the same magnitude.

INTRODUCTION

Specific strength and specific stiffness of rod shaped structures can be considerably improved by employing carbon fiber reinforced plastics (CFRP). However, it is particularly difficult to introduce axial loads to fiber reinforced rods. The large forces which can be carried by the rods themselves, can often be introduced into conventional structures only with additional laminate reinforcements and special metallic elements, so that the weight advantage of the fiber rods is partially compensated. By use of an optimal design, the load transmitting sections of the rods have low masses and are nevertheless able to carry high axial loads /1/.

I. EXAMPLES OF DIFFERENT LOAD TRANSMITTING CONSTRUCTIONS

The load transmitting areas of fiber reinforced rods differ in their way of working, which can be either form-locking or force-locking. With force-locking connection elements, the load is introduced by shear forces (bonding). A simple example of such a force-locking connection is a short piece of a tube bonded on the end of a rod. In this case the load-bearing capacity of the rod is extremely dependent on the bonding quality and the design of the bonded parts. Fig. 1 shows different possibilities of the design of bonded parts and shear stress curves referred to them /2/. According to Fig. 1, conical bonded parts lead to significantly lower stresses in the adhesive layer. As an example Fig. 2 depicts the load transmission into a fiber reinforced rod used in the AIRBUS /3/.

With form-locking load transmitting elements, tension and compression forces can be introduced directly into the load-carrying longitudinal fibers. In this case no substantial forces will be transmitted by the adhesive. A simple example of a form-locking transmitting element under compressive load is a metal tube with a cut-out, designed such that the end of a fiber reinforced rod is pressed against its vertical faces. Instead of utilizing the vertical faces of cut-outs, bolts and rivets can also press directly against the fibers of a rod and thus generate a form-locking connection for tension and compression loading.

Another method of producing form-locking connections is to bend fibers in the region of a metallic insert. First, there is the possibility of a fiber loop, where endless fibers are assembled and wrapped around a bolt /4/. Second, fibers can be bent in a radially symmetric fashion on a smaller diameter of a metal element, where they are fixed by hoop windings.

Fiber loops are wound as an integral part of the rod and in this simple form are only suitable for carrying tension forces. For compression loading, additional fittings are necessary. A typical fiber loop solution is shown in Fig. 3. There, endless fibers are wound around a metal pole cap at the end of the rod. While the pole cap is drawn into the fiber loops under tension load, compressive forces are introduced by a special metal compression cap screwed together with the wrapped tension cap.

In the case of the bent fibers, hoop windings act as locking rings on the metallic connection element, where its diameter is smaller. In this fashion, axial loads are introduced into the fibers. In contrast to the endless wound fiber loops, the longitudinal fibers of the rods are cut at the ends. Fig. 4 gives an example of such a version /5/. The fibers generate in the area of the load transmitting element a conical tube end, in which a metal cone is drawn. A hoop winding on the cone carries the high radial forces arising under tension loading. In the case of a compression load, a cap screwed together with the tension cone presses directly against the longitudinal fibers supported by the cone and the hoop winding.

The above-considered versions of load transmitting elements are representative of a lot of different variants and are suitable for transmitting high tension and compression loads in relative light weight and thin walled fiber reinforced rods. As a strength criterion for load transmission, the minimum load carrying capacity under compression loading should be the compressive strength of the rod itself, and under tension loading a value of the same magnitude should be achieved. The relative high requirements on strength and the low weight design goals often lead to a lot of work in production of the complete rods, which are mostly fabricated in a discontinous production process.

II. PROPOSAL OF TWO LOAD TRANSMITTING DESIGNS SUITABLE FOR FIBER REINFORCED RODS

In the following, two versions of load transmitting designs are represented, which have high strength combined with low weight and offer advantages in production /6/.

The first version uses a form-locking method of load transmission, where the principle of fiber bending on a smaller rod diameter is used. It consists of a metallic thread element with an annular groove turned in the element. Hoop windings force the load carrying unidirectional fibers into the annular groove and generate in that way a form-locking connection for tension and compression loading. The notched metal insert represents a double cone, such that both parts are drawn or pressed into the fiber ring (hoop windings) according to the force direction. The circumferential forces which arise are carried by the unidirectional fibers of the fiber ring. This construction, which entails only a slight amount of additional weight, leads to a strength which is higher than the compression strength of the rod itself. Special quality controls to ensure a satisfactory adhesion between the metal thread insert and the load carrying longitudinal fibers are dropped in this construction, because the load transmission is carried out even if the adhesion between the metal part and the longitudinal fibers does not exist. The production of the form-locking load transmission must occur "in situ" together with the fiber rod itself. A later assembly of the metal thread insert is no longer possible. In /6/ a special pultrusion process is described, which allows a continous production of CFRP-sandwich rods, where the above-mentioned form-locking load-transmitting elements are integrated.

A second version of load transmission is shown in Fig. 6 /6/. It consists of a metal thread insert, a CFRP-cap and a CFRP-wound ring. The CFRP-cap is built out of a C-fiber braid, the fibers of which generate small angles with respect to the longitudinal rod axis. Therefore, the cap has a high tensile and a low circumferential stiffness and has to be stiffened by hoop windings. In case of compression loading, the metal thread insert presses directly against the longitudinal fibers of the rod without considerable loading of the cap. Under tension load, the metal insert is drawn into the cap stiffened by the CFRP-hoop windings. Under compression, the load transmission occurs via form-locking and, under tension loading, via force-locking, because the CFRP-cap is bonded onto the rod end. The advantage of this load transmitting version is the possibility of a later assembly. Due to this, the fiber reinforced rods can be used as semi-finished material.

The proposed methods of load transmission were tested on short CFRP-sandwich rods, whose tension and compression strength amount to ca. 40 kN and ca. 24 kN respectively. The form-locking integrated load transmission already failed at a compression strength of only 70 % of the rod strength. This unacceptable failure follows from the very thin layer of the circumferential winding in the area of fiber bending (Fig. 7). However the desired load transmission strength of about 100 % can be achieved by some more hoop windings at this critical area of the load transmission. The installable load transmitting elements delivered compression strength values of ca. 100 %, where the failure occured in the rod, as expected. Under tension loading, all rods failed in the load transmission area, also as expected. The fracture of the form-locking load transmitting elements occured at ca. 140 % and that of the installables at ca. 160 % of the requested strength.

III. CONCLUSION

The above-mentioned strength criterion can be fulfilled by both of the described load transmitting methods. Both versions represent light weight load transmitting solutions with high potential for carrying axial loads, and both contain form-locking integrated metal thread inserts facilitating the connections with other structures. The capability of the load transmitting elements is not reduced by a lack of adhesion between metal and CFRP. Such a design is important with respect to thermal and corrosive incompatibility of the bonded parts. Force-locking exists only between the CFRP-cap and the rod end of the load transmitting element depicted in Fig. 6. But this represents a tube bonding of two compatible materials. Both load transmitting designs offer advantages in production and do not need much quality control.

REFERENCES

1 - G. Grüninger: "Möglichkeiten der Krafteinleitung in faserverstärkte Bauteile." Kohlenstoff- u. aramidfaserverstärkte Kunststoffe, VDI-Verlag, Düsseldorf 1977, 123-154.

2 - J. Franz, H. Laube: "Strength of Carbon Fiber Composite/Titanium Bonded Joints as Used for SPAS-Type Structures." ESA SP-243, (1986), 301-308.

3 - N.N.: "CFK/GFK-Stützen für den Airbus." "Commercial Newsletter", Felten + Guilleaume Energietechnik GmbH.

4 - U. Hütter: "Probleme der Krafteinleitung in Glasfaser-Kunststoff-Bauteile." Z. Kunststoffe, 56, (1966) 12.

5 - J. Glissmann, G. Hanselmann, H. Wurtinger: "Verbindungstechnologie für Struktur-Bauelemente in kohlefaserverstärktem Kunststoff." NT-Bericht 6/79 (1979) MAN-Neue Technologie.

6 - R. Schütze: "Untersuchungen an Krafteinleitungselementen für dünnwandige, zug- und druckbeanspruchte Faserverbundstäbe mit kreisförmigen Quer- schnitten." VDI-Fortschritt-Bericht Nr. 180, Reihe 1, Diss. RWTH Aachen.

FIGURES

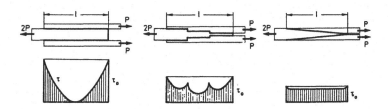

Fig. 1 - Shear stress curves for the adhesive layers of different joints (straight lap joint, multi step joint, tapered lap joint) /2/

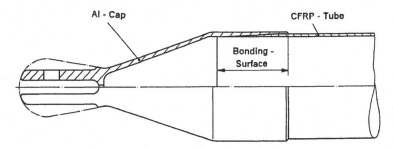

Fig. 2 - Joint between a CFRP-tube and a bonded aluminum cap (tapered lap joint) /3/

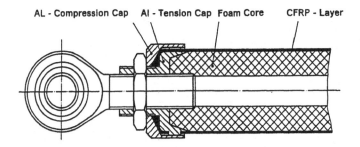

Fig. 3 - Form-locking by endless fiber wrapping around a metal tension cap; load transmission of the compression forces by a cap screwed on /5/

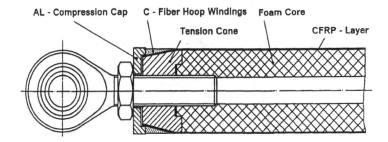

Fig. 4 - Form-locking by bending the fibers on a metal cone, where hoop windings act as a locking ring /5/

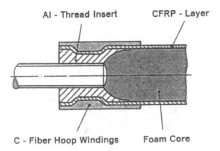

Fig. 5 - Form-locking by bending the fibers on a smaller diameter of the insert by means of hoop windings; the notched metal insert represents a double cone

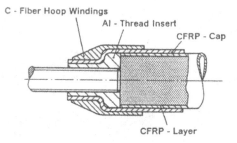

Fig. 6 - Bonded load transmitting element consisting of a braided CFRP-cap, an aluminum thread insert and C-fiber hoop windings; force-locking under tension and form-locking under compression forces

Fig. 7 - Failure of the load transmitting element shown in Fig. 5; a) under tension loading, b) under compression loading

Fig. 8 - Failure of the load transmitting element shown in Fig. 6; a) under tension loading, b) under compression loading

THE MONOLITHIC CFRP-SUBSTRUCTURE
WITH CORRUGATED V-SPARS
AN ALTERNATIVE TO SANDWICH CONSTRUCTIONS

T. SCHNEIDER, E. HENZE

Dornier Luftfahrt GmbH
Postfach 13 03 · 7990 Friedrichshafen · West-Germany

ABSTRACT

The classical substructure concept for airfoils consists of spar and rib elements. For thin airfoils, sandwich constructions are most commonly used. As an alternative, a monolithic CFRP-substructure was developed, featuring corrugated spars in a V-shaped arrangement. The corrugated CFRP-substructure (CSS) combines the spar and rib function in one element and allows an adaption of its dimensions to the structural loads without manufacturing limitations. The producibility and structural performance of this concept was verified in component tests.

I - CONSTRUCTION REQUIREMENTS

1.1 Basic requirements for a substructure.

In general, the requirements for a substructure in all kinds of air-foils (wings, stabilizers, flaps etc.) are:
- support of the skins
- internal shear load transfer
- connection of the skins to create sufficient bending and torsional stiffness.

The crucial dimensioning parameters are
- high strength and stability limits
- maximum weight savings.

1.2 Special requirements for an advanced monolithic CFRP-substructure

In addition to the basic requirements the present composite substructure was developed with the following objectives in view:
- combination of the spar and rib function in a single part (one shot bonding)
- local adaption of dimensions to the load distribution
- optimized skin buckling support.

II - REALIZATION OF THE CSS-CONCEPT

2.1 Background

Corrugated structures are very attractive for aircraft applications, as they offer superior buckling properties due to their geometrical shape. Therefore, additional stiffening elements or higher thickness are not required. The most famous example of this stiffening concept is the aircraft Ju 52, whose entire surface was built from corrugated aluminium sheets. These stability advantages are illustrated in fig. 1, where buckling loads of a corrugated and a flat web are compared.
The CSS-concept (corrugated substructure = CSS) combines the advantages of corrugated CFRP-spars (see /1/) with a V-shaped spar arrangement to obtain a monolithic structure. Former investigations at Dornier with metal corrugated V-spars using superplastic forming technique proved the load-bearing capability of this concept, but also the limitations concerning shape and dimensional variations from the manufacturing process. The use of composite material overcomes these problems and offers additional possibilities during the design process.

2.2 Dimensional parameters

Besides the laminate thickness and stacking sequence, the strength and stability of the CSS-structure is influenced by additional geometric parameters, which can be chosen according to structure requirements.
It is a unique property of the composite CSS, that in one monolithic part the laminate properties of the spars and flanges not only can be varied but that they are completely independent of each other. This enables the designer to adapt the dimensions of the structure to the internal loads only, without the manufacturing restrictions, in existance for metal parts with respect to forming processes and strains caused by deformation.
Another feature of the CSS-concept is the possibility of improving clamping conditions for the skins by an appropriate design of the flange width and shape. The reduction in the effective width of the buckling skin panel can offer considerable weight saving or increases in the stability limit of the skin. In addition, the corrugation parameters (see fig. 2) can be varied in different substructure areas.

2.3 Analysis

Four different buckling modes must be considered analytically in the layout procedure of a CSS: overall compression and shear buckling of the spar web as well as local compression and shear buckling of single corrugation waves. The analytical approach to the instability problem of anisotropic corrugated structures is described in refs. /2/ - /5/. The laminate strength was calculated with the Classical Laminate Theory (CLT) in conjunction with the Tsai-Wu failure criterion.

2.4 Test components

The producibility and the structural performance of the CSS-concept was proved by different test components, incl. V-spar elements with two different angles of corrugation and a complete bending-torsion box, part of which is shown in fig. 3.
The material was T300/1076E graphite/epoxy. For all parts, quasi-isotropic (-45/90/45/0)s-laminates from both unidirectional and fabric prepregs were used.

III - STRUCTURAL PERFORMANCE OF THE CSS

3.1 Spar function

The strength and stability of a V-spar element with a radius of corrugation of 25.4 mm, 1 mm spar and 2.5 mm flange thickness was tested using four components with different angles of corrugation and prepreg material.
Fig. 4 shows the component failure loads (pure shear) together with the theoretical strength and stability limit for this spar geometry. Obviously, the angle of corrugation as well as the prepreg type has only a minor influence on shear strength capability. Despite local bending effects, which lead to a superposition of normal and shear stresses, the failure load was only 10 - 25 % lower than the theoretical laminate shear strength.

3.2 Rib function

Rib function tests were limited to the determination of the component stiffness. Rib stiffness of the CSS (corrugation angle 46°, V-angle β = 25°, spar thickness 1 mm) is equivalent to conventional aluminium ribs with 1 mm thickness and 250 mm rib spacing.

3.3 Skin buckling support

The effect of the flange shape (as shown in figs. 2 and 3) on the skin compression buckling load also was investigated with component tests and compared to other flange geometries. As expected, the symmetry of the flange shape to the straight centreline (shape C in fig. 5) increases skin buckling stress by 30 % compared to the flange shape with symmetrical geometry to a corrugated centreline (shape B), and by 60 % related to a straight flange geometry (shape A), fig. 5.

IV - PROSPECTS

The present investigation confirmed the feasibility and capability of the CSS-concept. For future applications, improvements in the manufacturing process by using laminated short-fibre prepregs or injection-moulded composites may be envisaged. Another step to be done is the co-curing of a monolithic substructure with one of the skins, offering a substantial reduction of structural parts and assembly time.

REFERENCES

1. G. Porsche et.al., Report No. ET3-2175/82, VFW GmbH

2. S.P. Timoshenko and J.M. Gere, "Theory of Elastic Stability", McGraw-Hill Book Company, 2nd edition, 1961

3. D.H. Emero and L. Spunt, J. Aircraft, Vol. 3, No. 2, 1966

4. J.E. Ashton and J.M. Whitney, "Theory of Laminated Plates", Technomic Publishing Co., Inc., Stamford, 1970

5. J. Lemmer, Report No. TN-FE 212-22/77, MBB GmbH

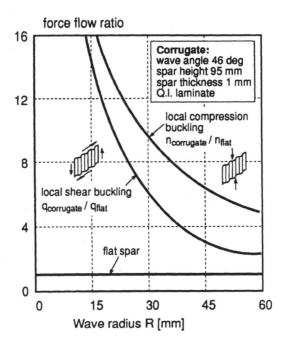

Fig. 1: Corrugated / flat web buckling load ratio

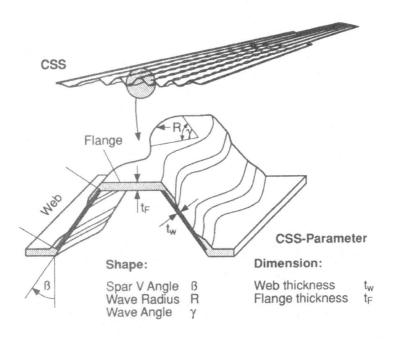

CSS

Flange

Web

R γ

t_F

t_w

β

CSS-Parameter

Shape:

Spar V Angle	β
Wave Radius	R
Wave Angle	γ

Dimension:

Web thickness	t_w
Flange thickness	t_F

Fig. 2: CSS design parameters

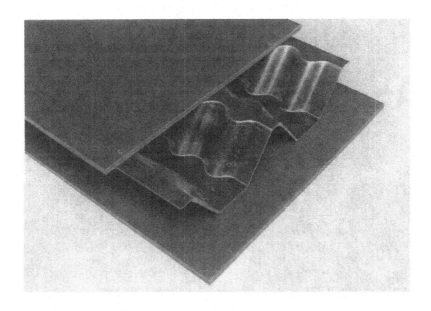

Fig. 3: CFRP Bending-Torsion Box (CSS + skins)

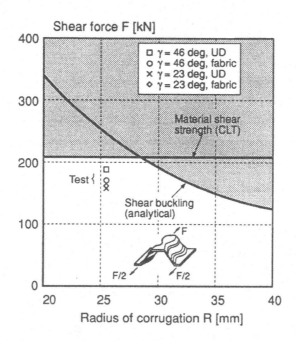

Fig. 4: Shear strength test results

Skin buckling stress [MPa]

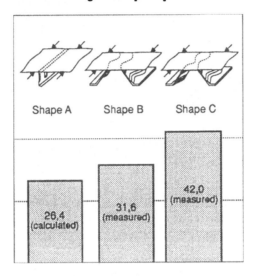

Fig. 5: Influence of flange shape on skin buckling stress

CHARACTERIZATION OF INTERLAMINAR MODE II FRACTURE USING BEAM SPECIMENS

J. WHITNEY, M. PINNELL*

Wright Research and Development Center
WRDC/MLBM - Wpafb Ohio 45433 - USA
*University of Dayton - Dayton Ohio - USA

ABSTRACT

Experimental data obtained in conjunction with the end notch flexure (ENF) test and the end notch cantilever (ENC) test are considered. These test methods are utilized to obtain critical interlaminar Mode II strain energy release rate on unidirectional graphite fiber reinforced composites. Data reduction is performed in conjunction with both classical beam theory and a polynomial approximation to a higher order beam theory derived from Reissner's principle.

INTRODUCTION

Mode II interlaminar fracture tests are often performed in conjunction with flexure specimens containing a midplane starter crack of desired length, a, at the end of the beam. The 3-point bend specimen configuration shown in Fig 1a is utilized in conjunction with a standard end notch flexure (ENF) test /1/. An alternate geometry in the form of an end notch cantilever (ENC) specimen is shown in Fig. 1b. A Mode II critical strain energy release rate, G_{IIC}, can be determined by measuring the load and deflection under the load nose at the instant the starter crack propagates. Classical beam theory can be utilized in deriving a relationship between deflection and load with the result

$$\delta = \frac{(c_1 L^3 + 3a^3)\ P}{c_2 E_1 bh^3} \qquad (1)$$

where

$$c_1 = 2 \text{ (ENF)}, \; 1 \text{ (ENC)}$$

$$c_2 = 8 \text{ (ENF)}, \; 2 \text{ (CNF)}$$

In addition, d and E_1 denote the deflection under the load nose and the effective bending modulus in the x- direction of a beam of thickness 2h, respectively. Combining eq. (1) with the definition of strain energy release rate for fixed load, we obtain the result

$$G_{II} = \frac{P}{2b} \frac{d\delta}{da} = \frac{9P^2a^2}{2c_2E_1b^2h^3} \tag{2}$$

Although these interlaminar Mode II beam tests have been utilized with both unidirectional and multidirectional laminates, difficulties have been encountered in the latter case /1/ due to the crack not being self-similar (crack not remaining in the mid-plane). This often results in high values of G_{IIC} as compared to unidirectional composites. Thus, these tests are currently being utilized in conjunction with unidirectional specimens only.

Finite element analysis of the ENF specimen based on either plane stress or plane strain theory of elasticity have been performed by a number of investigators /2-5/. In each of these studies significant departure from beam theory, eq. (2), was observed for a certain range of values of a/L. A recent analysis /6/, based on a higher beam theory developed in conjunction with Reissner's principle /7/, led to the following approximate solution for the ENF specimen:

$$\delta = \frac{P}{8E_1b} \left[2\overline{L}^3 + 3\overline{a}^3 + \frac{(39424\overline{L} + 6681\overline{a})}{425 \; \lambda^2} + \frac{9\overline{a}^2}{1} \right] \tag{3}$$

$$G_{II} = \frac{9P^2\overline{a}^2}{16E_1b^2h} \left[1 + \frac{1}{75\lambda^2\overline{a}^2} (131 + 1501\overline{a}) \right]$$

where

$$\lambda = 4 \sqrt{\frac{14}{5} \left(\frac{G_{13}}{E_1}\right)}, \quad \overline{L} = L/h, \quad \overline{a} = a/h$$

Equation (3) yields very accurate results compared to the finite element method utilized by Salpekar, Raju, and O'Brien /5/. A similar solution for the ENC specimen /6/ led to the following approximation:

$$\delta = \frac{P}{2E_1 b} \{ \overline{L}^3 + 3\overline{a}^3 + \frac{3}{25\lambda^2} [387\overline{L} + 75\overline{a} (1 + 1\overline{a})] \}$$

$$G_{II} = \frac{9P^2\overline{a}^2}{4E_1 b^2 h} [1 + \frac{1}{\lambda^2 \overline{a}^2} (1 + 21\overline{a})]$$

(4)

EXPERIMENTAL RESULTS

Experimental data is shown in Table 1 for AS4/3501-6 graphite/epoxy and APC2 graphite/thermoplastic matrix unidirectional composites. All specimens are either 16 or 24 plies thick. The number in the first column corresponds to the number of specimen replicates in the data set. Results obtained on different thickness specimens were averaged, as no significant difference in values of critical strain energy release rate was denoted between 16 and 24 ply specimens. Numerical results calculated from classical beam theory, eq. (2), is denoted by G_{IIB}, while values obtained from the polynomial approximations, eqs. (3) and (4), are denoted by G_{IIP}. All specimens were 25.4 mm wide. Crack lengths, a, and span lengths, L, as well as the ratio a/L are also indicated in Table 1. Starter cracks were in the form of teflon strips.

DISCUSSION

Similar values of G_{IIC} were obtained from both the ENC and ENF specimens. For APC2, however, somewhat lower values were obtained in conjunction with the ENC specimen compared to the ENF test. With the exception of the results for a/L = 0.3 in conjunction with the ENC specimen for APC2, G_{IIC} does not appear to depend on crack length. The increased toughness associated with thermoplastic matrix composites as compared to epoxy matrix composites is apparent from the data in Table 1.

ACKNOWLEDGEMENTS

The authors wish to acknowledge R. Cornwell and R. Esterline of the University of Dayton Research Institute for the fabrication and testing of composite specimens.

REFERENCES

1. A. J. Russell, and K. N. Street, in "Proceedings of the Fourth International Conference on Composite Materials" (1984) 279-286, ELSEVIER, North Holland.

2. S. Mall and N. K. Kochhar, J. Composites Technology and Research, 8 (1986) 54-57.

3. J. D. Barrett and R. O. Foschi, Engineering Fracture Mechanics, 9 (1977) 371-378.

4. J. W. Gillespie, Jr., L. A. Carlsson, and R. B. Pipes, Composites Science and Technology, 27 (1986)·177-197.

5. S. A.Salpekar, I. S. Raju, and T. K. O'Brien, J. Composites Research and Technology, 10 (1988) 133-139.

6. J. M.Whitney, J. Reinforced Plastics and Composites, 8 (1990) to be published.

7. E. Reissner, J. of Mathematics and Physics, 29 (1950) 90-97.

TABLE 1 EXPERIMENTAL VALUES OF G_{IIc} (J/m^2)

#	a(mm)	L(mm)	a/L	G_{IIB}	G_{IIP}
			AS4/3501-6		
			ENF		
3	15.2	50.8	0.3	310	367
4	25.4	50.8	0.5	351	385
3	35.6	50.8	0.7	295	318
			ENC		
5	15.2	50.8	0.3	267	308
4	25.4	50.8	0.5	344	376
5	35.6	50.8	0.7	334	356
			APC-2		
			ENF		
2	15.2	50.8	0.3	1309	1453
6	25.4	50.8	0.5	1587	1725
6	35.6	50.8	0.7	1576	1665
			ENC		
4	15.2	50.8	0.3	919	1042
1	30.5	101.6	0.3	961	1037
4	25.4	50.8	0.5	1237	1338
1	50.8	101.6	0.5	1307	1368
1	35.6	50.8	0.7	1366	1444
4	71.1	101.6	0.7	1261	1303

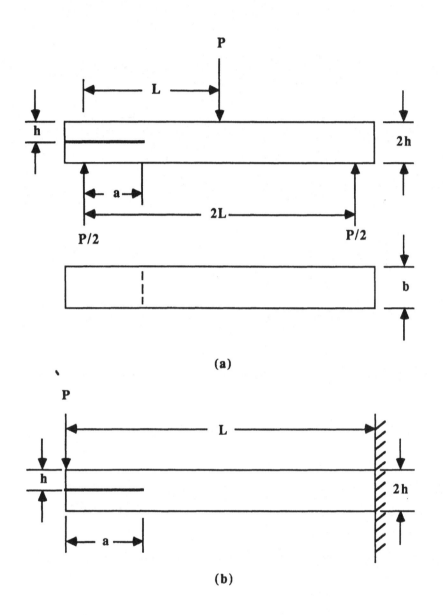

(a)

(b)

Fig. 1 - Geometry: (a) ENF specimen; (b) ENC specimen.

THROUGH - THICKNESS STRESS DISTRIBUTIONS IN TAPERED COMPOSITE BEAMS

A. MIRAVETE, R.I. KIM*, G. PIEDRAFITA, S. BASELGA

University of Zaragoza
Maria de Luna 3 - 50015 Zaragoza - Spain
*University of Dayton Research Institute
300 College Park - Dayton Ohio - USA

ABSTRACT

If the design of a laminated composite beam accounts for the concept of variable thickness, a remarkable weight saving can be obtained. In the present work, a theoretical study dealing with the stress distributions in tapered composite beams is presented. A finite element model has been used for calculating the stress components. A four point bending model has been applied. Three types of composite material are analyzed: Graphite/epoxy, Aramide/epoxy and Fiberglass/epoxy. Influence of material, laminate thickness, geometry, bending moment and aspect ratio on through-thickness stress distributions is studied.

INTRODUCTION

Composite materials have been increasingly used during the last decades, in order to light structures in fields like aeronautics and space. Two steps are essential in the process of taking advantage of this kind of materials: design and optimization.

In order to solve the problem of optimizing a composite structure, the variable thickness effect must be studied and understood properly.

Ochoa and Chan /1/ reported an outstanding work on delamination of tapered laminated subjected to in-plane loads. Quite a few references of failure modes in tapered laminates subjected to transverse loads have been found. All of them deal with applications to specific pieces /2/ to

/4/. Recently, Kim and Miravete /5/ studied the failure mechanisms of tapered laminated composite plates.

In the present work, a study of the stress distributions of variable thickness composite beams subjected to transverse loads is presented. A theoretical model based on a plane strain finite element theory is described.

I-MODEL ASSUMPTIONS AND METHOD OF ANALYSIS

The model used here presents the following assumptions:

- This study is limited to one-dimensional laminated composite plates, in order to avoid three-dimensional effects and to analyze properly the consequences of the variable thickness effect.

- The fiber orientation is longitudinal (x-direction). This is the optimal direction for one-dimensional laminated composite plates.

- A 2-D plane strain finite element model has been used.

- The mesh is composed by 1891 nodes.

- A three point bending model has been used for analyzing the failure mechanisms /5/ and mechanical behavior /6/ of tapered composite beams. However, a four point bending model seems more adequate for stress distribution studies because bending moment is constant and shear force vanishes at the middle of the span for constant thickness beams. Thus, for a tapered beam, interlaminar stress distributions in this area will be generated only by variable thickness effects. Figures 1 and 2 show the differences between both models. Smaller differences have been found between beam theory and stress distributions in 3 than in 4 point bending model /7/.

- Three materials have been used: Graphite/epoxy AS4/3501, Kevlar 39/epoxy and Glass E/epoxy. Mechanical properties are given in Table 1.

II-RESULTS OF THE STUDY

The definition of variables is shown in Figure 3. A 4 point bending model has been applied for the three types of materials mentioned above. Figures 4, 5 and 6 represent the longitudinal (σ_1), interlaminar normal

(σ_3) and interlaminar shear (σ_5) distributions through thickness in tapered area, respectively.

This study is focussed upon the interlaminar shear (σ_5) distribution because it is the critical component in this model. Experimental results show delamination failure modes in the tapered area. The peak of σ_5 (σ_{5max}) in Figure 6 is the critical parameter in the generation of this delamination.

This value (σ_{5max}) is function of the material and the ratio (t_1/t). Figure 7 shows the variation of σ_{5max} with respect to these two parameters. On the one hand, Fiberglass E/epoxy presents the maximum value and graphite/epoxy the minimum. On the other hand, variation of σ_{5max} in function of t_1/t presents a maximum, that is function of the material.

σ_{5max} is proportional to the bending moment at the middle of the span $P(L-L_1)/2$. And finally, σ_{5max} is not a function of the aspect ratio (L/t).

III-CONCLUSIONS

Distributions of longitudinal (σ_1), interlaminar normal (σ_3) and interlaminar shear (σ_5) distributions through thickness in the tapered area of a composite beam have been presented. A 4 point bending model has been used.

The peak of the distribution of σ_5 is the critical parameter in this study. This value, σ_{5max} is a function of the material, t_1/t and the bending moment.

REFERENCES

1. O.O. Ochoa and W.S. Chan, Proceedings of the American Society for Composites. Third Technical Conference, (1988) 1: 633-641.
2. W.J. Yu and H.C Kim, Composite Structures, (1988) 9: 279-300.
3. T. Goette, R. Jakobi and A. Puck, Kunststoffe, (1985) 75(2): 100-104.
4. C.J. Morris, Composite Structures, (1986) 5: 233-242.
5. R. Kim, and A. Miravete, J. Composite Materials (Submitted).
6. A. Miravete, AFWAL/MLBM, Technical report (In press).
7. N. Pagano, Interlaminar Response of Composite Materials, (Ed. Elsevier) (1989).

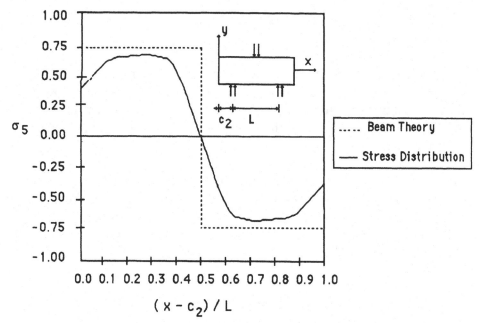

Figure 1. Distribution of normalized interlaminar shear stresses in the 3 point bending model for $[0]_{50}$ Graphite/epoxy, $L/t = 4$.

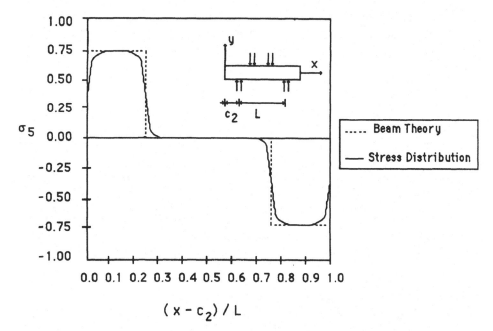

Figure 2. Distribution of normalized interlaminar shear stresses in the 4 point bending model for $[0]_{15}$ Graphite/epoxy, $L/t = 16$.

Material	AS4/3501	Kev39/epoxy	Vidrio-E/epoxi
E_x (GPa)	138.00	76.00	38.60
E_y (GPa)	8.96	5.50	8.27
v_x	0.30	0.34	0.26
E_s (GPa)	7.10	2.30	4.14
V_f	0.66	0.60	0.45

Table 1. Mechanical Properties of Composite Materials.

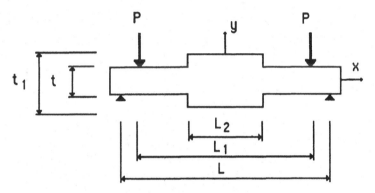

Figure 3. Definition of variables.

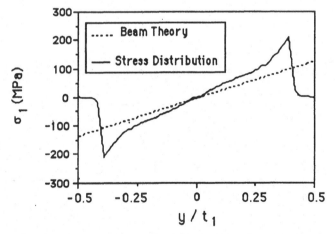

Figure 4. Distribution of longitudinal stresses for Kev39/epoxy and
P=1E5N/m, t=0.04 m, t_1=0.05 m, L_2=0.04 m, L_1=0.2 m and L=0.4 m.

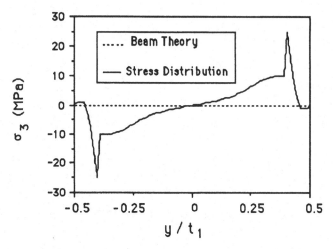

Figure 5. Distribution of interlaminar stresses for Kev39/epoxy and P=1E5N/m, t=0.04 m, t$_1$=0.05 m, L$_2$=0.04 m, L$_1$=0.2 m and L=0.4 m.

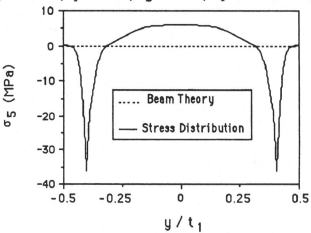

Figure 6. Distribution of interlaminar stresses for Kev39/epoxy and P=1E5N/m, t=0.04 m, t$_1$=0.05 m, L$_2$=0.04 m, L$_1$=0.2 m and L=0.4 m.

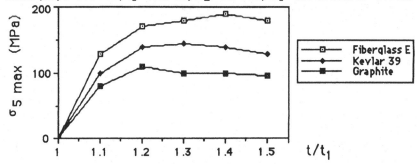

Figure 7. Distribution of σ_{5max} in function of t/t$_1$.

A MICROMECHANICAL PREDICTION OF MECHANICAL AND THERMAL PROPERTIES OF FRP WITH INTERFACIAL IMPERFECTIONS

G. PAPANICOLAOU, C. BAXEVANAKIS

University of Patras
Dept. of Mechanical Engineering - Applied Mechanics Laboratory
26110 Patras - Greece

ABSTRACT

A theoretical model approach for the prediction of the longitudinal elastic modulus, Poisson's Ratio and the thermal expansion coefficient in unidirectional fibre-reinforced composites was developed. The model takes into account the existence of the interphase which represents the third phase developed between the constituet phases of the composite. It was assumed that both fibre and matrix have well-defined mechanical properties while the interphase properties are varied following an exponential law of variation. Also the effect of the discontinuity of the properties at the fibre-matrix interface was investigated and interesting results have been found.

INTRODUCTION

The interphase represents the third phase developed between the constituent phases of the composite and it is characterized by mechanical imperfections, such as voids microcracks, etc., physicochemical interactions and limited mobility of macromolecules due to their absorption on the fibre surface. Interfacial conditions strongly affect the mechanicaland thermal properties of fibre-reinforced composites. In this type of composites, interfacial reaction is one of the main reasons for reducing composite strength.

Several mathematical models have been developed to predict the mechanical and/or thermal behaviour of FRP. A brief outline of the predictive theories is given in/1/. However it remains the problem since all these models assume interfaces as being perfect mathematical surfaces and that a perfect adhesion exists between matrix and fibre. In reality such conditions are hardly fulfilled and thus the above models may be considered as unrealistic.

A better approach has been made by models considering an RVE consisting of three phases i.e. the filler, the interphase and the matrix in the form of either concentric spheres in the case of particulates or in the form of coaxial cylinders for the fibe composites /2-10/.

In the present paper a theoretical model is developed for the prediction of the longitudinal modulus of elasticity, the logitudinal and the transverse coefficient of thermal expansion of a fibre-reinforced composite. The model takes into account the existence of the interphase as well as the mode of variation of the thermomechanical properties at the fibre-interphase common surface.

THE MODEL

Figure 1 shows the representative volume element (RVE) of the model considered in the present investigation.

If we denote by r_f, r_i and r_m the outer radii of the fibre, the interphase and the matrix respectively, then the fractions of the respective phases are given by:

$$U_f = r_f^2/r_m^2 \qquad U_i = (r_i^2 - r_f^2)/r_m^2 \qquad U_m = (r_m^2 - r_i^2)/r_m^2 \qquad (1)$$

with

$$U_f + U_i + U_m = 1 \qquad (2)$$

Next, we define the parameters α and β as follows

$$\alpha = E_i(r_f)/E_f \qquad \text{and} \qquad \beta = \nu_i(r_f)/\nu_f \qquad (3)$$

where E and ν are the elastic modulus and the Poisson's Ratio respectively while indices f, i and m state for the fibre, interphase and matrix respectively.

For $\alpha = \beta = 1$ a continuous variation of the elastic modulus and Poisson's Ratio is obtained at $r = r_f$. On the other hand, for $\alpha = E_m/E_f$ and $\beta = \nu_m/\nu_f$ the two-phase model is obtained.

The following relationship between α and β is assumed to exist:

$$\alpha = (k/\beta) + \lambda \qquad (4)$$

where k and λ are constants which can be determined from the following conditions:

At $r = r_f$, for $\alpha = 1$ $\qquad \beta = 1$

for $\alpha = E_m/E_f$ $\qquad \beta = \nu_m/\nu_f$ $\qquad (5)$

it follows that

$$k = (1-E_m/E_f)/(1-\nu_f/\nu_m)$$

and

(6)

$$\lambda = -(\nu_f/\nu_m-E_m/E_f)(1-\nu_f/\nu_m)$$

An exponential variation for $E_i(r)$, $\nu_i(r)$ and $a_i(r)$ where a denotes the coefficient of thermal expansion, is considered. The law of variation considered is of the form:

$$I(r) = A_I r e^{-C_I r} + B_I$$

(7)

where constants A_I, B_I and C_I can be evaluated from the boundary conditions:

At $r=r_i$ $dI(r)/dr=0$ and $I(r)=I_m$

(8)

At $r=r_f$ $E_i(r_f)=\alpha E_f$ $\nu_i(r_f)=\beta \nu_f$ and $a_i(r_f)=a_f$

Introducing now A_I, B_I and C_I (I=E,ν,a) in Equation 7 we obtain

$$E_i(r) = E_m + (\alpha E_f-E_m)R(r)$$

$$\nu_i(r) = \nu_m + (\beta \nu_f-\nu_m)R(r)$$

(9)

$$a_1(r) = a_m + (a_f-a_m)R(r)$$

where $R(r) = (1- \dfrac{r}{r_i} e^{(1-r/r_i)})/(1- \dfrac{r_f}{r} e^{(1-r_f/r_i)})$

(10)

Figure 2 shows the mode of variation of $E_i(r)$ modulus as a function of the polar radius r, according to relation (9). In a previous publication, by using the three-phase model, an expression relating the longitudinal modulus E_{CL} with $E_i(r)$ and $\nu_i(r)$ was derived. This relation is quite complex However, for the special case of perfect adhesion takes the form

$$E_{CL} = E_m U_m+E_f U_f+ \frac{2}{r_m} \int_{r_f}^{r_i} E_i(r)r \, dr$$

(11)

By introducing the expressions (9) for $E_i(r)$ and $\nu_i(r)$ into the above-mentioned relation, E_{CL} can be evaluated. Predicted E_{CL} values for $\alpha=\beta=1$ are plotted against U_f in Fig.3. In the same figure predicted values from other theories, as

879

well as experimental values, are plotted for comparison. It
can be seen that there is a good agreement between theory
and experiment. Also, application of the model for different α-values show that E_{CL}-values are insignificantly
influenced by the parameter α.

Following the same procedure, the model gives for the
longitudinal a_C^L and the transverse a_C^T thermal expansion
coefficients the following expressions

$$a_C^L = \frac{a_f E_f U_f + (2/r_m^2) \int_{r_f}^{r_i} a_i(r) E_i(r) dr + a_m E_m U_m}{E_f U_f + \frac{2}{r_m^2} \int_{r_f}^{r_i} E_i(r) dr + a_m E_m U_m} \qquad (12)$$

$$a_C^T = [(1+\nu_f)a_f - a_C^L \nu_f](r_f^2/r_m^2) + (2/r_m^2) \int_{r_f}^{r_i} (1+\nu_i(r))a_i(r)r \, dr +$$

$$(1+\nu_m)a_m U_m - (2a_C^L/r_m^2) \int_{r_f}^{r_i} \nu_i(r)r \, dr - a_C^L \nu_m U_m \qquad (13)$$

where $E_i(r)$, $\nu_i(r)$ and $a_i(r)$ are given by relations (9).
   The variation of a_C^L and a_C^T as a function of fibre-volume
fraction and for different values of the parameter α, is
shown in Figs.4 and 5 respectively. From these figures it
is clear that the coefficients of thermal expansion are
insensitive to the interfacial variations. However, a
strong effect of the mode of variation of the properties of
the interphase on the stress and strain fields developed
around the fibres is expected. This effect will be the
subject of future work.

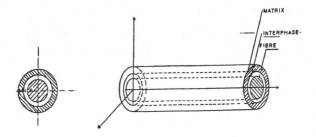

Fig.1   Representative Volume Element of the
        Three-Phase Model considered

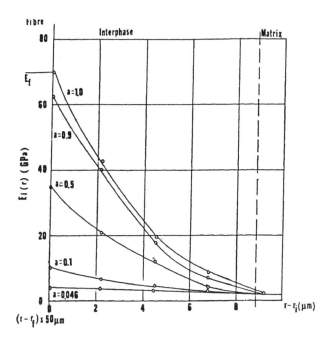

Fig.2 Variation of $E_i(r)$ with polar radius r

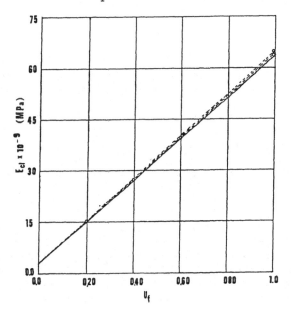

Fig.3 Predicted Values of E_C^L for $\alpha=\beta=1$ against U_f.

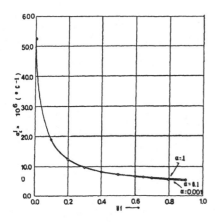

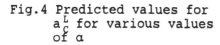

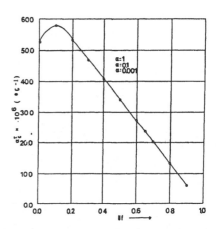

Fig.4 Predicted values for a_c^L for various values of α

Fig. 5 Predicted values for a_c^T for various values of α

REFERENCES

1- Paipetis S.A.,"Developments in Composite Materials 2", edited by G.S.Holister (Applied Science Publishers, London, 1981)

2- Papanicolaou G.C., Paipetis S.A., Theocaris P.S., Colloid and Polymer Science, 256 (1978) 625-630.

3- Papanicolaou G.C. and Theocaris P.S., Colloid and Polymer Science, 257 (1979) 239-246.

4- Theocaris P.S. and Papanicolaou G.C., Fibre Science and technology, 12 (1979) 421-433.

5- Theocaris P.S. and Papanicolaou G.C., Colloid and Polymer Science, 258 (1980) 1044-1051.

6- Papanicolaou G.C., Theocaris P.S. and Spathis G.D., Colloid and Polymer Science, 258 (1980) 1231-1237.

7- Theocaris P.S., Papanicolaou G.C. and Spathis G.D., Fibre Science and Technology, 15 (1981) 187-197.

8- Theocaris P.S., Sideridis E.P. and Papanicolaou G.C., Jnl. of Reinforced Plastics and Composites, 4 (1985) 396-418.

9- Sideridis E.P. and Papanicolaou G.C., Rheologica Acta, 27 (1988) 608-616.

10- Papanicolaou G.C., Messinis G.J. and Karakatsanidis S.S., Jnl. of Materials Science, 24 (1989) 395-401.

BEARING STRENGTH OF CARBON FIBRE/EPOXY COMPOSITES UNDER STATIC AND DYNAMIC LOADING

M. AKAY

University of Ulster at Jordanstown
Newtownabbey - Co Antrim - BT37 0QB - Northern Ireland

ABSTRACT

Pin bearing behaviour of a range of laminate systems consisting of uni-directional and woven carbon fibre reinforced epoxy matrices were examined under static and dynamic loading. Static bearing properties were determined under various levels of bolt constraint with particular attention to onset of failure and failure modes. Load-amplitude controlled fatigue tests were conducted at 3Hz to approximately 40% hole deformation. The properties depended mainly on the laminate type, ply stacking sequence and specimen geometry. Strength at failure initiation under static loading was found to be significant, particularly in relation to fatigue behaviour.

INTRODUCTION

Bolted joints can be a viable alternative for the assembly of laminate-type structures based on composite materials. They result in higher stress concentration than would be the case with bonded joints but enable easy inspection, facilitate assembly and provide reliability at an economic cost. Current applications include aerospace structures and the filament wound rocket booster case of the Challenger space shuttle, /1/.

Successful adoption of bolted joints in composites depends on a clear understanding of pin bearing strength. Limited information is available in review articles /2-4/ and specific research papers dealing with uni-directional carbon-fibre/epoxy laminates. The effect of ply lay-up in laminates has been well covered /5-9/. The influences of geometric features, viz specimen width (w) and thickness (t), end-distance (e), and hole-diameter (d) have been extensively covered, /3,6,10,11/. There is a general consensus that, for a given lay-up, minimum e/d and w/d ratios must be provided, otherwise the potential bearing strength will not be realised.

Work /5,6,9/ on the effects of lateral constraint has shown that bearing strength increases gradually and reaches a 'plateau' value with increasing clamping pressure via bolt-torque.

Various failure modes have been identified /2,4,6,12/ - end-section bearing, net-section tension, edge-section shear, end-section cleavage and 'tension-cleavage' combination. Some mixture of these modes may, of course, occur /10/. Failure can be defined as the maximum load sustained /6/ or as a deformation of the hole. The definition of bearing strength in relation to fixed diametric strain as in ASTM D953-87 is open to question. Clearly strength should be defined in relation to permanent damage within the joint, and the deformation to create damage is a function of both the tolerances of the hole and stress-deformation behaviour of a particular material system.

The volume of work to date has not however adequately covered woven carbon fibre composite systems and scarcely any work has been published on pin bearing strength under fatigue loading.

I - EXPERIMENTAL

1.1 Materials and specimens

Laminates were prepared from the following prepregs:

(a) Plain-weave (PW) fabric of 40% by weight resin content (wrc): carbon fibre (CF)/epoxy (Cycom 985 resin and T300 3k fibre).

(b) PW fabric of 40% wrc : CF/epoxy (Fiberite 934 resin and T300 3k fibre).

(c) 8-Harness (8HS) satin fabric of 37% wrc: CF/epoxy (Cycom 985 resin and T300 3k fibre).

(d) Uni-directional (UD) tape of 37% wrc: CF/epoxy (Cycom 985 resin and T300 3k fibre).

(e) UD tape of 37% wrc: CF/toughened epoxy (Cycom 1010 resin and T300 3k fibre).

(f) UD tape of 37% wrc : CF/epoxy (Fiberite 934 resin and T300 3k fibre).

The prepreg stacks were laminated as flat panels and, after debulking and bagging, autoclave cured at 180°C and 600 kP$_a$ for two hours. Lay-up details are presented in Table 1.

Coupons of approximately 194mm long and 30mm wide were cut and centrally-located holes of 4.84mm diameter and 35mm end-distance machined. The specimen-end to be gripped was bonded with aluminium alloy tabs.

1.2 Test procedures

1.2.1 Static pin bearing strength

Tests were conducted with reference to ASTM D953-87 and the CRAG test method. All of this work was conducted at room temperature, using an Instron floor model machine in tensile mode. Hole displacement was measured using an extensometer. Load was applied at 1mm/min via 4.76mm diameter pins.

Load displacement (1-d) curves were obtained for three specimens
from each laminate and the failure modes noted. Stress values,
load/(dt), were determined corresponding to: 4% hole-displacement (as
per ASTM D953-87); initial discontinuity in 1-d trace; and maximum
load.

All the specimens were prepared with finger-tight lateral constraint
and specimens 5 and 6 were further examined over a range of constraints,
corresponding to pin torque values of 0,2,4,6 and 8 Nm. The upper
limit of the applied torque was dictated by pin thread stripping. The
constraint pressure (σ press) has been identified elsewhere /6/ and
based on the pin torque (T) and the constrained area. It becomes:

$$\sigma \text{ press} = T / [(\pi/4) \ Kd_p \ (d_b^2 - d^2)]$$

where K = 0.2 is a torque coefficient /6/, and d_p, d_b, d are the
diameters of the pin, clamping bush and the hole, respectively.
Substituting average values of d_p = 4.76mm, d_b = 15.5mm and d = 4.84mm
gives σ_{press} (MPa) = 6.17[T(Nm)].

1.2.2 Fatigue test procedure

Sinusoidal load-amplitude tests on Instron servo-hydraulic machine
were conducted at 3Hz using a base load of 0.2kN and a range of pre-set
maximum loads. The tests were continued to a pin displacement of 2mm
(40% hole deformation); this value corresponds to the deformation
experienced under maximum static bearing load. Specimens which were
highly prone to tensile failure had fractured within this limit.

II - RESULTS AND DISCUSSION

2.1 Static bearing behaviour

Bearing resistance will be described as: the maximum stress (σ_m)
sustained by the specimen; the stress at initial failure (σ_1); and the
stress at which the bearing hole is deformed 4% of its diameter.

2,1,1 Influence of prepreg type and lay-up.

Figure 1 shows the maximum bearing stress for the laminate systems
constrained only by finger-tightening. In general woven laminates
supported higher stresses than UD laminates depending on the lay-up
and the geometric factors. The adoption of all 0°/90° fibre orientation
(as in Specimen A) generates high stress concentrations and causes
premature failure in tension.

The resin systems designated as Cycom 985 and Fiberite 934 are
considered to be similar, but Cycom 1010 is a toughened epoxy and
increases the ductility of the composite and hence reduces the
likelihood of high stress concentrations. Accordingly, specimen 2
produced the highest value for σ_m.

The failure modes are indicated on Fig.1. Most specimens underwent
significant degree of bearing failure, except the (0°,90°) PW specimen
A which failed in tension following limited bearing failure, and the 8H
specimen C which exhibited a mixture of bearing, cleavage and shear
type failures.

The woven laminates, particularly PW types, and, as would be
expected, the UD laminates with the higher number of plies showed a

greater resistance to the initiation of permanent damage (Fig.1). The low σ_i values experienced with the toughened epoxy system probably result from the greater ductility of the matrix. Toughened epoxies are excellent in containing damage but, they can suffer initial damage more readily under certain loading conditions.

In general the UD specimens exhibited slightly higher stress levels than the woven type specimens at 4% strain. This is as would be expected since UD laminates are inherently stiffer, suffer no weaving induced fibre distortion, and in this case, there was significant 0°-ply content, Table 1. At this low level of strain, the specimen suffered no visible damage or any damage which could be identified by X-radiography. It follows that in regard to material integrity the significance of the ASTM recommendation referred to earlier is open to question.

2.1.2 Influence of lateral constraint

Laminates 5 and 6 which were investigated over a range of lateral constraint, show, Fig.2, that the bearing stress gradually increased with increasing constraint pressure and levelled off above 25 MPa (specimen 5). The associated increases for specimens 5 and 6 respectively were approximately: 118% in σ_m; and 36% in σ_i (specimen 5) and 275% and 250% (specimen 6). These magnitudes are consistent with previously published work /5,9/. Thinner laminates benefit to a greater degree from the introduction of constraint pressures since they are more prone to local buckling.

A transition in the failure mode was noted and coincided with the change in bearing stress pattern, Fig.2. The thicker laminate, specimen 5, exhibited a bearing-type failure only up to a 25 MPa constraint; beyond this, limited initial bearing failure was followed by a final failure in the combined tension-cleavage mode. It would seem that effective inhibition of laminate 'brooming' requires a certain minimum constraint level. The transition in the failure mode occurred at 49 MPa for the thinner laminate (specimen 6) and the increase in bearing stress showed no levelling off up to this point. The failure mode changed from 100% bearing failure to limited bearing/ tension failure along 45° fibre alignment.

2.2 Fatigue bearing behaviour

An assessment was made of the correlation between static strengths, as represented by σ_m and σ_i, and fatigue behaviour. Figure 3 shows comparisons between specimens, of types 5 and 6, with similar σ_m but different σ_i values. Figure 4 shows comparisons between specimens A and B with different σ_m but similar σ_i. It can be seen that for $N \geq 10^5$, σ_i controls the fatigue performance; the fatigue strength approaches the value σ_i. Further work would be needed to confirm the correlation for $N > 10^6$.

CONCLUSIONS

Woven carbon fibre laminates exhibit higher initial bearing strengths (σ_m) compared with uni-directional (UD) laminates of similar lay-up sequence. In UD laminates the initial bearing strength increases with increasing number of plies.

Laminates with 90° fibres (PW) in the surface ply undergo limited bearing failure prior to fracturing in tension. Laminates with (0°,90°) fibres in an 8H weave in the surface ply exhibit a mixed mode of failure, comprising bearing, shear and cleavage.

A toughened epoxy resin matrix allows localised plastic deformation at the pin-hole and, accordingly, results in low initial bearing strengths and high maximum bearing strengths.

Laminates with uni-directional fibres provide an increase in stiffness compared to woven laminates and, accordingly, result in higher bearing strengths at 4% pin-hole displacement.

Lateral constraint results in higher initial bearing strengths and higher maximum bearing strengths - particularly for thinner laminates. The enhancement effect due to constraint is associated with a change in failure mode from 100% bearing to bearing cum tension for thinner laminates and bearing cum tension cum cleavage for thicker laminates.

Fatigue bearing behaviour exhibits a high correlation with initial static bearing strength; the residual ‹strength capacity, at a fatigue life $> 10^5$, approaches the initial bearing strength.

REFERENCES

1. Swanson S.R., and Burns J.S., 31st International SAMPE Symposium, (1986), 1070-1077.
2. Tsiang T-H., Composites Technical Review, 6 (1984), 74-77.
3. Godwin E.W., and Matthews F.L., Composites, July 1980, 155-160.
4. Vinson J.R., Polymer Engineering and Science, 29 (1989), 1333-1339.
5. Collings T.A., RAE Technical Report 75127 (1975).
6. Collings T.A., Composites, January 1977, 43-54.
7. Collings T.A., Composites, July 1982, 241-252.
8. Collings T.A., and Beauchamp M.J., Composites, 15 (1984), 33-38.
9. Smith P.A., and Pascoe K.J., Composite Structures, 6 (1986), 1-20.
10. Smith P.A., Pascoe K.J., Polak C., and Stroud D.O., Composite Structures, 6 (1986), 41-55.
11. Collings T.A., Joining Fibre-Reinforced Plastics (Ed. Matthews F.L.), Elsevier Applied Science (1987).
12. Ramkumar R.L., ASTM STP 734, 1981, 376-395.

Table 1

specimen	prepreg	ply nos	lay-up	% 0°-ply
A	a	14	(0°,90°)14	50
B	a	14	[(±45°)/(0°,90°)]7	25
C	c	8	[(0°,90°)/(±45°)/(0°,90°)/(±45°)]s	25
D	c	8	(±45°)2/(0°,90°)4/(±45°)2	25
E*	b	12	[(0°,90°)/(±45°)2/(0°,90°)/(±45°)2]s	16.7
1	d	18	[0°/45°/-45°/0°/45°/-45°/0°/45°/-45°]s	33.3
2	e	18	[0°/45°/-45°/0°/45°/-45°/0°/45°/-45°]s	33.3
3	d	12	[0°/45°/-45°/0°/45°/-45°]s	33.3
4	e	12	[0°/45°/-45°/0°/45°/-45°]s	33.3
5	f	18	[0°/45°/-45°/0°/45°/-45°/0°/45°/-45°]s	33.3
6	f	12	[0°/45°/-45°/0°/45°/-45°]s	33.3

* Specimen E was 35mm wide and the rest 30mm.

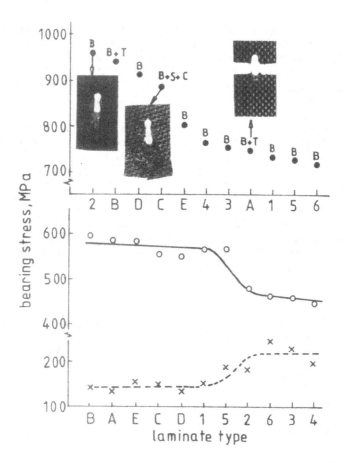

Fig. 1 Bearing stress of various laminate types: (●) σ_m; (o) σ_i;
(x) stress at 4% pin-hole deformation.

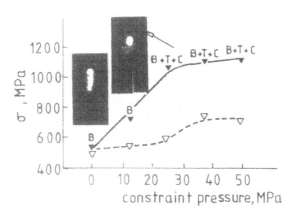

Fig. 2 Influence of constraint pressure on (▼) σ_m and (▽) σ_i.
(Types of failure: B-bearing, T-tension, S-shear, C-cleavage).

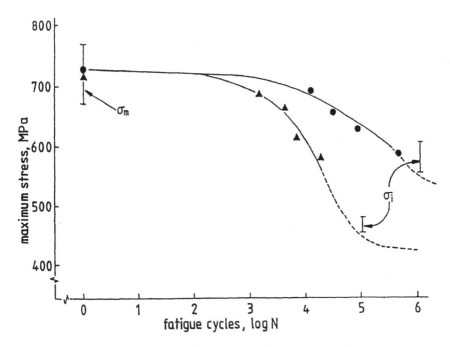

Fig. 3 S-N curve for laminates: (•) type 5; (▲) type 6

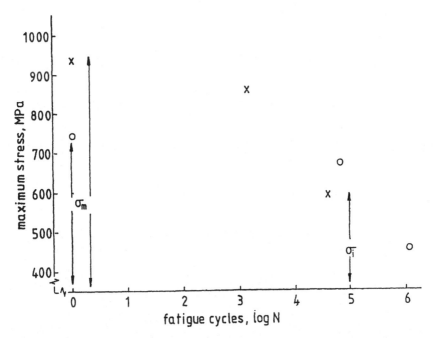

Fig. 4 S-N curve for laminates: (o) type A; (x) type B

MICROBUCKLING BEHAVIOUR OF A PLATE
WITH A HOLE USING IM6/3501, AS4-PEEK,
AND T300/976 COMPOSITE MATERIALS

L.B. LESSARD, F.-K. CHANG*

Dept of Mechanical Engineering
McGill University - H3A 2K6 Montreal - Canada
*Dept of Aeronautics & Astronautics
Stanford University - Stanford - California 94305 - USA

ABSTRACT

An investigation was performed to study the fiber buckling behaviour of three fiber-reinforced composite materials: IM6/3501, AS4-PEEK, and T300/976. The purpose was to analytically compare the response of these materials, using a plate with a hole subjected to compression loading. The results of this study, using computer model PDHOLEC, show inherent differences among the three material systems and how these differences affect the microbuckling behaviour. A parametric study is also presented for T300/976, showing the sensitivity of compression strength to micromechanical parameters

I - INTRODUCTION

One of the challenges for composite designers has been to predict composite behaviour in compression. Compression introduces some hard-to-predict failure modes such as sudden delamination, global buckling, and material microbuckling, otherwise known as fiber buckling or kinking failure. Global buckling has been predicted fairly accurately using finite element methods. Delamination growth has met with some success, while delamination initiation due to in-plane loading has been traditionally difficult to predict. The authors believe that microbuckling failure may initiate delaminations in many cases. In tests previously performed using a composite plate with a hole in compression /1/, rapidly propagating delaminations seem to initiate where microbuckling failure is predicted in cases where the composite laminate contains one or more 0-degree plies. Thus it would appear that microbuckling failure is a very important mode of failure, especially for structures that are not expected to fail by global buckling or other modes of failure.

The objective of this investigation is to examine some aspects of microbuckling failure. It is desired to know what factors may affect the failure at a micro-mechanical level (since this is a micro-mechanical phenomenon). This is accomplished by examining three composite materials and comparing the results in terms of microbuckling failure. Also, the sensitivity of microbuckling failure to individual micro-mechanical parameters is established by altering parameters and noting the effect on the microbuckling failure load.

II - PROBLEM STATEMENT

Consider a composite plate with a hole subjected to in-plane compressive loading (see Fig 1). Microbuckling failure may occur at stress concentrations (hole edge) at critical load levels. Find:

1. Load, and location of microbuckling failure occurrences.
2. Comparison of results from three composite materials.
3. Effect of various micro-mechanical parameters on failure.

III - METHOD OF APPROACH

This work is based on previous work /1-3/ where the formulation of a working microbuckling theory was developed. In the analysis a variational method is used based on the minimum potential energy principle. The composite problem shown in Fig 1 is examined layer by layer. For each layer, the stress concentration area is analyzed using the variational sub-model. Figure 2 shows how the sub-model idealizes the stress concentration area of a layer by a set of identical continuous fibers embedded uniformly in a matrix. Here the stresses on the layer are redistributed as equivalent point stresses on the individual fibers. The fibers are free to deform in an individual manner, and are modeled as beams of appropriate fiber property. The matrix between the fibers is modeled as an elastic medium of appropriate matrix property. The potential energy of the system is calculated, accounting for the following four terms: (1) strain energy due to fiber bending, (2) strain energy due to matrix extension, (3) strain energy due to matrix shearing, and (4) work done by the applied load. The total energy of the system is calculated and then minimized with respect to the fiber displacements in order to determine a critical loading situation. The initiation of buckling corresponds to this critical loading situation. Details of the fiber microbuckling sub-model are not presented here but may found in /3/.

IV - RESULTS

Theoretical microbuckling predictions are presented here for three composite materials: T300/976, IM6/3501, (both graphite/epoxy) and AS4-PEEK (graphite fiber thermoplastic), using the material properties found in Table 1 and geometry of Table 2.

T300/976 is a standard graphite/epoxy thermosetting material used in industry with a high modulus fiber (see Table 1 for properties). It is with this material that the microbuckling model was first developed /2,3/. Extensive tests have been performed for various geometries and ply layups /1/, and it was found that any layup containing 0 degree plies tended to fail in microbuckling mode rather than other types of failure. For the geometry chosen (see Table 2) the $[(0/90)_6]_s$ layup exhibited microbuckling failure at a theoretical compression strength of 433 MPa, confirmed by an experimental value of 426 Mpa /1/.

IM6/3501 is considered to be an "intermediate modulus" material which means that its fiber modulus is somewhere in-between the "high modulus" and "ultra-high modulus" materials. Thus IM6/3501 is a graphite/epoxy thermoset with stiffer properties than T300/976 (see Table 1). For a $[(0/90)_6]_s$ layup, and the geometry of Table 2, the predicted failure stress is 365 MPa.

AS4-PEEK is one of the most promising of the thermoplastic materials. Although its fiber modulus is of the same order as that of T300/976, its lower volume fraction contributes to its slightly lower ply modulus parameters (see Table 1). First a comparison was made with experimental results from Guynn & Bradley /4,5/. For two hole sizes of a 48 layer composite with the same W, and L as Table 2 and a ply layup of $[\pm45/0_2/\pm45/0_2/\pm45/0/90]_{2s}$, the paper /4/ shows experimental strengths of 355.7 and 422.6 MPa while the current model predicts 319 MPa (10% conservative) and 408 MPa (3% conservative). For a $[(0/90)_6]_s$ layup as modeled for the other two materials, the predicted failure strength is 304 MPa.

Here is a summary of the predicted compression strength of the three materials for the geometry of Table 2:

T300/976 : Compression Strength = 433 MPa
AS4-PEEK : Compression Strength = 304 MPa
IM6/3501 : Compression Strength = 365 MPa

Note that IM6/3501, which has the highest fiber strength (E_f) and Longitudinal Compression Property (X_c), does not have the highest open-hole compression strength, probably due to the fact that it has lowest fiber diameter and is thus susceptible to fiber microbuckling. In the fiber buckling submodel, a small fiber diameter leads to a more unstable geometry condition. T300/976 has the highest compression strength of three materials studied.

V - PARAMETRIC STUDY

The parametric study was conducted on the micro-mechanical level. Using T300/976, the parameters examined in the study were fiber modulus, fiber diameter, matrix modulus, volume fraction, and kink length ratio. The latter is the ratio between the length (along the fibers) of buckled fibers to the fiber diameter. This ratio is observed experimentally and gives an idea of the "scale" of the damaged area due to fiber microbuckling. Although parameters in the model were varied individually, it should be kept in mind that not all of the parameters can be considered independent of each other. For instance, the kink length is a complex phenomenon which is probably influenced by all of the other parameters, and also a change in fiber diameter would probably affect fiber modulus. For the sake of this sensitivity study, the parameters are varied one at a time and the results are shown graphically in Figure 3. The results show that microbuckling failure appears to depend more strongly on fiber diameter, kink length ratio and volume fraction, with a weak dependence on fiber and matrix modulus. This is an indication that the failure mechanism is highly geometry dependent rather than property dependent. This is confirmed by the low prediction for the IM6/3501 material

VI - CONCLUSIONS

Microbuckling failure, an important mode of compression failure, is a complex phenomenon which can be described by the current model. This model can be used to examine a composite plate with a hole in order to predict its behaviour under compressive loading. Of the three materials used here, two are accompanied by experimental data which support the model. Parametric studies are also presented which indicate the sensitivity of microbuckling failure to micromechanical geometry parameters. For the geometry tested, T300/976 seems best suited for compression loading.

REFERENCES

1 - Lessard, L. and Chang, F. K., "Damage Tolerance of Laminated Composites Containing an Open Hole and Subjected to Compressive Loadings: Part II--Experiment," J. of Composite Materials, (to appear June 1990).

2 - Chang, F. K. and Lessard, L., "Damage Tolerance of Laminated Composites Containing an Open Hole and Subjected to Compressive Loadings: Part I--Analysis," J. of Composite Materials, (to appear June 1990).

3 - Chang, F. K. and Lessard, L., "Effect of Load Distribution on the Fiber Buckling Strength of Unidirectional Composites," J. of Composite Materials, (to appear June 1990).

4 - Guynn, E. G., ASTM 2nd Symposium on Composite Materials: Fatigue & Fracture, Cincinnati (1987)

5 - Guynn, E. G. and Bradley, W. L., J. of Composite Materials, (May 1989) 479-504

Table 1. Material constants and parameters required for the use of PDHOLEC code with T300/976, AS4-PEEK and IM6/3501.

Moduli Parameters (units)	T300/976	AS4-PEEK	IM6/3501
Longitudinal modulus, E_x (GPa)	156	134	165
Transverse modulus, E_y (GPa)	13.0	8.9	9.6
Shear modulus, G_{xy} (GPa)	7.0·	5.1	4.9
Poisson's ratio, ν_x (–)	0.23	0.28	0.32
Strength Parameters (MPa)			
Longitudinal tension, X_t	1520	2130	2550
Longitudinal compression, X_c^o	1590	1180	1640
Transverse tension, Y_t^o	45	80	59
Transverse compression, Y_c	253	200	N/A
Longitudinal shear, cross ply, S_{cp}	107	160	124
Micromechanical Data (units)			
Fiber modulus, E_f (GPa)	238	231	300
Matrix modulus, E_m (GPa)	3.5	3.8	4.2
Fiber diameter, t_f (μm)	8	7	5
Fiber volume fraction, ν_f (–)	0.66	0.60	0.62
Kink length ratio (L/t_f), n (–)	10-30	14-42	N/A

Table 2. Microbuckling Comparison Geometry

D (cm)	H (cm)	W/D	L/D	Layers	Layup
0.64	0.34	4.0	8.0	24	$[(0/90)_6]_s$

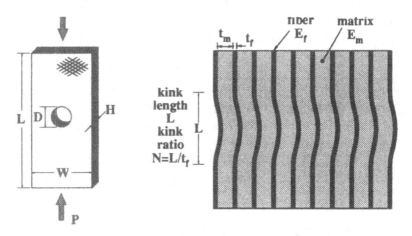

Fig. 1 Basic problem showing geometrical parameters.

Fig. 2 Fiber microbuckling sub-model showing parameters.

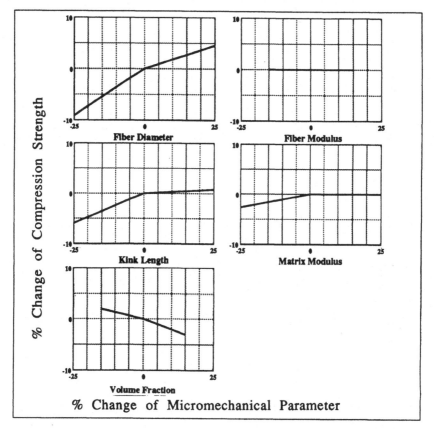

Fig. 3 Parametric study showing effect of micromechanical parameters on compression strength using T300/976.

THE INFLUENCE OF PRESTRESSING ON THE MECHANICAL
BEHAVIOUR OF UNI-DIRECTIONAL COMPOSITES

L.D.A. JORGE, A.T. MARQUES*, P.M.S.T. DE CASTRO*

*Universidade de Tras-os-Montes e Alto Douro
5000 Vila Real - Portugal
*Faculdade de Engenharia da Universidade do Porto
4099 Porto - Portugal*

ABSTRACT

The influence of pre-stressing the roving fibres upon the mechanical properties, namely the tensile strength and the modulus of an uni-directional composite was studied. The influence of strain rate upon the mentioned properties was also studied and a fractographic analysis was carried out seeking to caracterize the types of rupture and to assess the manual moulding technique used.

The composite material studied was a glass fibre reinforced plastic. The matrix was the Crystic 272 unsaturated polyester resin, and the reinforcement was fibre glass roving. A manual moulding process was developped, which made it possible to create the conditions for pre-stressing the material.

1   INTRODUCTION

The present work is part of an on-going effort being carried out by the authors in order to characterize the mechanical behaviour of polymeric matrix uni-directional composites.

The important applications of these materials, namely in the filament winding form, explain the need for a detailed study of the possible improvements of the perfomance of those materials in service. Among the several parameters with an influence upon the composite behaviour, it was decided to study here the effect of pre-stressing the rovings.

The influence of pre-stressing upon the mechanical properties, measured in the fibres direction, was studied using parallel bar type specimens of the composite system (272 Crystic matrix reinforced with fibre glass roving). The study was completed with the assessment of strain rate effects and a fractographic analysis.

The study of pre-stressing took into account the work of Manders and Chou [1] and Chi and Chou [2].

2   EXPERIMENTAL AND DISCUSSION

2.1   Material, specimens and testing conditions

The components of the uni-directional composite studied were the isophtalic polyester resin Crystic 272 and E type glass fibre ECR 144/2400 TEX. The cure system used consisted of Luperox GZN catalyst peroxide metil-etil ketone and a cobalt accelerator which consisted of a solution of cobalt octoate in styrene, with 1% Co.

Rectangular plates were made using a manual moulding technique. The composite plates were produced between two glass plates, separated from the composite by Melinex sheets, in order to facilitate the demoulding process. The rovings were applied through two comb like sets of steel pins, (Figure 1) in two parallel sides of the lower plate mould, producing a separation of aproximately 2mm between the rovings. Each roving made 6 to 8 consecutive longitudinal paths, and a good uniformity of pre-stressing was achieved.

Pre-stressing was carried out by applying suitable loads to the roving extremity, during and up to the complete cure of the composite. To improve the uniformity of pre-stressing, a suitable solid lubrification was applied to the contact of the rovings with the mentioned steel pins.

The resin was applied using two different procedures. One of the procedures consisted of applying first the resin to the Melinex sheet and then to the finished glass fibre assembly using rollers. The other procedure consisted of applying the resin to each layer separately. In both cases, weights were applied to the upper glass plate, seeking to produce a composite without any air bubles. The cure of 24 hours was followed by a 3 hours post-cure at 80oC.

Specimens were parallel bars of 290mm length, 25mm width and thicknesses betwen 3.5 and 5mm, according to the number of layers of roving.

Tests were carried out on an Instron 1125 testing machine at a crosshead speed of 2mm/min, except of course in the tests where the possible strain rate effect was assessed.

2.2   Pre-stressing influence

Manders and Chou /1/ and Chi and Chou /2/ concluded that the dynamic effects of a fibre rupture are more detrimental than the static stress concentration consequences of the preliminar rupture of weakened fibres during the composite manufacture. In particular, the possible rupture of weakened fibres during their pre-stressing may contribute to an increased reliability and strength in service. Unlike /2/, however, in the present experimental work the pre-stressing load was kept constant during the cure of the composite material.

The fibre content of the plates from which the specimens were cut was kept in the range 56% ±4%, since unfortunately it was not possible to ensure an uniform value of this parameter. Figure 2 gives the fibre content of the plates and the corresponding pre-stress value.

Figures 3 and 4 present some of the results obtained in the work, namely the influence of pre-stressing on the strength in the fibre direction (Figure 3) and the pre-stressing influence on the modulus in the same direction, (Figure 4). For the range of values of pre-stress tested, both the strength and the modulus display initially an increasing value with pre-stress, with a tendency to stabilise.

A fractographic analysis carried out on a scanning electron microscope, on the fracture surface of specimens of prestressed and non prestressed material did not display any influence of the pre-stress on the fracture surface morphology. Figure 5 gives a typical example of the fracture surface appearance.

Tests to assess possible strain rate effects were performed on specimens with fibres longitudinally oriented, cut from non prestressed plates. Figures 6 and 7 show the results of tests to measure strength (up to 50mm/min) and modulus (up to 800mm/min) respectively. The strength is almost constant (slightly increasing) for the range of speeds tested, whereas the modulus decreases sharply for speeds above 100mm/min.

3   CONCLUSIONS

For the range of pre-stress values used, pre-stressing increases modulus and strength up to a given level, after which these properties stabilize. It is suggested that this behaviour might be taken into account in filament winding manufacture of composite structures.

The composite displays a linear behaviour up to rupture. Scanning electron microscopy did not reveal differences between fracture surface appearance of pre-stresses and non pre-stressed material.

The manual moulding technique used proved acceptable for the manufacture of this composite.

REFERENCES

1   P. W. Manders, T.-W. Chou, "Enhancement of Strength in Composites Reinforced with   Previously Stressed Fibres", Journal of Composite Materials, Vol. 17, 1983, pp.26-44.
2   Z. Chi, T.-W. Chou, "An Experimental Study of the Effect of Pre-stressing Loose Carbon Strands on Composites Strength", Journal of Composite Materials, Vol.17, 1983,   pp.196-209.

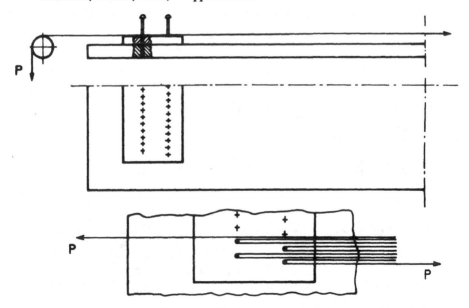

Fig 1 - Rig for the manufacture of the pre-stressed composite plates

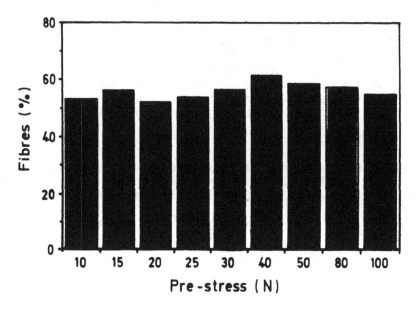

Fig 2 - Fibre content and pre-stress value

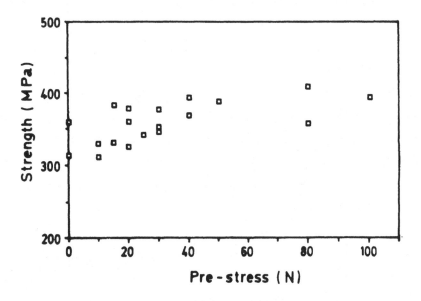

Fig 3 - Influence of pre-stressing on the strength

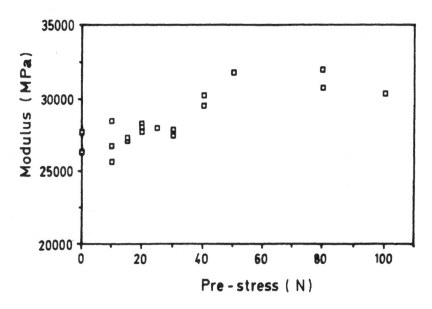

Fig 4 - Influence of pre-stressing on the modulus

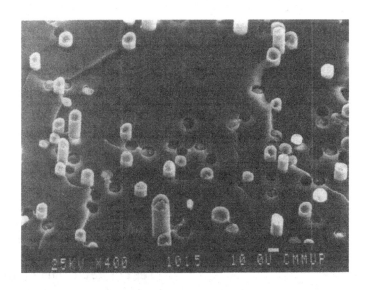

Fig 5 - Typical fracture surface appearance

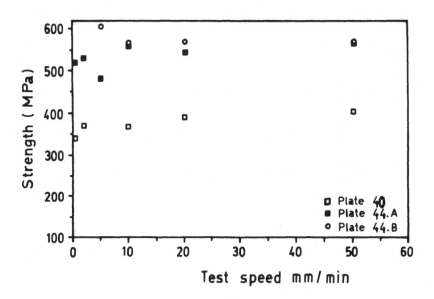

Fig 6 - Influence of test speed on strength (Fibre content: plate
40,56%; plate 44,72%)

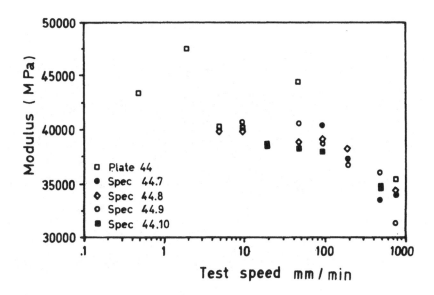

Fig 7 - Influence of test speed on modulus

THE INFLUENCE OF THE FIBRE TRANSVERSAL ARRANGEMENT ON THE TENSION AND FRACTURE OF UNIDIRECTIONAL FIBREGLASS PLASTIC

V. STAVROV, O. KRAVCHENKO*

BSSR Ministry of Public Education
220010 Minsk - USSR
*Polytechnical Institute
48 Oktober Av. - 246746 Gomel - USSR

ABSTRACT

The paper investigates tense and deformed states of glass fibre and bonding component in a unidirectional glas plastic in relation to the arrangement of fibres in th transversal section and applid external load. Regularitie in the distribution of random tensions and deformations i components at orderly and haphazard arrangement of fibre are established through computeraided modelling. The de pendence of the probability of microdeformation incidenc upon the structure of the material under a random ten sion is inferred. The paper also attempts to explain som of the pecularities of fracture of unidirectional fibe glass plastic at a microstructural level.

EINLEITUNG

Das Verformungs- und Bruchverhalten der Kompositen, wen: die Belastung kein Zug in der Faserrichtung ist, hängt voi der Faseranordnung im transversalen Schnitt ab. Diese: Umstand war noch auf den Anfangsstadien der Untersuchungei von GFK als Konstruktionswerkstoffe /1/, insbesnodere beii zweiachsigen Spannungszustand der Laminaten /2/ bekannt Durch die experimentellen Untersuchungen der makroskopi· schen Modellen /3/ wurde eine wesentliche Abhängigkeit dei Spannugskonzentration im Harz von der Faseranordnung bestä· tigt. Im nachfolgenden wurde von G.Gruninger eine sorgfälti· ge Untersuchung der RiBbildung in den Laminaten durchge· führt /4/. K.Stellbrink /5/ versuchte die zufällige Faser· anordnung zu berücksichtigen.Dabei wurden (auch von den an·

deren Autoren, s. z.B. /6/) beim Aufbau der GFK-Modelle die
wesentlichen Vereinfachungen der Struktur eingesetzt. Eine
direkte Untersuchung des Spannungszustandes der GFK-Kompo-
nenten ist technisch unmöglich.Die Berücksichtigung von Un-
regelmäßigkeiten ermöglichen die Methoden der statistischen
Mikromechanik /7,8/. Von einem der Autoren dieses Vortrages
wurde eine neue Methode vorgeschlagen, sog. Realisationsme-
thode. Mit Hilfe derer ist es möglich, Spannungs- und Deh-
nungszustand der Modellen sowohl mit regularen als auch mit
zufälligen Faseranordnungen zu errechnen. Es wird im vorliegenden Vortrag die Anwendung dieser Methode zur Analyse der strukturellen Effekte bei der Belastung eines unidirektionalen GFK beschrieben.

1 - MODELL UND METHODE

Der GFK wird durch ein Modell der unidirektionalen Fasern mit rundem Querschnitt vorgestellt. In der transversalen Ebene , die zur Faserrichtung senkrecht ist, werden die
Kreise der Faserschnitten zufällig oder regelmässig (z.B.
quadratich oder hexagonal) angeordnet. Die Struktur wird an
Computer simuliert. Der Algorithmus sichert bei der Simulation der zufälligen Anordnung der Kreise, die Übereinstimmung des Gesetzes der Durchmesserverteilung, der angegebene
Faservolumenanteil mit dem Modell und die Adäquanz im statistischen Sinne der Modellstruktur mit der Struktur der
realen GFK.
Die Glasfasern und das Harz werden als isotrope und
elastische Medien mit idealer Verbindung angenommen. In den
angeführten Beispielen wurde die Porosität nicht in Bet-
racht gezogen, obschon dies im Rahmen der eingesetzten Methode grundsätzlich möglich ist.
Das statistische Gleichungssystem wird bezuglich der Deformationen mit Hilfe der Funktionen von Green /7,8/ gelost.
Der Einfluß der Faseranordnung wird durch ein zufälliges
Funktional berüksichtigt, dessen Realisationswerte durch
die Integration nach dem repräsentativen Teil des Modells
berechnet werden.
Die Annahme von einem runden Querschnitt der Faser und
von der Isotropie der Komponenten führt zur Vereinfachungen
der Berechnungen, weist jedoch keine Besonderheit der Methode auf: diese Annahme kann nicht in Frage gezogen
werden, möglich sind die Untersuchungen des Einflüsses von
Faserschnittgeometrie sowie Anisotropie, was z.B. bei den
Kohlenstoff- und Aramidfaser wichtig ist.
Die Werte der Spannungen und Deformationen in Glasfasern
und Harz hängen nicht nur von ihren Eigenschaften und ihrer
Anordnung, sondern auch vom angegebenen Spannungszustand
der GFK-Bauteile ab.Darum wird bei den Berechnungen jeweils
ein Spannungstensor oder ein Deformationstensor eingeführt;
die Begrenzungen bei der Untersuchung des beliebigen Spannungszustandes sind nicht vorhanden.Die Berechnungen wurden
sowohl fur die einachsigen als auch für die zwei-und drei-

achsigen Spannungszustände vorgenommen.

Bei jedem Spannungszustand wurden Histogramme der Verteilungen von Spannungen und Deformationen in beiden Komponenten - Glasfasern und Harz - (sog. bedingte Verteilungen) sowohl in den mit der Faserrichtung verbundenen Koordinaten als auch in den Hauptachsen der Tensoren gezeichnet.

Die Adäquanz des Modells und die Genauigkeit der Methode wurden verschiedenartig ausgewertet: 1) durch den Vergleich Elastizitätsmodule, die bei dem verschiedenen Faseranteil rechnerisch und experimentell bekommen waren (es wurden dabei alle Komponente des Tensors errechnet); 2) durch den Vergleich der Konzentrationen von Spannungen im Harz bei quadratischer und hexagonaler Faseranordnung mit den bekannten Versuchsangaben (nach /3/); 3) durch den Vergleich der Mittelwerte der Spannungen und Deformationen in Faser und Harz, die nach verschiedenen Methoden errechnet wurden. Überzeugende Beweise bestätigen die Wirksamkeit der Methode.

2 - RECHNUNGSERGEBNISSE

Die meisten Berechnungen wurden für die GFK mit folgenden Elastizitätskonstanten von Komponenten durchgeführt: Glasfaser - E = 70 GPa, n =0,2; Harz - E = 3 GPa, n = 0,4 E - Moduli; n - Poissonsche Zahl). Variationskoeffizient für Faserdurchmesser - von 0 bis 50%; Glasgehalt nach Volumen - von 0,4 bis 0,7.

Die Berechnungen bestätigen den von einigen Autoren geäußerten Gedanken bezuglich der Unabhängigkeit von Elastizitätskonstanten eines unidirektionalen GFK (und nicht nur des E-Modul längs Fasern /1/) von der Faseranordnung im transversalen Schnitt. Einige Abweichungen sind nur für die Konstanten der Modelle mit quadratischer Faseranordnung im dem transversalen Schnitt erhalten. Daher ist die Benutzung dieses Modells, das besonders auf dem Gebiete der GFK-Mikromechanik /2/ gewöhnlich war, problematisch. Wie es sich herausstellte, hängen die effektiven Elastizitätsmoduli im transversalen Schnitt von der Verteilung der Faserdurchmesser und des Vorhandenseins (oder Fehlens) der garantierten Schichten des Harzes zwischen den Fasern nicht ab.

Ähnliche Gesetzmäßigkeiten sind auch für die Mittelwerte der Spannungen in den Komponenten beim beliebigen Spannungszustand festgestellt.

Zum erstenmal wurden Streuungsparameter, d.h. mittlere quadratische Abweichungen(Standartabweichungen) und die Variationskoeffizienten berechnet, die Histogramme der Verteilungen von Spannungen und Deformationen (in den Achsen X sowie in den Hauptachsen) im Harz und in den Glasfasern aufgebaut. Es wurden ein besonderer Asymmetrietyp und die Verteilungsinversion der Zustandsparameter beim Zug und Druck in den transversalen Richtungen bestimmt. Dabei unterscheidet sich das Verteilungsgesetz für Deformationen und Spannungen in den Komponenten von dem Gauss-Gesetz. Es wurde außerdem festgestellt, daß der Einfluß des Glasgehaltes, der Durchmesserstreuung und Faseranordnung auf das Verteilungsgesetz für die Deformationen und Spannungen in den Komponenten nicht zu hoch ist. Die Verteilungsparameter, vor allem die Variationskoeffizienten, hängen jedoch von der Faseranordnung ab. Nach den Histogrammen von Hauptspannungen und Deformationen im Harz wurden die Wahrscheinlichkeiten von Mikroschädigungen ausgewertet und die Konstanten der Microrißbildungsoberfläche berechnet.

Es wurden rechnerisch die Schädigungsgrenzen aufgebaut, die in der Regel durch die Versuchsmittel aufgezeichneten Mikroschädigungen kennzeichnen. Es wurde dabei angenommen, daß der Schädigungsgrenze nur dann erkannt werden kann, wenn etwa 5% des Harzes beschädigt worden sind.

Die Berechnungs- und Versuchsergebnisse zeigen, daß die Schädigung von etwa 10% des Harzvolumenanteils bei der Belastung in dem transversalen Schnitt in der Regel zu einem spontanen Wachstum des Risses und zum totalen Bruch der Probe und GFK-Bauteilen führt. Auf Grund dieser Voraussetzung wurden Festigkeitswerte berechnet und ihre Abhängigkeit von der Struktur untersucht.

Durch die Analyse der Wahrscheinlichkeiten von Rißbildung wurde festgestellt,daß die bei der Druckbelastung vorhandene Zerspaltung längs Faser der unidirektionalen GFK mit dem spröden Harz durch die Anweichungen der Faserrichtungen von der Druckrichtung verursacht ist. Die maximalen Abweichungswinkel sind bei diesem Bruchtyp berechnet.Es wurde auch die versuchsweise entdeckte Abhängigkeit der Schubfestigkeit in der Faserebene von der Belastungsart geklärt. Bekanntgemacht wurde die Rolle der Faseranordnung und anderer Strukturfaktoren für die erwähnten Besonderheiten des mechanischen Verhaltens von unidirektionalen GFK.

3 - ZUSAMMENFASSUNG

Die ausgearbeitete Methode gewährt neue Möglichkeiten für die Untersuchung der strukturellen Faktoren und Rißbildungen in den kompositen Materialien, für die Optimierung der Struktur und der Eigenschaften der Kompositen in den Bauteilen sowie fur die Prognostizierung des mechanischen Verhaltens von Bauteilen aus den kompositen Materialien unter Betriebsbelastung.

LITERATUR

1 - Kossira H., Diss. Stuttgart (1964)
2 - Puck A., Kunststoffe, 50 (1968) 886
3 - Roth S., Grüninger G., Kunststoffe, 59 (1969) 967
4 - Grüninger G., Diss. Univ. Stuttgart (1975)
5 - Stellbrink K., DLR - FB 77-56 (1977)
6 - Wanin G., Mikromechanik der Kompositen, Kiew (1985)
7 - Kröner E., Statistical Continuum Mechanics, (1971)
8 - Wolkow S.D., Stawrow W.P., Statistische Mechanik der kompositen Materialien, Minsk (1978)

TRANSVERSE CRACKING OF 0/90/0 CROSSPLY LAMINATES OF PMR-15 BASED CARBON FIBRE COMPOSITES

F.R. JONES, P.M. JACOBS, F.R. JONES, P.W.M. PETERS*

University of Sheffield - School of Materials
Elmfield - Northumberland Road - S10 2TZ Sheffield - Great Britain
*DLR - Institut für Werkstoff Forschung
Köln - West-Germany

ABSTRACT

Balanced 0/90/0 crossply laminates have been prepared from PMR-15 unidirectional prepreg using a novel "split laminate" technique. Transverse cracking has been studied in detail by acoustic emission, X-radiography and sensitive load measurements. Consistent values of first ply failure strain and the intrinsic transverse strength of the composite have been measured.

1-INTRODUCTION

Previous studies have highlighted the problem of thermal microcracking of the end-capped bismaleimide, PMR-15 based woven cloth composite. The relatively high curing temperatures have also led to thermal microcracking of crossply laminates on cooling from the curing temperature. Fabrication of good quality PMR-15 0/90/0 crossply laminates for the first time has enabled the transverse cracking phenomenon to be studied. Penetrant enhanced X-radiography has been used to show that microcrack free 0/90/0 laminates can be produced using the "split laminate" technique, and that subsequent post-curing does not result in microcracking of the laminate. Laminates produced in this way have been used to investigate the first transverse cracking strain of both cured and post-cured laminates, and in both cases consistent values for the first transverse cracking strain were obtained. The statistics of the multiple cracking has also been studied.

2-EXPERIMENTAL

It has been found that attempts to fabricate 0/90/0 laminates of PMR-15 by conventional techniques result in transverse cracking of the inner (90) ply of the laminate. To overcome this problem a "split laminate" technique was developed. In this technique the laminate of interest (e.g. 0/90/0 or 0/90) is sandwiched between two thin 0 plies. The three layers are separated by non-porous release film and the whole lay-up is cured in the conventional manner. After curing, the lay-up is removed from the mould prior to post-curing.

Tensile test specimens were cut from the laminate, and coupons from the ends of the specimens were removed for examination by penetrant enhanced X-radiography (PEXR). The penetrant used was a zinc iodide solution, which when applied to the edges of a coupon penetrates into any cracks or delaminations. Initially three configurations were selected for investigation ($0_2/90_4/0_2$, $0_2/90_8/0_2$ and $0_2/90_{12}/0_2$), however, it was found that transverse cracking occurred during curing in the $0_2/90_8/0_2$ and $0_2/90_{12}/0_2$ laminates and only laminates with a transverse ply thickness of $\leq 90_4$ could be used in the transverse cracking experiments. After ensuring that no microcracks were present in the laminates some of the specimens were post-cured. Coupons of the post-cured material were also examined by PEXR to ensure that no microcracks had developed during the post-cure. The laminates were dried to constant weight in an air circulating oven at $105^{\circ}C$ prior to testing.

Acoustic emission (AE) was used to identify the first transverse cracking strain in both cured and post-cured $0_2/90_4/0_2$ PMR-15 laminates. Tensile test specimens, with aluminium end-tags, were tested using a Mayes SM200 Universal Testing Machine. A strain rate of 0.2 mm/min was employed and the strain was measured by strain gauges attached to the specimens. AE was monitored using three transducers, which enabled the AE from a specific area of the specimen to be identified and any extraneous noise rejected.

Post-cured specimens for examination by PEXR were tested under identical conditions to those used for the acoustic emission study. The use of strain gauges enabled the strain in the material to be accurately monitored and the test terminated at a specified strain.

A complementary study, in which a piezo-electric transducer (PET) was used to monitor the transverse cracking behaviour of post-cured $0_2/90_4/0_2$ PMR-15 laminates was also undertaken. The PET detects the instantaneous load drop, which accompanies crack formation during a displacement controlled quasi-static test. This enables the accurate measurement of strain at which each sudden crack occurs, and allows for a more detailed analysis of the transverse cracking behaviour of the material. A complete description of this technique is presented elsewhere/1-2/.

3-RESULTS

AE was used to determine the strain at which the first transverse crack appears. Together with the average longitudinal thermal strain in the transverse plies (ε_{th}^l) determined from the curvature of 0/90 laminate beams/3-4/, the transverse ply failure strain can be estimated. This was found to be 0.89% for a $0_2/90_4/0_2$ PMR-15 laminate prior to post-curing, rising to 1.01% afterwards (Table1).

The PEXR investigation of the post-cured material showed that the first transverse cracks occurred between 0.4 and 0.5% strain (Fig. 1(a)). However, closer examination of the X-radiographs revealed that edge cracks began to develop at applied strains of between 0.3 and 0.4%. This is in good agreement with the value of 0.28% obtained by AE. It was also found that the maximum crack density for this material is 92 cracks per 100 mm, which corresponds to an average crack spacing of 1.09 mm (Fig. 1(b)). However, in areas of fibre misalignment the transverse cracking density increases, which may lead to premature failure of the laminate.

The piezo-electric transducer (PET) enabled the statistics of the distribution of transverse cracking strains to be monitored directly. Using the shear lag analysis, an element length (l_0) of 0.66 mm was calculated so that the 105 mm gauge length was considered to contain 177 elements.

A two-parameter Weibull distribution has been shown to describe transverse cracking data well[1-2], and the probability of failure (F) of the elements is given by

$$F = 1 - \exp - \left[\frac{e_{flo}}{\hat{e}_{flo}}\right]^a$$

where a is the shape parameter and $\hat{e}_{flo}$ is the characteristic fracture strain (at F = 0.632). The data for PMR-15, shown in Fig. 2, demonstrates the validity of this approach during the initial stages of transverse cracking. However, as the strain is increased the probability of failure is reduced, and the data is found to deviate from linearity. This occurs as the shear stress in the shear lag zone approaches either the shear yield stress of the matrix or the interlaminar shear strength.

4-DISCUSSION

It has been known for some time that multiple cracking of the inner ply of a 0/90/0 crossply laminate occurs when a tensile load is applied in the 0 direction. The transverse cracks increase in number with applied load. A simple multiple cracking theory based on the shear lag analysis in which the plies remain bonded was developed by Garrett and Bailey[5]. Decreasing the thickness of the transverse (90) ply of a 0/90/0 laminate resulted in an increase in the transverse cracking strain (e_{tu}) of the 90 ply, whereas the failure strain (ε_{tu}) for the unidirectional laminates were independent of thickness. Parvizi et al[6] defined a lower bound to the failure strain based on the thermodynamic requirement that the release of stored energy accompanying the formation of a crack must equal or exceed the energy of formation of the new crack surfaces. The constraint theory describes the influence of the transverse ply thickness on first ply failure and crack density reasonably effectively. However, this method cannot fully account for the measured distribution of crack spacings. It can be argued that constraint also reduces the severity of any flaws present and therefore probabilistic models have been examined[1,7,9]. The work of Manders et al[7] and Sheard[8] demonstrated that while the early stages of the process are strongly influenced by the presence of defects, once the shear transfer length approaches half the crack spacing the shear lag model is more appropriate. It is also likely that these defects arise from microstructural variations rather than gross laminate defects, as argued by Wang[9]. Peters[1-2] has attempted to predict the cracking phenomenon with a statistical model which employs the modified shear lag concept for the material and geometric factors. The PET data has been analysed according to this model.

The transverse cracking strain in carbon fibre composites with a thermoset matrix can be described with the aid of a two-parameter Weibull function[1]. Applying this analysis to the PET data, it was found that the first transverse cracking strain of 0.34% is, with compensation for ε_f^{tl}, equivalent to a ply failure strain 1.074% and is in good agreement with values from AE and PEXR, as shown in Table 1. Saturation in the multiple cracking can occur but this only occurred for two of the specimens in Fig. 2 ($\Diamond$, +) and for some of the PEXR specimens. This inconsistency probably arises from variations in thermal strain, which has been observed for this material/3/. This may also be responsible for the lower characteristic fracture strains of these specimens (Table 2).

It is evident that the first cracks in the transverse plies occur as a result of the largest defects. Thus, fracture in the lower part of the Weibull distribution is dominated by defect size, whereas in the upper part it is dominated by the intrinsic strength of the material, either the interface or matrix which ever is the lowest. If the last element to break is assumed to be defect free, then its fracture strain can be attributed to the failure of the fibre/matrix interface. This fracture strain (of the 177th element) can be determined by extrapolation, and is indicated in Table 2. The calculated values of interface strength are somewhat lower than for other CFRP systems: 1.75-1.92% for an epoxy composite/2,10/.

5-CONCLUSIONS

1. The first transverse cracking strain appears to be slightly higher for the post-cured material.

2. The decrease in residual thermal strain which occurs with increasing inner ply thickness is insufficient to compensate for the removal of constraint which causes the transverse failure strain to decrease, so that at large inner ply thickness, the 0/90/0 laminates can crack during manufacture. The presence of microstructural defects can also contribute to fabrication difficulties.

3. Areas of fibre misalignment in 0/90/0 laminates resulted in increased transverse cracking, illustrating the nature of the defects responsible for first ply failure, and this may result in premature failure of the laminate.

4. The "intrinsic transverse failure strength", which may occur by interfacial failure, has been estimated as 1.35-1.5%, using the two-parameter probabilistic model.

6-ACKNOWLEDGEMENTS

We acknowledge financial support from MOD and valuable discussions with Prof. G. Dorey (RAE).

REFERENCES

1. P.W.M. Peters, J. Comp. Mat., 18 (1984) 545.

2. P.W.M. Peters and S.I. Anderson, J. Comp. Mat., 23 (1989) 944.

3. F.R. Jones, M. Mulheron and J.E. Bailey, J. Mat. Sci., 18 (1983) 1522.

4. P.M. Jacobs and F.R. Jones, unpublished data.

5. K.W. Garrett and J.E. Bailey, J. Mat. Sci., 12 (1977) 157.

6. A. Parvizi, K.W. Garrett and J.E. Bailey, J. Mat. Sci., 13 (1978) 195.

7. P.W. Manders, T-W Chou, F.R. Jones, J.W. Rock, J. Mat. Sci., 18 (1983) 2876.

8. P.A. Sheard and F.R. Jones, "Computer Simulation of the Transverse Cracking Process in Glass Fibre Composites", in "ICCM-VI and ECCM-2" ,(F.L. Matthews et al, ed.) vol. 3, Elsevier, London (1987) 123.

9. A.S.D. Wang, P.C. Chou and S.C. Lei, J. Comp. Mat., 18 (1984) 239.

10. P.W.M. Peters, "A new method to determine the fibre matrix bond strength", in "IPCM-89"
(F.R. Jones, ed.) Sheffield (1989) 59.

Table 1

First ply failure strain of PMR-15 $O_2/9O_4/O_2$ crossply laminates determined by various methods. (see text for acronyms e.g. AE)

Cure temp. (C)	Post-cure temp. (C)	ε_{tf}^{th} (%)	ε_{tu} (%)	ε_{tlu} (%)		
				AE	PEXR	PET
290	-	0.65	0.24	0.89	-	-
290	316	0.73	0.28	1.01	1.03 - 1.13	1.07 ± 0.07

E_l = 110.8 GPa, E_t = 8.7 GPa, G_m = 1.43 GPa, width of shear transfer zone = 0.015 mm

Table 2

The transverse fracture data for PMR-15 with the Weibull shape parameter a and the characteristic fracture strain $\hat{\varepsilon}_{flo}$

Specimen number	Number of cracks	a	$\hat{\varepsilon}_{flo}$ (%)	Strain at interface failure (%)
1	57	25.45	1.321	1.410
2	78	25.11	1.405	1.506
3	52	22.62	1.263	1.358
4	51	19.56	1.301	1.416
5	86	25.70	1.262	1.346

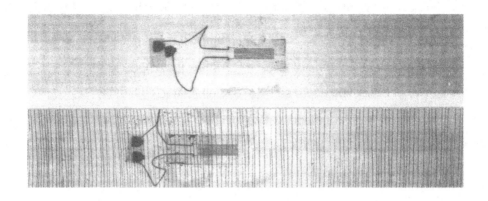

Figure 1. X-radiographs showing the build-up of transverse cracks with applied strain (a) 0.3% , (b) 0.8%.

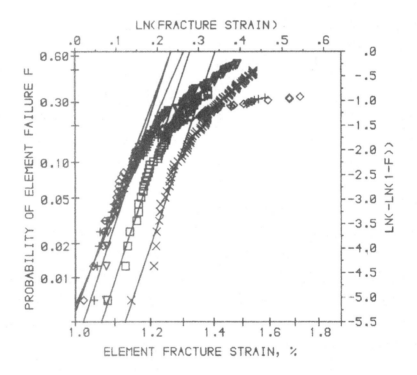

Figure 2. Two-parameter Weibull distributions based on the transverse cracking observed by PET.

COMPLIANCE/CRACK LENGTH MEASUREMENTS
FOR INDIVIDUAL TRANSVERSE PLY CRACKS
IN MODEL ARRAYS

L. BONIFACE, S.L. OGIN, P.A. SMITH

Dept of Materials Science & Engineering
University of Surrey - GU2 5XH Guildford - Great Britain

ABSTRACT

Closed-form expressions are presented for the compliance changes associated with the growth of individual matrix cracks in model arrays in cross-ply laminates. Comparisons are made with experimental results on GFRP cross-ply laminates with large transverse ply thicknesses such that it is possible to control the crack spacing and crack length, while the compliance changes are measurable using simple techniques. Reasonable agreement is achieved between predictions of the strain-energy-release-rate and the measurements based on compliance changes.

I - INTRODUCTION

The failure of composite laminates under tension-tension fatigue loading is not, in general, due to the initiation and growth of a single crack in a self-similar fashion, as in homogeneous materials, but to the growth and interaction of a number of different types of damage. Despite this complexity, isolating or considering particular types of damage, e.g. cracking between layers (delamination) or within layers (intralaminar cracking) enables fracture mechanics techniques to be applied, /see e.g. 1,2/. In this paper we concentrate on the growth of matrix cracks in simple cross-ply (0/90/0) laminates and the resulting compliance change with crack growth. Theoretical results derived for the strain energy release rate are also compared with experimental measurements.

II-ANALYSIS

2.1 Compliance/crack length relationships

The geometry of the cross-ply laminate is defined in Fig. 1. The laminate has a central 90-ply of thickness 2d, either side of which is a longitudinal ply of thickness b, and a width, W. The

913

plies have moduli E_1 and E_2 parallel and perpendicular to the fibres respectively and a transverse shear modulus G_T. The change in compliance when a transverse ply crack, which spans the transverse ply thickness, grows completely across the laminate width can be calculated as follows. A one-dimensional shear-lag analysis /3/ gives the normalised reduction in longitudinal modulus as a result of cracking to be:

$$E/E_0 = 1/[1 + \frac{dE_2}{bE_1} \frac{1}{\lambda s} \tanh (\lambda s)] \qquad \text{eqtn 1}$$

where 2s is the crack spacing, λ depends on the assumptions made in the shear-lag analysis (e.g. linear or parabolic longitudinal displacements in the transverse ply) and E_0 is the uncracked rule-of-mixtures laminate modulus parallel to the loading direction. Now, consider two cracks, A and B, at a distance 2s apart. The modulus of the section of the laminate between cracks A and B is given by eqtn 1. If a third crack appears instantaneously midway between the two initial cracks then it can be shown /4/ that the total compliance change, ΔC, for that portion of the laminate between cracks A and B is:

$$\Delta C = \frac{2s}{2W(b+d)} [\frac{1}{E_s} - \frac{1}{E_{2s}}] \qquad \text{eqtn 2}$$

where E_{2s} and E_s are the cracked laminate moduli, given by eqtn 1, for cracks with a spacing 2s and s respectively. We can attempt a further extension to this analysis by estimating the compliance as a function of crack length a (i.e. as a transverse ply crack advances incrementally across the width of the laminate, W; see Fig 2). In this case a simple "springs in parallel" model can be used to calculate the compliance, C, as a function of crack length giving

$$C = \frac{2s}{2W(b+d)} [\frac{1}{E_{2s} - \frac{a}{w} (E_{2s} - E_s)}] \qquad \text{eqtn 3}$$

2.2 Strain-energy-release-rate expression: compliance approach

The strain-energy-release-rate associated with matrix crack growth can be derived using the compliance expression:

$$G = \frac{P^2}{2t} \frac{dC}{da} \qquad \text{eqtn 4}$$

where P is the load on the laminate, t is the crack thickness (i.e. the transverse ply thickness in this case) and dC/da is the rate of change of laminate compliance with crack length. It is then easily shown that the expression for incremental crack growth is

$$G = (1 + \frac{b}{d}) \sigma^2 s \frac{(E_{2s} - E_s)}{[E_{2s} - \frac{a}{w} (E_{2s} - E_s)]^2} \qquad \text{eqtn 5}$$

where E_{2s} and E_s are the laminate moduli for crack spacings of 2s and s, respectively (given by eqtn 1). Alternatively, we can use the expression for the compliance change when a crack grows instantaneously across the laminate width to obtain an expression for G which is

$$G = \frac{P^2}{Wd} \frac{s}{2W(b+d)} [\frac{1}{E_s} - \frac{1}{E_{2s}}] \qquad \text{eqtn 6}$$

This approach may be generalised readily to include asymmetric arrays of three cracks. The drawback here is that internal balanced thermal stresses generated between the laminate layers during cure do not affect the laminate compliance. Hence, this approach is applicable when these stresses are non-existent or small, as in GFRP laminates with thick transverse plies. To include thermal stresses requires a return to first principles.

2.3 Strain-energy-release-rate expression: energy approach

To analyse the cracking problem incorporating thermal stresses is more complicated and requires consideration of the energy changes when cracking occurs under conditions of constant load. The overall energy balance for the instantaneous formation of crack C is

$$\delta W = \delta U_{el} + \delta T + G.2dW \qquad \text{eqtn 7}$$

where δW is the work done by the applied load and δU_{el} and δT are the change in the stored elastic energy associated with the longitudinal and shear stresses, respectively. G is the strain-energy-release-rate associated with the instantaneous crack growth across the laminate width. Two displacements need to be considered to calculate the work done -(i) the displacement as a result of the increased compliance of the laminate section between the two initial cracks and, (ii) the extra displacement as a result of the relaxation of the thermal stresses. Evaluation of eqtn 7 for the symmetric configuration enables G to be found /5/:

$$G = (1 + \frac{b}{d}) \ (\sigma \frac{E_2}{E_0} + \sigma^T)^2 \ \frac{E_0}{E_1 E_2} s \ [\frac{\tanh(\lambda s/2)}{\lambda s/2} - \frac{\tanh(\lambda s)}{\lambda s}] \qquad \text{eqtn 8}$$

Note that if the thermal stress is set to zero, then eqtns 6 and 8 are identical. Eqtn 8 can also be shown to be equivalent to the expression obtained by Laws and Dvorak /6/ who use a different form of shear lag analysis and method of calculation.

III-Experimental Method

Compliance/crack length measurements were made using cross-ply GFRP laminates with a thick transverse ply (about 3mm), giving a nominal $(0/90_5)s$ configuration. The laminates were fabricated using a simple filament winding technique in which glass rovings ('Silenka' E-glass 600-tex) were wound onto a 300 mm square steel frame in a 0/90/0 configuration and impregnated with the epoxy matrix (details are given elsewhere, see /4/). Plain rectangular test coupons 220 mm in length and 20 mm wide were cut from the laminates using a water-lubricated diamond saw with a 600-grade grit finish. Aluminium alloy end tabs approximately 50 mm long and 1.5 mm thick were bonded to the coupon using epoxy adhesive. Longitudinal laminate displacements were measured using an Instron dynamic extensometer with gauge lengths of 12.5 mm and 50 mm, depending on the spacing between cracks.

IV-RESULTS AND DISCUSSION

4.1 Compliance change due to crack growth

Examples of the compliance change with crack length are shown in Figure 3. In this diagram, AA', BB' and CC' show the progressive compliance change for the growth across the laminate width of cracks A, B and C respectively. (Note that the specimen compliance after the growth of crack A across the width is the same as the specimen compliance at the beginning of the growth of crack B, etc.) For non-interacting cracks (i.e. widely spaced cracks), as in Figure 3(a), the same compliance change occurs when each crack grows across the coupon width and the lines AA', BB' and CC' are parallel. For interacting cracks, the compliance changes are not the same. Hence, CC' in Fig 3b is not parallel to AA' and BB'.

We will first discuss the total compliance change when a crack grows across the laminate width before discussing the progressive change. Figure 4 shows the total compliance change plotted against the A-B crack spacing normalised by the 90-ply thickness. Above an A-B crack spacing of about 3 to 4 ply thicknesses the compliance change is independent of spacing and shows a plateau. This is the regime of non-interacting cracks. Below an A-B spacing of 2 ply thicknesses (i.e. an A-C spacing of 1 ply thickness), it is clear that crack interaction occurs and the total compliance change falls as the crack spacing decreases.

Superimposed on the diagram of total compliance change are

predictions using shear-lag theory (eqtn 6). The predictions for
linear shear-lag and parabolic shear-lag are shown. The predictions
follow the general trends of the experimental data, but none fits the
data well.

We now consider the progressive change in the measured specimen
compliance as a function of crack length for the third crack. Figure
5 shows an example of the compliance increment as a function of crack
length for initial spacings of cracks A and B of about 4.5mm. The
predicted curves of compliance increment as a function of crack
length for various values of the spacing between cracks A and B are
shown in Figure 6. These curves are based on the predicted
compliance/crack length relationship (eqtn 3).

The trends are similar to those shown by the data, but the
analysis is not sufficiently sophisticated to be applicable at the
extremes (near $a/W = 0$ and $a/W = 1$). Bearing in mind the experimental
scatter, the only quantitative measure of compliance change which we
can determine with confidence is the slope of the compliance/crack
length data at a given value of a/W for different crack spacings.
This enables experimental values of the strain energy release rate G
to be determined. Typical results are shown in Table 1, along with
the predicted values of G from eqtn 6 based on the parabolic shear
lag analysis. Agreement is reasonable.

V-CONCLUSIONS

1. Measurements of the compliance change when a crack grows in
   either a symmetric or an asymmetric model array show regimes of
   crack interaction and non-interaction, depending on the spacing
   of the cracks in the array.
2. Predictions of compliance changes based on shear lag analyses
   show the appropriate trends but do not give a good fit to the
   absolute values of the data.
3. Strain-energy-release rate measurements, on the other hand,
   which are based on the slope dC/da, are in better agreement
   with the predictions based on a one-dimensional parabolic shear
   lag expression.

ACKNOWLEDGEMENTS

We would like to thank the SERC (UK) for financial support and
our colleague Mr R Whattingham for technical assistance.

VI-REFERENCES

1. O'Brien T.K. ASTM STP 772 (1982) 140-167
2. Han Y.M., Hahn H.T. and Croman R.B., Comp. Sci. and Tech., 31
(1988) 165-177
3. Steif P., appendix to Ogin S.L., Smith P.A. and Beaumont P.W.R.,
Cambridge University Engineering Department Technical Report
CUED/C/MATS/TR.105 (1984)
4. Boniface L., Ogin S.L. and Smith P.A., presented at the ASTM
Symposium, Orlando, Florida, USA (1989) and to appear in the
associated ASTM STP
5. Boniface L., Ogin S.L. and Smith P.A., submitted to Proc. Roy.
Soc. Lond. (1990)
6. Laws N and Dvorak G.J., J. Comp. Mater., 22 (1988) 900-916

Table 1

G_{pred} calculated from eqtn 5 using a parabolic shear-lag analysis
G_{exp} calculated from eqtn 4 using dC/da determined experimentally at
$a/W=0.5$

Crack spacing 2s(mm)	5.0	12	50	
G_{pred}	7.2	10.8	11.0	Units of G are $\sigma^2 \times 10^{-3} J/m^2$
G_{exp}	5.9	12.0	13.1	with σ in MPa

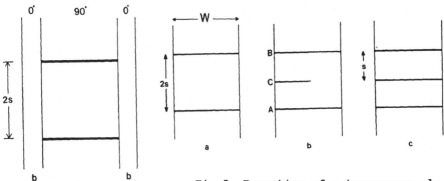

Fig.1 Geometry of crossply
laminate (edge view).

Fig.2 Formation of a transverse ply
crack between two existing cracks
A and B spaced 2s apart.

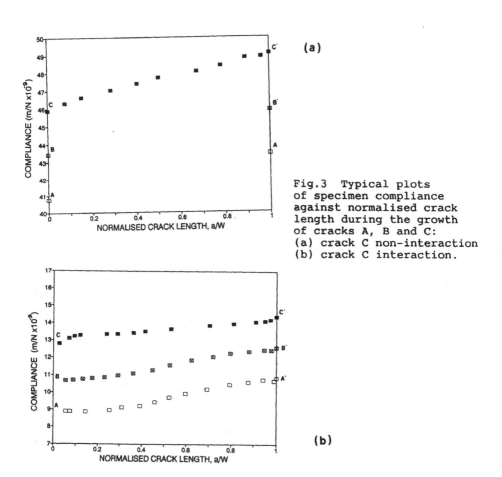

Fig.3 Typical plots
of specimen compliance
against normalised crack
length during the growth
of cracks A, B and C:
(a) crack C non-interaction
(b) crack C interaction.

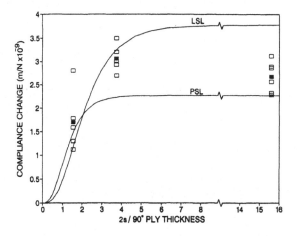

Fig.4 Total change in specimen compliance during growth of crack C from a/W=0 to a/W=1, including predictions from linear (LSL) and parabolic shear lag (PSL).

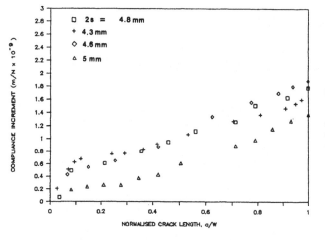

Fig.5 Compliance increment with crack length for the growth of crack C in specimens with cracks A and B spaced bewteen 4mm and 5mm apart.

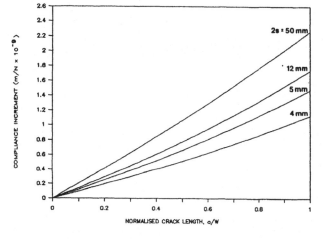

Fig.6 Predicted values of compliance increment with crack length for the growth of crack C at four different spacings of cracks A and B.

THE EFFECT OF POSTCURE ON TRANSVERSE CRACKING STRAIN IN TOUGHENED CARBON FIBER/EPOXY CROSS-PLY LAMINATES

J. YUAN, L. BERGLUND

Composite Group Dept of Mech. Engineering
Linköping University - 58183 Linköping - Sweden

ABSTRACT

The effect of postcure on the transverse cracking strain of HTA/6376 carbon fiber/epoxy cross-ply laminates was studied. Postcure at 200°C was found to reduce the transverse cracking strain due to increased residual thermal strains whereas the reduction in intralaminar G_{IC} due to postcure was negligible.

INTRODUCTION

Carbon fiber reinforced epoxy composites are widely used in aircraft structures. Since damage tolerance of these structures is important, toughened epoxies have been introduced as matrices to improve the composite toughness. Usually, thermoplastics such as polyethersulphone are added to improve the fracture toughness of the epoxy matrix [1]. A recent study of the cure kinetics of the toughened Fibredux 6376 carbon fiber/epoxy prepreg revealed some interesting facts [2]. After subjecting the prepreg to the cure cycle recommended by the prepreg supplier, the residual reactivity was as high as 15%. In other words, although the prepreg was cured at 175°C for two hours, 15% of the reactive groups had not participated in the crosslinking reaction. Incomplete cure is likely to lead to problems with the environmental stability of the materials [3]. For some applications it may therefore be desirable to postcure the epoxy at elevated temperature in order to increase the degree of cure of the matrix. However, postcuring has been shown to decrease the fracture toughness of postcured unidirectional composites /4/. An important question which is addressed in the present study is therefore how postcuring affects the toughness of the composite. The transverse cracking strain of cross-ply laminates was chosen as the primary toughness property to be measured.

I- EXPERIMENTAL

1.1 Materials and specimens fabrication

Fibredux 6376 toughened carbon fiber/epoxy prepreg from Ciba Geigy with HTA fiber from Toho was used. Ten laminates 300x300 mm were laid up, bagged and heated at 3°C/min in the hot press and cured at 175°C and 7 bar for 2 hours. Five of the laminates were postcured at 200°C for 1 hour. Lay ups were $[0_4/90_n]_s$, with n=0.5,1,2,4,8. The fiber volume fraction was 0.62 and the void content less then 0.5% as determined by a point-counting procedure in an optical microscope. Glass fiber/epoxy tab materials were adhesively bonded to the laminates. The specimens were cut with a diamond saw and edge polished /5/. The specimen width was 25.4 mm, length 235 mm with 155 mm gauge length. All laminates and specimens were stored under dry conditions in a desiccator.

In addition, two unidirectional $[0_8]_T$ laminates and two unsymmetric laminates $[0/90]_T$ were made using the same procedure. A total of sixteen compact tension specimens were cut from the two $[0_8]_T$ laminates. The width of the specimens was 50.8 mm and the other dimensions followed the recommendations in ref /6/. The precrack was first cut with the diamond saw and then sharpened with a razorblade.

1.2 Fracture testing and residual strain measurements

The transverse cracking strain of 90-degree layers in cross-ply laminates is difficult to detect, especially with thin 90-degree layers. A step-wise loading and unloading method was therefore used. The cross-head speed was 1.5 mm/min and the strain increase for each loading interval was 0.05%. After each loading interval, specimen edges were examined for cracks using a travelling optical microscope. A minimum of four specimens were tested from each material.

The chord-length and the beam displacement of the unsymmetic $[0|90]_T$ specimens were measured. The residual strain was calculated using the method described by Bailey /7/.

The critical strain energy release rate, G_{IC}, was determined on the compact tension specimens using the compliance calibration technique /8/. Typically five data were obtained from each specimen, loaded at a cross-head speed of 1.5 mm/min.

II- RESULTS AND DISCUSSION

The transverse cracking strain, ε_{cr}, in the 90-degree layers was measured for $[0_4/90_n]_s$ cross-ply laminates cured at 175°C (normally cured) and 200°C (postcured). Since we were also interested in the constraint effect /9/, ε_{cr} was measured as a function of n (90-degree layer thickness as defined by $[0_4/90_n]_s$). In this way we also obtained several independent data for each cure-condition.

The results are shown in Fig 1. For both cure-conditions, the constraining effect of the 0-degree layers is apparent for the laminates in the high constraint region (thin 90-degree layers, low values of n). ε_{cr} is significantly higher there than in the low constraint region (thick 90-degree layers, high values of n). But the most interesting observation is that postcure results in a decrease in ε_{cr}.

The decrease in ε_{cr} with postcure has two likely reasons. One is decreased critical strain energy release rate, G_{IC}, for intralaminar crack growth. This would be caused by decreased matrix G_{IC}, due to the higher crosslink density in the epoxy after postcure. Another possible reason is the increase of residual thermal strains due to the higher cure temperature.

In order to determine the relative influence of the two effects, the residual thermal strains were calculated from the curvature of unsymmetric laminates /7/. These residual thermal strains, ε_{th}, were added to the measured mechanical transverse cracking strains, ε_{cr}, to give the total strain at transverse cracking, ε_{tot}. All these strains are reported in Table I and Fig 2 for both normally cured and postcured laminates. The total strains, ε_{tot}, are similar for the differently cured laminates, except at n=4. These results suggest that the reduced transverse cracking strain resulting from postcure is entirely caused by increased residual thermal strains.

In order to conduct independent experiments to confirm this conclusion, the intralaminar composite G_{IC} was measured for normally cured and

921

postcured laminates. The results are given in Table II. As can be observed there is a relatively small influence of postcure on intralaminar composite G_{IC}. Using the shear lag analysis in ref /9/ it was found that the resulting difference in transverse cracking strain caused by the difference in G_{IC} would be less than 0.01%.

The postcuring effect on the residual thermal strain comes from the effect of postcure on the glass transition temperature, T_g, of thermosets. The residual thermal strains of epoxies have been shown to be linearly related to the temperature difference from T_g to room temperature /10/. T_g will be slightly above the maximum cure temperature /11/, 175°C for normally cured and 200°C for postcured laminates.

CONCLUSIONS

Postcure of HTA/6376 at 200°C was found to reduce the transverse cracking strain of cross-ply laminates. This was shown to be entirely due to increased residual thermal strains, whereas the effect of postcure on intralaminar G_{IC} was negligible. These results illustrate an advantage with the transverse cracking strain of cross-ply laminates as a measure of toughness. Residual thermal strains, which are also present in real composite structures, directly affect the measurements. This is not the case with toughness properties of unidirectional laminates.

ACKNOWLEDGEMENTS

We would like to thank Mr Per Tronarp and Mr Nils-Göran Persson at Saab Scania Aircraft Division in Linköping, Sweden for the donation of prepreg material and access to the results in ref /2/.

REFERENCES

1. M.S.Sefton, P.T.McGrail, P.Eustace, M.Chisolm, J.T.Carter, G.Almen, P.D.Mackenzie, M.Choate, Proc. of third annual international conference on crosslinked polymers, Luzern, Switzerland, May 30-June 1, (1990), p 221-229
2. L.Nicolais, J.M.Kenny, A.Trivisano, Processing of Fibredux 6376/HTA graphite epoxy prepreg, reported to the Saab Scania Aircraft Division, Linköping, Sweden.
3. A.Apicella, L.Nicolais, Adv. in Polymer Science, Vol 72, Springer-Verlag, (1985), p 69-79
4. E.R.Long, Jr., S.A.T.Long, W.D.Collions, S.L.Gray, J.G.Funk, SAMPE, Vol 20, No 3, April (1989), p 9-16

5. H.Toftegaard, Structure, Struers metallographic news, Vol 13, September, (1986), p 14-15
6. A.C.Garg Composites, Vol 17, No 2, April (1986), p 141-149
7. J.E.Bailey, P.T.Curtis, A.Parvizi, Proc Roy Soc, Lond, A366, (1979), p 599-623
8. M.Slepetz, L.Carlson, Fracture Mechanics of Composites, ASTM STP 593, (1975), p 143-16
9. M.G.Bader, J.E.Bailey, P.T.Curtis, A.Parvizi, Vol 3, ICM 3, Cambridge, England, August (1979), p 227-239
10. M.Shimbo, M.Ochi, Y.Shigeta, J. Applied Polymer Science, Vol 26, (1981), p 2265-2277
11. M.T.Aronhime, J.K.Gillham, Advances in Polymer Science, Vol 78, Springer-Verlag, (1986), p 83-114

Table I Transverse cracking strain ε_{cr}, residual thermal strain ε_{th}, and the total strain ε_{tot}, at different cure temperatures and values of n in $[0_4/90_n]_s$ cross-ply laminates of HTA/6376.

n	T_{cure} (°C)	ε_{tot} (%)	ε_{th} (%)	ε_{cr} (%)	$sd^*\varepsilon_{cr}$ (%)
0.5	175	0.88	0.42	0.46	0.05
1	175	0.85	0.41	0.44	0.05
2	175	0.80	0.41	0.39	0.02
4	175	0.75	0.40	0.35	0.09
8	175	0.68	0.38	0.30	0.08
0.5	200	0.90	0.48	0.42	0.06
1	200	0.87	0.48	0.39	0.06
2	200	0.77	0.47	0.30	0.02
4	200	0.67	0.46	0.23	0.08
8	200	0.67	0.44	0.23	0.14

*sd=standard deviation

Table II The intralaminar G_{IC} and the standard deviation of $[0_8]_T$ laminates cured at different cure temperatures.

Cure temperature (°C)	G_{IC} (J/m^2)
175	270 ±45
200	250 ±45

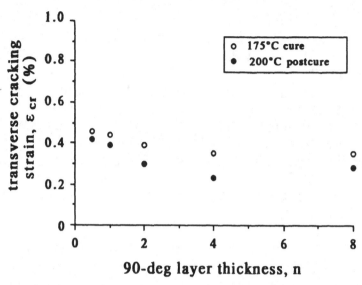

Figure 1. The mechanical strain at the onset of transverse matrix cracking as a function of 90 degree layer thickness for HTA/6376, $[0_4/90_n]_s$ cured at different temperatures.

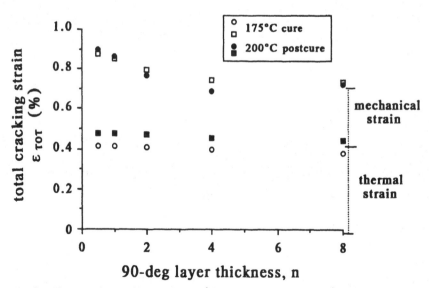

Figure 2. The total cracking strain (thermal+mechanical) as a function of 90 degree layer thickness for HTA/6376, $[0_4/90_n]_s$, cured at different temperatures.

DELAMINATIONS IN COMPOSITE PLATES
SUBJECTED TO TRANSVERSE LOADING

I. VERPOEST, L. LI, L. DOXSEE, M. SCHOLLE*

Dept Metallurgy & Mat. Engineering
Katholic University of Leuven
De Croylaan 2 - 3030 Leuven - Belgium
*DSM-Resins - PB 18 - 6160 MD Geleen - The Netherlands

ABSTRACT

Experiments were performed to study the growth of damage in fibre reinforced cross-ply carbon/epoxy composite plates subjected to transverse loads. The critical strain energy release rate for delamination growth was determined by relating the delamination area to the amount of energy dissipated by the formation of that delamination. This critical strain energy release rate was also compared with the values of the critical strain energy release rates for fracture in mode I and mode II.

INTRODUCTION

Even though the problem of laminated composites subjected to non-penetrating impact has received widespread and intensive investigation during the past several years /1-7/, because of its complexity, the impact event is still not completely understood. The impact event is complex because it consists of the elasto-dynamic response of the plate and impactor, coupled with the development and the growth of several different forms of damage within the composite. Much of the previous research on impact considered the impact process as a whole. In the present investigation, the dynamic aspects of impact are ignored and the attention is placed on just one aspect of the impact process, the propagation of a delamination.

One useful way to quantify the impact event is to characterize how the energy of the entire system is divided among the parts of the system at different times /6,7/. Thus, the goal of the present research was to determine experimentally the amount of energy which is required to propagate a single delamination crack in a laminated composite plate subjected to transverse loading.

APPROACH

The specimens were 55 mm square and 1.8 mm thick plates made from carbon-epoxy (Ciba-Geigy Fibredux 6376-T400-5-35%) with a $[0°_4/90°_4]s$ stacking sequence. This stacking sequence was chosen through trial and error so as to ensure the formation of only one delamination and to minimize matrix cracking. The specimens were manufactured by laminating the unidirectional preimpregated carbon-epoxy tape (width

925

75 mm) by hand. In each specimen two 20 mm diameter, 0.03 mm thick aluminium foils were placed in the centre of the laminate between one pair of 0° and 90° plies. The purpose of inserting the aluminium foils was to initiate a delamination crack, as in double cantilever beam specimens for fracture toughness tests /7,8/. Two pieces of the aluminium foil were sprayed with release agent (Ciba Geigy Transmittal QZ 11 Spray). The laminate was cured in an autoclave using the manufacturer's recommended cure cycle, resulting in a laminate having a 1.82±0.02 mm thickness. This laminate was cut to 55 mm square plate, so that the aluminium foil, which initiates the delamination, was kept in the centre of each plate. Each specimen was tested as follows:

a) The specimen was C-scanned to determine the extent of the initial delamination at the position of the aluminium foil.

b) The specimen was placed on an aluminium plate containing a 40 mm square hole. The specimens were placed in such a way that the initial delamination was centred over the hole and below the midplane of the plate, further from the point of loading. A loading was applied to the plate with a computer controlled servohydraulic test apparatus via an indentor with a 16 mm diameter hemispherical end. The loading was displacement controlled and applied at a rate of 0.2 mm/sec. The load and the displacement of the loading rod were measured and recorded with a computerized data acquisition system for later analysis. Initially, the specimen was loaded to a certain maximum displacement (either 0.2 mm or 0.4 mm) and then completely unloaded. The specimen was then reloaded to the same maximum displacement and completely unloaded again. These two load-unload cycles were called the initial load step.

c) The specimen was C-scanned again to measure the delamination.

d) In some cases, the specimen was then cut in half to either the 0° or 90° direction. The cut surfaces were polished, then examined and photographed by an optical microscope. Next, the cut specimen was soaked in a solution of penetrate dye for a period of 24 hours and finally radiographed.

e) In other cases, the specimens were reloaded to a maximum displacement greater than the previous maximum displacement. For most specimens, the increment in maximum displacement was equal to 0.4 mm but it sometimes was equal to 0.2 or 0.3 mm. The specimens were unloaded, reloaded to the same maximum displacement, and then unloaded again. These two load-unload cycles were called one load step.

f) Steps c through e were repeated several times. After each loading increment, some specimens were cut and examined as described in step d. The other specimens were reloaded. Some specimens were loaded and unloaded up to four times until a delamination was extended along the entire length of the specimen (corresponding to a maximum displacement of approximately 1.4 mm)

In addition to the quasi-static tests described above, drop weight impact tests, which employed the testing apparatus and method described in /6/, were performed on the same type of specimens as described above. After impact, the damage was assessed via C-scanning and optical microscopy.

Finally, double cantilever beam fracture tests were performed following the procedures described in /8-9/. From these tests, the values of the critical strain energy release rates for fracture in modes I and II were obtained.

RESULTS AND DISCUSSION

A typical load versus deflection curve from the quasi-static tests for specimens with aluminium initial delaminations is shown in figure 1. The point where the slope of the load-deflection curve in figure 1 suddenly decreases is the point at which delamination growth initiated during that particular load-unload cycle. Also cracking noise was heard from the specimen. The area within one load-deflection curve is the energy absorbed by the plate during that load-unload cycle and consist of two parts: the energy absorbed by delamintion growth (or other damage), and the energy absorbed by interlaminar friction and the friction between the specimen and the fixture. The area within the second load-deflection curve is the total friction energy. So the delamination energy is equal to the difference between the total absorbed energy and total friction energy.

From the results of the C-scan, microscopy and radiography observations the initiation and formation of damage were closely monitored. Three types of damage were observed: matrix cracking, delamination, and permanent indentation under the point of loading. No fibre breakage was observed. The schematic drawings of the damage recorded two different maximum displacement are shown in figures 2-3.

The relation between the delamination area A_d and the energy dissipated by the formation of that delamination (cumulative delamination energy, U_d) is shown in figure 4. In this plot, each symbol type corresponds to a different specimen and each data point corresponds to a given load step. The data in the plot indicate that the delamination energy is roughly linearly proportional to the measured delamination area. Thus the slope of the linear approximation to the test data is equal to $G_{dc} = U_d/A_d = 630$ J/m^2, and is approximately the amount of energy which was dissipated per unit area of the delamination growth under transverse loading condition.

In addition to the transverse loading tests described above, the fracture toughness of the material was measured by using unidirectional double cantilever beam specimens. The critical strain energy release rate for Mode I (G_{Ic}), calculated with the "Corrected (beam) displacement method", ranged from 250-300 J/m^2, depending on the crack length, in close agreement with the values reported in /8/. The critical strain energy release rate for crack growth initiation in mode II (G_{IIc}) was 1270±150 J/m^2. Thus, the value obtained for the energy dissipated by delamination (630 J/m^2) was within the range of the mode I and mode II critical strain energy release rates for the material, as might be expected.

Figure 5 is the schematic drawing of the damage observed in specimen subjected to drop weight impact. The development of damage in cross_ply composite plates subjected to a slow increase of transverse loads is qualitatively similar to the development of damage in identical plates subjected to impact loads.

CONCLUSION

Damage initiation and development in transversely loaded plates were closely monitored through C-scanning, optical microscopy, and radiography. All these three techniques were necessary to form a complete picture of the development of damage. The energy dissipated per unit area of delamination was determined to be between the critical strain energy release rate of mode I and mode II types of fracture. From the present data, we got a simple rule of mixture:

G_{dc} (600J/m^2) = 2/3 G_{Ic}(275J/m^2) + 1/3 G_{IIc} (1270J/m^2)

The value of the energy dissipated per unit area of (G_{dc}) delamination obtained in the present tests was determined for a specific material, stacking, specimen geometry, boundary condition, and loading condition. It is believed that all of these factors can influence the amount of energy dissipated by delamination growth. The simple rule of mixtures will hence have to be checked experimentally for different values of the above mentioned influencing factors. These data will then be compared with energy values, resulting from model calculations for interlaminar crack growth. Finally, the test method proposed in this paper may be useful for comparing the "delamination toughness" of different materials and stacking sequence, independence of the dynamic response to impact loading.

REFERENCES

1. Joshi, S.P. and C.T.Sun, J. Comp. Tech. and Resch., Vol. 9, 1987, 40.
2. Liu, D. and Malvern, L.E., J. Comp. Mats., Vol.21, 1989, 594
3. Sjoblom, P. and Hartness, J.T., J. Comp. Mats., Vol.22, 1988, 30.
4. Wu, H.Y. and Springer, G.S., J. Comps., Vol.22, 1988, 518.
5. Wu, H.Y. and Springer, G.S., J. Comps., Vol.22, 1988, 533.
6. Verpoest, I.J., Marien, J., Devos, J., Wevers, M., Proc. ECCM-II and ICCM-VI, ed F.L.Mathews et al., Elsevier Applied Scienc, London and Newyork, 1987, Vol.4, 485.
7. Verpoest, I.J., Marien, J., Devos, J., Wevers, M., Composites Evaluation, ed. J.Herriot, Butterworths Scientific Ltd., Seven Oaks, 1987, 69.
8. Hashemi, S., A. J.Williams, and J.G.Williams, J. Mats. Sci. Letters, 8 (1989), 125.
9. Hashemi,s.,A.J.Williams,and J.G.Williams, Proc. ECCM-II and ICCM-VI , ed.F.L.Mathews et al.,Elsevier Applied Science, London and New York, vol.3, (1987), 254.

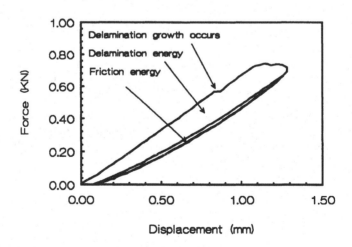

Absorbed Energy = Delamination Energy + Friction Energy

Fig. 1. The load verses deflection curve for two load-unload cycles of the quasi-static test.

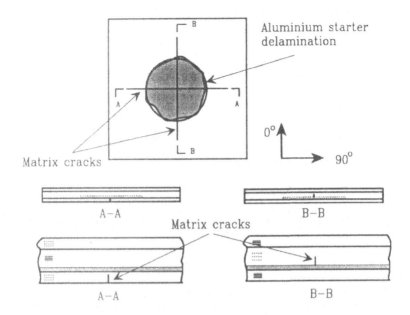

Fig. 2. Damage development after the specimen was loaded to deflection of 0.8 mm. Note that matrix cracks had formed but the delamination had not extended beyond the start delamination.(The bottom two drawings are enlarged cross sectional view on the specimen centre

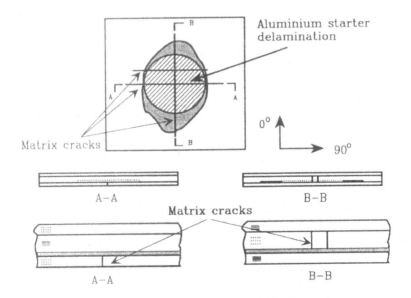

Fig. 3. Damage development after the specimen was loaded to a deflection of 1.2 mm.

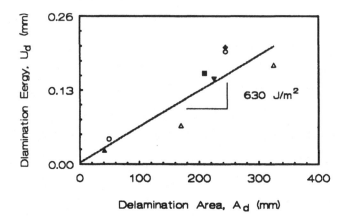

Fig. 4. Energy dissipated by the formation of delamination versus the area of that delamination. Each symbol type corresponds to a different specimen and each data point corresponds to a load step up to a given maximum displacement. The slope of the linear approximation (630J/m²) is the energy per unit area dissipated by delamination.

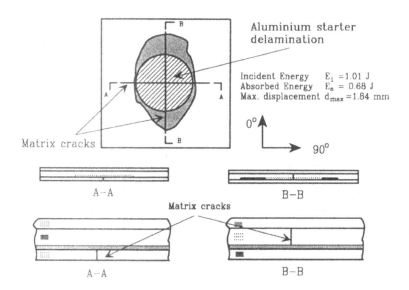

Fig. 5. Damage development in a specimen subjected to a drop weight impact. The mass of the impactor was 0.75 kg and it was dropped from a height of 0.15 mm. The following features were observed: single matrix cracks in the middle and bottom layers and a single delamination at the lower 0°/90° interface.

DEVELOPMENT OF STANDARDS
FOR ADVANCED COMPOSITES

G.D. SIMS

National Physical Laboratory
TW11 0LW Teddington - Great Britain

ABSTRACT

The increasing maturity of the industry and the imminent arrival of the Single European Market with associated European (EN) standards has significantly increased interest in standard test methods for composite materials. All the major worldwide initiatives concern polymer matrix composites (PMC's) but similar requirements can be expected to develop for metal (MMC) and ceramic (CMC) matrix composites. These current initiatives are reviewed in the paper, and a detailed analysis is made of an individual test method in order to assess the potential for harmonisation at an international level (eg EN or ISO).

I   INTRODUCTION

The increasing maturity of the advanced composite industry and the imminent arrival of the single European Market with associated EN standards has increased interest in standard test methods. Polymer matrix composites encompass a wide range of materials from injection moulded systems, through pultrusions to advance pre–preg laminates.

This range of materials has both common and individual requirements for standard test methods, validated data and material specifications. The requirements are wider than the purely mechanical properties that most initiatives cover (eg pultrusions:– product specifications, moulding materials:– shrinkage and mould finish, all:– fire resistance and smoke emission). At one end of the material range where properties are more isotropic "plastic" type test methods may be appropriate but characterisation is difficult due to the orientation and dispersion of the fibres that occurs during processing. For the highly anisotropic laminates the property requirements are more complex but the problem is potentially more tractable due to the greater regularity of the

931

structure arising from the use of pre–preg laminae and minimal flow during processing. Most initiatives relate to these latter materials.

Standards are required at several material levels:– fibre, matrix, interface, intermediate materials (eg pre–pregs), laminate and intermediate products (eg pultrusions); and for several different purposes:– material specifications, materials development, design data, quality assurance, non–destructive testing and research.

Section II of this paper reviews the current initiatives on standard test methods. The potential for harmonisation of test methods is discussed in Section III and examined for the 0° tensile test in Section IV. Concluding comments are given in Section V.

II   CURRENT INITIATIVES ON STANDARD TEST METHODS

2.1 The International Scene

The worldwide establishment of standards for composites is undertaken by the ISO TC 61 (Plastics)/SC 13 (Plastics – Composites and reinforcement fibres) committee. Although some existing ISO standards for plastics, such as ISO 1268 for flexural properties, includes unidirectional glass fibre/epoxy materials within a separate part; the need now is for a series of standards for advanced PMC's incuding carbon fibre and aramid fibre systems. The current list of work in progress shows no mechanical tests for carbon fibre laminates, although individual carbon fibres or carbon fibre strands are included. In addition, other standards such as manufacture of test panels could be revised to be applicable to fibres other than glass–fibre.

ISO TC61/SC13 is undertaking a strategy review of its work programme for the next few years. This review reflects the pressure on the international standards system to produce timely and required standards for relevant materials as shown by the existence of several well developed standards programmes driven by different "trade" groupings. The Japanese Industrial Standards (JIS) association is also likely to play an increasing role in standardisation, particularly for carbon fibres where it has an active trade association of fibre manufacturers.

2.2 The North American Scene

Although the ASTM D.30 committee has been responsible for long established test methods such as D.3039; "Tensile properties of fibre–resin composites", the major impetus is from the Suppliers of Advanced Composite Materials Association (SACMA). SACMA has published a well produced set of recommended methods for laminate tests. Where available the methods are based on existing ASTM standards and in some cases represent an improvement, as in the restriction to one specimen thickness in D.3039. In other cases where new methods are recommended, such as for the holed tension test of multidirectional laminates, the methods have been forwarded to ASTM. ASTM has organised round–robin trials on four test methods (compression of orientated PMC's, compression after impact, open–hole compression and open–hole tension) to obtain "precision and bias" data. Further test methods for carbon fibre properties now in draft form use S.I.

units and reflect the recently increased trend to "hard" metric versions of ASTM standards.

Other initiatives within North America include contacts between American and Canadian standard bodies, a U.S Department of Defence Programme Plan managed by the Army Watertown Materials Laboratory to harmonise defence standards using where possible ASTM standards, development of test methods for filament wound structures by a joint defence grouping (JANNAF) and the revisions of Military Handbook Vol 17 on polymer composites.

2.3 The European Scene

A similar trade initiated development of standards is also present in Europe except in this case the leading group is the European aircraft manufacturers (AECMA) and in particular the ACOTEG grouping of MBB, Aerospatiale and British Aerospace.

The importance of the European standardisation programme, which is a pre-requisite of the 1992 single market, is that the CEC members must publish the EN standard as their national standard and withdraw any existing national or international standard of the same scope. In fact, European EN standards only exist as these national standards. The CEN (Comite European de Normalisation) organisation, which includes both CEC and EFTA countries, where possible will adopt existing ISO standards for re-ballot using the normal weighted voting system. However, for advanced composites there are few suitable standards.

AECMA (Association Europeene des Constructures de Material Aerospace) has obtained equal status to national standards bodies and can propose standards to CEN for a European – Aerospace series for reinforced plastics. The first of these standards for interlaminar shear strength of glass reinforced plastics (EN 2377) has been published recently. The ACOTEG group quotes EN standards where available in its material specification document and will propose its own standards via AECMA to CEN where they do not exist. The European Trade Association of Advanced Composite Material Suppliers (ETAC) is negotiating a joint agreement to use these methods. Such an agreement would provide a powerful harmonisation drive covering aerospace requirements.

A "Plastics" work area has been set up in CEN which will include a sub-committee on composites. It remains to be seen whether the requirements of aerospace and general engineering can be satisfied by a single series of test methods; possibly with different tolerances if required for the two application areas.

2.4 The UK Scene

Meanwhile, in the UK the aerospace based CRAG (Composites Research Advisory Group) tests are most frequently requested in a quotable form. However, following their adoption and proposal to BSI as the basis of a set of standards for general use of composites by the Advanced Composites Group of the British Plastics Federation, no progress towards standards has been possible due to an absence of evidence for the validation and precision of the methods. These methods are now more likely to form the UK input into ISO and CEN standards.

III THE POTENTIAL FOR INTERNATIONAL HARMONISATION

There are many driving forces towards international harmonisation of test methods, such as the international nature of the supply and·user industries in the dominant aerospace industry, which increase the importance of ISO test methods, preferably with identical EN test methods.   The increased need for standards for use in many aspects of traceability, certification, free trade, product liability etc, increases the need for validation of test methods and that the validation should be at the highest possible level (ie ISO − International Standards Organisation).

The overriding need to harmonise the standards has been recognised within the USA by formation of an ad−hoc group consisting of representatives of many of the above organisations to consider the implementation of a national "Industrywide standardisation" plan.   Consequently. a paper on the "Potential for Rapid International Harmonisation of Advanced Composites" presented by the author was received sympathetically at both SACMA and ASTM meetings in October and November 1989 respectively.   This potential exists because many of the test methods proposed have common roots and reflect the dominant position on test methods and material specifications of the aerospace industry.   Now is an opportune moment to attempt this harmonsiation and ·this is one of the aims of current work at the NPL.

The VAMAS (Versailles Project on Advanced Materials and Standards) programme is a useful facility for international co−operation on new test methods where further research and validation is required.   The programme will enable more rapid international harmonisation of standards.   Current VAMAS projects for polymer matrix composites include round−robins on creep, fatigue (organised by NPL) and fracture toughness test methods.   Fracture toughness delamination tests are an area already involving international co−operation.   ASTM, JIS and EGF (European Group for Fracture) are co−operating on a joint protocol and round−robin programme for G_{IC} and G_{IIC} delamination.

IV COMPARISON OF TEST METHODS

In this paper it is only possible to use one of the test methods being assessed for their potential for harmonisation for illustration.   Ultimately experimental test results as part of the essential validation exercise are required, and these used to resolve remaining points of conflict.   A considerable step towards harmonisation may however be obtained purely from examination of the test specification.

Table 1 contains the essential points of several current procedures for the tensile strength of unidirectional material.   The methods are CRAG, draft aerospace series EN's, earlier non−mandatory EN equivalent to the ISO, ASTM and JIS.   Note that while some standards refer to all or a range of fibres (eg textiles−sized such as carbon, glass and aramid fibres), others such as the EN series refer to individual fibres.

The main similarity of these test methods is that they refer to plain, unwaisted strips of rectangular cross–section with the load applied through the tabbed ends of the specimen. Specimen thickness is normally 1 mm. The earlier preference for specimens waisted through the thickness using a large diameter profile, in order to keep the shear stresses below the shear strength, has been discontinued as the higher ultimate tensile strengths recorded were not obtained in the 0° plies of multidirectional coupon specimens or components. Further study is required to explain the difference but likely influences leading to the higher strength are the smaller volume of material, the lower chance of fibre ends within the much shorter gauge length and the absence of stress concentrations effects from the tabbing in the waisted specimens.

The straight specimen increases the importance of the shape and material used for tabbing. The stress concentration at the tab is influenced by a number of factors including the tabbing adhesive used. Finite element analysis has shown that a tapered tab gives the lowest stress concentrations.

Specimen width also varies, but all specimens fall within the range of 10 to 20 mm. While there is a need to ensure the specimen is a representative sample, and wider specimens may relate better to multi–directional specimens of similar width, there are also effects of alignment of both the grips and the specimen. Depending on the type of misalignment the effect can both increase and decrease with increasing specimen width. It would appear that a preferred size of say 12 or 15 mm, and an acceptable range of 10–20 mm would satisfy most requirements.

Differences also exist in the measurement of the elastic modulus. The current trend is to a chord value (also referred to as secant) using predefined points, normally in strain, but no consistency has yet been achieved. The most likely solution is for two strain related values, such as 0.1% and 0.3% which avoids the difficulties of defining the zero strain data point.

Given a constructive atmosphere it is clear that a harmonisation of this test method should be possible. Similar evaluations of other test methods show that there are several suitable for harmonisation. These possibilities should not camouflage the need for improved understanding of these tests (eg compression), new test methods (eg through–thickness properties), improved versions of some existing ones and better detailed specifications.

V CONCLUSIONS

There are several active initiatives on standards; also increasing acceptance of the need for rapid harmonisation. For several methods, as illustrated here for 0° tension, there is no large difference in the test philosophy but several detailed differences which are frequently historical rather than technically based. Equally other factors such as specimen preparation, edge finish, tabbing adhesive etc are not specified and may be more influential than the previously noted differences. The similarity of the methods confirms the potential for harmonisation. The cost of validation and the world–wide applicability of the tests suggests that the ISO is the appropriate level for this harmonisation.

935

TABLE 1 - Longitudinal tensile strength and modulus of unidirectional PMCs

Standard	CRAG 300(88)	prEN^ 2561(89)	prEN 2747(89)	EN*/ISO 61-71/3268-78	ASTM D3039-76	JIS 7073-88
Materials	Unid. C,G,A	Unid.C	Unid.G	GRPs	All FRPs Ef >20GPa	C
Specimen thickness, h (mm)	1	2 or 1*	1-3 3(to 1 waisted)	2-10	.5 to 2.5(C) .7 - 3.3(G)	1
Specimen width, b (mm)	10-20	10 or 16	10	25(50)	12.7	12.5
Specimen length, l (mm)	200-250	250	164	250	203	≥200
Distance between tabs (grips) (mm)	100-150 (-)	90-150 (76-136)	64 (50)	150 (170)	127 (-)	>100 (-)
Tab length (mm) /shape	> 50 straight	80-50 straight	> 50 straight	> 50 straight	> 38 tapered	- 50 tapered
Tab thickness (mm)	0.5 - 2	0.5 - 2	0.5-1	1-3h	1.5 - 4h	1-2
Tab material	Aluminium	± 45°	lower modulus	lower modulus	0/90°GRP	Aluminium or GRP
Modulus measurement	Secant 0 to 0.25% strain	Secant 0.1 to 0.5 of peak load	Initial tangent or secant 0 to 0.5% strain	Initial or secant 0 to 0.1% strain	Initial tangent	Initial tangent
Comment	Use CRAG 302 for fabrics	untabbed also	Also fabrics + injection moulded specimens	ISO Revision overdue	Also 90° and cross plies	Also 90° and cross plies

[NB. EN61 [28] published in UK as BS 2782 Method 1003] [^AFNOR T57-301(88) similar but tab specimens only.]
[* note indicates not necessarily suitable for unidirectional material.]
[SACMA as ASTM but h = 1 mm, l = 230 mm, tabs ~ 50 mm] [ACOTEG 0° as prEN 2561]

SHEAR IN CARBONE-EPOXY LAMINATES
AT VARIOUS STRAIN RATES

B. BOUETTE, C. CAZENEUVE, C. OYTANA*

Etablissement Technique Central de l'Armement
16 bis Avenue Prieur de la Côte d'Or
94114 Arcueil - France
*Laboratoire de Mécanique Appliquée
25030 Besançon - France

ABSTRACT

This paper describes interlaminar shear tests on T300/5208 carbon-epoxy laminates at high and low strain rates. The effect of strain rate is thus highlighted, particularly by the microfractographic investigations. The shear specimens used in the tests were subjected to a critical analysis based on finite element methods. These calculations have permitted the interpretation of experimental data, and the optimization of specimen geometry.

INTRODUCTION

Composite material damage is mainly in the form of shear and delamination and the behaviour of these materials seems to be more or less influenced by strain rate /1-2/. This is why the study of shear behaviour in composites, and more precisely of interlaminar shear, was undertaken.

I – EXPERIMENTAL

1.1. Material

The material tested was a carbon/epoxy composite (T300/5208), made by autoclave process, from prepregs. The elastic constants of the material, shown in Table 1, were determined through tensile tests on plane specimens /3/.

1.2. Specimen design

A large number of shear tests for composite materials are proposed in the literature /4-7/. The specimen we chose to generate interlaminar shear had to be useable under dynamic conditions ($\dot{\varepsilon} \approx 10^3 s^{-1}$: area explored by few authors /7-12/).

A first specimen, based on Harding's work /7/ was developed (figure 1). This specimen, of the double overlap type, consists of 3 x 16 plies of unidirectional T300/5208 material. The metal spacer required for specimen manufacture is not a source of stress, since its thermal expansion coefficient is close to that of the composite. This specimen can be used for failure strength tests.

For the determination of the interlaminar shear modulus (G_{13}), another type of specimen, with single overlap, was developed. Figure 2 shows this specimen, which consists of 24 plies of T300/5208 unidirectional composite flanked with 5 mm thick steel plates to reduce the normal stress component. The two steel plates were bonded at room temperature. Specimens were then machined from the plate thus obtained. The working zone is located between the two notches; the size of the shear zone can be varied by changing notch shape, depth and location. Strain is measured using an Althof extensometer /13/ placed on both sides of the shear zone. Under dynamic conditions, strain will be measured using gauges or an optical extensometer.

1.3. Testing systems

The quasi-static tests ($\dot{\varepsilon} \approx 10^{-3} s^{-1}$) were performed using a mechanical INSTRON 6025 machine with displacement control. At high strain rates ($\dot{\varepsilon} \approx 10^3 s^{-1}$), the test set-up operates on the following principle. The specimen is placed between an "input" bar and an "output" bar. A projectile, propelled by a carriage pulled by two elastic bands, slides on two rails and strikes an anvil, thus generating a tensile wave which propagates through the input bar. The output bar is instrumented with strain gauges. Knowing the caracteristics of the bars, the stress transmitted by the specimen can be determined from the gauge signal recorded by a NICOLET digital oscilloscope.

II - FINITE ELEMENT CALCULATION

A finite element analysis was performed on the two specimen types described above.

The first specimen was modelized using the ZEBULON isotropic finite element code. It was assumed that the specimen consists of a very thin resin layer (with modulus E = 3130 MPa) subjected to shear between isotropic supports with modulus E = 125 000 MPa. Calculations were made for two values of overlap length (5 mm and 10 mm). It can be seen that decreasing the overlap length tends to make shear stress uniform, and to greatly reduce the normal stress (the normal stress is the S_{yy} opening stress).

The second type of specimen was investigated in further details using the ABAQUS finite element code. The effects of overlap length and shear height were overviewed. Stress distribution in the shear zone was analyzed. In the center part of the shear zone, the S_{xy} stress is more or less uniform, and presents one or two maxima, depending on the overlap length ; The S_{yy} stress is not egal to zero all along the overlap. A plot of maximum reduced shear (S_{xy} max/S_{av}) and normal (S_{yy} max/S_{av}) stresses against overlap length (Figures 3 and 4) for a given shear height was made.(S_{av} is the average shear stress i.e. the strength divided by the overlap surface). It was pointed out that reducing the shear height is equivalent to increasing the overlap length. Further, it is found that clamping, i.e., preventing the normal stress by securing the bounds, is highly beneficial with long overlaps, but less so with shorter ones. It is found that S_{xy} max/S_{av} increases (the ideal case being = 1) when S_{yy} max increases. At short overlaps, e.g., 6 mm, there is little S_{yy} with or without clamping. The material under test being relatively stiff (approximately 6 GPa in the shear mode), the rise distance of the shear stress occurs over approximately 1 mm irrespective of overlap length. The ratio rise distance on overlap length ceases to be negligible at short overlaps. This accounts for the fact that, in this last case, S_{xy} max/S_{av} = 1.3, whereas S_{yy} max/S_{av} is close to 0.

III – RESULTS AND DISCUSSION

Double overlap specimens were tested at rates of $1.6\ 10^{-2}$ mm/s and 1.10^{4} mm/s. Failure stresses at these two rates were 38 MPa and 60 MPa respectively. Our values of failure stresses, although comparable to those found by others /7/ are relatively low. This is probably due to the S_{yy} stress, causing premature breaking of the specimen. Normal stress was pointed out through calculations, and its existence was confirmed by examinations of fracture surfaces using a scanning electron microscope. The fracture surfaces show resin hackles characteristic of resin shear /14/. There are fewer hackles at the overlap ends : therefore, this is no longer a pure mode II. However, the fracture surfaces of the specimens subjected to dynamic loading show a high density of resin fragments. This type of failure by fragmentation at high strain rates has been noticed in other brittle materials /15/.

The tests on the single overlap specimen, which was also used to determine the interlaminar shear modulus G_{13}, covered several notch sizes and shapes. It was noted that notch shape had a definite effect on failure stresses. For an overlap length of about 7 mm and a shear height of 1 mm, failure stresses were respectively 32 MPa, 36 MPa, and 38 MPa for square notches, notches with R = 0.5 radius and notches with R = 1 radius (the stress concentration being reduced when R = 1). Experimental and calculated data were fitted to determine G_{13}. Like the numerical simulation, mechanical testing is performed for a given displacement. For a given force, the experimental displacement, measured by the Althof extensometer, located on the overlap sides, is related to the theoretical displacement of the nodes occupying the same position. One finds G_{13} = 6.0 GPa. This value differs little from the "in-plane" shear modulus determined using a plane specimen (G_{12} = 5.6 GPa). To determine the interlaminar shear modulus experimentally, it is necessary to measure strain. For a given shear height adequate for such a measurement (h_{cis} = 3 mm), overlap length can be optimized. The curve of Figure 3 shows that the best overlap length is between 10 and 20 mm, with a minimum value of $S_{xy\ max}$ at 15 mm. (the ideal case being $S_{xy\ max}/S_{av}$ = 1). In contrast, the curve of Figure 4 shows a large increase in $S_{yy\ max}$ with length of overlap (the ideal case being 0). Therefore the best compromise seems to be L = 10 mm, the length selected for static and dynamic testing. The best results would be obtained with clamping.

CONCLUSION

The design of specimens for the determination of the interlaminar shear behaviour of composites at low and high shear rates was found to be a complex undertaking. The interlaminar shear failure stress was seen to increase by approximatively 50% with the strain rate. A change in fracture surfaces appearance was also noted in the case of the double overlap specimen. This specimen, while inadequate for modulus prediction, and showing a large normal stress component, permits the assessment of failure stress dependence on strain rate. To determine the shear modulus, it was found necessary to develop another specimen, which was the subject of a detailed numerical analysis. This development resulted in specimen optimization. Work is currently in progress to determine the fracture characteristics in pure modes, as well as the plane shear (torsional mode) behaviour.

REFERENCES

1. A. Valentin, C. Cazeneuve, Rap ETCA 88R101 (1988)
2. C. Cazeneuve, Rap ETCA 87R001 (1987)
3. S.W. Tsai, T. Massard, Composites Design 4th Ed. Think Composites (1988)
4. D.F. Adams, R.L. Thomas, 12th National SAMPE Symposium AC5
5. K. Kamimura, Thesis 10 juin 1981   Université Technologique de Compiègne
6. M. Kumosa, D. Hull, Elsevier Appl. Sc., 3 (1987) 3243-3253
7. J. Harding, 1ère Ecole d'Automne CODEMAC Guéthary (1988)
8. J.L. Lataillade, F. Collombet, J. Pouyet, Materiaux et Techniques,4, (1989) 43-49.
9. T.Parry, J. Harding, Univ. Oxford Rep OUEL 1635/81 (1981)
10. Z.G. Liu, C.Y. Chiem, 5ème Journéés Nationales sur les Composites Paris 9-11 (sept 1986)   741-755
11. P. Kumar, B. Rai, J. de Phys., Coll. C3 Supl 9 Tome 45   (sept 1988) 97-103
12. S.M. Werner, C.K.H. Dharan, J. of Comp. Mat. 20 (1986). 365-374
13. F. Schmit, R. Pauliard, J.C. Crasson, Rap ETCA 88R084 (1988)
14. M.F. Hibbs, N.L. Bradley, Fractography Modern Engineering Materials – Comp. & Metals –Nasville Nov 18-19 (1985) .68-100
15. B.R. Lawn, T.R. Wilshaw, Fracture of Brittle Solids – Cambridge Univ. Press (1975)

Vf (%vol)	E_1 (GPa)	E_2 (GPa)	G_{12} (GPa)	v_{12}	v_{21}	v_{23}^*
63	125	10.5	5.6	0.28	0.024	0.5

Table 1 : Elastic constants T300/5208 1 : fibres direction
2 : tranverse direction
* Ref /3/

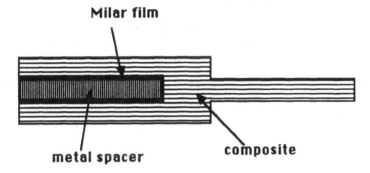

Figure 1 : double overlap specimen

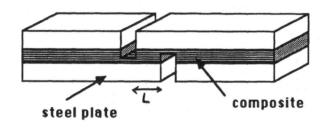

Figure 2 : simple overlap specimen

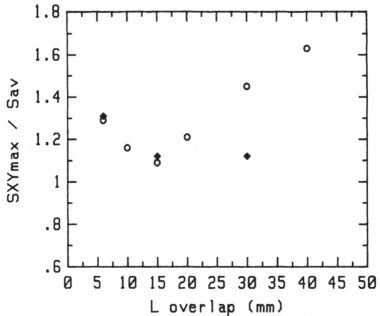

Figure 3 : reduced shear stress against overlap length

○ without clamp   ◆ with clamp

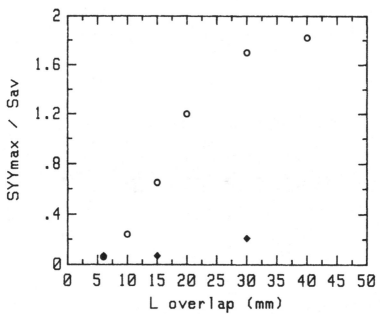

Figure 4 : reduced normal stress against overlap length

○ without clamp   ◆ with clamp

VISCOELASTICITY BEHAVIOUR
OF FIBRE REINFORCED COMPOSITE

B. MOUHAMATH

Commissariat à l'Energie Atomique
BP 12 - 91680 Bruyères-le-Chatel - France

ABSTRACT

Fibre reinforced composite materials such as carbon/epoxy can behave in a viscoelastic manner, and this should be taken into account when environmental conditions and the duration of applied loads are considered. To characterize the viscoelasticity of the T300/M10 composite, the dynamic mechanical technique has been used. Based on the classical time-temperature superposition principle (TTSP), accelerated tests results have been used to analyse the durability of the material. The creep function developed from the rheological model using parabolic elements has been evaluated for more than ten years.

INTRODUCTION

Fibre reinforced plastics (FRP), such as polymeric composites, are important candidate materials for many areas of engineering applications, especially when high weight efficiency and severe service environments limit the use of metals. However, it is well-known that while composite materials exhibit considerable potential properties, they also exhibit a few "undesirable" mechanical properties. Viscoelastic effects is one of their "undesirable" properties, they are affected by a variety of factors, including time, stress level, temperature, moisture conditions,..., and also their thermal history. The purpose of this investigation was to study the dynamic mechanical behaviour of viscoelastic composite material using an inversed torsion pendulum technique developed by METRAVIB (Société pour la mesure et le traitement des vibrations et du bruit). Prediction of the viscoelastic materials response can be a time consuming procedure, so that accelerated tests and the classical time-temperature superposition principle (TTSP) have been used. The rheological model with parabolic elements obtained by a transposition of the COLE-COLE dielectric model is presented for the evaluation of the creep of the material.

I - ELEMENTARY THEORY

The principle of the dynamic mechanical test method is to impart a small sinuoïdal mechanical strain onto a viscoelastic solid at a frequency ω. The sinuoïdal stress which results in response is resolved into real and imaginary components. The changes of these components correspond to the variation in the state of molecular motion in the material as temperatures and frequencies are scanned. The stress and the strain are not exactly in-phase (except when a stress is applied to a perfectly elastic solid) but out-of-phase with an angle δ ($0° < \delta < 90°$). In shear deformation, we get the complex dynamic shear modulus G^* :

$$G^* = \frac{\sigma}{\gamma} = \frac{\sigma_0 e^{i(\omega t + \delta)}}{\gamma_0 e^{i\omega t}} = \frac{\sigma_0}{\gamma_0} e^{i\delta} = \frac{\sigma_0}{\gamma_0}(\cos\delta + i\sin\delta)$$

$$= G' + i\, G''$$

G^* can be resolve into elastic and viscous components, G' and G''. They are called respectively the storage and the loss component.

$$G' = \frac{\sigma_0}{\gamma_0} \cos\delta \qquad \text{and} \qquad G'' = \frac{\sigma_0}{\gamma_0} \sin\delta$$

It is useful to refer at the dimensionless ratio, called the loss ratio :

$$\tan \delta = \frac{G''}{G'}$$

1.1 Rheological models

Linear viscoelastic can be described by the classical models MAXWELL or KELVIN formed with the simple elements, a spring (elastic behaviour) and a damper (viscous behaviour). These models have only one relaxation time. However it is found that the behaviour of polymers and hence, polymeric matrix composites, cannot be represented by one relaxation time. Therefore, MAXWELL or KELVIN assembly models are necessary to describe their behaviour. These models combine many parameters and a distribution of relaxation times and rapidly become complicated. To reduce the number of parameters, K.S.COLE and R.H.COLE / 1 / introduced parabolic elements in the study of dielectric problems. A parabolic element is definited by the creep function :

$$f(t) = A t^h \qquad \text{with} \quad 0 < h < 1$$

Later, in the polymer field, HUET / 2 / measured the dissymetric character of moduli curves $G'' = f(G')$ by using a model with two parabolic elements, called the biparabolic model. This model corresponds to the shear modulus equation :

$$G^* = G^0 + \frac{G^\infty - G^0}{1 + (i\omega\tau_1)^{-h} + (i\omega\tau_2)^{-k}}$$

where G^0 is the static modulus,

$G^\infty - G^0$ representes the mechanical dispersion,

τ_1, τ_2 are the relaxation times and h, k are the powers of the creep function.

Assuming that the time-temperature superposition principle / 3 / can be applied to the material, a variation of temperature corresponds to mutiplying all the relaxation times by the same factor :

$$\mu = \frac{\tau_1}{\tau_2}$$

This gives :

$$G^* = G^0 + \frac{G^\infty - G^0}{1 + \delta(i\omega\tau)^{-h} + (i\omega\tau)^{-k}}$$

and considering that : $G^0\left[\delta(i\omega\tau)^{-h} + (i\omega\tau)^{-k}\right] \ll G^\infty$

we get : $\quad G^*(i\omega) = \dfrac{G^\infty}{1 + \delta(i\omega\tau)^{-h} + (i\omega\tau)^{-k}}$ (1)

with $\delta = \mu^{-h}$

II - EXPERIMENTAL INVESTIGATION

2.1 Description of experimental arrangement

The apparatus is based on an inversed torsion pendulum but works under harmonic oscillations outside of its resonance at very low frequencies (1.0E-5 to 1Hz). This technique is described in more detail elsewhere / 4 /. The specimen was excited in strain with a relative amplitude between 5.0E-5 to 2.0E-3. The sample temperature at each relative amplitude was controlled in the range 20 to 160°C by a temperature programmer such that it was ramped up at a controlled rate (5°C/min for measurement). The dynamic shear modulus G* and the loss component tanδ were recorded by scanning frequency range from 0.001Hz to 1Hz at each temperature level.

2.2 Materials studied and specimen preparation

The material used in these experiments was known commercially as T300/M10. The carbon fibre told contained 3000 T300 filaments. Unidirectional carbon fibre reinforced epoxy was made by filament winding. All samples were cured at 125°C for 2 hours followed by poscuring at 150°C for 1 hour. The fibre volume fraction was maintained at 65% for most of the tests. The specimens used were in parallelepiped form 39-mm long, 5-mm wide and 1-mm thick.

III - RESULTS AND DISCUSSION

Using the experimental procedure outlined earlier, dynamic mechanical properties of the composite T300/M10 were evaluated for various orientations of specimens (θ = 0°, 15°, 30°, 45°, 75°, 90°). The experimental data for storage and loss modulus obtained at various temperatures and various relative amplitudes of the strain were reduced to a standard temperature of 60°C by using the time-temperature principle.

In the following analysis, the creep functions for the viscoelastic composite material T300/M10 are developed. It appears beneficial to outline some numerical process to obtain viscoelatic solutions. Using the LAPLACE transform inversion the constitutive relation between the compliance J*(iω) and the creep function F(t) may be written in the following form :

$$J^*(i\omega) = i\omega \int_0^\infty F(t)\, e^{-i\omega t}\, dt$$

or equivalently :

$$J^*(i\omega) = J^*(p) = \left[p \int_0^\infty F(t)\, e^{-pt}\, dt \right]_{i\omega = p}$$ (2)

The basic viscoelastic property J* can be determined from Equation (1) :

$$J^*(i\omega) = \frac{1}{G^*(i\omega)} = J^\infty\left[1 + \delta(i)^{-h} + (i)^{-k}\right]$$

with $J^\infty = \dfrac{1}{G^\infty}$ $\quad$ and $\quad \omega\tau = 1$

In the above, J*(p) is the LAPLACE transform of the time-dependent creep function F(t).

TEER HAAR / 5 / pointed out that the function t exp(-pt) resembles a delta function and therefore may be approximated by the function :

$$t \, e^{-pt} \approx \left[\delta (t - t_0) / p^2 \right]$$

where the factor p^2 has been introduced for the purpose of normalization. Making this substitution in Equation (2) and assuming the delta function location t_0 to be at the point where the intensity function is a maximum $(t = 1/p)$ allow Eq. (2) to be rewritten in the following form :

$$\left[J^*(p) \right]_{p = 1/t} = F(t)$$

The creep function is expressible in the terms of G^*, τ, h, k, δ parameters as :

$$F(t) = \left[J^*(p) \right]_{p = 1/t} = \frac{1}{\left[G^*(p) \right]_{p = 1/t}}$$

$$F(t) = \frac{1}{G^\infty} \left[1 + \delta \left(\frac{t}{\tau} \right)^h + \left(\frac{t}{\tau} \right)^k \right] \tag{3}$$

The dynamic tests result was used for calculing the parameters of the rheological models δ, h, k and τ by applying the procedure described in the fig.1. After that, the parameters values were introduced into Eq. (3) to obtain the creep function for each fibre orientation of the specimens. An examination of fig.2 indicates that the fibre orientation in the composite T300/M10 has almost no effect upon his creep response. The T300/M10 behaviour in the dynamic mechanical tests corresponds to the epoxy matrix behaviour. However, beyond the time range of 1.0E7 secondes, fibre orientation becomes important. The change in the character of the creep function is likely attributable to the fibre-matrix interface. This suggests that beyond this range of time, the fibre-matrix debonding and pull-out of broken fibres mechanisms begin to contribute to the creep phenomenon. It was also demonstrated (fig.3) that relative amplitude of the strain has not influence on the creep behaviour of T300/M10.

CONCLUSION
The results from the dynamic mechanical analysis of fibre reinforced plastics showed that this technique can be used to study the viscoelastic behaviour of the material and to get an analytical formulation to quantify it. In this experimental approach, we observed more than ten decades of time using the time-temperature superposition principle. The present method for characterizing viscoelastic T300/M10 material permits successful measurements and we can now introduce these results in the mechanical modelling of the global long-time behaviour of our material.

ACKNOWLEDGEMENTS
The authors would like to thank Mrs. C.TRENY (Research and Developement Service - METRAVIB) for realizing the experimental tests.

REFERENCES
1 - COLE K.S., COLE R.H.,
   J. Chem. Phys., 9 , 341 (141).

2 - HUET C.,
   Ann. Ponts et Chaussées, 6 , 5 , (1965).

3 - FERRY J, D.,
   "Viscoelastic Properties of Polymers" - 3rd Ed., Wiley, NEWYORK, (1980).

4 - SALVIA M., MERZEAU P.,
   Un nouvel outil d'analyse de spectrométrie mécanique : Le MICROANALYSER.
   Journées G3F " Rendez-vous des matériaux du futur " - LYON - 22/26 Juin 1987.

5 - TEER HAAR D.,
   J.Polymer Sci., 6 , 241, (1951).

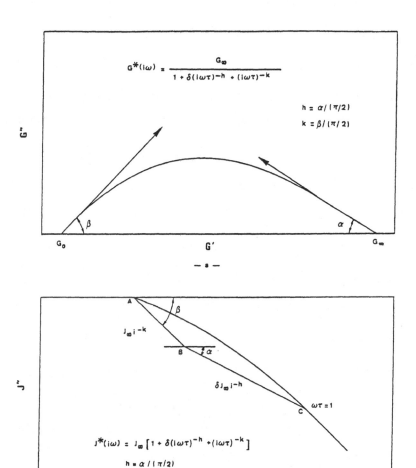

Figure 1 : The COLE-COLE diagram of the biparabolic model.
a - Calculation of the h, k parameters.
b - Calculation of the δ parameter.

947

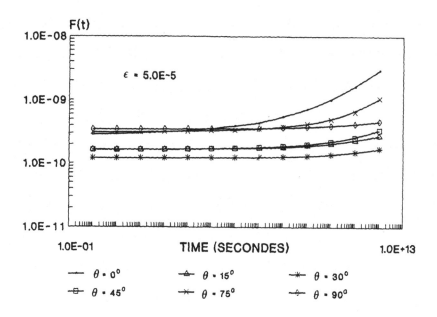

Figure 2 - The creep response of the T300/M10 for various fibre orientations of the specimens.

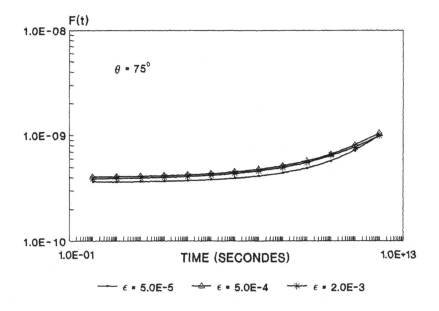

Figure 3 - The creep response of the T300/M10 for various relative amplitude of the strain.

EVIDENCING OF THE DAMAGE DUE TO
A SYMMETRICAL, ALTERNATE, CYCLICAL
AND HARMONIC STRESS, MEASURING THE INTERNAL
DAMPING IN A CARBON FIBER STRUCTURE
MADE OF EPOXY-RESIN

M. NOUILLANT, C.R. RAKOTOMANANA, P. MIOT, M. CHAPUIS

Université de Bordeaux I
Domaine Universitaire - 33405 Talence Cedex - France

Specimens consisting of thin tubes made of carbon-epoxy symmetrical laminate with various wiring orentations have been submitted to an harmonic torsional test for a wide range of stress level .As the shear stress τ exceeds a given threshold τmp, $\Delta w/w$. The so called damage depends on the maximum stress τmax experienced by the material (τ max> τmp).A decomposition in the laminate axes gives normal stresses σl(longitudinal), σt(transverse) and longitudinal shear σlt .The dependence of damage on σ l, σ t, σlt is discussed and seems to be a function of σt only , i.e : $\sigma t < (\sigma$ t)max

Stating the problem

The viscoelastic behaviour of the carbon fiber and resin composite elements is linear only within a reduce range of temperature and stress, compared with the mechanically and thermally admissible domain.

For instance, the diagram of figure 1 shows the evolution of damping as a function of temperature, and stress as the parameter according to torsion test realized with a tubular sample.

The sample is obtained winding 4 layers of wire with a +/- 75 degree angle respect to the revolution axis of the tube.

For these tests driven at a low stress level, the results obtained with a unique sample are independant from the way leading to one point of the characteristic (varying stress with a fixed temperature or vice versa).

In that case, the Δw/w variation seems to be perfectly determined given the temperature and stress variables

If we want to obtain those characteristics for a larger domain, a stress threshlod τ_{mp} has to be considered, above which some damage can be observed as an irreversible variation of Δw/w. This variation is connected to the difference between the cyclical stress applied, τ_{max}, and the stress threshlod τ_{mp}.

Various tests realized with initially nude samples, differently wire-winded (3 orientations : +/- 15°, +/- 45°, +/- 75°), seem to show that the stress threshlod is independant from the fiber orientation, apart from the experiment uncertainties.

	+/-15°	+/-45°	+/-75°
τ_{mp}	14 MPa	15 MPa	17 MPa

For instance, given the diagram of figure 2, obtained for +/- 75° orientation sample for given couple (T_0, τ_0), a steady increase of the damping value can be observed; this is due to the previous cycles applied to the sample, and to the evolution of their characteristics (increase in the maximal value of the stress).

Nevertheless, when (T_0, τ_0) are chosen close to the diagram origin, the damping comes back to its initial values.

In orders to determine the observed damage mechanism, we will analyze the contribution of $\sigma_l, \sigma_t, \sigma_{lt}$ in the fiber reference which is lying on a plane tangential to the tube.

The number of the parameters possibly involving the damage can be reduced because of the following points :

      - the +/-15° and +/-75° orientations lead to the same load of the fiber.

For these tests, the relationship between the stress components in the fiber reference, and the macroscopic stress τ_0, is given hereafter :

	σ_l/τ_0	σ_{lt}/τ_0	σ_t/τ_0
+/-15° +/-75°	2,8	0,3	0,1
+/-45°	2	0	0,1

- considering that the sample is a hollow and thin tube, the torsion and tension-compression tests can not induce relevant cross shear which can not be responsible for the observed damage.

Discussion

We can point out :

- the measured damping in a wire-winded composit structure made of carbon fiber and epoxy resin, is connected to the σ_{lt} stress component in the fiber reference. Indeed, the damping minimal values for +/- 45° orientation sample, are obtained for a nul value of σ_{lt}.

- defined as the increase of $\Delta w/w$ resulting from the application of a cyclical stress with an amplitude of σ_{max}, the damage seems to be independent of σ_{lt}, and therefor linked to σ_l or σ_t.

Considering the high numerical value of the ratio between the used fiber and matrix modulus :

R = fiber modulus / matrix modulus = 280/3 = 93

the low value of σ_l in the resin, called $(\sigma_i)_m$, let us think that, in all cases, the average stress applied to the composite fiber induces in the resin (along the 1 direction) $(\sigma_l)_m$ such that : $(\sigma_l)_m < \sigma_t$

Around the fiber, the σt stress implies a field bounded by a stress concentration coefficient.

If we call $(\sigma\ t)max$ this upper bound, we can admit, as a result of the tests :

$$(\sigma\ l)m < (\sigma\ t)max=10^-$$

Therefor, we can reasoably agree, with some other others, on the fact that the observed damage is connected to a mechanism induce by the σt stress acting on the matrix.

Finally, the explanation of the damage increase along with the macroscopic stress level, can be obtained considering the distance distribution between fibers. As a result, a specefic value, σt, is determined for each point of the matrix such that :

$$\sigma t<(\sigma\ t)max$$

$(\sigma\ t)max$ being the critical local damage threshold.

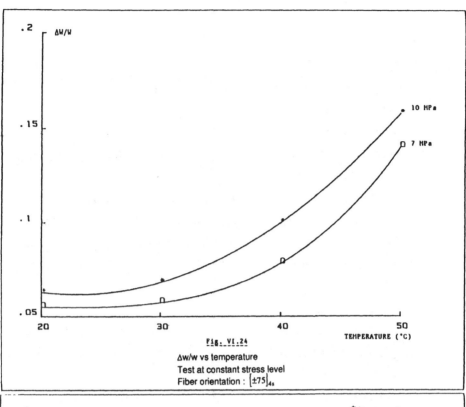

Fig. VI.24

Δw/w vs temperature

Test at constant stress level

Fiber orientation : $[\pm75]_{4s}$

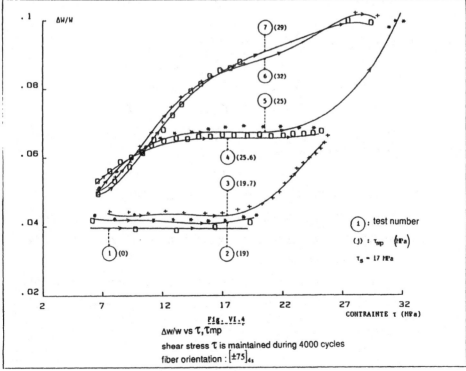

Fig. VI.4

Δw/w vs τ, τ_{mp}

shear stress τ is maintained during 4000 cycles

fiber orientation : $[\pm75]_{4s}$

(1) K. SCHULTE

Development of microdamage in composite laminates during fatigue loading.

International Conference on Testing, Evaluation and Quality Control of Composites, Univers. Surrey Guildfoerd England, Sept. 1983.

(2) R.D. JAMISSON, K. SCHULTE, K.L REIFSNIDER, W.W. STINCHCOMB

Characterization and analysis of mechanisms in fatigue of graphite/epoxy lamiates.- ASTM Symposium on the Effects of Defects in Composite Materials, San Francisco, California, Dec. 1982.

(3) I.M. DANIEL , J.W. LEE, G. YANIV

Damage mechanisms and stiffness degradation in graphite/epoxy composites.

Sixth International Conference on Composite Material ICCM/ECCM, LONDON, VOL. 4, Juillet 1987, pp. 4.129-4.138.

(4) J.J. NEVADUNSKY, J.J. LUCAS,M.J SALIND

Early fatigue damage detection in composite materials.

J. Composite Materials, vol.9,Oct.1975,pp.394-408.

(5) R.D. ADAMS, J.E. FLIT CROFT, H.T. HANCOX, W?N. REYNOLDS

Effects of shear damage on torsional behaviour of carbon fiber reinforced plastics

J. Composite materials, vol. 7, Janv. 1973, pp. 68-75.

(6) M. WEVERS, I. VERPOEST, E. AERNOUDT, P. DE MEESTER

Fatigue damage development in carbon fiber reinforced epoxy composites : correlation between the stiffness degration and the growth of different damage types.

Sixth International Conference onComposite materials ICCM/ECCM, London, vol. 4, Juillet 1987, pp. 4.114-4.127.

INFLUENCE OF MOISTURE ABSORPTION
ON THE MECHANICAL CHARACTERISTICS
OF POLYMER MATRIX COMPOSITES

A. CARDON, M. VERHEYDEN, Q. YANG, R. WILLEMS

Free University Brussels (V.U.B.)
TW (CONT - KB) Pleinlaan 2 - 1050 Brussels - Belgium

ABSTRACT

A long fibre reinforced polymeric matrix composite has an anisotropic viscoelastic behaviour. This implies dependent properties and sensitivity to environmental conditions such as temperature and moisture. For the durability analysis of a structural component accelerating factors must be considered in order to predict long term behaviour from short term tests. Such accelerating factors are the temperature and the stress level. Moisture absorption lowers the transition temperature of the composite and suggest to use it as accelerating factor. Before including the moisture absorption in a long termp prediction scheme it is necessary to investigate the different basic elements of the moisture diffusion in a polymeric matrix composite.

INTRODUCTION

H.Brinson and co-workers, [1],[2], demonstrated the applicability of a generalized time-temperature superposition principle to polymer matrix composites. The same research group proposed and tested the use of a time-stress superposition principle in order to use high stress levels during short time periods for the prediction of long term behaviour under normal stresses, [3],[4].

The influence of moisture level on the transition temperature of the polymer matrix and the composite suggest the use of moisture diffusion as accelerating factor for the analysis of long term behaviour.

A complete analysis of the influence of moisture absorption on the mechanical characteristics is necessary before the moisture diffusion can be included in a rational methodology for the durability analysis of the polymer matrix composite by a combined temperature-stress-moisture superposition principle. Especially the local effects of the moisture diffusion on the matrix, the fibres and the interphase region between matrix and fibres must be investigated for different composite systems. Moisture may induce damage and interactions such as those between moisture level and temperature or moisture diffusion and stress state must be analyzed.
Such interactions were studied by e.g. Y.Weitsman, [5] and S.Neumann and G.Marom, [6].

Experimental results obtained on graphite-epoxy composites are presented. The changes of the viscoelastic characteristics of the composite under different moisture conditions will be presented.
The possibility to use the Schapery model to take these effects into account, will be discussed.

VISCOELASTICITY OF UNIDIRECTIONAL REINFORCED POLYMER MATRIX COMPOSITE

For loadings in the fibre direction the behaviour of the composite laminate is elastic or elasto-plastic. If the tensile loading is applied transversely to the fibres the behaviour is viscoelastic. For any other loading direction, at angle θ with the reinforcement direction viscoelastic effects are observable function of the angle θ.
Under normal loading conditions at room temperature the linear anisotropic viscoelastic constitutive equation

$$\sigma_{ij} = \lambda_{ijkl}\, \varepsilon_{kl} \, , \tag{1}$$

where

$$f^{+} = p \int_{0}^{\infty} f(t)\, e^{-pt}\, dt \, ,$$

with the transverse isotropic reductions on λ_{ijkl}, is able to describe the mechanical behaviour.

At higher temperatures, higher stress levels and after moisture uptake the viscoelastic behaviour becomes nonlinear, the linearity limit being function of the angle θ.

A complete description of the mechanical behaviour under mechanical and environmental loading conditions needs a nonlinear viscoelastic model. Such a convenient model was proposed by R.Schapery and used by H.Brionson and co-workers, [4].

THE SCHAPERY MODEL

For simplicity and within this limited space we only mention the Schapery expression for the unidimensional case:

$$\varepsilon(t) = g_0[\sigma, T, \text{moisture uptake}] S_0 \sigma + g_1[\ldots] \int_0^\infty \Delta S(\psi - \psi') \frac{d[g_2(\ldots)\sigma]}{d\tau} d\tau \tag{2}$$

where

$$\psi = \int_0^t \frac{d\tau}{a_\sigma[\ldots]} \quad \text{and} \quad \psi' = \int_0^{t'} \frac{d\tau}{a_\sigma[\ldots]} \quad .$$

The nonlinearizing functions g_0, g_1, g_2, a_σ, depend on the stress level, the temperature and the moisture level. The limit values of the linear behaviour are different for the different nonlinearizing functions and for the different parameters.

The physical meaning of the nonlinearizing can be understood on the basis of a generalized standard solid model

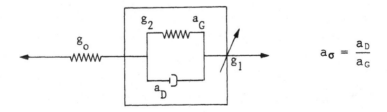

$$a_\sigma = \frac{a_D}{a_G}$$

957

g_0, g_1, a_G and g_1 are functions nonlinearizing conservative elements. a_D and thus a_σ is the nonlinearizing function of the dissipation and entropy related element.

For the analysis of the experimental results a general power law is used for the compliance :

$$S(t) = S_0 + \Delta S(t) = S_0 + \frac{S_\infty - S_0}{(1 + \frac{\tau}{t})^n} .$$

S_0 : initial compliance ; S_∞ : limiting rubber compliance ; τ : relaxation time.

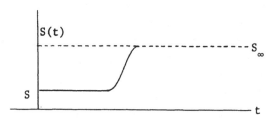

CREEP TESTS

The creep tests are performed under loading corresponding to $0.6 \, \sigma_u$.

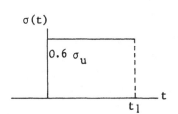 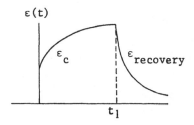

$$\varepsilon_c(t) = f_1(g_0, g_1, g_2, a_\sigma) \quad \text{for} \quad 0 < t < t_1$$
$$\varepsilon_r(t) = f_2(g_0, g_1, g_2, a_\sigma) \quad \text{for} \quad t \geq t_1.$$

In this part of the research the analysis is focused on the variation of g_0, g_1, g_2 and a_σ with the amount of moisture uptake in order to determine the limits of linear behaviour.

EXPERIMENTAL CONDITIONS AND RESULTS

The tests were performed on a unidirectional 8-layer lamiante FIBREDUX 914-C-TS(6K)-5-34%, at room temperature under two humidity levels :

- dry (25°C, silica)
- saturated (60°C,60 % RH) corresponding to 0.7 % moisture uptake.

Ultimate strengths	σ_u (N/mm^2) dry	σ_u (N/mm^2) wet
0°	1817	1825
15°	283	200
90°	18	16

The moisture uptake has practically no influence on σ_u in the 0°-direction but lowers the values for $\theta \neq 0$. The results on the 90°-directions are difficult to obtain and analyse because of the strong influence of the imperfections along the boundaries after the cutting of the specimen.

As example we give the parameters of the generalized power law :

	S_o (mm^2/N)	S_∞ (mm^2/N)	τ (sec)	n
90° wet	0.0066	0.0071	724.8	0.239
15° wet	0.0085	0.0088	33.6	0.408

The results on the nonlinearizing functions will be given during the poster presentation.

CONCLUSIONS

The different elements of the Schapery model can be obtained after moisture uptake from creep and creep recovery testing. The combination of these results with those at different temperatures and different stress levels opens the possibility to use the moisture uptake as an accelerating factor for the durability analysis of polymer matrix composite laminates.

ACKNOWLEDGEMENTS

This research was supported by the Belgian National Science Fund (NFWO-FNRS), by the Research Council of the University (OZR-VUB) and by the Belgian University of Science Policy.

REFERENCES

1 - Brinson H.F., Morris D.H. and Yeow Y.T., "A new experimental method for the accelerated characterisation of composite materials", Proc. of the 6th International Conference on Experimental Stress Analysis, München 395 (1978).

2 - Brinson H.F., "Viscoelastic behaviour and life-time (durability) predictions", in "Mechanical Characterisation of Load Bearing Fibre Composite Laminates" (ed. A.H.Cardon - G.Verchery), Elsevier Applied Science Publishers, pp. 3-20 (1985).

3 - Brinson H.F., Dillard D.A., "The prediction of long term viscoelastic properties of fibre reinforced plastics". 4th Int. Conference on Composite Materials. Progress in Science and Engineering of Composites 1 (ed. T.Hayashi, K.Kawata, S.Unekowa), Japan Society of Composite Materials, Tokyo (1982).

4 - Tuttle M., Brinson H.F., "Prediction of long term creep compliance of general composite laminates", Experimental Mechanicas, pp. 89-102, (March 1982).

5 - Weitsmann Y., "Stress assisted diffusion in elastic and viscoelastic materials", J. Mech. Phys. Solids 35(1), pp. 73-79 (1987).

6 - Neumann S. and Marom G., "Prediction of moisture diffusion parameters in composite materials under stress", Journal of Composite Materials, vol. 21, pp. 68)80 (January 1987).

PROPERTY VARIATIONS OF GLASS FIBRE REINFORCED POLYMERS DUE TO CORROSIVE INFLUENCE

A. SCHMIEMANN, A.K. BLEDZKI*, G.W. EHRENSTEIN**

BAYER AG - Bayerwerk - 5090 Leverkusen - West-Germany
*GH - Kassel - Wilhelmshöher Allee 73 - 3500 Kassel - West Germany
**Frierich-Alexander-Uni - Schlossplatz 4 - 8520 Erlangen - West-Germany

ABSTRACT

Mechanical properties of glass-fibre reinforced plastics show often a big scatter in results. This is applicable to virgin material and specially to composites, which have been exposed to a corrosive media under mechanical load. With an increasing damage stage not only the bearable forces but also the reliability of the characteristic data decreases. The uniformity and accuracy of mechanical data of glass fibre reinforced thermosets are determined and criticized using definite type laminates. Important influence factors are discussed in detail.

ZIELSETZUNG

Ein Ziel der vorgestellten Arbeit ist die Erfassung und Beurteilung von Gleichmäßigkeit und Genauigkeit mechanischer Kennwerte bei Glasfaser-Kunststoff-Verbunden. Dazu wurden verschiedene Laminattypen: Mattenlaminate - Gewebelaminate - UD-Laminate (Rundprofile) untersucht.

I - GLEICHMÄßIGKEIT UND GENAUIGKEIT MECHANISCHER KENNWERTE VON GFK

Die Ergebnisse der Zug- und Biegeversuche zeigen, daß in der genannten Reihenfolge nicht nur die mechanischen Eigenschaften, sondern auch die Gleichmäßigkeit der Kurzzeit-Kennwerte zunehmen.

Der Matrixeinfluß und somit der Einfluß derjenigen Komponente, die erst während der Verarbeitung ihre Eigenschaften erhält, ist bei Mattenlaminaten noch sehr groß, d.h. die Eigenschaften der Mattenlaminate sind matrixdominant. Aber auch bei den Gewebelaminaten kann ein Matrixeinfluß auf die Kennwertstreuungen nachgewiesen werden. Bei diesen orthotropen Laminaten beherrscht das Querschichtversagen deutlich das Eigenschaftsbild. Mischlaminate bilden eine Zwischenstufe. Prinzipiell gilt für alle Laminate, daß mit zunehmendem Schädigungsfortschritt die Kennwerte ungleichmäßiger werden (Bild 1).

Immer dann, wenn örtliche Spannungsüberhöhungen auftreten, sinken nicht nur die Kennwerte, sondern auch die Streuungen ab. Es kann aufgrund geringer Meßwertstreuungen folglich nicht darauf geschlossen werden, daß optimale Kennwerte erreicht wurden. Beschädigungen hingegen führen zur Abnahme der Kennwerte und zur Zunahme der Streuungen. Weiterhin kann davon ausgegangen werden, daß mit zunehmendem Probekörpervolumen bei Langfaserverbunden und auch bei Einzelfasern die Kennwerte und deren Streuungen sinken, da die Wahrscheinlichkeit für Fehlstellen zunimmt.

Die Genauigkeit oder Reproduzierbarkeit mechanischer Kennwerte von Laminaten mit Wirrfaserverstärkung ist geringer als bei Laminaten mit Langfaserverstärkung. Die Eigenschaften von UD-Verbunden sind sehr genau vorhersehbar und berechenbar. Hingegen weisen nach gleichen Angaben gefertigte Hand-, Faserspritz- und Preßlaminate von Fertigungs- zu Fertigungscharge Mittelwertabweichungen auf, die sich nur zum Teil auf unterschiedliche Glasgehalte zurückführen lassen. Weitere, wichtige Ursachen dafür sind:

- Abweichungen bei der Einhaltung der Nachhärtebedingungen
- örtliche Spannungsüberhöhungen aufgrund einer ungünstigen Krafteinleitung/Probengeometrie oder großer Lufteinschlüsse
- frühzeitige, starke Grenzflächenablösungen aufgrund hoher Eigenspannungen im Harz oder eines sehr spröden Harzsystems oder schlechter Faserbenetzung /1/.

Aus Untersuchungen an zahlreichen Wirrfaser-Laminatchargen kann abgeleitet werden, daß sich die Kennwertstreuungen in zwei Komponenten aufgliedern lassen: Einer inhärenten oder inneren, mittelwertunabhängigen Grund- oder Basisstreuung, die laminatspezifisch ist, überlagert sich eine zweite Komponente, die mit zunehmendem Mittelwert ansteigt. Nur diese zweite Komponente unterliegt Fertigungs- und Prüfeinflüssen. Mit einer minimalen Standardabweichung von nur 4 MPa, einer typischen von 7 MPa und einer oberen Grenze von 13 MPa (σ_B = 100 MPa) für wirrfaserverstärkte Laminate (Hand/-, Faserspritz- und Preßlaminate) ergeben sich nach dem Fraktilwertkonzept für die Abminderung von σ, um zu

962

einem Dimensionierungskennwert zu kommen, maximal k•S = 23
MPa für einen Probenumfang von n = 20 (5%-Fraktile bei W =
90%). Ein Probenumfang von n = 20 ist zwar hoch und für
einen zulässigen Fehler von 5% reichen n = 10 Proben oft
aus (Aussagewahrscheinlichkeit W = 90 %), ein größerer
Stichprobenumfang und damit größerer Informationsumfang
hat aber zur Folge, daß der k-Wert abnimmt. n = 20 stellt
dabei ein Optimum dar, da eine weitere Erhöhung des Stich-
probenumfangs dann nur noch unwesentliche Änderungen zur
Folge hat /2,3/.

II - KORROSIVE EINFLÜSSE

Korrosive Beanspruchungen haben einen starken Einfluß auf
die Eigenschaften von Glasfaser-Kunststoff-Verbunden, wenn
sie unter mechanischer Gebrauchslast stehen. Einerseits
werden die ertragbaren Lasten kleiner und andererseits
vergrößert sich die Streubreite. Da der Angriffsmechanis-
mus von Säuren und Laugen grundsätzlich verschieden ist,
müssen getrennte Betrachtungen durchgeführt werden.

2.1. Korrosion in Säuren

Das Verhalten von GFK in Säuren ist deshalb so gefährlich,
weil sehr plötzliches, katastrophales Versagen eintreten
kann. Dabei erweisen sich die Glasfasern als spannungsriß-
korrosionsempfindlich und Spannungsrißkorrosion von GFK
bezieht sich primär auf die Glasfaser. Gelangt das korrosive Medium (verdünnte Säure) z.B. durch einen Matrixriß
an belastete Glasfasern, so werden diese in einem örtlich
sehr begrenztem Bereich - der Rißspitze -geschädigt (kor-
rodiert), und es kommt zum Versagen. Die Standzeiten der
einzelnen Faser sinken dabei exponentiell mit steigender
Belastung. Beispielsweise ergibt sich für eine Belastung
von $\sigma_B/10$ eine Standzeit von ca. 30 h für eine 17 μm-E-
Glasfaser (Medium: 1 N H_2SO_4, Temperatur: 23°C). Dies ist
genau die Zeitspanne, die nötig ist, damit unter gleichen
Bedingungen E-Glasfasern ohne Einwirkung einer äußeren
Last erste, nennenswerte Rißbildungsphänomene nach Säure-
einwirkung und anschließender Trocknung zeigen /4/. Verantwortlich dafür sind die im äußeren Fasermantel konzen-
trierten Eigenspannungen, die durch die extremen Bedingungen beim Faserziehprozeß eingefroren werden.

Der korrosive Angriff auf Einzelfasern besteht darin, daß
sie regelrecht ausgelaugt werden; bei unbelasteten Glasfasern kann nach genügend langen Zeiten nur noch ein
nacktes SiO_2-Gerüst nachgewiesen werden. Dieses ist so
stark geschwächt, daß die eingefrorenen Eigenspannungen
Radial-, Spiral- oder Axialrißbildung in den Glasfasern
auslösen können. Eine Zunahme der Abzugsgeschwindigkeit

und Abnahme der Schmelztemperatur führt zu einer starken Deformation des Glasfasernetzwerks /5,6/.

Ein korrosiver Angriff erfolgt ungleichmäßig; da von außen einwirkende Säure an bestimmten, exponierten Stellen zuerst an die tragenden Fasern gelangt und bei diesen - je nach Belastung - in vergleichsweise kurzen Zeiten Spannungsrißkorrosion auslöst. Nach Säurebehandlung kommt es daher zu einer bimodalen Verteilung der "Rest"-Kennwerte, wobei das untere Maximum gleich Null und das obere Maximum fast dem Ausgangskennwert entspricht. Akzeptable Standzeiten bei korrosiver Beanspruchung in saurer Umgebung erfordern folglich eine drastische Reduzierung der Gebrauchslasten, so daß sowohl frühzeitige Rißbildung als auch aufgrund der mechanischen Belastung stark erhöhte Diffusionskoeffizienten vermieden werden, oder es werden geeignete Maßnahmen zur Verbesserung der Beständigkeit ergriffen. Möglichkeiten dazu zeigt zusammenfassend Bild 2 auf.

2.2. Korrosion in Alkalien

Das Erreichen optimaler Beständigkeiten von Verbunden in Alkalien muß vor allem bei Polykondensat-Matrixes als schwieriger eingestuft werden als in saurer Umgebung. Einerseits ist die Matrix in den Korrosionsprozeß einbezogen (Hydrolyse) und andererseits sind speziell für Säuren geeignete Fasern (z.B. ECR-Glasfasern) in Säuren wesentlich beständiger als alkaliresistente Glasfasern (z.B. CemFil-Fasern) in alkalischen Medien. Der korrosive Angriff von alkalischen Medien folgt streng den Diffusionsgesetzen. Auch die Streuungen der mechanischen Kennwerte werden nicht oder nur unwesentlich größer, da ein intakter Kernbereich vorliegt. Es kann dabei beobachtet werden, daß z.B. bei biegebelasteten Probekörpern die Zug- und Druckzonen zuerst und stärker korrosiv beeinflußt werden. Der Zugbereich unterliegt dabei dem stärksten korrosiven Angriff. Bei der Alkalikorrosion kann von einer Spannungskorrosion gesprochen werden. Unterschiede zwischen den wenig aufwendigen Zeitstand-Biegeversuchen und Zeitstand-Zugversuchen, die dem Anwendungsfall GFK im Spannbetonbau näher kommen, sind darauf zurückzuführen, daß bei einer Biegebelastung nur die äußeren Schichten so hoch belastet werden wie der gesamte Querschnitt bei reinem Zug. Versuche bei erhöhten Temperaturen belegen, daß eine klare Temperatur-Zeit-Analogie vorliegt, die dem Arrhenius-Prinzip folgt. Die Zugabe vom Aluminiumionen zur Lauge bringt einen nicht unerheblichen Festigkeitsanstieg im Fall der Einzelfasern und hemmt die Korrosion von GFK in alkalischen Lösungen mit hohem pH-Wert und kann daher als eine geeignete Maßnahme zur Verbesserung der Alkalibeständigkeit von GFK gelten.

Insgesamt läuft die Korrosion im alkalischen Medium bei gleichen Laminaten sehr viel langsamer ab als im sauren Medium.

LITERATUR

/1/ Seiler U., Thesis, RWTH-Aachen (1987)

/2/ Ehrenstein G.W., Spaude, R., Kunststoffe 72 (1982) 479-483

/3/ Schmiemann A., Ehrenstein G.W., DFG-Forschungsbericht Eh 60/12-1, Univ. Kassel (1987)

/4/ Spaude R., Univ. Kassel (1984)

/5/ Kobayashi M., Fujinowa T., Hashimoto K., Yosomiya R. Glastechnische Berichte 55 (1982) 75-80

/6/ Schmiemann A., Gehde M., Ehrenstein G.W. Proc. 44th Annual Conference 1989, RP/CI, SPI, Section 2-C

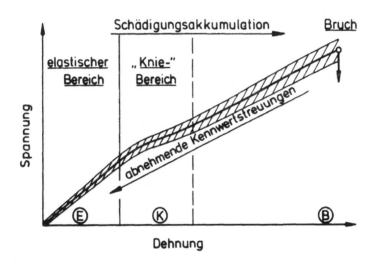

Bild 1: Bereiche im Spannungs-Dehnungs-Diagramm

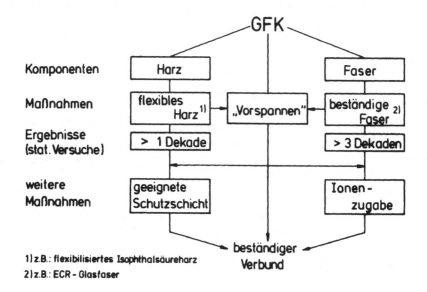

Komponenten

Maßnahmen

Ergebnisse
(stat. Versuche)

weitere
Maßnahmen

1) z.B.: flexibilisiertes Isophthalsäureharz
2) z.B.: ECR - Glasfaser

Bild 2: Möglichkeiten zur Verbesserung der Korrosionsbeständigkeit in sauren Medien

THE EFFECT OF STRAIN RATE ON THE INTERLAMINAR SHEAR STRENGTH OF WOVEN REINFORCED LAMINATES

J. HARDING, Y. LI*, M. TAYLOR

Dept of Engineering Science - University of Oxford
Parks Road - OX1 3PJ Oxford - Great Britain
*Northwestern Polytechnical University
Xi'An - China

ABSTRACT

A recently developed testing technique for determining the interlaminar shear strength of laminated composites at impact rates of strain /1/ is described. Results are presented for the interlaminar shear strength at the interface between a) two woven glass plies, b) two woven carbon plies and c) a woven glass and a woven carbon ply at both a quasi-static and an impact rate of loading. In each case a significant increase in interlaminar strength at the impact rate is observed.

INTRODUCTION

A change in the failure process with increasing strain rate has been seen in tensile tests on both woven /2/ and unidirectional /3/ glass reinforced composites loaded in the direction of reinforcement. Fibre fracture dominates the failure process at quasi-static rates while at impact rates failure is almost entirely by either interlaminar shear, in the woven laminates, or by fibre pull-out, in the unidirectionally reinforced specimens. Such behaviour is consistent with the hypothesis that the tensile strength of the fibres increases more rapidly with strain rate than either the interlaminar shear strength or the interfacial bond strength. For this hypothesis to be checked it is necessary to devise techniques which allow the effect of strain rate on each of these various properties to be independently determined. The present paper specifically considers the interlaminar shear strength and reports on an attempt to determine its rate dependence at the interfaces between a) two woven glass plies, b) two woven carbon plies and c) a woven glass and a woven carbon ply. In each case a significant increase in the shear strength with strain rate is observed.

SPECIMEN DESIGN

A double-lap design of shear specimen, as shown in Figs. 1 and 2, was used in both quasi-static and impact tests. A tensile load applied across the ends of the specimen gives a shear failure on the two pairs of overlapping interlaminar planes. At the quasi-static rate specimens were tested in an Instron loading machine and failed in about 10s, while in impact tests the specimen was fixed between the input and output bars of a split Hopkinson bar apparatus and failure occurred after about 30μs. The lay-up for an all-carbon specimen, as used to determine the ILSS between neighbouring carbon reinforced plies, is shown in Fig. 1. Similar lay-ups for determining the ILSS on the interface between a carbon ply and a glass ply or between two glass plies are shown in Figs. 2a and b.

A major problem in most designs of shear specimen is the non-uniform stress and strain distribution on the failure plane. This is shown in Fig. 2, which gives the variation of direct, normal and shear strain, as derived from a finite element analysis, along the overlap zone of the double-lap shear specimen for the hybrid and the all-glass lay-ups. At each end of the failure plane, points X in Fig. 2, large shear strain concentrations may be seen. These are minimised by increasing the bending stiffness of the specimen, i.e. by using carbon for the outer plies of the hybrid lay-up. Even so it is clear that the experimentally determined shear failure load will only give an average value of the ILSS at failure.

To allow direct comparison with earlier experimental work in impact tension /1,4/ and compression /1/, specimens were prepared by hand lay-up from dry plain weave fabric using the Ciba-Geigy XD 927 epoxy resin system and cured under an external pressure of 90psi. This led to a loss of resin and some variation in the final fibre weight fraction between different specimen lay-ups. Full details are given elsewhere /5/.

RESULTS

Figs. 3a and b, respectively, show typical load-time signals for tests on all-carbon specimens at a quasi-static and an impact rate. Macrographs of a failed hybrid specimen after quasi-static loading are shown in Fig. 4. Assuming failure occurs at the peak load and that there is an equal loading on the two failure planes, the average shear stress on the interlaminar plane at failure was determined for the three types of specimen at the two loading rates. The results are summarised in Table I. In the quasi-static tests all specimens failed in the manner shown in Fig. 4. In the impact tests, however, 1 of the 5 hybrid specimens also failed in tension, at section (a) in Fig. 5, while 3 of the 6 all-carbon specimens showed similar behaviour, 2 failing at section (b) and one at section (c).

DISCUSSION

As is clear from Table I the quasi-static interlaminar shear strength is the same, about 26MPa, for all three types of interface and in each case shows a very significant increase with rate of loading. The highest increase, by about 70%, is for the carbon/carbon interface and the smallest, about 42%, for the glass/glass interface. This suggests that at the quasi-static rate the ILSS is controlled primarily by the shear strength of the resin system. Under impact loading, however, where the resin shear strength is likely to be much higher, there would appear to be an additional effect, probably arising from a stronger mechanical interaction at the coarse-weave carbon/carbon interface, under rapid shear displacement, than at the much finer weave glass/glass interface.

The difference in weave geometry for the carbon and glass reinforcements is apparent on the failure surface of the hybrid specimen in Fig. 4, implying that there is a step across the failure surface. This is modelled in Fig. 6, which assumes a thin layer of epoxy between neighbouring reinforcing plies. Failure is predicted to initiate at the two points of maximum shear strain and to propagate along the interface between the resin and either the carbon or the glass reinforced ply before finally joining up across the resin. A step of this type was seen in tests at both loading rates on all three types of specimen. Fig. 7 shows an SEM of the step region on a glass/glass interface after impact.

It is interesting to note that only in the all-glass specimens were simultaneous tensile failures entirely absent under impact loading, in accord with the earlier suggestion that for glass/epoxy laminates the increase in the tensile strength of the fibres with loading rate is at least as great as the increase in ILSS. This is not true for the carbon/carbon and carbon/glass interfaces where tensile failures were observed and where the increase in ILSS was more marked. The tensile strength of carbon fibres, however, is known to be rate insensitive /6/. Nevertheless, while results such as those reported in Table I may be used in a qualitative discussion, anomalies arise /7/ if attempts are made to use them directly in a more detailed analytical or numerical modelling of composite mechanical behaviour under impact loading.

CONCLUSIONS

The double-lap shear specimen has been used to determine the effect of loading rate on the ILSS in woven reinforced carbon/epoxy, glass/epoxy and hybrid carbon-glass/epoxy laminates. An increase in loading rate of about 6 orders of magnitude increases the ILSS by about 42%, for the glass/glass interface, 56% for the carbon/glass interface and 72% for the carbon/carbon interface. While these results are of qualitative significance, in view of the large strain concentrations on the failure plane the specific values obtained should be used with caution.

ACKNOWLEDGMENTS

This research was sponsored by the Air Force Office of Scientific Research, Air Force Systems Command, USAF, under Grant No. AFOSR-87-0129.

969

REFERENCES

1. J. Harding, Y. L. Li, K. Saka and M. E. C. Taylor, <u>4th Oxford Int.
    Conf. on Mech. Props. of Materials at High Rates of Strain</u>, Inst.
    Phys. Conf. Series No. 102, Institute of Physics, London and Bristol
    (1989), 403-410.

2. L. M. Welsh and J. Harding, Proc. DYMAT 85, <u>Int. Conf. on Mechanical
    and Physical Behaviour of Materials under Dynamic Loading</u>, Jour, de
    Physique, Colloque C5, (1985), 405-414.

3. L. M. Welsh and J. Harding, <u>Proc. 5th. Int. Conf. on Composite Mater-
    ials</u>, ICCM V, TMS-AIME, (1985), 1517-1531.

4. J. Harding, K. Saka and M. E..C. Taylor, Proc. IMPACT 87, <u>Impact
    Loading and Dynamic Behaviour of Materials</u>, eds. C. Y. Chiem, H.-D.
    Kunze and L. W. Meyer, DGM Informationsgesellschaft mbH, Oberursel,
    (1988), <u>1</u>, 515-522.

5. Y. L. Li, J. Harding and M. E. C. Taylor, Oxford University Engineer-
    ing Laboratory Report, (in preparation).

6. J. Harding and L. M. Welsh, <u>J. Mater. Sci.</u>, <u>18</u>, (1983), 1810-1826.

7. Y. L. Li, C. Ruiz and J. Harding, Composites Science and Technology
    (in press).

Table I <u>Effect of Loading Rate on Interlaminar Shear Strength</u>
(MPa)

Interface type	Quasi-Static Tests	Impact Tests
Carbon/carbon	26.4 ± 1.7 (4 tests)	45.0 ± 4.2 (6 tests)
Carbon/glass	26.6 ± 3.9 (5 tests)	41.4 ± 4.7 (5 tests)
Glass/glass	26.3 ± 3.2 (5 tests)	37.4 ± 0.2 (3 tests)

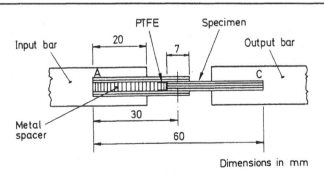

Fig. 1 <u>Double-Lap Shear Specimen</u>

Fig. 2   Specimen Lay-Up and Strain Distribution on Interlaminar Plane

a) Hybrid Specimen   b) All-Glass Specimen

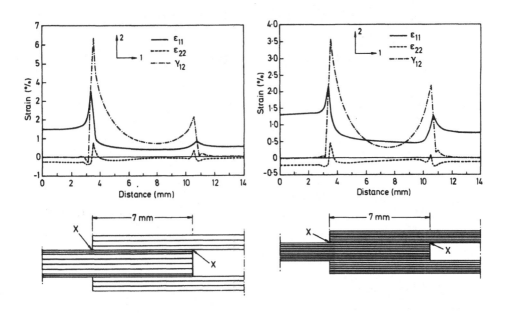

Fig. 3   Load-Time Signals for Tests on All-Carbon Specimens

a) Quasi-Static Test   b) Impact Test

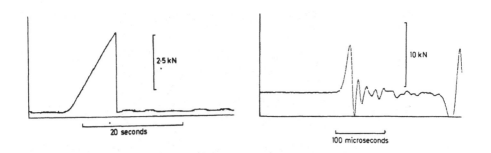

Fig. 4   Macrographs of Quasi-Statically Tested Hybrid Specimen (x1.2)

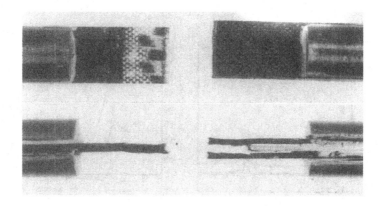

Fig. 5 Types of failure in Impact Tests

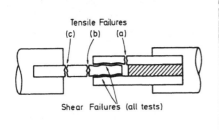

Fig. 6 Schematic Model of Failure Process

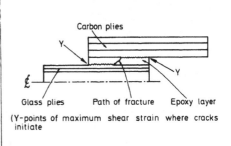

Fig. 7   Scanning Electron Micrograph of Step Region
(All-glass specimen after impact loading, x670)

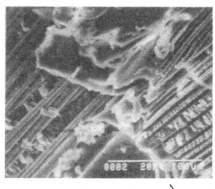

972

MECHANICAL PROPERTIES AND FRACTURE
OF HYBRID PARTICULATE COMPOSITES

G. LEVITA, A. MARCHETTI, A. LAZZERI, S. DE PETRIS

University of Pisa - Dept Chemical Engineering
Via Diotisalvi 2 - 56100 Pisa - Italy

ABSTRACT

A rigid epoxy resin was modified by acrylic core-shell particles and glass beads (with and without surface treatment). Glass transition temperature, elastic modulus and fracture toughness were determined for all formulations. The fracture toughness increased linearly with the rubber content (from 0.82 MPa√m for the neat resin to 2.6 MPa√m for 20% rubber). Toughness also increased on adding glass particles (for 50% glass up to 1.9 and 2.3 MPa√m depending on the surface treatment). The stress-strain behaviour in compression depended on glass content and surface sizing. The fracture behaviour markedly depended on the radius of curvature of the crack tip. The toughness of sharply cracked samples increased on increasing the glass content. On the contrary samples containing blunt cracks were embrittled by the hard particles. This was attributed to crack tip sharpening caused by detached beads.

INTRODUCTION

The fracture behaviour of brittle polymers such as epoxies can be substantially modified by incorporating small particles capable of interacting with the crack as it propagates. These particulate composites have received much attention by researchers because they are good models for two component materials and for their technical importance. Depending on the particle to matrix modulus ratio, interface strength, particle cohesive strength and intrinsic toughness of the matrix the particulate composite can be either tougher or more fragile than the matrix polymer. The experimental evidence is that soft particles are better toughener than hard ones. The latter on the other hand have positive effects on other important properties as HDT, wear, hardness and dimensional stability. Both soft and hard particles can be simultaneously incorporated in order to combine the merit of either two types of filler. Such particulate materials are known as hybrid composite /1-6/. In this work we have investigated the effects of preformed rubber particles and glass beads on the fracture toughness of a medium T_g epoxy resin.

EXPERIMENTAL

A low molecular weight epoxy prepolymer (Epon 828EL by Shell Co) crosslinked with the stoichiometric amount of isophorone-diamine was used as the matrix (cured at 60°C and postcured at 180°C). The soft component was a fully acrylic modifier (KM334 by Rohm & Haas) with a core and shell structure. Glass beads (size range 4÷44μm, mean size ≈15μm) were used with no surface treatment and af-ter being treated with a water soluble silane coupling agent (A187 by Union Carbide). Fracture testing was carried out at room temperature on prismatic bars in 3-point bending at 1 mm/min. The critical parameter K_{Ic} was used as the measure of fracture resistance. Thermomechanical characterization was carried out using a Polymer Laboratory analyzer at the frequency of 1 cps. All compositions are by volume.

RESULTS AND DISCUSSION

Resin/Rubber Formulations. All formulations behave as incompatible blends. In all cases two main transitions (T_{g1}=-50°C, T_{g2}=130°C) are observed at temperatures corresponding to those of pure constituents, fig-1. This is likely due to the lack of reactive groups in the acrylic modifier which cannot react with the epoxy resin. Widely used tougheners like CTBNs or ATBNs on the contrary plasticize to a certain extent the matrix with the consequence of lowering its distortion temperature. The brittleness of the pure matrix notably reduces on adding the soft component. Up to 20% of rubber K_{Ic} linearly depends on composition, fig-2. The efficiency of preformed particles is apparently higher than that of reactive liquid rubbers if comparison is made at the same cure tempera-ture. ATBN/DGEBA formulations cured in the same conditions as in this case (60°C) do not show any appreciable toughening /7/. The use of pre-formed particles is therefore highly effective for low cure temperature formulations (for instance adhesives, casting and potting applications).
In the case of reactive rubbers the soft phase is generated by phase se-paration. The final morphology is therefore largely affected by several parameters such as temperature, viscosity and solubility parameters. Ty-pically the particle size is in the range 1÷10μm. Preformed particles are much smaller (≈0.3μm) and relatively uniform in size and their dimen-sions and amount do not depend on cure temperature.
Depending on sample geometry a stress whitened zone develops in the crack tip area indicating extensive plastic damage primarily due to ca-vitation around the soft particles. It has been suggested /8-11/ that a major toughening mechanism in rubber toughened epoxies is internal ca-vitation of the particles. In the present case the crosslinked nature of the particle core and the lack of chemical links at the interface suggest that voiding, as revealed by SEM examinations, mainly takes place at the particle-matrix interface. Since the particles are very small the surface to volume ratio is very high and the stress relieve at the crack tip, caused by the interface failure, is therefore significant.

974

Resin/Beads Formulations. The material becomes stiffer on adding the rigid filler. The Ishai and Cohen lower bound satisfactorily models the elastic properties of the composites, fig-3. It has been frequently observed that the toughness of brittle polymers increases on adding glass beads or microfibers. The effects can be accounted for in terms of several crack tip processes such as pinning, blunting, deflection or shielding. As shown in fig-4 the resin is toughened by the rigid filler. A plateau value is reached at about 50% glass. K_{IC} doubles for 10% acrylic rubber and increases by a factor of 1.5 for the same amount of glass. By treating the glass surface with the coupling agent the toughening efficiency of the filler further increases. Evans /12/ has modeled the advancement of a crack through an array of rigid particles. He calculated lower and upper bounds for the dependence of toughness on filler content, fig-5. The lower bound represent the case in which the secondary cracks elastically interact whereas the upper one refers to not interacting cracks. At low filler concentration no interactions are expected for particles are largely apart. On increasing the particle density the toughening efficiency decreases and the experimental data move to the interaction line.

The toughness largely depends on the crack tip geometry. As expected the K_{IC} of the neat resin is much higher when blunt cracks are used. In this case, however, the glass beads promote brittleness. The toughness from sharp and blunt crack fracture tests becomes independent of tip geometry at high glass concentrations. In the crack tip region a triaxial state of stress develops on loading. The negative hydrostatic stress generated in this region causes the matrix to detach from the particles. The matrix is highly deformed and rapidly brought to the ultimate elongation. Secondary microcracks, with the natural radius of curvature, thus generate with the effect of rendering sharper the principal blunt crack. The sharpening effect is more effective as the number of particles in the process zone increases and toughness consequently deteriorates. At sufficiently high filler content the network of sharp microcracks in the tip region makes the principal blunt crack mechanically equivalent to a sharp one. The sharpening mechanism can be opposed by strengthening the interface in order to minimize the number of secondary cracks. As shown in fig-4 the toughness from blunt-crack determinations is constantly higher than that from sharp-crack determinations in the case of treated beads.

Hybrid formulations. To a first approximation the effects of rubber of rubber and beads are additive when the crack is sharp, fig-6. Bluntly cracked hybrid composites behave differently. The toughness in fact does not worsen by adding the glass filler as for the simple resin/bead materials. This indicate that in the tip region of blunt cracks sharp secondary microcracks do no form. The secondary cracks would in fact undergo blunting and K_I would to a letter extent depend on filler content. Since the preformed particles are much smaller than the glass beads the rubber modified matrix can be regarded as homogeneous as far as its behaviour around the glass particles is concerned. The rubber causes a major change in the crack propagation mechanism. In the case of resin/bead materials a large number of beads appear on the fracture surface (crack pinning is the active mechanism). On the contrary, a rather limited number of particles appear on the fracture surface of hybrid composites, irrespective of the glass surface treatment. In these materials fracture propagates at constant CTOD so that the increase of K_I depends on the effect of the rigid beads on both the elastic modulus and the yield strength. Because the glass particles show little interaction with the crack tip, the secondary microcracks that might form are less armful to toughness and K_I is consequently less influenced by the glass content in bluntly cracked materials.

References
1) S.Bandyopadhyay, V.M.Silva, *Proc. 6th Int.Conf.Fracture,vol.4,* Pergamon Press,Oxford 1984,page 2971.
2) A.J.Kinlock, D.L.Maxwell, R.J.Young, *J.Mat.Sci.,*20(1985)4169.
3) R.J.Young, D.L.Maxwell, A.J.Kinlock, *ibid* 21(1986)380.
4) I.M.Low, Y.M.May, S.Bandyopadhyay, V.M.Silva, *Mat.Forum,*10(1987)241
5) A.C.Garg, Y.M.May, *Comp.Sci.Techn.,*31(1988)193.
6) S.Bandyopadhyay, V.M.Silva, Y.M.May, *Plast.Rub.Proc.Appl.,*10(1988)193.
7) G.Levita, A.Marchetti, E.Butta, *Polymer,*26(1985)1110.
8) A.J.Kinlock, S.J.Shaw, D.A.Tod, *ibid,*24(1983)1341.
9) A.J.Kinlock, S.J.Shaw, D.L. Hunston, *ibid,*24(1983)1355.
10) A.F.Yee, R.A.Pearson, *J.Mat.Sci.,*21(1986)2462.
11) A.F.Yee, R.A.Pearson, *ibid,*21(1986)2475.
12) A.G.Evans, *Phil.Mag.,*26(1972)1327.

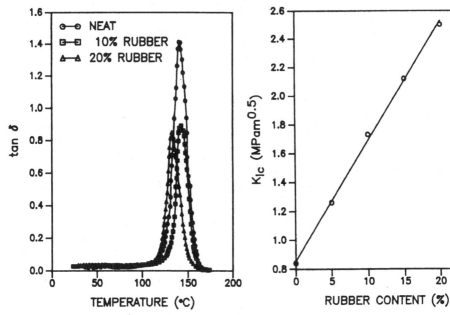

Fig.1 — LOSS FACTOR VS TEMPERATURE FOR RUBBER MODIFIED RESINS.

Fig.2 — FRACTURE TOUGHNESS VS RUBBER CONTENT.

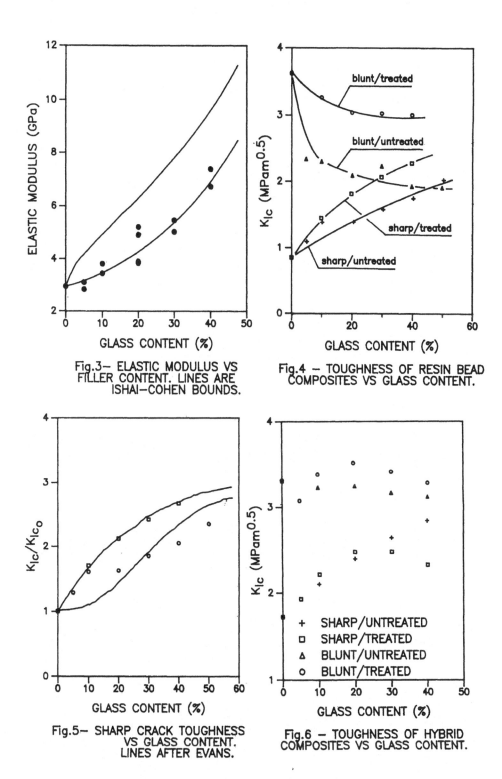

Fig.3– ELASTIC MODULUS VS FILLER CONTENT. LINES ARE ISHAI–COHEN BOUNDS.

Fig.4 – TOUGHNESS OF RESIN BEAD COMPOSITES VS GLASS CONTENT.

Fig.5– SHARP CRACK TOUGHNESS VS GLASS CONTENT. LINES AFTER EVANS.

Fig.6 – TOUGHNESS OF HYBRID COMPOSITES VS GLASS CONTENT.

CHARACTERIZATION OF THE CURING AND HIGH-TEMPERATURE DECOMPOSITION CHARACTERISTICS OF A GLASS-FILLED POLYMER COMPOSITE USED IN HIGH-TEMPERATURE APPLICATIONS

J. HENDERSON, E. POST

Netzsch - Gerätebau GmbH
Wittelsbacherstr. 42 - Postfach 14 60 - 8672 Selb - West-Germany

ABSTRACT

The complete characterization of a thermosetting polymer composite has been carried out during both the curing and high-temperature decomposition reactions using various thermal analysis techniques combined with mass spectrometry.

INTRODUCTION

Thermosetting polymers and their composites are being used in an increasing number of new. high-temperature applications. The temperature range of many of these applications has increased to the point that improved thermosets are required to meet the design specifications. These improvements, in many cases, can be accomplished through alteration of the curing cycle, and/or the use of coupling agents, the inducement of specific morphological features during processing, the orientation and type of reinforcement, etc. One of the most powerful tools for characterizing the effect of these various changes on the performance of the polymer is thermal analysis combined with mass spectrometry.

BACKGROUND

A great deal of research has been directed at characterizing the curing behavior and evaluating its effects on the properties of thermosetting polymers and their composites. Some excellent work has.

been conducted in this area by Gillham /1, 2/, Enns and Gillham /3/ and Prime /4/, among others.

The majority of curing studies have been conducted isothermally using thermal analysis techniques other than thermogravimetric analysis (TGA). A notable exception is the work performed by Greenberg and Kamel /5/ in which the kinetics of anhydride formation in a poly(acrylic acid)-alumina composite were studied using a combination of isothermal and dynamic mass loss (TG) and mass spectrometry data. In addition, Cassel /6/ used isothermal TG data to determine the degree of cure of a phenolic resin. Finally, Lee and Levi /7/ used TGA to evaluate the effects of curing temperature on the degradation of an epoxide resin.

In this work a combination of TGA, differential scanning calorimetry (DSC), and mass spectrometry were used to study both the curing and decomposition reactions in a glass-filled polymer composite under dynamic heating conditions. The effects of curing rate on the high-temperature stability were studied along with the energetics of curing and decomposition using TGA and DSC. Mass spectrometry data were then used to clarify the difference in behavior.

EXPERIMENTAL

Material: The polymer composite studied consisted of approximately 39.5% phenol-formaldehyde resin and 60.5% glass and talc filler. The polymer was stored in a sealed container in a freezer maintained at approximately 0°C. All tests were conducted within 2 weeks of the receipt of the polymer.

Instrumentation: The study was carried out using a Netzsch model 409 simultaneous thermal analyzer (STA) coupled with a Balzers model QMA-120 quadrupole mass spectrometer. The STA unit is equipped with a platinum furnace capable of operation to 1500°C at heating rates ranging from 0.1 to 100°C/min. The Balzers control unit is a QMS-420 capable of analyzing 16 mass numbers simultaneously. Data acquisition and instrument control were accomplished using a 16/32 bit computer system with peripheral units and the appropriate Netzsch software. Data evaluation, i.e. specific heat, kinetic analysis, mass loss, and gas analysis was accomplished using standard Netzsch software.

Procedure: Approximately 40 mg of the uncured polymer was heated from room temperature to about 1500°C at rates of 80, 60, 40, 20, 10, 5, 2, 1, and 0.5°C/min in an inert atmosphere. This temperature range provided data for 3 separate and distinct sets of reactions. These are the curing, pyrolysis, and carbon-silica reactions which occur in the temperature range of 100-250°C, 300-800°C, and 1200-1600°C, respectively. The temperature, mass loss, rate of mass loss, and differential temperature were monitored continuously throughout the experiments. In addition, 16 mass numbers between 16 and 94 amu were continuously monitored. It should be pointed out here that the pyrolysis reactions for a phenolic resin yield molecules with molecular weights greater than 100 were not monitored.

RESULTS AND DISCUSSION

The fraction of mass remaining, m/m_0, as a function of temperature and heating rate for the curing cycle is shown in figure 1. As can be seen, the curing mass loss for this particular polymer is complete by 250°C for all of the heating rates studied. This indicates 100% cure and was confirmed by the DSC and mass spectrometry data which showed no additional curing energetics or release of volatile components above 250°C. Also, as expected, the curing kinetics are a strong function of the heating rate. For example, at $m/m_0 = 0.95$ the temperature difference between the 0.5 and 80°C/min heating rates is over 75°C. In addition, it is clear from figure 1 that the curing kinetics change with temperature for each of the heating rates studied. Finally, it is important to note that there is a clear and consistent increase in the curing mass loss with heating rate. As can be seen, the mass loss at 250°C increases from about 8.12% for the 0.5°C/min heating rate to approximately 9.9% for the 80°C/min heating rate. This results in a different microstructure, and hence different decomposition kinetics, for each heating rate, even though in each case the polymer is fully cured. The reason for the increased mass loss with heating rate will be clarified when mass spectrometry data are discussed.

Figure 2 depicts the decomposition mass loss as a function of temperature at a heating rate of 40°C/min for the samples cured at rates of 80, 1, and 0.5°C/min. As shown, decreasing the curing rate results in a consistent shift in the mass loss curves to higher temperatures. In addition, the mass loss at 1000°C increases with the curing rate. For example, the mass loss increases by about 0.7% as the curing heating rate is increased from 0.5 to 80°C/min. The increase in material performance which accompanies the slower curing rates is a direct result of the higher degree of cross-linking. Again, the reason for the change in stability with curing rate will be clarified when the mass spectrometry data are discussed.

The results of the specific heat and curing energetics measurements at a heating rate of 20°C/min for the uncured material are shown in figure 2. Figure 2 shows that the specific heat of the uncured and cured (virgin) components are essentially linear functions of temperature. Also, as shown, the curing energetics for this material are endothermic in the temperature range of approximately 100 to 180°C and exothermic from approximately 180 to 220°C. The magnitude of the endothermic and exothermic reactions is 25.4 and 19.8 J/g, respecitvely. Mass spectrometry measurements show that the endothermic activity is a result of the vaporization of water, formaldehyde, etc. This, of course, is not a part of the actual curing, but it is impossible to eliminate this vaporization without pressurizing the sample. A more complete discussion will be given later.

The results of the specific heat and reaction energetics measurements for heating rates of 20°C/min for the pyrolysis reactions are shown in figure 3. As can be seen, the pyrolysis reactions are endothermic, with a value of 214 J/g. Also, it can be shown that the

peaks in the apparent specific heat curve generally correspond to
those in the rate of mass loss curve over the temperature range of
the pyrolysis reactions. This is the expected behavior since an increase in the rate of mass loss is generally accompanied by higher
endothermic activity. Also of interest in figure 3 is the fact that
at 1050°C the magnitude of the apparent specific heat is in close
agreement with the char specific heat. Again this is the expected behavior because the material used in the char test was conditioned to
1050°C. Finally, the techniques for computing the specific heat and
reaction energetics have been discussed in detail by Henderson and
Emmerich /8/ and in the interest of space will not be discussed here.

The results of the mass spectrometry measurements are shown in
figures 4 through 6. Figure 5 shows the evaluation of carbon monoxide
and formaldehyde for heating rates of 20 and 60°C/min for all three
sets of reactions. The curves numbered 2 and 4 show the evolution of
formaldehyde for the heating rates of 60 and 20°C/min, respectively.
As can be seen, the evolution of formaldehyde begins at approximately
100°C and continues to 200°C for the 20°C/min heating rate and 250°C
for the 60°C/min heating rate. Comparison of figures 3 and 5 shows
that the onset of endothermic activity at approximately 100°C corresponds to the release of the formaldehyde. It is also clear that there
is a much greater quantity of formaldehyde released for the 60°C/min
heating rate. The reason for this is that the curing reactions are
rate limited and therefore at the higher heating rate larger quantities of formaldehyde are driven off before the reactions can occur.
This, of course, is one of the major reasons for the increased curing
mass loss with increased curing rate. Also, it is speculated that at
the higher heating rates formaldehyde that is needed for curing is
driven off, thereby reducing the cross-linking density and thus the
high temperature stability as shown in figure 2. Also of interest in
figure 4 is the release of free formaldehyde during the pyrolysis
reactions at temperatures between 500 and 600°C for the 60°C/min
heating rate. Clearly, this is one of the reasons for the increased
mass loss with curing rate during these reactions. Finally, curves 1
and 3 in figure 4 depict the intensity of carbon monoxide released
for the 60 and 20°C/min heating rates respectively. As can be seen,
the largest quantities of CO are released during the high-temperature
carbon-silica reactions at temperatures above 1200°C. This is consistent with previous work in which it was speculated that the major
components released during these reactions are CO and CO_2.

Figure 5 shows the release of carbon dioxide, phenol, and the
fragment with mass number 66 for a heating rate of 60°C/min. As can
be seen, the onset of the release of phenol during the curing reactions occurs at approximately 150°C. Comparison of figures 3 and 6
reveals that this temperature corresponds to the point at which the
apparent specific heat curve starts in the exothermic direction, indicating the onset of strong condensation reactions. Also of interest
is the release of phenol during the pyrolysis reactions between 500
and 700°C. Although not shown in this figure, the release of phenol
during the pyrolysis reactions for the 20°C/min heating rate is much
lower than for the 60°C/min heating rate. This, of course, also

accounts for the increased mass loss with curing rate which is shown in figure 1. Finally, the release of CO_2 during both the pyrolysis and carbon-silica reactions is clearly shown. This agrees well with the findings of Sykes /9/ for a similar phenolic resin.

Shown in figure 6 is the release of water vapor and formaldehyde for the 60°C/min heating rate. Note that the onset temperature of the release of water vapor (curve 2) corresponds closely to that for formaldehyde. This confirms that the release of water vapor at temperatures between 100 and 150°C contributes to the endothermic nature of the curing shown in figure 2. Also of interest is the inflection point in the water vapor curve at approximately 150°C indicating the onset of strong curing. It should be noted that this temperature also corresponds to the onset of the release of phenol as shown in figure 5. Finally, the release of large quantities of water vapor over the entire range of the pyrolysis reactions should be noted. Again, this corresponds with the findings of Sykes /9/.

CONCLUDING COMMENTS

Combined STA/mass spectrometry is a powerful tool which can be used to unravel the details of complicated reactions which occur in polymeric materials. STA data yield general information regarding the reactions, while mass spectrometry data provide specific details regarding the characteristics of these reactions.

With regard to the specific polymer studied here, it has been shown that curing rate influences not only the high temperature stability, but the energetics of curing and decomposition as well. The mass spectrometry data were used to clarify why these changes occur. Finally, it should be pointed out that while mass spectrometry data for only 6 mass numbers were presented, significant quantities of several other gases were released. These data, however, were not presented due to space limitations.

REFERENCES

1. J.K. Gillham, Polymer Engineering and Science, Vol. 16, No. 5 (1976) 353
2. J.K. Gillham, Polymer Engineering and Science, Vol. 26, No. 20 (1986) 1429
3. J.B. Enns and J.K. Gillham, Journal of Applied Polymer Science, Vol. 28 (1983) 2567
4. R.B. Prime, Polymer Engineering and Science, Vol. 13, No. 5 (1973) 365
5. A.R. Greenberg and I. Kamel, Journal of Polymer Science, Polym. Chem. Edition, Vol. 15 (1977) 2137
6. B. Cassel, Perkin-Elmer Report No. MA-29 (1976)
7. H.T. Lee and D.W. Levi, Journal of Applied Polymer Science, Vol. 13 (1969) 1703
8. J.B. Henderson and W.-D. Emmerich, Thermochim. Acta, Vol. 131 (1988) 7
9. G.F. Sykes, Jr., NASA Technical Report No. TN D-3810 (1967)

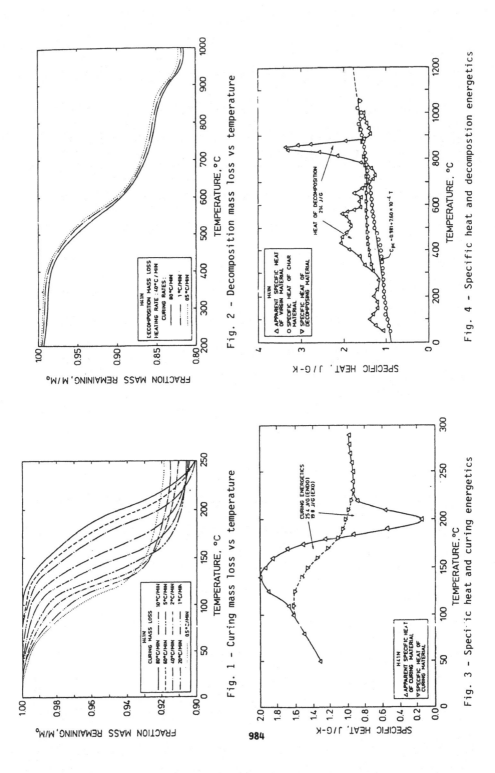

Fig. 1 - Curing mass loss vs temperature

Fig. 2 - Decomposition mass loss vs temperature

Fig. 3 - Specific heat and curing energetics

Fig. 4 - Specific heat and decompostion energetics

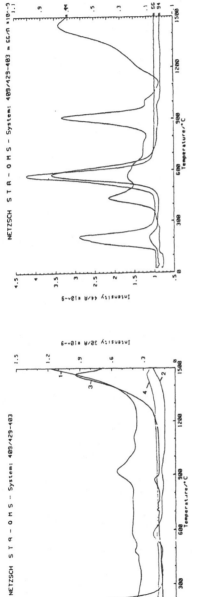

Fig. 5 – Intensity for mass numbers 28 and 30

Fig. 7 – Intensity for mass numbers 18 and 30

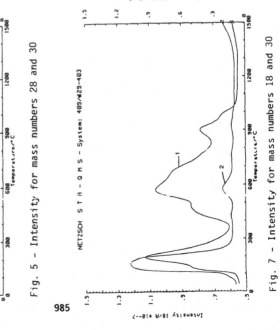

Fig. 6 – Intensity for mass numbers 44, 66 and 94

IMPACT BEHAVIOUR OF LOADED COMPOSITES

H. MORLO, J. KUNZ

Wehrwissenschaftliches Institut für Mat. Untersg.
Landshuter Strasse 70 - 8058 Erding - West-Germany

Abstract

This article concerns tests on the impact behaviour of
loaded carbon fibre reinforced plastics. In these
investigations both stress-free specimens and specimens
loaded uniaxially with compression stress of 25 % and 50 %
of the compression strength (undamaged specimens) are
tested. Following this, the damage area is determined
ultrasonically and the residual compression strength by a
quasi-static method. Results achieved are presented and
discussed.

Introduction

Carbon fibre reinforced plastics (CFRP) show a high
susceptibility to short-time impact loads as compared to
isotropic materials. Thereby primarily delaminations occur
which lead to a reduction in component strength. Type and
extent of this impact damage are influenced by a number of
parameters. Among these are the material-specific
characteristics like resin and fibre properties, lay-up,
moisture content and temperature as well as external
influences like impact energy, velocity, shape, size and
material of the impactor. Another factor influencing the
impact behaviour is represented by the stress condition of
the component. On this subject, however, only very few,
mainly theoretical, investigations are available /1,2/. In
practical use, however, components are in many cases

987

subject to static and dynamic stresses which may affect their impact behaviour.

The aim of these tests therefore is to systematically evaluate the influence of uniaxial compression stresses on the impact behaviour and residual compression strength of CFRP.

Material

Specimens measuring 100 mm x 100 mm x 3 mm made of Narmco T300/5208 with the quasi-isotropic lay-up [(45°,0°,-45°,90°)$_4$]s are to be tested.

Test method

Impact tests are conducted on the basis of Boeing Specification BSS 7260, using an instrumented drop weight impact device, Fig.1. This intrumentation permits measuring of the force-time and deflection-time curves, thus making a calculation of the energy balance possible. The test fixture support is square in shape with a length of 90 mm. The front end of the impactor is ball-shaped with a diameter of 11 mm. During the impact test specimen can be loaded with a uniaxial compression stress. The design of the loading fixture is such that denting of the specimen is prevented.

Following the impact test the resulting damage is determined by means of a highfrequency ultrasound C-Scan equipment. Determination of residual compression strength is also performed in accordance with BSS 7260.

Results

Three impact energy steps, namely 4 J, 6.4 J and 13.2 J have been selected for test purposes. Compression stresses used for the impact test are 25 % and 50 % of the compression strength of the undamaged specimen (≈456 N/mm²). Figures 2 and 3 illustrate the force-time and deflection-time curves for the compression stress of 0 % and 50 % and impact energy of 13.2 J. A comparison reveals that under stress-free condition the maximum force is greater than under stress condition. The energy curve, too, is different. This leads to the assumption of different damage mechanisms. In addition to that, the impact tested specimen with compression stress reveals a greater deflection value, so that in both cases the same energy is absorbed. This also holds true for the compression stress of 25 %. With increasing compression stress however, the damage area also increases, Fig. 4. The externally visible damage, on the other hand, decreases. Under stress-free conditions the visible damage is largest on the rear surface of the specimen. Thus, under stress-free conditions a large amount of the energy absorbed is used for residual deformations of the components, whereas under compression stress conditions the energy absorbed results in more delaminations.

Figure 5 illustrates the residual compression strength and its dependence on impact energy and compression stress. At 4 J the impact damage with and without compression stress only results in a damage area of approx. 2 mm². Thus the residual compression strength values only differ marginally. With increasing impact energy and compression stress a greater difference between residual compression strength values can be observed.

Conclusions

The tests presented reveal that the impact behaviour of CFRP is influenced by external compression stress. Changes can be observed in both the morphology of the specimen and the damage size. This also affects the residual compression strenth of the specimen.

References

1-Sun C.T. and Chattopadhyay S., Journal of Applied Mechanics, 42 (1975) 693-698

2-Sun C.T. and Chen K., Journal of Composite Material, 19 (1985) 490-504

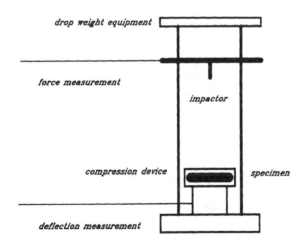

Fig.1 - Instrumented impact test equipment

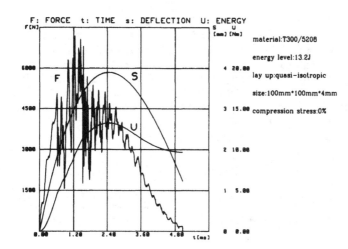

Fig.2 - Force, deflection and energy in dependence of time
Without stress

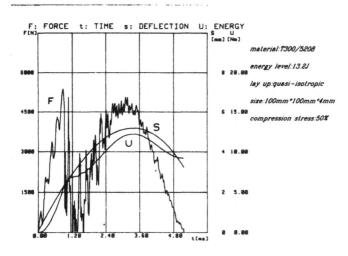

Fig.3 - Force, deflection and energy in dependence of time
50 % compression stress

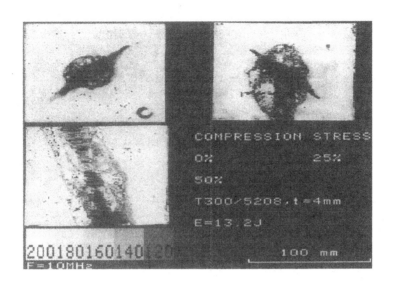

Fig.4 - Ultrasonic images of the damaged specimen

material:T300/5208

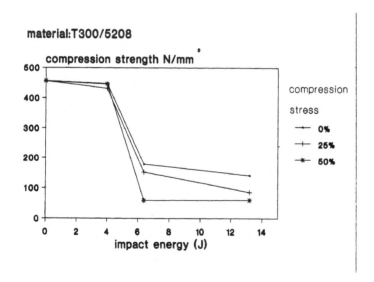

Fig.5 - Compression strength in dependence of energy level
and compression stress

IMPACT STRESS WAVES
IN FIBRE COMPOSITE LAMINATES

W.A. GREEN, E.R. GREEN*

Nottingham University - University Park
Nottingham NG7 2RD - Great Britain
*Leicester University
Engineering Dept - The University - Leicester LE1 7RH - Great Britain

ABSTRACT

The propagator matrix method together with the integral
transform formalism is employed to derive the transient
response in a fibre composite plate due to a normal
impulsive line load acting on its upper surface. Each
layer of the composite is modelled as a transversely
isotropic elastic material which is inextensible along the
fibre direction. Detailed plots are presented showing the
stress variation with distance from the impact line at two
different times and two orientations of the line load.

INTRODUCTION

The problem under consideration is the determination of
the transient stress waves in a four-ply laminate of fibre
reinforced material due to an impulsive line load acting on
the upper surface. The laminate is of infinite lateral
extent and each of the plies is modelled as a transversely
isotropic elastic continuum with the axis of transverse
isotropy lying in the plane of the ply and parallel to the
fibre direction. The system of differential equations
governing the motion is simplified by treating the material
as inextensible in the fibre direction. This idealisation
reflects the fact that the extensional modulus of the
composite along the fibre direction can be of the order of
fifty times that in the cross fibre direction. A
discussion of the consequences of this assumption is
contained in references [1] and [2].

METHOD OF SOLUTION

The layers forming the laminate are all of the same
material and of uniform thickness h. They are arranged in
a symmetric cross-ply configuration with the fibre

directions in the two inner plies being parallel to each other and at right angles to the fibre directions in the two outer plies. A system of Cartesian coordinates is chosen so that the middle surface of the plate is the plane $z = 0$, with the y-axis parallel to the fibre direction in the two outer layers and the x-axis parallel to the fibre direction in the two inner layers. An impulsive load having uniform magnitude along its length is imposed along the line $s = x \cos a + y \sin a = 0$ on the upper surface. This sets up a motion of the plate in which the components of stress and displacement in each layer are functions of z, s and the time t only. Taking Laplace transforms with respect to t and Fourier transforms with respect to s the equations of motion in the individual layers are solved to give the transformed components as functions of the thickness variable z only. Imposing the interface continuity conditions and the surface boundary conditions then yields the analytic solutions for the transforms throughout the plate. The stress and displacement components as functions of s and t are recovered by inverting the transforms. Full details of the technique are given in [1].

RESULTS

The results presented in Figs.1-4 relate to the orientations $a = 30$ (Figs 1 & 3) and $a = 60$ (Figs 2 & 4). Each figure consists of five curves which show the stress at the upper surface, the upper interface, the mid-surface, the lower interface and the lower surface, as a function of distance s from the impact line. The curves in Figs.1 & 2 are for $t = 40h/c$ and those in Figs.3 & 4 for $t = 200h/c$ where c is the speed of the fastest body wave in the continuum.

The most striking feature of these results is the existence of the large spike in the upper surface stress in Figs.1 & 3 which does not exist at any of the other levels. The magnitude of this spike stays constant with time, showing that the associated disturbance propagates along the upper surface without attenuation. No such surface wave is set up at $a = 60$ (Figs.2 & 4) since the spikes appearing at both the upper and lower surfaces in Fig.2 are reduced in amplitude by a factor between 1/2 and 1/3 in Fig.4. These results are in agreement with the conclusions reached concerning the existence of the surface wave in reference [2].

ACKNOWLEDGEMENT

This work is supported by the U.S.A.F. Office of Scientific Research under research grant number AFOSR-88-0353.

REFERENCES

1. E. Rhian Baylis and W. A. Green, in "Proc. 3rd Int. Conf. on Recent Advances in Structural Dynamics." AFWAL-TR-88-3034 Vol 1 (1988) 171-183.

2. W. A. Green and E. R. Baylis, in "Wave Propagation in Structural Composites" (T.C.Ting and A.K.Mal, eds) ASME-AMD Vol 90 (1988) 53-67.

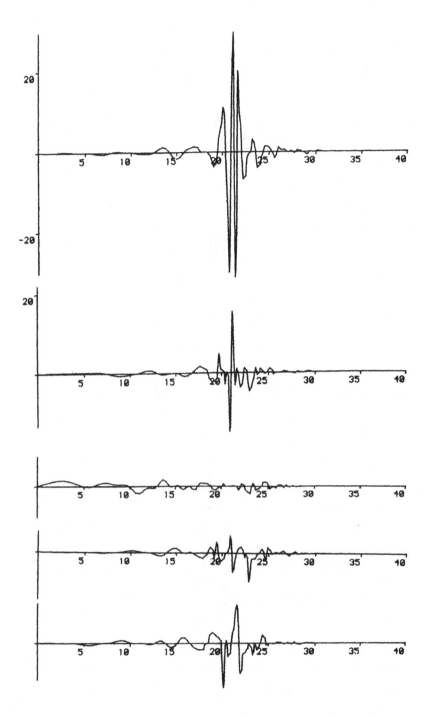

FIG.1 Stress versus distance from impact line: $t=40h/c$, $a=30$.
(a) $z = 2h$, (b) $z = h$, (c) $z = 0$, (d) $z = -h$, (e) $z = -2h$.

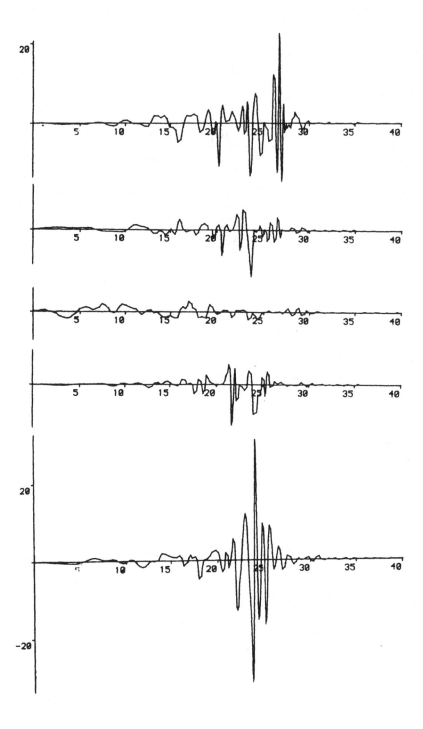

FIG.2 Stress versus distance from impact line: $t=40h/c$, $a=60$.
(a) $z = 2h$, (b) $z = h$, (c) $z = 0$, (d) $z = -h$, (e) $z = -2h$.

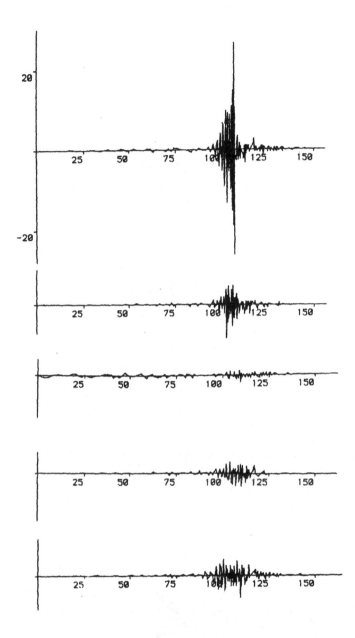

FIG.3 Stress versus distance from impact line: $t = 200h/c$, $a = 30$.
(a) $z = 2h$, (b) $z = h$, (c) $z = 0$, (d) $z = -h$, (e) $z = -2h$.

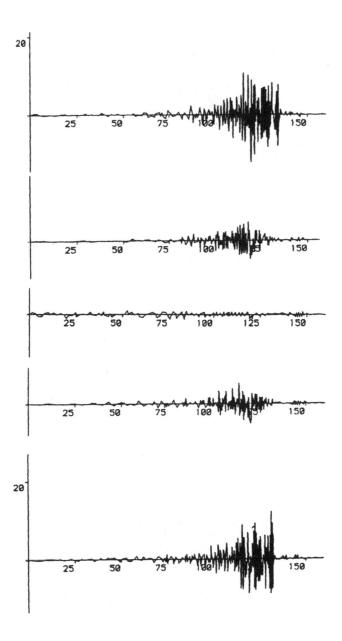

FIG.4 Stress versus distance from impact line:$t=200h/c$,a=60.
(a) $z = 2h$,(b) $z = h$,(c) $z = 0$,(d) $z = -h$,(e) $z = -2h$.

THERMOPLASTICS

Chairmen : **Dr. J.A.N. SCOTT**
Shell Research B.V.
Prof. K. FRIEDRICH
Technical University of
Hamburg-Harburg
now University of
Kaiserslautern

EXPERIMENTAL BACKGROUND FOR FINITE ELEMENT ANALYSIS OF THE INTERPLY-SLIP PROCESS DURING THERMOFORMING OF THERMOPLASTIC COMPOSITES

R. SCHERER, K. FRIEDRICH

Technical University of Hamburg-Harburg
Polymer and Composites Group - Harburger Schloßstraße 20
2100 Hamburg 90 - West-Germany

ABSTRACT

One of the main advantages using thermoplastic polymers as matrices for continuous fibre reinforced composites is their feature to melt under heat and pressure. During a thermoforming process, e.g. over a single curvature tool, the fibre rich layers must slip relative to each other, in order to reach the new, desired shape without fibre breakage or wrinkling effects. To characterize this flow process, a single prepreg layer has been pulled-out of a laminate to measure the resistance against slip in between the layers as a function of lay-up, temperature, slip velocity and pressure applied normal to the slipping direction. The results will act as input parameters for a Finite Element Analysis of the forming process.

INTRODUCTION

The main feature of thermoforming continuous fibre-reinforced thermoplastics is the fluid state of the matrix when being heated above melting temperature. For the production of thermoplastic composites, preimpregnated layers are laid-up in a flat mould and consolidated under heat and pressure to a laminated sheet. This sheet can be reheated whenever being thermoformed into the final shape. Several flow processes of the fibres and the polymer have been classified in the literature /1-3/.

After consolidation, the laminate is built up of alternating fibre rich layers and resin interlayers. The interply-slip process is desired because otherwise an initially flat laminate cannot be formed satisfactorily to a single-curvature part This slippage mainly takes place in the resin interlayer. The aim of this report is to describe the basic concepts of the interply-slip investigations. Experimental results and computer modelling will act together to determine the resistance against sliding as a function of 1)lay-up, 2)slip velocity, 3)temperature and 4)pressure applied normal to the pull-out direction.

EXPERIMENTAL SET-UP

The experiments are based on experiences obtained by pulling out a single, unidirectional reinforced layer out of a stack of several layers by dead-weight loading /1/. Figure 1 presents the experimental set-up for the pull-out experiments. Test specimens were fixed in a hydraulic press with heatable platens. The tension mode of a universal testing machine was used to apply the pulling-out load, inclusive force and displacement recording devices. A thin steel wire was first attached to the centre ply of the specimen, then turned round by a free rotatable roll and finally fixed to the cross-head of the universal testing machine. Additionally, a displacement transducer was attached next to the centre ply of the specimen to measure its movement only. This avoided an incorrect displacement measurement because the universal testing machine would have measured the extension of the steel wire and the displacement of the centre ply simultaneously.

Before starting the experiment, the laminate was at first preconsolidated (T=180°C,t=6 min.,p=0.71 MPa). The pressure was then released down to a point that the specimen and the heat platens were in slight contact only. The specimens were wrapped in thin steel foil to avoid tranverse flow when being subjected to presssure and temperature. Thus, the main flow occured only in the pull-out direction. Easily, a 0°-layer could be pulled out as the unidirectionally orientated fibres allowed a load tranfer, but no traction load could be applied to differently orientated layers (e.g. 45 and 90°-layers). Therefore, the polymeric film on one side of the prepreg was removed by grinding, and this side was then adhesively bonded to a metal foil. As the prepreg and this load transfer foil expend differently under heat, both sides of the metal foil were prepared to avoid thermally induced warping of the specimen. Figure 2 presents schematically the force-displacement-time relationship. No displacement could be measured until the pulling-out force reached a critical value. Above this threshold, the force slightly increased before it reached a stationary level which was dependent on the pull-out velocity. Both, the yield level and the stationary values at different velocities mainly describe the materials behaviour under shear traction.

RESULTS AND DISCUSSSION

First, the influence of the number of layers was measured. If we consider a Newtonion fluid as a first approximation, the shear stress is proportional to the applied shear rate (Fig. 3). This shear rate is dependent on the number of resin interlayers with the individual height h. For a multiple lay-up, the shear stress decreases proportional to the number of layers n. Even though the assumption of a linear-Newtonion relationship of shear stress (pulling-out force) and shear strain rate (pulling-out velocity) was not confirmed by experiment, the linear decrease of the pulling-out force by stepwise increasing the number of layers was measured. Thus, the following results were obtained by testing the slippage behaviour only of one centre ply against one outer ply on each side. A better theoretical description would be a power-law $\tau = \tau_{yield} + \eta \gamma^{n-1}$ of the molten resin interlayer under shear traction.

Secondly, the pressure normal to the sample was gradually increased (Fig.4a). For low velocities, hardly any influence on the slip resistance was measured. For higher velocities, the pulling-out force, which represents the resistance against slipping at a certain velocity level, increases. This slip rate dependence of the interply-slip indicates that the resistance cannot be described as a frictional slip. The inner friction, namely the materials viscosity, is more accurate. If we consider the temperature T as a parameter which mainly influences the materials viscosity (Fig. 4b). It should be recommended to thermoform parts at higher temperatures because the interply-slip resistance decreases. On the other hand, other flow processes, like fibre washing and misalignment, are more likely to occur. Furtheron, the experiments were carried out at 180°C.

The lay-up of a composite also influences the interply-slip behaviour significantly. The resistance increases from left to right:

Outer Ply [0]:   [0,45,0]   [0,90,0]   [0,0,0]
Outer Ply [+45]: [45,0,45] [45,-45,45] [45,90,45] [45,45,45]
Outer Ply [-45]: [45,0,45] [-45,45,-45] [45,90,45] [45,45,45]
Outer Ply [90]: [90,0,90]   [90,45,90] [90,90,90]

In futhergoing experiments, a flat, preconsolidated laminate was thermoformed to a 90°-angle. Figs. 6a and 6b present two cross-sections of a ten layer PP/CF laminate, 40 Vol.% (ICI´s Plytron). If the slipping of adjacent plies does not occur, disasterous fibre buckling and wrinkling are the consequence of inadequate processing conditions.

CONCLUDING REMARKS

Experimental Results have been presented considering the influence of various parameters on the interply-slip process during thermoforming of continuous fibre-reinforced

composites. These results are planned to act as input
parameters for a Finite Element Modelling, focussing the
prediction of interply-slip for different part geometries,
forming speeds, and lay-ups. An optimum processing window
should be determined to produce good quality components.

ACKNOWLEDGEMENT

The authors would like to thank Mr H. Hargarter and Mr. K.
Meyer for their experimental work. Thanks are also due to
Imperial Chemical Industries, ICI Wilton, UK, for the
material supply. The financial support of the DFG
(FR 675-7-1), Bonn, West Germany, is greatly acknowledged.

REFERENCES

1 F.N. Cogswell
  The Processing of Thermoplastics Containing High Loading
  of Long and Continuous Reinforcing Fibers, Proc. of The
  Polymer Society,Freund Publ. House, London 1986, 345-361

2 J.A. Barnes, F.N. Cogswell
  Tranverse Flow Processes in Continuous Fibre-Reinforced
  Thermoplastic Composites,Composites Vol.20 (1) 1989, 38-4

3 R. Scherer, N. Zahlan, K. Friedrich
  Modelling the Interply-Slip Process during Thermoforming
  of Thermoplastic Composites using Finite Element Analysis
  Proc. CADCOMP 90, Brussels,Belgium, April 1990

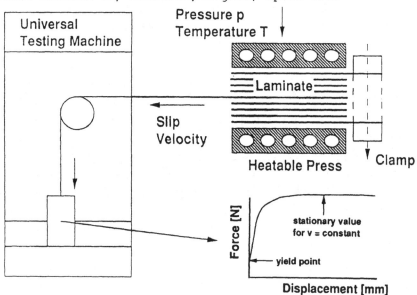

Figure 1   Pull-out Experiment - Test Apparatus

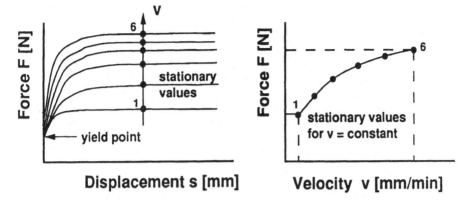

Figure 2 Pull-out Experiment - Recorded Curves

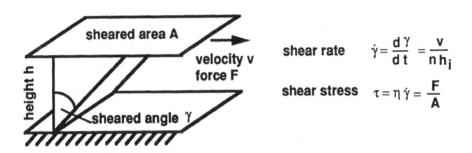

shear rate $\quad \dot{\gamma} = \dfrac{d\gamma}{dt} = \dfrac{v}{n\,h_i}$

shear stress $\quad \tau = \eta\,\dot{\gamma} = \dfrac{F}{A}$

Figure 3 Shear Stress - Shear Strain Relationship

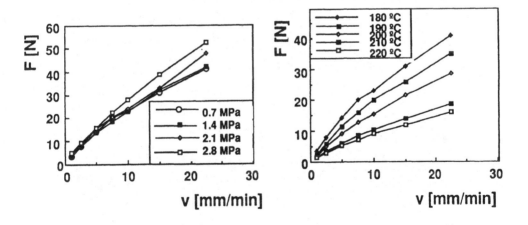

Fig. 4a Parameter Pressure p   Fig. 4b Parameter Temperature T
(T=180°C=const.)   (p=slight contact=const.)

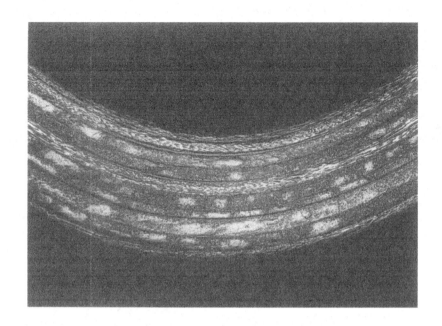

Figure 6a Thermoformed Laminate - Interply-Slip

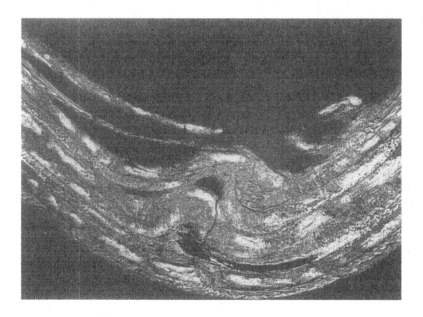

Figure 6b Thermoformed Laminate - No Slip, Buckling of Plies

INFLUENCE OF PRODUCTION PARAMETERS ON THE MECHANICAL BEHAVIOUR OF CF/PEEK

R. WEIß

Schunk Kohlenstofftechnik GmbH
6300 Gießen - West-Germany

ABSTRACT

The influence of processing parameters and raw materials on the resulting mechanical properties of CF/PEEK-composites has been investigated. The mechanical properties of the PEEK-composites have been compared with samples made of APC2. Comparable composite properties can be obtained independent of the processing techniques like film-stacking or compression moulding of prepregs (APC2) or comingled PEEK yarn materials. The mechanical behaviour of CF/PEEK-composites is mainly controlled by the cooling rates and/or incompatible fibre-finish-systems.

INTRODUCTION

Carbon fibre reinforced polyetheretherketones are a candidate material for aircraft as well as industrial applications. From industrial viewpoints the influence of raw materials and processing techniques on the final mechanical properties is most important to result in reproducable composite properties. According to literature the best composite properties should be obtained by using the commercial PEEK-system APC2 from ICI.
Therefore the aim of the present study was to determine the limitations in strength, stiffness and transverse properties of PEEK-composites depending on the raw materials and the processing techniques. All composites have been tested in comparison to APC2.

MATERIALS

As reinforcing fibres Hercules AS4 (without finish) and Tenax-J HTA7 (from Akzo) with a standard epoxy finish (1,3 %) have been used. Furthermore a comingled yarn (hybride) from BASF Structural Materials based on Hercules AS4 6K C-fibres (without finish) and PEEK-fibres (150 G-type) with a fibre volume fraction of 62,7 % was selected. Unidirectional composites based on PEEK foils from ICI, Stabar XK300 (crystalline) and Stabar K200 (amorphous) have been investigated in comparison to APC2 (ICI) and the comingled material (BASF).

EXPERIMENTAL

All UD-composites have been compression moulded in a closed metal die. The maximum temperature have been 420 °C in case of hybrided (comingled material) as well as the film stacked material, whereas in case of APC2 only 400 °C have been applied for 15 minutes at a pressure of 97 bar. The following two different cooling rates have been selected to obtain different degrees of crystallinity:

low cooling rate   -   20 °/h
high cooling rate   -   22 - 23 °/min

All mechanical properties were determined according DIN 29971. The degree of crystallinity was measured by DSC.

RESULTS

The mechanical properties of various CF/PEEK-composites are given in table 1 and 2. As can be seen, the flexural strength (table 1) is improved in all cases by increased crystallinity (low cooling rates), whereas the modulus shows no clear correlation in respect to the two different cooling rates.
The composites prepared by film-stacking (AS4/crystalline and AS4/amorph) or compression moulding of comingled yarns (hybride) are comparable or even superior in longitudinal flexural strength and modulus compared to the composites with APC2 (see table 1).
However, in case of matrix controlled properties like transverse flexural strength or interlaminar shear strength the film-stacking-technique is always combined with a slight decrease of the transverse properties (see table 2). The composites based on comingled yarns (hybrides) show the same transverse flexural strength (173 MPa) and ILSS (121 MPa) as the composites made with APC2 (172 MPa or 170 MPA, respectively).
This behaviour of matrix controlled properties can be explained by the fibre/matrix distribution of the four different types of PEEK-composites (fig. 1 - 4).
APC2 composites show an absolutely homogeneous fibre/matrix distribution (fig. 1). The typical fibre distribution of composites with comingled yarns is given in fig. 2. The distribution is comparable to APC2 with typical circular PEEK reach areas due to the comingling process. In all cases of composites made by film-stacking-techniques the fibre matrix layers can be recognized. Composites made with amorphous PEEK foils (fig. 3) show thin intermediate PEEK layers, which are increased in case of a more crystalline foil (fig. 4). Therefore the transverse properties of AS4/amorph-composites are superior compared to the AS4/crystalline type (see table 2) due to the improved fibre/matrix distribution.
Furthermore it should be pointed out that increased crystallinity improves only the transverse flexural strength of composites with APC2 and comingled yarns. All polyetheretherketone composites made by film-stacking-processes show a remarkable decrease in transverse strength with increasing crystallinity (table 2). In case of the

reinforcement with unsized Hercules AS4 C-fibres (AS4/crystalline and AS4/amorph) there exists no explanation for this mechanical behaviour. Further investigations are necessary to clear this point. In case of film-stacked materials with epoxy sized HTA7-C-fibres the decrease of transverse strength with increasing cooling time (low cooling rate) is caused by the thermal inducesd decomposition of the epoxy finish. Whereas in case of reinforcement of polysulfone and polyethersulfone with pressing temperatures of at about 350 °C no incompatibility of the epoxy finish could be observed, thermal decomposition occurs in HTA7/PEEK-composites (pressing temperatures 420 °C) resulting in the formation of internal voids and pores. This decomposition of the finish is responsible for the tremendous decrease in transverse strength of the HTA7/crystalline composites at low cooling rates (from 80 MPa to 15 MPa, see table 2). The more homogenous composites with amorphous PEEK foils (HTA7/amorph) show only a loss of flexural strength of 12 MPA due to the increased decomposition time by low cooling rates.

Furthermore it should be mentioned that the longitudinal flexural strength datas of the composites with HTA7 C-fibres (table 1) are in contradiction to the transverse flexural strength in respect to the cooling rates. In case of longitudinal flexural load the increased crystallinity seems to be dominant over the composite damage by void formation. The transverse strength properties are more sensitive against faults like voids and pores than the longitudinal flexural strength and the ILSS, too.

Although in case of HTA7/PEEK-composites the lowest ILSS datas have been obtained, the shear strength reduction is lower than the reduction in transverse strength in comparison to the composites with the compatible unsized Hercules AS4 C-fibres (table 2).

Table 3 shows the degree of crystallinity of CF/PEEK-composites. The crystallinity increases with decreasing cooling rate. However, there exists no direct correlation between the absolute degree of crystallinity and the longitudinal flexural strength.

SUMMARY

Same or even superiour mechanical properties of CF/PEEK-composites can be obtained with comingled materials or by film-stacking-techniques in comparison to APC2-samples. The longitudinal flexural strength of PEEK-composites increases with increasing crystallinity. The transverse flexural strength is more sensitive to matrix modifications by the processing technique than the ILSS. Thermal decompositions of incompatible finish systems results in a tremendous decrease in transverse flexural strength.

ACKNOWLEDGEMENT

This study was supported by BMFT under the grant No 03M1017A6

Raw Material	Cooling Rate	Flexural Strength (MPa)	Young's Modulus (GPa)
APC2	high	2169	123,5
APC2	low	2284	131,2
Hybride	high	2049	119,6
Hybride	low	2301	128,3
AS4/cryst.	high	2002	111,1
AS4/cryst.	low	2066	101,6
AS4/amorph	high	2162	106,4
AS4/amorph	low	2426	128,0
HTA7/cryst.	high	1861	118,2
HTA7/cryst.	low	1949	106,6
HTA7/amorph	high	1597	115,0
HTA7/amorph	low	1913	105,7

Table 1 -
Influence of the cooling rate on the longitudinal flexural strength and modulus of various PEEK-composites

Raw Material	Cooling Rate	Transverse Flexural Strength (MPa)	ILSS (MPa)
APC2	high	154	108
APC2	low	172	117
Hybride	high	144	113
Hybride	low	173	121
AS4/cryst.	high	146	101
AS4/cryst.	low	114	103
AS4/amorph	high	169	107
AS4/amorph	low	131	108
HTA7/cryst.	high	80	94
HTA7/cryst.	low	15	80
HTA7/amorph	high	73	90
HTA7/amorph	low	61	75

Table 2 -
Influence of cooling rate on the matrix dominant properties of various PEEK-composites

Crystallinity of PEEK

Table 3 -
Crystallinity of PEEK-composites

Raw Material	Cooling Rate	Crystallinity (%)	Melting energy (J/g)
APC2	high	25,72	40,59
APC2	low	30,53	48,18
Hybride	high	22,56	35,59
Hybride	low	32,87	51,86
AS4/cryst.	high	25,77	40,66
AS4/cryst.	low	36,04	56,87
AS4/amorph	high	20,77	32,77
AS4/amorph	low	25,26	39,87
HTA7/cryst.	high	27,12	42,80
HTA7/cryst.	low	30,25	47,72
HTA7/amorph	high	26,16	41,28
HTA7/amorph	low	28,96	45,70

Figure 1 -
Fibre /matrix distri-
bution in APC2-
composites

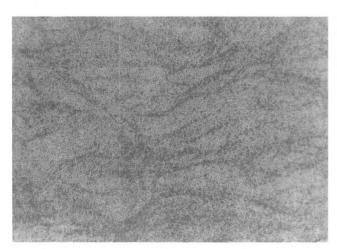

Figure 2 -
Fibre/matrix distribution in composites made with comingled materials

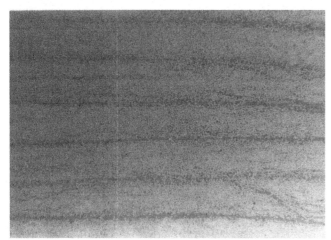

Figure 3 -
Fibre/matrix distribution of PEEK-composites made by film-stacking-techniques with amorphous PEEK foils (Hercules AS4/amorph)

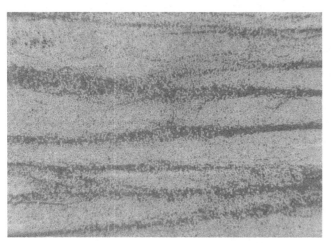

Figure 4 -
Fibre/matrix distribution of PEEK-composites made by film-stacking-techniques with crystalline PEEK foils (Hercules AS4/ crystalline)

MATRIX MORPHOLOGY EFFECTS ON FRACTURE TOUGHNESS OF UNIDIRECTIONAL THERMOPLASTIC COMPOSITES (POLYAMIDE + GLASS FIBRES)

B. GOFFAUX, I. VERPOEST

Katholieke Universiteit Leuven
De Croylaan 2 - 3030 Leuven - Belgium

ABSTRACT

The toughness transfer from the matrix to the composite is studied for a glass fibre reinforced thermoplastic (polyamide) composite. Fracture toughness of the composite and of the pure matrix as measured experimentally. Control of the morphology by Differential Scanning Calorimetry and optical microscopy explains the toughness values obtained for different systems. Finally a correlation between the toughness of the matrix and the composite toughness is discussed.

INTRODUCTION

It has now been admitted that toughness in general, and interlaminar fracture more specifically are the major critical composite properties for structures subjected to service loads. Different solutions have been presented to improve the interlaminar fracture toughness, and hence the delamination resistance. One of these solutions is the use of a toughened polymer as matrix material. Nevertheless the resulting composite toughness is lower than one could expect. This fact has been related to the restricted crack-tip deformation zone due to the presence of the fibres[1]. Moreover the use of a toughened polymer can lead to a decrease of the properties in a hot and wet environment.

Finally, a basic understanding of the way the matrix toughness is transferred to the composite is still not available.

In order to study more fundamentally these transfer mechanisms, a polymer/fibre system has been selected, for which different matrix toughness values can be obtained by changing the morphology of the matrix. This is achieved by controlling the cooling rate during the lamination process and by further carrying out several thermal treatments.

A further variation of the fracture toughness values could be achieved by using polymers of the same type but with different physico-chemical properties (i.e. the molecular weight). The results of that study will be the subject of another publication.

EXPERIMENTAL PROCEDURE

1. The materials

The polymer, used in this study, is an aromatic polyamide (PA) supplied by SOLVAY. The glass transition temperature and the melting temperature measured at 10°C/min are respectively 75°C and 234°C [2].

The pure polymer specimens are injection molded and either cooled slowly (to obtain crystalline PA) or quenched to ensure an amorphous state.

To manufacture the composite prepregs, glass fibres are impregnated with polyamide using the powder impregnation method developed by FLEXLINE b.v.. Then, the prepregs are laminated into unidirectional plates of 200 mm width and 2.5 mm thickness and consolidated in a hot press (at 260 °C during 5 min.). The cooling rate is regulated to obtain 5°C/min and 2000°C/min. The cooling rates ensure to have respectively a crystalline and an amorphous laminate as shown in the Differential Scanning Calorimetry (DSC) scans (figure 1). Specimens of 20 mm width were then were from the plates. Some amorphous specimens were isothermally crystallized at different temperatures; the controlling parameter is the "surfusion" (ΔT = T melting - T crystallization) which was respectively 140°C and 85°C.

2. Fracture toughness tests

Composite specimens of 200 mm length by 20 mm width were cut and tested in Mode I following the Double Contilever Beam test procedure described elsewhere[3]. R-curves (crack-resistance versus crack length) were calculated from the raw data following the displacement method.

Polymer Compact Tension Specimens were cut following the procedure described elsewhere[4]. Amorphous and crystalline specimens were tested. The toughness of the polymer was determined by measuring the critical stress intensity factor (Kc).

3. Morphology observations

Thermal characterization of the samples was carried out using the Dupont DSC Differential Scanning Colorimetry. Specimens were heated under argon flow at 10°C/min from room temperature to 300°C.

Optical microscopy observations were done in transmission on cross-sections. This last technique requires to make specimens of only 10 μm thick. These thin specimens were obtained by mechanical polishing.

EXPERIMENTAL RESULTS

1. Composites

During mode I fracture toughness tests, it was observed that the crack propagation can be stable or unstable, depending on the morphology of the composite (figure 2).

For amorphous specimens, the crack propagation is unstable. The toughness at the onset of instability is about 1350 J/m²; the crack stops at a toughness value of 700 J/m² (figure 2a - table 1).

On the contrary, the crack propagation is stable for specimens crystallized in the press (Xc) during the consolidation of the laminate (figure 2b). The toughness for the crack initiation is 220 J/m² and for the crack propagation 300 J/m². Microscopical observations on these directly crystallized specimens revealed that the matrix exhibits a Maltese Cross pattern. The polyamide spherulites are 7-8 μm in diameter.

Finally, specimens which have been crystallized from amorphous specimens also exhibit an instability during the crack propagation like the amorphous specimens do (figure 2c). Nevertheless, those specimens are fully crystalline as shown by the DSC-scans (figure 3). When the degree of surfusion is 140°C (crystallization temperature of 95°C), the toughness values are the same as for the amorphous specimens. For the lower degree of surfusion ($\Delta T = 85°$, Tc = 150°C), the G_{IC} for the onset of the instability decreases to 900 J/m².

The crystalline morphology of both these specimens is very fine; for the higher degree of surfusion, the spherulites are so fine that the difference with the amorphous specimens is very small. When the degree of surfusion decreases, we observe an increase in the spherulite diameter.

2. Polymer

As shown in table 2, the fracture toughness as measured by the critical stress intensity factor, is for the crystalline PA only 2/3 of the value for the amorphous PA. The fracture toughness G_{IC}, calculated from K_{IC}, show an even larger difference.

DISCUSSION

The experimental results of this study on thermoplastic polymer and composites, clearly show that the correlation between the polymer fracture toughness and the composite interlaminar fracture toughness is complex.

If one compares the amorphous and crystalline pure polymer with the amorphous and crystalline composite, obtained during fast or slow cooling after consolidation, a simple, linear relation could be found (figure 4).

The experimental results however showed that the composite fracture toughness can change drastically by changing the morphology of the crystalline phase.

If the spherulites are very fine (specimen XA_{95}), the G_{IC} - value approaches the value for the amorphous material (1400 J/m²). By increasing the spherulite dimensions over 1 μm (specimen XA_{150}) to 8 μm (specimen X_c), the fracture toughness gradually decreases over 900 to 300 J/m². It is hence clear that the crystalline matrix morphology has a definite influence on the fracture toughness of this thermoplastic composite. Experiments are currently carried out to create the same crystalline morphology in pure PA - polymer, in order to compare their fracture toughness values with the ones found for the thermoplastic composites in the actual study.

Finally, one has to emphasize that the morphology, and hence the intrinsic toughness of the matrix, can drastically change the fracture mode. We however believe that the transition from a stable to an unstable crack growth mode, when the toughness increases, can be explained on the basis of energy balance calculations for the DCB-specimen used. This will be presented in a future paper.

CONCLUSION

The present study confirms, for mode I fracture toughness tests, the low transfer of toughness from the polymer to the composite for unidirectional glass-fibre-polyamide composites. Moreover it shows the opportunity to modulate the transfer of toughness by applying thermal treatments to the composite. The morphology dependance of the pure polymer and of composite toughness is clearly demonstrated.

This study shows that the toughness transfert is very complex; more remains to be done theoretically as well as experimentally.

ACKNOWLEDGEMENT

We would like to thank dr. DECROLY and dr. DELIMOY (SOLVAY S.A.), and mr. De Jager and mr Merkelbach (FLEXLINE B.V.) for their collaboration, and for providing the test materials.

BIBLIOGRAPHY

(1) HUNSTON, D.L., MOULTON, R.J., JOHNSTON, N.J., BASCOM, W.D., "Matrix Resin Effects in Composite Delamination: Mode I Fracture Aspects", Toughened Composites, ASTM STP 937, Norman J. Johnston, Ed. , American Society for Testing and Materials, Philadelphia, 1987, pp. 74-94.

(2) GOFFAUX, B., Internal report.

(3) DAVIES, P, EGF Task Group on Polymers and Composites, A protocol for interlaminar Fracture Testing, Switzerland, 1989.

(4) HALE, G.E., A testing protocol for conducting J-R curve tests on plastics, EGF task Group on Polymers and Composites, February 1990.

$$G_{IC} \; (J/m^2)$$

	initiation	propagation	instability onset	instability stop
A	--	--	1350	700
X_c	220	300	--	--
XA_{95}	--	--	1400	605
XA_{150}	--	--	900	600

Table 1   Fracture toughness values for unidirectional G/PA laminates A = amorphous specimen, X_c = crystallized in the press, XA_{95} = crystallized by annealing at 95° from amorphous specimens XA_{150} = crystallized by annealing at 150° from amorphous specimens

K_{IC} (MPa √m)	E (MPa)	G_{IC} (J/m²)
A 6.8 ± 0.2	4.04	8600 ± 400
X_c 4.2 ± 0.8	5.06	2700 ± 1000

Table 2 Fracture toughness values for amorphous (A) and crystalline (X_c) polyamide.

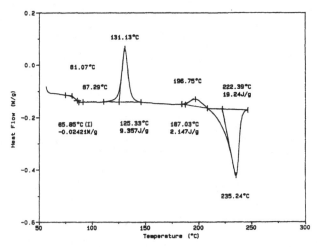

Fig. 1. DSC-scan on heating (10°C/min) for G/PA laminate rapidly cooled (2000°C/min.) under pressure from the lamination temperature 260°C.

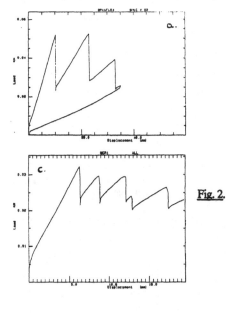

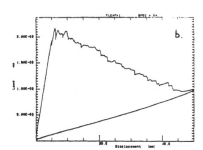

Fig. 2. Load displacement curves for continuous crack propagation. Crosshead rate = 2 mm/min.
a. Amorphous G/PA = A
b. Crystallized in the press = X_c
c. Crystallized by annealing amorphous specimen at 150°C = XA_{150}

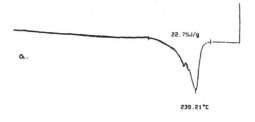

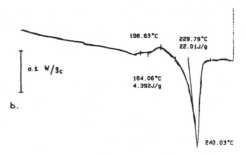

Fig. 3. DSC-scans of G/PA specimens crystallized from amorphous specimens.
a. Tc = 95 °C   b. Tc = 150°C

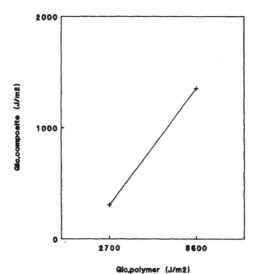

Fig. 4. Correlation between the matrix fracture toughness and the composite interlaminar fracture toughness; $G_{IC,comp}(J/m^2) = 0.18\ G_{IC,poly}(J/m^2) - 180\ (J/m^2)$

MICROSTRUCTURAL INVESTIGATION OF THE MATRIX CRYSTALLISATION IN PPS/CARBON FIBRE COMPOSITES

A. BOUDET, L. CARAMARO*, B. CHABERT*, J. CHAUCHARD*

CEMES-LOE - CNRS
BP 4347 - 31055 Toulouse - France
*Laboratoire d'Etudes des Matériaux Plastiques et Biomatériaux
UA-CNRS 507 - Université Lyon I
43 Bvd du 11 novembre 1918 - 69622 Villeurbanne - France

ABSTRACT

The mechanical properties of a composite with a thermoplastic matrix reinforced by carbon fibres are directly related to the morphology of the bulk polymer matrix (spherulitic semicrystalline structure) and that in the vicinity of the fibre surface (existence of a transcrystalline phase).

Investigations were carried out to understand the influence of the elaboration conditions (melting-crystallization cycle) on the microstructure and superstructure of the PPS matrix. In particular, the conditions to obtain a transcrystalline phase in the vicinity of the fibres were defined. The techniques of investigation were DSC, polarized light microscopy, X ray and electron diffraction transmission and scanning electron microscopy.

INTRODUCTION

Polyphenylene sulfide (PPS) - the chemical formula of which is $(-C_6H_4-S-)_n$ - is a semi crystalline thermoplastic polymer with a melting temperature of about 280° C. It possesses interesting mechanical properties and especially thermal and chemical stability, resistance to fire. However at high temperature, PPS can undergo damage processes like chaine scissions, recombination, branching and crosslinking /1/. It is then important to know and control the melting conditions in order to manage :
- the chemical stability of the material,
- the number of nucleation sites still present in the melt state before the cooling stage.

The final crystalline superstructure of the matrix depends on the density of nucleation sites and on their growing during the cooling stage. Furthermore the reinforcement fibres can have an influence as nucleating agents on the crystallization kinetics as well as on the crystalline morphology at the fibre/matrix interface /3,4,5,6,7,8,9/.

The study of thermodynamical and cinetical aspects of the crystallization of PPS in contact with carbon fibre was performed taking into account either the molecular weight, or the mineral impurity rate (NaCl) or the surface treatment of the fibre, or especially the thermal history of the sample. The study was conducted in the case of anisothermal cooling in order to be as close as possible to the industrial processing conditions.

I - MATERIALS.

The PPS used was provided by Phillips Petroleum Society. It contains a low percentage of trifunctionnal branching introduced by the presence of trichlorobenzene during the polymerization reaction. The molecular weight is $M_p = 65\ 000$ g.mol^{-1}.

The reinforcing fibres were provided by Soficar Society. They consist in oxidized T300 fibres. A bundle of fibres contains 6000 monofilaments of diameter 7μm.

II - APPARATUS.

The study of melting and crystallization conditions was conducted in differential scanning calorimetry with a Perkin Elmer DSC4 device equipped with a data acquisition module TADS.

The observations of the crystalline superstructure in the bulk polymer and in the vicinity of the fibres was performed with an optical microscope Zeiss in polarized light equipped with a Mettler heating stage and a Leica photographic set.

The electronic microscopy was performed on a JEOL 200CX microscope at an accelerating voltage of 200kV and scanning electronic microscopy on a CAMBRIDGE INSTRUMENT S250. X ray diffraction was operated using θ/ω goniometer with graphite monochromator and X ray localisation detector.

III - EXPERIMENTAL RESULTS :

3.1. Melting-crystallization :

The polymer was placed in the DSC cell and rapidly heated to the melt state at a temperature T_i during a time t_i then cooled at a rate of $10K.min^{-1}$.The exothermic peak was recorded and the temperature T_C of the peak maximum was noted.

In order to avoid a shift in the results due to some degradation of the polymer, each test was performed on a new sample of the same polymer. For a fixed value of t_i, the evolution of $T_C = f(T_i)$ can be depicted. On the curve shown in fig. 1, where t_i=6mn, three areas can be observed :

In the first one, T_C decreases with T_i as the preexisting crystalline nuclei become more and more destroyed. In the second one, T_C remains constant as the preexisting crystalline nuclei have totally disappeared. In the third one, the T_C increase is due to the lowering of M_n in relation with degradation. So two temperatures were chosen for structural investigations : one in the first area (T_i= 300°C), one in the second area (T_i = 350°C).

To understand the influence of the carbon fibre, preliminary observations on PPS alone have been carried out in the same thermal conditions of crystallization. In light or electron microscopy, numerous spherulites can be seen, with a size comprised between 5 and 60 µm, the larger ones being obtained at the highest T_i (350°C).

X ray diffraction showed that the sample are pretty well crystallized with an orthorhombic unit cell. Electron diffraction patterns confirmed that growth directions visible in the electron images lie along the b axis of the unit cell.

3.2 Transcrystallization

No sample with T_i = 300°C exhibits any transcrystallization phase. A spherulitic superstructure was observed up to the vicinity of the carbon fibre which seems to play no particular role in the crystallization of the PPS.

Transcrystallization areas have been observed along the carbon fibre in samples with T_i = 350°C (fig. 2). In polarized light microscopy, the transcrystalline lamellae have an extension similar to the radius of matrix spherulites. When the carbon fibres are very close, the matrix in between is made of transcrystalline phase only. Note that with a PPS of low molecular weight and rich in sodium chloride, the existence of heterogenous nuclei strongly diminishes the nucleating role of the fibre and the importance of transcrystallization. These results were confirmed by the observation in TEM (fig. 3).

The local structure could be obtained by electron diffraction patterns on small selected areas : there again, b is parallel to the growth direction, perpendicular to the fibre surface. Observations are going on to precise this point just on the surface of the fibre, with the help of dark field images. The properties of the transcrystallinity are different of the other morphologies, then it should be of importance in the global mechanical properties of the composite and this point is under investigation.

REFERENCES

1- T. HAWKINS, Macromolecules, 9, n°2, (1976), 189-1.94

2 - CAMPBELL D., QAYYUM M.M., J. Pol. Sci, Polym. Phys. Ed. 18, (1980), 83-93

3 - TUNG C.M., DYNES P.J., J. Appl. Polym. Sci., 33, (1987), 505-520

4 - BURTON R.H., FOLKES, PLast. Rub. Proc. App., 3, (1983), 129

5 - BURTON R.H., DAY T.M., FOLKES M.J., Polym. Communications, 25, (1984), 361

6 - ZENG C., HO G., Die Angew Makrom Chem. 127, (1984), 103-114

7 - CHABERT B., CHAUCHARD J., CINQUIN J., Mak. Chem., Macr. Symp., 9, (1987), 99-111

8 - CHABERT B., BOUYAHYA O.I., GUILLET J., NEMOZ G., Composites, 5, (1984), 49-55

9 - CARAMARO L., CHABERT B., CHAUCHARD J., C.R. Acad. Sci. PARIS, 306, (1988), 887-890

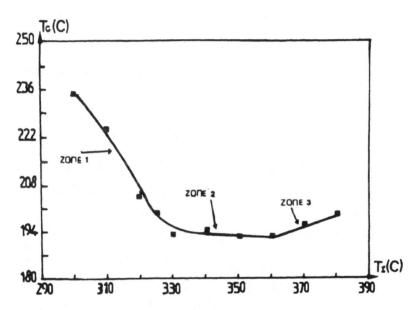

Fig.1 - Crystallization temperature determined in DSC, as a function of the temperature of annealing in the melt state.

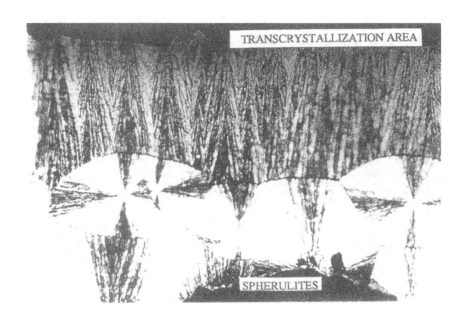

Fig.2 - Transcrystallization area close to the carbon fibre surface. Polarized light microscopy.

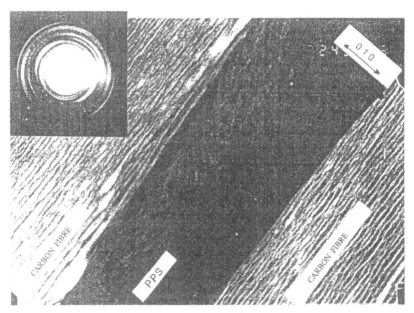

Fig.3 - Transcrystallization area between two carbon fibres and electron diffraction of PPS. Transmission electron microscopy.

EFFECTS OF ACCELERATED AGEING ON TRANSITION TEMPERATURES AND MECHANICAL PROPERTIES OF A GLASS FIBRE REINFORCED THERMOPLASTIC

C. ALLIE, D. VALENTIN

Ecole des Mines - Centre des Matériaux
BP 87 - 91003 Evry Cedex - France

ABSTRACT

The behaviour of a thermoplastic polymer, reinforced or not when subjected to accelerated ageing has been determined. This material is a polyamide 66 either bulk or reinforced with unidirectional continuous E glass fibre. Various ageing conditions were imposed on the samples. Then physical and mechanical characterization was carried out by differential scanning calorimetry and viscoelastic analyses. The principal effect of moisture is the decrease of glass-transition temperature towards 0°C, the drop of modulus from this temperature is reached during static or dynamic tests. More over, the appearance of other relaxations, for dynamic analysis above the T_g, is emphasized for the aged composite.

INTRODUCTION

At present, research for large, low cost industrial applications of composites shows a trend towards the use of glass fibre reinforced thermoplastics, in order to avoid the manufacturing and salvage problems associated with thermosets. The studied material is unidirectional continuous E glass fibre reinforced polyamide 66. The volume percentage of fibre is 55%. This polymer is well-known for its water absorbing. So PA66 sheets, reinforced or not were subjected to accelerated ageing. The results are briefly reported here.The characterization of the various materials was carried out by differential scanning calorimetry, dynamic mechanical analysis and some static 3 point bend tests.

I - AGEING METHODS AND OBSERVATIONS

Nylon 66 absorbs water. The three possible ways of fixation of H_2O molecules are the following /1/ : H_2O molecules can fix between two neighbouring carbonyl groups, or between - CO and HN - groups, or by condensation on a molecule already fixed. It is usually assumed that water is fixed in amorphous regions of the polymer. In order to determine the speed and the rate of moisture diffusion in the composite and in the polymer, ten hygrothermal conditions were imposed to both materials in environmental chambers. Relative humidities were varied from 0 to 100% while temperatures ranged from 40°C to 90°C. Extended details were reported previously /2/. Results of weighings produced absorption curves showing a shape near Fickian pattern ; the first part is linear and then the curve tends to a limit value, equal to the saturation percentage. However, for a given humidity, this percentage depends upon the ageing temperature ; for the composite at 90°C and 100% RH it is 0,8 % while at 40°C and 100% RH it is 1,2 %. The same behaviour has been observed concerning the polymer by itself.

These phenomena can be explained by the following facts : the glass transition temperatures are 70°C for the polymer and 75°C for the composite. At the beginning of ageing at 70°C or 90°C, the materials are in a state either of transition or rubbery behaviour while at 40°C they are in a glassy state. It has been proved by other authors /3/ and by further analyses, that the T_g (glass transition temperature) decreases during ageing. This infers that at the end of the test, i.e. when the sample reached saturation, all samples are in rubbery state. It is difficult to determine a diffusion law because of this change of state, which probably provokes a change of mechanisms. The difficulty is increased as no evolution can be observed on absorption curves. The effects of fibre presence on absorbed water quantity have been calculated, and we can assume that the amorphous part of nylon 66 absorbs more water than that of reinforced PA66. This is probably due to a problem of bulkiness, where fibres, by changing the polymer structure, become an obstacle to diffusion.

II - CHARACTERIZATION

The calorimetric and viscoelastic studies allowed us to determine the following properties : transition temperatures, modulus and damping evolutions.

2.1. Calorimetry

The D.S.C. melting thermograms of the bulk polymer and of the composite show that T_m remains constant whatever the absorbed moisture content of the material (T_m = 265°C). The volume fraction crystallinity increases with ageing, but melting curves present a double endothermic peak varying with rate of increasing temperature and with the ageing temperature and humidity imposed (figure n° 1). This phenomenon is attributed /4/ to the Brill transition ; the usual crystalline form of nylon 66 at ambient temperature is triclinic, and this Brill transition corresponds to a pseudo-hexagonal form which appears at higher temperatures. But it was proved, by observing dry and wetted cuts of polymer with an optical transmission microscope, that crystals are of different sizes and that they grow with ageing ; it seems that crystallization induces the pseudo-hexagonal crystal phase.

The effects of ageing on the glass transition temperature consist in a shift towards lower temperatures. The materials aged at 40°C and 100% RH present a larger transition region and this may indicate the presence of a second transition temperature.

2.2. Viscoanalysis

2.2.1. Method and apparatus

Experiments were carried out on PA66, reinforced or not, using the PL DMTA and the METRAVIB viscoanalyser. The study is based on bending tests, but in order to compare results, compression-compression tests were also accomplished. Multifrequence tests were imposed with the PL apparatus, with temperature scanning at a 1°C/mn rate from -100 to 150°C, and at 5 frequencies : 30 , 10 , 3 , 1 and 0,33 Hz. The results of these tests allowed us to study the evolution of the loss modulus through tg δ and of the elastic modulus (E') - not corresponding exactly to bending modulus because of the testing mode ; single cantilever. Using the METRAVIB instrument, 12 frequencies were used during successive temperature steps around T_g . Two transition temperatures appear on the bulk and reinforced polyamide graphs. β and α transitions evolve, with ageing and frequency, in range and localization. The figure n° 2 represents the spectra of tg δ versus temperature for a dry composite and a dry polymer sheet.

The β relaxation

β transition is at approximately - 50°C. A third transition can be supposed to be below - 100°C. The damping rate is considerably lower for the composite as it consists of only 45% resin. The transition temperatures, corresponding to peak maxima increase with frequency.

When the non-reinforced polyamide is aged at 100% of relative humidity, the β peak is shifted to a slightly lower temperature with an attenuation of the signal. The shifting phenomenon is lower for composite.

Various authors have tried to explain the β relaxation, but the phenomenon is not yet clearly understood /5/. This secondary relaxation involves local motion mechanisms. An interpretation is that β is ascribed to the motion of neighbouring chains with non hydrogen-bonded amide groups, while another is that β is due to the motion of carbonyl groups with water molecules which are attached by hydrogen-bonds. These mechanisms can not be verified, but it seems obvious that β involves the motion of a water-polymer complex.

The α relaxation

The α transition is attributed to large scale chain motion in the disordered (amorphous) regions, which involves the failure of hydrogen bonds and related to the glass transition temperature. And α peak temperatures are shifted towards lower values with water absorption.

The bulk polymer and the composite sheets show the same shifting T_g effect but for the composite, the magnitude seems to be less and another transition region appears at approximately 100°C (figure n° 3). This phenomenon can be related to one of the following possibilities:
1) Water diffusion was not homogeneous ; the transition at the higher temperature of aged sample is near α transition of the dried sample.
2) The temperature increasing involves a crystallization of polymer during the dynamic tests.
3) The ageing produced growth of another crystal with the nylon 66 pseudo-hexagonal form.

To clarify this situation, some compression-compression tests were carried out on the METRAVIB instrument. These showed the presence of a similar peak even for the dry, bulk polymer. It seems therefore that water is not the only reason for this relaxation. Further study of the microstructure is necessary to determine it, but several authors /5/ have mentionned the possibility of relaxations above the glass-transition temperature being due to recrystallization.

Another feature of Nylon 66 has been determined by these tests ; the real (or elastic) modulus. This modulus drops considerably with increasing temperature, from 0°C upwards for the aged materials. The value of E at low temperature (-100°C) is constant.

CONCLUSION

The great effects of water absorption have been established and show that even a reinforced polyamide 66 is damaged. This confirms the static 3 point bending test results on moduli and failure stress ; moisture induces a drop of properties from a room temperature test.

Moreover the presence of fibres combined with ageing, seems to emphazise crystallization phenomenon which appears near α relaxation modifying tg δ peak form, or above α through another relaxation.

ACKNOWLEDGEMENTS

The authors would like to thank IRCHA for the sample manufacture and the use of the PL DMTA.

REFERENCES

1 - Soulier J.P., Thèse de l'Université Claude Bernard, Lyon, (1975).

2 - Allié C., Valentin D., JNC6, (1988), 321-332.

3 - Cinquin J., Thèse de l'Université Claude Bernard, Lyon, Bernard, Lyon, (1988).

4 - Bunn C.W., Garner E.V., Proc. Roy. Soc. London, (1947), 189, 39-68.

5 - Mc Crun N.G., Read B.E. and Williams G., Anelastic and Dielectric Effects in Polymer Solids, Wiley, New York, (1967), CHAPTER 12.

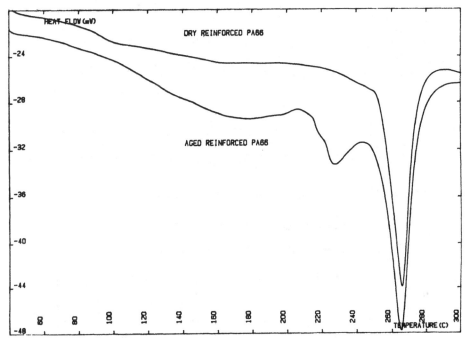

Fig. 1 : D.S.C. thermograms for a dry and a 100 RH, 90°C aged reinforced Nylon 66.

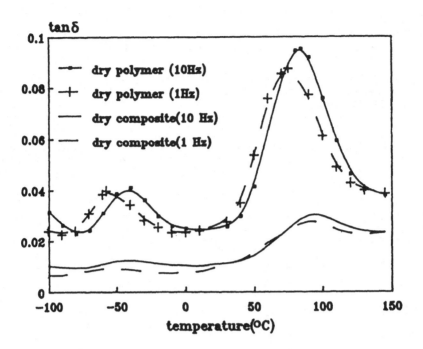

Fig. 2 : Evolution of tg δ with temperature obtained with PL DMTA for dry bulk and reinforced Nylon 66.

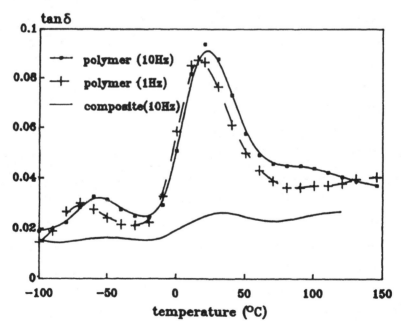

Fig. 3 : Evolution of tg δ with temperature for 100 RH aged bulk and reinforced Nylon 66.

THE EFFECT OF STACKING SEQUENCE
UPON THE MECHANICAL PROPERTIES
OF THERMOPLASTIC COMPOSITE LAMINATES

A.J. BALL, R. YOUNG, A. CERVENKA*

UMIST - Materials Science Centre - Grosvenor Street
M60 1QD Manchester - Great Britain
*Shell Research - Arnhem - The Netherlands

ABSTRACT

The effect of the laminate stacking sequence upon the tensile
modulus, the ultimate tensile strength and the failure strain was
determined for symmetrical 8-ply laminates of APC-2 with ±45°
orientations. Classical lamination theory predicts that the stacking
sequence will have no effect on these properties. However, results
showed that both the strength and the failure strain increased as the
ply number was increased. Fractography suggested that the stacking
sequence effect was due to an increase in the amount of interlaminar
and intralaminar shear with ply number.

INTRODUCTION

A major disadvantage of unidirectional composite materials is that
they are highly anisotropic. One way around this is to form
laminates by stacking two or more laminae on top of each other with
various orientations of the fibre direction and the mechanical
properties are dependent upon the way the individual layers are
stacked i.e. the stacking sequence.

Predictions for cross-ply and angle-ply laminate mechanical
properties are based upon classical lamination theory /1/ which is
well-established and has been in existence for over 50 years. In
order to predict the stress-strain behaviour of a laminate it is
assumed that the individual laminae are perfectly bonded together,
with the bond being infinitesimally thin and the laminae unable to
slip over one another. Firstly the lamina properties are calculated

using micromechanics and macromechanics /2/ to give the stiffness matrix $|Q|$ and the transformed stiffness matrix $|\bar{Q}|$. Application of a stress to the laminate results in the production of forces and moments which can be obtained by the integration of the stresses in each lamina through the laminate thickness /1/. The resultant can be represented by three more stiffness matrices; $|A|$ the extensional, $|B|$ the coupling and $|D|$ the bending stiffnesses. The laminate stiffness is the appropriate element, a_{11}, of the inverse extensional stiffness matrix, $|a| = |A|^{-1}$, /3/.

For angle-ply laminates the tensile strength is calculated by carrying out two additional steps /4/. Firstly the force distribution matrix $|f|$ is derived which relates the stress level for a given lamina to the resulting force, N, on a symmetrical laminate under a tensile load. Then the force, Ni, acting on an individual lamina, i, is calculated using the Tsai-Hill failure criterion /1/.

The object of this investigation was to examine the effect of the laminate stacking sequence on the tensile properties of symmetrical angle-ply laminates of APC-2 and its relation to the theory. For 8-ply laminates, assuming that there are equal numbers of plies in each direction (4 +45° and 4 -45°), there are 3 possible symmetrical stacking sequences as follows:

+ - + -	- + - +	ply no.=6
+ - - +	+ - - +	ply no.=4
+ + - -	- - + +	ply no.=2
		(| = line of symmetry)

For each type there are a number of interfaces between plies of different orientation; this is the ply number. APC-2 with unidirectional AS4 carbon fibres has a longitudinal modulus of 134 GPa, a transverse modulus of 9.2 GPa and a shear modulus of 4.9 GPa /5/. From this theory predicts that each of the above sequences will have a modulus of 17.4 GPa and a strength of 162 MPa, i.e. the laminate properties are independent of the stacking sequence. Previous work /2,4/ has shown that experiment/theory correlation is good for the modulus. However the fit for the ultimate properties is poor. It has also been shown that the thickness of individual plies can affect the laminate strength /6/.

EXPERIMENTAL

I - COMPRESSION MOULDING OF LAMINATES

The APC-2(AS4) carbon fibre reinforced PEEK pre-preg was supplied as a 0.17mm thick, 210mm wide strip wound on to a reel. From this strip eight 210 x 210mm squares were cut and stacked into a picture frame mould for compression moulding into a 1mm thick laminate sheet. The squares were spot welded together by the use of a high powered soldering iron to prevent undesirable movement of the material during moulding. The mould was compressed between two 270 x 270 x 10mm aluminium platens coated with a release agent (Frekote 700).

Compression moulding was carried out using a heated press with a water cooling facility. The mould was placed in the press at 380°C for 10 minutes under a force of 70 kN to 'contact' the plies. The load was then increased to 200 kN for a further 15 minutes at 380°C. Next the moulding was rapidly water cooled to 190°C and held at this temperature for 15 minutes at a load of 300 kN. Finally the laminate was released from the mould and allowed to cool to room temperature.

II - MECHANICAL TESTING

Tensile tests were carried out using an Instron mechanical testing machine (model 4505) following the ASTM standard (D3039-74) /7/ modified with respect to the specimen dimensions. The specimens used for mechanical testing had a length of 180mm and a width of 20mm and were cut from the laminate sheets by the use of a 'wet' diamond cut off wheel. Tapered aluminium tabs were fixed to the composite using a high shear strength adhesive (Araldite 2005). Strain was measured by an extensometer (Instron model 2620-601) with a gauge length of 25mm which was attached to the middle of the specimens by two rubber 'O-rings'. A cross-head jaw separation speed of 2mm/min was employed. For each specimen the following information was obtained from the stress-strain data; Youngs modulus (E), the ultimate tensile strength (6) and the failure strain (6).

RESULTS

Typical stress-strain curves for the three different laminate types are shown in figures 1-3. The initial part of the curve is the same for each one, however as the ply number is increased the latter section, after the 'knee', becomes much more extended, and thus the area under the curve is increased.

The modulus was found to remain constant with the ply number which was as expected (figure 4). However the measured values for the modulus were consistently higher than the theoretical value of 17.4 GPa (shown by the dashed line). Studies involving specimens with various lengths and widths indicated that one of the reasons for this discrepancy was that the specimens used were too short and subject to a constraint at the grips.

It was also found that as the ply number is increased the composite strength is increased (figure 5), a similar trend was found for the failure strain. This result was in contradiction with the theory. The measured strengths were substantially higher than the predicted value of 162 MPa (again shown by a dashed line) in fact for a ply number of 6 a strength of 300 MPa was achieved. This figure is comparable to that found by other workers /4/.

THE STACKING SEQUENCE EFFECT

This research has shown that for ±45° angle-ply laminates the stacking sequence has a marked effect on the ultimate tensile properties. One possible reason for this contradiction with the theory is the contribution of shear to the failure process. When a specimen of a ±45° laminate is placed under a tensile load the fibres tend to rotate in such a way as to align themselves with the loading direction. This movement of fibres gives rise to shear between fibres in the same ply (intralaminar shear) and shear between plies of different orientations (interlaminar shear) /8/.

Scanning electron microscopy (SEM) of fracture surfaces of the tensile specimens clearly gave evidence for both of these shearing processes. Figure 6 shows an SEM micrograph of interlaminar shear. As the ply number increased the number of interfaces undergoing interlaminar shear are increased. SEM also showed that as the ply number was increased the amount of intralaminar shear also increased.

Obviously the more shearing of the matrix that takes place the greater the energy absorbed by the system, shown by the increase in the area under the stress-strain curve with ply number (figures 1-3). Hence, as the ply number is increased the material becomes stronger.

CONCLUSIONS

For APC-2 ±45° angle-ply laminates it has been found that:-

1) The modulus is independent of the stacking sequence.

2) As the ply number is increased both the tensile strength and the failure strain are increased.

3) The stacking sequence effect is at least partly caused by the contribution of shearing mechanisms to the failure process.

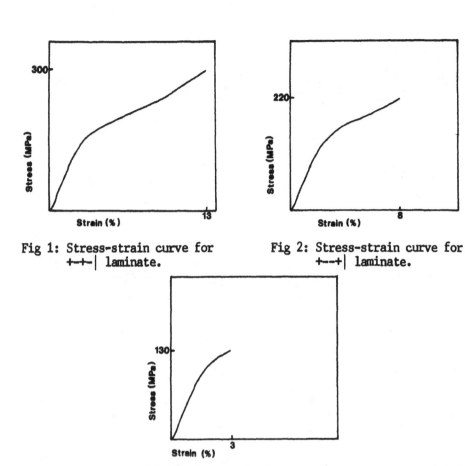

Fig 1: Stress-strain curve for
+-+-| laminate.

Fig 2: Stress-strain curve for
+--+| laminate.

Fig 3: Stress-strain curve for
++--| laminate.

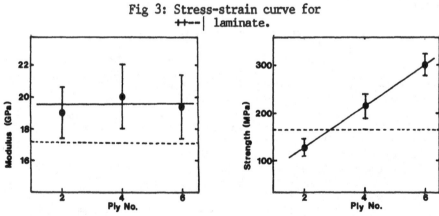

Fig 4: Modulus vs Ply No.

Fig 5: Strength vs Ply No.

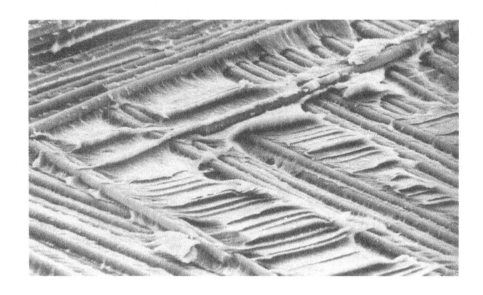

Figure 6: SEM micrograph of interlaminar shear (X460).

ACKNOWLEDGEMENTS

The authors wish to thank Shell Research Limited for making this work possible through their sponsorship and provision of materials. Dr. A. Cervenka is thanked for his valuable contributions and discussion.

REFERENCES

1- Jones, R.M., "Mechanics of Composite Materials", Scripta Book Co. Ltd., Washington DC., 1975
2- Cervenka, A., Polymer Composites, 9, No.4, p263,1988
3- Cervenka, A., "Composite Modelling", Proceeds. "Rolduc Polymer Meeting-2", Limburg, The Netherlands, April 1987, (being published).
4- Cervenka, A. and Sheard, P., "Mechanical Behaviour of Angle Ply PEEK/Carbon Fibre Laminates: Theory/Experiment Correlation", Shell Research Ltd., Chester, Sept 1988.
5- "Victrex" PEEK/Carbon fibre, provisional property data of APC-2, ICI plc, Welwyn Garden City, 1983.
6- Curtis, P.T., J. Mat. Sci., 19, p167,1985.
7- ASTM Standard D 3039-74.
8- Hull, D., "An Introduction To Composite Materials", Cambridge University Press, 1981.

LONG FIBRE REINFORCED THERMOPLASTICS FOR INJECTION MOULDING : THE RELATIONSHIP BETWEEN IMPREGNATION AND PROPERTIES

R.S. BAILEY, W. LEHR, D.R. MOORE, I.M. ROBINSON, P.M. RUTTER

ICI Advanced Materials - Wilton Materials Research Centre
TS6 8JE Wilton Middlesbrough - Great Britain

ABSTRACT

Long fibre reinforced thermoplastics have become widely available through the use of pultrusion technology. The degree of impregnation between the fibres and the thermoplastic matrix in such compounds can be varied, for example, by changing the processing route. In an attempt to explore the relationship between impregnation and properties, three types of long fibre compound have been studied; these are classified as being Ideal, Intermediate and Poorly impregnated. The microstructure for each of these compounds has been measured by a variety of methods including ultrasonic C-scanning, microscopy and the determination of the fibre length and orientation distribution functions. Mechanical properties have been characterised in terms of stiffness, toughness and strength.

INTRODUCTION

Long fibre reinforced thermoplastics are becoming increasingly accepted in load bearing engineering applications and serve to extend the property range of traditional extrusion compounded materials /1/. These materials are pultrusion compounded and it is possible to impregnate the continuous fibre rovings by varying extents. It has long been assumed that a minimum degree of impregnation is required to achieve the property potential of enhanced fibre lengths attainable in moulded components. Recently, there has been evidence that an encapsulation process would give adequate properties, and that a high impregnation level was unnecessary.

This work aims to highlight that a high level of impregnation is important in service, and while materials produced by encapsulation offer attractive short term properties, there is some doubt about their application to load bearing situations.

I - PROCESSING AND MICROSTRUCTURE

1.1 Materials and Techniques

Three 50% w/w glass fibre reinforced nylon 6.6 compounds were prepared by pultrusion compounding and were chopped to 10mm granule lengths for subsequent injection moulding. The feedstock materials, incorporating differing levels of impregnation, were characterised by two techniques:-

i) Ultrasonic C-scanning of consolidated pultruded sections.
ii) Fibre dispersion by image analysis.

The materials exhibited different qualities of impregnation and fibre dispersion and can be considered as suitable simulation materials for the purposes of this study. The materials are identified as follows:-

(1) Ideal - well impregnated and dispersed fibres.
(2) Intermediate
(3) Poor - low impregnation, simulating encapsulated materials.

These materials were subsequently injection moulded under identical conditions into 160 mm² - coathanger gated plaques. The moulding conditions were selected in order that the test pieces would retain desirable fibre length distributions by minimising the screw back pressure during the preplasticising stage of the cycle. The plaques were also Ultrasonic C-scanned.

It is clear from these measurements that the moulding process induces further impregnation of the fibres. Materials (1) and (2) appear to be very similar in quality, once moulded. However Material (3) retains a dramatically different level of fibre impregnation. The test piece morphologies were subsequently characterised by studying material taken from the exact centre of the plaque. This location coincides with the region used in the mechanical test program.

1.2 Characterisation

1.2.1 Fibre length distributions

These were measured using the sieve/image analysis method as described by Bailey /2/. The results are given below:-

Material	Skin (Θ_{100})		Core (Θ_{700})	
Material	number average (mm)	weight average (mm)	number average (mm)	weight average (mm)
(1) Ideal	3.93	5.71	2.98	5.69
(2) Intermediate	3.67	6.66	3.71	6.13
(3) Poor	2.88	5.88	3.16	5.81

1.2.2 Fibre orientation structures

Skin-core ratios

The orientation structure can be considered in two parts, firstly the extent of the skin-core structure, and secondly the orientation with these régimes. For the purposes of simplification and later modelling this can be represented by sixteen layers.

impact speed of 5 m/sec. 10 tests were made on each grade of compound, and the resultant means and standard deviations for the energy used to initiate cracking and total energy absorbed during impact are given below.

Compound Type	Initiation Energy (J)	Total Energy (J)
Ideal	0.59 (0.07)	10.78 (1.00)
Intermediate	0.55 (0.08)	10.50 (0.76)
Poor	0.51 (0.05)	9.96 (0.71)

The data shows that there is a consistent improvement in impact performance with better impregnation. However this particular mode of deformation (bending through the plane of reinforcement) is fibre dominated so large differences in the toughness for the compounds are not expected.

2.3 Fatigue

Tension-tension fatigue tests were performed on parallel sided specimens machined from the centre of the coathanger plaque mouldings. The specimen geometry was given the label 90° (with the stress aligned perpendicular to the mould fill direction). Each specimen had a width and thickness of approximately 12.5 mm by 3 mm and a gauge length of 100 mm. The fatigue tests were carried out on a servohydraulic 1341 Instron, with the applied load having a sinusoidal wave form. A computer system recorded the stiffness reduction data during the test, with the modulus being defined by a tangent method /5/. The data for the normalised stiffness reduction for two 90° specimens, one an Ideal impregnated compound and one a Poorly impregnated compound are shown in Figure (1). As can be seen, the Ideal compound retains it's stiffness out to 10,000 cycles, whereas the Poorly impregnated compound shows a significant stiffness loss almost immediately. Such a stiffness loss can be attributed to the presence of microcracking in the skin-core structure of the compound due to debonding between the fibres and matrix. For this particular orientation, the major damage occurred in the skin layers.

2.4 Deformational Properties

The measurement of tensile modulus and volume strain were made by tensile deformation of parallel sided specimens machined from the centre of the coathanger plaque mouldings. Two types of specimen geometry were tested, labelled 0° (with the stress aligned with the mould fill direction) and 90° (with the stress aligned perpendicular to the mould fill direction). Each specimen had a width and thickness of approximately 12.5 mm by 3 mm and a gauge length of 150 mm. The deformation was performed using a screw driven 6025 Instron, with strain measurements being monitored in three mutually perpendicular directions together with the applied stress. The equipment and method for calculating the volume strain are described more fully in Reference /6/.

The measurement of volume strain during a tensile test provides an insight into the various deformational mechanisms that govern material behaviour, as well as monitoring the usual stress-strain, modulus-strain functions and measuring the lateral contraction ratios for the material. The basic shape of the volume strain vs axial strain function gives a qualitative method for distinguishing the dominant deformation mechanism. These are either volume increasing (voiding, crazing, cracking, debonding) or shear banding processes (yielding) where deformation occurs without major volume increase. Debonding mechanisms in composites are strongly related to the amount of adhesion between the fibre and the polymer matrix and are consequently dependent on the level of impregnation achieved in a moulding. The measured volume strain and tensile modulus at an applied axial strain of 1% for the 0° and 90° specimens of each compound are shown beneath.

Material	% Core (mm)
(1) ideal	24
(2) Intermediate	25
(3) Poor	23-26

Fibre orientations within the layered structures can be measured by image analysis. Each layer contains a three dimensional fibre orientation distribution, and a symmetry is assumed about a section through the centre. Therefore, the orientation in only eight layers requires definition: $\Theta_{1 to 6}$ are taken as "skin" layers and $\Theta_{7,8}$ as "core layers" in

accord with the relative "skin/core" dimensions. The fibres are generally confined to in-plane orientations as a consequence of the mould fill process. These results give approximate average values for each layer. The angles are relative to the axis of injection into the mould.

Layer	Average orientation angle w.r.t injection
Outer surface Θ_1	30°
Θ_2	20°
Θ_3	20°
Θ_4	10°
Θ_5	20°
Θ_6	30°
Θ_7	50°
Θ_8	65°
Centre of section	

These results indicate that the fibre length and skin-core structures are approximately the same. It may be assumed that the orientation structures are similar for each material. However, local fibre fractions (or fibre dispersion) are evidently different between the compounds.

II - MECHANICAL PROPERTY MEASUREMENTS

2.1 Mechanical Tests

Deformation, toughness and strength studies were performed using a number of mechanical property techniques. These principally involved modulus and volume strain vs axial strain measurements, impact energy analysis and fatigue measurements of strength and modulus. In recognizing the possible influence of moisture on these compounds, all specimens were stored dry prior to testing. All measurements were made at 23° C in a temperature and humidity controlled laboratory.

2.2 Impact Measurements

Instrumented Falling Weight Impact (IFWI) analysis of panels of long fibre reinforced compounds gave information on the energy required to initiate a crack and also the total energy absorbed during the impact process /3,4/. All tests used a ring support of 40 mm internal diameter with a hemispherical striker of diameter 10 mm and were performed at an

Compound Type	Volume Strain at 1.0% Axial Strain		Tensile Modulus (GPa) at 1.0% Axial Strain	
	0°	90°	0°	90°
Ideal	0.24 (0.03)	0.29 (0.10)	12.2 (0.4)	7.3 (0.3)
Intermediate	0.16 (0.02)	0.19 (0.08)	11.1 (0.2)	7.0 (0.3)
Poor	0.27 (0.01)	0.33 (0.07)	12.1 (0.4)	7.0 (0.3)

As expected the 0° moduli for all three compounds are greater than the 90° moduli, reflecting the relative stiffness contributions from each layer of the skin-core microstructure. Little difference can be seen for the values of the modulus at each orientation for all three compounds, indicating that the short term stiffness is dominated by the fibre orientation distribution. However the volume strain measurements indicate that improved impregnation gave smaller amounts of microdebonding, due to the improved adhesion between the fibre and matrix as expected. This can especially be seen in the results for the 90° specimens, where the mode of deformation encourages the largest amount of fibre/matrix debonding.

III - CONCLUSIONS

This study has shown clear differences in impregnation levels between three types of long fibre reinforced thermoplastics and described fully the resultant microstructures. Mechanical tests that are dominated by a fibre response (such as IFWI and short term stiffness) have shown little difference in properties with impregnation. However, tests designed to measure the interfacial response (such as stiffness in fatigue and volume strain measurements) have shown that improved properties are seen with compounds that have better levels of impregnation. More tests are currently being performed to explore further the relationship between properties and impregnation in this class of composite material.

ACKNOWLEDGEMENTS

The authors wish to thank NJ Burgoyne, RS Prediger and G von Bradsky for their contributions to this work.

REFERENCES

1 - C.R Gore, Composite Polymers, 1 (1988) p 280
2 - R.S Bailey and H Kraft, Int. Polymer Processing, 2 (1987) p 94
3 - P.A Gutteridge et al, Kunstoffe, 72 (1982) p 9
4 - D.C Leach and D.R Moore, Composites, 16 (1985) p 113
5 - S.A Hitchen and S.L Ogin, "FRC '90", Inst. Mech. E. (1990) p 215, London
6 - I.T Barrie, D.R Moore and S. Turner, Plastics and Rubber Proc. & Appl, 3 (1983) p 293

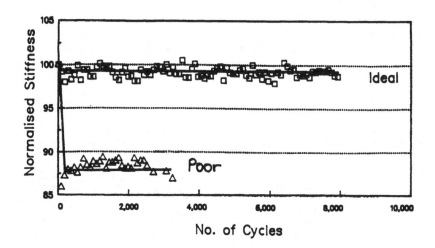

Figure 1:- Normalised stiffness data for 90° specimens of Ideal and Poorly impregnated compounds, measured at a mean applied stress of 50 MPa

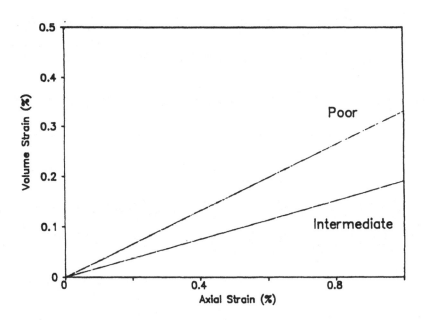

Figure 2:- Average volume strain versus Axial Strain for the Intermediate and Poorly impregnated compounds

FRACTURE PERFORMANCE OF RANDOM, CONTINUOUS GLASS STRAND REINFORCED POLYAMIDE-RIM COMPOSITE

J. KARGER-KOCSIS, K. FRIEDRICH

Technical University of Hamburg-Harburg
2103 Hamburg-Harburg - West-Germany

ABSTRACT
The fracture behavior of reaction injection molded (RIM) polyamide block copolymer reinforced with randomly arranged continuous glass fiber bundles (glass mats) was studied under monotonic static and dynamic loading conditions.Static fracture measurements carried out with compact tension specimens at different crosshead speeds (1 and 1000 mm/min) and temperatures (-40 to 80 $^{\circ}$C).Fracture toughness values were determined from maximum loads.Dynamic fracture toughness and fracture energy values were derived from high-speed impact bending tests carried out on different Charpy and Izod specimens, respectively.The fracture mechanical data were compared with each other and with those of conventional short fiber reinforced polyamides. Failure mode of the composites was characterized by fractography and possible failure events were summarized.Recommendations were given for further material improvements.

INTRODUCTION
Reinforced reaction injection molded polyamide block copolymers reinforced with continuous swirl glass fiber mat (PA-RRIM) are promising candidates to replace metals in the automobile industry.The RIM technology allows a cost-effective production of large-area panels of sophisticated geometry which can hardly be achieved by injection molding or thermoforming from thermoplastic matrix composites.The target application field requires a profound fracture mechanical characterization of this kind of composites.The aim of this contribution is to deliver static and dynamic fracture mechanical data from measurements carried out with monotonc increased load at different testing conditions.

FRACTURE MECHANICAL CHARACTERIZATION OF THE PA-RRIM COMPOSITES

Materials

The plates investigated (150x150x4 mm^3) were molded by RAPRA Technology Ltd (Shawbury,GB) using the components of the Nyrim^R 2000 system of DSM (Maastricht,NL).This PA-RIM material was a block copolymer containing 20 wt.% elastomeric polyether soft segments.

Fiber loading ($\approx$20-50 wt.%) were produced by stacking continuous E-glass strand mats (Unifilo U816 swirl mat from Vetrotex International,Aix-les-Bains,F).This GF mat of 450 g/m^2 surface weight was sized by silane and bonded by a polyester binder.The strands with a diameter of about 150 μm were composed from GF monofilaments of 15 μm diameter and 25 tex strength.

Methods

For the static fracture measurements razor blade notched compact tension (CT) specimens of 35x36 mm^2 dimension were used.Their exact form and size can be taken from references /1-2/.Static fracture was carried out at two crosshead speeds, v= 1 and =1000 mm/min, at different temparatures, T=-40, 20 and 80 ^{o}C.The stress intensity factor or fracture toughness, K_c, was determined according to the ASTM E 399 standard.

For the dynamic fracture characterization instrumented high-speed impact bending techniques were used.Charpy specimens of different size and notching were cut from the plaques according to Figure 1 and impacted at v=3.6 m/s striker speed (DIN 53 455 standard).The dynamic fracture characteriation was extended to Izod specimens (cf. Figure 1) too,fractured at v=3.46 m/s (ASTM D 256 standard).Measurements were followed in both cases at different temperatures (T=-40 and 20 ^{o}C) using an instrumented impact pendulum device.This device was connected to an AFS-MK3 fractoscope (Ceast,Torino,I) and additionally to a Hewlett-Packard computer for data acqusition and storage by means of a home-made program.

From the data registered and evaluated both the dynamic fracture toughness, K_d, and the dynamic strain energy release rate or fracture energy, G_d, were determined.K_d was got as the slope of a straight line through the origo giving the best fit for the $\sigma.Y$ vs. $1/\sqrt{a}$ poit pairs, where "σ" is the maximum stress,"Y" is the shape factor and "a" is the notch length.Since the shape factor of the Izod specimens is unknown a related value from the Charpy range has been used.G_d was determined as the slope of a straigth line runing through the origo by plotting the E vs. BWϕ values, where "E" is the energy absorbed, "B" and "W" are the specimen thickness and width, respectively (cf. Figure 1), and "ϕ" is the energy calibration factor.Detailed

informations about the evaluation methods used can be taken from the literature (e.g. /3-5/).By the selection of the specimens to be impacted and differing from each other in type,size and notching answers for the following questions were expected:
i- which is the effect of specimen size on facture mechanical data (cutting results from the initial continuous fiber reinforced composite a "chopped" fiber reinforced version)
ii- is there any anisotropy in this preplaced GF mat reinforced composites
iii-wether machining (sawing) additional flaws originates

Static Fracture
    Figure 2 indicates that increasing GF mat loading improves the fracture toughness considerably.The PA-RIM matrix possesses a relative low K_C, it behaves very ductilely due to the "intrinsic plastification" by the polyether segments in block form.Incorporation of about 50 wt.% GF mat yields a 4-6 fold increment in K_C depending on the crosshead speed and temperature.It should be noticed here, that higher loading than 50 wt% can not be achieved using this swirl mat.The scatter in the K_C values is the lower the higher the GF mat loading is.This should be connected therefore with the local arrangement of the mat at the crack tip.
    Increasing crosshead speed (≈strain rate) resulted in higher fracture toughness values for both the matrix and composites (cf. Figure 2).Such a matrix behavior is rather charac-teristic for impact-modified unreinforced thermoplastics into the group of which the PA-RIM studied also belongs.On the other hand, the observed trend for the composites is rather peculiar, since increasing strain rate is generally accompanied with a drop in K_C for chopped fiber filled composites.It is supposed to be an effect of multiple brittle matrix cracking initiated by stress concentration on and among the fibers in the frequency embrittled matrix /6/.
    Increasing temperature generally lowers the K_C of thermoplastic matrices and related composites /6-7/, this trend is followed in this case, too (cf. Figure 2b).At low crosshead speed there was practically no difference between the K_C values at T=20 and 80 °C.At high v on the other hand, a monotonic decrease in K_C with temperature could be observed (Figure 2b).It should be connected with heating up of the crack tip due to the suddenly introduced mechanical work at loading.

Dynamic Fracture
    Figure 3 shows the K_d and $G_{d,i}$ evaluation for the matrix at different specimens.$G_{d,i}$ denotes the initial fracture energy, representing the onset of fracture at the maximum stress.This term was used since complete separation

of the specimen not always occurred.The K_d values of the matrix were found in the range of 2.8 to 3.6 MPa$\sqrt{m}$ for the different specimens.This range agrees reasonably with the K_C value got by static fracture at high crosshead speed (cf. Figure 2).The $G_{d,i}$ values established on different specimens agreed very well with each other (from 9.2 to 11.2 kJ/m^2; cf.Figure 3b) with the only exception of the related value for the C-SN specimens ($\approx$5 kJ/m^2).This can be explained by the molding-induced skin-core morphology of the matrix, which showed flow line contours based on light microscopic investigations.

Both the K_d and $G_{d,i}$ values of the PA-RRIM composites increased according to the following sequence:

C-SN$\approx$ C-S < I < C-B.

This trend suggests that there is an effect of specimen size and thus of an effective fiber length.Since the width of the Izod (I) and big Charpy (C-B) specimens was the same (cf. Figure 1) one expected similar values.In spite of this a slight but reproducible difference was found according to the above ranking.It may be connected with the difference in the clamping of the specimens and thus with the related stress state upon bending.

The evaluation method of $G_{d,i}$ allowed an estimation of the flaw size induced by machining (Figure 4).In this case the "ineffective" notch size (causing fracture away from the razor blade notch) agreed very well with the value derived by the method of Akay and Barkley /9/.This machining induced flaw size proved to be in the range of 0.6 to 1.2 mm depending on the test conditions,specimen type and size. Investigations on C-SN specimens resulted in a much smaller value (0.2-0.4 mm) which can be treated as a molding induced flaw size due to the notching position of this specimen (cf. Figure 1).

The effect of fiber loading on K_d and $G_{d,i}$ is demonstrated in Figure 5.According to this figure both K_d and $G_{d,i}$ increase with V_f, the increment is , however, the smaller the higher V_f lays.

Based on the results in Figure 5 the effect of temperature is negligible in the temperature range studied.This statement relates especially for the PA-RRIM composites.

FAILURE BEHAVIOR

The failure events of injection molded traditional short fiber composites can be divided into the following groups: a) matrix-related events comprising cleavage, crazing,voiding and shear yielding and b) fiber-related ones including fiber debonding, pull-out and bridging phenomena /6,10/.The matrix fails (micro)ductilely depending upon the testing temperature in this case.On the fracture surface characteristic markings of secondary cracking could be resolved.It was proved by energy dispersive X-ray scattering

(EDAX) that their initiation sites are not dissolved catalyst rests.

The incorporation of swirl GF mat alters the relative proportion of the fiber related failure events.Fiber bridging due to a "mesh deformation" of the mat becomes very important, this process is namely responsible why complete separation of the specimens is very seldom.Among other pecularities of PA-RRIM the intraroving breakage and fibrillization of the strands should be mentioned.These events contribute to the energy dissipation in the damage zone, the exact size of which can hardly be determined.It is worth mentioning that all PA-RRIM composites exhibited a stable fracture propagation stage, their final failure never occurred by catastrophic manner.The energy values in the fracture initiation (ended up by E_{peak} used in $G_{d,i}$ evaluation) and fracture propagation range can be related to each other.Such a ratio may deliver informations about the probability of onset of catastrophic failure /11/.

RECOMMENDATIONS FOR MATERIAL IMPROVEMENT

The presence of catalyst rests is rather characteristic for all PA-RIM matrices.Since they act as stress concentrators and induce undesired secondary cracks the dissolution of the catalyst should be improved.It is strongly recommended to increase the "mesh deformability" of the GF mat by the "construction" of an adequate mat network. Use of coupling and sizing agents seems to be also beneficial as it is the case in all polyamide matrix systems /12/. From the point of view of molding the main problem is associated with the escape of the air during wetting of the preplaced GF mat by the molten RIM system.Adequate tool and processing modifications might reduce the surface flaws, caused by the air compressed, considerably.

CONCLUSIONS

Improved fracture mechanical data can be noticed for the PA-RRIM composites with GF mat reinforcement when one compares the corresponding values of chopped and mat reinforced composites /7/.The upgrading is more obvious at high-frequency than at low frequency loadings, so thus PA-RRIM with GF mat reinforcement is a material of outstanding impact resistance.

ACKNOWLEDGEMENT

The financial support of this study by the EURAM program of the European Community (MA 1E/0043/C) is gratefully acknowledged.

REFERENCES
1   Karger-Kocsis, J and Friedrich,K.:Plast.Rubb.Process. Appl., 8 (1987), 91-104
2   Idem:Compos.Sci.Technol., 32 (1988), 293-325

3   Williams,J.G.:Fracture Mechanics of Polymers, Ellis
    Horwood Ltd, Chichester, 1984
4   Savadori,A.:Polym.Test., 5 (1985), 209-241
5   Plati,E. and Wiliams,J.G.:Polym.Eng.Sci., 15 (1975),
    470-477
6   Friedrich,K. and Karger-Kocsis,J.:Fractography and
    failure mechanisms of unfilled and short fibre
    reinforced semicrystalline thermoplastics,Ch.11 in
    "Fractography and Failure Mechanisms of Polymers and
    Composites" Ed.:Roulin-Moloney,A.C.),Elsevier Appl.Sci.,
    Barking, 1989,pp.437-494
7   Karger-Kocsis,J.:Fracture of short-fibre reinforced
    thermoplastics,Ch.6 in "Applicaton of Fracture Mechanics
    to Composite Materials" (Ed.:Friedrich,K.),Elsevier,
    Amsterdam,1989,pp.189-247
8   Idem:Structure and fracture mechanics of injection
    molded composites in "Encyclopedia of Composites"
    (Ed.:Lee,S.M.),VCH Publ., N.Y. (to be appeared)
9   Akay,M. and Barkley,D.:Polym.Test., 7 (1987), 391-404
10 Friedrich,K:Fractographic analysis of polymer composites
    Ch.11 in "Application of Fracture Mechanics to Composite
    Materials"(Ed.:Friedrich,K.),Elsevier,Amsterdam,1989,pp.
    425-487
11 Otaigbe,J.U.and Harland,W.G.:J.Appl.Polym.Sci.,37 (1989)
    77-89
12 Friedrich,K. and Karger-Kocsis,J.:Fracture and fatigue
    of unfilled and reinforced polyamides and polyesters in
    "Solid State Behavior of Linear Polyesters and
    Polyamides" (Eds.:Schultz,J.M and Fakirov,S.),Prentice
    Hall,Englewood Cliff (to be appeared)

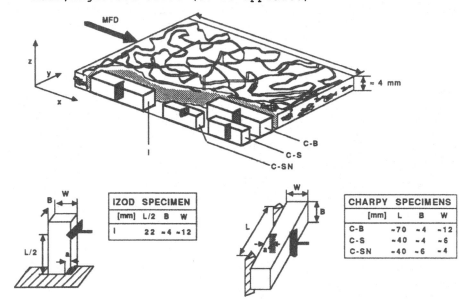

Figure 1 Machining and impact mode of the specimens used

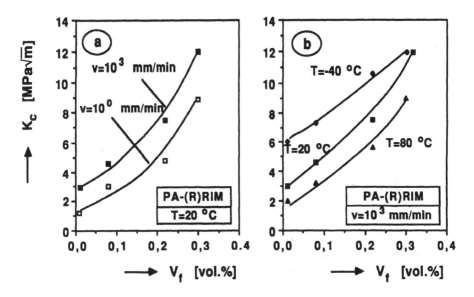

<u>Figure 2</u> K_C values in function of the GF mat volume fraction (V_f) at different temperatures and crosshead speeds

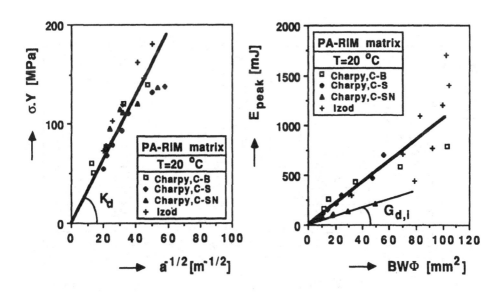

<u>Figure 3</u> Evaluation of K_d (a) and $G_{d,i}$ of the PA-RIM matrix (b) on different specimens

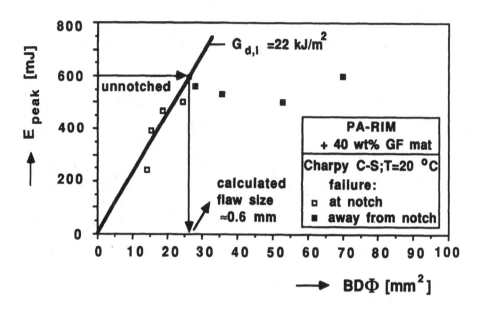

Figure 4 Determination of the sawing-induced flaw size

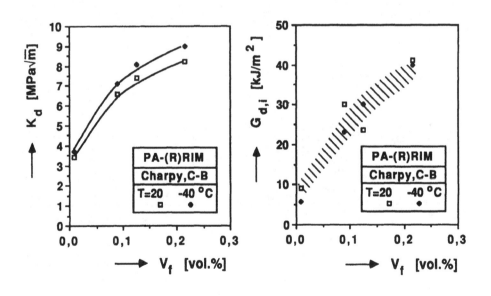

Figure 5 Effect of temperature and fiber mat loading on the K_d (a) and $G_{d,i}$ values (b) of the PA-(R)RIM materials

FAILURE MECHANISMS OF SHORT GLASS FIBER REINFORCED POLY(ETHYLENE TEREPHTHALATE)

K. TAKAHASHI, N.-S. CHOI

Research Institute for Applied Mechanics
Kyushu University - 816 Kasuga-Shi - Japan

ABSTRACT

Failure mechanisms of short glass fiber reinforced poly(ethylene terephthalate) were investigated with particular attention to the effects of fiber volume fraction (V_f). A fracture morphology study was carried out for the surface and the interior of uniaxial tensile specimens. On the surface, tensile cracks occurring mostly at the fiber ends seemed to be more influential in catastrophic fracture initiation with decreasing V_f. In the interior, however, shear failure along the fiber-matrix interface and/or between fibers was most influential for all the fiber volume fractions. A "specific layer" was formed in the matrix along the interface.

1. INTRODUCTION

The tensile deformation and fracture processes of short fiber rein-forced thermoplastics generally involve microcracking, microvoiding, shear band formation and/or crazing in the matrix and at the interface. Fiber breakage and fiber pull-out are also involved. Those micromecha-nistic failure processes seem to be associated with the occurrence of catastrophic fracture initiation. Several authors investigated the failure mechanisms using notched /1-3/ or unnotched /4-6/ specimens. Fractographic observation in these studies, however, has been made mostly on fracture surfaces or unbroken specimen surfaces.

In this paper, failure mechanisms of short glass fiber reinforced poly(ethylene terephthalate) were examined on the surface as well as in the interior of specimens under uniaxial tensile loading. The study also considered the effects of fiber volume fraction on the various failure mechanisms which lead to .catastrophic fracture initiation.

2. EXPERIMENTAL

2.1. Composites: Dumbbell-type short glass fiber reinforced poly(ethylene terephthalate)[SGFR-PET] tensile specimens were prepared for this study. The E-glass fibers 10 μm in diameter and 3mm in length were treated with an epoxy sizing agent and an amino-silane coupling agent. The weight fractions (w/o) of short glass fibers (SGF) embedded in the PET resin were 1%, 30% and 60%. The compounds were melted and were injected into end-gated dumbbell-shaped molds at 30°C. The molded specimen geometry and mold fill direction are shown in Fig.1. The gage portion was 70mm by 12.5mm with a thickness of 3mm. Short fibers in the specimens became oriented more to the mold fill direction with increasing fiber volume fraction (V_f). Average fiber lengths became shorter as V_f increased: 194μm for SGF 1 w/o, 123μm for 30 w/o and 99 μm for 60 w/o. The resin was in a mostly amorphous state. The interfacial coupling state between fibers and matrix was reasonably good.

2.2. Microscopic failure observation: All specimens were uniaxially loaded with a strain rate of 7×10^{-3}/min at a temperature of 23°C. Unreinforced and SGF 1 w/o-PET specimens were unloaded from the loaded state, and were examined under a polarized optical microscope. For specimens of SGF 30 and 60 w/o-PET, in situ scanning electron microscopic (SEM) observation was carried out similarly as done by Sato et al./4/ to examine failure mechanisms on the specimen surface. In order to study the mechanisms in the interior for the SGF 1, 30 and 60 w/o-PET specimens, we examined a thin section of a failed specimen under a polarized optical microscope in transmission. We prepared the section by polishing the gage portion to a thickness of about 250μm.

3. RESULTS AND DISCUSSION

3.1 Stress-strain curves: Figure 2 shows typical stress-strain curves for SGF 1 w/o, SGF 30 w/o and SGF 60 w/o-PET specimens. With increasing fiber volume fraction, fracture strength and elastic modulus of the specimens became larger while the fracture strain became smaller.

3.2 Failure processes on the surface: Many tensile microcracks occurred and grew on the surface of unreinforced PET specimen. This microcracking may have resulted from some inhomogeneities in the PET resin. Shear bands were formed at sites of the microcracks. Growth direction of the shear bands was about 52 from the tensile direction.

Figure 3 shows failure processes optically observed on the surface of a SGF 1 w/o-PET specimen. With an increase of load, tensile microcracking and shear banding occurred in the matrix resin around the fiber ends. The tensile microcracking was caused by the tensile stress concentration at the fiber ends. The high interfacial shear strength between fibers and matrix must have affected the concentration /1/. With further increase of the load to a strain of about 3% corresponding to the maximum stress level, other tensile microcracks were formed in the matrix between fibers. These microcracks grew and the grown cracks were then bridged together by shear bands which had been formed at both tensile microcrack tips. Tensile cracks (arrow A) occurred following the microcrack growth. The shear bands grew into the matrix (arrow B) and were joined together by different shear bands. Shear yielding proceeded along the fiber-matrix interface in the fiber length direction. Shear cracking followed this shear yielding, which caused the fibers to pull out of the matrix (arrow C). Some fibers were broken, which induced other tensile microcracks and also shear bands in the matrix. The most important factor in the fracture processes above is considered to be the microcrack-induced tensile cracks at the fiber ends which grew into the matrix. Those tensile cracks were very influential in the catastrophic fracture initiation on the specimen surface. Based on the above result, the failure mechanisms on the surface of SGF 1 w/o-PET are schematically presented in Fig.4.

In situ SEM photographs were obtained to study the surface morphology of a SGF 30 w/o-PET specimen as shown in Fig.5. Microcrack-induced tensile cracks occurred at the broken or unbroken fiber ends during loading to a strain of about 2.6% corresponding to the maximum stress level, and they grew into the matrix (arrow A in Fig.5). Shear yielding took place around the fiber ends and it was followed by shear cracking along the interface in the fiber length direction (arrow B). The shear crack induced voiding as well as fiber pull-out from the matrix at the fiber ends. The shear yielding and shear cracking along the interface may have proceeded in parallel with the shear band growth in the matrix (see Fig.3). With further loading until catastrophic fracture, the tensile cracks and the shear cracks above described joined together to initiate a macroscopic fracture. Thus, both microcrack-induced tensile cracks and shear band-induced cracks were considered to have a cooperative effect on catastrophic fracture initiation.

For SGF 60 w/o-PET specimen, growth of the tensile cracks at the fiber ends was suppressed by other closely neighboring fibers. Instead, shear failure proceeded in the matrix between fibers. Shear cracks propagated on the matrix side near the interface like partial delamination. Shear stress interaction may have been greatly enhanced due to the closeness between fibers and also due to the orientation of the fibers toward the tensile direction /7/. The enhanced interaction must have an effect on the delamination-like failure. Thus, the shear cracks strongly influenced catastrophic fracture initiation on the surface of SGF 60 w/o-PET specimens.

3.3 Failure processes in the interior: Shear bands appeared at some probable inclusion sites in the interior of an unreinforced specimen. The growth direction of those shear bands were almost the same as that on the surface. In addition, debonding took place along the interface between the inclusions and resin. No discernable tensile cracking occurred in this case. It is worth noting that the failure processes in the interior of unreinforced specimens were different from those on the surface.

Failure processes in the interior of a SGF 1 w/o-PET specimen were also somewhat different from those on the surface. Polarized optical photograph in Fig.6 was taken in the interior of the specimen in Fig.3. Shear bands were formed with an increase of load around the fiber ends. Shear cracking occurred at the fiber ends as the shear bands grew and proceeded on the matrix side near the interface. As the load increased further to a strain of about 3%, the shear bands grew much more and joined with other shear bands. The shear cracks at the fiber end grew further in the fiber length direction. Some fibers were broken, which caused tensile microcracking and another shear banding around the broken ends. During loading to a strain of about 160%, the shear crack propagation (arrow D) induced pull-out of the fibers from the matrix as well as voiding at the fiber ends (arrow E). The voids grew into the matrix and joined together, which caused catastrophic fracture.

It should be noted that a "specific layer" appeared at the foot of the shear bands during the deformation (arrow F). The layer was not vacant but contained matrix material. Very few macrovoids or discrete shear cracks were observed in the layer. This type of layer has also been observed in specimens of SGF 30 w/o and 60 w/o-PET. Friedrich /1/ suggested the sequence of the shear crack formation in case of cohesive failure between fibers and matrix could be : 1. microcracking, 2. formation of shear crazes, 3. shear macrovoiding and 4. shear fracture between the macrovoids. Lhymn /3/ reported that discrete shear cracks were formed similar to the above processes 1. and 2., but that the cracks are connected by shear deformation, thus the fiber is pulled out with the residual matrix phase adhering to the fiber surface. The specific layer, on the other hand, was observed on the matrix side near the interface. This layer is thought to be a kind of plastic deformation layer which was transformed from severely deformed shear bands around the fiber end. It is thus considered to be a transition layer from the shear bands to the shear cracks near the interface. The above observed failure mechanisms in the interior of SGF 1 w/o-PET specimens are schematically presented in Fig.7. Growth and joining-together of the shear band-induced voids at the fiber ends exhibited strong effects on the catastrophic fracture initiation, which was different from the failure processes (see Fig.3) on the surface.

Figure 8 shows failure processes optically observed in the interior of a SGF 30 w/o-PET specimen. With an increase of load to a strain of about 2.6%, shear cracks (arrow A') near the interface propagated in the fiber length direction and the shear bands also grew around the fiber ends (arrow A). The specific layer was also observed to form in the matrix near the interface (arrow B). The shear cracks induced matrix macrocracking and caused pull-out of the fibers from matrix. Thus, all of the shear-band induced cracks seemed strongly influential in the initiation of catastrophic fracture. The failure mechanims were also different from those on the surface (see Fig.5), where tensile cracks at the fiber ends had an effect on the fracture initiation.

Delamination-like shear cracks near the interface were also recognized in the interior of a SGF 60 w/o-PET specimen similar to the case on the surface. The shear cracking was also very important in the initiation of catastrophic fracture.

SUMMARY: Table 1 summarizes the factors influencing catastrophic fracture initiation in unreinforced and short glass fiber reinforced PET specimens. The failure mechanisms on the specimen surface were different from those in the interior and were greatly affected by the fiber volume fractions.

ACKNOWLEDGEMENTS: The authors are indebted to Dr. A.Kato of Idemitsu Petrochemical Co. who kindly provided specimens for this study.

REFERENCES
1. K.Friedrich, Fortschr.-Ber. VDI-Zeitschr. Series 18 No.18, VDI-Verlag, Düsseldorf (1984).
2. C.LHYMN and J.M.SCHULTZ, Polym. Eng. Sci., 24(1984)1064.
3. C.LHYMN, J. Mater. Sci. Letters, 4(1985)1323.

4. N.SATO, T.KURAUCHI, S.SATO and O.KAMIGAITO, J. Comp. Mater., 22(1988)850.
5. P.T.CURTIS, M.G.BADER and J.E.BAILEY, J. Mater. Sci., 13(1978)377.
6. J.YUAN, A.HILTNER, E.BAER and D.RAHRIG, J.Mater.Sci. 20(1985)4377.
7. T.F.MACLAUGHLIN, J. Comp. Mater., 2(1968)44.

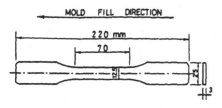

Fig.1   Specimen geometry and
        mold fill direction.

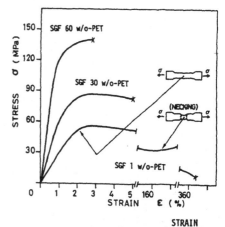

Fig.2
Stress-strain curves of
SGFR-PET specimens.

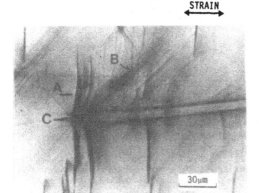

Fig.3
Polarized optical microscopic observation
of failure processes on the surface of a
SGF 1 w/o -PET specimen.

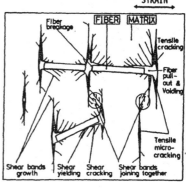

Fig.4
A schematic drawing of failure
mechanisms on the surface of
SGF 1 w/o-PET specimens.

Fig.5
In situ scanning electron microscopic
(SEM) observation of failure processes
on the surface of a SGF 30 w/o-PET
specimen.

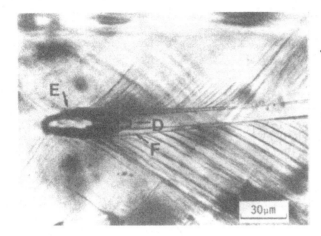

STRAIN

Fig.6
Polarized optical microscopic observation of failure processes in the interior of a SGF 1 w/o -PET specimen.

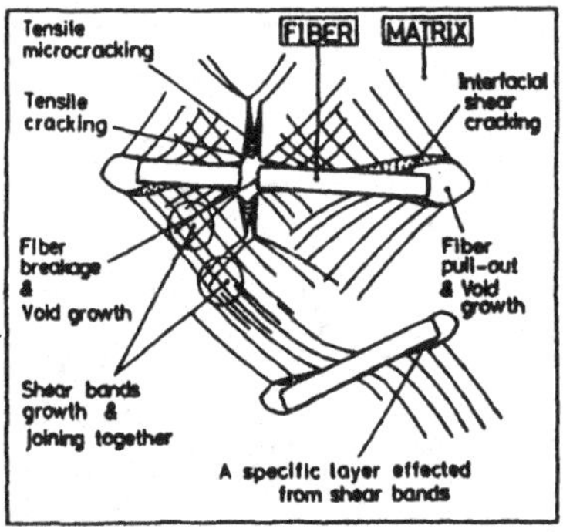

Fig.7
A schematic drawing of failure mechanisms in the interior of SGF 1 w/o-PET specimens.

STRAIN

Tensile microcracking
FIBER MATRIX
Tensile cracking
Interfacial shear cracking
Fiber breakage & Void growth
Fiber pull-out & Void growth
Shear bands growth & joining together
A specific layer effected from shear bands

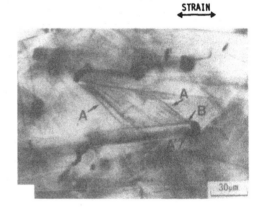

STRAIN

Fig.8
Polarized optical microscopic observation of failure processes in the interior of a SGF 30 w/o -PET specimen.

w/o	SURFACE	INTERIOR
Unreinforced PET	▲Tensile microcracking	•Shear band formation
SGF 1 w/o-PET	▲Tensile cracks growing at fiber ends	•Shear band-induced voids growing at fiber ends
SGF 30 w/o-PET	▲Tensile cracks growing at fiber ends •Shear crack propagation along fiber-matrix interface and/or between fibers	•Shear crack propagation along fiber-matrix interface and/or between fibers
SGF 60 w/o-PET	•Shear crack propagation along fiber-matrix interface and/or between fibers	

Table 1

Influential factors in catastrophic fracture initiation in unrein forced and short fiber reinforced PET specimen

1056

THERMO-MECHANICAL PERFORMANCES
OF POLYETHYLENE FIBRES/POLYETHYLENE
MATRIX COMPOSITES

C. MARAIS

ONERA - 29 Avenue de la Division Leclerc
92320 Chatillon - France

ABSTRACT

The present investigation is concerned with polyethylene films/ polyethylene fibres (PE/PE) composites. After a short description of the prepreg and laminates manufacturing process and a recalling of the mechanical properties of the fibres and the polymer matrix, the mechanical performances of unidirectional PE/PE laminates (as moulded or after irradition by electron beams) are presented. Some important physical properties as the cristallinity and the electromagnetic properties in radar frequency ranges are also discussed. The results are compared to those of glass or aramid-reinforced epoxy resins, demonstrating that PE/PE composites can be considered as a category of material which could be turned to the best account in the near future.

I-INTRODUCTION

Among the mechanical properties of composites, both strength and lightness are always put forward as advantages, mainly over light alloys. This assertion is still quite right with the composites made of high modulus polyethylene fibres.

Since polyethylene films were introduced, composites have been manufactured by using conventional thermoset resins as the matrix (polyesters, epoxies). Unfortunately, there is a limited compatibility between these thermoset resins and thermoplastic fibres. So, the fibre/ matrix bonding strength remains very poor despite extensive works on fibre surface treatments tried to increase the interface characteristics.

In order to meet the compatibility requirements, a polyethylene matrix less crystalline than the fibres has been used, resulting in polyethylene/polyethylene (PE/PE) composites. Using polyethylene as the matrix allows also to match both fibres and matrix use temperature ranges.

As an additional advantage, polyethylene provides some inherent properties for technical applications as, for example, lightness, chemical resistance, dielectrical properties, toughness, impact resistance, hydrophobicity, thermoformability.

II-MATERIALS

2.1 Polyethylene fibres

To-day, three fibres have entered the market: Dyneema SK60 (DSM), Spectra 900 and 1000 (Allied Fibers) and Tekmilon (Mitsui Petrochemical). In fact, the European market is entirely covered by DSM, which explains why the study has been based mainly on Dyneema SK60 fibres. However, some information about the other fibres will be introduced below for comparison.

Due to the high level of macromolecular orientation during the manufacturing process /1,2/, Dyneema SK60 polyethylene fibres have very good mechanical properties: dried tows of Dyneema SK60 exhibit a Young's modulus of about 85 GPa and a tensile strength of 2 GPa /3/.

As to the Dyneema SK60 filament, we have tried to determine its mechanical properties by tensile testing. It is very difficult to obtain the Young's modulus or the breaking strength because there is some doubt about what does the filament consist in, and thus measuring the actual cross-section. In order to illustrate this difficulty, the cross-sections analysis by optical micrography of transversely cutted filaments encapsulated in an epoxy resin has been accomplished on all the above fibres.

Fig. 1 shows the different cross-sections at the same scale. The Dyneema SK60 filaments do not exhibit circular shapes as Tekmilon filaments do. In fact, they tend to a pronounced decomposition into smaller and smaller fibrillar entities. However, all the fibres defibrillate more or less as illustrated, for example, in Fig. 2 where the defibrillation of the Tekmilon filaments is shown. This kind of damage is very similar to what is observed for Kevlar fibres and, accordingly, polythyelene fibres are expected to behave as Kevlar fibres, i.e. good mechanical along the fibre axis but poor properties across it. As far as Dyneema SK60 is concerned, the pronounced defibrillation is not expected to affect the mechanical properties more than for the other polyethylene fibres.

2.2 Polyethylene matrix

A polyethylene matrix has been defined to produce the PE/PE laminates by compression moulding. That process imposes especially a moulding temperature in the interval between the melting points of the fibres and the matrix respectively. Five polymers were selected as the first choice: low density polyethylene, low linear density polyethylene, irradiated low density polyethylene, high density polyethylene and polypropylene.

The final matrix, selected on the basis of DSC thermograms, is a high-density polyethylene (Hostalen GM9240HT) with a melting point of 128°C slightly below the fibre melting point (145°C). With this high melting point, the physical properties are maintained at higher temperature than for the three other low density polyethylenes. As to polypropylene, its melting point is slightly higher than the fibres one, and it is obvious that laminates cannot be manufactured.

III-COMPOSITE FABRICATION

First, prepregs have been fabricated with Dyneema SK60 fibres and Hostalen films by filament winding. Full details of the preparation of prepregs are given elsewhere /4/.

The laminates, made of a lay-up of PE/PE prepreg sheets, are moulded under pressure up to 9.6 MPa and at a temperature of 132°C. That a uniform, closely controlled, temperature has to be arrived at throughout the lay-up, is a critical point of the moulding process.

Fibre volume contents as high as 75 % can be achieved with no porosity. Lower pressures may be applied, but there is a risk of getting a lot of voids in the laminates: undoubtedly, selecting a high density polymer to make the composites imposes these high pressures.

As reported in the literature /5/, a moderate irradation leads to the formation of intermolecular cross-links in the PE polymers, the main benefit being the improvement of creep behaviour. According to the differences of cristallinity of the fibres and the matrix (see next section), irradiation will likely modify only the matrix of the PE/PE composites.

IV-MAIN PHYSICAL PROPERTIES

4.1 Crystallinity

Crystallinity has been determined by DSC (Fig. 3) and the respective value for the fibre and the matrix calculated with respect to their relative content in the composite. Typical results are: matrix 60 %, fibres 81 %, composite 75 %.

It must be pointed out that the Hostalen matrix has a slightly higher crystallinity than Hostalen films before moulding (55 %). The microscopic analysis of PE/PE composites in polarized light reveals a transcrystallinity of the matrix around fibres (anyway, with a fibre volume content of 75 %, there is no opportunity for spherulitic structures to form) This transcrystallinity confers on the matrix and the fibres/matrix interface their mechanical performances.

4.2 Electromagnetic properties

As a verification, the complex permittivity of PE/PE laminates has been determined by means of an open resonator. The laminates had a thickness of 165 µm and a fibre volume content of Ca. 75 %. As expected, the dielectric constant ε' values are very low, namely 2 and 2.2 for fibres parallel or perpendicular to E respectively at 10 GHz.

The dielectric constant of polyethylene has practically the same value for the fibres and the matrix. The PE/PE composites present the lowest dielectric constant in the composites family, in all the radar frequency range. It is also the same for the loss tangents, which are about 3.10^{-4} at 10 GHz.

V-MECHANICAL PROPERTIES OF PE/PE COMPOSITES

In- and off-axis tensile tests on unidirectional laminates have been carried out. Fig. 4 shows the elastic modulus for various orientations with regard to the fibre direction.

The variation of the Young's modulus as a function of Θ is consistent with the usual law of orthotropic materials, with E varying from 73 GPa in the fibre direction down to 1.6 in the transverse direction. The latter is higher ($\approx$ 2x) than the modulus of the neat Hostalen film. As to the modulus at 0°C, it is similar to that of Kevlar/epoxy composites. Obviously, due to the low density of the polyethylene, the specific modulus is much better than that of Kevlar/epoxy composites. Moreover, while these results are obtained at room temperature where the polyethylene is still in a viscoelastic state, a much better value is expected at the glass transition temperature (Ca. −110°C): an elastic modulus of 100 GPa was determined at −70°C.

On the other hand, the analysis of the mechanical properties of composites irradiated by electron beams has been performed. The neat Hostalen matrix exhibits an elongation at break of Ca. 700 % before irradiation. It falls off to a few per cent after exposure to radiation doses of approximately 40 Mrad, which corresponds to the elongation at break of Dyneema SK60 fibres. Therefore, this radiation dose of 40 Mrad has been taken as the limit in the PE/PE composites. Fig. 5 shows the influence of the radiation dose and temperature on the elastic tensile modulus of unidirectional composites. For testing along the fibres axis, there is no effect of the irradiation on the modulus, which proves that the fibres are not damaged and that the overall quality of the composite is not affected. More tests are still in progress, since what we expect actually from irradiation is an improvement of viscoelastic and viscoplastic properties of the matrix, and thereby of the related composite characteristics as creep, shear, etc...

Moreover, the elastic modulus decreases with temperature whether the composite is irradiated or not: at 100°C, there is a 50 % drop of the modulus from the initial value at room temperature.

A similar study has been performed for the ultimate tensile strength (Fig. 6). For tests in the fibre direction, the strength is quite comparable to that of Kevlar/epoxy composites. For off-axis specimens, with increasing Θ, the strength falls off dramatically. The decline becomes moderate beyond $\Theta = 30°$.

VI-CONCLUSION

In summary, three main aspects have been emphasized in this study:

1) PE/PE prepregs can be made by filament winding PE fibres on PE thin films and laminates manufactured by compression moulding.

2) The mechanical properties of the resultant laminates are quite comparable with those of other, more traditional, composites. Furthermore, the specific properties of PE/PE composites transcend those of glass/thermoset materials... For various applications, PE/PE composites could even replace polyaramide epoxy composites.

   Finally, because the glass transition temperature is very low, the mechanical properties are better when the temperature decreases. This gives access to applications such as cold and even cryogenic structures.

3) After manufacturing, the inherent properties of polyethylene, namely chemical resistance, toughness, damping, thermoformability, soldability, electromagnetic properties, etc... are preserved, so that various applications may be anticipated in technological structures for ballistic protection or radomes, as well as in sporting or biological goods.

ACKNOWLEDGEMENTS

Financial support for this work has been received from the Direction des Recherches et Etudes Techniques (DRET) for which grateful thanks are addressed. The author is also indebted to G. Désarmot for helpful discussions and assistance.

REFERENCES

1. P. Smith, P.J. Lemstra, B. Kalb and A.J. Pennings, Polymer Bull., 1 (1979) p. 733

2. P. Smith and P.J. Lemstra, "Ultra-high strength polyethylene filaments by solution spinning/drawing", Journal of Materials Science, 15 (1980) p. 505

3. C. Cazeneuve, A. Valentin, D. Degout, "Les polyéthylènes", Rapport ETCA 90-R014, février 1990

4. C. Marais, G. Désarmot, "Procédé de fabrication d'un matériau composite constitué d'une matrice thermoplastique renforcée par des fibres longues et matériau composite obtenu par ce procédé, Demande de brevet n° 89-04327, mars 1989

5. A. Chapiro, "Radiation chemistry of polymeric systems", Wiley Interscience, New York (1962)

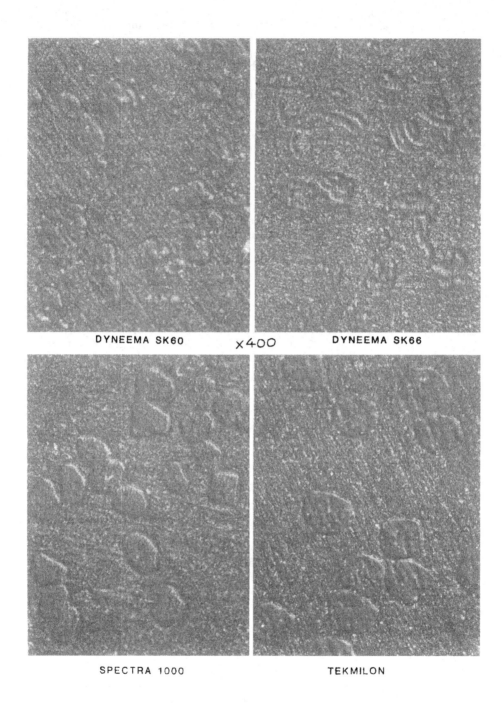

DYNEEMA SK60   ×400   DYNEEMA SK66

SPECTRA 1000   TEKMILON

Figure 1 - Cross-sections of polyethylene fibres

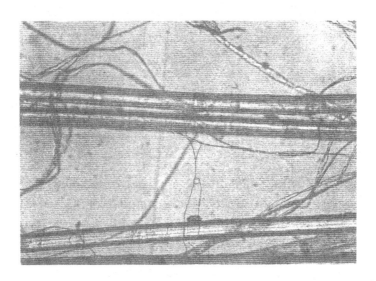

Figure 2 – Defibrillation of polyethylene fibres
(Tekmilon)

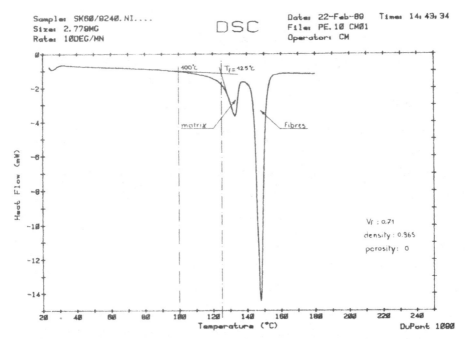

Figure 3 – DSC of Dyneema SK60 polyethylene fibres/
Hostalen polyethylene films composite

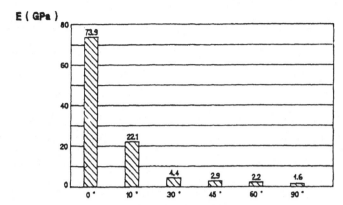

Figure 4 – Elastic modulus vs. fibre orientation Θ
for unidirectional composites (Vf ≈ 0.75)

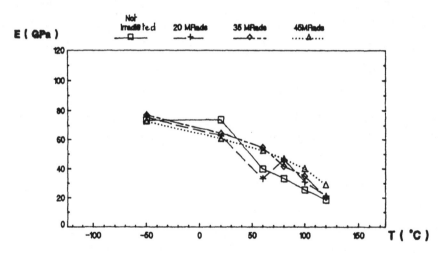

Figure 5 – Variation of elastic modulus as a function of
the temperature for various irradiation doses

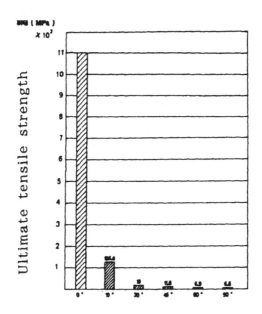

Figure 6 –

Ultimate strength vs. fibre
orientation Θ for unidirectional
PE/PE composites

STRUCTURAL EVOLUTION OF A STAMPING
REINFORCED THERMOPLASTIC (SRT) IN BOTH
EXTENSIONAL AND SHEARING FLOWS

Y. LETERRIER, C. G'SELL, A. GERARD*

Laboratoire de Physique du Solide (URA CNRS DO 155)
Parc des Mines de Nancy - Parc de Saurupt
54042 Nancy Cedex - France
*Developpement Plastique et Composite
PSA DAT/PL - BP 79 - 25400 Audincourt - France

ABSTRACT

The axisymmetric flow of a polypropylene/long glass fibers SRT is analysed during isothermal or non-isothermal compression. SEM observations, uniaxial tension tests and measurements of the local weight fraction of glass reveal a circumferential orientation of the deep fibers under the effects of extensional strains, while at the surface shearing strains induce fiber rotation and breakage. Particular attention is paid to the shearing effects by means of a novel plane shearing device, showing strong shear-thinning of the material. A good agreement is found between the experimental data and a rheological calculation ; a thermal transfert law shows a surface layer of congealed PP which depth grows as the square root of the compression rate.

I-INTRODUCTION

Processing of SRT materials has known a large expansion in the automobile industry during the last decade. This flexible technology is really suitable to realise complex shapes or large area parts under high production rates, and is less time and energy consuming than others compression molded composites such as SMC. During its flow in the mold cavity, the SRT undergoes considerable strains and high strain rates, up to 100 s^{-1} at typical closing speeds of 1 to 10 mm/s. The occurence of both extensional and shearing effects induces changes in the fiber orientation state. Consequently, the stamped parts exhibit significants amounts of mechanical anisotropy.

Most of the studies on flow kinetics in compression molding were dedicated to the SMC process [1]. Their aim was mainly to simulate the mold filling procedure [2,3] and to analyse the orientation of the fibers in real parts, thanks to various orientation techniques [4-6]. Nevertheless those rheological calculations need physical characterizations and our purpose was more fundamentally devoted to the structural evolution of axisymmetric parts stamped according to isothermal or non-isothermal paths. In that scope, a particular attention was paid to the local fiber orientation following on shearing and extensional strains.

II-PROCESSING AND INVESTIGATION DEVICES

The material involved in this study have been the object of preliminary analyses [7,8]. It concerns a composite whose polypropylene (PP) matrix is reinforced by 35 wt% long E-Glass fibers. Originally, the SRT is provided in the form of a 3 to 4 mm thick sheet. It is preheated at 200 °C in the hot air stream of an environmental chamber. The hot blanks are loaded in the mold at 200 °C (experimental isothermal compression) or at 50 °C (industrial non-isothermal compression). In the non-isothermal procedure, the parts are cupel-shaped with a diameter of 20 cm, while in the isothermal stamping it concerns 10 cm diameter disks.

In addition to the experimental techniques previously described [7] we have developed a novel plane shearing device equipped with transparent plates in order to visualize the SRT behaviour under large strains. With these features, it constitutes a sliding plate rheometer and allows the measure of the viscosity as a function of the applied shear, shear rate and temperature.

III-EXPERIMENTAL RESULTS

On the grounds of the isotropy of the SRT sheet characterized by an in-plane random fiber orientation with light curvatures, we have analysed any deviation from this distribution as a consequence of the flow process. SEM obervations reveal specific incidence of shearing or extensional strains on local fiber orientation state, whereas tension tests provide accurate information about global fiber distribution.

3.1 Configuration of the superficial fibers in stamped parts

Figure 1 shows the slip-stream left by a fiber end at the surface of the solidified polymer. Underlying fibers with strong curvatures appear on figure 2, involving fiber wear and damage. Therefore one will remain the large fiber rotation with a preferential orientation in the flow direction.

3.2 Configuration of the deep fibers in stamped parts

The in-core fibers are characterized by a circumferential orientation, as revealed on figure 3 by the uniaxial tension tests performed on radial and circumferential samples, cut out of isothermal or non-isothermal stamped axisymmetric parts. One notice the light increase of the stress in the circumferential samples, due to the rotation of the fibers perpendicularly to the radial flow. Furthermore, radial samples break by delamination, whereas circumferential samples snap at higher strains levels. This stress improvement increases when flow is isothermal as the fibers are free to move in the molten PP during the whole compression period.

3.3 Viscosity measurements

The viscosity of molten SRT has been measured by means of two complementary experiments (figure 4). Plane shearing results show a smooth decrease (i. e. strong shear-thinning) of the apparent shear viscosity with increasing shear rate $\dot{\gamma}$, reaching almost 100 Pa.s at typical rates of industrial processing. By contrast, the viscosity deduced from compression load measurement during the course of isothermal stamping [9] exhibits two distinct types of behaviours : at low shear rates (i. e. slow compression speed) the viscosity is roughly of the same order of magnitude ; at higher shear rates the viscosity shows a steep fall which lessens with increasing $\dot{\gamma}$.

IV-RHEOLOGICAL ANALYSIS

4.1 Kinematic analysis

The elongational rate $\dot{\varepsilon}$ and shear rate $\dot{\gamma}$ have been calculated on the basis of an isothermal HELE-SHAW model flow [1] for the axi-symmetric case. Figure 5 gives the maps of the evolution of both strain rates at the end of the compression period, for a closing speed of 10 mm/s, a final thickness of 4 mm and a final diameter of 200 mm. This calculation shows that the extensional strain rate, $\dot{\varepsilon}$, goes through a maximum in the mid-plane layer and cancels out at the mold contact ; it is constant throughout a given plane parallel to the mold surface. On the other hand, the shear rate, $\dot{\gamma}$, is higher in the surface layers while it decreases downwards the center of the SRT charge, its distribution is more complex than that of the $\dot{\varepsilon}$ rate, and its intensity is about 100 to 1000 times higher.

The relative influence of both rates has been examined previously in the case of injection molded reinforced thermoplastics [10] where the configuration of the strain field is similar to that of stamping flow. The theoretical model usually used [11], despite restrictive hypotheses, provides good information on fiber behaviour. Shear flow induces rotation of the fibers which may eventually be tipped over in the direction of the flow, while extension, divergent in axisymmetric flows orients the fibers perpendicularly to the flow direction.

These noteworthy configurations are actually present in stamped parts, and as in every point of the SRT charge both kinds of strains occur with variable intensity, the interpretation of the orientation of a given fiber is difficult. Nevertheless the surface heteroge-neities would be explained with the combination of the two mecanisms (elongation and shearing) : under the effects of shearing an under-lying fiber tips up, one of its ends reaches the surface while the other one moves off with the elongational flow, and orients per-pendicularly, from which result strong curvatures and fiber damage. Deeper in the SRT charge, on the contrary, shearing is overtaken by elongation and the gradual orientation of the fiber does not involve fiber breakage.

4.2 Thermal analysis

The cooling of the SRT during its stamping flow has been modelled through a simplified thermic analysis by assuming that the in-plane heat flows are negligible versus the conduction flow normal to the sheet surface. At the mold contact, isothermal boundary conditions were chosen for simplicity. The evolution of the temperature profile is given on figure 6, for five steps of the SRT flow period. It is evident that the cooling takes place in a thin layer, which temperature rapidly rises from 50-80 °C up to 130-160 °C. From this we have calculated the thickening rate of the congealed layer, and this "cooling rate" was found to increase roughly about the square root of the compression speed. For example a 10 mm/s stamping implies a 3 mm/s cooling rate, giving a final congealed thickness of about 60 % of the whole charge thickness, which is considerable.

CONCLUSIONS

The rheological parameters and the orientation distribution of the fibers of isothermal or non-isothermal stamped axisymmetric parts of an isotropic SRT sheet have been determined by means of SEM observations and uniaxial tension tests in addition to previous investigations. The following remarks were found :

i) during the flow, the shearing stains are concentrated in the superficial layers, while the elongational strains are predominant in the deep layers with 100 to 1000 times lower intensities ;

ii) the connection of the deformation mecanisms by shearing and extension induces strong curvatures and breaking of the underlying fibers. The elongational flow of the deep layers induces a circumferential orientation of the fibers without damage ;

iii) the apparent shear viscosity decreases slightly with increasing shear rate whereas the apparent compression viscosity undergoes a large fall for shear rates over 0.3 s^{-1} ;

iv) the cooling rate of the non-isothermal stamped charge increases as the square root of the compression speed.

REFERENCES

1 - C. L. Tucker III, in "Injection and compression molding fundamentals" (A. E. Isayev, ed.) (1987) 481-565 (New York and Basel)

2 - W. C. Jackson, S. G. Advani, C. L. Tucker, J. Compos. Mater. 20 (1986) 539-557

3 - M. R. Barone, T. A. Osswald, J. Non-Newt. Fluid Mech. , 26 (1987) 185-207

4 - A. Vaxman, M. Narkis, A. Siegman, S. Kenig, J. Mater. Sci. Letters, 7 (1988) 25-30

5 - G. Fischer, P. Eyerer, Polym. Compos. , 9 (1988) 297-304

6 - M. W. Darlington, P. L. McGinley, J. Mater. Sci. , 10 (1975) 906-910

7 - Y. Leterrier, C. G'sell, in "Matériaux composites à matrice thermoplastique" (C. G'sell and J. P. Favre, PLURALIS, ed.) (1990) St.Etienne

8 - Y. Leterrier, C. G'sell, A. Gérard, in "Matrices et fibres polymères" (AMAC, ed.) (1990) Colloque GFP-AMAC, Sophia-Antipolis

9 - J. R. Scott, Trans. Inst. Rubber Ind. , 7 (1931) 169

10 - M. Vincent, J. F. Agassant, in "Initiation à la chimie et physico-chimie macromoléculaires. Volume 7. Matériaux composites à matrice polymère" (GFP, ed.) (1989) 285-314 (Strasbourg)

11 - G. B. Jeffery, Proc. Roy. Soc. , London A102 (1922) 161

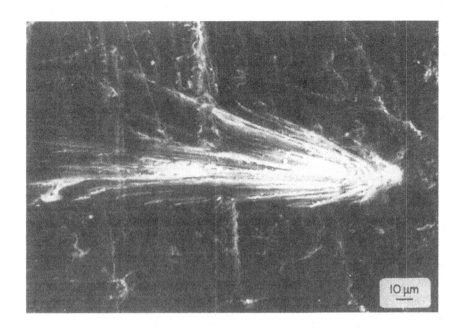

figure 1 : Slip-stream left by a fiber end at the surface of a non-isothermal stamped SRT ; the flow direction is from left to right.

figure 2 : Effects of the combination of shearing and extension on the curvature and damage of underlying fibers ; the flow direction is from left to right.

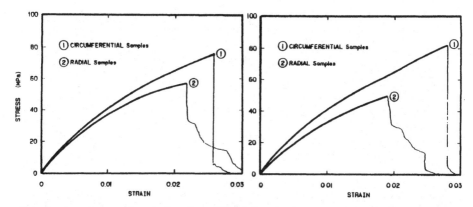

figure 3 : Comparison of the uniaxial tension properties for circumferential and radial samples cut off on non-isothermal (a) and isothermal (b) axisymmetric stamped parts.

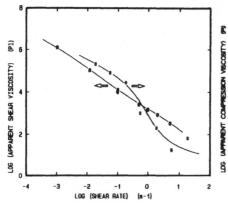

figure 4 : Evolution of the apparent viscosities in shear and compression of a SRT at 200 °C as a function of the shear rate.

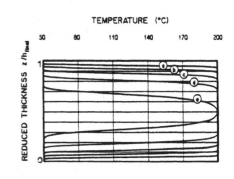

figure 6 : Cooling profiles at 5 steps of the non-isothermal SRT compression (a : 20 % of the flow time, b : 40 %, c : 60 %, d : 80 % e : end of the compression).

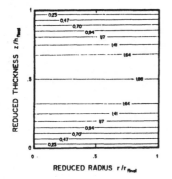

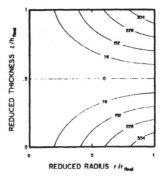

figure 5 : Maps of the strain rates distribution in the SRT charge computerized at the end of the isothermal compression stage. The values are given in s^{-1} for a 10 mm/s compression speed.

DATA FOR DESIGNING WITH CONTINUOUS
GLASS FIBRE REINFORCED POLYPROPYLENE

A.C. LOWE, D.R. MOORE, I.M. ROBINSON

ICI Advanced Materials
Wilton Materials Research Centre
TS6 8JE Wilton Middlesbrough - Great Britain

ABSTRACT

Engineering materials are used in a wide range of load-bearing applications. The process by which a material might find selection for a
particular application normally involves a design feasibility study.
The purpose of this paper is to discuss the requirements of engineering
design properties for a new type of continuous fibre reinforced thermoplastic composite. The paper will concentrate on a 35% by volume
fraction V_f glass fibre / polypropylene system. The stiffness, strength

and toughness of various lay-ups of 35% V_f GL/PP have been measured and

discussed in terms of simplified design criteria. This has allowed
predictions of the properties of more lay-ups to be made, with some
success.

INTRODUCTION

Considerable literature exists on the properties of continuous carbon
fibre reinforced materials [1], including those based on high performance
thermoplastic matrices [2,3]. These materials are used in high technology
components, for example in the aerospace industry. In contemplating
design with continuous fibre reinforced composites that need not include
carbon fibres, there is an interest in applying continuous fibre thermoplastic technology to a much wider area of load-bearing components.
Components manufactured with continuous glass fibre/polypropylene tape
can be fabricated by a wide range of routes such as laminate formation,
tape laying, braiding, filament winding, weaving, hybrids and rubber
block stamping. There is therefore a special need to describe the
relationship between material properties and the performance of a complex
component made from such composites. In contemplating a new material
for a load-bearing component, it is likely that a design feasibility
study will require definition of the stiffness, toughness and strength.
The paper is divided into three main sections. The first section will
deal with the measurement of stiffness, particularly in terms of the
anisotropy these composites display. The second section will briefly
deal with strength, where some simple statements on the static tensile

properties of a few lay-ups will be given. The third section will concentrate on the impact performance of various lay-ups. Impact toughness is defined in two ways; in terms of resistance to impact damage and in terms of the total energy absorption capacity of a material.

1 MATERIALS AND EXPERIMENTAL

The material used in the study is designated 35% V_f GL/PP and is manufactured by ICI Advanced Materials under the trade name "Plytron". It is available in the form of continuous unidirectional sheet or tape.

The measurements of tensile modulus were made on an Instron 6025 screw machine. Tensile coupon specimens were machined from specially prepared laminates, and were subjected to ramp tests to obtain the stress-strain data. For the determination of the pseudo-elastic constants, two transverse extensometers were used in addition to the axial extensometer. The measurements of shear modulus were conducted by an isochronous (100 seconds) torsion test of a rectangular prismatic specimen. Full details of these experimental techniques have been reported elsewhere [4,5]. Impact toughness was measured on an instrumented falling weight impact (IFWI) apparatus where photography during impact enabled interpretation of the force-time signals. Full details of this method have been reported elsewhere [6,7].

2 STIFFNESS

2.1 Engineering Constants and the prediction of stiffness

Unidirectional continuous fibre composite laminates can be assumed to exhibit transverse isotropy [4]. For this particular class of isotropy there are 5 independent pseudo-elastic constants that fully describe the anisotropy of the material. Consequently, these 5 parameters can be used to predict the stiffness of other laminate lay-ups eg [±45°], [+45°,0°,-45°,90°] and so on [4] through the use of analytical methods such as Classical Laminate Theory. The 5 pseudoelastic constants can be calculated from separate tensile and torsion tests conducted on 0° and 90° specimens machined from the unidirectional laminate. The tests generate a total of 8 engineering properties from which the 5 pseudo-elastic constants are calculated and the results are given below:-

E_{11}	28.7 GPa
E_{22}	4.1 GPa
ν_{12}	0.40
ν_{13}	0.39
ν_{21}	0.04
ν_{23}	0.62
G_{12}	1.35 GPa
G_{23}	1.26 GPa

Engineering constants measured for 35% V_f GL/PP at 23° C and 0.2% strain, and at a displacement rate of 1 mm/min

1074

From transverse isotropy, it is known that certain of the engineering constants are related. In particular, the following constants should be equivalent:-

$$\nu_{12} = \nu_{13}$$

$$\nu_{21} = \frac{E_{22} \cdot \nu_{12}}{E_{11}}$$

$$G_{23} = \frac{E_{22}}{2(1 + \nu_{23})}$$

The values given above are in agreement with these predictions within experimental limits, confirming that the assumption of transverse isotropy is valid for this material.

The values for the derived pseudo-elastic constants are given below:-

Pseudoelastic Constants	
S_{11}	= 34.843 pm²/N
S_{12}	= - 13.936 pm²/N
S_{22}	= 243.902 pm²/N
S_{23}	= -151.220 pm²/N
S_{55}	= 740.740 pm²/N

The pseudoelastic constants for 35% V, GL/PP expressed in picometres² (10^{-12})m² per Newton measured at 23° C and at an applied strain of 0.2%.

Classical Laminate Theory (CLT) provides a method for predicting the modulus of plates fabricated from unidirectional lamina, with varying stacking sequences. The theory assumes that the in-plane properties are the only important ones, consequently the through-plane properties can be ignored. CLT can be seen to be a simplification of the conditions described by transverse isotropy. Thus the engineering constants involving through-plane properties (e.g. ν_{13}, ν_{23} and G_{23}) are neglected in calculations of more complicated lay-ups, and other engineering constants can be directly calculated from the assumptions of transverse isotropy.

For 35% V, GL/PP, the engineering constants required for CLT predictions reduce to:-

E_{11}	28.7 GPa
E_{22}	4.1 GPa
ν_{12}	0.40
G_{12}	1.35 GPa

Using CLT, predictions for stiffness have been made for a range of lay-ups. These were within 5% of the measured values, which given the assumptions made in the use of CLT, are good.

Lay-up	Prediction GPa	Measured GPa
$(\pm 45)_{2S}$	4.7	4.4 (0.5)
$(0/90)_{2S}$	16.6	15.8 (0.5)
$(+45/0/-45/90)_s$	12.0	11.2 (0.5)
$(\pm 23)_{2S}$	17.4	16.7 (0.8)

3 STRENGTH

Strength will be considered using the concepts of principal stress theory. This assumes that the maximum stress that any component can sustain is related to the maximum principal stress in the structure, which in turn is related to the tensile strength of that laminate. The tensile strength for the laminates studied are given beneath:-

Lay-up	Tensile Strength MPa
$(0)_8$	426
$(+45/0/-45/90)_s$	178
$(\pm 45)_{2S}$	50
$(0/90)_{2S}$	280

Design with these values in mind should allow for high enough factors of safety, so that an applied load on a structure will produce a principle stress on the laminates that do not exceed the tensile strength values given above, and should ideally remain at some stress level well beneath those shown above. Under the action of dynamic loading, these strengths are expected to reduce as a function of number of cycles under load in the normal manner for continuous fibre composites.

4 IMPACT

Photographic IFWI provides a means for understanding the toughness and failure mechanisms in impact [8]. Careful examination of the photographs during impact defines the position on the force-displacement trace at which crack damage initiates in the panel specimen. The energy absorbed by the specimen associated with this event can also be determined. The data are summarised below using the mean values for various parameters associated with the impact tests. These parameters are, U_I, the energy absorbed by the specimen at initial crack damage on the tension surface, F_I, the force at this point, $U_P \& U_T$, which are the energies absorbed at the "main peak" in the force - displacement curve and the total energy involved in the impact event, respectively.

Lay-up Configuration	$U_I(J)$	$F_I(kN)$	$U_P(J)$	$U_T(J)$
$(\pm 22.5)_{2S}$	1.10 (0.5)	1.41 (0.24)	35 (16)	106
$(0)_8$	0.92 (0.5)	1.14 (0.28)	8.4 (1)	16
$(0/90)_{2S}$	0.88 (0.3)	1.37 (0.13)	50 (5)	84
$(\pm 45)_{2S}$	0.75 (0.1)	1.53 (0.07)	38 (8)	60
$(+45/-45/0/90)_S$	0.69 (0.2)	1.41 (0.07)	25 (2)	52
$(+45/0/-45/90)_S$	0.69 (0.2)	1.22 (0.2)	26 (2)	49

The specimens at the end of the impact test were all intact, with the exception of the unidirectional laminate. However, it was apparent from inspection that considerable crack damage occurred well before the end point of the force - displacement curve. Therefore the term designated "total energy" is a combination of crack initiation, crack propagation and a term associated with bringing the impactor to rest on the specimen. "Total energy" is therefore not a material property alone. The initiation of crack damage can be seen to occur at an early part of the force-displacement curve; generally associated with the second peak in the impact signal. Therefore it is apparent that full fracture damage (initiation and propagation crack damage) must be at some energy absorption between initiation energy and total energy. The definition of full fracture energy is somewhat elusive, but could be equated with the energy associated at the main peak in the force-displacement curve. The energy to create initial crack damage is a far more precisely defined parameter, and is identified as the point where tension surface splitting occurs. It is therefore reasonable to suggest, for example, that the energy to initiate such an event for the $(\pm 22.5)_{2S}$ laminate is higher than that for the quasisotropic lay-up.

CONCLUSIONS

The stiffness, strength and toughness of various lay-ups of 35% V_f GL/PP have been measured and discussed in terms of simplified design criteria. The pseudoelastic constant data for the unidirectional composite, measured at 23° C, has been used via Classical Laminate Theory (available in the form of a commercial software package) to predict the stiffness of a range of various plate lay-up configurations. The strength of the various lay-ups under static and fatigue conditions are reported. It is hoped that the data, combined with the simplified design criteria, will be of use to the engineering community.

ACKNOWLEDGEMENTS

The authors wish to acknowledge the contributions of R.S Prediger, B Slater, P.M Rutter, N.J Burgoyne, G Cuff and N Walker in this paper.

[1] D Hull "An introduction to composite materials"
(Cambridge University Press, 1981)

[2] L Carlsson (ed) "Thermoplastic Composites"
to be published by Elsevier, 1990.

[3] Cowan G, Measuria U, Turner R.M, Proc. I. Mech. Eng. Conf. on "Fibre
Reinforced Composites", 1986, p 105

[4] Zahlan N, Moore D.R and Robinson I.M, SAMPE, July 1989, Birmingham,
U.K

[5] Carsile D.R, Leach D.C, Moore D.R, Zahlan N, "Mechanical Properties
of the Carbon Fibre/PEEK Composite APC-2/AS4 for Structural
Applications"
To be published by ASTM

[6] Moore D.R and Prediger R.S, Poly. Comp, **9**, (1988), p 330

[7] Chivers R.A and Moore D.R, Meas. Sci. Tech, **1** (1990) P 313

AUTHORS' INDEX

KEY-WORD INDEX

COMMITTEE E.A.C.M.

STANDING COMMITTEE :

FRANCE	A.R. BUNSELL, Chairman
	A. MASSIAH, General Secretary
GREAT-BRITAIN	A. KELLY
ITALY	I. CRIVELLI-VISCONTI
DENMARK	H. LILHOLT
GERMANY	K. SCHULTE

ADVISORY COMMITTEE :

FRANCE	F.X. de CHARENTENAY
	P. LAMICQ, R. NASLAIN
GREAT-BRITAIN	M.G. BADER, J.H. GREENWOOD
	D.C.PHILLIPS
WEST-GERMANY	H. KELLERER, K. FRIEDRICH
ITALY	M. AGNETTI, G. CONNI
SWEDEN	Th. JOHANNESSON
BELGIUM	N. SPRECHER
SWITZERLAND	P. MEIER

CORRESPONDENTS :

U.S.A.	B. PIPES, J.C. SEFERIS
	S.W. TSAI
JAPAN	T. HAYASHI, A. KOBAYASHI
	M. UEMERA
CHINA	H. GU

EACM STAFF :

Comité d'Expansion Aquitaine
2 Place de la Bourse
33076 Bordeaux Cedex - France
Tel. : 56.52.65.47
Telex : 572651 F
Fax : 56 44 32 69

A.R. BUNSELL, Chairman
A. MASSIAH, General Secretary
D. DOUMEINGTS, Public Relations
H. BENEDIC, Secretary

MEMBERSHIP

FOR YOUR SUBSCRIPTION

Whether you are in Europe, the United-States, Japan or elsewhere you have a need to know what is happening in all the areas of the composite material fields.

In this aim EACM has set up delegates and formed different committees in the whole world.

ADVANTAGES OF MEMBERSHIP

For company membership :

- Reduced registration fees for employees to ECCM conference
- Reduced price for the European Directory
- Provide a phone-in information service
- Free subscription to the European Composite Materials Newsletter

For Laboratory and individual membership :

- Reduced registration at the ECCM conference
- Reduced price for the European Directory
- Reduced group travel to major meetings such as ICCM
- Access to the information service
- Free subscription to the European Composite Materials Newsletter

MEMBERSHIP APPLICATION

To be returned to :
EACM
2 Place de la Bourse
33076 Bordeaux Cedex
France
Phone : (33) 56 52 98 94 - 56 52 65 47
Fax : (33) 56 44 32 69

NAME :...................................... ADDRESS :.....................................

FORENAME :.................................. ....

POSITION :.................................. POSTCODE :...................................

COMPANY :.................................. COUNTRY :...................................

BUSINESS ACTIVITIES :..

PHONE :.................................. TELEX :...................................
FAX :..................................

Check type of membership for which you qualify :

• COMPANY MEMBERSHIP : /__/

- Large sized firms	10 000 FF.
- Small and medium sized firms	1 500 FF.
- For each additionnal person of the same firm	350 FF.

• RESEARCH LABORATORY MEMBERSHIP /__/   1 000 FF.

• INDIVIDUAL MEMBER (*) /__/   350 FF.

Please make out your cheque to : EACM

(*) Members of national associations (AMAC, CODEMAC -France-, CMC -Italy-, BCS -UK-...) are considered as being associate members of EACM.

LA NEF

Dépôt légal septembre 1990 - N° imprimeur 6705
Imprimé en France

Printed in the United States
By Bookmasters